Radiation Detection and Measurement

Third Edition

Glenn F. Knoll
Professor of Nuclear Engineering and Radiological Sciences
University of Michigan
Ann Arbor, Michigan

John Wiley & Sons, Inc.

ACQUISITIONS EDITOR	Bill Zobrist
MARKETING MANAGER	Katherine Hepburn
SENIOR PRODUCTION EDITOR	Robin Factor
ILLUSTRATION EDITOR	Sigmund Malinowski
DESIGN DIRECTOR	Madelyn Lesure

This book was set in New Times Roman by Argosy, and printed and bound by Quebecor Printing. The cover was printed by Phoenix Color

This book is printed on acid-free paper.

To order books or for customer service please, call 1(800)-CALL-WILEY (225-5945).

Library of Congress Cataloging in Publication Data:

Knoll, Glenn F.
    Radiation detection and measurement/Glenn F. Knoll. — 3rd ed.
       p.   cm.
    Includes index.
    ISBN 0-471-07338-5 (cloth: alk. paper)
      1. Nuclear Counters.   2. Radiation—Measurement. I. Title.
QC787.C6K56   1999
539.7'7—dc21
                                                      99-34621
                                                        CIP

Printed in the United States of America

10 9 8 7 6 5

About the Author

GLENN FREDERICK KNOLL is Professor of Nuclear Engineering and Radiological Sciences in the College of Engineering at the University of Michigan. Following his undergraduate education at Case Institute of Technology, he earned a Master's degree from Stanford University and a doctorate in Nuclear Engineering from the University of Michigan. During his graduate work, he held national fellowships from the Atomic Energy Commission and the National Science Foundation.

He joined the Michigan faculty in 1962, and served as Chairman of the Department of Nuclear Engineering from 1979 to 1990 and as Interim Dean of the College of Engineering from 1995-96. He held appointments as Visiting Scientist at the Nuclear Research Center in Karlsruhe, Germany and as Senior Fellow in the Department of Physics at the University of Surrey, U.K. His research interests have centered on radiation measurements, nuclear instrumentation, and radiation imaging. He is author or co-author of over 140 technical publications, 8 patents, and 2 textbooks.

He has been elected a Fellow of the American Institute for Medical and Biological Engineering, the American Nuclear Society, and the Institute of Electrical and Electronics Engineers. He has been selected to receive three national awards given annually to a single recipient for achievements in engineering and education: the 1979 Glenn Murphy Award from the American Society for Engineering Education, the 1991 Arthur Holly Compton Award of the American Nuclear Society, and the 1996 Merit Award of the IEEE/Nuclear and Plasma Sciences Society. He is one of five receiving editors of *Nuclear Instruments and Methods in Physics Research, Part A,* and a past or present member of the Editorial Boards for *Nuclear Science and Engineering, IEEE Transactions on Medical Imaging,* and *Physica Medica.* In 1999, he was elected to membership in the National Academy of Engineering. He has served as consultant to 25 industrial and government organizations in technical areas related to radiation measurements, and is a Registered Professional Engineer in the State of Michigan.

Dedicated in memory of my parents,
Oswald Herman Knoll and Clara Bernthal Knoll

Preface

In the 20 years since the first edition of this book was published, the methods for the detection and measurement of ionizing radiation have undergone significant evolution. Techniques that were unknown several decades ago, such as the cryogenic devices described in Chapter 19, have provided new alternatives as sensors of radiation. In contrast, some common detectors such as the sodium iodide scintillators discussed in Chapters 8 through 10 are quite similar to those that first appeared following their widespread introduction in the 1950s. The changes that are included in this third edition reflect some of the recent additions to the field, and only a few topics have been dropped from the second edition. As a result the book has grown somewhat longer, and it is even more important that instructors using it as a textbook make selective reading assignments.

Some of the more substantial additions are the inclusion of fluence-to-dose data in Ch. 2, the concept of minimum detectable activity in Ch. 3, tissue equivalent proportional counters and microstrip gas chambers in Ch. 6, the "time-to-first-count" technique in Ch. 7, new scintillation materials and fiber scintillators in Ch. 8, position-sensing and hybrid photomultiplier tubes in Ch. 9, use of silicon diodes for X-ray spectroscopy and dose measurement in Ch. 11, new sections on pixel detectors and charge coupled devices, together with an expanded discussion of room-temperature semiconductor materials in Ch. 13, the capture-gated neutron spectrometer in Ch. 15, a new section on digital pulse processing in Ch. 17, discussion of high-pressure xenon ion chambers, recently-developed cryogenic spectrometers, optically-stimulated luminescence dosimeters, and superheated drop detectors in Ch. 19, and new appendices on Poisson statistics and the Shockley-Ramo theorem for induced charge calculations. There are also many smaller changes in each chapter to update individual topics. The references to current literature have been supplemented with more recent publications, and nearly a hundred figures have been added or revised. A small number of new problems appear at the end of chapters, largely limited to material that was added following the second edition.

I am grateful for the assistance provided in the preparation of the manuscript by many individuals. Significant technical contributions and suggestions were provided by the reviewers of the manuscript and their students, John Valentine of Georgia Tech and Walter Gilboy of the University of Surrey. Faculty colleagues David Wehe, Zhong He, and Doug McGregor also were of great assistance in providing valuable technical input. Graduate students Jim LeBlanc, Eric Smith, and Yanfeng Du produced graphical and tabular data that appear in several of the chapters. Valentin Jordanov, Ron Keyser, and Richard Helmer helped in supplying descriptions and/or figures that added greatly to the timeliness of the coverage. Pat Moore, Beth Seibert, and Peter Skrabis from the University of Michigan, together with Cecelia Musselman and Susannah Green of Argosy, were very helpful in the manuscript preparation. My wife Gladys provided countless hours of reading and checking the proofs without complaint. To all, you have my sincere thanks.

This is also an opportunity to acknowledge some of those individuals who have had a strong personal influence early in my career on understanding of the broad topic covered in this text. They include John King, William Parkinson, Jacob Trombka, Joachim Janecke, Geza Gyorey, Karl Beckurts, Walter Gilboy, and Les Rogers. Others whose published work has been instrumental to me in clarifying and extending the state-of-the-art in radiation measurements include W. J. Price, P. R. Bell, Kai Siegbahn, J. B. Birks, Emilio Gatti,

D. H. Wilkinson, Velko Redeka, Fred Goulding, John Walter, Ed Fairstein, Paul Siffert, J. W. Müller, Josef Kemmer, Marek Moszynski, Steve Derenzo, Hobie Kraner, Fabio Sauli, and Amos Breskin. It is also axiomatic at universities that one usually learns more from graduate students than vice versa. I would like to acknowledge the following former Ph.D. students whose research and intellectual exchanges were particularly relevant to topics in this book: Bernie Snyder, Ken Wanio, Don Oliver, Marty Dresser, Dave Gilliam, Mike Flynn, Bill Stephany, Larry Emmons, Mark Davis, Hadi Bozorgmanesh, John Engdahl, Jay Williams, Warren Snapp, Dan Grady, George Baldwin, Merzhad Mahdavi, Yung Lai, Ken Zasadny, Eiping Quang, Doug McGregor, Valentin Jordanov, Jeff Martin, Nestor Tsirliganis, Manos Christodoulou, Jun Miyamoto, and David Stuenkel.

Glenn F. Knoll

Ann Arbor, Michigan
October 1999

Preface *To The Second Edition*

$\mathbf{I}$n this second edition, I have maintained the dual objective of the original: to serve as a textbook for those new to the field and to provide sufficient substance so that the book may also be helpful to those actively involved in radiation measurements. In the decade since the first edition was published, there has been significant development in most areas of the subject matter. As a result, those familiar with the original will notice modifications, additions, and updates throughout this version. Inevitably, there have been more additions than deletions, reflecting the growing breadth of topics important in radiation measurements. Therefore, it is even more important that the instructor using this book as a text provide guidance to the students on which sections are most essential for the purposes of the course. The illustrative problems at the ends of the chapters have been doubled, and the companion solutions manual should be helpful in the tutorial use of the book.

The organization by chapters has remained unchanged, with one exception. The sequencing of the chapters on statistics and detector general properties (Chapters 3 and 4) is reversed from the first edition. This change was made to facilitate the use of statistical concepts in the discussions throughout Chapter 4.

The most extensive changes will be found in Chapter 12, reflecting the replacement of Ge(Li) detectors by the newer HPGe type. Developments in scintillation materials and in the use of photodiodes to convert the scintillation light have led to significant revisions in Chapters 8 and 9. Elsewhere in this edition, pulse-type ion chambers, long in dormancy, are now described more fully because of their enhanced importance in heavy ion measurements. A section has been added on the self-quenched streamer mode of operation in gas detectors. The description of silicon detectors now emphasizes the fully depleted configurations, and the discussions of CdTe and HgI_2 detectors have been expanded. Activation counters used in pulsed fast neutron measurements are now described, and a section has been added on developmental cryogenic and superconducting detectors.

In the areas that are related to pulse processing, full derivations are now provided for the pulse shape from coaxial germanium detectors and from proportional counters. Reflecting a growing interest in high counting rate applications, discussions of reset preamplifiers, gated integrators, and fast pulse-processing methods are now included. A section on the pile-up contribution to recorded spectra also is new. Because of the increasing interest in position-sensitive detectors in many applications, more attention has been given to techniques for determining the location of the radiation interaction in the major detector types. Finally, many small updates and refinements were made throughout the book.

References to the literature have been updated where needed to keep up with current practice. The citations are intended to lead to more detailed descriptions than are possible in this text and to provide starting points for literature searches. No attempt has been made to compile a comprehensive bibliography.

I acknowledge a number of individuals who provided significant assistance in the preparation of the second edition. Don Miller of Ohio State University contributed substantially to an update of the discussion of reactor instrumentation in Chapter 14, and also to Chapter 4. Dennis Persyk of Siemens Gammasonics and Mario Martini of EG&G ORTEC provided critical reviews of early drafts of Chapters 9 and 12, respectively, and made many suggestions that have now been incorporated. The Japanese translators of the first edition, Itsuru Kimura of Kyoto University and the late Eiji Sakai of JAERI, corrected a number of errors and provided valuable points of clarification throughout the text. Detailed reviews of the manuscript by Stephen Binney of Oregon State University, Gary Catchen of Penn State, and Bradley Micklich of Illinois caught many errors and made helpful suggestions for improvements. Faculty colleague David Wehe, departmental students Tim DeVol, Alison Stolle, Yuji Fujii and Richard Kruger, and my son Tom Knoll were also extremely helpful during stages of the manuscript preparation. Finally, many authors have been kind enough to provide original art for figures that have been taken from previously published articles. To all, I express sincere thanks.

The loving support of my wife Gladys throughout this endeavor has been essential and greatly appreciated.

Glenn F. Knoll

Ann Arbor, Michigan
July 1988

Preface *To The First Edition*

This book serves two purposes. First, as a textbook, it is appropriate for use in a first course in nuclear instrumentation or radiation measurements. Such courses are taught at levels ranging from the junior undergraduate year through the first year of graduate programs and are an important part of most curricula in nuclear engineering or radiation physics. Students in health physics, radiation biology, and nuclear chemistry often will also include a similar course in their program. Substantially more material is included, however, than can possibly be covered in the usual one-term course. I have intentionally done so in order that the book remain useful to the student after completion of the course, and so that it may also serve its second purpose as a general review of radiation detection techniques for scientists and engineers actively involved in radiation measurements. The instructor using this book as a text will therefore need to select only those portions deemed most relevant to the purposes of the course. I feel that this inconvenience is offset by the larger scope and more lasting value of the book. Problems intended as student exercises are provided at the end of most chapters.

The level of the discussions assumes an elementary background in radioactivity, radiation properties, and basic electrical circuits. Some topics from these categories are reviewed in the first two chapters, but only in the limited context of laboratory radiation sources and the more important interaction mechanisms. Readers who would like information beyond the scope of the text are referred to the current scientific literature that is cited at the end of each chapter. These references have largely been limited to fairly recent publications, and I apologize in advance to my colleagues whose important but older work may not be referenced.

The important detection techniques for ionizing radiations with energies below about 20 MeV are covered in various chapters of the book. These are the radiations of primary interest in fission and fusion energy systems, as well as in medical, environmental, and industrial applications of radioisotopes. I have concentrated on the basic detector configurations most frequently encountered by the typical user and excluded more complex or specialized detection systems that may be found in many research laboratories. Also not included are the instruments such as bubble chambers, spark chambers, and calorimeters of principal use in high-energy particle physics research. The sections on electronic components and pulse processing aspects of detector signals are based on functional descriptions rather than detailed circuit analyses. This approach reflects the usual interests of the user rather than those of the circuit designer.

Although illustrative applications are included, the discussions emphasize the principles of operation and basic characteristics of the various detector systems. Other publications are available to the reader who seeks more detailed and complete description of specific applications of these instruments. A good example is NCRP Report No. 58, "A Handbook of Radioactivity Measurement Procedures," published by the National Council on Radiation Protection and Measurements, Washington, DC, in 1978 [a second edition of this report published in 1985 is now available].

The SI system of units is used throughout the text. Many traditional radiation units familiar to experienced users are destined to be phased out, so that those not familiar with the newer SI units of activity, gamma-ray exposure, absorbed dose, and dose equivalent should review the definitions as introduced in Chapters 1 and 2.

I have been primarily responsible for teaching nuclear instrumentation courses at the University of Michigan since 1962. Parts of the manuscript have evolved from lecture notes developed over this time, and a preliminary version of the text has been in use for several semesters. This book therefore reflects considerable student feedback that has been essential in improving the clarity of presentation in many areas. I particularly thank present and former graduate students George Baldwin, Hadi Bozorgmanesh, John Engdahl, Dan Grady, Bill Halsey, Bill Martin, Warren Snapp, and Jay Williams, who provided valuable input in many forms. I also thank those faculty colleagues who reviewed portions of the manuscript and offered many helpful comments and suggestions: David Bach, Chihiro Kikuchi, John Lee, Craig Robertson, and Dieter Vincent, and also Lou Costrell and Ron Fleming of the National Bureau of Standards. Pam Hale carried out the formidable task of typing the manuscript with great skill and patience.

I owe a special debt to Jim Duderstadt who provided the initial encouragement that transformed good intentions into a definite commitment toward this project. The steadfast support, understanding, and help of my wife Gladys and sons Tom, John, and Peter throughout the several years required for its completion have been an essential contribution for which I will always be grateful.

Glenn F. Knoll

Ann Arbor, Michigan
September 1978

Credits

Many figures and tables in this text have been reproduced from previously published and copyrighted sources. The cooperation of the publishers in granting permission for the use of this material has been a major contribution.

Except for those sources directly acknowledged in captions and footnotes, most are identified by citing a reference in a scientific journal where the table or figure first appeared. A compilation of these citations, arranged by publication, is given below.

Nuclear Instruments and Methods in Physics Research
Copyright © by Elsevier Science, Amsterdam. Figures 1.6, 1.12, 2.3, 2.4, 2.7, 2.9, 2.12, 2.14, 2.15, 2.16, 2.17, 5.4, 5.13, 6.10, 6.11, 6.12, 6.14, 6.15, 6.18, 6.19, 6.23, 6.25, 8.3, 8.4, 8.11, 8.12, 8.13, 8.15, 8.18, 8.19, 8.21, 8.23, 9.6, 9.15, 9.16, 9.21, 9.22, 9.24, 10.14, 10.15, 10.16, 10.17, 10.20, 10.26, 10.27, 10.30, 10.31, 11.14, 11.16, 11.18, 12.4, 12.5, 12.8, 12.10, 12.12, 12.13, 12.17, 12.22, 12.23, 12.25, 12.26, 12.28, 12.31, 12.33, 12.34, 13.2, 13.3, 13.4, 13.5, 13.6, 13.7, 13.9, 13.10, 13.11, 13.12, 13.13, 13.14, 13.15, 13.18, 13.19, 13.20, 13.23, 13.26, 13.27, 13.28, 13.29, 13.31, 13.32, 13.33, 13.34, 13.35, 13.37, 13.39, 15.5, 15.7, 15.8, 15.18, 15.19, 15.21, 15.25, 16.14, 17.19, 17.28, 17.35, 18.14, 19.1, 19.2, 19.3, 19.7, 19.25, 20.1, 20.2, 20.3, 20.4, 20.5, 20.6, 20.7, 20.9, 20.10, C.2. Tables 1.6, 6.1, 6.2, 8.2, 10.1, 11.2, 12.3, 12.4, 13.1, 13.2, 13.3, 15.1, 19.1, 19.4, 20.1, 20.2.

IEEE Transactions on Nuclear Science
Copyright © by The Institute of Electrical and Electronics Engineers, Inc., New York. Figures 6.5, 8.8, 8.10, 9.3, 9.4, 9.11, 9.12, 9.23, 10.18, 10.21, 11.2, 11.10, 11.19, 11.20, 12.7, 12.9, 12.14, 12.16, 12.18, 12.27, 12.29, 13.8, 13.17, 13.22, 13.25, 13.30, 13.36, 16.13, 17.20, 17.24, 17.25, 17.26, 17.29, 19.4, 19.5. Table 13.3.

International Journal of Applied Radiation and Isotopes
Copyright © by Pergamon Press, Elsevier Science, Amsterdam. Figures 1.10, 1.13, 1.14*a*, 1.14*b*, 1.14*c*. Table 1.7.

Health Physics
Copyright © by Pergamon Press, Elsevier Science, Amsterdam. Figures 1.7, 1.14*d*, 5.12, 7.11.

Annals of the ICRP
Copyright © 1991 by Pergamon Press, Elsevier Science, Amsterdam. Tables 2.2 and 2.3.

Nuclear Science and Engineering
Copyright © by the American Nuclear Society, LaGrange Park, Illinois. Figures 1.11, 14.7.

American National Standard ANSI/ANS-6.1.1-1991
Copyright © 1992 by the American Nuclear Society, La Grange Park, Illinois. Figures 2.22*a* and 2.22*b*.

Transactions of the American Nuclear Society
Copyright © by the American Nuclear Society, LaGrange Park, Illinois. Figure 15.2.

Review of Scientific Instruments
Copyright © by the American Institute of Physics, New York. Figures 5.2, 8.5.

Physical Review
Copyright © by the American Physical Society, New York. Figures 1.4*a*, 1.4*b*.

Reviews of Modern Physics
Copyright © by the American Physical Society, New York. Figure 2.2.

Journal of Applied Physics
Copyright © by the American Institute of Physics, New York. Figure 13.21.

Nuclear Data Tables
Copyright © by Academic Press, Inc., Orlando. Figure 2.11.

Atomic and Nuclear Data Tables
Copyright © by Academic Press, Inc., Orlando. Table 1.3.

In addition, a number of figures and tables were obtained from publications of various government laboratories or agencies. These include Figures 1.3, 1.9, 6.6, 6.20, 10.5, 10.11, 10.19, 10.28, 11.15, 12.20, 12.21, 12.24, 14.6, 14.13, 14.15, 14.17, 15.3, 15.4, A.1, A.2. Tables 1.4, 1.5, 12.1, and 12.2.

Contents

Chapter 1

Radiation Sources

The radiations of primary concern in this text originate in atomic or nuclear processes. They are conveniently categorized into four general types as follows:

Charged particulate radiation   { Fast electrons
                                                { Heavy charged particles

Uncharged radiation   { Electromagnetic radiation
                                                { Neutrons

Fast electrons include beta particles (positive or negative) emitted in nuclear decay, as well as energetic electrons produced by any other process. *Heavy charged particles* denote a category that encompasses all energetic ions with mass of one atomic mass unit or greater, such as alpha particles, protons, fission products, or the products of many nuclear reactions. The *electromagnetic radiation* of interest includes X-rays emitted in the rearrangement of electron shells of atoms, and gamma rays that originate from transitions within the nucleus itself. *Neutrons* generated in various nuclear processes constitute the final major category, which is often further divided into *slow neutron* and *fast neutron* subcategories (see Chapter 14).

The energy range of interest spans over six decades, ranging from about 10 eV to 20 MeV. (Slow neutrons are technically an exception but are included because of their technological importance.) The lower energy bound is set by the minimum energy required to produce ionization in typical materials by the radiation or the secondary products of its interaction. Radiations with energy greater than this minimum are classified as *ionizing radiations*. The upper bound is chosen to limit the topics in this coverage to those of primary concern in nuclear science and technology.

The main emphasis in this chapter will be the laboratory-scale sources of these radiations, which are likely to be of interest either in the calibration and testing of radiation detectors described in the following chapters, or as objects of the measurements themselves. Natural background radiation is an important additional source and is discussed separately in Chapter 20.

The radiations of interest differ in their "hardness" or ability to penetrate thicknesses of material. Although this property is discussed in greater detail in Chapter 2, it is also of considerable concern in determining the physical form of radiation sources. Soft radiations, such as alpha particles or low-energy X-rays, penetrate only small thicknesses of material. Radioisotope sources must therefore be deposited in very thin layers if a large fraction of these radiations is to escape from the source itself. Sources that are physically thicker are subject to "self-absorption," which is likely to affect both the number and the energy spectrum of the radiations that emerge from its surface. Typical thicknesses for such sources are therefore measured in micrometers. Beta particles are generally more penetrating, and sources up to a few tenths of a millimeter in thickness can usually be tolerated. Harder radiations, such as gamma rays or neutrons, are much less affected by self-absorption and sources can be millimeters or centimeters in dimension without seriously affecting the radiation properties.

I. UNITS AND DEFINITIONS

A. Radioactivity

The *activity* of a radioisotope source is defined as its rate of decay and is given by the fundamental law of radioactive decay

$$\frac{dN}{dt}\bigg|_{\text{decay}} = -\lambda N \tag{1.1}$$

where N is the number of radioactive nuclei and λ is defined as the *decay constant*.[†] The historical unit of activity has been the *curie* (Ci), defined as exactly 3.7×10^{10} disintegrations/second, which owes its definition to its origin as the best available estimate of the activity of 1 gram of pure ^{226}Ra. Its submultiples, the millicurie (mCi) or microcurie (μCi), generally are more suitable units for laboratory-scale radioisotope sources.

Although still widely used in the literature, the curie is destined to be replaced gradually by its SI equivalent, the becquerel (Bq). At its 1975 meeting, the General Conference of Weights and Measures (GCPM) adopted a resolution declaring that the becquerel, defined as one disintegration per second, has become the standard unit of activity. Thus

$$1 \text{ Bq} = 2.703 \times 10^{-11} \text{ Ci}$$

Radioactive sources of convenient size in the laboratory are most reasonably measured in kilobecquerels (kBq) or megabecquerels (MBq).

It should be emphasized that activity measures the source disintegration rate, which is not synonymous with the emission rate of radiation produced in its decay. Frequently, a given radiation will be emitted in only a fraction of all the decays, so a knowledge of the decay scheme of the particular isotope is necessary to infer a radiation emission rate from its activity. Also, the decay of a given radioisotope may lead to a daughter product whose activity also contributes to the radiation yield from the source. A complete listing of radioisotope decay schemes is tabulated in Ref. 1.

The *specific activity* of a radioactive source is defined as the activity per unit mass of the radioisotope sample. If a pure or "carrier-free" sample is obtained that is unmixed with any other nuclear species, its specific activity can be calculated from

$$\text{specific activity} \equiv \frac{\text{activity}}{\text{mass}} = \frac{\lambda N}{NM/A_v} = \frac{\lambda A_v}{M} \tag{1.2}$$

where

M = molecular weight of sample

A_v = Avogadro's number ($= 6.02 \times 10^{23}$ nuclei/mole)

λ = radioisotope decay constant ($= \ln 2/\text{half-life}$)

[†]One should be aware that Eq.(1.1) represents the decay rate only, and the *net* value of dN/dt may be altered by other production or disappearance mechanisms. As one example, the radioisotope may be produced as the daughter product of the decay of a parent species also present in the sample. Then a production term is present for the daughter that is given by the decay rate of the parent multiplied by the fraction of such decays that lead to the daughter species. If the half-life of the parent is very long, the number of daughter nuclei increases until the daughter activity reaches an equilibrium value (after many daughter half-lives have passed) when the production and decay rates are equal, and $dN/dt = 0$ for the number of daughter nuclei.

Radioisotopes are seldom obtained in carrier-free form, however, and are usually diluted in a much larger concentration of stable nuclei of the same element. Also, if not prepared in pure elemental form, additional stable nuclei may be included from other elements that are chemically combined with those of the source. For sources in which self-absorption is a problem, there is a premium on obtaining a sample with high specific activity to maximize the number of radioactive nuclei within a given thickness. From Eq. (1.2), high specific activity is most readily obtained using radionuclides with large λ (or small half-life).

B. Energy

The traditional unit for measurement of radiation energy is the *electron volt* or eV, defined as the kinetic energy gained by an electron by its acceleration through a potential difference of 1 volt. The multiples of kiloelectron volt (keV) and megaelectron volt (MeV) are more common in the measurement of energies for ionizing radiation. The electron volt is a convenient unit when dealing with particulate radiation because the energy gained from an electric field can easily be obtained by multiplying the potential difference by the number of electronic charges carried by the particle. For example, an alpha particle that carries an electron charge of +2 will gain an energy of 2 keV when accelerated by a potential difference of 1000 volts.

The SI unit of energy is the *joule* (J). When dealing with radiation energies, the submultiple femtojoule (fJ) is more convenient and is related to the electron volt by the conversion

$$1 \text{ eV} = 1.602 \times 10^{-19} \text{ J}$$

or
$$1 \text{ fJ} (= 10^{-15} \text{ J}) = 6.241 \times 10^3 \text{ eV}$$

It is not clear to what extent the electron volt will be phased out in future usage because its physical basis and universal use in the literature are strong arguments for its continued application to radiation measurements.

The energy of an X- or gamma-ray photon is related to the radiation frequency by

$$E = h\nu \tag{1.3}$$

where
$$h = \text{Planck's constant } (6.626 \times 10^{-34} \text{ J} \cdot \text{s, or } 4.135 \times 10^{-15} \text{ eV} \cdot \text{s})$$
$$\nu = \text{frequency}$$

The wavelength λ is related to the photon energy by

$$\lambda = \frac{1.240 \times 10^{-6}}{E}$$

where λ is in meters and E in eV.

II. FAST ELECTRON SOURCES

A. Beta Decay

The most common source of fast electrons in radiation measurements is a radioisotope that decays by beta-minus emission. The process is written schematically

$$^A_Z\text{X} \rightarrow \, ^A_{Z+1}\text{Y} + \beta^- + \bar{\nu} \tag{1.4}$$

where X and Y are the initial and final nuclear species, and $\bar{\nu}$ is the antineutrino. Because neutrinos and antineutrinos have an extremely small interaction probability with matter, they are undetectable for all practical purposes. The recoil nucleus Y appears with a very

Table 1.1 Some "Pure" Beta-Minus Sources

Nuclide	Half-Life	Endpoint Energy (MeV)
^{3}H	12.26 y	0.0186
^{14}C	5730 y	0.156
^{32}P	14.28 d	1.710
^{33}P	24.4 d	0.248
^{35}S	87.9 d	0.167
^{36}Cl	3.08×10^5 y	0.714
^{45}Ca	165 d	0.252
^{63}Ni	92 y	0.067
^{90}Sr/^{90}Y	27.7 y/64 h	0.546/2.27
^{99}Tc	2.12×10^5 y	0.292
^{147}Pm	2.62 y	0.224
^{204}Tl	3.81 y	0.766

Data from Lederer and Shirley.[1]

small recoil energy, which is ordinarily below the ionization threshold, and therefore it cannot be detected by conventional means. Thus, the only significant ionizing radiation produced by beta decay is the fast electron or beta particle itself.

Because most radionuclides produced by neutron bombardment of stable materials are beta-active, a large assortment of beta emitters are readily available through production in a reactor flux. Species with many different half-lives can be obtained, ranging from thousands of years down to as short a half-life as is practical in the application. Most beta decays populate an excited state of the product nucleus, so that the subsequent de-excitation gamma rays are emitted together with beta particles in many common beta sources. Some examples of nuclides that decay directly to the ground state of the product and are therefore "pure beta emitters" are shown in Table 1.1.

Each specific beta decay transition is characterized by a fixed decay energy or Q-value. Because the energy of the recoil nucleus is virtually zero, this energy is shared between the beta particle and the "invisible" neutrino. The beta particle thus appears with an energy that varies from decay to decay and can range from zero to the "beta endpoint energy," which is numerically equal to the Q-value. A representative beta energy spectrum is illustrated in Fig. 1.1. The Q-value for a given decay is normally quoted assuming that the

Figure 1.1 The decay scheme of ^{36}Cl and the resulting beta particle energy distribution.

transition takes place between the ground states of both the parent and daughter nuclei. If the transition involves an excited state of either the parent or daughter, the endpoint energy of the corresponding beta spectrum will be changed by the difference in excitation energies. Since several excited states can be populated in some decay schemes, the measured beta particle spectrum may then consist of several components with different endpoint energies.

B. Internal Conversion

The continuum of energies produced by any beta source is inappropriate for some applications. For example, if an energy calibration is to be carried out for an electron detector, it is much more convenient to use a source of monoenergetic electrons. The nuclear process of *internal conversion* can be the source of *conversion electrons*, which are, under some circumstances, nearly monoenergetic.

The internal conversion process begins with an excited nuclear state, which may be formed by a preceding process—often beta decay of a parent species. The common method of de-excitation is through emission of a gamma ray-photon. For some excited states, gamma emission may be somewhat inhibited and the alternative of internal conversion can become significant. Here the nuclear excitation energy E_{ex} is transferred directly to one of the orbital electrons of the atom. This electron then appears with an energy given by

$$E_{e^-} = E_{ex} - E_b \qquad (1.5)$$

where E_b is its binding energy in the original electron shell.

An example of a conversion electron spectrum is shown in Fig. 1.2. Because the conversion electron can originate from any one of a number of different electron shells within the atom, a single nuclear excitation level generally leads to several groups of electrons with different energies. The spectrum may be further complicated in those cases in which more than one excited state within the nucleus is converted. Furthermore, the electron energy spectrum may also be superimposed on a continuum consisting of the beta spectrum of the parent nucleus that leads to the excited state. Despite these shortcomings, conversion electrons are the only practical laboratory-scale source of monoenergetic electron groups in the high keV to MeV energy range. Several useful radioisotope sources of conversion electrons are compiled in Table 1.2.

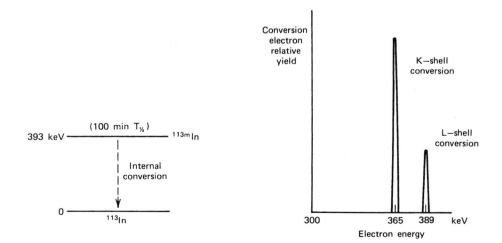

Figure 1.2 The conversion electron spectrum expected from internal conversion of the isomeric level at 393 keV in ^{113m}In.

Table 1.2 Some Common Conversion Electron Sources

Parent Nuclide	Parent Half-Life	Decay Mode	Decay Product	Transition Energy of Decay Product (keV)	Conversion Electron Energy (keV)
^{109}Cd	453 d	EC	^{109m}Ag	88	62 84
^{113}Sn	115 d	EC	^{113m}In	393	365 389
^{137}Cs	30.2 y	β^-	^{137m}Ba	662	624 656
^{139}Ce	137 d	EC	^{139m}La	166	126 159
^{207}Bi	38 y	EC	^{207m}Pb	570 1064	482 554 976 1048

Data from Lederer and Shirley.[1]

C. Auger Electrons

Auger electrons are roughly the analogue of internal conversion electrons when the excitation energy originates in the atom rather than in the nucleus. A preceding process (such as electron capture) may leave the atom with a vacancy in a normally complete electron shell. This vacancy is often filled by an electron from one of the outer shells of the atom with the emission of a characteristic X-ray photon. Alternatively, the excitation energy of the atom may be transferred directly to one of the outer electrons, causing it to be ejected from the atom. This electron is called an Auger electron and appears with an energy given by the difference between the original atomic excitation energy and the binding energy of the shell from which the electron was ejected. Auger electrons therefore produce a discrete energy spectrum, with different groups corresponding to different initial and final states. In all cases, their energy is relatively low compared with beta particles or conversion electrons, particularly because Auger electron emission is favored only in low-Z elements for which electron binding energies are small. Typical Auger electrons with a few keV initial energy are subject to pronounced self-absorption within the source and are easily stopped by very thin source covers or detector entrance windows.

III. HEAVY CHARGED PARTICLE SOURCES

A. Alpha Decay

Heavy nuclei are energetically unstable against the spontaneous emission of an alpha particle (or ^{4}He nucleus). The probability of decay is governed by the *barrier penetration* mechanism described in most texts on nuclear physics, and the half-life of useful sources varies from days to many thousands of years. The decay process is written schematically as

$$^A_Z X \rightarrow {}^{A-4}_{Z-2} Y + {}^4_2 \alpha$$

where X and Y are the initial and final nuclear species. A representative alpha decay scheme is shown in Fig. 1.3, together with the expected energy spectrum of the corresponding alpha particles emitted in the decay.

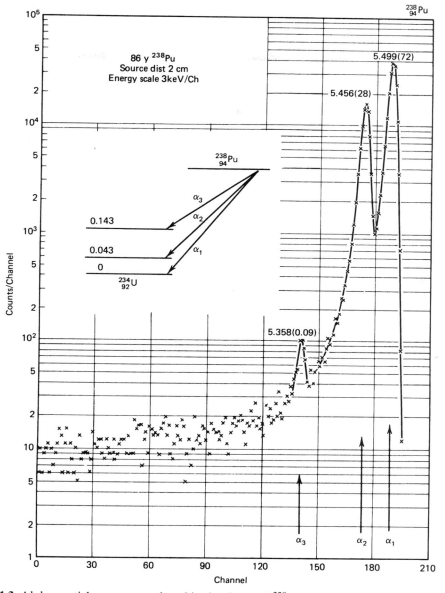

Figure 1.3 Alpha particle groups produced in the decay of ^{238}Pu. The pulse height spectrum shows the three groups as measured by a silicon surface barrier detector. Each peak is identified by its energy in MeV and percent abundance (in parentheses). The insert shows the decay scheme, with energy levels in the product nucleus labeled in MeV. (Spectrum from Chanda and Deal.[2])

The alpha particles appear in one or more energy groups that are, for all practical purposes, monoenergetic. For each distinct transition between initial and final nucleus (e.g., between ground state and ground state), a fixed energy difference or Q-value characterizes the decay. This energy is shared between the alpha particle and the recoil nucleus in a unique way, so that each alpha particle appears with the same energy given by $Q(A-4)/A$. There are many practical instances in which only one such transition is involved and for which the alpha particles are therefore emitted with a unique single energy. Other examples, such as that shown in Fig. 1.3, may involve more than one transition energy so that the alpha particles appear in groups with differing relative intensities.

Table 1.3 lists some properties of the more common radioisotope sources of alpha particles. It is no accident that most alpha particle energies are limited to between about 4 and 6 MeV. There is a very strong correlation between alpha particle energy and half-life of the parent isotope, and those with the highest energies are those with the shortest half-life.

Table 1.3 Common Alpha-Emitting Radioisotope Sources

Source	Half-Life	Alpha Particle Kinetic Energy (with Uncertainty) in MeV		Percent Branching
^{148}Gd	93 y	3.182787	±0.000024	100
^{232}Th	1.4×10^{10} y	4.012	±0.005	77
		3.953	±0.008	23
^{238}U	4.5×10^9 y	4.196	±0.004	77
		4.149	±0.005	23
^{235}U	7.1×10^8 y	4.598	±0.002	4.6
		4.401	±0.002	56
		4.374	±0.002	6
		4.365	±0.002	12
		4.219	±0.002	6
^{236}U	2.4×10^7 y	4.494	±0.003	74
		4.445	±0.005	26
^{230}Th	7.7×10^4 y	4.6875	±0.0015	76.3
		4.6210	±0.0015	23.4
^{234}U	2.5×10^5 y	4.7739	±0.0009	72
		4.7220	±0.0009	28
^{231}Pa	3.2×10^4 y	5.0590	±0.0008	11
		5.0297	±0.0008	20
		5.0141	±0.0008	25.4
		4.9517	±0.0008	22.8
^{239}Pu	2.4×10^4 y	5.1554	±0.0007	73.3
		5.1429	±0.0008	15.1
		5.1046	±0.0008	11.5
^{240}Pu	6.5×10^3 y	5.16830	±0.00015	76
		5.12382	±0.00023	24
^{243}Am	7.4×10^3 y	5.2754	±0.0010	87.4
		5.2335	±0.0010	11
^{210}Po	138 d	5.30451	±0.00007	99+
^{241}Am	433 y	5.48574	±0.00012	85.2
		5.44298	±0.00013	12.8
^{238}Pu	88 y	5.49921	±0.00020	71.1
		5.4565	±0.0004	28.7
^{244}Cm	18 y	5.80496	±0.00005	76.4
		5.762835	±0.000030	23.6
^{243}Cm	30 y	6.067	±0.003	1.5
		5.992	±0.002	5.7
		5.7847	±0.0009	73.2
		5.7415	±0.0009	11.5
^{242}Cm	163 d	6.11292	±0.00008	74
		6.06963	±0.00012	26
^{254m}Es	276 d	6.4288	±0.0015	93
^{253}Es	20.5 d	6.63273	±0.00005	90
		6.5916	±0.0002	6.6

Data from Rytz.[3]

Beyond about 6.5 MeV, the half-life can be expected to be less than a few days, and therefore the source is of very limited utility. On the other hand, if the energy drops below 4 MeV, the barrier penetration probability becomes very small and the half-life of the isotope is very large. If the half-life is exceedingly long, the specific activity attainable in a practical sample of the material becomes very small and the source is of no interest because its intensity is too low. Probably the most common calibration source for alpha particles is ^{241}Am, and an example of its application to the calibration of silicon solid-state detectors is shown in Fig. 11.15.

Because alpha particles lose energy rapidly in materials, alpha particle sources that are to be nearly monoenergetic must be prepared in very thin layers. In order to contain the radioactive material, typical sources are covered with a metallic foil or other material that must also be kept very thin if the original energy and monoenergetic nature of the alpha emission are to be preserved.

B. Spontaneous Fission

The fission process is the only spontaneous source of energetic heavy charged particles with mass greater than that of the alpha particle. Fission fragments are therefore widely used in the calibration and testing of detectors intended for general application to heavy ion measurements.

All heavy nuclei are, in principle, unstable against spontaneous fission into two lighter fragments. For all but the extremely heavy nuclei, however, the process is inhibited by the large potential barrier that must be overcome in the distortion of the nucleus from its original near-spherical shape. Spontaneous fission is therefore not a significant process except for some transuranic isotopes of very large mass number. The most widely used example is ^{252}Cf, which undergoes spontaneous fission with a half-life (if it were the only decay process) of 85 years. However, most transuranic elements also undergo alpha decay, and in ^{252}Cf the probability for alpha emission is considerably higher than that for spontaneous fission. Therefore, the actual half-life for this isotope is 2.65 years, and a sample of 1 microgram of ^{252}Cf will emit 1.92×10^7 alpha particles and undergo 6.14×10^5 spontaneous fissions per second.

Each fission gives rise to two fission fragments, which, by the conservation of momentum, are emitted in opposite directions. Because the normal physical form for a spontaneous fission source is a thin deposit on a flat backing, only one fragment per fission can escape from the surface, whereas the other is lost by absorption within the backing. As described later in this chapter, each spontaneous fission in ^{252}Cf also liberates a number of fast neutrons.

The fission fragments are medium-weight positive ions with a mass distribution illustrated in Fig. 1.4a. The fission is predominantly asymmetric so that the fragments are clustered into a "light group" and "heavy group," with average mass numbers of 108 and 143. The fragments appear initially as positive ions for which the net charge approaches the atomic number of the fragment. As the fragment slows down by interacting with the matter through which it passes, additional electrons are picked up by the ion, reducing its effective charge.

The energy shared by the two fragments averages about 185 MeV. The distribution of this energy is also asymmetric with the light fragment receiving the greater fraction. A plot of their initial energy distribution is shown in Fig. 1.4b. Because they also lose energy readily in solid materials, self-absorption and energy loss of the fragments are important considerations unless the source is prepared in a very thin layer. The type of degradation of a fission fragment energy spectrum observed from thicker sources (for the case of neutron-induced ^{235}U fission) is illustrated in Fig. 14.7.

Figure 1.4a The mass distribution of ^{252}Cf spontaneous fission fragments. Also shown is the corresponding distribution from fission of ^{235}U induced by thermal neutrons. (From Nervik.[4])

Figure 1.4b The distribution in kinetic energy of the ^{252}Cf spontaneous fission fragments. The peak on the left corresponds to the heavy fragments, and that on the right to the light fragments. (From Whetstone.[5])

IV. SOURCES OF ELECTROMAGNETIC RADIATION

A. Gamma Rays Following Beta Decay

Gamma radiation is emitted by excited nuclei in their transition to lower-lying nuclear levels. In most practical laboratory sources, the excited nuclear states are created in the decay of a parent radionuclide. Four common examples widely used as gamma-ray calibration sources are illustrated in the decay schemes in Fig. 1.5. In each case, a form of beta decay leads to the population of the excited state in the daughter nucleus. For the examples shown, the beta decay is a relatively slow process characterized by a half-life of hundreds of days or greater, whereas the excited states in the daughter nucleus have a much shorter average lifetime (typically of the order of picoseconds or less). De-excitation takes place through the emission of a gamma-ray photon whose energy is essentially equal to the difference in energy between the initial and final nuclear states. The gamma rays therefore appear with a half-life characteristic of the parent beta decay but with an energy that reflects the energy level structure of the daughter nucleus. For example, although "^{60}Co gamma rays" decrease in intensity with the 5.26-year half-life characteristic of ^{60}Co, they actually arise from transitions in the ^{60}Ni nucleus. Decay schemes of the type shown in Fig. 1.5 are compiled for all radioactive nuclei in Ref. 1. From the probabilities of various de-excitation transitions (or "branching ratios") given in these decay schemes, the number of gamma-ray photons per disintegration of the parent nucleus can be deduced. Some

Figure 1.5 Decay schemes for some common gamma reference sources. Only major transitions are shown. The energies and yields per disintegration of X- and gamma rays emitted in each decay are listed below the diagram. (Data from Lederer and Shirley.[1])

specific radionuclide gamma-ray sources useful in the precise energy calibration and efficiency calibration of gamma-ray detectors are listed in Tables 12.1 and 12.2.

Because nuclear states have very well-defined energies, the energies of gamma rays emitted in state-to-state transitions are also very specific. The gamma rays from any one transition are nearly monoenergetic, and the inherent line width of the photon energy distribution is nearly always small compared with the energy resolution of any of the detectors described later in this text. A measurement of the detector response is therefore indicative of its own limiting resolution rather than any variation in the incident gamma-ray energy.[†]

The common gamma-ray sources based on beta decay are generally limited to energies below about 2.8 MeV. One nuclide that can be useful[6] as a potential source for gamma rays of higher energy is ^{56}Co. The decay scheme of this isotope, which includes both electron capture and β^+ decay, gives rise to a complex spectrum of gamma rays whose energies extend to as high as 3.55 MeV. The short half-life of ^{56}Co (77 days), however, largely limits its use to facilities with access to accelerators necessary to carry out its production through the ^{56}Fe(p,n) reaction. Another possible radioisotope for high-energy calibrations is ^{16}N, with gamma rays of 6.13 and 7.11 MeV emitted through its β^- decay to ^{16}O. Again, the very short half-life of just over 7 seconds requires the local production of this isotope in reactors or accelerators, which can be employed with a continuous circulation loop[7] to provide a steady-state source of this radionuclide.

Gamma-ray reference sources are an essential accessory in any radiation measurements laboratory in which gamma-ray measurements are carried out. They normally consist of samples of radioisotopes of a few microcuries (around 10^5 Bq) encased in plastic disks or rods. The thickness of the encapsulation is generally large enough to stop any particulate radiation from the decay of the parent nucleus, and the only primary radiation emerging from the surface is the gamma radiation produced in the daughter decay. However, secondary radiations such as annihilation photons or bremsstrahlung can be significant at times (see below). Although the radiation hazard of such sources is minimal, the gamma-ray emission rate is sufficiently high to permit ready energy calibration of most types of gamma-ray detectors.

If the sources are to be used to carry out accurate efficiency calibration, their absolute activity must also be known. In these cases, the radioisotope deposits are generally prepared on much thinner backings with a minimum of overlying cover to reduce gamma-ray attenuation and scattering within the source structure. External absorbers must then be used to eliminate any particulate emission if its presence will interfere with the application.

B. Annihilation Radiation

When the parent nucleus undergoes β^+ decay, additional electromagnetic radiation is generated. The origin lies in the fate of the positrons emitted in the primary decay process. Because they generally travel only a few millimeters before losing their kinetic energy (see Chapter 2), the inherent encapsulation around the source is often sufficiently thick to fully stop the positrons. When their energy is very low, near the end of their range, they combine with normal negative electrons in the absorbing materials in the process of *annihilation*.[††] The original positron and electron disappear and are replaced by two oppositely directed 0.511 MeV electromagnetic photons known as *annihilation radiation*. This radiation is then superimposed on whatever gamma radiation may be emitted in the subsequent decay of

---

[†]A rare exception may occur if the emitting nuclei have large velocities. The Doppler effect can then introduce an energy spread that may be significant in detectors with excellent energy resolution. An example is given later in this chapter in which gamma rays are emitted from nuclei that are still moving after being formed in a nuclear reaction.

[††]This step can take place either directly or through an intermediate stage in which the positron and electron form a quasistable combination, known as positronium, which may exist for a few nanoseconds.

the daughter product. For example, in the decay of ^{22}Na, shown in Fig. 1.5, photons of both 0.511 and 1.274 MeV energy are emitted from encapsulated sources.

C. Gamma Rays Following Nuclear Reactions

If gamma rays with energies higher than those available from beta-active isotopes are needed, some other process must lead to the population of higher-lying nuclear states. One possibility is the nuclear reaction

$$\,^4_2\alpha + \,^9_4\text{Be} \rightarrow \,^{12}_{6}\text{C}^* + \,^1_0\text{n}$$

where the product nucleus ^{12}C is left in an excited state. Its decay gives rise to a gamma-ray photon of 4.44 MeV energy. Unfortunately, the average lifetime of this state is so short (61 fs) that the recoil carbon atom does not have time to come to rest before the gamma ray is emitted. The resulting photon energies are therefore broadened by Doppler effects, depending on the relative orientation of the recoil atom velocity and the photon direction, and there is an inherent spread of about 1% in the gamma-ray energies. This line width is adequate for many calibration purposes, but it is too large for detectors with very good energy resolution (such as the germanium detectors of Chapter 12). Neutrons are also emitted in this reaction, and the number of 4.44 MeV gamma rays per neutron has been measured[8] to be about 0.59 for typical sources based on radioisotope alpha particle emitters mixed with beryllium.

Another possibility is the reaction

$$\,^4_2\alpha + \,^{13}_6\text{C} \rightarrow \,^{16}_8\text{O}^* + \,^1_0\text{n}$$

Here the product nucleus ^{16}O can be formed in an excited state at 6.130 MeV above the ground state and with a lifetime of about 2×10^{-11} s. This lifetime is sufficiently long to eliminate almost all Doppler effects, and the resulting 6.130 MeV gamma-ray photons are essentially monoenergetic.

Both the above reactions can be exploited by combining a radioisotope that decays by alpha emission with the appropriate target material (either ^{9}Be or ^{13}C). Because sources of this type are more commonly used to produce neutrons, further discussion of the choice of alpha emitter and other aspects of the source design will be postponed until the following section on neutron sources. Because most alpha particles do not lead to a reaction before losing their energy in the target material, large activities of the alpha emitter must be used to produce a gamma-ray source of practical intensity. For example, a typical source[9] fabricated from 6×10^9 Bq of ^{238}PuO$_2$ and 200 mg of isotopically separated ^{13}C produces a 6.130 MeV gamma-ray yield of 770 photons/s.

Gamma rays are also commonly emitted following the absorption of thermal neutrons by typical nuclei. Sources of thermal neutrons used in the production of gamma rays can be intense beams from nuclear reactors[10] or accelerator facilities, or weaker fluxes from moderated radioisotope sources of neutrons.[11,12] These "neutron-capture gamma rays" typically have energy ranging as high as 9 MeV, and data on the probabilities of emission per neutron capture that can be useful for detector calibrations are available from various sources (e.g., Ref. 13).

D. Bremsstrahlung

When fast electrons interact in matter, part of their energy is converted into electromagnetic radiation in the form of *bremsstrahlung*. (This process is discussed in somewhat more detail in Chapter 2.) The fraction of the electron energy converted into bremsstrahlung increases with increasing electron energy and is largest for absorbing materials of high atomic number. The process is important in the production of X-rays from conventional X-ray tubes.

For monoenergetic electrons that slow down and stop in a given material, the bremsstrahlung energy spectrum is a continuum with photon energies that extend as high as the electron energy itself. The shape of a typical spectrum produced by monoenergetic electrons is shown in Fig. 1.6. The emission of low-energy photons predominates, and the average photon energy is a small fraction of the incident electron energy. Because these spectra are continua, they cannot be applied directly to the energy calibration of radiation detectors.

The shape of the energy spectrum from an X-ray tube can be beneficially altered by *filtration* or passage through appropriate absorber materials. Through the use of absorbers that preferentially remove the lower-energy photons, a peaked spectrum can be produced that, although far from monoenergetic, can be useful in the energy calibration of detectors whose response changes only gradually with energy. Some examples of filtered spectra from X-ray tubes are plotted in Fig. 1.7. At lower energies, the abrupt change in filter transmission at its *K*-absorption edge (see Chapter 2) can produce a prominent peak at the corresponding energy in the filtered spectrum.[16]

Bremsstrahlung is also produced by other sources of fast electrons, including beta particles. Therefore, some bremsstrahlung photons are generated by any beta-active isotope encapsulated to stop the beta particles. Some examples of bremsstrahlung spectra from specific isotopes are plotted in Fig. 10.5.

In addition to bremsstrahlung, characteristic X-rays (see following section) are also produced when fast electrons pass through an absorber. Therefore, the spectra from X-ray tubes or other bremsstrahlung sources also show characteristic X-ray emission lines superimposed on the continuous bremsstrahlung spectrum.

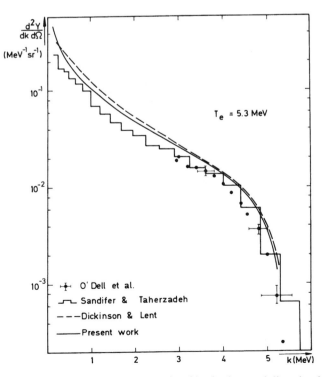

Figure 1.6 The bremsstrahlung energy spectrum emitted in the forward direction by 5.3 MeV electrons incident on a Au-W target. A 7.72 g/cm^2 aluminum filter also was present. (From Ferdinande et al.[14])

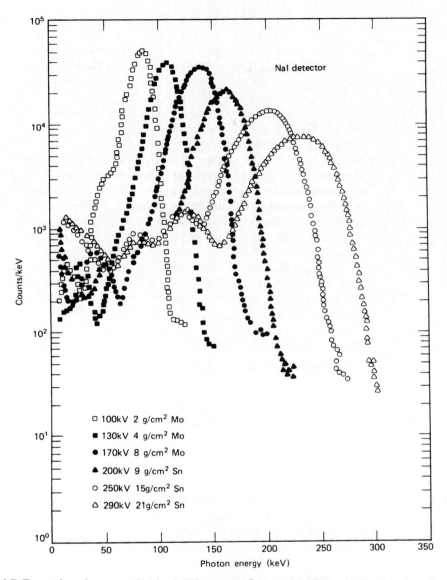

Figure 1.7 Examples of measured pulse height spectra [using a NaI(Tl) scintillator] after filtration of an X-ray tube output using the indicated absorbers and tube voltages. (From Storm et al.[15])

E. Characteristic X-Rays

If the orbital electrons in an atom are disrupted from their normal configuration by some excitation process, the atom may exist in an excited state for a short period of time. There is a natural tendency for the electrons to rearrange themselves to return the atom to its lowest energy or ground state within a time that is characteristically a nanosecond or less in a solid material. The energy liberated in the transition from the excited to the ground state takes the form of a *characteristic X-ray* photon whose energy is given by the energy difference between the initial and final states. For example, if the vacancy is temporarily created in the K shell of an atom, then a characteristic K X-ray is liberated when that vacancy is subsequently filled. If that electron comes from the L shell, then a K_α photon is produced whose energy is equal to the difference in binding energies between the K and L shells. If the filling electron originated in the M shell instead, then a K_β photon is produced with slightly larger energy, and so on. The maximum K-series photon is produced when the vacancy is filled by a free or unbound electron, and the corresponding photon energy is then simply given by the K shell binding energy. Vacancies created in outer shells by the filling of a K shell vacancy are subsequently filled with the emission of L-, M-, . . . series characteristic X-rays.

Because their energy is greatest, the K-series X-rays are generally of most practical significance. Their energy increases regularly with atomic number of the element and is, for example, about 1 keV for sodium with $Z = 11$, 10 keV for gallium with $Z = 31$, and 100 keV for radium with $Z = 88$. The L series X-rays do not reach 1 keV until $Z = 28$ and 10 keV at $Z = 74$. Extensive tables of precise characteristic X-rays can be found in Ref. 1. Because the energy of the characteristic X-ray is unique to each individual element, they are often used in the elemental analysis of unknown samples (see Fig. 13.17).

For an atom in an excited state, the ejection of an Auger electron is a competitive process to the emission of characteristic X-rays. The *fluorescent yield* is defined as the fraction of all cases in which the excited atom emits a characteristic X-ray photon in its de-excitation. Values for the fluorescent yield are often tabulated as part of spectroscopic data.

A large number of different physical processes can lead to the population of excited atomic states from which characteristic X-rays originate. In general, the relative yields of the K, L, and subsequent series will depend on the excitation method, but the energy of the characteristic photons is fixed by the basic atomic binding energies. We list below those excitation mechanisms that are of most practical importance for compact laboratory sources of characteristic X-rays.

1. EXCITATION BY RADIOACTIVE DECAY

In the nuclear decay process of electron capture, the nuclear charge is decreased by one unit by the capture of an orbital electron, most often a K-electron. The resulting atom still has the right number of orbital electrons, but the capture process has created a vacancy in one of the inner shells. When this vacancy is subsequently filled, X-rays are generated that are characteristic of the product element. The decay may populate either the ground state or an excited state in the product nucleus, so that the characteristic X-rays may also be accompanied by gamma rays from subsequent nuclear de-excitation.[†]

Internal conversion is another nuclear process that can lead to characteristic X-rays. As defined earlier in this chapter, internal conversion results in the ejection of an orbital electron from the atom leaving behind a vacancy. Again, it is the K-electrons that are most readily converted, and therefore the K-series characteristic X-rays are the most prominent. Because gamma-ray de-excitation of the nuclear state is always a competing process to internal conversion, radioisotope sources of this type usually emit gamma rays in addition to the characteristic X-rays. The conversion electrons may also lead to a measurable bremsstrahlung continuum, particularly when their energy is high.

In Table 13.2, some examples are given of radioisotopes that involve either electron capture or internal conversion and are possible sources of characteristic X-rays. In all these examples, the yield of high-energy gamma rays from a nuclear transition is large compared with that of the characteristic X-rays. If a pure X-ray source free of gamma-ray contamination is required, a radioisotope that decays by electron capture leading directly to the

---

[†]Electron capture can lead to another form of continuous electromagnetic radiation known as *inner bremsstrahlung*. In the decay process, an orbital electron is captured by the nucleus and therefore must undergo some acceleration. From classical theory, an accelerated charge must emit electromagnetic radiation. Because the acceleration may vary over a wide range depending on the specifics of the capture process, the resulting emission spectrum is a continuum ranging up to a maximum photon energy given by the Q-value of the electron capture decay (the maximum energy available in the nuclear transition). Inner bremsstrahlung therefore adds a continuous electromagnetic spectrum to the other radiations normally expected as a product of the electron capture decay, although in many cases the intensity of this spectrum may be negligibly small.

Table 1.4 Some Radioisotope Sources of Low-Energy X-Rays

Nuclide	Half-Life	Weighted K_α X-Ray Energy	Fluorescent Yield	Other Radiations
^{37}Ar	35.1 d	2.957 keV	0.086	Some IB[a]
^{41}Ca	8×10^4 y	3.690	0.129	Pure
^{44}Ti	48 y	4.508	0.174	γ Rays at 68 and 78 keV
^{49}V	330 d	4.949	0.200	IB
^{55}Fe	2.60 y	5.895	0.282	Weak IB

[a]IB represents inner bremsstrahlung.

Data from Amlauer and Tuohy.[17]

ground nuclear state of the daughter must be chosen. Table 1.4 gives some examples for the lower energy range. Of these, ^{55}Fe is most widely used because of its convenient half-life and high available specific activity. It is very nearly a pure source of manganese K-series X-rays at about 5.9 keV, and the inner bremsstrahlung associated with the electron capture process is very weak.

Self-absorption is a significant technical problem in the preparation of radioisotope X-ray sources. As the thickness of the radioisotope deposit is increased, the X-ray flux emerging from its surface approaches a limiting value because only those atoms near the surface can contribute photons that escape. The number of emitting atoms within a given distance of the deposit surface is maximized by increasing the specific activity of the radioisotope, and carrier-free samples will exhibit the maximum attainable source intensity per unit area.

2. EXCITATION BY EXTERNAL RADIATION

The general scheme sketched in Fig. 1.8 may also be used to generate characteristic X-rays. Here an external source of radiation (X-rays, electrons, alpha particles, etc.) is caused to strike a target, creating excited or ionized atoms in the target through the processes detailed in Chapter 2. Because many excited atoms or ions in the target subsequently de-excite to the ground state through the emission of characteristic X-rays, the target can serve as a localized source of these X-rays.

The energy of the X-rays emitted depends on the choice of target material. Targets with low atomic number result in soft characteristic X-rays, and high-Z targets produce harder or higher energy X-rays. The incident radiation must have a higher energy than the maximum photon energy expected from the target, because the excited states leading to the corresponding atomic transition must be populated by the incident radiation.

Figure 1.8 General method for the generation of characteristic X-rays from a specific target element. The exciting radiation can be X-rays, electrons, alpha particles, or any other ionizing radiation.

As one example, the incident radiation may consist of X-rays generated in a conventional X-ray tube. These X-rays may then interact in the target through photoelectric absorption, and the subsequent de-excitation of the target ions creates their characteristic X-ray spectrum. In this case, the process is called *X-ray fluorescence*. Although the characteristic X-ray spectrum can be contaminated by scattered photons from the incident X-ray beam, this component can be kept below about 10 or 20% of the total photon yield with proper choice of target and geometry.[15]

As an alternative to bulky X-ray tubes, radioisotopes that emit low-energy photons may also be used as the source of the excitation. An example is ^{241}Am, in which gamma rays of 60 keV energy are emitted in 36% of the decays. Characteristic X-ray sources using this isotope for excitation of targets in the geometry shown in Fig. 1.9 are currently available using up to 10 mCi of ^{241}Am activity.

Another method of exciting the target is through the use of an external electron beam. For targets of low atomic number, acceleration potentials of only a few thousand volts are required so that relatively compact electron sources can be devised. In the case of electron excitation, the characteristic X-ray spectrum from the target will be contaminated by the continuous bremsstrahlung spectrum also generated by interactions of the incident electrons in the target. For thin targets, however, the bremsstrahlung photons are preferentially emitted in the forward direction, whereas the characteristic X-rays are emitted isotropically.[18] Placing the exit window at a large angle (120–180°) with respect to the incident electron direction will therefore minimize the bremsstrahlung contamination.

The incident radiation in Fig. 1.8 can also consist of heavy charged particles. Again, the interactions of these particles in the target will give rise to the excited atoms necessary for the emission of characteristic X-rays. For compact and portable sources, alpha particles emitted by radioisotope sources are the most convenient source of the incident particles. Of the various potential alpha emitters, the most useful are ^{210}Po and ^{244}Cm because of their convenient half-lives and relative freedom from contaminant electromagnetic radiation (see Table 1.5). Alpha particle excitation avoids the complication of bremsstrahlung associated with electron excitation and is therefore capable of generating a relatively "clean" characteristic X-ray spectrum. A cross-sectional diagram of a typical source of this type is shown in Fig. 1.9. The X-ray yields into one steradian solid angle per mCi (37 MBq) of ^{244}Cm ranges from 1.7×10^5 photons/second for a beryllium target to about 10^4 photons/second for targets of higher Z (Ref. 17).

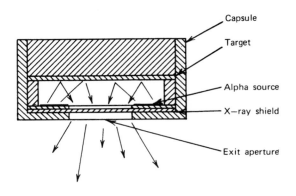

Figure 1.9 Cross-sectional view of a compact source of characteristic X-rays which utilizes alpha particle excitation of a target. (From Amlauer and Tuohy.[17])

Table 1.5 Alpha Particle Sources Useful for Excitation of Characteristic X-rays

	^{210}Po	^{244}Cm
Half-life	138 d	17.6 y
Alpha emissions	5.305 MeV (100%)	5.81, 5.77 MeV
Gamma-rays	803 keV (0.0011%)	43 keV (0.02%) 100 keV (0.0015%) 150 keV (0.0013%) 262 keV (1.4×10^{-4} %) 590 keV (2.5×10^{-4} %) 820 keV (7×10^{-5} %)
X-rays	Pb characteristic L and M (trace)	Pu characteristic L and M

Data from Amlauer and Tuohy.[17]

F. Synchrotron Radiation

Another form of radiation is produced when a beam of energetic electrons is bent into a circular orbit. From electromagnetic theory, a small fraction of the beam energy is radiated away during each cycle of the beam. Originally considered to be something of a nuisance by high-energy accelerator designers, this *synchrotron radiation* has turned out over the past several decades to be a very useful form of electromagnetic radiation. When extracted from the accelerator in a tangential direction to the beam orbit, the radiation appears as an intense and highly directional beam of photons with energy that can span the range from visible light (a few eV) through X-ray energies ($\sim 10^4$ eV). Monochromators can be used to produce nearly monoenergetic photon beams with a very high intensity. A number of high-energy accelerators (typically producing electrons with energy of 500 MeV or more) have since been built that are dedicated to the production of synchrotron radiation. Although limited to large-scale centralized user facilities, this unique form of electromagnetic radiation has proven to be very useful because of the high intensity and tunable energy of the available source.

V. NEUTRON SOURCES

Although nuclei created with excitation energy greater than the neutron binding energy can decay by neutron emission, these highly excited states are not produced as a result of any convenient radioactive decay process. Consequently, practical isotope sources of neutrons do not exist in the same sense that gamma-ray sources are available from many different nuclei populated by beta decay.[†] The possible choices for radioisotope neutron sources are much more limited and are based on either spontaneous fission or on nuclear reactions for which the incident particle is the product of a conventional decay process.

[†]The specific beta decay with the longest half-life that leads to an excited state that does de-excite by neutron emission is

Because the half-life of this beta decay is only 55 s, ^{87}Br is not practical as a laboratory neutron source.

A. Spontaneous Fission

Many of the transuranic heavy nuclides have an appreciable spontaneous fission decay probability. Several fast neutrons are promptly emitted in each fission event, so a sample of such a radionuclide can be a simple and convenient isotopic neutron source. Other products of the fission process are the heavy fission products described earlier, prompt fission gamma rays, and the beta and gamma activity of the fission products accumulated within the sample. When used as a neutron source, the isotope is generally encapsulated in a sufficiently thick container so that only the fast neutrons and gamma rays emerge from the source.

The most common spontaneous fission source is ^{252}Cf. Its half-life of 2.65 years is long enough to be reasonably convenient, and the isotope is one of the most widely produced of all the transuranics. The dominant decay mechanism is alpha decay, and the alpha emission rate is about 32 times that for spontaneous fission. The neutron yield is 0.116 n/s per Bq, where the activity is the combined alpha and spontaneous fission decay rate. On a unit mass basis, 2.30×10^6 n/s are produced per microgram of the sample. Compared with the other isotopic neutron sources described below, ^{252}Cf sources involve very small amounts of active material (normally of the order of micrograms) and can therefore be made in very small sizes dictated only by the encapsulation requirements.

The energy spectrum of the neutrons is plotted in Fig. 1.10. The spectrum is peaked between 0.5 and 1 MeV, although a significant yield of neutrons extends to as high as 8 or 10 MeV. The shape of a typical fission spectrum is approximated by the expression

$$\frac{dN}{dE} = E^{1/2} e^{-E/T} \tag{1.6}$$

For the spontaneous fission of ^{252}Cf, the constant T in Eq. (1.6) has a value of 1.3 MeV (Ref. 19). In addition to about 3.8 neutrons on average, each fission in ^{252}Cf also results in an average of about 9.7 emitted gamma-ray photons.[20] Most of these (> 85%) are relatively high-energy prompt gamma rays that are emitted within the first nanosecond following the fission event.[21]

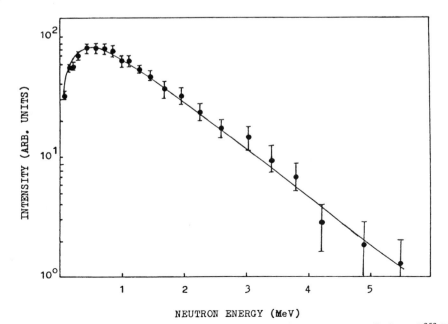

Figure 1.10 Measured neutron energy spectrum from the spontaneous fission of ^{252}Cf. (From Batenkov et al.[18])

B. Radioisotope (α, n) Sources

Because energetic alpha particles are available from the direct decay of a number of convenient radionuclides, it is possible to fabricate a small self-contained neutron source by mixing an alpha-emitting isotope with a suitable target material. Several different target materials can lead to (α, n) reactions for the alpha particle energies that are readily available in radioactive decay. The maximum neutron yield is obtained when beryllium is chosen as the target, and neutrons are produced through the reaction

$$\tfrac{4}{2}\alpha + \tfrac{9}{4}\text{Be} \rightarrow \tfrac{12}{6}\text{C} + \tfrac{1}{0}\text{n}$$

which has a Q-value of $+5.71$ MeV.

The neutron yield from this reaction when a beam of alpha particles strikes a target that is thick compared with their range is plotted in Fig. 1.11. Most of the alpha particles simply are stopped in the target, and only 1 in about 10^4 reacts with a beryllium nucleus. Virtually the same yield can be obtained from an intimate mixture of the alpha particle emitter and beryllium, provided the alpha emitter is homogeneously distributed throughout the beryllium in a small relative concentration. All the alpha emitters of practical interest are actinide elements, and investigations have shown that a stable alloy can be formed between the actinides and beryllium of the form MBe_{13}, where M represents the actinide metal. Most of the sources described below therefore are metallurgically prepared in the form of this alloy, and each alpha particle has an opportunity to interact with beryllium nuclei without any intermediate energy loss.

Some of the common choices for alpha emitters and properties of the resulting neutron sources are listed in Table 1.6. Several of these isotopes, notably ^{226}Ra and ^{227}Ac, lead to long chains of daughter products that, although adding to the alpha particle yield, also contribute a large gamma-ray background. These sources are therefore inappropriate for some applications in which the intense gamma-ray background interferes with the measurement. Also, these Ra-Be and Ac-Be sources require more elaborate handling procedures because of the added biological hazard of the gamma radiation.

Figure 1.11 Thick target yield of neutrons for alpha particles on beryllium. (From Anderson and Hertz.[22])

Table 1.6 Characteristics of Be(α, n) Neutron Sources

Source	Half-Life	E_α (MeV)	Neutron Yield per 10^6 Primary Alpha Particles		Percent Yield with $E_n < 1.5$ MeV	
			Calculated	Experimental	Calculated	Experimental
^{239}Pu/Be	24000 y	5.14	65	57	11	9−33
^{210}Po/Be	138 d	5.30	73	69	13	12
^{238}Pu/Be	87.4 y	5.48	79[a]	—	—	—
^{241}Am/Be	433 y	5.48	82	70	14	15−23
^{244}Cm/Be	18 y	5.79	100[b]	—	18	29
^{242}Cm/Be	162 d	6.10	118	106	22	26
^{226}Ra/Be + daughters	1602 y	Multiple	502	—	26	33−38
^{227}Ac/Be +daughters	21.6 y	Multiple	702	—	28	38

[a]From Anderson and Hertz.[22] All other data as calculated or cited in Geiger and Van der Zwan.[23]

[b]Does not include a 4% contribution from spontaneous fission of ^{244}Cm.

The remaining radioisotopes in Table 1.6 involve simpler alpha decays and the gamma-ray background is much lower. The choice between these alternatives is made primarily on the basis of availability, cost, and half-life. Because the physical size of the sources is no longer negligible, one would like the half-life to be as short as possible, consistent with the application, so that the specific activity of the emitter is high.

The ^{239}Pu/Be source is probably the most widely used of the (α, n) isotopic neutron sources. However, because about 16 g of the material is required for 1 Ci (3.7×10^{10} Bq) of activity, sources of this type of a few centimeters in dimension are limited to about 10^7 n/s. In order to increase the neutron yield without increasing the physical source size, alpha emitters with higher specific activities must be substituted. Therefore, sources incorporating ^{241}Am (half-life of 433 years) and ^{238}Pu (half-life of 87.4 years) are also widely used if high neutron yields are needed. Sources utilizing ^{244}Cm (half-life of 18 years) represent a near ideal compromise between specific activity and source lifetime, but the isotope is not always widely available.

The neutron energy spectra from all such alpha/Be sources are similar, and any differences reflect only the small variations in the primary alpha energies. A plot of the spectrum from a ^{239}Pu/Be source is shown in Fig. 1.12. The various peaks and valleys in this energy distribution can be analyzed in terms of the excitation state in which the ^{12}C product nucleus is left.[23,24] The alpha particles lose a variable amount of energy before reacting with a beryllium nucleus, however, and their continuous energy distribution washes out much of the structure that would be observed if the alpha particles were monoenergetic. For sources that contain only a few grams of material, the spectrum of neutrons that emerges from the source surface is essentially the same as that created in the (α, n) reactions. For larger sources, the secondary processes of neutron scattering within the source, (n, 2n) reactions in beryllium, and (n, fission) events within the plutonium or other actinide can introduce some dependence of the energy spectrum on the source size.[25]

Because large activities of the actinide isotope are involved in these neutron sources, special precautions must be taken in their fabrication to ensure that the material remains safely encapsulated. The actinide–beryllium alloy is usually sealed within two individually

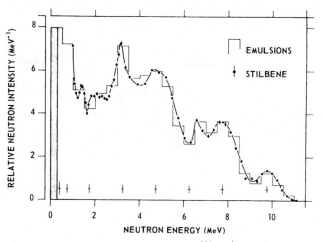

Figure 1.12 Measured energy spectra for neutrons from a ^{239}Pu/Be source containing 80 g of the isotope. (From Anderson and Neff.[25])

welded stainless steel cylinders in the arrangement shown in Fig. 1.13. Some expansion space must be allowed within the inner cylinder to accommodate the slow evolution of helium gas formed when the alpha particles are stopped and neutralized.

When applied to the efficiency calibration of detectors, some caution must be used in assuming that the neutron yield for these sources decays exactly as the half-life of the principal actinide alpha emitter. Small amounts of contaminant alpha activity, present in either the original radioisotope sample or produced through the decay of a precursor, can influence the overall neutron yield. For example, many ^{239}Pu/Be sources have been prepared from plutonium containing small amounts of other plutonium isotopes. The isotope ^{241}Pu is particularly significant, because it beta decays with a half-life of 13.2 years to form ^{241}Am, an alpha emitter. The neutron yield of these sources can therefore gradually increase with time as the ^{241}Am accumulates in the source. An original ^{241}Pu isotopic fraction of 0.7% will result in an initial growth rate of the neutron yield of 2% per year.[26]

A number of other alpha-particle-induced reactions have occasionally been employed as neutron sources, but all have a substantially lower neutron yield per unit alpha activity compared with the beryllium reaction. Some of the potential useful reactions are listed in Table 1.7. Because all the Q-values of these reactions are less than that of the beryllium reaction, the resulting neutron spectra shown in Fig 1.14 have a somewhat lower average energy. In particular, the ^{7}Li (α, n) reaction with its highly negative Q-value leads to a neutron spectrum with a low 0.5 MeV average energy that is especially useful in some applications. Details of this spectrum shape are given in Ref. 28.

Figure 1.13 Typical double-walled construction for Be (α, n) sources. (From Lorch.[19])

Table 1.7 Alternative (α, n) Isotopic Neutron Sources

Target	Reaction	Q-Value	Neutron Yield per 10^6 Alpha Particles
Natural B	^{10}B (α, n)	+1.07 MeV	13 for ^{241}Am alpha particles
	^{11}B(α, n)	+0.158 MeV	
F	^{19}F(α, n)	−1.93 MeV	4.1 for ^{241}Am alpha particles
Isotopically separated ^{13}C	^{13}C(α, n)	+2.2 MeV	11 for ^{238}Pu alpha particles
Natural Li	^{7}Li(α, n)	−2.79 MeV	
Be (for comparison)	^{9}Be(α, n)	+5.71 MeV	70 for ^{241}Am alpha particles

Data from Lorch[19] and Geiger and Van der Zwan.[27]

Figure 1.14 Neutron energy spectra from alternative (α, n) sources. (^{7}Li data from Geiger and Van der Zwan,[27] remainder from Lorch.[19])

C. Photoneutron Sources

Some radioisotope gamma-ray emitters can also be used to produce neutrons when combined with an appropriate target material. The resulting *photoneutron sources* are based on supplying sufficient excitation energy to a target nucleus by absorption of a gamma-ray photon to allow the emission of a free neutron. Only two target nuclei, ^{9}Be and ^{2}H, are of any practical significance for radioisotope photoneutron sources. The corresponding reactions can be written

$$Q$$
$$^9_4\text{Be} + h\nu \rightarrow {}^8_4\text{Be} + {}^1_0\text{n} \qquad -1.666 \text{ MeV}$$
$$^2_1\text{H} + h\nu \rightarrow {}^1_1\text{H} + {}^1_0\text{n} \qquad -2.226 \text{ MeV}$$

A gamma-ray photon with an energy of at least the negative of the Q-value is required to make the reactions energetically possible, so that only relatively high-energy gamma rays can be applied. For gamma-ray energies that exceed this minimum, the corresponding neutron energy can be calculated from

$$E_\text{n}(\theta) \cong \frac{M(E_\gamma + Q)}{m + M} + \frac{E_\gamma[(2mM)(m + M)(E_\gamma + Q)]^{1/2}}{(m + M)^2} \cos \theta \qquad (1.7)$$

where θ = angle between gamma photon and neutron direction

E_γ = gamma energy (assumed $\ll$ 931 MeV)

M = mass of recoil nucleus $\times c^2$

m = mass of neutron $\times c^2$

One decided advantage of photoneutron sources is that if the gamma rays are monoenergetic, the neutrons are also nearly monoenergetic. The relatively small kinematic spread obtained from Eq. (1.7) by letting the angle θ vary between 0 and π broadens the neutron energy spectrum by only a few percent. For large sources, the spectrum is also somewhat degraded by the scattering of some neutrons within the source before their escape.

The main disadvantage of photoneutron sources is the fact that very large gamma-ray activities must be used in order to produce neutron sources of attractive intensity. For the type of source sketched in Fig. 1.15, only about 1 gamma ray in 10^5 or 10^6 interacts to produce a neutron, and therefore the neutrons appear in a much more intense gamma-ray background. Some of the more common gamma-ray emitters are ^{226}Ra, ^{124}Sb, ^{72}Ga, ^{140}La, and ^{24}Na. Some properties and energy spectra of corresponding photoneutron sources are shown in Table 1.8 and Fig. 1.16. For many of these sources, the half-lives of the gamma emitters are short enough to require their reactivation in a nuclear reactor between uses.

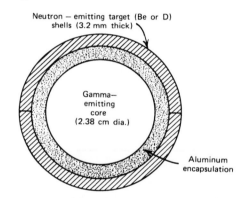

Figure 1.15 Construction of a simple spherical photoneutron source.

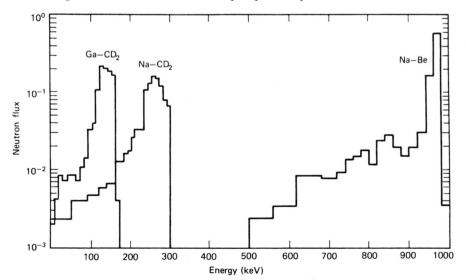

Figure 1.16 Neutron spectra calculated for the photoneutron source dimensions shown in Fig. 1.15. The gamma emitters are either ^{72}Ga or ^{24}Na. The outer shells are either deuterated polyethylene (CD_2) or beryllium (Be).

Table 1.8 Photoneutron Source Characteristics

Gamma-Ray Emitter	Half-Life[a]	Gamma Energy[a] (MeV)	Target	Neutron Energy[b] (keV)	Neutron Yield (n/s) for 10^{10} Bq Activity[c]
^{24}Na	15.0 h	2.7541	Be	967	340,000
		2.7541	D	263	330,000
^{28}Al	2.24 min	1.7787	Be	101	32,600
^{38}Cl	37.3 min	2.1676	Be	446	43,100
^{56}Mn	2.58 h	1.8107	Be	129 ⎫	
		2.1131		398 ⎬	91,500
		2.9598		1,149 ⎭	
		2.9598	D	365	162
^{72}Ga	14.1 h	1.8611	Be	174 ⎫	
		2.2016		476 ⎬	64,900
		2.5077		748 ⎭	
		2.5077	D	140	25,100
^{76}As	26.3 h	1.7877	Be	109 ⎫	3,050
		2.0963		383 ⎭	
^{88}Y	107 d	1.8361	Be	152 ⎫	229,000
		2.7340		949 ⎭	
		2.7340	D	253	160
^{116m}In	54.1 min	2.1121	Be	397	15,600
^{124}Sb	60.2 d	1.6910	Be	23	210,000
^{140}La	40.3 h	2.5217	Be	760	10,200
		2.5217	D	147	6,600
^{144}Pr	17.3 min	2.1856	Be	462	690

[a]Decay data from Ref. 1.

[b]Calculated for $\theta = \pi/2$, approximate midpoint of primary spectrum.

[c]Monte Carlo calculations for the source dimensions given in Fig. 1.15. Outer target shells are either metallic Be or deuterated polyethylene. Core materials assumed to be NaF, Al, CCl_4, MnO_2, Ga_2O_3, As_2O_3, Y_2O_3, In, Sb, La_2O_3, and Pr_2O_3.

Source: G. F. Knoll, "Radioisotope Neutron Sources," Chap. 2 in *Neutron Sources for Basic Physics and Applications*, Pergamon Press, New York, 1983.

D. Reactions from Accelerated Charged Particles

Because alpha particles are the only heavy charged particles with low Z conveniently available from radioisotopes, reactions involving incident protons, deuterons, and so on must rely on artificially accelerated particles. Two of the most common reactions of this type used to produce neutrons are

$$Q$$

		Q
The D-D reaction	$^{2}_{1}\text{H} + ^{2}_{1}\text{H} \rightarrow ^{3}_{2}\text{He} + ^{1}_{0}\text{n}$	+3.26 MeV
The D-T reaction	$^{2}_{1}\text{H} + ^{3}_{1}\text{H} \rightarrow ^{4}_{2}\text{He} + ^{1}_{0}\text{n}$	+17.6 MeV

Because the coulomb barrier between the incident deuteron and the light target nucleus is relatively small, the deuterons need not be accelerated to a very high energy in order to create a significant neutron yield. These reactions are widely exploited in "neutron generators" in which deuterium ions are accelerated by a potential of about 100–300 kV. Because the incident particle energy is then small compared with the Q-value of either reaction, all the neutrons produced have about the same energy (near 3 MeV for the D-D reaction and 14 MeV for the D-T reaction). A 1 mA beam of deuterons will produce about 10^9 n/s from a thick deuterium target, and about 10^{11} n/s from a tritium target. Somewhat smaller yields are produced in compact neutron generators consisting of a sealed tube containing the ion source and target, together with a portable high-voltage generator.

A number of other charged-particle-induced reactions that involve either a negative Q-value or a target with higher atomic number are also applied to neutron generation. Some common examples are $^9Be(d, n)$, $^7Li(p, n)$, and $^3H(p, n)$. In these cases, a higher incident particle energy is required, and large accelerator facilities such as cyclotrons or Van de Graaf accelerators are needed to produce the incident particle beam.

PROBLEMS

1.1 Radiation energy spectra can be categorized into two main groups: those that consist of one or more discrete energies (line spectra) and those that consist of a broad distribution of energies (continuous spectra). For each of the radiation sources listed below, indicate whether "line" or "continuous" is a better description:

 (a) Alpha particles.

 (b) Beta particles.

 (c) Gamma rays.

 (d) Characteristic X-rays.

 (e) Conversion electrons.

 (f) Auger electrons.

 (g) Fission fragments.

 (h) Bremsstrahlung.

 (i) Annihilation radiation.

1.2 Which has the higher energy: a conversion electron from the L shell or from the M shell, if both arise from the same nuclear excitation energy?

1.3 By simultaneously conserving energy and momentum, find the alpha-particle energy emitted in the decay of a nucleus with mass number 210 if the Q-value of the decay is 5.50 MeV.

1.4 What is the lowest wavelength limit of the X-rays emitted by a tube operating at a potential of 195 kV?

1.5 From a table of atomic mass values, find the approximate energy released by the spontaneous fission of ^{235}U into two equal-mass fragments.

1.6 Calculate the specific activity of pure tritium (^3H) with a half-life of 12.26 years.

1.7 What is the highest energy to which doubly ionized helium atoms (alpha particles) can be accelerated in a dc accelerator with 3 MV maximum voltage?

1.8 What is the minimum gamma-ray energy required to produce photoneutrons in water from the trace heavy water content?

1.9 By simultaneously conserving energy and momentum, calculate the energy of the neutron emitted in the forward direction by a beam of 150 keV deuterons undergoing the D-T reaction in a tritium target.

REFERENCES

 1. C. M. Lederer and V. S. Shirley, *Table of Isotopes,* 7th ed., Wiley-Interscience, New York, 1978.
 2. R. N. Chanda and R. A. Deal, *Catalog of Semiconductor Alpha-Particle Spectra*, IN-1261, (1970).
 3. A. Rytz, *Atomic Data and Nuclear Data Tables* **12**, 479 (1973).
 4. W. E. Nervik, *Phys. Rev.* **119**, 1685 (1960).
 5. S. L. Whetstone, *Phys. Rev.* **131**, 1232 (1963).
 6. R. J. Gehrke, J. E. Cline, and R. L. Heath, *Nucl. Instrum. Meth.* **91**, 349 (1971).
 7. E. L. Hull et al., *Nucl. Instrum. Meth.* **A385**, 489 (1997).
 8. S. Croft, *Nucl. Instrum. Meth.* **A281**, 103 (1989).
 9. J. P. Mason, *Nucl. Instrum. Meth. Phys. Res.* **A241**, 207 (1985).
 10. C. Coceva, A. Brusegan, and C. van der Vorst, *Nucl. Instrum. Meth.* **A378**, 511 (1996).
 11. T. Ikuta et al., *Nucl. Instrum. Meth.* **A323**, 697 (1992).
 12. J. G. Rogers et al., *Nucl. Instrum. Meth.* **A413**, 249 (1998).
 13. *X-Ray and Gamma-Ray Standards for Detector Calibration*, IAEA-TECDOC-619, IAEA, Vienna (1991).
 14. H. Ferdinande, G. Knuyt, R. Van De Vijver, and R. Jacobs, *Nucl. Instrum. Meth.* **91**, 135 (1971).
 15. E. Storm, D. W. Lier, and H. I. Israel, *Health Phys.* **26**, 179 (1974).
 16. J. L.-H Chan and A. Macovski, IEEE *Trans. Nucl. Sci.* **NS-24**, No. 4, 1968 (1977).
 17. K. Amlauer and I. Tuohy, *Proceedings, ERDA X- and Gamma Ray Symposium*, Ann Arbor CONF-760539, p. 19 (1976).
 18. O. I. Batenkov et al., INDC (NDS)-146, (1983).
 19. E. A. Lorch, *Int. J. Appl. Radiat. Isotopes* **24**, 585 (1973).

20. K. Skarsvaq, *Phys. Rev.* **C22(2)**, 638 (1980).

21. H. Maier-Leibnitz et al., *Proc. Symp. Phys. and Chem. of Fission* **2**, 113 (1965).

22. M. E. Anderson and M. R. Hertz, *Nucl. Sci. Eng.* **44**, 437 (1971).

23. K. W. Geiger and L. Van der Zwan, *Nucl. Instrum. Meth.* **131**, 315 (1975).

24. A. Kumar and P. S. Nagarajan, *Nucl. Instrum. Meth.* **140**, 175 (1977).

25. M. E. Anderson and R. A. Neff, *Nucl. Instrum. Meth.* **99**, 231 (1972).

26. M. E. Anderson, *Nucl. Appl.* **4**, 142 (1968).

27. K. W. Geiger and L. Van der Zwan, *Health Phys.* **21**, 120 (1971).

28. D. R. Weaver, J. G. Owen, and J. Walker, *Nucl. Instrum. Meth.* **198**, 599 (1982).

Chapter **2**

Radiation Interactions

The operation of any radiation detector basically depends on the manner in which the radiation to be detected interacts with the material of the detector itself. An understanding of the response of a specific type of detector must therefore be based on a familiarity with the fundamental mechanisms by which radiations interact and lose their energy in matter. Many general reference works are available concerning this broad topic; the classic text by Evans,[1] to mention only one, has served as a standard reference for several decades.

To organize the discussions that follow, it is convenient to arrange the four major categories of radiations introduced in Chapter 1 into the following matrix:

Charged Particulate Radiations		Uncharged Radiations
Heavy charged particles (characteristic distance $\cong 10^{-5}$ m)	$\Leftarrow$	Neutrons (characteristic length $\cong 10^{-1}$ m)
Fast electrons (characteristic distance $\cong 10^{-3}$ m)	$\Leftarrow$	X-rays and gamma rays (characteristic length $\cong 10^{-1}$ m)

The entries in the left column represent the charged particulate radiations that, because of the electric charge carried by the particle, continuously interact through the coulomb force with the electrons present in any medium through which they pass. The radiations in the right column are uncharged and therefore are not subject to the coulomb force. Instead, these radiations must first undergo a "catastrophic" interaction (often involving the nucleus of constituent atoms) that radically alters the properties of the incident radiation in a single encounter. In all cases of practical interest, the interaction results in the full or partial transfer of energy of the incident radiation to electrons or nuclei of the constituent atoms, or to charged particle products of nuclear reactions. If the interaction does not occur within the detector, these uncharged radiations (e.g., neutrons or gamma rays) can pass completely through the detector volume without revealing the slightest hint that they were ever there.

The horizontal arrows shown in the diagram illustrate the results of such catastrophic interactions. An X- or gamma ray, through the processes described in this chapter, can transfer all or part of its energy to electrons within the medium. The resulting *secondary electrons* bear a close similarity to the fast electron radiations (such as the beta particle) discussed in Chapter 1. Devices designed to detect gamma rays are tailored to promote such interactions and to fully stop the resulting secondary electrons so that their entire energy may contribute to the output signal. In contrast, neutrons may interact in such a way as to produce secondary heavy charged particles, which then serve as the basis of the detector signal.

Also listed in the diagram are order-of-magnitude numbers for the characteristic distance of penetration or average path length (range or mean free path) in solids for typical energy radiations in each category.

I. INTERACTION OF HEAVY CHARGED PARTICLES

A. Nature of the Interaction

Heavy charged particles, such as the alpha particle, interact with matter primarily through coulomb forces between their positive charge and the negative charge of the orbital electrons within the absorber atoms. Although interactions of the particle with nuclei (as in Rutherford scattering or alpha-particle-induced reactions) are also possible, such encounters occur only rarely and they are not normally significant in the response of radiation detectors. Instead, charged particle detectors must rely on the results of interactions with electrons for their response.

Upon entering any absorbing medium, the charged particle immediately interacts simultaneously with many electrons. In any one such encounter, the electron feels an impulse from the attractive coulomb force as the particle passes its vicinity. Depending on the proximity of the encounter, this impulse may be sufficient either to raise the electron to a higher-lying shell within the absorber atom (*excitation*) or to remove completely the electron from the atom (*ionization*). The energy that is transferred to the electron must come at the expense of the charged particle, and its velocity is therefore decreased as a result of the encounter. The maximum energy that can be transferred from a charged particle of mass m with kinetic energy E to an electron of mass m_0 in a single collision is $4Em_0/m$, or about $1/500$ of the particle energy per nucleon. Because this is a small fraction of the total energy, the primary particle must lose its energy in many such interactions during its passage through an absorber. At any given time, the particle is interacting with many electrons, so the net effect is to decrease its velocity continuously until the particle is stopped.

Representative paths taken by heavy charged particles in their slowing down process are schematically represented in the sketch below. Except at their very end, the tracks tend to be quite straight because the particle is not greatly deflected by any one encounter, and interactions occur in all directions simultaneously. Charged particles are therefore characterized by a definite *range* in a given absorber material. The range, to be defined more precisely below, represents a distance beyond which no particles will penetrate.

The products of these encounters in the absorber are either excited atoms or *ion pairs*. Each ion pair is made up of a free electron and the corresponding positive ion of an absorber atom from which an electron has been totally removed. The ion pairs have a natural tendency to recombine to form neutral atoms, but in some types of detectors, this recombination is suppressed so that the ion pairs may be used as the basis of the detector response.

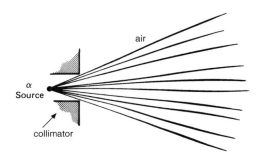

In particularly close encounters, an electron may undergo a large enough impulse that after having left its parent atom, it still may have sufficient kinetic energy to create further ions. These energetic electrons are sometimes called *delta rays* and represent an indirect means by which the charged particle energy is transferred to the absorbing medium. Under typical conditions, the majority of the energy loss of the charged particle occurs via these delta rays. The range of the delta rays is always small compared with the range of the incident energetic particle, so the ionization is still formed close to the primary track. On a microscopic scale, one effect of this process is that the ion pairs normally do not appear as randomly spaced single ionizations, but there is a tendency to form many "clusters" of multiple ion pairs distributed along the track of the particle.

B. Stopping Power

The *linear stopping power S* for charged particles in a given absorber is simply defined as the differential energy loss for that particle within the material divided by the corresponding differential path length:

$$S = -\frac{dE}{dx} \tag{2.1}$$

The value of $-dE/dx$ along a particle track is also called its *specific energy loss* or, more casually, its "rate" of energy loss.

For particles with a given charge state, S increases as the particle velocity is decreased. The classical expression that describes the specific energy loss is known as the *Bethe formula* and is written

$$-\frac{dE}{dx} = \frac{4\pi e^4 z^2}{m_0 v^2} NB \tag{2.2}$$

where

$$B \equiv Z\left[\ln \frac{2m_0 v^2}{I} - \ln\left(1 - \frac{v^2}{c^2}\right) - \frac{v^2}{c^2} \right]$$

In these expressions, v and ze are the velocity and charge of the primary particle, N and Z are the number density and atomic number of the absorber atoms, m_0 is the electron rest mass, and e is the electronic charge. The parameter I represents the average excitation and ionization potential of the absorber and is normally treated as an experimentally determined parameter for each element. For nonrelativistic charged particles ($v \ll c$), only the first term in B is significant. Equation (2.2) is generally valid for different types of charged particles provided their velocity remains large compared with the velocities of the orbital electrons in the absorbing atoms.

The expression for B in Eq. (2.2) varies slowly with particle energy. Thus, the general behavior of dE/dx can be inferred from the behavior of the multiplicative factor. For a given nonrelativistic particle, dE/dx therefore varies as $1/v^2$, or inversely with particle energy. This behavior can be heuristically explained by noting that because the charged particle spends a greater time in the vicinity of any given electron when its velocity is low, the impulse felt by the electron, and hence the energy transfer, is largest. When comparing different charged particles of the same velocity, the only factor that may change outside the logarithmic term in Eq. (2.2) is z^2, which occurs in the numerator of the expression. Therefore, particles with the greatest charge will have the largest specific energy loss. Alpha particles, for example, will lose energy at a rate that is greater than protons of the same velocity but less than that of more highly charged ions. In comparing different materials as absorbers, dE/dx depends primarily on the product NZ, which is outside the logarithmic term. This product NZ represents

Figure 2.1 Variation of the specific energy loss in air versus energy of the charged particles shown. (From Beiser.[2])

the electron density of the absorber. High atomic number, high-density materials will consequently result in the greatest linear stopping power.

The variation of the specific energy loss for a number of different charged particles is shown in Fig. 2.1 over a wide energy range. This figure shows that the value of dE/dx for many different types of charged particles approaches a near-constant broad minimum value at energies above several hundred MeV, where their velocity approaches the velocity of light. This specific energy loss corresponds to about 2 MeV per g/cm^2 in light materials. Because of their similar energy loss behavior, such relativistic particles are sometimes referred to as "minimum ionizing particles." Fast electrons also fall into this category at energies as low as about 1 MeV because their much lower mass results in relativistic velocities even at such modest energy.

The Bethe formula begins to fail at low particle energies where charge exchange between the particle and absorber becomes important. The positively charged particle will then tend to pick up electrons from the absorber, which effectively reduce its charge and consequent linear energy loss. At the end of its track, the particle has accumulated z electrons and becomes a neutral atom.

C. Energy Loss Characteristics

1. THE BRAGG CURVE

A plot of the specific energy loss along the track of a charged particle such as that shown in Fig. 2.2 is known as a *Bragg curve*. The example is shown for an alpha particle of several MeV initial energy. For most of the track, the charge on the alpha particle is two electronic charges, and the specific energy loss increases roughly as $1/E$ as predicted by Eq. (2.2). Near the end of the track, the charge is reduced through electron pickup and the curve falls off. Plots are shown both for a single alpha particle track and for the average behavior of a parallel beam of alpha particles of the same initial energy. The two curves differ somewhat due to the effects of straggling, to be discussed below.

Figure 2.2 The specific energy loss along an alpha track.

Related plots showing $-dE/dx$ versus particle energy for a number of different heavy charged particles are given in Fig. 2.3. These examples illustrate the energy at which charge pickup by the ion becomes significant. Charged particles with the greatest number of nuclear charges begin to pick up electrons early in their slowing-down process. Note that in an aluminum absorber, singly charged hydrogen ions (protons) show strong effects of charge pickup below about 100 keV, but doubly charged ^{3}He ions show equivalent effects at about 400 keV.

2. ENERGY STRAGGLING

Because the details of the microscopic interactions undergone by any specific particle vary somewhat randomly, its energy loss is a statistical or stochastic process. Therefore, a spread in energies always results after a beam of monoenergetic charged particles has passed through a given thickness of absorber. The width of this energy distribution is a measure of energy straggling, which varies with the distance along the particle track.

Figure 2.4 shows a schematic presentation of the energy distribution of a beam of initially monoenergetic particles at various points along its range. Over the first portion, the distribution becomes wider (and more skewed) with penetration distance, showing the

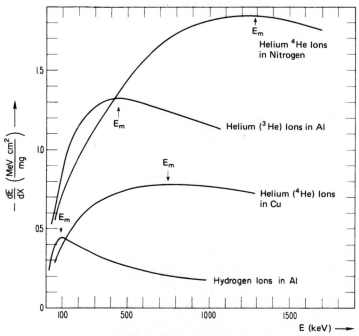

Figure 2.3 Specific energy loss as a function of energy for hydrogen and helium ions. E_m indicates the energy at which dE/dx is maximized. (From Wilken and Fritz.[3])

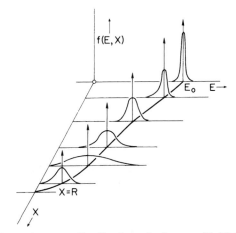

Figure 2.4 Plots of energy distribution of a beam of initially monoenergetic charged particles at various penetration distances. E is the particle energy and X is the distance along the track. (From Wilken and Fritz.[3])

increasing importance of energy straggling. Near the end of the range, the distribution narrows again because the mean particle energy has greatly been reduced.

D. Particle Range

1. DEFINITIONS OF RANGE

In order to quantify the definition of particle range, we refer to the conceptual experiment sketched in Fig. 2.5. Here a collimated source of monoenergetic alpha particles is counted by a detector after passing through an absorber of variable thickness. (We later contrast the behavior of other types of radiation when observed under similar conditions.) For alpha particles, the results are also plotted in Fig. 2.5. For small values of the absorber thickness, the only effect is to cause an energy loss of the alpha particles in the absorber as they pass through. Because the tracks through the absorber are quite straight, the total number that reach the detector remains the same. No attenuation in the number of alpha particles takes place until the absorber thickness approaches the length of the shortest track in the absorbing material. Increasing the thickness then stops more and more of the alpha particles, and the intensity of the detected beam drops rapidly to zero.

The range of the alpha particles in the absorber material can be determined from this curve in several ways. The *mean range* is defined as the absorber thickness that reduces the alpha particle count to exactly one-half of its value in the absence of the absorber.

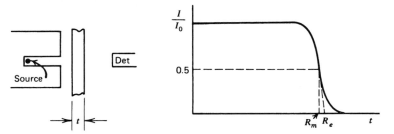

Figure 2.5 An alpha particle transmission experiment. I is the detected number of alpha particles through an absorber thickness t, whereas I_0 is the number detected without the absorber. The mean range R_m and extrapolated range R_e are indicated.

This definition is most commonly used in tables of numerical range values. Another version that appears in the literature is the *extrapolated range*, which is obtained by extrapolating the linear portion of the end of the transmission curve to zero.

The range of charged particles of a given energy is thus a fairly unique quantity in a specific absorber material. In the early days of radiation measurement, experiments of the type sketched in Fig. 2.5 were widely used to measure the energy of alpha particles indirectly by determining the absorber thickness equivalent to their mean range. With the availability of detectors that provide an output signal directly related to the alpha particle energy, such indirect measurements are no longer necessary.

Some graphs of the mean range of various charged particles in materials of interest in detectors are given in Figs. 2.6 through 2.8. As one obvious application of these curves, any detector that is to measure the full incident energy of a charged particle must have an active thickness that is greater than the range of that particle in the detector material.

2. RANGE STRAGGLING

Charged particles are also subject to *range straggling*, defined as the fluctuation in path length for individual particles of the same initial energy. The same stochastic factors that lead to energy straggling at a given penetration distance also result in slightly different

Figure 2.6   Range–energy plot for alpha particles in air at 15°C and 760 mm Hg pressure. (From *Radiological Health Handbook*, U.S. Department of Health, Education and Welfare, Washington, DC, 1970.)

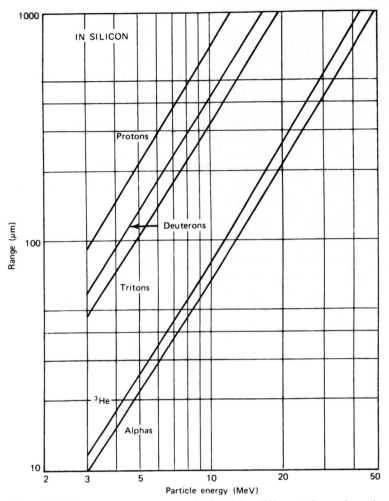

Figure 2.7 Range–energy curves calculated for different particles in silicon. The near-linear behavior of the log–log plot over the energy range shown suggests an empirical relation to the form $R = aE^b$, where the slope-related parameter b is not greatly different for the various particles. (From Skyrme.[4])

total path lengths for each particle. For heavy charged particles such as protons or alphas, the straggling amounts to a few percent of the mean range. The degree of straggling is evidenced by the sharpness of the cutoff at the end of the average transmission curve plotted in Fig. 2.2. Differentiating this curve leads to a peak whose width is often taken as a quantitative measure of the importance of range straggling for the particles and absorber used in the measurement.

3. STOPPING TIME

The time required to stop a charged particle in an absorber can be deduced from its range and average velocity. For nonrelativistic particles of mass m and kinetic energy E, the velocity is

$$v = \sqrt{\frac{2E}{m}} = c\sqrt{\frac{2E}{mc^2}} = \left(3.00 \times 10^8 \frac{\text{m}}{\text{s}}\right)\sqrt{\frac{2E}{(931 \text{ MeV/amu})m_A}}$$

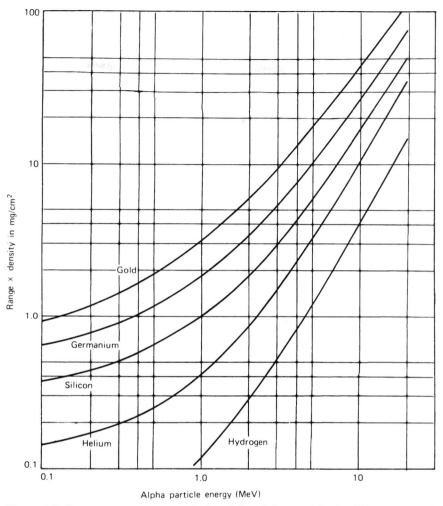

Figure 2.8 Range–energy curves calculated for alpha particles in different materials. Units of the range are given in mass thickness (see p. 54) to minimize the differences in these curves. (Data from Williamson et al.[5])

where m_A is the particle mass in atomic mass units and E is the particle energy in MeV. If we assume that the average particle velocity as it slows down is $\langle v \rangle = Kv$, where v is evaluated at the initial energy, then the stopping time T can be calculated from the range R as

$$T = \frac{R}{\langle v \rangle} = \frac{R}{Kc}\sqrt{\frac{mc^2}{2E}} = \frac{R}{K(3.00 \times 10^8\,\text{m/s})}\sqrt{\frac{931\ \text{MeV/amu}}{2}}\sqrt{\frac{m_A}{E}}$$

If the particle were uniformly decelerated, then $\langle v \rangle$ would be given by $v/2$ and K would be 0.5. However, charged particles generally lose energy at a greater rate near the end of their range, and K should be a somewhat higher fraction. By assuming $K = 0.60$, the stopping time can be estimated as

$$T \cong 1.2 \times 10^{-7}\,R\,\sqrt{\frac{m_A}{E}} \tag{2.3}$$

where T is in seconds, R in meters, m_A in amu, and E in MeV. This approximation is expected to be reasonably accurate for light charged particles (protons, alpha particles, etc.) over much of the energy range of interest here. It is not, however, to be used for relativistic particles such as fast electrons.

Using typical range values, stopping times calculated from Eq. (2.3) for charged particles are a few picoseconds in solids or liquids and a few nanoseconds in gases. These times are generally small enough to be neglected for all but the fastest-responding radiation detectors.

E. Energy Loss in Thin Absorbers

For thin absorbers (or detectors) that are penetrated by a given charged particle, the energy deposited within the absorber can be calculated from

$$\Delta E = -\left(\frac{dE}{dx}\right)_{\text{avg}} t \qquad (2.4)$$

Figure 2.9 The specific energy loss calculated for different charged particles in silicon. (From Skyrme.[4])

where t is the absorber thickness and $(-dE/dx)_{avg}$ is the linear stopping power averaged over the energy of the particle while in the absorber. If the energy loss is small, the stopping power does not change much and it can be approximated by its value at the incident particle energy. Tabular values for dE/dx for a number of different charged particles in a variety of absorbing media are given in Refs. 5–10. Some graphs for materials of interest are shown in Figs. 2.9 through 2.11.

For absorber thicknesses through which the energy loss is not small, it is not simple to obtain a properly weighted $(-dE/dx)_{avg}$ value directly from such data. In these cases, it is easier to obtain the deposited energy in a way that makes use of range–energy data of the type plotted in Figs. 2.6 through 2.8. The basis of the method is as follows: Let R_1 represent the full range of the incident particle with energy E_0 in the absorber material. By subtracting the physical thickness of the absorber t from R_1, a value R_2 is obtained that represents the range of those alpha particles that emerge from the opposite surface of the absorber. By finding the energy corresponding to R_2, the energy of the transmitted charged particles

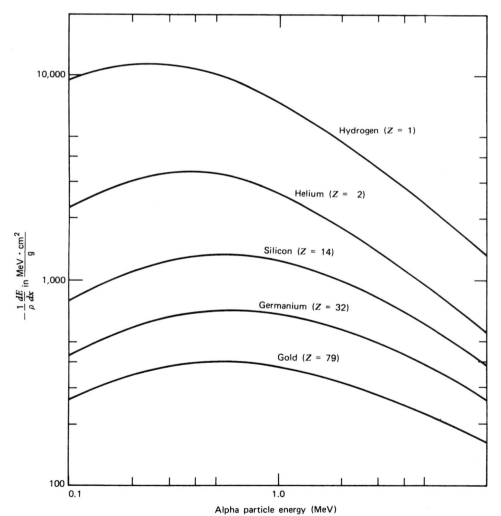

Figure 2.10 The specific energy loss calculated for alpha particles in different materials. Values are normalized by the density of the absorber material. (Data from Williamson et al.[5])

Figure 2.11 Plots showing the specific energy loss of various heavy ions in aluminum. The abscissa is the ion energy divided by its mass, and the ordinate is $-dE/dx$ divided by the density of aluminum and the square of the ion atomic number. Typical fission fragments (e.g., iodine) show a continuously decreasing $-dE/dx$ while slowing from their initial energy (~1 MeV/amu). (From Northcliffe and Schilling.[8])

E_t is obtained. The deposited energy ΔE is then given simply by $E_0 - E_t$. These steps are illustrated below:

The procedure is based on the assumption that the charged particle tracks are perfectly linear in the absorber, and the method does not apply in situations where the particle can be significantly deflected (such as for fast electrons).

The combined effects of particle range and the decrease in dE/dx with increasing energy are illustrated in Fig. 2.12. Here the energy loss of protons in a thin detector is plotted versus the incident proton energy. For low energies, the proton range is less than the detector thickness. Therefore, as the energy is increased, the energy deposited in the detector (which is just equal to the incident energy) increases linearly. At a proton energy of 425 keV, the range is exactly equal to the detector thickness. For higher energies, only a portion of incident energy is deposited, and the transmitted proton carries off the remainder. Under these conditions, the energy deposited in the detector is given by Eq. (2.4), or simply the product of the detector thickness and the average linear stopping power. Because the stopping power continuously decreases with increasing energy in this region (see Fig. 2.3), the deposited energy therefore *decreases* with further increases in the

Figure 2.12 Energy loss of protons of initial energy E_0 in a silicon detector of 4.6 μm thickness (shown as dots). The transmitted energy for penetrating protons is also shown (as crosses). (From Wilken and Fritz.[3])

incident proton energy. The second curve in Fig. 2.12 plots the transmitted energy (E_t on the diagram on the preceding page) as recorded by a second thick detector.

F. Scaling Laws

Sometimes data are not available on the range or energy loss characteristics of precisely the same particle-absorber combination needed in a given experiment. Recourse must then be made to various approximations, most of which are derived based on the Bethe formula [Eq. (2.2)] and on the assumption that the stopping power per atom of compounds or mixtures is additive. This latter assumption, known as the *Bragg–Kleeman rule*, may be written

$$\frac{1}{N_c}\left(\frac{dE}{dx}\right)_c = \sum_i W_i \frac{1}{N_i}\left(\frac{dE}{dx}\right)_i \tag{2.5}$$

In this expression, N is the atomic density, dE/dx is the linear stopping power, and W_i represents the atom fraction of the ith component in the compound (subscript c). As an example of the application of Eq. (2.5), the linear stopping power of alpha particles in a metallic oxide could be obtained from separate data on the stopping power in both the pure metal and in oxygen. Some caution should be used in applying such results, however, since some measurements[12–15] for compounds have indicated a stopping power differing by as much as 10–20% from that calculated from Eq. (2.5).

It can be shown[11] that the range of a charged particle in a compound material can also be estimated provided its range is known in all the constituent elements. In this derivation, it is necessary to assume that the shape of the dE/dx curve is independent of the stopping medium. Under these conditions, the range in the compound is given by

$$R_c = \frac{M_c}{\sum_i n_i \left(A_i / R_i\right)} \tag{2.6}$$

where R_i is the range in element i, n_i is the number of atoms of element i in the molecule, A_i is the atomic weight of element i, and M_c is the molecular weight of the compound.

If range data are not available for all the constituent elements, estimates can be made based on a semiempirical formula (commonly called the Bragg–Kleeman rule as well)

$$\frac{R_1}{R_0} \cong \frac{\rho_0 \sqrt{A_1}}{\rho_1 \sqrt{A_0}} \qquad (2.7)$$

where ρ and A represent density and atomic weight, and subscripts 0 and 1 refer to different absorbing materials. The accuracy of this estimate diminishes when the two materials are of widely different atomic weights, so it is always best to use range data from a material that is as close as possible in A to the absorber of interest.

Range data can also be generalized to different charged particles within a given absorber material. By integration of Eq. (2.2), it can be shown that the range of a particle of mass m and charge z can be represented by

$$R(v) = \frac{m}{z^2} F(v) \qquad (2.8)$$

where $F(v)$ represents a unique function of the particle initial velocity v. For particles of the same initial velocity, this factor will be identical and therefore we can write

$$R_a(v) = \frac{m_a z_b^2}{m_b z_a^2} R_b(v) \qquad (2.9)$$

where the subscripts a and b refer to different charged particles. Thus, the range of a particle for which data are not available can be estimated by calculating its initial velocity, finding the range of any other particle of the same initial velocity in the same material, and applying Eq. (2.9). It should be emphasized that these estimates are only approximate, because no account is taken of the change in charge state of the particle as it nears the end of its path. Correction factors necessary to compensate for this effect and predict the range more accurately are presented by Evans.[1]

G. Behavior of Fission Fragments

The heavy fragments produced as a result of neutron-induced or spontaneous fission of heavy nuclei are energetic charged particles with properties somewhat different from those discussed up to this point. Because the fragments start out stripped of many electrons, their very large effective charge results in a specific energy loss greater than that encountered with any other radiations discussed in this text. Because the initial energy is also very high (see Fig. 1.4*b*), however, the range of a typical fission fragment is approximately half that of a 5 MeV alpha particle.

An important feature of a fission fragment track is the fact that the specific energy loss $(-dE/dx)$ decreases as the particle loses energy in the absorber. This behavior is in marked contrast to the lighter particles, such as alpha particles or protons, and is a result of the continuous decrease in the effective charge carried by the fragment as its velocity is reduced. The pickup of electrons begins immediately at the start of the track, and therefore the factor z in the numerator of Eq. (2.2) continuously drops. The resulting decrease in $-dE/dx$ is large enough to overcome the increase that normally accompanies a reduction in velocity. For particles with much lower initial charge state, such as the alpha particle, electron pickup does not become significant until near the end of the range.

H. Secondary Electron Emission from Surfaces

As charged particles lose their kinetic energy during the slowing-down process, many electrons from the absorber are given a sufficient impulse to travel a short distance from the original track of the particle. These include the delta rays mentioned earlier, whose energy is high enough to ionize other absorber atoms, and also electrons whose kinetic energy of only a few eV is below the minimum energy for further ionization. If an energetic charged particle is incident on or emerges from a solid surface, some of these electrons may migrate to the surface and have sufficient energy to escape. The term *secondary electrons* is conventionally applied to these escaping low-energy electrons. (*Note*: The same term is often used to characterize the much more energetic electrons formed in the gamma-ray interactions described later in this chapter—a very different type of secondary electron.)

A single heavy ion such as a fission fragment can produce hundreds of these escaping secondaries, whereas lighter alpha particles might typically result in 10 or fewer secondaries per particle.[16] Fast electrons such as beta particles are much less likely to create escaping secondaries, and only a few percent will result in secondary emission. To a first approximation, the yield of secondaries will be proportional to the energy lost within a near-surface layer whose thickness represents the maximum distance an electron will tend to migrate from its point of origin and still retain enough energy to escape from the surface. Thus, the dE/dx value of the particle is a reasonable predictor of secondary yield from a given material. Those materials with low work function and large escape distance will have the largest yield. Thin films of alkali halides, and cesium iodide in particular, are observed to produce secondaries with high yield. Models of the electron transport in alkali halides[17] confirm their suitability as prolific sources of secondary electrons.

The energy spectrum of these escaping secondary electrons is a continuum with a mean value that is very low compared with the primary particle energy. For example, the energy of secondaries from a carbon surface average 60–100 eV for alpha particles and 290 eV for fission fragments.[16] The energy of secondary electrons produced by lighter particles such as fast electrons is even lower. Thus it is normally difficult or impossible to observe the secondaries since they are readily reabsorbed even in air. If they are emitted into a vacuum or low-pressure gas, however, they can be accelerated and easily guided by electric fields because of their low initial velocity. For example, an electric field created by the application of 1000 V with respect to the surface would have a strong influence on the trajectory of a 100 eV electron, but almost no effect on a 1 MeV fast electron. This property has led to the successful use of secondary electron emission from surfaces as a means of detecting the positions at which beta particles emerge from a surface[18] and in the imaging of X-rays and fast electrons.[19] Secondary electrons emitted from a thin foil and directed to an external electron detector such as a microchannel plate have also served as the basis for fast timing measurements with heavy ions.[20,21] One special application of the secondary electron emission process will be detailed in Chapter 9, where a typical electron in a photomultiplier tube is accelerated to a few hundred eV before striking a solid surface, causing the emission of 5–10 secondaries in a charge multiplication process.

II. INTERACTION OF FAST ELECTRONS

When compared with heavy charged particles, fast electrons lose their energy at a lower rate and follow a much more tortuous path through absorbing materials. A series of tracks from a source of monoenergetic electrons might appear as in the sketch below:

Large deviations in the electron path are now possible because its mass is equal to that of the orbital electrons with which it is interacting, and a much larger fraction of its energy can be lost in a single encounter. In addition, electron–nuclear interactions, which can abruptly change the electron direction, sometimes occur.

A. Specific Energy Loss

An expression similar to that of Eq. (2.2) has also been derived by Bethe to describe the specific energy loss due to ionization and excitation (the "collisional losses") for fast electrons:

$$-\left(\frac{dE}{dx}\right)_c = \frac{2\pi e^4 N Z}{m_0 v^2} \left(\ln \frac{m_0 v^2 E}{2I^2(1-\beta^2)} - (\ln 2)(2\sqrt{1-\beta^2} - 1 + \beta^2) \right.$$
$$\left. + (1-\beta^2) + \frac{1}{8}\left(1 - \sqrt{1-\beta^2}\right)^2 \right) \quad (2.10)$$

where the symbols have the same meaning as in Eq. (2.2) and $\beta \equiv v/c$.

Electrons also differ from heavy charged particles in that energy may be lost by radiative processes as well as by coulomb interactions. These radiative losses take the form of *bremsstrahlung* or electromagnetic radiation, which can emanate from any position along the electron track. From classical theory, any charge must radiate energy when accelerated, and the deflections of the electron in its interactions with the absorber correspond to such acceleration. The linear specific energy loss through this radiative process is

$$-\left(\frac{dE}{dx}\right)_r = \frac{NEZ(Z+1)e^4}{137 m_0^2 c^4} \left(4 \ln \frac{2E}{m_0 c^2} - \frac{4}{3} \right) \quad (2.11)$$

For the particle types and energy ranges of interest in this text, only fast electrons can have a significant yield of bremsstrahlung. The yield from heavy charged particles is negligible as indicated by the presence of the m_0^2 factor in the denominator of the multiplicative term in Eq. (2.11). The factors of E and Z^2 in the numerator of Eq. (2.11) show that radiative losses are most important for high electron energies and for absorber materials of large atomic number. For typical electron energies, the average bremsstrahlung photon energy is quite low (see Fig. 1.6) and is therefore normally reabsorbed fairly close to its point of origin. In some cases, however, the escape of bremsstrahlung can influence the response of small detectors.

The total linear stopping power for electrons is the sum of the collisional and radiative losses:

$$\frac{dE}{dx} = \left(\frac{dE}{dx}\right)_c + \left(\frac{dE}{dx}\right)_r \quad (2.12)$$

The ratio of the specific energy losses is given approximately by

$$\frac{(dE/dx)_r}{(dE/dx)_c} \cong \frac{EZ}{700} \quad (2.13)$$

where E is in units of MeV. For the electrons of interest here (such as beta particles or secondary electrons from gamma-ray interactions), typical energies are less than a few MeV. Therefore, radiative losses are always a small fraction of the energy losses due to ionization and excitation and are significant only in absorber materials of high atomic number.

B. Electron Range and Transmission Curves

1. ABSORPTION OF MONOENERGETIC ELECTRONS

An attenuation experiment of the type discussed earlier for alpha particles is sketched in Fig. 2.13 for a source of monoenergetic fast electrons. Even small values of the absorber thickness lead to the loss of some electrons from the detected beam because scattering of the electron effectively removes it from the flux striking the detector. A plot of the detected number of electrons versus absorber thickness therefore begins to drop immediately and gradually approaches zero for large absorber thicknesses. Those electrons that penetrate

Figure 2.13 Transmission curve for monoenergetic electrons. R_e is the extrapolated range.

the greatest absorber thickness will be the ones whose initial direction has changed least in their path through the absorber.

The concept of range is less definite for fast electrons than for heavy charged particles, because the electron total path length is considerably greater than the distance of penetration along the initial velocity vector. Normally, the electron range is taken from a transmission plot, such as that given in Fig. 2.13, by extrapolation of the linear portion of the

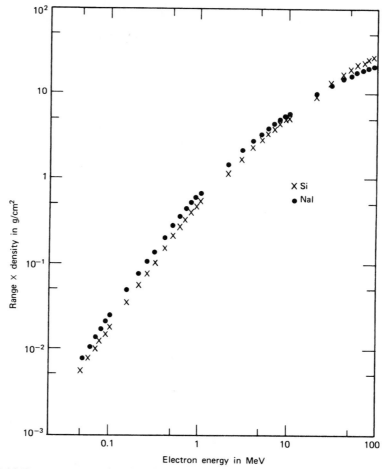

Figure 2.14 Range–energy plots for electrons in silicon and sodium iodide. If units of mass thickness (distance × density) are used for the range as shown, values at the same electron energy are similar even for materials with widely different physical properties or atomic number. (Data from Mukoyama.[24])

curve to zero and represents the absorber thickness required to ensure that almost no electrons can penetrate the entire thickness.

For equivalent energy, the specific energy loss of electrons is much lower than that of heavy charged particles, so their path length in typical absorbers is hundreds of times greater. As a very crude estimate, electron ranges tend to be about 2 mm per MeV in low-density materials, or about 1 mm per MeV in materials of moderate density.

Tabular data are given in Refs. 22 and 23 for the stopping power and range of electrons and positrons in elements and compounds, covering a large region of energy. To a fair degree of approximation, the product of the range times the density of the absorber is a constant for different materials for electrons of equal initial energy. Plots of the range of electrons in two common detector materials are given in Fig. 2.14.

2. ABSORPTION OF BETA PARTICLES

The transmission curve for beta particles emitted by a radioisotope source, because of the continuous distribution in their energy, differs significantly from that sketched in Fig. 2.13 for monoenergetic electrons. The "soft" or low-energy beta particles are rapidly absorbed even in small thicknesses of the absorber, so that the initial slope on the attenuation curve is much greater. For the majority of beta spectra, the curve happens to have a near-exponential shape and is therefore nearly linear on the semilog plot of the type shown in Fig. 2.15. This exponential behavior is only an empirical approximation and does not have a fundamental basis as does the exponential attenuation of gamma rays [see Eq. (2.20)]. An *absorption coefficient n* is sometimes defined by

$$\frac{I}{I_0} = e^{-nt} \tag{2.14}$$

where

I_0 = counting rate without absorber

I = counting rate with absorber

t = absorber thickness in g/cm^2

The coefficient n correlates well with the endpoint energy of the beta emitter for a specific absorbing material. This dependence is shown in Fig. 2.16 for aluminum. Through the use of such data, attenuation measurements can be used to identify indirectly endpoint

Figure 2.15 Transmission curves for beta particles from
^{185}W (endpoint energy of 0.43 MeV). (From Baltakmens.[25])

Figure 2.16 Beta particle absorption coefficient n in aluminum as a function of the endpoint energy E_m, average energy E_{av}, and $E' \equiv 0.5\,(E_m + E_{av})$ of different beta emitters. (From Baltakmens.[26])

energies of unknown beta emitters, although direct energy measurements are more common.

3. BACKSCATTERING

The fact that electrons often undergo large-angle deflections along their tracks leads to the phenomenon of *backscattering*. An electron entering one surface of an absorber may undergo sufficient deflection so that it re-emerges from the surface through which it entered. These backscattered electrons do not deposit all their energy in the absorbing medium and therefore can have a significant effect on the response of detectors designed to measure the energy of externally incident electrons. Electrons that backscatter in the detector "entrance window" or dead layer will escape detection entirely.

Backscattering is most pronounced for electrons with low incident energy and absorbers with high atomic number. Figure 2.17 shows the fraction of monoenergetic electrons that are backscattered when normally incident on the surface of various absorbers. Additional data for materials commonly used as electron detectors are given in Table 10.1.

Backscattering can also influence the apparent yield from radioisotope sources of beta particles or conversion electrons. If the source is deposited on a thick backing, electrons that are emitted initially into this backing may backscatter and re-emerge from the surface of the source.

C. Positron Interactions

The coulomb forces that constitute the major mechanism of energy loss for both electrons and heavy charged particles are present for either positive or negative charge on the particle. Whether the interaction involves a repulsive or attractive force between the incident particle and orbital electron, the impulse and energy transfer for particles of equal mass are about the same. Therefore, the tracks of positrons in an absorber are similar to those of normal negative electrons, and their specific energy loss and range are about the same for equal initial energies.

Positrons differ significantly, however, in that the annihilation radiation described in Chapter 1 is generated at the end of the positron track. Because these 0.511 MeV photons are very penetrating compared with the range of the positron, they can lead to the deposition of energy far from the original positron track.

Figure 2.17 Fraction η of normally incident electrons that are backscattered from thick slabs of various materials, as a function of incident energy E. (From Tabata et al.[27])

III. INTERACTION OF GAMMA RAYS

Although a large number of possible interaction mechanisms are known for gamma rays in matter, only three major types play an important role in radiation measurements: *photoelectric absorption, Compton scattering,* and *pair production.* All these processes lead to the partial or complete transfer of the gamma-ray photon energy to electron energy. They result in sudden and abrupt changes in the gamma-ray photon history, in that the photon either disappears entirely or is scattered through a significant angle. This behavior is in marked contrast to the charged particles discussed earlier in this chapter, which slow down gradually through continuous, simultaneous interactions with many absorber atoms. The fundamentals of the gamma-ray interaction mechanisms are introduced here but are again reviewed at the beginning of Chapter 10 in the context of their influence on the response of gamma-ray detectors.

A. Interaction Mechanisms

1. PHOTOELECTRIC ABSORPTION

In the photoelectric absorption process, a photon undergoes an interaction with an absorber atom in which the photon completely disappears. In its place, an energetic *photoelectron* is ejected by the atom from one of its bound shells. The interaction is with the atom as a whole and cannot take place with free electrons. For gamma rays of sufficient energy, the most probable origin of the photoelectron is the most tightly bound or K shell of the atom. The photoelectron appears with an energy given by

$$E_{e^-} = h\nu - E_b \tag{2.15}$$

where E_b represents the binding energy of the photoelectron in its original shell. For gamma-ray energies of more than a few hundred keV, the photoelectron carries off the majority of the original photon energy.

In addition to the photoelectron, the interaction also creates an ionized absorber atom with a vacancy in one of its bound shells. This vacancy is quickly filled through capture of a free electron from the medium and/or rearrangement of electrons from other shells of the atom. Therefore, one or more characteristic X-ray photons may also be generated. Although in most cases these X-rays are reabsorbed close to the original site through photoelectric absorption involving less tightly bound shells, their migration and possible escape from radiation detectors can influence their response (see Chapter 10). In some fraction of the cases, the emission of an Auger electron may substitute for the characteristic X-ray in carrying away the atomic excitation energy.

As an example of the complexity of these interactions, consider incident photons with energy above 30 keV that undergo photoelectric absorption in xenon. About 86% interact through K-shell absorptions in the xenon atom.[28] Of these, 87.5% result in K-shell characteristic (or "fluorescent") X-rays (a mixture of K-α and K-β) and 12.5% de-excite through the emission of Auger electrons. The remaining 14% of the incident photons that do not undergo K-shell interactions are absorbed through photoelectric interaction with the L or M shells. These result in much lower energy characteristic X-rays or Auger electrons that are very short range and, to first approximation, are reabsorbed very near the site of the original interaction. Another example of such a "cascade sequence" is shown in Chapter 10 as Fig. 10.20.

The photoelectric process is the predominant mode of interaction for gamma rays (or X-rays) of relatively low energy. The process is also enhanced for absorber materials of high atomic number Z. No single analytic expression is valid for the probability of photoelectric absorption per atom over all ranges of E_γ and Z, but a rough approximation is

$$\tau \cong \text{constant} \times \frac{Z^n}{E_\gamma^{3.5}} \tag{2.16}$$

where the exponent n varies between 4 and 5 over the gamma-ray energy region of interest.[1] This severe dependence of the photoelectric absorption probability on the atomic number of the absorber is a primary reason for the preponderance of high-Z materials (such as lead) in gamma-ray shields. As further detailed in Chapter 10, many detectors used for gamma-ray spectroscopy are chosen from high-Z constituents for the same reason.

A plot of the photoelectric absorption cross section for a popular gamma-ray detection material, sodium iodide, is shown in Fig. 2.18. In the low-energy region, discontinuities in the curve or "absorption edges" appear at gamma-ray energies that correspond to the binding energies of electrons in the various shells of the absorber atom. The edge lying highest in energy therefore corresponds to the binding energy of the K-shell electron. For gamma-ray energies slightly above the edge, the photon energy is just sufficient to undergo a photoelectric interaction in which a K-electron is ejected from the atom. For gamma-ray

Figure 2.18 Energy dependence of the various gamma–ray interaction processes in sodium iodide. (From *The Atomic Nucleus* by R. D. Evans. Copyright 1955 by the McGraw-Hill Book Company. Used with permission.)

energies slightly below the edge, this process is no longer energetically possible and therefore the interaction probability drops abruptly. Similar absorption edges occur at lower energies for the $L, M, \ldots$ electron shells of the atom.

2. COMPTON SCATTERING

The interaction process of Compton scattering takes place between the incident gamma-ray photon and an electron in the absorbing material. It is most often the predominant interaction mechanism for gamma-ray energies typical of radioisotope sources.

In Compton scattering, the incoming gamma-ray photon is deflected through an angle θ with respect to its original direction. The photon transfers a portion of its energy to the electron (assumed to be initially at rest), which is then known as a *recoil electron*. Because all angles of scattering are possible, the energy transferred to the electron can vary from zero to a large fraction of the gamma-ray energy.

The expression that relates the energy transfer and the scattering angle for any given interaction can simply be derived by writing simultaneous equations for the conservation of energy and momentum. Using the symbols defined in the sketch below

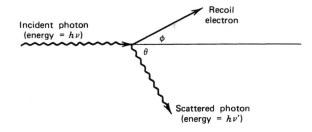

we can show[1] that

$$hv' = \frac{hv}{1 + \dfrac{hv}{m_0 c^2}(1 - \cos \theta)}$$ (2.17)

where $m_0 c^2$ is the rest-mass energy of the electron (0.511 MeV). For small scattering angles θ, very little energy is transferred. Some of the original energy is always retained by the incident photon, even in the extreme of $\theta = \pi$. Equations (10.3) through (10.6) describe some properties of the energy transfer for limiting cases. A plot of the scattered photon energy predicted from Eq. (2.17) is also shown in Fig. 10.7.[†]

The probability of Compton scattering per atom of the absorber depends on the number of electrons available as scattering targets and therefore increases linearly with Z. The dependence on gamma-ray energy is illustrated in Fig. 2.18 for the case of sodium iodide and generally falls off gradually with increasing energy.

The angular distribution of scattered gamma rays is predicted by the *Klein–Nishina formula* for the differential scattering cross section $d\sigma/d\Omega$:

$$\frac{d\sigma}{d\Omega} = Zr_0^2 \left(\frac{1}{1 + \alpha(1 - \cos \theta)} \right)^2 \left(\frac{1 + \cos^2 \theta}{2} \right) \left(1 + \frac{\alpha^2 (1 - \cos \theta)^2}{(1 + \cos^2 \theta)[1 + \alpha(1 - \cos \theta)]} \right)$$ (2.18)

where $\alpha \equiv hv/m_0 c^2$ and r_0 is the classical electron radius. The distribution is shown graphically in Fig. 2.19 and illustrates the strong tendency for forward scattering at high values of the gamma-ray energy.

3. PAIR PRODUCTION

If the gamma-ray energy exceeds twice the rest-mass energy of an electron (1.02 MeV), the process of pair production is energetically possible. As a practical matter, the probability of this interaction remains very low until the gamma-ray energy approaches several MeV and therefore pair production is predominantly confined to high-energy gamma rays. In the interaction (which must take place in the coulomb field of a nucleus), the gamma-ray photon disappears and is replaced by an electron–positron pair. All the excess energy carried in by the photon above the 1.02 MeV required to create the pair goes into kinetic energy shared by the positron and the electron. Because the positron will subsequently annihilate

[†]The simple analysis here neglects the atomic binding of the electron and assumes that the gamma-ray photon interacts with a free electron. If the small binding energy is taken into account, the unique energy of the scattered photon at a fixed angle predicted by Eq. 2.17 is spread into a narrow distribution centered about that energy (see Fig. 13.9).

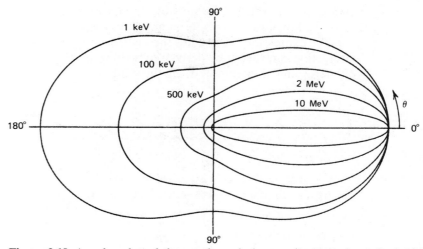

Figure 2.19 A polar plot of the number of photons (incident from the left) Compton scattered into a unit solid angle at the scattering angle θ. The curves are shown for the indicated initial energies.

after slowing down in the absorbing medium, two annihilation photons are normally produced as secondary products of the interaction. The subsequent fate of this annihilation radiation has an important effect on the response of gamma-ray detectors, as described in Chapter 10.

No simple expression exists for the probability of pair production per nucleus, but its magnitude varies approximately as the square of the absorber atomic number.[1] The importance of pair production rises sharply with energy, as indicated in Fig. 2.18.

The relative importance of the three processes described above for different absorber materials and gamma-ray energies is conveniently illustrated in Fig. 2.20. The line at the left represents the energy at which photoelectric absorption and Compton scattering are equally probable as a function of the absorber atomic number. The line at the right represents the energy at which Compton scattering and pair production are equally probable. Three areas are thus defined on the plot within which photoelectric absorption, Compton scattering, and pair production each predominate.

Figure 2.20 The relative importance of the three major types of gamma-ray interaction. The lines show the values of Z and $h\nu$ for which the two neighboring effects are just equal. (From *The Atomic Nucleus* by R. D. Evans. Copyright 1955 by the McGraw-Hill Book Company. Used with permission.)

4. COHERENT SCATTERING

In addition to Compton scattering, another type of scattering can occur in which the gamma-ray photon interacts coherently with all the electrons of an absorber atom. This *coherent scattering* or *Rayleigh scattering* process[1] neither excites nor ionizes the atom, and the gamma-ray photon retains its original energy after the scattering event. Because virtually no energy is transferred, this process is often neglected in basic discussions of gamma-ray interactions, and we will also ignore it in the discussions that follow. However, the direction of the photon is changed in coherent scattering, and complete models of gamma-ray transport must take it into account. The probability of coherent scattering is significant only for low photon energies (typically below a few hundred keV for common materials) and is most prominent in high-Z absorbers. The average deflection angle decreases with increasing energy, further restricting the practical importance of coherent scattering to low energies.

B. Gamma-Ray Attenuation

1. ATTENUATION COEFFICIENTS

If we again picture a transmission experiment as in Fig. 2.21, where monoenergetic gamma rays are collimated into a narrow beam and allowed to strike a detector after passing through an absorber of variable thickness, the result should be simple exponential attenuation of the gamma rays as also shown in Fig. 2.21. Each of the interaction processes removes the gamma-ray photon from the beam either by absorption or by scattering away from the detector direction and can be characterized by a fixed probability of occurrence per unit path length in the absorber. The sum of these probabilities is simply the probability per unit path length that the gamma-ray photon is removed from the beam:

$$\mu = \tau(\text{photoelectric}) + \sigma(\text{Compton}) + \kappa(\text{pair}) \tag{2.19}$$

and is called the *linear attenuation coefficient*. The number of transmitted photons I is then given in terms of the number without an absorber I_0 as

$$\frac{I}{I_0} = e^{-\mu t} \tag{2.20}$$

The gamma-ray photons can also be characterized by their *mean free path* λ, defined as the average distance traveled in the absorber before an interaction takes place. Its value can be obtained from

$$\lambda = \frac{\displaystyle\int_0^\infty x e^{-\mu x}\, dx}{\displaystyle\int_0^\infty e^{-\mu x}\, dx} = \frac{1}{\mu} \tag{2.21}$$

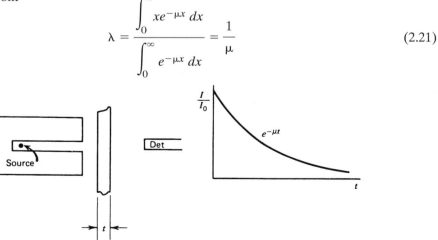

Figure 2.21 The exponential transmission curve for gamma rays measured under "good geometry" conditions.

and is simply the reciprocal of the linear attenuation coefficient. Typical values of λ range from a few mm to tens of cm in solids for common gamma-ray energies.

Use of the linear attenuation coefficient is limited by the fact that it varies with the density of the absorber, even though the absorber material is the same. Therefore, the *mass attenuation coefficient* is much more widely used and is defined as

$$\text{mass attenuation coefficient} = \frac{\mu}{\rho} \tag{2.22}$$

where ρ represents the density of the medium. For a given gamma-ray energy, the mass attenuation coefficient does not change with the physical state of a given absorber. For example, it is the same for water whether present in liquid or vapor form. The mass attenuation coefficient of a compound or mixture of elements can be calculated from

$$\left(\frac{\mu}{\rho}\right)_c = \sum_i w_i \left(\frac{\mu}{\rho}\right)_i \tag{2.23}$$

where the w_i factors represent the weight fraction of element i in the compound or mixture.

2. ABSORBER MASS THICKNESS

In terms of the mass attenuation coefficient, the attenuation law for gamma rays now takes the form

$$\frac{I}{I_0} = e^{-(\mu/\rho)\rho t} \tag{2.24}$$

The product ρt, known as the *mass thickness* of the absorber, is now the significant parameter that determines its degree of attenuation. Units of mass thickness have historically been mg/cm^2, and this convention is retained in this text. The thickness of absorbers used in radiation measurements is therefore often measured in mass thickness rather than physical thickness, because it is a more fundamental physical quantity.

The mass thickness is also a useful concept when discussing the energy loss of charged particles and fast electrons. For absorber materials with similar neutron/proton ratios, a particle will encounter about the same number of electrons passing through absorbers of equal mass thickness. Therefore, the stopping power and range, when expressed in units of ρt, are roughly the same for materials that do not differ greatly in Z.

3. BUILDUP

The gamma-ray attenuation experiment of Fig. 2.21, in which the gamma rays are collimated to a narrow beam before striking the absorber, is sometimes characterized as a "narrow beam" or "good geometry" measurement. The essential characteristic is that only gamma rays from the source that escape interaction in the absorber can be counted by the detector. Real measurements are often carried out under different circumstances (as sketched below) in which the severe collimation of the gamma rays is absent.

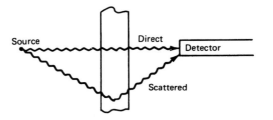

Now the detector can respond either to gamma rays directly from the source, to gamma rays that reach the detector after having scattered in the absorber, or to other types of secondary photon radiation. Many types of detectors will be unable to distinguish between these possibilities, so that the measured detector signal will be larger than that recorded under equivalent "good geometry" conditions. The conditions that lead to the simple exponential attenuation of Eq. (2.20) are therefore violated in this "broad beam" or "bad geometry" measurement because of the additional contribution of the secondary gamma rays. This situation is usually handled by replacing Eq. (2.20) by the following:

$$\frac{I}{I_0} = B(t, E_\gamma)e^{-\mu t} \tag{2.25}$$

where the factor $B(t, E_\gamma)$ is called the *buildup factor*. The exponential term is retained to describe the major variation of the gamma-ray counting rate with absorber thickness, and the buildup factor is introduced as a simple multiplicative correction. The magnitude of the buildup factor depends on the type of gamma-ray detector used, because this will affect the relative weight given to the direct and secondary gamma rays. (With a detector that responds only to the direct gamma rays, the buildup factor is unity.) The buildup also depends on the specific geometry of the experiment. As a rough rule of thumb, the buildup factor for thick slab absorbers tends to be about equal to the thickness of the absorber measured in units of mean free path of the incident gamma rays, provided the detector responds to a broad range of gamma-ray energies.

IV. INTERACTION OF NEUTRONS

A. General Properties

In common with gamma rays, neutrons carry no charge and therefore cannot interact in matter by means of the coulomb force, which dominates the energy loss mechanisms for charged particles and electrons. Neutrons can also travel through many centimeters of matter without any type of interaction and thus can be totally invisible to a detector of common size. When a neutron does undergo interaction, it is with a nucleus of the absorbing material. As a result of the interaction, the neutron may either totally disappear and be replaced by one or more secondary radiations, or else the energy or direction of the neutron is changed significantly.

In contrast to gamma rays, the secondary radiations resulting from neutron interactions are almost always heavy charged particles. These particles may be produced either as a result of neutron-induced nuclear reactions or they may be the nuclei of the absorbing material itself, which have gained energy as a result of neutron collisions. Most neutron detectors utilize some type of conversion of the incident neutron into secondary charged particles, which can then be detected directly. Specific examples of the most useful conversion processes are detailed in Chapters 14 and 15.

The relative probabilities of the various types of neutron interactions change dramatically with neutron energy. In somewhat of an oversimplification, we will divide neutrons into two categories on the basis of their energy, either "fast neutrons" or "slow neutrons," and discuss their interaction properties separately. The dividing line will be at about 0.5 eV, or about the energy of the abrupt drop in absorption cross section in cadmium (the *cadmium cutoff* energy).

B. Slow Neutron Interactions

For slow neutrons, the significant interactions include elastic scattering with absorber nuclei and a large set of neutron-induced nuclear reactions. Because of the small kinetic energy of slow neutrons, very little energy can be transferred to the nucleus in elastic scattering. Consequently, this is not an interaction on which detectors of slow neutrons can be based. Elastic collisions tend to be very probable, however, and often serve to bring the slow neutron into thermal equilibrium with the absorber medium before a different type of interaction takes place. Much of the population in the slow neutron energy range will therefore be found among these *thermal neutrons*, which, at room temperature, have an average energy of about 0.025 eV.

The slow neutron interactions of real importance are neutron-induced reactions that can create secondary radiations of sufficient energy to be detected directly. Because the incoming neutron energy is so low, all such reactions must have a positive Q-value to be energetically possible. In most materials, the radiative capture reaction [or (n, γ) reaction] is the most probable and plays an important part in the attenuation or shielding of neutrons. Radiative capture reactions can be useful in the indirect detection of neutrons using activation foils as described in Chapter 19, but they are not widely applied in active neutron detectors because the secondary radiation takes the form of gamma rays, which are also difficult to detect. Instead, reactions such as (n, α), (n, p), and (n, fission) are much more attractive because the secondary radiations are charged particles. A number of specific reactions of this type are detailed in Chapter 14.

C. Fast Neutron Interactions

The probability of most neutron–induced reactions potentially useful in detectors drops off rapidly with increasing neutron energy. The importance of scattering becomes greater, however, because the neutron can transfer an appreciable amount of energy in one collision. The secondary radiations in this case are *recoil nuclei*, which have picked up a detectable amount of energy from neutron collisions. At each scattering site, the neutron loses energy and is thereby *moderated* or slowed to lower energy. The most efficient moderator is hydrogen because the neutron can lose up to all its energy in a single collision with a hydrogen nucleus. For heavier nuclei, only a partial energy transfer is possible [see Eq.(15.4) and the associated discussion].

If the energy of the fast neutron is sufficiently high, *inelastic scattering* with nuclei can take place in which the recoil nucleus is elevated to one of its excited states during the collision. The nucleus quickly de-excites, emitting a gamma ray, and the neutron loses a greater fraction of its energy than it would in an equivalent elastic collision. Inelastic scattering and the subsequent secondary gamma rays play an important role in the shielding of high-energy neutrons but are an unwanted complication in the response of most fast neutron detectors based on elastic scattering.

D. Neutron Cross Sections

For neutrons of a fixed energy, the probability per unit path length is a constant for any one of the interaction mechanisms. It is conventional to express this probability in terms of the *cross section* σ per nucleus for each type of interaction. The cross section has units of area and has traditionally been measured in units of the *barn* (10^{-28} m^2). For example, each nuclear species will have an elastic scattering cross section, a radiative capture cross section, and so on, each of which will be a function of the neutron energy. Some examples of cross section plots are given in Figs. 14.1 and 14.2.

When multiplied by the number of nuclei N per unit volume, the cross section σ is converted into the macroscopic cross section Σ

$$\Sigma = N\sigma \qquad (2.26)$$

which now has dimensions of inverse length. Σ has the physical interpretation of the probability per unit path length for the specific process described by the "microscopic" cross section σ. When all processes are combined by adding together the cross sections for each individual interaction

$$\Sigma_{\text{tot}} = \Sigma_{\text{scatter}} + \Sigma_{\text{rad. capture}} + \cdots \tag{2.27}$$

the resulting Σ_{tot} is the probability per unit path length that any type of interaction will occur. This quantity has the same significance for neutrons as the linear absorption coefficient for gamma rays defined earlier. If a narrow beam attenuation experiment is carried out for neutrons, as sketched earlier for gamma rays, the result will be the same: The number of detected neutrons will fall off exponentially with absorber thickness. In this case the attenuation relation is written

$$\frac{I}{I_0} = e^{-\Sigma_{\text{tot}}t} \tag{2.28}$$

The neutron mean free path λ is, by analogy with the gamma-ray case, given by $1/\Sigma_{\text{tot}}$. In solid materials, λ for slow neutrons may be of the order of a centimeter or less, whereas for fast neutrons, it is normally tens of centimeters.

Under most circumstances, neutrons are not narrowly collimated so that typical shielding situations involve broad beam or "bad geometry" conditions. Just as in the case of gamma rays, the exponential attenuation of Eq. (2.28) is no longer an adequate description because of the added importance of scattered neutrons reaching the detector. A more complex neutron transport computation is then required to predict the number of transmitted neutrons and their distribution in energy.

When discussing the rate of reactions induced by neutrons, it is convenient to introduce the concept of neutron flux. If we first consider neutrons with a single energy or fixed velocity v, the product $v\Sigma$ gives the interaction frequency for the process for which Σ is the macroscopic cross section. The reaction rate density (reactions per unit time and volume) is then given by $n(r)v\Sigma$, where $n(r)$ is the neutron number density at the vector position r, and $n(r)v$ is defined as the *neutron flux* $\varphi(r)$ with dimensions of length^{-2} time^{-1}. Thus, the reaction rate density is given by the product of the neutron flux and the macroscopic cross section for the reaction of interest:

$$\text{reaction rate density} = \varphi(r)\Sigma \tag{2.29}$$

This relation can be generalized to include an energy-dependent neutron flux $\varphi(r, E)$ and cross section $\Sigma(E)$:

$$\text{reaction rate density} = \int_0^\infty \varphi(r, E)\,\Sigma(E)\,dE \tag{2.30}$$

V. RADIATION EXPOSURE AND DOSE

Because of their importance in personnel protection at radiation-producing facilities and in the medical applications of radiation, the concepts of radiation exposure and dose play prominent roles in radiation measurements. In the following sections, we introduce the fundamental concepts that underlie the quantities and units of importance in this area. Precise definitions of quantities related to radiation protection are continually evolving, and the reader is referred to publications of the International Commission on Radiological Units and Measurements (ICRU), the International Commission on Radiological Protection (ICRP), and the U.S. National Commission on Radiation Protection (NCRP) for up-to-date information. As examples, Refs. 29 and 30 are comprehensive reviews of the units and detailed issues that arise in the monitoring of radiation exposures.

A. Gamma-Ray Exposure

The concept of gamma-ray exposure was introduced early in the history of radioisotope research. Defined only for sources of X- or gamma rays, a fixed exposure rate exists at every point in space surrounding a source of fixed intensity. The exposure is linear, in that doubling the source intensity also doubles the exposure rate everywhere around the source.

The basic unit of gamma-ray exposure is defined in terms of the charge dQ due to ionization created by the secondary electrons (negative electrons and positrons) formed within a volume element of air and mass dm, when these secondary electrons are completely stopped in air. The exposure value X is then given by dQ/dm. The SI unit of gamma-ray exposure is thus the coulomb per kilogram (C/kg), which has not been given a special name. The historical unit has been the *roentgen* (R), defined as the exposure that results in the generation of one electrostatic unit of charge (about 2.08×10^9 ion pairs) per 0.001293 g (1 cm^3 at STP) of air. The two units are related by

$$1 \text{ R} = 2.58 \times 10^{-4} \text{ C/kg}$$

The exposure is therefore defined in terms of the effect of a given flux of gamma rays on a test volume of air and is a function only of the intensity of the source, the geometry between the source and test volume, and any attenuation of the gamma rays that may take place between the two. Its measurement fundamentally requires the determination of the charge due to ionization produced in air under specific conditions. Inherent in the above definition is the obligation to track each secondary electron created by primary gamma-ray interactions in the test volume under consideration, and to add up the ionization charges formed by that secondary electron until it reaches the end of its path. This requirement is often difficult or impossible to achieve in actual practice, and therefore instruments that are designed to measure gamma-ray exposure usually employ approximations that involve the principle of compensation introduced in Chapter 5.

The gamma-ray exposure, although not directly tied to physical phenomena, is often of interest in gamma-ray dosimetry. It is therefore often convenient to be able to calculate the exposure rate at a known distance from a point radioisotope source. If we assume that the yield per disintegration of X- and gamma rays is accurately known for the radioisotope of interest, the exposure per unit activity of the source at a known distance can simply be expressed under the following conditions:

1. The source is sufficiently small so that spherical geometry holds (i.e., the photon flux diminishes as $1/d^2$, where d is the distance to the source).
2. No attenuation of the X- or gamma rays takes place in the air or other material between the source and measuring point.
3. Only photons passing directly from the source to the measuring point contribute to the exposure, and any gamma rays scattered in surrounding materials may be neglected.

The exposure rate $\dot{X}$ is then

$$\dot{X} = \Gamma_\delta \frac{\alpha}{d^2} \tag{2.31}$$

where α is the activity of the source, and Γ_δ is defined as the *exposure rate constant* for the specific radioisotope of interest. The subscript δ implies that the assumption has been made that all X- and gamma rays emitted by the source above an energy δ contribute to the dose, whereas those below this energy are not sufficiently penetrating to be of practical interest. The value of Γ_δ for a particular radioisotope can be calculated from its gamma-ray yield and the energy-dependent absorption properties of air. Some particular values for $\delta = 0$ are listed in Table 2.1.

Table 2.1 Exposure Rate Constant for Some
Common Radioisotope Gamma-Ray Sources

Nuclide	Γ^a
Antimony-124	9.8
Cesium-137	3.3
Cobalt-57	0.9
Cobalt-60	13.2
Iodine-125	~0.7
Iodine-131	2.2
Manganese-54	4.7
Radium-226	8.25
Sodium-22	12.0
Sodium-24	18.4
Technetium-99m	1.2
Zinc-65	2.7

[a]The exposure rate constant Γ is in units of $(R \cdot cm^2)/(hr \cdot mCi)$.

Source: The Health Physics and Radiological Health Handbook,
Nucleon Lectern Associates, Olney, MD, 1984.

B. Absorbed Dose

Two different materials, if subjected to the same gamma-ray exposure, will in general absorb different amounts of energy. Because many important phenomena, including changes in physical properties or induced chemical reactions, would be expected to scale as the energy absorbed per unit mass of the material, a unit that measures this quantity is of fundamental interest. The energy absorbed from any type of radiation per unit mass of the absorber is defined as the *absorbed dose*. The historical unit of absorbed dose has been the *rad*, defined as 100 ergs/gram. As with other historical radiation units, the rad is being gradually replaced by its SI equivalent, the *gray* (Gy) defined as 1 joule/kilogram. The two units are therefore simply related by:

$$1 \text{ Gy} = 100 \text{ rad}$$

The absorbed dose should be a reasonable measure of the chemical or physical effects created by a given radiation exposure in an absorbing material. Careful measurements have shown that the absorbed dose in air corresponding to a gamma-ray exposure of 1 coulomb/kilogram amounts to 33.8 joules/kilogram, or 33.8 Gy. If water is substituted for the air, its absorption properties per unit mass do not differ greatly because the average atomic number of water is much the same as that of air. For absorbing materials of greatly dissimilar atomic numbers, however, the interaction mechanisms have different relative importance, and therefore the absorbed dose per unit exposure would show greater differences.

In order to measure absorbed dose in a fundamental manner, some type of energy measurement must be carried out. One possibility is a calorimetric measurement in which the rate of rise of the temperature in a sample of absorber is used to calculate the rate of energy deposition per unit mass. Because of their difficulty, these measurements are not commonplace for routine application since the thermal effects created even by large doses of radiation are very small. Instead, indirect measurements of absorbed dose are much more common, in which its magnitude is inferred from ionization measurements carried out under proper conditions (see Chapter 5).

C. Dose Equivalent

When the effects of radiation on living organisms are evaluated, the absorption of equal amounts of energy per unit mass under different irradiation conditions does not guarantee the same biological effect. In fact, the extent of the effects can differ by as much as an order of magnitude depending on whether the energy is deposited in the form of heavy charged particles or electrons.

The biological damage created by ionizing radiation is traceable to the chemical alteration of the biological molecules that are influenced by the ionization or excitation caused by the radiation. The severity and permanence of these changes are directly related to the local rate of energy deposition along the particle track, known as the *linear energy transfer*[†] *L*. Those radiations with large values of *L* (such as heavy charged particles) tend to result in greater biological damage than those with lower *L* (such as electrons), even though the total energy deposited per unit mass is the same.

The concept of *dose equivalent* has therefore been introduced to more adequately quantify the probable biological effect of the given radiation exposure. A unit of dose equivalent is defined as that amount of any type of radiation that, when absorbed in a biological system, results in the same biological effect as one unit of absorbed dose delivered in the form of low-*L* radiation. The dose equivalent *H* is the product of the absorbed dose *D* and the *quality factor Q* that characterizes the specific radiation:

$$H = DQ \qquad (2.32)$$

The quality factor increases with linear energy transfer *L* as shown in Table 2.2.

For the fast electron radiations of interest in this text, *L* is sufficiently low so that *Q* is essentially unity in all applications. Therefore, the dose equivalent is numerically equal to the absorbed dose for beta particles or other fast electrons. The same is true for X-rays and gamma rays because their energy is also delivered in the form of fast secondary electrons. Charged particles have a much higher linear energy transfer and the dose equivalent is larger than the absorbed dose. For example, *Q* is approximately 20 for alpha particles of typical energies. Because neutrons deliver most of their energy in the form of heavy

Table 2.2 Quality Factors for Different Radiations	
L in Water (keV/μm)	*Q*
< 10	1
10–100	$0.32L - 2.2$
> 100	$300/\sqrt{L}$

From ICRP Publication 60.

[†]The linear energy transfer is nearly identical to the specific energy loss $(-dE/dx)$ defined earlier. The only differences arise when a substantial portion of the radiation energy is liberated in the form of bremsstrahlung, which may travel a substantial distance from the particle track before depositing its energy. The specific energy loss includes the bremsstrahlung as part of the energy loss of the particle, but the linear energy transfer *L* counts only the energy that is deposited along the track and therefore excludes the bremsstrahlung.

charged particles, their effective quality factor is also considerably greater than unity and varies significantly with the neutron energy.

The units used for dose equivalent H depend on the corresponding units of absorbed dose D in Eq. (2.32). If D is expressed in the historical unit of the rad, H is defined to be in units of the *rem*. Historically, the rem (or millirem) was universally used to quantify dose equivalent. Under the SI convention, D is instead expressed in grays, and a corresponding unit of dose equivalent called the *sievert* (Sv) has been introduced. For example, an absorbed dose of 2 Gy delivered by radiation with Q of 5 will result in a dose equivalent of 10 Sv. Because 1 Gy = 100 rad, the units are interrelated by

$$1 \text{ Sv} = 100 \text{ rem}$$

Guidelines for radiation exposure limits to personnel are quoted in units of dose equivalent in order to place exposures to different types and energies of radiation on a common basis.

D. Fluence-to-Dose Conversion

For neutrons and gamma rays that are relatively penetrating radiations, it is sometimes convenient to be able to estimate the dose equivalent to exposed personnel that results from a given *fluence* of the radiation. The fluence is generally defined as $\Phi = dN/da$ where dN represents the differential number of gamma-ray photons or neutrons that are incident on a sphere with differential cross-sectional area da. In terms of the neutron flux defined on p. 57, the fluence is the time integral of the flux over the exposure duration. For a monodirectional beam, the fluence is simply the number of photons or neutrons per unit area, where the area is perpendicular to the direction of the beam. The number of counts recorded from typical radiation detectors is most easily interpreted in terms of the number of radiation-induced interactions in the detector, and therefore is more closely related to fluence than to dose. A conversion between fluence and dose can therefore be helpful in interpreting measurements made using specific detectors.

Because the attenuation of gamma rays or fast neutrons in air is negligible over small distances, the direct (or unscattered) fluence often can be estimated from point sources by assuming that $\Phi = N/4\pi d^2$, where N is the number of photons or neutrons emitted by the source and d is the distance from the source. More complicated geometries require the use of radiation transport codes to predict the fluence and energy spectrum at a given location.

The conversion from fluence to dose must take into account the energy-dependent probabilities of producing ionizing secondary particles from gamma-ray and/or neutron interactions, the kinetic energy that is deposited by these particles, and the quality factor Q that should be applied to convert deposited energy per unit mass to dose equivalent. By assuming a physical model for the human body, gamma-ray and neutron transport codes can be applied to calculate energy deposition and dose for the various tissues and organ systems. These calculations will be sensitive to the direction of incidence of the radiation because of self-shielding and attenuation effects within the body. Taking into account the differing radiosensitivity of various organs and tissues through the tissue-weighting factors recommended by the ICRP (see following section), the individual dose components can then be combined into an *effective dose equivalent* H_E (see Ref. 31) to represent an estimate of the overall biological effect of a uniform, whole body exposure to the assumed fluence. The effective dose equivalent is then written as

$$H_E = h_E \Phi \tag{2.33}$$

where h_E is the fluence-to-dose factor. Values of h_E for gamma rays and neutrons are plotted as a function of energy in Fig. 2.22.

(a)

(b)

Figure 2.22 Fluence-to-dose conversion factors h_E for uniform whole body exposures to (*a*) gamma rays and (*b*) neutrons. The labels refer to different assumed directions of the incident particle flux: AP for frontal exposure of the body, PA for rear exposure, LAT for exposure from the side, ROT for uniform rotation of the body about its axis, perpendicular to the directional flux, and ISO for an isotropic incident flux. Tabulated data and additional details are given in Ref. 31, which is the source of these data. (Copyright 1992 by the American Nuclear Society, LaGrange Park, Illinois.)

E. ICRP Dose Units

In its report ICRP Publication 60, the International Commission on Radiation Protection has introduced a somewhat different set of definitions that are also used in the radiation protection literature. A new quantity is named the *equivalent dose* and given the symbol $H_{T,R}$. It is obtained from the absorbed dose $D_{T,R}$ averaged over a tissue or organ T due to radiation R, multiplied by a radiation weighting factor w_R that accounts for the different biological effects of various radiations:

$$H_{T,R} = w_R \cdot D_{T,R} \tag{2.34}$$

Table 2.3 Radiation Weighting Factors

Type and Energy Range	Radiation Weighting Factor, w_R
Photons, all energies	1
Electrons and muons, all energies	1
Neutrons, energy < 10 keV	5
10 keV to 100 keV	10
> 100 keV to 2 MeV	20
> 2 MeV to 20 MeV	10
> 20 Me V	5
Protons, other than recoil protons, energy > 2 MeV	5
Alpha particles, fission fragments, heavy nuclei	20

From ICRP Publication 60.

Values of w_R are shown in Table 2.3. Units for equivalent dose are sieverts when the absorbed dose is expressed in grays. If there is a mix of radiations, then the total equivalent dose H_T is given by a sum over all radiation types:

$$H_T = \sum_R H_{T,R} = \sum_R w_R \cdot D_{T,R} \tag{2.35}$$

To account for the differing radiosensitivity of various organs and tissues, a set of tissue-weighting factors w_T are introduced to allow calculation of what is now called the *effective dose E:*

$$E = \sum_T w_T \cdot H_T \tag{2.36}$$

This quantity is intended to fill the same role as the effective dose equivalent H_E described in the previous section, in that it is an estimate of the overall effect of a given exposure to radiation. The tissue-weighting factors are chosen to give greater weight to those organs and tissues that are most sensitive to radiation. The factors are normalized in that their sum is unity, so that a uniform equivalent dose H_T to the whole body of one sievert will result in an effective dose E of one sievert. Recommended values for the tissue-weighting factors can be found in ICRP Publication 60.

It should be emphasized that the estimation of the effect of a given exposure to ionizing radiation is by its very nature an inexact science.[32] Biological effects are not absolute physical quantities that can be measured with high precision. The concepts of effective dose or effective dose equivalent are intended only to provide guidance in approximating the potential effects of a given exposure to radiation and should not be regarded as highly accurate or exactly reproducible quantities.

PROBLEMS

2.1 Estimate the time required for a 5 MeV alpha particle to slow down and stop in silicon. Repeat for the same particle in hydrogen gas.

2.2 With the aid of Fig. 2.7, estimate the energy remaining in a beam of 5 MeV protons after passing through 100 μm of silicon.

2.3 Using Fig. 2.10, find the approximate energy loss of 1 MeV alpha particles in a thickness of 5 μm of gold.

2.4 Estimate the range of 1 MeV electrons in aluminum with the aid of Fig. 2.14.

2.5 Calculate the energy of a 1 MeV gamma-ray photon after Compton scattering through 90°.

2.6 Give a rough estimate of the ratio of the probability per atom for photoelectric absorption in silicon to that in germanium.

2.7 Indicate which of the three major interaction processes (photoelectric absorption, Compton scattering, pair production) is dominant in the following situations:

(a) 1 MeV gamma rays in aluminum.

(b) 100 keV gamma rays in hydrogen.

(c) 100 keV gamma rays in iron.

(d) 10 MeV gamma rays in carbon.

(e) 10 MeV gamma rays in lead.

2.8 (a) From Fig. 2.18, calculate the mean free path of 1 MeV gamma rays in sodium iodide (specific gravity = 3.67).

(b) What is the probability that a 600 keV gamma ray undergoes photoelectric absorption in 1 cm of sodium iodide?

2.9 Define the following terms:

(a) Absorber mass thickness.

(b) Buildup.

(c) Neutron inelastic scattering.

(d) Macroscopic cross section.

(e) Neutron flux.

2.10 For 140 keV gamma rays, the mass attenuation coefficients for hydrogen and oxygen are 0.26 and 0.14 cm²/g, respectively. What is the mean free path in water at this energy?

2.11 How many 5 MeV alpha particles are required to deposit a total energy of 1 J?

2.12 A beam of 1 MeV electrons strikes a thick target. For a beam current of 100 microamperes (μA), find the power dissipated in the target.

2.13 Using the data in Table 2.1, estimate the exposure rate 5 m from a 1-Ci source of ^{60}Co.

2.14 Calculate the rate of temperature rise in a sample of liquid water adiabatically exposed to radiation that results in an absorbed dose rate of 10 mrad/h.

2.15 Using the data in Fig. 2.22b, estimate the effective dose equivalent to an individual who spends an 8-hour day working at an average distance of 5 m from a 3 μg ^{252}Cf fast neutron source (see discussion of spontaneous fission sources in Chapter 1).

REFERENCES

1. R. D. Evans, *The Atomic Nucleus*, Krieger, New York, 1982.
2. A. Beiser, *Revs. Mod. Phys.* **24**, 273 (1952).
3. B. Wilken and T. A. Fritz, *Nucl. Instrum. Meth.* **138**, 331 (1976).
4. D. J. Skyrme, *Nucl. Instrum. Meth.* **57**, 61 (1967).
5. C.F. Williamson, J. P. Boujot, and J. Picard, CEA-R3042 (1966).
6. C. Hanke and J. Laursen, *Nucl. Instrum. Meth.* **151**, 253 (1978).
7. W. H. Barkas and M. J. Berger, National Academy of Sciences, National Research Council, Publications 1133, 103 (1964).
8. L. C. Northcliffe and R. F. Schilling, *Nuclear Data Tables*, **A7**, 233 (1970).
9. J. F. Ziegler, *The Stopping and Ranges of Ions in Matter*, Pergamon Press, New York, 1977.
10. "Stopping Powers and Ranges for Protons and Alpha Particles," ICRU Report 49, ICRU, Bethesda (1993).
11. G. Cesini, G. Lucarini, and F. Rustichelli, *Nucl. Instrum. Meth.* **127**, 579 (1975).
12. P. D. Bourland and D. Powers, *Phys. Rev. B* **3**, 3635 (1971).
13. J. S.-Y. Feng, W. K. Chu, and M-A. Nicolet, *Phys. Rev. B* **10**, 3781 (1974).
14. S. Matteson, E. K. L. Chau, and D. Powers, *Phys. Rev. A* **14**, 169 (1976).
15. D. I. Thwaites, *Nucl. Instrum. Meth.* **B69**, 53 (1992).
16. S. Saro et al., *Nucl. Instrum. Meth.* **A381**, 520 (1996).
17. A. Akkerman, et al., *J. Appl. Phys.* **76** (8), 4656 (1994).
18. Y. K. Dewaraja, R. F. Fleming, M. A. Ludington, and R. H. Fleming, *IEEE Trans. Nucl. Sci.* **41 (4)**, 871 (1994).
19. A. Breskin, *Nucl. Phys. B (Proc. Suppl.)* **44**, 351 (1995).
20. T. Odenweller, et al., *Nucl. Instrum. Meth.* **198**, 263 (1982).
21. W. Starzecki, A. M. Stefanini, S. Lunardi, and C. Signorini, *Nucl. Instrum. Meth.* **193**, 499 (1982).
22. M. J. Berger and S. M. Seltzer, National Bureau Of Standards Publication NBSIR 82-2550-A (1982).
23. L. Pages et al., *Atomic Data* **4**, 1 (1972).
24. T. Mukoyama, *Nucl. Instrum. Meth.* **134**, 125 (1976).
25. T. Baltakmens, *Nucl. Instrum. Meth.* **82**, 264 (1970).
26. T. Baltakmens, *Nucl. Instrum. Meth.* **142**, 535 (1977).
27. T. Tabata, R. Ito, and S. Okabe, *Nucl. Instrum. Meth.* **94**, 509 (1971).
28. G. Zavattini et al., *Nucl. Instrum. Meth.* **A401**, 206 (1997).
29. "1990 Recommendations of the International Commission on Radiological Protection," ICRP Publication 60, Pergamon Press, Oxford (1991).
30. "Quantities and Units in Radiation Protection Dosimetry," ICRU Report 51, ICRU, Bethesda (1993).
31. "Neutron and Gamma-Ray Fluence-to-Dose Factors," American National Standard ANSI/ANS-6.1.1-1991.
32. W. K. Sinclair, *Health Phys.* **70(6)**, 781 (1996).

Chapter 3

Counting Statistics and Error Prediction

Radioactive decay is a random process. Consequently, any measurement based on observing the radiation emitted in nuclear decay is subject to some degree of statistical fluctuation. These inherent fluctuations represent an unavoidable source of uncertainty in all nuclear measurements and often can be the predominant source of imprecision or error. The term *counting statistics* includes the framework of statistical analysis required to process the results of nuclear counting experiments and to make predictions about the expected precision of quantities derived from these measurements.

The value of counting statistics falls into two general categories. The first is to serve as a check on the normal functioning of a piece of nuclear counting equipment. Here a set of measurements is recorded under conditions in which all aspects of the experiment are held as constant as possible. Because of the influence of statistical fluctuations, these measurements will not all be the same but will show some degree of internal variation. The amount of this fluctuation can be quantified and compared with predictions of statistical models. If the amount of observed fluctuation is not consistent with predictions, one can conclude that some abnormality exists in the counting system. The second application is generally more valuable and deals with the situation in which we have only one measurement. We can then use counting statistics to predict its inherent statistical uncertainty and thus estimate an accuracy that should be associated with that single measurement.

The distinctions made in the organization of this chapter are a critical part of the topic. The confusion that often arises when the student is first introduced to counting statistics originates more from a failure to keep separate the concepts presented in Sections I and II than from any other single cause. In Section I we are careful to limit the discussion to methods used in the characterization or organization of experimental data. We are not particularly concerned where these data come from but rather are interested only in presenting the formal methods by which we can describe the amount of fluctuation displayed by the data. In Section II, we discuss the separate topic of probabilistic mathematical models, which can sometimes represent real measurement systems. For purposes of the discussion in Section II, however, we are concerned only with the structure and predictions of these models as mathematical entities. We reserve, until Section III, the demonstration of how the statistical models can be matched to experimental data, resulting in the two common applications of counting statistics mentioned above. In Section IV, we examine how the predicted statistical uncertainties propagate through the calculations typically needed to produce a quoted final result that is calculated from counting data. The final three sections of the chapter illustrate some further examples of applications of statistical principles in radiation measurements.

I. CHARACTERIZATION OF DATA

We begin by assuming that we have a collection of N independent measurements of the same physical quantity:

$$x_1, x_2, x_3, \ldots, x_i, \ldots, x_N$$

We further assume that a single typical value x_i from this set can only assume integer values so that the data might represent, for example, a number of successive readings from a radiation counter for repeated time intervals of equal length. Two elementary properties of this data set are

$$\text{Sum} \qquad \Sigma \equiv \sum_{i=1}^{N} x_i \qquad (3.1)$$

$$\text{Experimental mean} \qquad \bar{x}_e \equiv \Sigma/N \qquad (3.2)$$

The experimental mean is written with the subscript to distinguish it from the mean of a particular statistical model that will be introduced later.

It is often convenient to represent the data set by a corresponding *frequency distribution function* $F(x)$. The value of $F(x)$ is the relative frequency with which the number appears in the collection of data. By definition

$$F(x) \equiv \frac{\text{number of occurrences of the value } x}{\text{number of measurements } (= N)} \qquad (3.3)$$

The distribution is automatically normalized, that is,

$$\sum_{x=0}^{\infty} F(x) = 1 \qquad (3.4)$$

As long as we do not care about the specific sequence of the numbers, the complete data distribution function $F(x)$ represents all the information contained in the original data set.

Table 3.1 Example of Data Distribution Function

Data		Frequency Distribution Function	
8	14	$F(3) = 1/20$	$= 0.05$
5	8	$F(4)$	$= 0$
12	8	$F(5)$	$= 0.05$
		$F(6)$	$= 0.10$
10	3	$F(7)$	$= 0.10$
13	9	$F(8)$	$= 0.20$
7	12	$F(9)$	$= 0.10$
		$F(10)$	$= 0.15$
9	6	$F(11)$	$= 0.05$
10	10	$F(12)$	$= 0.10$
6	8	$F(13)$	$= 0.05$
		$F(14)$	$= 0.05$
11	7	$\sum_{x=0}^{\infty} F(x)$	$= 1.00$

For purposes of illustration, Table 3.1 gives a hypothetical set of data consisting of 20 entries. Because these entries range from a minimum of 3 to maximum of 14, the data distribution function will have nonzero values only between these extreme values of the argument x. The corresponding values of $F(x)$ are also shown in Table 3.1.

A plot of the data distribution function for the example is given in Fig. 3.1. Also shown directly above the plot is a horizontal bar graph of the original 20 numbers from which the distribution was derived. These data show an experimental mean of 8.8, and the distribution function is in some sense centered about that value. Furthermore, the relative shape of the distribution function indicates qualitatively the amount of internal fluctuation in the data set. For example, Fig. 3.2 shows the shape of the distribution functions corresponding to two extreme sets of data: one with large amounts of scatter about the mean and one with little scatter. An obvious conclusion is that the width of the distribution function is a relative measure of the amount of fluctuation or scattering about the mean inherent in a given set of data.

It is possible to calculate the experimental mean by using the data distribution function, because the mean of any distribution is simply its first moment

$$\bar{x}_e = \sum_{x=0}^{\infty} x \, F(x) \tag{3.5}$$

It is also possible to derive another parameter, called the *sample variance*, which will serve to quantify the amount of internal fluctuation in the data set. The first step is to define the

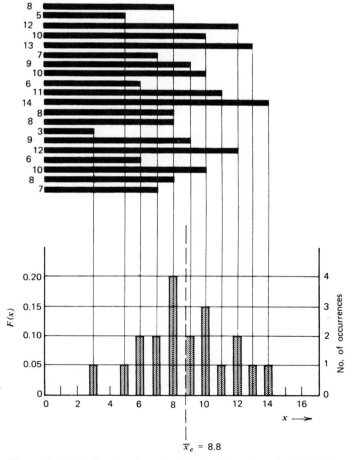

Figure 3.1 Distribution function for the data given in Table 3.1.

Figure 3.2 Distribution functions for two sets of data with differing amounts of internal fluctuation.

residual of any data point as the amount by which it differs from the experimental mean value

$$d_i \equiv x_i - \bar{x}_e \tag{3.6}$$

To illustrate, the example of the 20 numbers given in Table 3.1 is shown as the bar graph of Fig. 3.3*a*. The individual residuals of these values have been separately plotted in part (*b*) of the figure. There must be an equal contribution of positive and negative residuals, so that

$$\sum_{i=1}^{N} d_i = 0$$

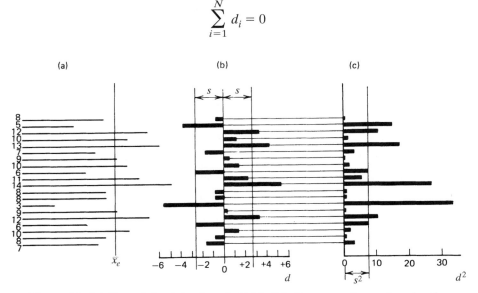

Figure 3.3 (*a*) Plot of the data given in Table 3.1. Corresponding values for the residuals *d* and for d^2 are shown in parts (*b*) and (*c*).

If we take the square of each residual, however, a positive number will always result. These are plotted for the example in Fig. 3.3c.

We next define the *deviation* of a given data point as the amount by which it differs from the true mean value $\overline{x}$

$$\epsilon_i \equiv x_i - \overline{x} \tag{3.7}$$

The deviation defined in this way is similar to the residual introduced above, except that the distance from the true mean value $\overline{x}$ appears in the definition rather than the experimental mean $\overline{x}_e$. We now can introduce the definition of the sample variance as the average value of each of these deviations after squaring

$$s^2 \equiv \overline{\epsilon^2} = \frac{1}{N} \sum_{i=1}^{N} (x_i - \overline{x})^2 \tag{3.8}$$

The sample variance is a useful index of the degree of the internal scatter in the data or as a measure of how different a typical number is from another.

This definition presents a practical difficulty, since we can never know the exact value of the true mean $\overline{x}$ without collecting an infinite number of data points. The best we can do is to use the experimental mean value $\overline{x}_e$ that we have measured, and thus use residuals rather than deviations. But the process of using the experimental rather than the true mean value will affect the calculated value of the sample variance, and we cannot simply substitute $\overline{x}_e$ into Eq. 3.8. Instead, the analysis given in Appendix B shows that the alternative expression

$$s^2 = \frac{1}{N-1} \sum_{i=1}^{N} (x_i - \overline{x}_e)^2 \tag{3.9}$$

is now valid when the experimental mean is used. The sum of squared residuals in the above equation is divided by $N - 1$ rather than by N as in Eq. 3.8, a distinction that is significant only when the number of measurements N is small. For large data sets, therefore, the sample variance can be thought of as the mean squared value of either the residuals or the deviations.

The sample variance s^2 for the example of 20 numbers is shown graphically in Fig. 3.3c. Because it is essentially a measure of the average value of the squared deviations of each point, s^2 is an effective measure of the amount of fluctuation in the original data. A data set with a narrow distribution will have a small typical deviation from the mean, and therefore the value for the sample variance will be small. On the other hand, data with a large amount of fluctuation will have a wide distribution and a large value for typical deviations, and the corresponding sample variance will also be large. It is important to note that the sample variance is an absolute measure of the amount of internal scatter in the data and does not, to first approximation, depend on the number of values in the data set. For example, if the data shown in Fig. 3.3 were extended by simply collecting an additional 20 values by the same process, we would not expect the sample variance calculated for the extended collection of 40 numbers to be substantially different from that shown in Fig 3.3.

We can also calculate the sample variance directly from the data distribution function $F(x)$. Because Eq. (3.8) indicates that s^2 is simply the average of $(x - \overline{x})^2$, we can write that same average as

$$s^2 = \sum_{x=0}^{\infty} (x - \overline{x})^2 \, F(x) \tag{3.10}$$

Equation (3.10) is not introduced so much for its usefulness in computation as for the parallel it provides to a similar expression, Eq. (3.17), which will be introduced in a later discussion of statistical models. An expansion of Eq. (3.10) will yield the well-known result

$$s^2 = \overline{x^2} - (\overline{x})^2 \tag{3.11}$$

We now end our discussion of the organization of experimental data with two important conclusions:

1. Any set of data can be completely described by its frequency distribution function $F(x)$.
2. Two properties of this frequency distribution function are of particular interest: the experimental mean and the sample variance.

The experimental mean is given by Eq. (3.5) and is the value about which the distribution is centered. The sample variance is given by Eq. (3.10) and is a measure of the width of the distribution, or the amount of internal fluctuation in the data.

II. STATISTICAL MODELS

Under certain conditions, we can predict the distribution function that will describe the results of many repetitions of a given measurement. We define a measurement as counting the number of *successes* resulting from a given number of *trials*. Each trial is assumed to be a *binary* process in that only two results are possible: Either the trial is a success or it is not a success. For everything that follows, we also assume that the probability of success is a constant for all trials.

To show how these conditions apply to real situations, Table 3.2 gives three separate examples. The third example indicates the basis for applying the theoretical framework that follows to the case of counting nuclear radiation events. In this case a trial consists of observing a given radioactive nucleus for a period of time t, the number of trials is equivalent to the number of nuclei in the sample under observation, and the measurement consists of counting those nuclei that undergo decay. We identify the probability of success of any one trial as p. In the case of radioactive decay, that probability is equal to $(1 - e^{-\lambda t})$, where λ is the decay constant of the radioactive sample.

Three specific statistical models are introduced:

1. *The Binomial Distribution.* This is the most general model and is widely applicable to all constant-p processes. It is, unfortunately, computationally cumbersome in radioactive decay, where the number of nuclei is always very large, and is used only rarely in nuclear counting applications.
2. *The Poisson Distribution.* This model is a direct mathematical simplification of the binomial distribution under conditions that the success probability p is small and constant. In practical terms, this condition implies that we have chosen an observation time that is small compared with the half-life of the source. Then the number of radioactive nuclei remains essentially constant during the observation, and the probability of recording a count from a given nucleus in the sample is small.
3. *The Gaussian or Normal Distribution.* The third important distribution is the Gaussian, which is a further simplification if the average number of successes is relatively large (say greater than 20 or 30). This condition will apply for any situation in which we accumulate more than a few counts during the course of the measurement.

Table 3.2 Examples of Binary Processes

Trial	Definition of Success	Probability of Success $\equiv p$
Tossing a coin	Heads	$1/2$
Rolling a die	A six	$1/6$
Observing a given radioactive nucleus for a time t	The nucleus decays during the observation	$1 - e^{-\lambda t}$

This is most often the case so that the Gaussian model is widely applicable to many problems in counting statistics.

It should be emphasized that all the above models become identical for processes with a small individual success probability p but with a large enough number of trials so that the expected mean number of successes is large.

A. The Binomial Distribution

The binomial distribution is the most general of the statistical models discussed. If n is the number of trials for which each trial has a success probability p, then the predicted probability of counting exactly x successes can be shown[1] to be

$$P(x) = \frac{n!}{(n-x)! \, x!} \, p^x (1-p)^{n-x}$$

(3.12)

$P(x)$ is the predicted probability distribution function, as given by the binomial distribution, and is defined only for integer values of n and x.

We show one example of an application of the binomial distribution. Imagine that we have an honest die so that the numbers 1 through 6 are all equally probable. Let us define a successful roll as one in which any of the numbers 3, 4, 5, or 6 appear. Because these are four of the six possible results, the individual probability of success p is equal to $\frac{4}{6}$ or 0.667. We now roll a die a total of 10 times and record the number of rolls that result in success as defined above. The binomial distribution now allows us to calculate the probability that exactly x out of the 10 trials will be successful, where x can vary between 0 and 10. Table 3.3 gives the values of the predicted probability distribution from Eq. (3.12) for the parameters $p = \frac{2}{3}$ and $n = 10$. The results are also plotted in Fig. 3.4 . We see that 7 is the most probable number of successes from 10 rolls of the die, with a probability of occurrence slightly greater than one out of four. From the value of $P(0)$ we see that only twice out of 100,000 tests would we expect to see no successes from 10 rolls of the die.

Table 3.3 Values of the Binomial Distribution for the Parameters $p = \frac{4}{6}$ or $\frac{2}{3}, n = 10$

x	$P(x)$
0	0.00002
1	0.00034
2	0.00305
3	0.01626
4	0.05690
5	0.13656
6	0.22761
7	0.26012
8	0.19509
9	0.08671
10	0.01734

$$\sum_{x=0}^{10} P(x) = 1.00000$$

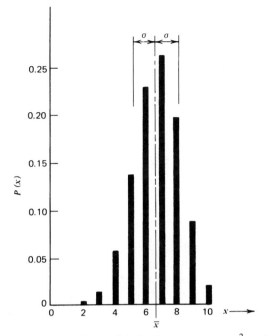

Figure 3.4 A plot of the binomial distribution for $p = \frac{2}{3}$ and $n = 10$.

Some properties of the binomial distribution are important. First, the distribution is normalized:

$$\sum_{x=0}^{n} P(x) = 1 \tag{3.13}$$

Also, we know that the average or mean value of the distribution is given by

$$\bar{x} = \sum_{x=0}^{n} xP(x) \tag{3.14}$$

If we now substitute Eq. (3.12) for $P(x)$ and carry out the summation, a remarkably simple result is derived:

$$\bar{x} = pn \tag{3.15}$$

Thus, we can calculate the expected average number of successes by multiplying the number of trials n by the probability p that any one trial will result in a success. In the example just discussed, we calculate an average number of successes as

$$\bar{x} = pn = \left(\tfrac{2}{3}\right)(10) = 6.67 \tag{3.16}$$

The mean value is obviously a very fundamental and important property of any predicted distribution.

It is also important to derive a single parameter that can describe the amount of fluctuation predicted by a given distribution. We have already defined such a parameter, called the sample variance, for a set of experimental data as defined in Eq. (3.10). By analogy we now define a *predicted variance* σ^2, which is a measure of the scatter about the mean predicted by a specific statistical model $P(x)$:

$$\sigma^2 \equiv \sum_{x=0}^{n} (x - \bar{x})^2 \, P(x) \tag{3.17}$$

Conventionally, σ^2 is called the *variance*, and we emphasize the fact that it is associated with a predicted probability distribution function by calling it a *predicted variance*. It is also conventional to define the *standard deviation* as the square root of σ^2. Recall that the variance is in some sense a typical value of the squared deviation from the mean. Therefore, σ represents a typical value for the deviation itself, hence the name "standard deviation."

Now if we carry out the summation indicated in Eq. (3.17) for the specific case of $P(x)$ given by the binomial distribution, the following result is obtained:

$$\sigma^2 = np(1 - p) \tag{3.18}$$

Because $\bar{x} = np$, we can also write

$$\sigma^2 = \bar{x}(1 - p) \tag{3.19}$$

$$\sigma = \sqrt{\bar{x}(1 - p)} \tag{3.20}$$

We now have an expression that gives an immediate prediction of the amount of fluctuation inherent in a given binomial distribution in terms of the basic parameters of the distribution, n and p, where $\bar{x} = np$.

To return to the example of rolling a die, we defined success in such a way that $p = \frac{2}{3}$. We also assumed 10 rolls of the die for each measurement so that $n = 10$. For this example, the predicted mean number of successes is 6.67 and we can proceed to calculate the predicted variance

$$\sigma^2 = np(1 - p) = (10)(0.667)(0.333) = 2.22 \tag{3.21}$$

By taking the square root we get the predicted standard deviation:

$$\sigma = \sqrt{\sigma^2} = \sqrt{2.22} = 1.49 \tag{3.22}$$

The significance of the standard deviation is illustrated in Fig. 3.4 .The mean value of the distribution is shown as the dashed line, and one value of the standard deviation is shown on either side of this mean. Because σ is a typical value for the difference between a given measurement and the true value of the mean, wide distributions will have large values for σ and narrow distributions will correspond to small values. The plot illustrates that the association of σ with the width of the distribution is not inconsistent with the example shown in Fig. 3.4.

B. The Poisson Distribution

Many categories of binary processes can be characterized by a constant and small probability of success for each individual trial. Included are most nuclear counting experiments in which the number of nuclei in the sample is large and the observation time is short compared with the half-life of the radioactive species. Similarly, in a particle beam experiment, many particles from an accelerator might strike a target for every recorded reaction product. Under these conditions, the approximation holds that the success probability is small and constant, and the binomial distribution reduces to the Poisson form shown below. For circumstances in which the observation is not short compared with the half-life of the source, the simple Poisson shown below no longer holds and a modified Poisson distribution, outlined in Appendix C, will apply.

It can be shown[2] that for a constant and small probability of success, the binomial distribution reduces to the form

$$P(x) = \frac{(pn)^x e^{-pn}}{x!} \tag{3.23}$$

Because $pn = \bar{x}$ holds for this distribution as well as for the parent binomial distribution,

$$P(x) = \frac{(\bar{x})^x e^{-\bar{x}}}{x!} \tag{3.24}$$

which is now the familiar form of the Poisson distribution.

Recall that the binomial distribution requires values for two parameters: the number of trials n and the individual success probability p. We note from Eq. (3.24) that a significant simplification has occurred in deriving the Poisson distribution—only one parameter is required, which is the product of n and p. This is a very useful simplification because now we need only know the mean value of the distribution in order to reconstruct its amplitude at all other values of the argument. That is a great help for processes in which we can in some way measure or estimate the mean value, but for which we have no idea of either the individual probability or the size of the sample. Such is usually the case in nuclear measurements.

Some properties of the Poisson distribution follow directly. First, it is also a normalized distribution, or

$$\sum_{x=0}^{n} P(x) = 1 \tag{3.25}$$

We can also calculate the first moment or mean value of the distribution:

$$\bar{x} = \sum_{x=0}^{n} x P(x) = pn \tag{3.26}$$

which is the intuitively obvious result also obtained for the binomial distribution. The predicted variance of the distribution, however, differs from that of the binomial and can be evaluated from our prior definition

$$\sigma^2 \equiv \sum_{x=0}^{n} (x - \bar{x})^2\, P(x) = pn \tag{3.27}$$

or noting the result from Eq. (3.26)

$$\sigma^2 = \bar{x} \tag{3.28}$$

The predicted standard deviation is just the square root of the predicted variance, or

$$\sigma = \sqrt{\bar{x}} \tag{3.29}$$

Thus, we see that the predicted standard deviation of any Poisson distribution is just the square root of the mean value that characterizes that same distribution. Note that the corresponding result obtained earlier for the binomial distribution [Eq. (3.20)] reduces to the above result in the limit of $p \ll 1$ already incorporated into the Poisson assumptions.

We again illustrate with an example. Suppose we randomly select a group of 1000 people and define our measurement as counting the number of current birthdays found among all members of that group. The measurement then consists of 1000 trials, each of which is a success only if a particular individual has his or her birthday today. If we assume a random distribution of birthdays, the probability of success p is equal to $1/365$. Because p is much less than one in this example, we can immediately turn to the Poisson distribution to evaluate the probability distribution function that will describe the expected results from many such samplings of 1000 people. Thus, for our example,

$$p = 1/365 = 0.00274 \qquad\qquad \bar{x} = pn = 2.74$$

$$n = 1000 \qquad\qquad \sigma = \sqrt{\bar{x}} = 1.66$$

$$P(x) = \frac{\bar{x}^x e^{-\bar{x}}}{x!} = \frac{(2.74)^x e^{-2.74}}{x!}$$

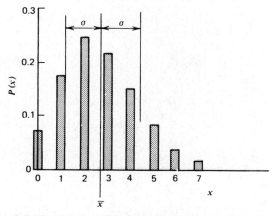

Figure 3.5 The Poisson distribution for a mean value $\bar{x} = 2.74$.

x	$P(x)$
0	0.064
1	0.177
2	0.242
3	0.221
4	0.152
5	0.083
6	0.038
7	0.014
⋮	⋮

Recall that $P(x)$ gives the predicted probability that exactly x birthdays will be observed from a random sampling of 1000 people. The numerical values are plotted in Fig. 3.5 and show that $x = 2$ is the most probable result. The mean value of 2.74 is also shown in the figure, together with one value of the standard deviation of 1.66 on either side of the mean. The distribution is roughly centered about the mean value, although considerable asymmetry is evident for this low value of the mean. Again the size of the standard deviation gives some indication of the width of the distribution or the amount of scatter predicted by the distribution.

C. The Gaussian or Normal Distribution

The Poisson distribution holds as a mathematical simplification to the binomial distribution in the limit $p \ll 1$. If, in addition, the mean value of the distribution is large (say greater than 20), additional simplifications can generally be carried out[2] which lead to the *Gaussian* distribution:

$$P(x) = \frac{1}{\sqrt{2\pi\bar{x}}} \exp\left(-\frac{(x - \bar{x})^2}{2\bar{x}}\right) \tag{3.30}$$

This is again a pointwise distribution function defined only for integer values of x. It shares the following properties with the Poisson distribution:

1. It is normalized: $\displaystyle\sum_{x=0}^{\infty} P(x) = 1$.

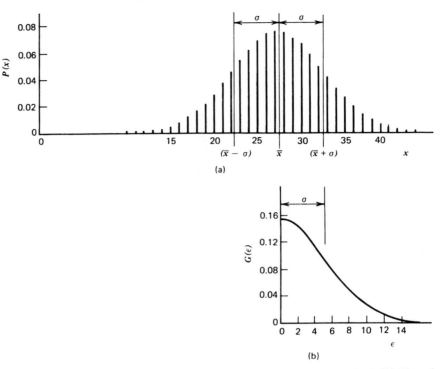

Figure 3.6 (*a*) Discrete Gaussian distribution for a mean value $x = 27.4$. (*b*) Plot of the corresponding continuous form of the Gaussian.

2. The distribution is characterized by a single parameter $\bar{x}$, which is given by the product np.

3. The predicted variance σ^2 as defined in Eq. (3.17) is again equal to the mean value $\bar{x}$.

We can again illustrate an example of a physical situation in which the Gaussian distribution is applicable. Suppose we return to the previous example of counting birthdays out of a group of randomly selected individuals, but now consider a much larger group of 10,000 people. For this example, $p = \frac{1}{365}$ and $n = 10{,}000$, so the predicted mean value of the distribution $\bar{x} = np = 27.4$. Because the predicted mean is larger than 20, we can turn to the Gaussian distribution for the predicted distribution of the results of many measurements, each of which consists of counting the number of birthdays found in a different group of 10,000 people. The predicted probability of observing a specific count x is then given by

$$P(x) = \frac{1}{\sqrt{2\pi \cdot 27.4}} \exp\left(-\frac{(x - 27.4)^2}{2 \cdot 27.4} \right) \tag{3.31}$$

and the predicted standard deviation for the example is

$$\sigma = \sqrt{\bar{x}} = \sqrt{27.4} = 5.23 \tag{3.32}$$

The results are shown graphically in Fig. 3.6*a*.

Two important observations can be made at this point about the Gaussian distribution:

1. The distribution is symmetric about the mean value $\bar{x}$. Therefore, $P(x)$ depends only on the absolute value of the deviation of any value x from the mean, defined as $\epsilon \equiv |x - \bar{x}|$.

2. Because the mean value $\bar{x}$ is large, values of $P(x)$ for adjacent values of x are not greatly different from each other. In other words, the distribution is slowly varying.

These two observations suggest a recasting of the distribution as an explicit function of the deviation ϵ (rather than of x) and as a continuous function (rather than a pointwise discrete function). These changes are accomplished by rewriting the Gaussian distribution as

$$G(\epsilon) = \sqrt{\frac{2}{\pi\bar{x}}}\, e^{-\epsilon^2/2\bar{x}} \qquad (3.33)$$

where $G(\epsilon)\, d\epsilon$ is now defined as the differential probability of observing a deviation in $d\epsilon$ about ϵ. Comparing Eq. (3.33) with Eq. (3.30), we note a factor of 2 that has entered in $G(\epsilon)$ because there are two values of x for every value of the deviation ϵ.

Figure 3.6*b* shows the continuous form of the Gaussian distribution for the same example chosen to illustrate the discrete case. Comparing Fig. 3.6*a* and 3.6*b*, the scale factors for each abscissa are the same but the origin for Fig. 3.6*b* has been shifted to illustrate that a value of zero for the deviation ϵ corresponds to the position of the mean value $\bar{x}$ on Fig. 3.6*a*. If a factor of 2 difference in the relative ordinate scale is included as shown, then the continuous distribution $G(\epsilon)$ represents the smooth curve that connects the pointwise values plotted in Fig. 3.6*a*.

Because we are now dealing with a continuous function, we must redefine some properties of the distribution as shown in Fig. 3.7. It should be particularly noted that quantities of physical interest now involve integrals of the distribution between set limits, or *areas* under the curve, rather than sums of discrete values.

Equation (3.33) can be rewritten in a more general form by incorporating several observations. We have already seen that the standard deviation σ of a Gaussian distribution is given by $\sigma = \sqrt{\bar{x}}$, or $\bar{x} = \sigma^2$. With this substitution in Eq. (3.33), the value of the exponential factor now depends only on the ratio of ϵ to σ. Formally defining this ratio as

$$t \equiv \frac{\epsilon}{\sigma}$$

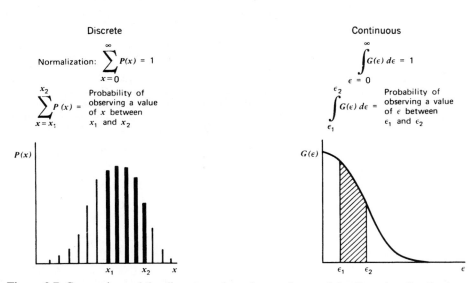

Figure 3.7 Comparison of the discrete and continuous forms of the Gaussian distribution .

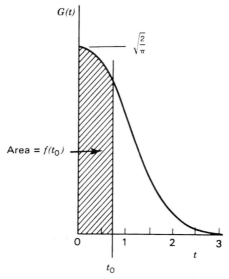

Figure 3.8 A plot of the general Gaussian curve.

the Gaussian distribution can be rewritten in terms of this new variable t:

$$G(t) = G(\epsilon) \frac{d\epsilon}{dt} = G(\epsilon)\sigma$$

$$G(t) = \sqrt{\frac{2}{\pi}}\; e^{-t^2/2} \qquad (3.34)$$

where $0 \leq t \leq \infty$. We now have a universal form of the Gaussian distribution, shown in Fig. 3.8, that is valid for all values of the mean $\bar{x}$. Recall that t is just the observed deviation $\epsilon \equiv |x - \bar{x}|$ normalized in units of the standard deviation σ.

From the definitions illustrated in Fig. 3.7, the probability that a typical normalized deviation t predicted by a Gaussian distribution will be less than a specific value t_0 is given by the integral

$$\int_0^{t_0} G(t)\, dt \equiv f(t_0)$$

Table 3.4 Probability of Occurrence of Given Deviations Predicted by the Gaussian Distribution

ϵ_0	t_0	$f(t_0)$
0	0	0
0.674 σ	0.674	0.500
σ	1.00	0.683
1.64 σ	1.64	0.900
1.96 σ	1.96	0.950
2.58 σ	2.58	0.990
3.00 σ	3.00	0.997

Tabulated values of this function can be found in most collections of statistical tables. Some selected values are given in Table 3.4. The value of $f(t_0)$ gives the probability that a random sample from a Gaussian distribution will show a normalized deviation $t(\equiv \epsilon/\sigma)$ that is less than the assumed value t_0. For example, we can conclude that about 68% of all samples will deviate from the true mean by less than one value of the standard deviation.

III. APPLICATIONS OF STATISTICAL MODELS

The first two sections of this chapter dealt with independent topics: the organization of experimental data in Section I, and the structure of certain statistical models in Section II. The practical uses of statistical analysis are now illustrated by bringing together these two separate topics.

There are two major applications of counting statistics in nuclear measurements. The first of these, "Application A," involves the use of statistical analysis to determine whether a set of multiple measurements of the same physical quantity shows an amount of internal fluctuation that is consistent with statistical predictions. This application usually is used to determine whether a particular counting system is functioning normally. Although this is a useful application, a far more valuable contribution of counting statistics arises in situations in which we have only a single experimental measurement. In "Application B" we examine the methods available to make a prediction about the uncertainty one should associate with that single measurement to account for the unavoidable effects of statistical fluctuations.

Application A: Checkout of the Counting System to See if Observed Fluctuations Are Consistent with Expected Statistical Fluctuation

A common quality control procedure in many counting laboratories is to periodically (perhaps once a month) record a series of 20–50 successive counts from the detector system while keeping all experimental conditions as constant as possible. By applying the analytical procedures to be described here, it can be determined whether the internal fluctuation shown by these multiple measurements is consistent with the amount of fluctuation expected if statistical fluctuations were the only origin. In this way abnormal amounts of fluctuation can be detected which could indicate malfunctioning of some portion of the counting system.

Fig. 3.9 shows the chain of events that characterizes this application of counting statistics. Properties of the experimental data are confined to the left half of the figure, whereas on the right side are listed properties of an appropriate statistical model. We start in the upper-left corner with the collection of N independent measurements of the same physical quantity. These might be, for example, successive 1 min counts from a detector. Using the methods outlined in Section I, we can characterize the data in several ways. The data distribution function $F(x)$ as defined in Eq. (3.3) can be compiled. From this distribution, the mean value $\bar{x}_e$ and the sample variance s^2 can be computed by the formulas given in Eqs. (3.5) and (3.9). Recall that the mean value $\bar{x}_e$ gives the value about which the distribution is centered, whereas the sample variance s^2 is a quantitative measure of the amount of fluctuation present in the collection of data.

We now are faced with the task of matching these experimental data with an appropriate statistical model. Almost universally we will want to match to either a Poisson or Gaussian distribution (depending on how large the mean value is), either of which is fully specified by its own mean value $\bar{x}$. What should we choose for $\bar{x}$? We would be rather foolish if we chose any value other than $\bar{x}_e$, which is our only estimate of the mean value for the distribution from which the data have been drawn. Setting $\bar{x} = \bar{x}_e$ then provides the bridge from left to right in the figure, so that we now have a fully specified statistical model. If we let $P(x)$ represent the Poisson or Gaussian distribution with $\bar{x} = \bar{x}_e$, then the measured

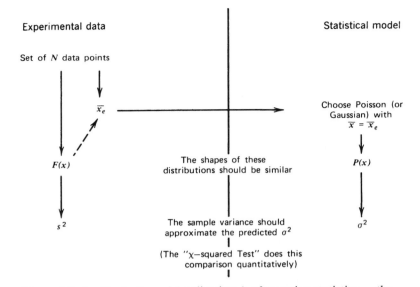

Figure 3.9 An illustration of Application A of counting statistics — the inspection of a set of data for consistency with a statistical model.

data distribution function $F(x)$ should be an approximation to $P(x)$ provided the statistical model accurately describes the distribution from which the data have arisen. One method of carrying out a comparison at this level is simply to make a superimposed plot of $F(x)$ and $P(x)$ and then to compare the shape and amplitude of the two distributions.

But such a comparison of two functions is, as yet, only qualitative. It is desirable to extract a single parameter from each distribution so that they can be compared quantitatively. The most fundamental parameter is the mean value, but these have already been matched and are the same by definition. A second parameter of each distribution is the variance, and we can carry out the desired quantitative comparison by noting the predicted variance σ^2 of the statistical model and comparing with the measured sample variance s^2 of the collection of data. If the data are actually characterized by the statistical model and show a degree of internal fluctuation that is consistent with statistical prediction, these two variance values should be about the same.

To illustrate the direct comparison of the data distribution function with the predicted probability distribution function, we return to the example of data given in Table 3.1. In Fig. 3.10 data distribution function has been replotted as the solid vertical bars. The mean value for these data was calculated to be $\bar{x}_e = 8.8$, so the transition to the appropriate statistical model is made by assuming its mean value to be $\bar{x} = 8.8$. Because the mean value is not large, we are prohibited from using the Gaussian distribution and we therefore use the Poisson as the assumed statistical model. The points on Fig. 3.10 are the values of the predicted distribution function of the Poisson distribution for a mean value of 8.8. Because the Poisson is defined only for discrete values of x, the continuous curve is drawn only to connect the points for visual reference.

At this point a comparison of the two distributions is difficult. Because relatively few data points were obtained (20 measurements) the value of $F(x)$ at each point is subject to rather large fluctuations. One would expect that, if more data were gathered, the fluctuations would diminish and the data distribution function $F(x)$ would adhere more and more closely to the predicted probability distribution function $P(x)$, provided the data are indeed a true sample from the predicted statistical model. From Fig. 3.10 we can only say that the experimental data are not grossly at variance with the prediction.

Figure 3.10 A direct comparison of experimental data (from Table 3.1) with predictions of a statistical model (the Poisson distribution for $\bar{x} = 8.8$).

To take the comparison one step further, we would now like to compare the value of the sample variance and the predicted variance from the statistical model. The sample variance calculated from Eq. (3.8) for the same set of data is found to be

$$s^2 = 7.36$$

Because the assumed statistical model is the Poisson distribution, the predicted variance is given by

$$\sigma^2 = \bar{x} = 8.80$$

These two results show that there is less fluctuation in the data than would be predicted if the data were a perfect sample from a Poisson distribution of the same mean. With a limited sample size, however, one would not expect these two parameters to be precisely the same and a more quantitative test is required to determine whether the observed difference is really significant. This function is provided by the "chi-squared test."

Chi-squared is simply another parameter of the experimental data distribution and is defined as

$$\chi^2 \equiv \frac{1}{\bar{x}_e} \sum_{i=1}^{N} (x_i - \bar{x}_e)^2 \tag{3.35}$$

where the summation is taken over each individual data point x_i. Chi-squared is closely related to the sample variance and the two are related by

$$\chi^2 = \frac{(N-1)s^2}{\bar{x}_e} \tag{3.36}$$

Now if the amount of fluctuation present in the data is closely modeled by the Poisson distribution, then $s^2 \cong \sigma^2$. But we know that for the Poisson distribution, $\sigma^2 = \bar{x}$. Furthermore, we have chosen $\bar{x}$ to be equal to $\bar{x}_e$. Therefore, the degree to which the ratio $s^2/\bar{x}_e$ deviates from unity is a direct measure of the extent to which the observed sample variance differs from the predicted variance. Now referring to Eq. (3.36), the degree to which χ^2 differs from $N - 1$ is a corresponding measure of the departure of the data from predictions of the

Table 3.5 Portion of a Chi-Squared Distribution Table

Statistical Degrees of Freedom	Number of Measurements N	$p = 0.8$	0.7	0.6	0.5
18	19	12.85	14.44	15.89	17.33
19	20	13.72	15.35	16.85	18.33
20	21	14.58	16.26	17.80	19.34

Poisson distribution. Chi-squared distribution tables may be found (e.g. Ref. 3) that are generally cast in the form shown in Table 3.5. The column on the left indicates the number of statistical degrees of freedom in the system. (This is one less than the number of independent measurements used to derive the value of χ^2 because $\bar{x}_e$ has been calculated from the same set of data.) Each column in the table is headed by a specific value of p, defined as the probability that a random sample from a true Poisson distribution would have a larger value of χ^2 than the specific value shown in the table. Very low probabilities (say less than 0.02) indicate abnormally large fluctuations in the data, whereas very high probabilities (greater than 0.98) indicate abnormally small fluctuations. A perfect fit to the Poisson distribution for large samples would yield a probability of 0.50, whereas the somewhat arbitrary limits listed above indicate situations in which the counting system may be displaying either abnormally large fluctuations (which is the usual type of malfunction) or data that are too regular and show abnormally small fluctuations. Figure 3.11 gives a plot of the χ^2 distribution for a wider range of the parameters involved.

For the illustrative example given above, we calculate a χ^2 value of 15.89. From Table 3.5 for $N = 20$ we find (by interpolation) a value of $p = 0.66$. Because that probability is neither

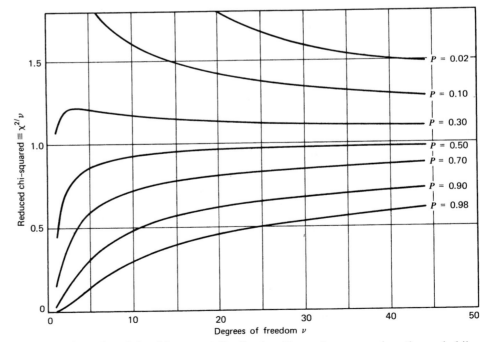

Figure 3.11 A plot of the chi-squared distribution. For each curve, p gives the probability that a random sample of N numbers from a true Poisson distribution would have a larger value χ^2/ν than that of the ordinate. For data for which the experimental mean is used to calculate χ^2, the number of degrees of freedom $\nu = N - 1$.

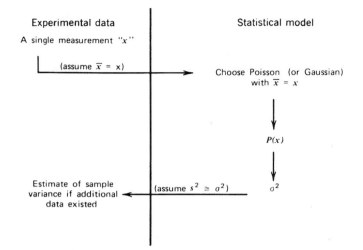

Figure 3.12 An illustration of Application B of counting statistics—prediction of the precision to be associated with a single measurement.

very large nor very small, we would conclude that the equipment used to generate the set of numbers originally shown does not give rise to abnormal fluctuations.

Application B: Estimation of the Precision of a Single Measurement

A more valuable application of counting statistics applies to the case in which we have only a single measurement of a particular quantity and wish to associate a given degree of uncertainty with that measurement. To state the objective in another way, we would like to have some estimate of the sample variance to be expected if we were to repeat the measurement many times. The square root of the sample variance should be a measure of the typical deviation of any one measurement from the true mean value and thus will serve as a single index of the degree of precision one should associate with a typical measurement from that set. Because we have only a single measurement, however, the sample variance cannot be calculated directly but must be estimated by analogy with an appropriate statistical model.

The process is illustrated in Fig. 3.12. Again, the left half of the figure deals only with experimental data, whereas the right half deals only with the statistical model. We start in the upper-left corner with a single measurement, x. If we make the assumption that the measurement has been drawn from a population whose theoretical distribution function is predicted by either a Poisson or Gaussian distribution, then we must match an appropriate theoretical distribution to the available data. For either model we must start with a value for the mean $\bar{x}$ of the distribution. Because the value of our single measurement x is the only information we have about the theoretical distribution from which it has been drawn, we have no real choice other than to assume that the mean of the distribution is equal to the single measurement, or $\bar{x} = x$. Having now obtained an assumed value for $\bar{x}$, the entire predicted probability distribution function $P(x)$ is defined for all values of x. We can also immediately find a value for the predicted variance σ^2 of that distribution. We can then use the association that, if the data are drawn from the same distribution, an estimate of the sample variance s^2 of a collection of such data should be given by σ^2. Through this process we have therefore obtained an estimate for the sample variance of a repeated set of measurements that do not exist but that represent the expected results if the single measurement were to be repeated many times.

The conclusion we reach can then be stated as follows:

The expected sample variance	$s^2 \cong \sigma^2$	of the statistical model from which we think the measurement x is drawn
	$= \bar{x}$	provided the model is either Poisson or Gaussian
	$\cong x$	because x is our only measurement on which to base an estimate of $\bar{x}$

We therefore conclude that

$$\sqrt{s^2} \cong \sigma = \sqrt{x}$$

is our best estimate of the deviation from the true mean that should typify our single measurement x.

This conclusion can be stated somewhat more quantitatively provided the assumed probability distribution function is a Gaussian (x is large). Then the range of values $x \pm \sigma$ or $x \pm \sqrt{x}$ will contain the true mean $\bar{x}$ with 68% probability. This conclusion follows directly from earlier statements about the shape of the Gaussian curve.[†] It is conventional to quote the uncertainty or "error" of a single measurement as simply one value of the standard deviation σ. If we quote a larger uncertainty, then the probability of including the true mean within the quoted interval is increased, and vice versa.

To illustrate, assume we have a single measurement $x = 100$. Then

$$\sigma = \sqrt{x} = \sqrt{100} = 10$$

Because our best estimate of the mean value of the distribution from which this measurement was drawn (the measurement itself) is large, we can assume that the parent distribution is a Gaussian. From the shape of the Gaussian curve (see Table 3.4) we can then construct Table 3.6 for the specific example. The table gives various options available in quoting the uncertainty to be associated with our single measurement. The conventional choice is to quote the measurement plus or minus one value of the standard deviation σ, or 100 ± 10. This interval is expected to contain the true mean value $\bar{x}$ with a probability of 68%. If we

Table 3.6 Examples of Error Intervals for a Single Measurement $x = 100$

Interval		Probability that the True Mean $\bar{x}$ Is Included
$x \pm 0.67\sigma$	93.3–106.7	50%
$x \pm \sigma$	90 –110	68%
$x \pm 1.64\sigma$	83.6–116.4	90%
$x \pm 2.58\sigma$	74.2–125.8	99%

[†]Technically, we can only say that the interval $\bar{x} \pm \sqrt{\bar{x}}$ has a 68% probability of containing our single measurement x. However, since the mean value has already been assumed to be large, we do not expect a large difference between $\bar{x}$ and any typical measurement x. Thus, we can interchange $\bar{x}$ and x in the statement above without seriously affecting the general conclusions.

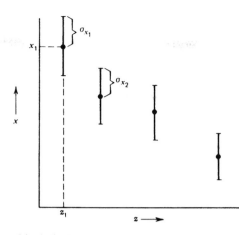

Figure 3.13 A graphical display of error bars associated with experimental data.

wish to increase the probability that the true mean is included, we can do so only by expanding the interval or error associated with the measurement. For example, to achieve a 99% probability that the true mean is included, the interval must be expanded to 2.58σ, or the range 100 ± 25.8 for our example. Unless otherwise stated, the errors quoted with a particular nuclear measurement normally represent one standard deviation.

The *fractional standard deviation*, defined as σ/x, of a simple counting measurement is given by $\sqrt{x}/x$, or $1/\sqrt{x}$. Thus, the total number of recorded counts x completely determines the fractional error to be associated with that measurement. If 100 counts are recorded, the fractional standard deviation is 10%, whereas it can be reduced to 1% only by increasing the total counts recorded to 10,000. For events occurring at a constant rate, this relation implies that the time required to achieve a given fractional error will increase as the inverse square of the desired statistical precision.

When a set of measurements is presented graphically, the estimated errors associated with each measurement are often displayed on the same graph. Figure 3.13 gives a hypothetical set of measurements of a quantity x as a function of some other variable or parameter z. The measured data are presented as points, whereas the uncertainty associated with each point is indicated by the length of the "error bar" drawn around each point. It is conventional to show the length of the error bar equal to one value of σ on either side of the point or the total length of the error bar equal to 2σ. Under these conditions, if one were to attempt a fit of an assumed functional behavior $x = f(z)$, the fitted function should at best pass through 68% (or roughly two-thirds) of all the error bars associated with the data.

Caution

All the conclusions we have drawn apply *only* to a measurement of a number of successes (number of heads in coin tossing, number of birthdays, etc.) In radioactive decay or nuclear counting, we may directly apply $\sigma = \sqrt{x}$ *only* if x represents a counted number of successes, that is, a number of events over a given observation time recorded from a detector. The vast majority of mistakes made in the use of counting statistics results from the misapplication of the above relation.

One *cannot* associate the standard deviation σ with the square root of any quantity that is not a directly measured number of counts. For example, the association does *not* apply to

   1. Counting *rates*.

   2. *Sums* or *differences* of counts.

3. *Averages* of independent counts.
4. Any *derived* quantity.

In all these cases the quantity is *calculated* as a function of the number of counts recorded in a given experiment. The error to be associated with that quantity must then be calculated according to the methods outlined in the next section.

IV. ERROR PROPAGATION

The distributions we have treated so far represent the direct results of counting experiments. In typical applications, one is seldom interested in the unprocessed data consisting of the number of counts over a particular time interval. More often the data are processed through multiplication, addition, or other functional manipulation to arrive at a derived number of more immediate interest. The values that are produced by these processing steps will be distributed in a way that is dependent on both the original distribution and the types of operations carried out.

To illustrate, assume that measurements are made under conditions in which it is expected that the "raw" counts should follow the Gaussian distribution of Eq. (3.30). Only one parameter, the mean value, is needed to fully specify the distribution, and the standard deviation is just the square root of the mean. Now assume that these counts were recorded using a background-free detector system that records only half of the quanta emitted by the source of the radiation, or, as will be formally introduced in Chapter 4, the system has an *absolute detection efficiency* of 50%. We would then multiply the number of counts by a factor of 2 to obtain a value for of the number of quanta emitted by the source. If identical measurements were then repeated many times, the multiplication process will produce a distribution of calculated emission numbers that will follow a new distribution. Clearly its mean will be twice that of the average number of counts, but what will be its shape and standard deviation?

The answer to this question is that a multiplication by a constant as in this example will produce a new distribution that also has the *shape* of a Gaussian, but for which the standard deviation σ is no longer given by the square root of the mean value. Now it is necessary to represent the distribution with *two* parameters, its mean and standard deviation. This form of the continuous Gaussian distribution analogous to Eq. (3.30) is conventionally written as

$$G(x)\,dx = \frac{1}{\sigma}\frac{1}{\sqrt{2\pi}}\exp\left(-\frac{(x-\bar{x})^2}{2\sigma^2}\right)dx$$

where $G(x)dx$ is the probability of observing a value between x and $x + dx$. The shape of the distribution is still described by Eq. (3.34):

$$G(t)dt = \sqrt{\frac{2}{\pi}}\,e^{-t^2/2}\,dt$$

where

$$t \equiv \frac{\left|x-\bar{x}\right|}{\sigma}$$

and the properties of the distribution listed in Table 3.4 remain valid.

It turns out that many of the common manipulations when carried out on counting data that were originally Gaussian distributed will produce derived values that also follow a Gaussian shape. Such operations include multiplying or dividing the data by a constant, combining two Gaussian-distributed variables through addition, subtraction, or multiplication, or calculating the average of a series of independent measurements. While the mean value of the derived distribution is simply predicted from the means of the original distributions, the standard deviation values are not so obvious. This section describes the process that can be used to predict these standard deviations. In the examples shown, the derived distribution will be of Gaussian shape provided the distribution of the original data is also Gaussian. It can be shown[3] that if the errors are individually small and symmetric about zero, a general result can be obtained for the expected error to be associated with any quantity that is calculated as a function of any number of independent variables. If x, y, z, ... are directly measured counts or related variables for which we know $\sigma_x, \sigma_y, \sigma_z, \ldots$ then the standard deviation for any quantity u derived from these counts can be calculated from

$$\sigma_u^2 = \left(\frac{\partial u}{\partial x}\right)^2 \sigma_x^2 + \left(\frac{\partial u}{\partial y}\right)^2 \sigma_y^2 + \left(\frac{\partial u}{\partial z}\right)^2 \sigma_z^2 + \ldots \tag{3.37}$$

where $u = u(x, y, z, \ldots)$ represents the derived quantity. Equation (3.37) is generally known as the *error propagation formula* and is applicable to almost all situations in nuclear measurements. The variables $x, y, z, \ldots$, however, must be chosen so that they are truly independent in order to avoid the effects of correlation. The same specific count should not contribute to the value of more than one such variable. The use of Eq. (3.37) can be illustrated by application to some simple cases.

Case 1. Sums or Differences of Counts
If we define

$$u = x + y \quad \text{or} \quad u = x - y$$

then

$$\frac{\partial u}{\partial x} = 1 \quad \text{and} \quad \frac{\partial u}{\partial y} = \pm 1$$

Application of Eq. (3.37) yields

$$\sigma_u^2 = (1)^2 \sigma_x^2 + (\pm 1)^2 \sigma_y^2$$

or

$$\boxed{\sigma_u = \sqrt{\sigma_x^2 + \sigma_y^2}} \tag{3.38}$$

In words, combining the standard distributions in this manner is frequently called "combining in quadrature" or taking their "quadrature sum."

A common application of this case arises when counts resulting from a radioactive source must be corrected by subtracting an appropriate background count. If we assume equal counting times, then

$$\text{net counts} = \text{total counts} - \text{background counts}$$

or

$$u = x - y$$

Because both x and y are directly measured numbers of counts (or successes), the expected standard deviation of each is known to be its own square root. The object is to deduce the expected standard deviation of the net counts, a derived number. Because a simple difference is involved, the answer will be given by Eq. (3.38).

To illustrate by example, suppose we have recorded the following data for equal counting times

$$\text{total counts} = x = 1071$$

$$\text{background counts} = y = \underline{521}$$

then

$$\text{net counts} = u = 550$$

We know a priori

$$\sigma_x = \sqrt{x} = \sqrt{1071}$$
$$\sigma_y = \sqrt{y} = \sqrt{521}$$

Thus

$$\sigma_u = \sqrt{\sigma_x^2 + \sigma_y^2} = \sqrt{x + y} = \sqrt{1592} = 39.9$$

We would then quote the result plus or minus one standard deviation as

$$\text{net counts} = 550 \pm 39.9$$

Case 2. Multiplication or Division by a Constant
If we define

$$u = Ax$$

where A is a constant (no associated uncertainty), then

$$\frac{\partial u}{\partial x} = A$$

and application of Eq. (3.37) gives

$$\boxed{\sigma_u = A\sigma_x} \tag{3.39}$$

Similarly, if

$$v = \frac{x}{B}$$

where B is also a constant, then

$$\boxed{\sigma_v = \frac{\sigma_x}{B}} \tag{3.40}$$

Note that, in either case, the final "fractional error" (σ_u/u or σ_v/v) is the same as the original fractional error (σ_x/x). As we would expect intuitively, multiplying or dividing a value by a constant does not change its relative error.

A familiar example of the above case is the calculation of a counting rate. If x counts are recorded over a time t, then

$$\text{counting rate} \equiv r = \frac{x}{t}$$

The usual assumption is that the time is measured with very small uncertainty, so that t can be considered a constant. Then Eq. (3.40) can be used to calculate the expected standard deviation in r corresponding to the known standard deviation in the number of counts x.

As an example, suppose

$$x = 1120 \text{ counts and } t = 5 \text{ s}$$

Then

$$r = \frac{1120}{5 \text{ s}} = 224 \text{ s}^{-1}$$

The associated standard deviation is

$$\sigma_r = \frac{\sigma_x}{t} = \frac{\sqrt{1120}}{5\text{ s}} = 6.7\text{ s}^{-1}$$

Therefore, the counting rate is

$$r = 224 \pm 6.7 \text{ counts}/\text{s}$$

Case 3. Multiplication or Division of Counts
For the case

$$u = xy, \quad \frac{\partial u}{\partial x} = y \quad \frac{\partial u}{\partial y} = x$$

$$\sigma_u^2 = y^2\sigma_x^2 + x^2\sigma_y^2$$

Dividing both sides by $u^2 = x^2y^2$

$$\left(\frac{\sigma_u}{u}\right)^2 = \left(\frac{\sigma_x}{x}\right)^2 + \left(\frac{\sigma_y}{y}\right)^2 \tag{3.41}$$

Similarly, if

$$u = \frac{x}{y}, \quad \frac{\partial u}{\partial x} = \frac{1}{y} \quad \frac{\partial u}{\partial y} = -\frac{x}{y^2}$$

$$\sigma_u^2 = \left(\frac{1}{y}\right)^2 \sigma_x^2 + \left(-\frac{x}{y^2}\right)^2 \sigma_y^2$$

Again, dividing both sides by $u^2 = x^2/y^2$

$$\left(\frac{\sigma_u}{u}\right)^2 = \left(\frac{\sigma_x}{x}\right)^2 + \left(\frac{\sigma_y}{y}\right)^2 \tag{3.41'}$$

Thus, for either $u = xy$ or $u = x/y$, the *fractional errors* in x and y (σ_x/x and σ_y/y) combine in quadrature sum to give the *fractional error* in u.

As an example, suppose we wish to calculate the ratio of two source activities from independent counts taken for equal counting times (background is neglected). Assume

$$\text{counts from source 1} \equiv N_1 = 16{,}265$$
$$\text{counts from source 2} \equiv N_2 = 8192$$
$$\text{activity ratio: } R \equiv \frac{N_1}{N_2} = \frac{16{,}265}{8192} = 1.985$$

From Eq. (3.41')

$$\left(\frac{\sigma_R}{R}\right)^2 = \left(\frac{\sigma_{N_1}}{N_1}\right)^2 + \left(\frac{\sigma_{N_2}}{N_2}\right)^2 = \frac{N_1}{N_1^2} + \frac{N_2}{N_2^2}$$

$$= 1.835 \times 10^{-4}$$

$$\frac{\sigma_R}{R} = 0.0135$$

and multiplying by the value of R

$$\sigma_R = 0.027$$

Therefore, the reported result would be

$$R = 1.985 \pm 0.027$$

Case 4. Mean Value of Multiple Independent Counts

Suppose we have recorded N repeated counts from the same source for equal counting times. Let the results of these multiple counts be designated $x_1, x_2, \ldots, x_n$ and their sum be Σ. Then

$$\Sigma = x_1 + x_2 + \cdots + x_N$$

If we formally apply the error propagation formula [Eq. (3.37)] to find the expected error in Σ, we find $\partial \Sigma / \partial x_i = 1$ for all independent counts x_i, and therefore

$$\sigma_\Sigma^2 = \sigma_{x_i}^2 + \sigma_{x_2}^2 + \cdots + \sigma_{x_N}^2$$

But because $\sigma_{x_i} = \sqrt{x_i}$ for each independent count,

$$\sigma_\Sigma^2 = x_1 + x_2 + \cdots + x_N = \Sigma$$

$$\sigma_\Sigma = \sqrt{\Sigma} \tag{3.42}$$

This result shows that the standard deviation expected for the sum of all the counts is the same as if the measurement had been carried out by performing a single count, extending over the entire period represented by all the independent counts.

Now if we proceed to calculate a mean value from these N independent measurements,

$$\bar{x} = \frac{\Sigma}{N} \tag{3.43}$$

Equation (3.43) is an example of dividing an error-associated quantity (Σ) by a constant (N). Therefore, Eq. (3.40) applies and the expected standard deviation of this mean value is given by

$$\sigma_{\bar{x}} = \frac{\sigma_\Sigma}{N} = \frac{\sqrt{\Sigma}}{N} = \frac{\sqrt{N\bar{x}}}{N}$$

$$\boxed{\sigma_{\bar{x}} = \sqrt{\frac{\bar{x}}{N}}} \tag{3.44}$$

Note that the expected standard deviation of any single measurement x_i is

$$\sigma_{x_i} = \sqrt{x_i}$$

Because any typical count will not differ greatly from the mean, $x_i \cong \bar{x}$, and we therefore conclude that the mean value based on N independent counts will have an expected error that is smaller by a factor $\sqrt{N}$ compared with any single measurement on which the mean is based. A general conclusion is that, if we wish to improve the statistical precision of a given measurement by a factor of 2, we must invest four times the initial counting time.

Case 5. Combination of Independent Measurements with Unequal Errors

If N independent measurements of the same quantity have been carried out and they do not all have nearly the same associated precision, then a simple average (as discussed in Case 4) no longer is the optimal way to calculate a single "best value." We instead want to give more weight to those measurements with small values for σ_{x_i} (the standard deviation associated with x_i) and less weight to measurements for which this estimated error is large.

Let each individual measurement x_i be given a weighting factor a_i and the best value $\langle x \rangle$ computed from the linear combination

$$\langle x \rangle = \frac{\sum\limits_{i=1}^{N} a_i x_i}{\sum\limits_{i=1}^{N} a_i} \tag{3.45}$$

We now seek a criterion by which the weighting factors a_i should be chosen in order to minimize the expected error in $\langle x \rangle$.

For brevity, we write

$$\alpha \equiv \sum_{i=1}^{N} a_i$$

so that

$$\langle x \rangle = \frac{1}{\alpha} \sum_{i=1}^{N} a_i x_i$$

Now apply the error propagation formula [Eq. (3.37)] to this case:

$$\sigma_{\langle x \rangle}^2 = \sum_{i=1}^{N} \left(\frac{\partial \langle x \rangle}{\partial x_i} \right)^2 \sigma_{x_i}^2$$

$$= \sum_{i=1}^{N} \left(\frac{a_i}{\alpha} \right)^2 \sigma_{x_i}^2$$

$$= \frac{1}{\alpha^2} \sum_{i=1}^{N} a_i^2 \sigma_{x_i}^2$$

$$\sigma_{\langle x \rangle}^2 = \frac{\beta}{\alpha^2} \tag{3.46}$$

where

$$\beta \equiv \sum_{i=1}^{N} a_i^2 \sigma_{x_i}^2$$

In order to minimize $\sigma_{\langle x \rangle}$, we must minimize $\sigma_{\langle x \rangle}^2$ from Eq. (3.46) with respect to a typical weighting factor a_j:

$$0 = \frac{\partial \sigma_{\langle x \rangle}^2}{\partial a_j} = \frac{\alpha^2 \dfrac{\partial \beta}{\partial a_j} - 2\alpha\beta \dfrac{\partial \alpha}{\partial a_j}}{\alpha^4} \tag{3.47}$$

Note that

$$\frac{\partial \alpha}{\partial a_j} = 1 \qquad \frac{\partial \beta}{\partial a_j} = 2 a_j \sigma_{x_j}^2$$

Putting these results into Eq. (3.47), we obtain

$$\frac{1}{\alpha^4} \left(2\alpha^2 a_j \sigma_{x_j}^2 - 2\alpha\beta \right) = 0$$

and solving for a_j, we find

$$a_j = \frac{\beta}{\alpha} \cdot \frac{1}{\sigma_{x_j}^2}$$

(3.48)

If we choose to normalize the weighting coefficients,

$$\sum_{i=1}^{N} a_i \equiv \alpha = 1$$

$$a_j = \frac{\beta}{\sigma_{x_j}^2}$$

Putting this into the definition of β, we obtain

$$\beta = \sum_{i=1}^{N} a_i^2 \sigma_{x_i}^2 = \sum_{i=1}^{N} \left(\frac{\beta}{\sigma_{x_i}^2}\right)^2 \sigma_{x_i}^2$$

or

$$\beta = \left(\sum_{i=1}^{N} \frac{1}{\sigma_{x_i}^2}\right)^{-1}$$

(3.49)

Therefore, the proper choice for the normalized weighting coefficient for x_j, is

$$a_j = \frac{1}{\sigma_{x_j}^2} \left(\sum_{i=1}^{N} \frac{1}{\sigma_{x_i}^2}\right)^{-1}$$

(3.50)

We therefore see that *each data point should be weighted inversely as the square of its own error.*

Assuming that this optimal weighting is followed, what will be the resultant (minimum) error in $\langle x \rangle$? Because we have chosen $\alpha = 1$ for normalization, Eq. (3.46) becomes

$$\sigma_{\langle x \rangle}^2 = \beta$$

In the case of optimal weighting, β is given by Eq. (3.49). Therefore,

$$\frac{1}{\sigma_{\langle x \rangle}^2} = \sum_{i=1}^{N} \frac{1}{\sigma_{x_i}^2}$$

(3.51)

From Eq. (3.51), the expected standard deviation $\sigma_{\langle x \rangle}$ can be calculated from the standard deviations σ_{x_i} associated with each individual measurement.

V. OPTIMIZATION OF COUNTING EXPERIMENTS

The principle of error propagation can be applied in the design of counting experiments to minimize the associated statistical uncertainty. To illustrate, consider the simple case of measurement of the net counting rate from a long-lived radioactive source in the presence of a steady-state background. Define the following:

$S \equiv$ counting rate due to the source alone without background

$B \equiv$ counting rate due to background

The measurement of S is normally carried out by counting the source plus background (at an average rate of $S + B$) for a time T_{S+B} and then counting background alone for a time T_B. The net rate due to the source alone is then

$$S = \frac{N_1}{T_{S+B}} - \frac{N_2}{T_B}$$

(3.52)

where N_1 and N_2 are the total counts in each measurement.

Applying the results of error propagation analysis to Eq. (3.52), we obtain

$$\sigma_S = \left[\left(\frac{\sigma_{N_1}}{T_{S+B}}\right)^2 + \left(\frac{\sigma_{N_2}}{T_B}\right)^2\right]^{1/2}$$

$$\sigma_S = \left(\frac{N_1}{T_{S+B}^2} + \frac{N_2}{T_B^2}\right)^{1/2}$$

$$\sigma_S = \left(\frac{S+B}{T_{S+B}} + \frac{B}{T_B}\right)^{1/2} \tag{3.53}$$

If we now assume that a fixed total time $T = T_{S+B} + T_B$ is available to carry out both measurements, the above uncertainty can be minimized by optimally choosing the fraction of T allocated to T_{S+B} (or T_B). We square Eq. (3.53) and differentiate

$$2\sigma_S \, d\sigma_S = -\frac{S+B}{T_{S+B}^2} \, dT_{S+B} - \frac{B}{T_B^2} \, dT_B$$

and set $d\sigma_S = 0$ to find the optimum condition. Also, because T is a constant, $dT_{S+B} + dT_B = 0$. The optimum division of time is then obtained by meeting the condition

$$\left.\frac{T_{S+B}}{T_B}\right|_{\text{opt}} = \sqrt{\frac{S+B}{B}} \tag{3.54}$$

A figure of merit that can be used to characterize this type of counting experiment is the inverse of the total time, or $1/T$, required to determine S to within a given statistical accuracy. If certain parameters of the experiment (such as detector size and pulse acceptance criteria) can be varied, the optimal choice should correspond to maximizing this figure of merit.

In the following analysis, we assume that the optimal division of counting times given by Eq. (3.54) is chosen. Then we can combine Eqs. (3.53) and (3.54) to obtain an expression for the figure of merit in terms of the fractional standard deviation of the source rate, defined as $\epsilon \equiv \sigma_S/S$

$$\frac{1}{T} = \epsilon^2 \frac{S^2}{(\sqrt{S+B} + \sqrt{B})^2} \tag{3.55}$$

Equation (3.55) is a useful result that can be applied to analyze the large category of radiation measurements in which a signal rate S is to be measured in the presence of a steady-state background rate B. For example, it predicts the attainable statistical accuracy (in terms of the fractional standard deviation ϵ) when a total time T is available to measure the signal plus background and the background alone. The assumption has been made that this time is subdivided optimally between the two counts. Note that, in common with simple counting measurements, the time required varies as the inverse square of the fractional standard deviation desired for the net signal rate precision. Cutting the predicted statistical error of a measurement in half requires increasing the available time by a factor of 4.

It is instructive to examine two extreme cases in the application of Eq. (3.55). If the source-induced rate is much greater than the background, $S \gg B$ and Eq. (3.55) reduces to

$$\frac{1}{T} \cong \epsilon^2 S \tag{3.56}$$

In this limit, the statistical influence of background is negligible. The figure of merit $1/T$ is maximized simply by choosing all experiment parameters to maximize S, or the rate due to the source alone.

The opposite extreme of a small source rate in a much larger background ($S \ll B$) is typical of low-level radioactivity measurements. In this case, Eq. (3.55) reduces to

$$\frac{1}{T} \cong \epsilon^2 \frac{S^2}{4B} \tag{3.57}$$

For such applications, the figure of merit is maximized by choosing experimental conditions so that the ratio S^2/B is maximized. As an example of the application of Eq. (3.57), assume that changing the detector configuration in a low-level counting experiment increases the rate due to the source alone by a factor of 1.5, but also increases the background by a factor of 2.0. The ratio S^2/B is then $(1.5)^2/2.0 = 1.125$ times its former value. Because this ratio exceeds unity, the change will slightly improve the overall statistical accuracy of the net source rate determination if the total measurement time is held constant.

VI. LIMITS OF DETECTABILITY

There are some circumstances under which it is convenient to estimate the smallest signal that can be detected reliably in order to set a "detection limit" for the counting system. Some regulatory agencies require that a certain *minimum detectable amount (MDA)* of activity be measurable while monitoring for the possible presence of radioactive contaminants. One of the most widely used definitions of MDA was first introduced by Currie.[4] It is based on a binary (yes or no) decision whether a given sample contains activity.

In the simplest case, a counting system is set up to detect the radiation of interest and the total number of counts recorded for equal periods of time as different samples are put in place. Let N_T be the number of counts recorded with an unknown sample and N_B be the number of recorded counts when a blank sample is substituted to determine the background level. (In the analysis of the previous section, it was shown that equal allocation of time to the sample and background is the proper choice to minimize the uncertainty in the net counts when the source is weak compared with background.) The net counts resulting from the unknown are then calculated as

$$N_S = N_T - N_B \tag{3.58}$$

To make a decision whether the sample contains activity, N_S is then compared with a *critical level* L_C that will be determined in the analysis that follows. A simple protocol is followed: If N_S is less than L_C, it is concluded that the sample does not contain activity, whereas if N_S exceeds L_C, it is assumed that some real activity is present.

In the absence of statistical fluctuations and other instrumental variations, L_C could be set at zero, and any net positive counts could be interpreted as evidence of real activity. With the statistical fluctuations that are inevitable in any counting measurement, however, there will be many instances of positive N_S that will be observed even for samples with no activity. One would therefore like to choose a value of L_C that is high enough to minimize the likelihood of such *false positives*, while keeping it low to reduce the possibility of missing real activity when some is actually present (*false negatives*).

If we assume that the counting time is long enough so that the total number of counts recorded in each of these measurements is large (say > 30), then both N_T and N_B should follow Gaussian distributions. When these two variables are subtracted to determine N_S, the net counts should also follow a Gaussian whose mean is the true net number of counts, and whose standard deviation can be predicted using Eq. (3.38):

$$\sigma_{N_S}^2 = \sigma_{N_T}^2 + \sigma_{N_B}^2 \tag{3.59}$$

Case 1. No Real Activity Is Present

If there is no actual activity present in the sample, then the true mean values of N_T and N_B are the same, and the mean value of N_S is zero. The expected distribution of N_S values is

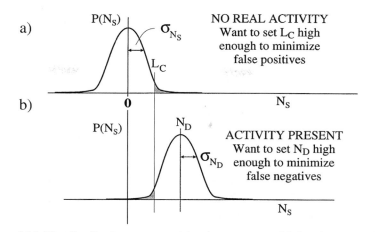

Figure 3.14 The distributions expected for the net counts N_S for the cases of (*a*) no activity present, and (*b*) a real activity present. L_C represents the critical level or "trigger point" of the counting system.

shown in Fig. 3.14*a*. Since under these conditions $\sigma_{N_T} = \sigma_{N_B}$, the standard deviation of the Gaussian distribution for N_S, from Eq. (3.59) is just

$$\sigma_{N_S} = \sqrt{2\sigma_{N_B}^2} = \sqrt{2}\ \sigma_{N_B} \tag{3.60}$$

If the only significant fluctuations entering the measurement are from counting statistics, then $\sigma_{N_B} = \sqrt{N_B}$ and $\sigma_{N_S} = \sqrt{2N_B}$. Since no true activity is present, any positive indication will be a false positive. We therefore want to set L_C high enough to ensure that, under these conditions, the probability that a particular measurement of N_S exceeds L_C is acceptably small. From Table 3.4, we note that the probability that a random sample from a Gaussian distribution having a value within the interval of the mean $\pm 1.645\sigma$ is 90%. Since we are only concerned with positive deviations from the mean (negative values of N_S are always interpreted as no source being present), there is a 95% probability that a random sample will lie below the mean plus 1.645σ. Thus setting

$$L_C = 1.645\ \sigma_{N_S} = 1.645\sqrt{2}\sigma_{N_B}$$
$$L_C = 2.326\ \sigma_{N_B} \tag{3.61}$$

ensures that a false-positive probability will be no larger than 5%. This choice has become a common basis for setting L_C.

Case 2. Real Activity Is Present

We now presume that the true mean value of N_S is a positive value, as shown in Fig. 3.14*b*. Any conclusions that no activity is present are now false negatives. The question we address is: How large must this mean value be so that false negatives are highly unlikely? Clearly, for mean values of N_S equal to L_C, the false-negative rate will be 50% because the Gaussian distribution is symmetric about its mean. We will define the MDA of activity to be the source strength necessary to produce a mean value of N_S that is high enough to reduce the false-negative rate to an acceptable level. Again, a 5% false-negative probability is commonly chosen as the basis.

Let N_D represent the minimum net value of N_S that meets this criterion. N_D will therefore become a numerical estimate of the mean number of net counts that correspond to the MDA. To ensure that 95% of the area under the N_S distribution lies above L_C, we require that

$$N_D = L_C + 1.645\sigma_{N_D} \tag{3.62}$$

As a first approximation, we assume that all signals are small compared with background, and therefore $N_D \ll N_B$. Then $\sigma_{N_D} \cong \sqrt{2}\sigma_{N_B}$ and

$$N_D = L_C + 1.645\sqrt{2}\,\sigma_{N_B}$$

$$N_D = L_C + 2.326\,\sigma_{N_B} \tag{3.63}$$

Using Eq. (3.61) for L_C

$$N_D = 2.326\,\sigma_{N_B} + 2.326\,\sigma_{N_B}$$

$$N_D = 4.653\,\sigma_{N_B} \tag{3.64}$$

Again, if only statistical fluctuations are considered, $\sigma_{N_B} = \sqrt{N_B}$. The value of N_D given by Eq. (3.64) now can serve as a first approximation to the net counts associated with the MDA, assuming that a level L_C has been set as the criteria on net counts for declaring the presence or absence of activity.

Implicit in the above derivation is the assumption that the width of the net distribution of N_S with activity is the same as the width of the net distribution without activity. Because of the small real signal, these widths are not exactly the same, and a more rigorous derivation is $\sigma_{N_D} = \sqrt{2N_B + N_D}$. Expanding for $N_D \ll N_B$:

$$\sigma_{N_D} \cong \sqrt{2N_B}\left(1 + \frac{N_D}{4N_B}\right)$$

Using the previous approximation of Eq. (3.64) that $N_D \cong 4.653\,\sigma_{N_B}$

$$\sigma_{N_D} = \sqrt{2N_B}\left(1 + \frac{4.653\,\sigma_{N_B}}{4N_B}\right) \tag{3.65}$$

Assuming all variances are from counting statistics, we can put $\sigma_{N_B} = \sqrt{N_B}$ into Eq. (3.65) to get

$$\sigma_{N_D} = \sqrt{2}\,\sigma_{N_B} + 1.645 \tag{3.66}$$

From Eq. (3.62), we require that

$$N_D = L_C + 1.645\,\sigma_{N_D}$$

Using the values for σ_{N_D} specified by Eq. (3.66) and L_C from Eq. (3.61)

$$N_D = 2.326\,\sigma_{N_B} + 1.645(\sqrt{2}\,\sigma_{N_B} + 1.645)$$

$$\boxed{N_D = 4.653\,\sigma_{N_B} + 2.706} \tag{3.67}$$

This is the frequently-quoted "Currie Equation."[4]

N_D as defined above can now be interpreted as the minimum number of counts needed from the source to ensure a false-negative rate no larger than 5% when the system is operated with a critical level (or "trigger point") of L_C that, in turn, ensures a false-positive rate of no greater than 5%. The critical level was previously chosen from Eq. (3.61) to be $L_C = 2.326\,\sigma_{N_B}$.

To convert N_D in counts to minimum detectable activity α, the additional factors of radiation yield per disintegration (f) and absolute detection efficiency (ϵ) from Chapter 4 must be taken into account:

$$\alpha = \frac{N_D}{f\epsilon T} \tag{3.68}$$

where T is the counting time per sample.

VII. DISTRIBUTION OF TIME INTERVALS

The time intervals separating random events are often of practical interest in radiation measurements. We present some results that apply to any random process characterized by a constant probability of occurrence per unit time. In statistical language, any such series of events is called a *Poisson random process*. One of the characteristics of such a process is that the system has no memory and the probability per unit time remains constant regardless of past behavior. Neglecting "dead time" (see Chapter 4), this model will usually be adequate to describe the behavior of a radiation detector undergoing irradiation at a reasonable distance from a steady-state or long-lived source of radiation. (The distance stipulation minimizes possible perturbations caused by any parent–daughter correlated events originating from the same nucleus.) No known measurements have ever been substantiated that contradict the fundamental assumption that nuclear decay is a Poisson random process.[5] By definition, the differential probability dp of occurrence of an event within a differential time interval dt is given by rdt, where r is the average rate of occurrence. For a finite time interval T, the average number of events occurring will simply be rT.

A. Intervals between Successive Events

In order to derive a distribution function to describe the time intervals between adjacent random events, first assume that an event has occurred at time $t = 0$. What is the differential probability that the next event will take place within a differential time dt after a time interval of length t? Two independent processes must take place: No events may occur within the time interval from 0 to t, but an event must take place in the next differential time increment dt. The overall probability will then be given by the product of the probabilities characterizing the two processes, or

$$
\begin{array}{ccccc}
\text{probability of next} & & \text{probability of} & & \text{probability} \\
\text{event taking place in} & = & no \text{ events during} & \times & \text{of an event} \\
dt \text{ after delay of } t & & \text{time from 0 to } t & & \text{during } dt \\
I_1(t)dt & = & P(0) & \times & r\,dt
\end{array}
\tag{3.69}
$$

The first factor on the right side follows directly from the earlier discussion of the Poisson distribution. We seek the probability that no events will be recorded over an interval of length t for which the average number of recorded events should be rt. From Eq. (3.24)

$$
P(0) = \frac{(rt)^0 e^{-rt}}{0!}
$$
$$
P(0) = e^{-rt}
\tag{3.70}
$$

Substituting Eq. (3.70) into Eq. (3.69) leads to

$$
\boxed{I_1(t)dt = re^{-rt}\,dt}
\tag{3.71}
$$

$I_1(t)$ is now the distribution function for intervals between adjacent random events. The plot below shows the simple exponential shape of this distribution.

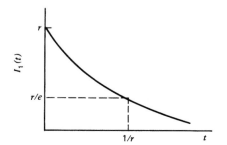

Note that the *most probable* interval is zero. The *average* interval length is calculated by

$$\bar{t} = \frac{\int_0^\infty tI_1(t)\, dt}{\int_0^\infty I_1(t)\, dt} = \frac{\int_0^\infty tre^{-rt}\, dt}{1} = \frac{1}{r} \tag{3.72}$$

which is the expected result.

In some radiation applications, the counting rate is low enough so that each individual count can be visually observed as the data are being collected. Experienced observers soon learn what a true exponential interval distribution looks like and occasionally can spot a malfunctioning detector by noting a deviation from an expected random input signal.

B. Time-to-Next-Event

In the above derivation of the exponential distribution of intervals, we started with the statement: "first assume that an event has occurred at time $t = 0$." Thus the result applies to time intervals that begin and end with true events. But we could have started with the alternative initial assumption: "select a random point in time," and the logic of Eq. (3.69) would not change. Therefore the time from a random start to the next true event will follow the same exponential distribution as in Eq. (3.71) and will have the same average value of $1/r$. This fact is exploited in the "time-to-next-event" method of measuring r described on p. 214.

At first glance, the equivalence of these two time interval distributions appears puzzling.[6] A random point in time must always fall somewhere within an interval between true events that has already begun, and it is tempting to think that therefore the average length of time to the next true event must be shorter than the average length of intervals that both begin and end with true events. Intuitively, this difference might be expected to be a factor of 2, since a randomly selected starting time would tend to fall, on the average, near the midpoint of the interval between two events. This erroneous conclusion might be further defended by correctly noting that the average time from the *previous* true event for the same random point in time is also $1/r$, since the logic of Eq. (3.69) also applies to this case.

The answer to this apparent dilemma is to note that selecting a random point in time is *not* equivalent to randomly selecting an interval from the distribution of Eq. (3.71). A randomly selected time will more likely fall within a long interval than a short one, so there is a bias toward the selection of longer intervals. The distribution describing these selected intervals can be obtained by noting that the intervals described by Eq. (3.71) are chosen with a weighting factor that is proportional to their length, and given by $t/\bar{t} = rt$. Thus the selected interval distribution $I_s(t)$ is given by:

$$I_s(t) = rtI_1(t) = r^2te^{-rt} \tag{3.73}$$

Its average value is given by the first moment of this distribution:

$$\bar{t}_s = \frac{\int_0^\infty tI_s(t)\, dt}{\int_0^\infty I_s(t)\, dt} = \frac{2}{r} \tag{3.74}$$

Thus the average selected interval is twice as long as the average random interval, and the apparent dilemma is resolved.

C. Intervals Between Scaled Events

There are some occasions in which a digital "scaler" may be employed to reduce the rate at which data are recorded from a detector system. A scaler functions as a data buffer by producing an output pulse only when N input pulses have been accumulated.

A general form for the distribution that describes scaled intervals can be derived using arguments parallel to those given earlier for unscaled intervals. Again, two independent processes must occur: A time interval of length t must be observed over which exactly $N - 1$ events are presented to the scaler, and an additional event must occur in the increment dt following this time interval. Under these conditions, a scaled interval of length within dt about t will take place. The parallel expression to Eq. (3.69) then becomes

$$I_N(t)\, dt = P(N - 1)r\, dt \tag{3.75}$$

Again using the Poisson form for $P(N - 1)$, Eq. (3.75) becomes

$$I_N(t)\, dt = \frac{(rt)^{N-1}e^{-rt}}{(N - 1)!}\, r\, dt \tag{3.76}$$

$I_N(t)$ is the interval distribution for N-scaled intervals. A plot is given in Fig. 3.15 for various scaling factors and shows the more uniform intervals that accompany larger values of N. The average interval is

$$\bar{t} = \frac{\displaystyle\int_0^\infty t I_N(t)\, dt}{\displaystyle\int_0^\infty I_N(t)\, dt} = \frac{N}{r} \tag{3.77}$$

whereas the most probable interval is evaluated by setting

$$\frac{dI_n(t)}{dt} = 0$$

and leads to

$$t\Big|_{\text{most probable}} = \frac{N - 1}{r} \tag{3.78}$$

as the most probable length of an N-scaled interval.

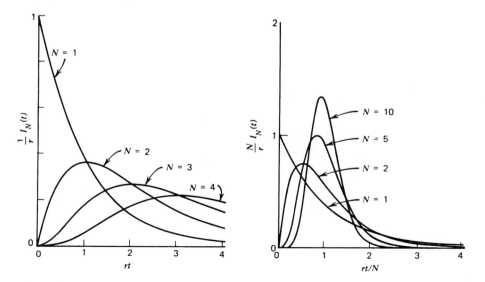

Figure 3.15 Graphical representation of the scaled interval distribution $I_N(t)$. (*a*) Four distributions for scaling factors of 1, 2, 3 and 4. (*b*) Interval distributions for $N = 1$ through $N = 10$ normalized to the same average interval N/r.

PROBLEMS

3.1 A series of 100 measurements of a physical quantity have been made that show a random fluctuation characterized by a sample variance of 2% of the mean value. If the series is lengthened to 1000 measurements made under the same conditions, estimate the sample variance of the larger set of data.

3.2 Find the probability that exactly 8 heads will occur in 12 tosses of a coin.

3.3 A given large population consists of 75% males. Random samples of 15 people are taken from this population and the number of males tallied for each sample. Find the predicted mean, variance, and standard deviation of the expected results.

3.4 Find the probability that no sixes turn up in 10 rolls of a die.

3.5 A computer programmer averages one error per 60 program statements.

  (a) Find the mean and standard deviation of the number of errors expected in a 250-statement program.

  (b) Find the probability that a 100-statement program is free of errors.

3.6 From the following list, single out those measurements for which the square root of a typical measurement is a proper estimate of the standard deviation of the distribution from which the measurement is drawn:

  (a) A 1-min count from a detector.

  (b) A 5-min count from a detector.

  (c) The net counts from a detector over a 1-min period after background subtraction.

  (d) The counting rate expressed as counts per second based on a 100-s measurement.

  (e) The average of five sequential 1-min counts.

  (f) The sum of five sequential 1-min counts.

3.7 A source is counted for 1 min and gives 561 counts. The source is removed and a 1-min background count gives 410 counts. What is the net count due to the source alone and its associated standard deviation?

3.8 A 10-min count of a source + background gives a total of 846 counts. Background alone counted for 10 min gives a total of 73 counts. What is the net counting rate due to source alone, and what is its associated standard deviation?

3.9 The measurement described in Problem 3.8 is to be repeated, but in this case the available 20 min is to be subdivided optimally between the two separate counts. Find the optimal allocation of time that minimizes the expected standard deviation in the net source counting rate. By what factor has the expected statistical error been reduced from the situation of Problem 3.8?

3.10 In a given application, a 10-min measurement resulted in a statistical uncertainty of 2.8%. How much additional time must be allocated to reduce the statistical uncertainty to 1.0%?

3.11 A designer has the choice in a specific application of either doubling the signal from the source or reducing the background by a factor of 2. From the standpoint of counting statistics, which should be chosen under the following conditions:

  (a) The signal is large compared with background.

  (b) The signal is small compared with background.

3.12 A flow counter shows an average background rate of 2.87 counts/min. What is the probability that a given 2-min count will contain (a) exactly 5 counts and, (b) at least 1 count? What length of counting time is required to ensure with > 99% probability that at least one count is recorded?

3.13 The following data are obtained from sources A and B of the same isotope:

		Timing Period
Source A + background	251 counts	5 min
Source B + background	717 counts	2 min
Background	51 counts	10 min

What is the ratio of the activity of source B to source A, and what is the percent standard deviation in this ratio?

3.14 The background count from a detector was measured to be 845 over a 30-min period. A source to be measured increases the total counting rate to about 80 counts/min. Estimate the time the source should be counted to determine the counting rate due to the source alone to within a fractional standard deviation of 3%.

3.15 Thirty different students have measured the background counting rate with the same apparatus. Each used the same procedure, consisting of recording the number of counts in five 1-min intervals and taking their average. A set of numbers from a typical student is shown below:

$$25 = \text{count in first minute}$$
$$35 = \text{count in second minute}$$
$$30 = \text{count in third minute}$$
$$23 = \text{count in fourth minute}$$
$$27 = \text{count in fifth minute}$$
$$\text{total} = \overline{140}$$

$$\text{mean} = \frac{140}{5} = 28.0 \; \frac{\text{counts}}{\text{min}}$$

  (a) Do these data seem reasonable assuming all the fluctuations are statistical? Substantiate your conclusion quantitatively.

  (b) Based on the above data, what is the expected standard deviation of the mean?

  (c) Estimate the sample variance of the 30 numbers representing a similar calculation of the mean background rate by each of the 30 students.

(d) Again assuming only statistical variations, estimate the standard deviation of the final answer for the mean obtained by averaging all 30 independent values.

3.16 The following set of 25 counts was recorded under identical detector conditions and counting times. Apply the chi-squared test to determine whether the observed fluctuations are consistent with expectations from Poisson statistics.

3626	3711	3677	3678	3465
3731	3617	3630	3624	3574
3572	3572	3615	3652	3601
3689	3578	3605	3595	3540
3625	3569	3591	3636	3629

3.17 An average of five sequential 2-min counts of a constant source by Lab Group A gave a resulting value of 2162.4 counts/min. Lab Group B then used the same source and detector in identical conditions and arrived at a value of 2081.5 counts/min based on four sequential 5-min counts. Is the difference between these two results statistically significant?

3.18 You are asked to calibrate the activity of a Cs-137 gamma-ray source by comparison with a standard Cs-137 reference source of approximately the same activity. The standard source has a quoted activity of 3.50 ± 0.05 μCi ($\pm$ one standard deviation) and either source alone gives rise to a counting rate of about 1000 per second in the available counter. Background rates are negligible. Assuming that each source is counted separately for equal counting times, how much *total* time will be required to determine the unknown activity to within a 2% expected standard deviation?

3.19 A particular counting system has a stable average background rate (measured over a long time) of 50 counts/min. A decaying radioisotope source was introduced and a 10-min count showed a total of 1683 counts. After a delay of 24 h, the 10-min count was repeated, this time giving a total of 914 counts.

(a) What is the half-life of the source?

(b) What is the expected standard deviation of the half-life value due to counting statistics?

3.20 An engine wear test is to be carried out in which the weight of radioactive piston ring particles in an oil sample is to be determined. A sample of the used oil gives 13,834 counts over a 3-min period. A standard has been prepared using exactly 100 μg of the same activity material which gives 91,396 counts over a 10-min period. Background for the detector has been determined to be 281 counts/min, measured over a very long counting period (~24 h). Find the weight of the particles in the sample and its expected fractional standard deviation.

3.21 The decay constant λ of a radioisotope sample is to be determined by counting in a detector system with negligible

background. An approximate value λ' is already known. The procedure will be to count for a short time τ at $t = 0$, wait a time Δt, and then count again for the same time τ. Assuming that $\tau \ll 1/\lambda'$, what value of the waiting time Δt will minimize the expected statistical error in the value of λ derived from these measurements?

3.22 The thickness of nominal 1-cm sheet aluminum is to be monitored by noting the attenuation of a gamma-ray parallel beam passing perpendicularly through the sheet. The source and the detector are well shielded, so background and scattering into the detector are negligible. Any given sample will spend 1 s in the beam. The detector counting rate with no sheet in place has a mean value (measured over a long time) of 10,000 per second.

(a) Find the optimum value of the linear attenuation coefficient μ that will minimize the uncertainty in the derived sheet thickness value due to statistical fluctuations. (What is the corresponding gamma-ray energy?)

(b) What is the lowest attainable fractional standard deviation under these conditions?

3.23 At an average rate of 100 per second, what fraction of the intervals between randomly occurring events are shorter than 10 ms?

3.24 A measurement of possible ^{137}Cs contamination on an air filter is made on a daily basis. The measurement consists of placing the filter in a counting system with an absolute gamma ray counting efficiency of 15% at 662 keV for a period of 30 min. A new uncontaminated filter is then substituted in the counting system and another 30-min background count recorded. The background count averages 100 counts per minute.

(a) Where should the critical level L_C be set (in units of counts per 30-min measurement) to declare the positive presence of contamination following the criteria outlined in the MDA discussion?

(b) Under these conditions, find the minimum detectable amount (MDA) of ^{137}Cs on the filter as defined in the same discussion.

3.25 (a) A municipal bus system, when operating perfectly on schedule, provides a bus at any given bus stop every 30 min. You arrive at a random time at a bus stop. What is the average waiting time until the next bus arrives?

(b) The busses become totally disorganized so that the spacing between busses is made random (assume that their arrival is a Poisson random process.) The same number of busses are still in operation. You again arrive at a bus stop at an arbitrary time. What is the average waiting time under these circumstances?

REFERENCES

1. R. D. Evans, *The Atomic Nucleus*, Krieger, New York, 1982, Chaps. 26–28.
2. W. Feller, *An Introduction to Probability Theory and Its Applications*, 2nd ed., Wiley, New York, 1957.
3. P. R. Bevington, *Data Reduction and Error Analysis for the Physical Sciences*, McGraw-Hill, New York, 1969.
4. L. A. Currie, *Anal. Chem.* **40 (3)**, 586 (1968).
5. "Particle Counting in Radioactivity Measurements," ICRU Report 52, ICRU, Bethesda (1994).
6. J. W. Müller, *Nucl. Instrum. Meth.* **A309**, 555 (1991).

Chapter **4**

General Properties of Radiation Detectors

Before discussing the different types of radiation detectors individually, we first outline some general properties that apply to all types. Included will be some basic definitions of detector properties, such as efficiency and energy resolution, together with some general modes of operation and methods of recording data that will be helpful in categorizing detector applications.

I. SIMPLIFIED DETECTOR MODEL

We begin with a hypothetical detector that is subject to some type of irradiation. Attention is first focused on the interaction of a *single* particle or quantum of radiation in the detector, which might, for example, be a single alpha particle or an individual gamma-ray photon. In order for the detector to respond at all, the radiation must undergo interaction through one of the mechanisms discussed in Chapter 2. As indicated by Eq. (2.3), the interaction or stopping time is very small (typically a few nanoseconds in gases or a few picoseconds in solids). In most practical situations, these times are so short that the deposition of the radiation energy can be considered instantaneous.

The net result of the radiation interaction in a wide category of detectors is the appearance of a given amount of electric charge within the detector active volume.[†] Our simplified detector model thus assumes that a charge Q appears within the detector at time $t = 0$ resulting from the interaction of a single particle or quantum of radiation. Next, this charge must be collected to form the basic electrical signal. Typically, collection of the charge is accomplished through the imposition of an electric field within the detector, which causes the positive and negative charges created by the radiation to flow in opposite directions. The time required to fully collect the charge varies greatly from one detector to another. For example, in ion chambers the collection time can be as long as a few milliseconds, whereas in semiconductor diode detectors the time is a few nanoseconds. These times reflect both the mobility of the charge carriers within the detector active volume and the average distance that must be traveled before arrival at the collection electrodes.

[†]Strictly true only for detectors such as ion chambers, proportional tubes, G-M tubes, or semiconductor diode detectors. The discussion is also useful for detector types in which the charge is formed indirectly, as from a photomultiplier tube used with a scintillation crystal.

We therefore begin with a model of a prototypical detector whose response to a single particle or quantum of radiation will be a current that flows for a time equal to the charge collection time. The sketch below illustrates one example for the time dependence the detector current might assume, where t_c represents the charge collection time.

$$\int_0^{t_c} i(t)\, dt = Q$$

The time integral over the duration of the current must simply be equal to Q, the total amount of charge generated in that specific interaction.

In any real situation, many quanta of radiation will interact over a period of time. If the irradiation rate is high, situations can arise in which current is flowing in the detector from more than one interaction at a given time. For purposes of the present discussion, we assume that the rate is low enough so that each individual interaction gives rise to a current that is distinguishable from all others. The magnitude and duration of each current pulse may vary depending on the type of interaction, and a sketch of the instantaneous current flowing in the detector might then appear as shown in the sketch below.

It is important to recall that, because the arrival of radiation quanta is a random phenomenon governed by Poisson statistics, the time intervals between successive current pulses are also randomly distributed.

II. MODES OF DETECTOR OPERATION

We can now introduce a fundamental distinction between three general modes of operation of radiation detectors. The three modes are called *pulse mode, current mode,* and *mean square voltage mode* (abbreviated MSV mode, or sometimes called *Campbelling mode*). Pulse mode is easily the most commonly applied of these, but current mode also finds many applications. MSV mode is limited to some specialized applications that make use of its unique characteristics. Although the three modes are operationally distinct, they are interrelated through their common dependence on the sequence of current pulses that are the output of our simplified detector model.

In pulse mode operation, the measurement instrumentation is designed to record each individual quantum of radiation that interacts in the detector. In most common applications, the time integral of each burst of current, or the total charge Q, is recorded since the energy deposited in the detector is directly related to Q. All detectors used to measure the energy of individual radiation quanta must be operated in pulse mode. Such applications are categorized as *radiation spectroscopy* and are the subject of much of the remainder of this text.

In other circumstances, a simpler approach may suit the needs of the measurement: All pulses above a low-level threshold are registered from the detector, regardless of the value of Q. This approach is often called *pulse counting*, and we will show various examples later in this text. It can be useful in many applications in which only the intensity of the radiation is of interest, rather than sensing any changes in or information about the incident energy distribution of the radiation.

At very high event rates, pulse mode operation becomes impractical or even impossible. The time between adjacent events may become too short to carry out an adequate analysis, or the current pulses from successive events may overlap in time. In such cases, one can revert to alternative measurement techniques that respond to the time average taken over many individual events. This approach leads to the remaining two modes of operation: current mode and MSV mode.

A. Current Mode

In the sketch below, we show a current-measuring device (an ammeter or, more practically, a picoammeter) connected across the output terminals of a radiation detector.

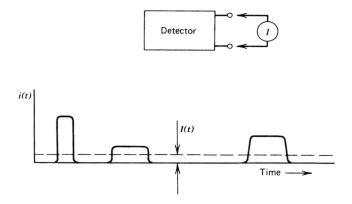

If we assume that the measuring device has a fixed response time T, then the recorded signal from a sequence of events will be a time-dependent current given by

$$I(t) = \frac{1}{T} \int_{t-T}^{t} i(t')\, dt' \tag{4.1}$$

Because the response time T is typically long compared with the average time between individual current pulses from the detector, the effect is to average out many of the fluctuations in the intervals between individual radiation interactions and to record an average current that depends on the product of the interaction rate and the charge per interaction. In current mode, this time average of the individual current bursts serves as the basic signal that is recorded.

At any instant of time, however, there is a statistical uncertainty in this signal due to the random fluctuations in the arrival time of the event. In many ways, the integration time T is analogous to the measurement time discussed in the statistical analysis of the previous chapter. Thus, the choice of large T will minimize statistical fluctuations in the signal but will also slow the response to rapid changes in the rate or nature of the radiation interactions.

The *average* current is given by the product of the average event rate and the charge produced per event.

$$I_0 = rQ = r\,\frac{E}{W}\,q \tag{4.2}$$

where r = event rate

$Q = Eq/W$ = charge produced for each event

E = average energy deposited per event

$W = \dfrac{\text{average energy required to produce a unit}}{\text{charge pair (e.g., electron-ion pair)}}$

$q = 1.6 \times 10^{-19}$ C

For steady-state irradiation of the detector, this average current can also be rewritten as the sum of a constant current I_0 plus a time-dependent fluctuating component $\sigma_i(t)$, as sketched below.

Here $\sigma_i(t)$ is a random time-dependent variable that occurs as a consequence of the random nature of the radiation events interacting within the detector.

A statistical measure of this random component is the variance or mean square value, defined as the time average of the square of the difference between the fluctuating current $I(t)$ and the average current I_0. This mean square value is given by

$$\overline{\sigma_I^2(t)} = \frac{1}{T}\int_{t-T}^{t} [I(t') - I_0]^2 \, dt' = \frac{1}{T}\int_{t-T}^{t} \sigma_i^2(t') \, dt' \tag{4.3}$$

and the standard deviation follows as

$$\overline{\sigma_I(t)} = \sqrt{\overline{\sigma_I^2(t)}} \tag{4.4}$$

Recall from Poisson statistics that the standard deviation in the number of recorded events n over a given observation period is expected to be

$$\sigma_n = \sqrt{n} \tag{4.5}$$

Therefore, the standard deviation in the number of events occurring at a rate r in an effective measurement time T is simply

$$\sigma_n = \sqrt{rT} \tag{4.6}$$

If each pulse contributes the same charge, the *fractional* standard deviation in the measured signal due to random fluctuations in pulse arrival time is given by

$$\frac{\overline{\sigma_I(t)}}{I_0} = \frac{\sigma_n}{n} = \frac{1}{\sqrt{rT}} \tag{4.7}$$

Here $\overline{\sigma_I(t)}$ is the time average of the standard deviation in the measured current, where T is the response time of the picoammeter and I_0 is the average current read on the meter. This result is useful in estimating the uncertainty associated with a given current mode measurement.

It should be noted that, in the derivation of Eq. (4.7), the charge produced in each event *(Q)* is assumed to be constant. Therefore, the result accounts for only the random fluctuations in pulse arrival time, but not for fluctuations in pulse amplitude. In some applications, however, this second source of variance in the signal is small in comparison with the first, and the general character of the results given remains applicable.[1]

B. Mean Square Voltage Mode

An extension of this discussion of the statistical properties of the signal in current mode leads us to the next general mode of operation: the mean square voltage (MSV) mode. Suppose that we send the current signal through a circuit element that blocks the average current I_0 and only passes the fluctuating component $\sigma_i(t)$. By providing additional signal-processing elements, we now compute the time average of the squared amplitude of $\sigma_i(t)$. (The details of these circuits are not important here; for further discussion see Ref. 2.) The processing steps are illustrated below:

The result corresponds to the quantity $\overline{\sigma_I^2(t)}$ defined previously in Eq. (4.3). Combining Eqs. (4.2) and (4.7), we predict the magnitude of the signal derived in this way to be

$$\overline{\sigma_I^2(t)} = \frac{rQ^2}{T} \tag{4.8}$$

We see that this mean square signal is directly proportional to the event rate *r* and, more significantly, proportional to the *square of the charge Q produced in each event.* An analysis of this mode of operation was first given by Campbell,[3] and the term *Campbelling mode* is therefore substituted for *MSV mode* in some usage.

The MSV mode of operation is most useful when making measurements in mixed radiation environments when the charge produced by one type of radiation is much different than that from the second type. If simple current mode operation is chosen, the measured current will linearly reflect the charges contributed by each type. In MSV mode, however, the derived signal is proportional to the *square* of the charge per event. This operational mode will therefore further weight the detector response in favor of the type of radiation giving the larger average charge per event. As one example of the useful application of the MSV mode, in Chapter 14 we describe its use with neutron detectors in reactor instrumentation to enhance the neutron signal compared with the response due to smaller-amplitude gamma-ray events.

C. Pulse Mode

In reviewing various applications of radiation detectors, we find that current mode operation is used with many detectors when event rates are very high. Detectors that are applied to radiation dosimetry are also normally operated in current mode for reasons that will be discussed in Chapter 5. MSV mode is useful in enhancing the relative response to large-amplitude events and finds widespread application in reactor instrumentation. Most applications, however, are better served by preserving information on the amplitude and timing of individual events that only pulse mode can provide. Consequently, the remainder of this chapter deals with various aspects of pulse mode operation.

The nature of the signal pulse produced from a single event depends on the input characteristics of the circuit to which the detector is connected (usually a preamplifier). The equivalent circuit can often be represented as shown below.

Here R represents the input resistance of the circuit, and C represents the equivalent capacitance of both the detector itself and the measuring circuit. If, for example, a preamplifier is attached to the detector, then R is its input resistance and C is the summed capacitance of the detector, the cable used to connect the detector to the preamplifier, and the input capacitance of the preamplifier itself. In most cases, the time-dependent voltage $V(t)$ across the load resistance is the fundamental signal voltage on which pulse mode operation is based. Two separate extremes of operation can be identified that depend on the relative value of the time constant of the measuring circuit. From simple circuit analysis, this time constant is given by the product of R and C, or $\tau = RC$.

CASE 1. SMALL RC ($\tau \ll t_c$)

In this extreme the time constant of the external circuit is kept small compared with the charge collection time, so that the current flowing through the load resistance R is essentially equal to the instantaneous value of the current flowing in the detector. The signal voltage $V(t)$ produced under these conditions has a shape nearly identical to the time dependence of the current produced within the detector as illustrated in Fig. 4.1b. Radiation detectors are sometimes operated under these conditions when high event rates or timing information is more important than accurate energy information.

CASE 2. LARGE RC ($\tau \gg t_c$)

It is generally more common to operate detectors in the opposite extreme in which the time constant of the external circuit is much larger than the detector charge collection time. In this case, very little current will flow in the load resistance during the charge collection time and the detector current is momentarily integrated on the capacitance. If we assume that the time between pulses is sufficiently large, the capacitance will then discharge through the resistance, returning the voltage across the load resistance to zero. The corresponding signal voltage $V(t)$ is illustrated in Fig. 4.1c.

Because the latter case is by far the most common means of pulse-type operation of detectors, it is important to draw some general conclusions. First, the time required for the signal pulse to reach its maximum value is determined by the charge collection time within the detector itself. No properties of the external or load circuit influence the rise time of the pulses. On the other hand, the decay time of the pulses, or the time required to restore the signal voltage to zero, is determined only by the time constant of the load circuit. The conclusion that the leading edge is detector dependent and the trailing edge circuit dependent is a generality that will hold for a wide variety of radiation detectors operated under the conditions in which $RC \gg t_c$. Second, the amplitude of a signal pulse shown as V_{max} in Fig. 4.1c is determined simply by the ratio of the total charge Q created within the detector during one radiation interaction divided by the capacitance C of the equivalent load circuit. Because this capacitance is normally fixed, *the amplitude of the signal pulse is directly proportional to the corresponding charge generated within the detector* and is given by the simple expression

$$V_{max} = Q/C \tag{4.9}$$

Thus, the output of a detector operated in pulse mode normally consists of a sequence of individual signal pulses, each representing the results of the interaction of a single quan-

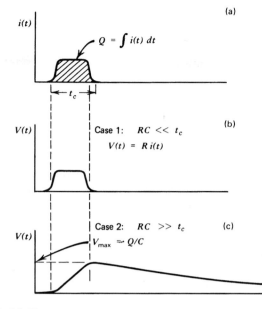

Figure 4.1 (*a*) The assumed current output from a hypothetical detector. (*b*) The signal voltage *V(t)* for the case of a small time constant load circuit. (*c*) The signal voltage *V(t)* for the case of a large time constant load circuit.

tum of radiation within the detector. A measurement of the rate at which such pulses occur will give the corresponding rate of radiation interactions within the detector. Furthermore, the amplitude of each individual pulse reflects the amount of charge generated due to each individual interaction. We shall see that a very common analytical method is to record the distribution of these amplitudes from which some information can often be inferred about the incident radiation. An example is that set of conditions in which the charge Q is directly proportional to the energy of the incident quantum of radiation. Then, a recorded distribution of pulse amplitudes will reflect the corresponding distribution in energy of the incident radiation.

As shown by Eq.(4.9), the proportionality between V_{max} and Q holds only if the capacitance C remains constant. In most detectors, the inherent capacitance is set by its size and shape, and the assumption of constancy is fully warranted. In other types (notably the semiconductor diode detector), the capacitance may change with variations in normal operating parameters. In such cases, voltage pulses of different amplitude may result from events with the same Q. In order to preserve the basic information carried by the magnitude of Q, a type of preamplifier circuit known as a *charge-sensitive* configuration has come into widespread use. As described in Chapter 17, this type of circuit uses feedback to largely eliminate the dependence of the output amplitude on the value of C and restores proportionality to the charge Q even in cases in which C may change.

Pulse mode operation is the more common choice for most radiation detector applications because of several inherent advantages over current mode. First, the sensitivity that is achievable is often many factors greater than when using current or MSV mode because each individual quantum of radiation can be detected as a distinct pulse. Lower limits of detectability are then normally set by background radiation levels. In current mode, the minimum detectable current may represent an average interaction rate in the detector that is many times greater. The second and more important advantage is that each pulse amplitude carries some information that is often a useful or even necessary part of a particular application. In both current and MSV mode operations, this information on individual pulse amplitudes is lost and all interactions, regardless of amplitude, contribute to the average measured current. Because of these inherent advantages of pulse mode, the emphasis in nuclear instrumentation is largely in pulse circuits and pulse-processing techniques.

III. PULSE HEIGHT SPECTRA

When operating a radiation detector in pulse mode, each individual pulse amplitude carries important information regarding the charge generated by that particular radiation interaction in the detector. If we examine a large number of such pulses, their amplitudes will not all be the same. Variations may be due either to differences in the radiation energy or to fluctuations in the inherent response of the detector to monoenergetic radiation. The pulse amplitude distribution is a fundamental property of the detector output that is routinely used to deduce information about the incident radiation or the operation of the detector itself.

The most common way of displaying pulse amplitude information is through the *differential pulse height distribution*. Figure 4.2a gives a hypothetical distribution for purposes of example. The abscissa is a linear pulse amplitude scale that runs from zero to a value larger than the amplitude of any pulse observed from the source. The ordinate is the differential number dN of pulses observed with an amplitude within the differential amplitude increment dH, divided by that increment, or dN/dH. The horizontal scale then has units of pulse amplitude (volts), whereas the vertical scale has units of inverse amplitude (volts^{-1}). The number of pulses whose amplitude lies between two specific values, H_1 and H_2, can be obtained by integrating the area under the distribution between those two limits, as shown by the cross-hatched area in Fig. 4.2a:

$$\text{number of pulses with amplitude between } H_1 \text{ and } H_2 = \int_{H_1}^{H_2} \frac{dN}{dH}\, dH \qquad (4.10)$$

The total number of pulses N_0 represented by the distribution can be obtained by integrating the area under the entire spectrum:

$$N_0 = \int_0^{\infty} \frac{dN}{dH}\, dH \qquad (4.11)$$

Most users of radiation instrumentation are accustomed to looking at the shape of the differential pulse height distribution to display significant features about the source of the pulses. The maximum pulse height observed (H_5) is simply the point along the abscissa at which the distribution goes to zero. Peaks in the distribution, such as at H_4, indicate pulse amplitudes about which a large number of pulses may be found. On the other hand, valleys or low points in the spectrum, such as at pulse height H_3, indicate values of the pulse amplitude around which relatively few pulses occur. The physical interpretation of differential pulse height spectra always involves *areas* under the spectrum between two given limits of pulse height. The value of the ordinate itself (dN/dH) has no physical significance until multiplied by an increment of the abscissa H.

A less common way of displaying the same information about the distribution of pulse amplitudes is through the *integral pulse height distribution*. Figure 4.2b shows the integral distribution for the same pulse source displayed as a differential spectrum in Fig. 4.2a. The abscissa in the integral case is the same pulse height scale shown for the differential distribution. The ordinate now represents the number of pulses whose amplitude exceeds that of a given value of the abscissa H. The ordinate N must always be a monotonically decreasing function of H because fewer and fewer pulses will lie above an amplitude H that is allowed to increase from zero. Because all pulses have some finite amplitude, the value of the integral spectrum at $H = 0$ must be the total number of pulses observed (N_0). The value of the integral distribution must decrease to zero at the maximum observed pulse height (H_5).

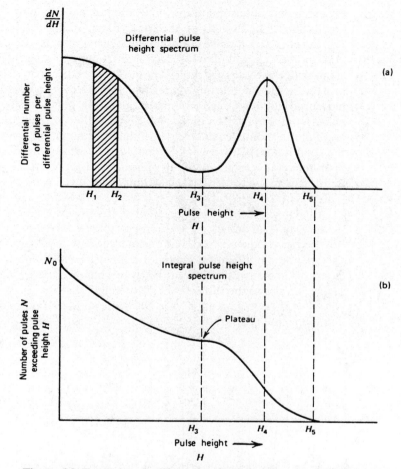

Figure 4.2 Examples of differential and integral pulse height spectra for an assumed source of pulses.

The differential and integral distributions convey exactly the same information and one can be derived from the other. The amplitude of the differential distribution at any pulse height H is given by the absolute value of the slope of the integral distribution at the same value. Where peaks appear in the differential distribution, such as H_4, local maxima will occur in the magnitude of the slope of the integral distribution. On the other hand, where minima appear in the differential spectrum, such as H_3, regions of minimum magnitude of the slope are observed in the integral distribution. Because it is easier to display subtle differences by using the differential distribution, it has become the predominant means of displaying pulse height distribution information.

IV. COUNTING CURVES AND PLATEAUS

When radiation detectors are operated in pulse counting mode, a common situation often arises in which the pulses from the detector are fed to a counting device with a fixed discrimination level. Signal pulses must exceed a given level H_d in order to be registered by the counting circuit. Sometimes it is possible to vary the level H_d during the course of the measurement to provide information about the amplitude distribution of the pulses. Assuming that H_d can be varied between 0 and H_5 in Fig. 4.2, a series of measurements can be carried out in which the number of pulses N per unit time is measured as H_d is changed through a sequence of values between 0 and H_5. This series of measurements is just an experimental determination of the integral pulse height distribution, and the measured counts should lie directly on the curve shown in Fig. 4.2*b*.

In setting up a pulse counting measurement, it is often desirable to establish an operating point that will provide maximum stability over long periods of time. For example, small drifts in the value of H_d could be expected in any real application, and one would like to establish conditions under which these drifts would have minimal influence on the measured counts. One such stable operating point can be achieved at a discrimination point set at the level H_3 in Fig. 4.2. Because the slope of the integral distribution is a minimum at that point, small changes in the discrimination level will have minimum impact on the total number of pulses recorded. In general, regions of minimum slope on the integral distribution are called *counting plateaus* and represent areas of operation in which minimum sensitivity to drifts in discrimination level are achieved. It should be noted that plateaus in the integral spectrum correspond to valleys in the differential distribution.

Plateaus in counting data can also be observed with a different procedure. For a particular radiation detector it is often possible to vary the gain or amplification provided for the charge produced in radiation interactions. This variation could be accomplished by varying the amplification factor of a linear amplifier between the detector and counting circuit, or in many cases more directly by changing the applied voltage to the detector itself. Figure 4.3 shows the differential pulse height distribution corresponding to three different values of voltage gain applied to the same source of pulses. Here the value of gain can be defined as the ratio of the voltage amplitude for a given event in the detector to the same

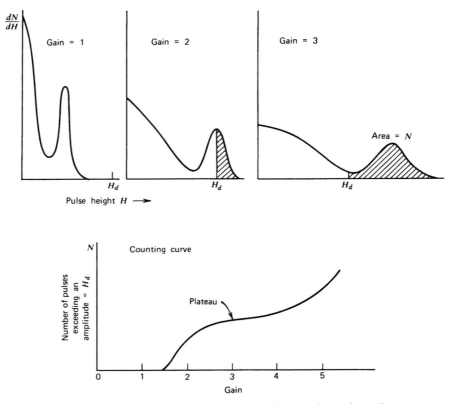

Figure 4.3 Example of a counting curve generated by varying gain under constant source conditions. The three plots at the top give the corresponding differential pulse height spectra.

amplitude before some parameter (such as amplification or detector voltage) was changed. The highest voltage gain will result in the largest maximum pulse height, but in all cases the area under the differential distribution will be a constant. In the example shown in Fig. 4.3, no counts will be recorded for a gain $G = 1$ because under those conditions all pulses will be smaller than H_d. Pulses will begin to be recorded somewhere between a gain $G = 1$ and $G = 2$. An experiment can be carried out in which the number of pulses recorded is measured as a function of the gain applied, sometimes called the *counting curve*. Such a plot is also shown in Fig. 4.3 and in many ways resembles an integral pulse height distribution. We now have a mirror image of the integral distribution, however, because small values of the gain will record no pulses, whereas large values will result in counting nearly all the pulses. Again, plateaus can be anticipated in this counting curve for values of the gain in which the effective discrimination pulse height H_d passes through minima in the differential pulse height distribution. In the example shown in Fig. 4.3, the minimum slope in the counting curve should correspond to a gain of about 3, in which case the discrimination point is near the minimum of the valley in the differential pulse height distribution.

In some types of radiation detectors, such as Geiger–Mueller tubes or scintillation counters, the gain can conveniently be varied by changing the applied voltage to the detector. Although the gain may not change linearly with voltage, the qualitative features of the counting curve can be traced by a simple measurement of the detector counting rate as a function of voltage. In order to select an operating point of maximum stability, plateaus are again sought in the counting curve that results, and the voltage is often selected to lie at a point of minimum slope on this counting curve. We shall discuss these plateau measurements more specifically in Chapters 6 and 7 in connection with proportional counters and Geiger–Mueller detectors.

V. ENERGY RESOLUTION

In many applications of radiation detectors, the object is to measure the energy distribution of the incident radiation. These efforts are classified under the general term *radiation spectroscopy*, and later chapters give examples of the use of specific detectors for spectroscopy involving alpha particles, gamma rays, and other types of nuclear radiation. At this point we discuss some general properties of detectors when applied to radiation spectroscopy and introduce some definitions that will be useful in these discussions.

One important property of a detector in radiation spectroscopy can be examined by noting its response to a monoenergetic source of that radiation. Figure 4.4 illustrates the differential pulse height distribution that might be produced by a detector under these

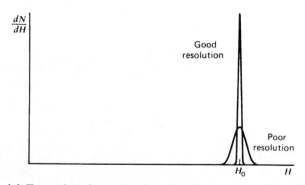

Figure 4.4 Examples of response functions for detectors with relatively good resolution and relatively poor resolution.

conditions. This distribution is called the *response function* of the detector for the energy used in the determination. The curve labeled "Good resolution" illustrates one possible distribution around an average pulse height H_0. The second curve, labeled "Poor resolution," illustrates the response of a detector with inferior performance. Provided the same number of pulses are recorded in both cases, the areas under each peak are equal. Although both distributions are centered at the same average value H_0, the width of the distribution in the poor resolution case is much greater. This width reflects the fact that a large amount of fluctuation was recorded from pulse to pulse even though the same energy was deposited in the detector for each event. If the amount of these fluctuations is made smaller, the width of the corresponding distribution will also become smaller and the peak will approach a sharp spike or a mathematical delta function. The ability of a given measurement to resolve fine detail in the incident energy of the radiation is obviously improved as the width of the response function (illustrated in Fig. 4.4) becomes smaller and smaller.

A formal definition of detector *energy resolution* is shown in Fig. 4.5. The differential pulse height distribution for a hypothetical detector is shown under the same assumption that only radiation for a single energy is being recorded. The full width at half maximum (FWHM) is illustrated in the figure and is defined as the width of the distribution at a level that is just half the maximum ordinate of the peak. This definition assumes that any background or continuum on which the peak may be superimposed is negligible or has been subtracted away. The energy resolution of the detector is conventionally defined as the FWHM divided by the location of the peak centroid H_0. The energy resolution R is thus a dimensionless fraction conventionally expressed as a percentage. Semiconductor diode detectors used in alpha spectroscopy can have an energy resolution less than 1%, whereas scintillation detectors used in gamma-ray spectroscopy normally show an energy resolution in the range of 5–10%. It should be clear that the smaller the figure for the energy resolution, the better the detector will be able to distinguish between two radiations whose energies lie near each other. An approximate rule of thumb is that one should be able to resolve two energies that are separated by more than one value of the detector FWHM.

There are a number of potential sources of fluctuation in the response of a given detector that result in imperfect energy resolution. These include any drift of the operating characteristics of the detector during the course of the measurements, sources of random noise within the detector and instrumentation system, and statistical noise arising from the discrete nature of the measured signal itself. The third source is in some sense the most important because it represents an irreducible minimum amount of fluctuation that will always be present in the detector signal no matter how perfect the remainder of the system is made. In a wide category of detector applications, the statistical noise represents the dominant source of fluctuation in the signal and thus sets an important limit on detector performance.

The statistical noise arises from the fact that the charge Q generated within the detector by a quantum of radiation is not a continuous variable but instead represents a discrete number of charge carriers. For example, in an ion chamber the charge carriers are the ion pairs produced by the passage of the charged particle through the chamber, whereas in a scintillation counter they are the number of electrons collected from the photocathode of the photomultiplier tube. In all cases the number of carriers is discrete and subject to random fluctuation from event to event even though exactly the same amount of energy is deposited in the detector.

An estimate can be made of the amount of inherent fluctuation by assuming that the formation of each charge carrier is a Poisson process. Under this assumption, if a total number N of charge carriers is generated on the average, one would expect a standard deviation of $\sqrt{N}$ to characterize the inherent statistical fluctuations in that number [see Eq. (3.29)]. If this were the only source of fluctuation in the signal, the response function, as shown in Fig. 4.5, should have a Gaussian shape, because N is typically a large number. In

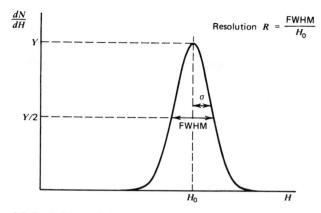

Figure 4.5 Definition of detector resolution. For peaks whose shape is Gaussian with standard deviation σ, the FWHM is given by 2.35σ.

this case, the Gaussian function introduced in the previous chapter is most conveniently written

$$G(H) = \frac{A}{\sigma \sqrt{2\pi}} \exp\left(-\frac{(H - H_0)^2}{2\sigma^2}\right) \tag{4.12}$$

The width parameter σ determines the FWHM of any Gaussian through the relation FWHM $= 2.35\sigma$. (The remaining two parameters, H_0 and A, represent the centroid and area, respectively.)

The response of many detectors is approximately linear, so that the average pulse amplitude $H_0 = KN$, where K is a proportionality constant. The standard deviation σ of the peak in the pulse height spectrum is then $\sigma = K\sqrt{N}$ and its FWHM is $2.35K\sqrt{N}$. We then would calculate a limiting resolution R due only to statistical fluctuations in the number of charge carriers as

$$R\Big|_{\text{Poisson limit}} \equiv \frac{\text{FWHM}}{H_0} = \frac{2.35K\sqrt{N}}{KN} = \frac{2.35}{\sqrt{N}} \tag{4.13}$$

Note that this limiting value of R depends only on the number of charge carriers N, and the resolution improves (R will decrease) as N is increased. From Eq. (4.13) we see that in order to achieve an energy resolution better than 1%, one must have N greater than 55,000. An ideal detector would have as many charge carriers generated per event as possible, so that this limiting resolution would be as small a percentage as possible. The great popularity of semiconductor detectors stems from the fact that a very large number of charge carriers are generated in these devices per unit energy lost by the incident radiation.

Careful measurements of the energy resolution of some types of radiation detectors have shown that the achievable values for R can be lower by a factor as large as 3 or 4 than the minimum predicted by the statistical arguments given above. These results would indicate that the processes that give rise to the formation of each individual charge carrier are not independent, and therefore the total number of charge carriers cannot be described by simple Poisson statistics. The *Fano factor* has been introduced in an attempt to quantify the departure of the observed statistical fluctuations in the number of charge carriers from pure Poisson statistics and is defined as

$$F \equiv \frac{\text{observed variance in } N}{\text{Poisson predicted variance } (= N)} \tag{4.14}$$

Because the variance is given by σ^2, the equivalent expression to Eq. (4.13) is now

$$R\big|_{\text{Statistical limit}} = \frac{2.35K\sqrt{N}\sqrt{F}}{KN} = 2.35\sqrt{\frac{F}{N}} \qquad (4.15)$$

Although the Fano factor is substantially less than unity for semiconductor diode detectors and proportional counters, other types such as scintillation detectors appear to show a limiting resolution consistent with Poisson statistics and the Fano factor would, in these cases, be unity.

Any other source of fluctuations in the signal chain will combine with the inherent statistical fluctuations from the detector to give the overall energy resolution of the measuring system. It is sometimes possible to measure the contribution to the overall FWHM due to a single component alone. For example, if the detector is replaced by a stable pulse generator, the measured response of the remainder of the system will show a fluctuation due primarily to electronic noise. If there are several sources of fluctuation present and each is symmetric and independent, statistical theory predicts that the overall response function will always tend toward a Gaussian shape, even if the individual sources are characterized by distributions of different shape. As a result, the Gaussian function given in Eq. (4.12) is widely used to represent the response function of detector systems in which many different factors may contribute to the overall energy resolution. Then the total FWHM will be the quadrature sum of the FWHM values for each individual source of fluctuation:

$$(\text{FWHM})^2_{\text{overall}} = (\text{FWHM})^2_{\text{statistical}} + (\text{FWHM})^2_{\text{noise}} + (\text{FWHM})^2_{\text{drift}} + \dots$$

Each term on the right is the square of the FWHM that would be observed if all other sources of fluctuation were zero.

VI. DETECTION EFFICIENCY

All radiation detectors will, in principle, give rise to an output pulse for each quantum of radiation that interacts within its active volume. For primary charged radiation such as alpha or beta particles, interaction in the form of ionization or excitation will take place immediately upon entry of the particle into the active volume. After traveling a small fraction of its range, a typical particle will form enough ion pairs along its path to ensure that the resulting pulse is large enough to be recorded. Thus, it is often easy to arrange a situation in which a detector will see every alpha or beta particle that enters its active volume. Under these conditions, the detector is said to have a counting efficiency of 100%.

On the other hand, uncharged radiations such as gamma rays or neutrons must first undergo a significant interaction in the detector before detection is possible. Because these radiations can travel large distances between interactions, detectors are often less than 100% efficient. It then becomes necessary to have a precise figure for the detector efficiency in order to relate the number of pulses counted to the number of neutrons or photons incident on the detector.

It is convenient to subdivide counting efficiencies into two classes: *absolute* and *intrinsic*. Absolute efficiencies are defined as

$$\epsilon_{\text{abs}} = \frac{\text{number of pulses recorded}}{\text{number of radiation quanta emitted by source}} \qquad (4.16)$$

and are dependent not only on detector properties but also on the details of the counting geometry (primarily the distance from the source to the detector). The intrinsic efficiency is defined as

$$\epsilon_{\text{int}} = \frac{\text{number of pulses recorded}}{\text{number of radiation quanta incident on detector}} \qquad (4.17)$$

and no longer includes the solid angle subtended by the detector as an implicit factor. The two efficiencies are simply related for isotropic sources by $\epsilon_{int} = \epsilon_{abs} \cdot (4\pi/\Omega)$, where Ω is the solid angle of the detector seen from the actual source position. It is much more convenient to tabulate values of intrinsic rather than absolute efficiencies because the geometric dependence is much milder for the former. The intrinsic efficiency of a detector usually depends primarily on the detector material, the radiation energy, and the physical thickness of the detector in the direction of the incident radiation. A slight dependence on distance between the source and the detector does remain, however, because the average path length of the radiation through the detector will change somewhat with this spacing.

Counting efficiencies are also categorized by the nature of the event recorded. If we accept all pulses from the detector, then it is appropriate to use *total* efficiencies. In this case all interactions, no matter how low in energy, are assumed to be counted. In terms of a hypothetical differential pulse height distribution shown in Fig. 4.6, the entire area under the spectrum is a measure of the number of all pulses that are recorded, regardless of amplitude, and would be counted in defining the total efficiency. In practice, any measurement system always imposes a requirement that pulses be larger than some finite threshold level set to discriminate against very small pulses from electronic noise sources. Thus, one can only approach the theoretical total efficiency by setting this threshold level as low as possible. The *peak* efficiency, however, assumes that only those interactions that deposit the full energy of the incident radiation are counted. In a differential pulse height distribution, these full energy events are normally evidenced by a peak that appears at the highest end of the spectrum. Events that deposit only part of the incident radiation energy then will appear farther to the left in the spectrum. The number of full energy events can be obtained by simply integrating the total area under the peak, which is shown as the cross-hatched area in Fig. 4.6. The total and peak efficiencies are related by the *peak-to-total* ratio r

$$r = \frac{\epsilon_{peak}}{\epsilon_{total}} \tag{4.18}$$

which is sometimes tabulated separately. It is often preferable from an experimental standpoint to use only peak efficiencies, because the number of full energy events is not sensitive to some perturbing effects such as scattering from surrounding objects or spurious noise. Therefore, values for the peak efficiency can be compiled and universally applied to a wide variety of laboratory conditions, whereas total efficiency values may be influenced by variable conditions.

To be complete, a detector efficiency should be specified according to both of the above criteria. For example, the most common type of efficiency tabulated for gamma-ray detectors is the *intrinsic peak efficiency*.

Figure 4.6 Example of the full-energy peak in a differential pulse height spectrum.

A detector with known efficiency can be used to measure the absolute activity of a radioactive source. In the following discussion, we assume that a detector with an intrinsic peak efficiency ϵ_{ip} has been used to record N events under the full energy peak in the detector spectrum. For simplicity, we also assume that the source emits radiation isotropically and that no attenuation takes place between the source and detector. From the definition of intrinsic peak efficiency, the number of radiation quanta S emitted by the source over the measurement period is then given by

$$S = N\frac{4\pi}{\epsilon_{ip}\Omega} \tag{4.19}$$

where Ω represent the solid angle (in steradians) subtended by the detector at the source position. The solid angle is defined by an integral over the detector surface that faces the source, of the form

$$\Omega = \int_A \frac{\cos\alpha}{r^2}\,dA \tag{4.20}$$

where r represents the distance between the source and a surface element dA, and α is the angle between the normal to the surface element and the source direction. If the volume of the source is not negligible, then a second integration must be carried out over all volume elements of the source. For the common case of a point source located along the axis of a right circular cylindrical detector, Ω is given by

$$\Omega = 2\pi\left(1 - \frac{d}{\sqrt{d^2 + a^2}}\right) \tag{4.21}$$

where the source–detector distance d and detector radius a are shown in the sketch below:

For $d \gg a$, the solid angle reduces to the ratio of the detector plane frontal area A visible at the source to the square of the distance:

$$\Omega \cong \frac{A}{d^2} = \frac{\pi a^2}{d^2} \tag{4.22}$$

Another commonly encountered circumstance, shown in the sketch below, involves a uniform circular disk source emitting isotropic radiation aligned with a circular disk detector, both positioned perpendicular to a common axis passing through their centers:

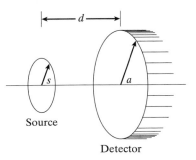

Source

Detector

In terms of the dimensions shown on the sketch, it can be shown[4] that the effective solid angle averaged over the surface of the source is given by solving the integral

$$\Omega = \frac{4\pi a}{s} \int_0^\infty \frac{\exp(-dk)J_1(sk)J_1(ak)}{k} \, dk$$

where the $J_1(x)$ are Bessel functions of x. This integral does not have an analytic solution, so it can only be solved using numerical techniques. A useful approximate solution is

$$\Omega \cong 2\pi \left[1 - \frac{1}{(1+\beta)^{1/2}} - \frac{3}{8} \frac{\alpha\beta}{(1+\beta)^{5/2}} + \alpha^2[F1] - \alpha^3[F2] \right]$$

$$F1 = \frac{5}{16} \frac{\beta}{(1+\beta)^{7/2}} - \frac{35}{16} \frac{\beta^2}{(1+\beta)^{9/2}}$$

$$F2 = \frac{35}{128} \frac{\beta}{(1+\beta)^{9/2}} - \frac{315}{256} \frac{\beta^2}{(1+\beta)^{11/2}} + \frac{1155}{1024} \frac{\beta^3}{(1+\beta)^{13/2}}$$

where

$$\alpha \equiv \left(\frac{a}{d}\right)^2 \qquad \beta \equiv \left(\frac{s}{d}\right)^2$$

This approximation becomes inaccurate when the source or detector diameters become too large compared with their spacing, but it has been shown[5] to give very accurate results over a wide range of dimensions.

Published values for Ω can sometimes be found for more complicated geometric arrangements involving off-axis or volumetric sources, or detectors with more complex shapes. Some specific examples of data or descriptions of algorithms useful in solid angle computations are given in Refs. 6–15.

VII. DEAD TIME

In nearly all detector systems, there will be a minimum amount of time that must separate two events in order that they be recorded as two separate pulses. In some cases the limiting time may be set by processes in the detector itself, and in other cases the limit may arise in the associated electronics. This minimum time separation is usually called the *dead time* of the counting system. Because of the random nature of radioactive decay, there is always some probability that a true event will be lost because it occurs too quickly following a preceding event. These "dead time losses" can become rather severe when high counting rates are encountered, and any accurate counting measurements made under these conditions must include some correction for these losses. In this section we discuss some simple models of dead time behavior of counting systems, together with two experimental methods of determining system dead time. Ref. 16 is an excellent presentation of more detailed analyses of related topics that are beyond the present scope.

A. Models for Dead Time Behavior

Two models of dead time behavior of counting systems have come into common usage: *paralyzable* and *nonparalyzable* response. These models represent idealized behavior, one or the other of which often adequately resembles the response of a real counting system. The fundamental assumptions of the models are illustrated in Fig. 4.7. At the center of the figure, a time scale is shown on which six randomly spaced events in the detector are indicated. At the bottom of the figure is the corresponding dead time behavior of a detector assumed to be nonparalyzable. A fixed time τ is assumed to follow each true event that

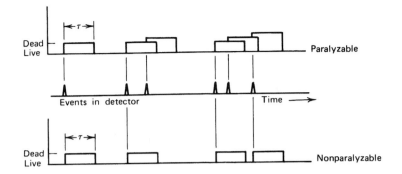

Figure 4.7 Illustration of two assumed models of dead time behavior for radiation detectors.

occurs during the "live period" of the detector. True events that occur during the dead period are lost and assumed to have no effect whatsoever on the behavior of the detector. In the example shown, the nonparalyzable detector would record four counts from the six true interactions. In contrast, the behavior of a paralyzable detector is shown along the top line of Fig. 4.7. The same dead time τ is assumed to follow each true interaction that occurs during the live period of the detector. True events that occur during the dead period, however, although still not recorded as counts, are assumed to extend the dead time by another period τ following the lost event. In the example shown, only three counts are recorded for the six true events.

The two models predict the same first-order losses and differ only when true event rates are high. They are in some sense two extremes of idealized system behavior, and real counting systems will often display a behavior that is intermediate between these extremes. The detailed behavior of a specific counting system may depend on the physical processes taking place in the detector itself or on delays introduced by the pulse processing and recording electronics.

In the discussion that follows, we examine the response of a detector system to a steady-state source of radiation, and we adopt the following definitions:

$$n = \text{true interaction rate}$$

$$m = \text{recorded count rate}$$

$$\tau = \text{system dead time}$$

We assume that the counting time is long so that both n and m may be regarded as average rates. In general, we would like to obtain an expression for the true interaction rate n as a function of the measured rate m and the system dead time τ, so that appropriate corrections can be made to measured data to account for the dead time losses.

In the nonparalyzable case, the fraction of all time that the detector is dead is given simply by the product $m\tau$. Therefore, the rate at which true events are lost is simply $nm\tau$. But because $n - m$ is another expression for the rate of losses,

$$n - m = nm\tau \tag{4.23}$$

Solving for n, we obtain

$$\boxed{n = \frac{m}{1 - m\tau}} \quad \begin{array}{c}\text{Nonparalyzable} \\ \text{model}\end{array} \tag{4.24}$$

In the paralyzable case, dead periods are not always of fixed length, so we cannot apply the same argument. Instead, we note that rate m is identical to the rate of occurrences of time intervals between true events which exceed τ. The distribution of intervals between random events occurring at an average rate n was previously shown [Eq. (3.71)] to be

$$P_1(T)\, dT = ne^{-nT}\, dT \qquad (4.25)$$

where $P_1(T)\, dT$ is the probability of observing an interval whose length lies within dT about T. The probability of intervals larger than τ can be obtained by integrating this distribution between τ and ∞

$$P_2(\tau) = \int_{\tau}^{\infty} P_1(T)\, dT = e^{-n\tau} \qquad (4.26)$$

The rate of occurrence of such intervals is then obtained by simply multiplying the above expression by the true rate n

$$\boxed{m = ne^{-n\tau}} \quad \begin{array}{l}\text{Paralyzable}\\ \text{model}\end{array} \qquad (4.27)$$

The paralyzable model leads to a more cumbersome result because we cannot solve explicitly for the true rate n. Instead, Eq. (4.27) must be solved iteratively if n is to be calculated from measurements of m and knowledge of τ.

A plot of the observed rate m versus the true rate n is given in Fig. 4.8 for both models. When the rates are low the two models give virtually the same result, but the behavior at high rates is markedly different. A nonparalyzable system will approach an asymptotic value for the observed rate of $1/\tau$, which represents the situation in which the counter barely has time to finish one dead period before starting another. For paralyzable behavior, the observed rate is seen to go through a maximum. Very high true interaction rates result in a multiple extension of the dead period following an initial recorded count, and very few true events can be recorded. One must always be careful when using a counting system that may be paralyzable to ensure that ostensibly low observed rates actually correspond to low interaction rates rather than very high rates on the opposite side of the maximum. Mistakes in the interpretation of nuclear counting data from paralyzable systems have occurred in the past by overlooking the fact that there are always two possible true interaction rates corresponding to a given observed rate. As shown in Fig. 4.8, the observed rate m_1 can correspond

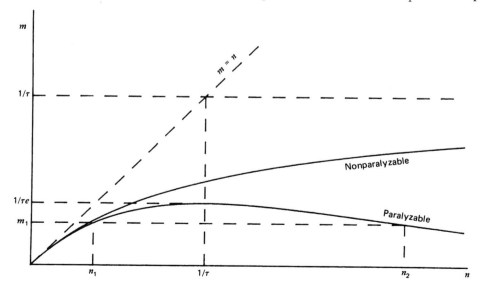

Figure 4.8 Variation of the observed rate m as a function of the true rate n for two models of dead time losses.

to either true rates n_1 or n_2. The ambiguity can be resolved only by changing the true rate in a known direction while observing whether the observed rate increases or decreases.

For low rates ($n \ll 1/\tau$) the following approximations can be written :

Nonparalyzable
$$m = \frac{n}{1 + n\tau} \cong n(1 - n\tau) \tag{4.28}$$

Paralyzable
$$m = ne^{-n\tau} \cong n(1 - n\tau) \tag{4.29}$$

Thus, the two models lead to identical results in the limit of small dead time losses.

If possible, one should avoid measurement conditions under which dead time losses are high because of the errors that inevitably occur in making corrections for the losses. The value of τ may be uncertain or subject to variation, and the system behavior may not follow exactly either of the models described above. When losses are greater than 30 or 40%, the calculated true rate becomes very sensitive to small changes in the measured rate and the assumed system behavior. Instead, the user should seek to reduce the losses by changing the conditions of the measurement or by choosing a counting system with smaller dead time.

B. Methods of Dead Time Measurement

In order to make dead time corrections using either model, prior knowledge of the dead time τ is required. Sometimes this dead time can be associated with a known limiting property of the counting system (e.g., a fixed resolving time of an electronic circuit). More often, the dead time will not be known or may vary with operating conditions and must therefore be measured directly. Common measurement techniques are based on the fact that the observed rate varies nonlinearly with the true rate. Therefore, by assuming that one of the specific models is applicable, and by measuring the observed rate for at least two different true rates that differ by a known ratio, the dead time can be calculated.

The common example is the *two-source method*. The method is based on observing the counting rate from two sources individually and in combination. Because the counting losses are nonlinear, the observed rate due to the combined sources will be less than the sum of the rates due to the two sources counted individually, and the dead time can be calculated from the discrepancy.

To illustrate the method, let n_1, n_2, and n_{12} be the true counting rates (sample plus background) with source 1, source 2, and the combined sources, respectively, in place. Let m_1, m_2, and m_{12} represent the corresponding observed rates. Also, let n_b and m_b be the true and measured background rates with both sources removed. Then

$$n_{12} - n_b = (n_1 - n_b) + (n_2 - n_b) \tag{4.30}$$
$$n_{12} + n_b = n_1 + n_2$$

Now assuming the nonparalyzable model [Eq. (4.24)] and substituting, we obtain

$$\frac{m_{12}}{1 - m_{12}\tau} + \frac{m_b}{1 - m_b\tau} = \frac{m_1}{1 - m_1\tau} + \frac{m_2}{1 - m_2\tau} \tag{4.31}$$

Solving this equation explicitly for τ gives the following result:

$$\tau = \frac{X(1 - \sqrt{1 - Z})}{Y} \tag{4.32}$$

where

$$X \equiv m_1 m_2 - m_b m_{12}$$
$$Y \equiv m_1 m_2 (m_{12} + m_b) - m_b m_{12}(m_1 + m_2)$$
$$Z \equiv \frac{Y(m_1 + m_2 - m_{12} - m_b)}{X^2}$$

A number of approximations to this general solution are often recommended in textbooks. For example, in the case of zero background $(m_b = 0)$

$$\tau = \frac{m_1 m_2 - [m_1 m_2 (m_{12} - m_1)(m_{12} - m_2)]^{1/2}}{m_1 m_2 m_{12}} \tag{4.33}$$

Other simplifications of Eq. (4.32) have appeared that are based on various mathematical approximations. However, the use of any type of approximation should be discouraged because significant errors can be introduced under typical experimental conditions.[17] Because the two-source method involves a substantial amount of experimental time and effort, it is difficult to justify the use of any expression other than Eq. (4.32) in analyzing the results.

Because the method is essentially based on observing the difference between two nearly equal large numbers, careful measurements are required in order to get reliable values for the dead time. The measurement is usually carried out by counting source 1, placing source 2 nearby and measuring the combined rate, and then removing source 1 to measure the rate caused by source 2 alone. During this operation, care must be exercised not to move the source already in place, and consideration must be given to the possibility that the presence of a second source will scatter radiation into the detector that would not ordinarily be counted from the first source alone. In order to keep the scattering unchanged, a dummy second source without activity is normally put in place when the sources are counted individually. Best results are obtained by using sources active enough to result in a fractional dead time $m_{12}\tau$ of at least 20%.

A second method can be carried out if a short-lived radioisotope source is available.[†] In this case the departure of the observed counting rate from the known exponential decay of the source can be used to calculate the dead time. The technique, known as the *decaying source method*, is based on the known behavior of the true rate n:

$$n = n_0 e^{-\lambda t} + n_b \tag{4.34}$$

where n_0 is the true rate at the beginning of the measurement and λ is the decay constant of the particular isotope used for the measurement.

In the limit of negligible background, a simple graphical procedure can be applied to analyze the resulting data. Then Eq. (4.34) becomes

$$n \cong n_0 e^{-\lambda t} \tag{4.35}$$

By inserting Eq. (4.35) into Eq. (4.24) and carrying out some algebra, we get the following relation for the nonparalyzable model:

$$m e^{\lambda t} = -n_0 \tau m + n_0 \tag{4.36}$$

If we identify, as in Fig. 4.9a, the abscissa as m and the ordinate as the product $m e^{\lambda t}$, then Eq. (4.36) is that of a straight line. The experimental procedure consists of measuring the observed rate m as a function of time t, and thus defining points that should lie on this line starting from the right and moving left as the source decays. By fitting the best straight line to the data, the intercept will give n_0, the true rate at the beginning of the measurement, and the negative slope will give the product of $n_0\tau$. The dead time τ then follows directly from the ratio of the slope to the intercept.

[†]For laboratories with access to neutron irradiation facilities, a convenient isotope is ^{116m}In (half-life of 54.0 min), which is readily produced by neutron absorption in an indium foil.

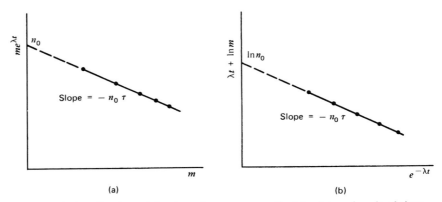

Figure 4.9 Application of the decaying source method to determine dead time.

For the paralyzable model, inserting Eq. (4.35) into Eq. (4.27) gives the following result:

$$\lambda t + \ln m = -n_0 \tau e^{-\lambda t} + \ln n_0 \qquad (4.37)$$

Again by choosing the abscissa and ordinate as shown in Fig. 4.9*b*, the equation of a straight line results. In this case the intercept gives the value $\ln n_0$, whereas the slope again gives the negative of the product $n_0 \tau$. The dead time can simply be derived from these two values. The decaying source method offers the advantage of not only being able to measure the value of the dead time but also being able to test the validity of the assumed models. If a nonparalyzable model best describes the counting system, the data will most closely fit a straight line for the format shown in Fig. 4.9*a*. On the other hand, if a paralyzable model is more appropriate, the format shown in Fig. 4.9*b* will result in a more nearly linear plot of the data. In order to be effective, the measurements should be carried out for a time period at least equal to the half-life of the decaying radioisotope, and the initial loss fraction $m\tau$ should be at least 20%.

If the background is more than a few percent of the smallest measured rate, the graphical procedure can lead to significant errors. Although some improvement will result from subtracting the observed background rate from all the measured m values, this correction is not rigorous and an exact analysis can only be made by reverting back to Eq. (4.34). It then becomes necessary to use numerical computation techniques to derive values for n_0 and τ which, when inserted in an assumed model of the dead time behavior, result in a best fit to the measured data.

The methods described above are applicable to determine the dead time of simple counting systems in which all pulses above a threshold amplitude are recorded. For spectroscopic systems in which the pulse height spectrum is recorded from the detector, other methods are available to deal with counting losses that are based on the mixture of artificial pulses from a pulse generator into the signal processing chain. These pulser-based techniques are treated later in Chapter 17.

C. Statistics of Dead Time Losses

When measuring radiation from steady-state sources, we normally assume that the true events occurring in the detector follow Poisson statistics in which the probability of an event occurring per unit time is a constant. The effect of system dead time is to remove selectively some of the events before they are recorded as counts. Specifically, events occurring after short time intervals following preceding events are preferentially discarded, and the interval distribution [Eq. 3.71)] is modified from its normal exponential shape. Dead time losses therefore distort the statistics of the recorded counts away from a true Poisson behavior. If the losses are small ($n\tau$ less than 10 or 20%), however, this distortion has little practical effect on the validity of the statistical formulas for the prediction of statistical counting uncertainties developed in Chapter 3.

If the dead time losses are not small, the deviations from Poisson statistics become more significant. The discarding of events that occur after short time intervals causes the sequence of recorded counts to become somewhat more regular, and the variance expected in repeated measurements is reduced. Detailed analyses of the statistics of counts distorted by dead time are beyond the present scope, but can be found in Refs. 18–21.

With either paralyzable or nonparalyzable behavior, there is some chance that more than one true event is lost per dead period. A recorded count therefore can correspond to the occurrence of any number of true events, from one to many. The relative probability that multiple true events are contained in a typical dead period will increase as the true event rate becomes higher. Because the true events still obey Poisson statistics, relatively simple analyses can be made to predict the probability that a typical recorded count results from the combination of exactly x true events. These analyses are given later in this text, beginning on p. 636 in connection with the closely related topic of *pulse pile-up*.

D. Dead Time Losses from Pulsed Sources

The analysis of counting losses due to detector dead time in the previous sections assumed that the detector was irradiated by a steady-state source of constant intensity. There are many applications in which the source of radiation is not continuous but instead consists of short pulses repeated at a constant frequency. For example, electron linear accelerators used to generate high-energy X-rays can be operated to produce pulses of a few microsecond width with a repetition frequency of several kilohertz. In such cases, the results given previously may not be applicable to correct properly for the effects of dead time losses. Substitute analytical methods must now be applied that make use of some of the statistical principles introduced in Chapter 3.

We confine our analysis to a radiation source that can be represented by the time-dependent intensity sketched below:

It is assumed that the radiation intensity is constant throughout the duration T of each pulse, and that the pulses occur at a constant frequency f. Depending on the relative value of the detector dead time τ, several conditions may apply:

1. If τ is much smaller than T, the fact that the source is pulsed has little effect, and the results given earlier in this section for steady-state sources may be applied with reasonable accuracy.

2. If τ is less than T but not by a large factor, only a small number of counts may be registered by the detector during a single pulse. This is the most complicated circumstance and is beyond the scope of the present discussion. Detailed analysis of this case may be found in Refs. 22 and 23.

3. If τ is larger than T but less than the "off" time between pulses (given by $1/f - T$), the following analysis applies. Note that under these conditions, we can have a maximum of only one detector count per source pulse. Also, the detector will be fully recovered at the start of each pulse.

For the analysis that follows, it is no longer necessary that the radiation be constant over the pulse length T, but it is required that each radiation pulse be of the same intensity. We carry through the following definitions:

$$\tau = \text{dead time of the detector system}$$

$$m = \text{observed counting rate}$$

$$n = \text{true counting rate (if } \tau \text{ were 0)}$$

$$T = \text{source pulse length}$$

$$f = \text{source pulse frequency}$$

Since there can be at most a single count per pulse, the probability of an observed count per source pulse is given by m/f.

The average number of true events per source pulse is by definition equal to n/f. (Note that this average can be greater than 1.) We can apply the Poisson distribution [Eq. (3.24)] to predict the probability that *at least one* true event occurs per source pulse:

$$
\begin{aligned}
P(> 0) &= 1 - P(0) \\
&= 1 - e^{-\bar{x}} \\
&= 1 - e^{-n/f}
\end{aligned}
\tag{4.38}
$$

Since the detector is "live" at the start of each pulse, a count will be recorded if at least one true event occurs during the pulse. Only one such count can be recorded, so the above expression is also the probability of recording a count per source pulse. Equating the two expressions for this probability, we obtain

$$\frac{m}{f} = 1 - e^{-n/f}$$

or

$$m = f(1 - e^{-n/f}) \tag{4.39}$$

A plot of this behavior is shown below:

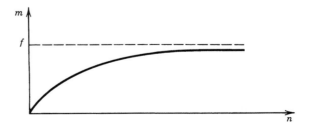

Under these conditions, the maximum observable counting rate is just the pulse repetition frequency, since no more than a single count can be recorded per pulse. Also, neither the specific length of the dead time nor the detailed dead time behavior of the system (whether it is paralyzable or nonparalyzable) has any influence on the losses.

Under normal circumstances, we are more interested in a correction formula to predict the true rate from the measured rate and the system dead time. Solving Eq. (4.39) for n, we derive

$$n = f \ln\left(\frac{f}{f - m}\right) \tag{4.40}$$

Recall that this correction is valid only under the conditions $T < \tau < (1/f - T)$.

In this case, the dead time losses are small under the condition $m \ll f$. Expanding the logarithmic term above for this limit, we find that a first-order correction is then given by

$$n \cong \frac{m}{1 - m/2f} \tag{4.41}$$

This result, because of its similarity to Eq. (4.24), can be viewed as predicting an *effective dead time* value of $1/2f$ in this low-loss limit. Since this value is now one-half the source pulsing period, it can be many times larger than the actual physical dead time of the detector system.

PROBLEMS

4.1 Calculate the amplitude of the voltage pulse produced by collecting a charge equal to that carried by 10^6 electrons on a capacitance of 100 pF. ($e = 1.602 \times 10^{-19}$ C).

4.2 Compare the characteristics of pulse, MSV, and current mode operations as they are applied in radiation measurement systems. Include a table that lists the advantages and disadvantages of each.

4.3 Derive Eq. (4.8).

4.4 A detector with charge collection time of 150 ns is used with a preamplifier whose input circuit can be represented by the parallel combination of 300 pF and 10,000 ohms. Does this situation fall in the category of small or large collection circuit time constant?

4.5 A scintillation counter operated at a given voltage produces a differential pulse height spectrum as sketched below:

(a) Draw the corresponding integral pulse height spectrum.

(b) Sketch the expected counting curve obtained by varying the voltage to the detector while counting above a fixed threshold.

4.6 Sketch both the differential and integral pulse height spectra (using the same horizontal scale) for the following cases:

(a) Pulses with single amplitude of 1 V.

(b) Pulses uniformly distributed in amplitude between 0 and 1 V.

(c) Pulses distributed around an average amplitude of 1.5 V with a pulse height resolution of 8%.

4.7 A gamma-ray spectrometer records peaks corresponding to two different gamma-ray energies of 435 and 490 keV. What must be the energy resolution of the system (expressed as a percentage) in order just to distinguish these two peaks?

4.8 In a detector with a Fano factor of 0.1 what should be the minimum number of charge carriers per pulse to achieve a statistical energy resolution limit of 0.5%?

4.9 A pulse-processing system operated over a long period of time shows a typical drift that broadens single-amplitude pulses into a distribution with pulse height resolution of 2%. If this system is used with a detector with an intrinsic pulse height resolution of 4%, what will be the expected overall pulse height resolution?

4.10 Find the solid angle subtended by the circular end surface of a cylindrical detector (diameter of 10 cm) for a point source located 20 cm from the surface along the cylindrical axis.

4.11 The diameter of the moon as seen from earth subtends an angle of about 0.5°. Find the probability that a laser beam aimed in a random direction from the earth's surface will stike the moon.

4.12 The detector of Problem 4.10 has an intrinsic peak efficiency at 1 MeV of 12%. The point source emits a 1 MeV gamma ray in 80% of its decays and has an activity of 20 kBq. Neglecting attenuation between the source and detector, calculate the number of counts that will appear under the 1 MeV full-energy peak in the pulse height spectrum from the detector over a 100 s count.

4.13 A source of ^{116m}In (half-life = 54.0 min.) is counted using a G-M tube. Successive 1-min observations gave 131,340 counts at 12:00 noon and 93,384 counts at 12:40. Neglecting background and using a reasonable model for dead time losses, calculate the true interaction rate in the G-M tube at 12:00.

4.14 Counters A and B are nonparalyzable with dead time of 30 and 100 μs, respectively. At what *true* event rate will dead time losses in counter B be twice as great as those for counter A?

4.15 A counter with negligible background gives exactly 10,000 counts in a 1-s period when a standard source is in place. An identical source is placed beside the first, and the counter now records 19,000 counts in 1 s. What is the counter dead time?

4.16 A paralyzable detector system has a dead time of 1.5 μs. If a counting rate of 10^5 per second is recorded, find the two possible values for the true interaction rate.

4.17 As a source is brought progressively closer to a paralyzable detector, the measured counting rate rises to a maximum and then decreases. If a maximum counting rate of 50,000 per second is reached, find the dead time of the detector.

REFERENCES

1. L. Purghel and N. Valcov, *Nucl. Instrum. Meth.* **B95**, 7 (1995).
2. J. M. Harrer and J. G. Beckerley, *Nuclear Power Reactor Instrumentation Systems Handbook,* Vol. 1, Chap. 5, TID-25952-PI (1973).
3. N. R. Campbell and V. J. Francis, *IEEE 93*, Part III (1946).
4. L. Ruby, *Nucl. Instrum. Meth.* **A337**, 531 (1994).
5. H. R. V. Carrillo, *Nucl. Instrum. Meth.* **A371**, 535 (1996).
6. H. Gotoh and H. Yagi, *Nucl. Instrum. Meth.* **96,** 485 (1971).
7. P. Oblozinsky and I. Ribansky, *Nucl Instrum. Meth.* **94,** 187 (1971).
8. K. Verghese, R. P. Gardner, and R. M. Felder, *Nucl. Instrum. Meth.* **101,** 391 (1972).
9. M. Belluscio, R. De Leo, A. Pantaleo, and A. Vox, *Nucl. Instrum. Meth.* **114,** 145 (1974).
10. M. V. Green, R. L. Aamodt, and G. S. Johnston, *Nucl. Instrum. Meth.* **117,** 409 (1974).
11. R. Carchon et al., *Nucl. Instrum. Meth.* **128,** 195 (1975).
12. L. Wielopolski, *Nucl. Instrum. Meth.* **143,** 577 (1977).
13. J. Cook, *Nucl. Instrum. Meth.* **178,** 561 (1980).
14. L. Wielopolski, *Nucl. Instrum. Meth.* **226,** 436 (1984).
15. R. A. Rizk, A. M. Hathout, and A.-R. Z. Hussein, *Nucl. Instrum. Meth.* **A245,** 162 (1986).
16. "Particle Counting in Radioactivity Measurements," ICRU Report 52, ICRU, Bethesda (1994).
17. W. S. Diethorn, *Int. J. Appl. Rad. Isotopes* **25,** 55 (1974).
18. J. W. Müller, *Nucl. Instrum. Meth.* **112,** 47 (1973).
19. J. W. Müller, *Nucl. Instrum. Meth.* **117,** 401 (1974).
20. J. Libert, *Nucl. Instrum. Meth.* **151,** 555 (1978).
21. G. Faraci and A. R. Pennisi, *Nucl. Instrum. Meth.* **212,** 307 (1983).
22. A. M. Cormack, *Nucl. Instrum. Meth.* **15,** 268 (1962).
23. C. H. Westcott, *Proc. R. Soc. London Ser. A* **194,** 508 (1948).

Chapter 5

Ionization Chambers

Several of the oldest and most widely used types of radiation detectors are based on the effects produced when a charged particle passes through a gas. The primary modes of interaction involve ionization and excitation of gas molecules along the particle track. Although the excited molecules can at times be used to derive an appropriate signal (as in the gas scintillators discussed in Chapter 8), the majority of gas-filled detectors are based on sensing the direct ionization created by the passage of the radiation. The detectors that are the topic of the following three chapters (ion chambers, proportional counters, Geiger tubes) all derive, in somewhat different ways, an electronic output signal that originates with the ion pairs formed within the gas filling the detector.

Ion chambers in principle are the simplest of all gas-filled detectors. Their normal operation is based on collection of all the charges created by direct ionization within the gas through the application of an electric field. As with other detectors, ion chambers can be operated in current or pulse mode. In most common applications, ion chambers are used in current mode as dc devices, although some examples of pulse mode applications will be given at the end of this chapter. In contrast, proportional counters or Geiger tubes are almost always used in pulse mode.

The term *ionization chamber* has conventionally come to be used exclusively for the type of detector in which ion pairs are collected from gases. The corresponding process in solids is the collection of electron–hole pairs in the semiconductor detectors described in Chapters 11–13. Direct ionization is only rarely exploited in liquids, although some developments of this type are described in Chapter 19.

Many details that are omitted in the following discussions can be found in the classic books on ionization chambers by Rossi and Staub[1] and by Wilkinson.[2] More specific descriptions of chamber design and construction are included in other books, of which Refs. 3–5 are representative examples.

I. THE IONIZATION PROCESS IN GASES

As a fast charged particle passes through a gas, the types of interaction detailed in Chapter 2 create both excited molecules and ionized molecules along its path. After a neutral molecule is ionized, the resulting positive ion and free electron are called an *ion pair*, and it serves as the basic constituent of the electrical signal developed by the ion chamber. Ions can be formed either by direct interaction with the incident particle, or through a secondary process in which some of the particle energy is first transferred to an energetic electron or "delta ray" (see Chapter 2). Regardless of the detailed mechanisms involved, the practical quantity of interest is the total number of ion pairs created along the track of the radiation.

Table 5.1 Values of the Energy Dissipation per Ion Pair (the W-Value) for Different Gases[a]

Gas	First Ionization Potential (eV)	W-Value (eV/ion pair)	
		Fast Electrons	Alpha Particles
Ar	15.7	26.4	26.3
He	24.5	41.3	42.7
H_2	15.6	36.5	36.4
N_2	15.5	34.8	36.4
Air		33.8	35.1
O_2	12.5	30.8	32.2
CH_4	14.5	27.3	29.1

[a] Values for W from ICRU Report 31, "Average Energy Required to Produce an Ion Pair," International Commission on Radiation Units and Measurements, Washington, DC, 1979.

A. Number of Ion Pairs Formed

At a minimum, the particle must transfer an amount of energy equal to the ionization energy of the gas molecule to permit the ionization process to occur. In most gases of interest for radiation detectors, the ionization energy for the least tightly bound electron shells is between 10 and 25 eV. However, there are other mechanisms by which the incident particle may lose energy within the gas that do not create ions. Examples are excitation processes in which an electron may be elevated to a higher bound state in the molecule without being completely removed. Therefore, the average energy lost by the incident particle per ion pair formed (defined as the *W-value*) is always substantially greater than the ionization energy. The *W*-value is in principle a function of the species of gas involved, the type of radiation, and its energy. Empirical observations, however, show that it is not a strong function of any of these variables and is a remarkably constant parameter for many gases and different types of radiation. Some specific data are shown in Table 5.1, and a typical value is 25–35 eV/ion pair. Therefore, an incident 1 MeV particle, if it is fully stopped within the gas, will create about 30,000 ion pairs. Assuming that W is constant for a given type of radiation, the deposited energy will be proportional to the number of ion pairs formed and can be determined if a corresponding measurement of the number of ion pairs is carried out.

B. The Fano Factor

In addition to the mean number of ion pairs formed by each incident particle, the fluctuation in their number for incident particles of identical energy is also of interest. These fluctuations will set a fundamental limit on the energy resolution that can be achieved in any detector based on collection of the ions. In the simplest model, the formation of each ion pair will be considered a Poisson process. The total number of ion pairs formed should therefore be subject to fluctuations characterized by a standard deviation equal to the square root of the average number formed. As discussed in Chapter 4, many radiation detectors show an inherent fluctuation that is less than predicted by this simplified model. The *Fano factor* is introduced as an empirical constant by which the predicted variance must be multiplied to give the experimentally observed variance [see Eq. (4.14)].

The Fano factor reflects to some degree the fraction of all the incident particle energy that is converted into information carriers within the detector. If the entire energy of the incident radiation were always converted into ion pairs, the number of pairs produced would always be precisely the same and there would be no statistical fluctuation. Under these conditions the Fano factor would be zero. If only a very small fraction of the incident radiation is converted, however, then the ion pairs would be formed far apart and with a relatively low probability, and there would be a good reason to expect that the distribution in their number should follow a Poisson distribution. In gases, the Fano factor is empirically observed to be less than 1 so that the fluctuations are smaller than would be predicted based on Poisson statistics alone.

The Fano factor has significance only when the detector is operated in pulse mode. We therefore postpone further discussions of its magnitude in gases to the following chapter on proportional counters, where pulse mode operation and good energy resolution are more important considerations.

C. Diffusion, Charge Transfer, and Recombination

The neutral atoms or molecules of the gas are in constant thermal motion, characterized by a mean free path for typical gases under standard conditions of about 10^{-6}–10^{-8} m. Positive ions or free electrons created within the gas also take part in the random thermal motion and therefore have some tendency to diffuse away from regions of high density. This diffusion process is much more pronounced for free electrons than for ions since their average thermal velocity is much greater. A point-like collection of free electrons will spread about the original point into a Gaussian spatial distribution whose width will increase with time. If we let σ be the standard deviation of this distribution as projected onto an arbitrary orthogonal axis $(x, y, \text{or } z)$, and t be the elapsed time, then it can be shown that

$$\sigma = \sqrt{2Dt} \qquad (5.1)$$

The value of the diffusion coefficient D in simple cases can be predicted from kinetic gas theory, but in general, a more complex transport model is required to accurately model experimental observations.[6]

Of the many types of collisions that will normally take place between free electrons, ions, and neutral gas molecules, several that are important in understanding the behavior of gas-filled detectors are shown schematically in Fig. 5.1. *Charge transfer collisions* can

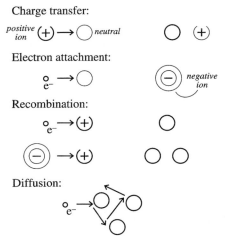

Figure 5.1 Shown are some types of interactions of charged species in a gas that can influence the behavior of gas-filled detectors. In the first four illustrations, the interacting species are on the left and the products of the interaction are on the right. Neutral atoms or molecules are shown as simple circles.

occur when a positive ion encounters another neutral gas molecule. In such a collision, an electron is transferred from the neutral molecule to the ion, thereby reversing the roles of each. This charge transfer is particularly significant in gas mixtures containing several different molecular species. There will then be a tendency to transfer the net positive charge to the gas with the lowest ionization energy because energy is liberated in collisions which leave that species as the positive ion.

The free electron member of the original ion pair also undergoes many collisions in its normal diffusion. In some species of gas, there may be a tendency to form *negative ions* by the attachment of the free electron to a neutral gas molecule. This negative ion then shares many properties with the original positive ion formed in the ionization process, but with opposite electric charge. Oxygen is an example of a gas that readily attaches electrons, so that free electrons diffusing in air are rapidly converted to negative ions. In contrast, nitrogen, hydrogen, hydrocarbon gases, and noble gases all are characterized by relatively low electron attachment coefficients, and therefore the electron continues to migrate in these gases as a free electron under normal conditions.

Collisions between positive ions and free electrons may result in *recombination* in which the electron is captured by the positive ion and returns it to a state of charge neutrality. Alternatively, the positive ion may undergo a collision with a negative ion in which the extra electron is transferred to the positive ion and both ions are neutralized. In either case, the charge represented by the original pair is lost and cannot contribute further to the signal in detectors based on collection of the ionization charge.

Because the collision frequency is proportional to the product of the concentrations of the two species involved, the recombination rate can be written

$$\frac{dn^+}{dt} = \frac{dn^-}{dt} = -\alpha n^+ n^- \tag{5.2}$$

where
n^+ = number density of positive species

n^- = number density of negative species

α = recombination coefficient

The recombination coefficient is normally orders of magnitude larger between positive ions and negative ions compared with that between positive ions and free electrons. In gases that readily form negative ions through electron attachment, virtually all the recombination takes place between positive and negative ions.

There are two general types of recombination loss: *columnar* recombination and *volume* recombination. The first type (sometimes also called *initial* recombination) arises from the fact that ion pairs are first formed in a column along the track of the ionizing particle. The local density of ion pairs is therefore high along the track until the ion pairs are caused to drift or diffuse away from their point of formation. Columnar recombination is most severe for densely ionizing particles such as alpha particles or fission fragments compared with fast electrons that deposit their energy over a much longer track. This loss mechanism is dependent only on the local conditions along individual tracks and does not depend on the rate at which such tracks are formed within the detector volume. In contrast, volume recombination is due to encounters between ions and/or electrons after they have left the immediate location of the track. Since many tracks are typically formed over the time it takes for ions to drift all the way to the collecting electrodes, it is possible for ions and/or electrons from independent tracks to collide and recombine. Volume recombination therefore increases in importance with irradiation rate. Thus charge separation and collection should be as rapid as possible in order to minimize recombination, and high electric fields are indicated for this purpose.

II. CHARGE MIGRATION AND COLLECTION

A. Charge Mobility

If an external electric field is applied to the region in which ions or electrons exist in the gas, electrostatic forces will tend to move the charges away from their point of origin. The net motion consists of a superposition of a random thermal velocity together with a net drift velocity in a given direction. The drift velocity for positive ions is in the direction of the conventional electric field, whereas free electrons and negative ions drift in the opposite direction.

For ions in a gas, the drift velocity can be fairly accurately predicted from the relation

$$v = \frac{\mu \mathscr{E}}{p} \tag{5.3}$$

where

v = drift velocity

$\mathscr{E}$ = electric field strength

p = gas pressure

The *mobility* μ tends to remain fairly constant over wide ranges of electric field and gas pressure and does not differ greatly for either positive or negative ions in the same gas. Tabulated values for the mobility of ions in various gases can be found in Ref. 2. Typical values are between 1 and 1.5×10^{-4} m^2 atm/V · s for detector gases of medium atomic number. Therefore, at 1 atm pressure, a typical electric field of 10^4 V/m will result in a drift velocity of the order of 1 m/s. Ion transit times over typical detector dimensions of a centimeter will therefore be approximately 10 ms. By most standards, this is a very long time.

Free electrons behave quite differently. Their much lower mass allows a greater acceleration between encounters with neutral gas molecules, and the value of the mobility in Eq. (5.3) is typically 1000 times greater than that for ions. Typical collection times for electrons are therefore of the order of microseconds, rather than milliseconds. In some gases, for example, hydrocarbons and argon–hydrocarbon mixtures, there is a saturation effect in the electric drift velocity (see Fig. 5.2). Its value approaches a maximum for high values of the electric field and may even decrease slightly if the field is further increased. In many

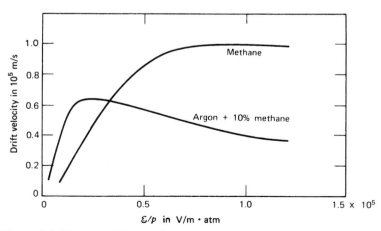

Figure 5.2 Electron drift velocity as a function of electric field $\mathscr{E}$ divided by gas pressure p. (Data from Bortner et al.[7])

other gases (see Fig. 6.15), the electron drift velocity continues to increase for the largest $\mathscr{E}/p$ values likely to be used in gas-filled counters.

As the electrons are drifting through the gas under the influence of the electric field, they will to first approximation follow the path of the electric field line that passes through their point of origin (but in the reverse direction to the electric field vector). Random diffusion of the electrons will still be taking place, however, causing each individual electron to follow a slightly different path. For strong electric field values, the increased average energy given the electron in the direction of the field results in different values of the diffusion coefficient [D in Eq. (5.1)] for the directions that are parallel to or transverse to the field. Over the few microseconds that are typically required for the electrons to reach a collecting electrode, the diffusion in either direction might be of the order of a millimeter or less. While in most standard detectors this charge spreading will have little practical effect, it does play a potential role in limiting the position resolution attainable in "position-sensitive" gas detectors that deduce the location of the ionizing event through the position of arrival of the electrons at the anode, or through measuring the electron drift time.

B. The Ionization Current

In the presence of an electric field, the drift of the positive and negative charges represented by the ions and electrons constitutes an electric current. If a given volume of gas is undergoing steady-state irradiation, the rate of formation of ion pairs is constant. For any small test volume of the gas, this rate of formation will be exactly balanced by the rate at which ion pairs are lost from the volume, either through recombination or by diffusion or migration from the volume. Under the conditions that recombination is negligible and all the charges are efficiently collected, the steady-state current produced is an accurate measure of the rate at which ion pairs are formed within the volume. Measurement of this *ionization current* is the basic principle of the dc ion chamber.

Figure 5.3 illustrates the basic elements of a rudimentary ion chamber. A volume of gas is enclosed within a region in which an electric field can be created by the application of an

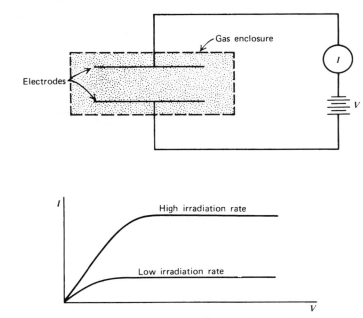

Figure 5.3 The basic components of an ion chamber and the corresponding current–voltage characteristics.

external voltage. At equilibrium, the current flowing in the external circuit will be equal to the ionization current collected at the electrodes, and a sensitive ammeter placed in the external circuit can therefore measure the ionization current.

The current–voltage characteristics of such a chamber are also sketched in Fig. 5.3. Neglecting some subtle effects related to differences in diffusion characteristics between ions and electrons, no net current should flow in the absence of an applied voltage because no electric field will then exist within the gas. Ions and electrons that are created ultimately disappear either by recombination or by diffusion from the active volume. As the voltage is increased, the resulting electric field begins to separate the ion pairs more rapidly, and columnar recombination is diminished. The positive and negative charges are also swept toward the respective electrodes with increasing drift velocity, reducing the equilibrium concentration of ions within the gas and therefore further suppressing volume recombination between the point of origin and the collecting electrodes. The measured current thus increases with applied voltage as these effects reduce the amount of the original charge that is lost. At a sufficiently high applied voltage, the electric field is large enough to effectively suppress recombination to a negligible level, and all the original charges created through the ionization process contribute to the ion current. Increasing the voltage further cannot increase the current because all charges are already collected and their rate of formation is constant. This is the region of *ion saturation* in which ion chambers are conventionally operated. Under these conditions, the current measured in the external circuit is a true indication of the rate of formation of all charges due to ionization within the active volume of the chamber.

C. Factors Affecting the Saturation Current

Several factors can detract from reaching saturation current in an ion chamber. The most important of these is recombination, which is minimized by ensuring that a large value of the electric field exists everywhere within the ion chamber volume. Columnar recombination along the track of heavy charged particles (such as alpha particles or fission fragments) is particularly significant, so that larger values of the applied voltage are required in these cases to achieve saturation, compared with electron or gamma-ray irradiation. Also, the effects of volume recombination are more important for higher intensity of the irradiation (and thus higher ion currents), as illustrated in Fig. 5.4. Here the percentage by which the measured current falls short of true saturation is seen to increase as a function of the measured ion current. At the lower irradiation levels, the density of ions and electrons (or negative ions) is correspondingly low, and from Eq. (5.2), the recombination rate is less

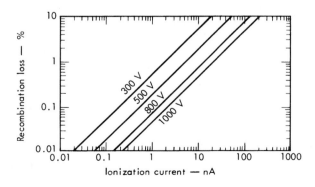

Figure 5.4 Losses due to recombination in an ion chamber filled with argon at 1 atm. These losses are minimized at high values of the applied voltage as shown. (From Colmenares.[8])

significant than at high rates. Thus higher voltages are required at high irradiation rates to approach the true saturated ion current. This dependence is evident in the current–voltage characteristics shown in Fig. 5.3 and also is apparent in the example of an ionization chamber used for neutron monitoring shown in Fig. 14.14. In chambers operated with ambient air as the fill gas, increases in the humidity level will also raise the volume recombination rate.[9]

If the production of ion pairs is uniform throughout the volume of the ion chamber, the drift of positive charges toward the cathode and negative charges toward the anode will create some imbalance in the steady-state concentrations of the two charge carriers. The concentration of positive ions will be greatest near the cathode; the opposite is true for the negative charges. Wherever a gradient exists for a species that is free to migrate, some net diffusion must take place in the direction of decreasing concentration. The direction of diffusion is therefore opposite that of the charge carrier flow caused by the electric field, and the effect can be to reduce the measured ion current. From an analysis given by Rossi and Staub[1] the perturbation in the measured ion current is given by

$$-\frac{\Delta I}{I} = \frac{\epsilon k T}{eV} \tag{5.4}$$

where parallel planar electrodes have been assumed. In this expression,

ϵ = ratio of the average energy of charge carrier with electric field present to that without the electric field

k = Boltzmann constant

T = absolute temperature

e = electronic charge

V = applied voltage between electrodes

The size of this loss in the saturation current depends primarily on the applied voltage and the size of the quantity ϵ defined above. At room temperature, the factor kT/e is approximately 2.5×10^{-2} V. For ions, ϵ is not much larger than unity, and Eq. (5.4) therefore predicts that the loss due to diffusion is negligible even for low values of the applied voltage. When the negative charge is carried by free electrons, however, ϵ can be of the order of several hundred and a significant loss in saturation current due to electron diffusion is possible. Because ϵ will tend toward a limiting saturated value as the voltage V is raised to high values, the loss is minimized by operating the chamber at high values of the applied voltage.

In many cases, losses due to diffusion and volume recombination can be reduced to insignificant levels at reasonable values of the applied voltage. Columnar recombination is more difficult to eliminate completely, and small losses may persist even at the highest available voltages. It is then helpful to make a number of measurements of the ionization current as a function of voltage in order to determine the true saturated current. It has been shown[10] that under these conditions a plot of $1/I$ versus $1/V$ can be used to extrapolate the measurements to zero on the $1/V$ axis (infinite electric field) to make an accurate prediction of the saturated value of the current.

III. DESIGN AND OPERATION OF DC ION CHAMBERS

A. General Considerations

When an ion chamber is used in direct current mode, it is possible to collect the negative charges either as free electrons or as negative ions. Therefore, virtually any fill gas can be used for the chamber, including those that have high electron attachment coefficients. Although recombination is more significant when negative ions are formed, diffusion losses are less important. Conditions of saturation can normally be achieved over dimensions of a few centimeters using applied voltages that are no greater than tens or hundreds of volts. Air is the most common fill gas and is one in which negative ions are readily formed. Air is required in those chambers designed for the measurement of gamma-ray exposure (see p. 140). Denser gases such as argon are sometimes chosen in other applications to increase the ionization density within a given volume. The fill gas pressure is often 1 atmosphere, although higher pressures are sometimes used to increase the sensitivity.

The geometry chosen for an ion chamber can be varied greatly to suit the application, provided the electric field throughout the active volume can be maintained high enough to lead to saturation of the ion current. Parallel plate or planar geometry leads to a uniform electric field between the plates. Also common is a cylindrical geometry in which the outer shell of the cylinder is operated at ground potential and an axial conducting rod carries the applied voltage. In this case, a field that varies inversely with radius is created. Analytical methods that can be used to predict the current–voltage characteristics of chambers of various geometries are described in Ref. 11.

B. Insulators and Guard Rings

With any design, some sort of supporting insulator must be provided between the two electrodes. Because typical ionization currents in most applications are extremely small (of the order of 10^{-12} A or less), the leakage current through these insulators must be kept very small. In the simple scheme of Fig. 5.3, any leakage through the insulator will add to the measured ionization current and cause an unwanted component of the signal. In order to keep this component below 1% of an ionization current of 10^{-12} A for an applied voltage of 100 V, the resistance of the insulator would need to be greater than 10^{16} ohms. Although it may be possible to find materials with bulk resistivity sufficiently high to meet this criterion, leakage paths across the surface of the insulator due to absorbed moisture or other contaminants almost always represent a lower resistance. A different design is therefore often used in low-current applications of ion chambers in which a *guard ring* is employed to reduce the effects of insulator leakage.

A diagram illustrating the use of a guard ring is shown in Fig. 5.5. The insulator is now segmented into two parts, one part separating the conducting guard ring from the negative electrode and the other part separating it from the positive electrode. Most of the voltage drop occurs across the outer segment in which the resulting leakage current does not pass through the measuring instrument. The voltage drop across the inner segment is only the

Figure 5.5 Cross-sectional view of one end of a cylindrical ion chamber that utilizes guard ring construction. Most of the applied voltage V appears across the outer insulator, for which the resulting leakage current does not contribute to the measured current I.

voltage difference across the ammeter terminals and can be very small. Therefore, the component of leakage current which is added to the signal is greatly reduced compared with the case without the guard ring.

Insulators for applications that do not involve high radiation fields are normally manufactured from one of the high-resistivity synthetic plastics. Care is taken to keep the surface smooth and as free of defects as possible in order to minimize the amount of moisture absorption. Radiation damage in these materials in high radiation applications (such as in reactor instruments) can lead to rapid deterioration of the insulation properties. In such cases, inorganic materials such as ceramics are preferred because of their higher resistance to radiation damage.

C. Measurement of the Ion Current

The magnitude of the ionization current under typical conditions is much too small to be measured using standard galvanometer techniques. Instead, some active amplification of the current must be carried out to allow its indirect measurement. An *electrometer* indirectly measures the current by sensing the voltage drop across a series resistance placed in the measuring circuit (see Fig. 5.6). The voltage developed across the resistor (typically with a value of 10^9–10^{12} ohms) can be amplified and serves as the basis for the measured signal. One weakness of standard electrometer circuits is that they must be dc coupled throughout. Any small drift or gradual change in component values therefore results in a corresponding change in the measured output current. Thus, circuits of this type must frequently be balanced by shorting the input and resetting the scale to zero.

An alternative approach is to convert the signal from dc to ac at an early stage, which then allows a more stable amplification of the ac signal in subsequent stages. This conversion is accomplished in the *dynamic-capacitor* or *vibrating reed* electrometer by collecting the ion current across an RC circuit with long time constant, as shown in Fig. 5.7. At equilibrium, a constant voltage is developed across this circuit, which is given by

$$V = IR \tag{5.5}$$

where I is the steady-state ionization current. A charge Q is stored on the capacitance, which is given by

$$Q = CV \tag{5.6}$$

If the capacitance is now caused to change rapidly compared with the time constant of the circuit, a corresponding change will be induced in the voltage across C given by

$$\Delta V = \frac{Q}{C^2}\Delta C$$

$$\Delta V = I\frac{R}{C}\Delta C \tag{5.7}$$

Figure 5.6 Measurement of small ion currents through the use of a series resistance R and an electrometer to record the resulting dc voltage V_R. The chamber capacitance plus any parallel stray capacitance is represented by C. Provided the ion current does not change for several values of the measurement time constant RC, its steady-state value is given by $I = V_R/R$.

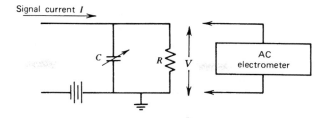

Signal current *I*

Figure 5.7 Principle of the vibrating reed electrometer. Oscillations of the capacitance induce an ac voltage that is proportional to the steady-state signal current *I* [see Eq. (5.7)].

If the value of the capacitance is varied sinusoidally about an average value *C*, the amplitude of the ac voltage that is induced is therefore proportional to the ionization current.

The average ionization current can also be measured over finite periods of time by integration methods. If the value of *R* in Fig. 5.6 is made infinite, any ionization current from the chamber is simply integrated across the capacitance *C*. By noting the change in the voltage across the capacitance over the measurement period, the total integrated ionization current or ionization charge can be deduced. If the amount of natural leakage across the capacitance can be kept small, this integration technique has the potential of being able to measure much smaller ionization currents than through direct dc current measurement.

To illustrate, suppose that the ion chamber of Fig. 5.8 is originally charged to a voltage V_0. If the leakage across the chamber insulators and external capacitor is negligible, this voltage would be maintained at its original value indefinitely in the absence of ionizing radiation. If radiation is present, the ions will act to partially discharge the capacitance and reduce the voltage from its original value. If a charge ΔQ (defined either as the positive charge of the positive ions or the negative charge of the electrons) is created by the radiation, then the total charge stored on the capacitance will be reduced by ΔQ. The voltage will therefore drop from its original value of V_0 by an amount ΔV given by

$$\Delta V = \frac{\Delta Q}{C} \tag{5.8}$$

A measurement of ΔV thus gives the total ionization charge or the integrated ionization current over the period of the measurement.

D. The Electret

The previous section described the measurement of ionization current integrated over a period of time by noting the drop in voltage measured across the chamber capacitance after an initial charging to a known voltage. A similar measurement can be carried out by

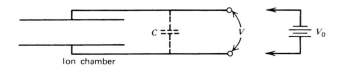

Ion chamber

Figure 5.8 Measurement of the total charge due to ionization ΔQ over a finite period of time by integration of the current. The voltage across the chamber is initially set to V_0 and the voltage source is disconnected. By measuring V at a later time, the ionization charge is given by $\Delta Q = C \Delta V$, where $\Delta V \equiv V_0 - V$, and C is the capacitance of the chamber plus any parallel capacitance.

noting the drop in surface voltage on an element known as an *electret*[12-14] that is placed inside an ion chamber. An electret is a sample of dielectric material (commonly Teflon) that carries a quasi-permanent electrical charge. A disk of the material of approximately 1-cm diameter and 1-mm thickness is fabricated by heating in the presence of an electric field and then cooling to "freeze" electric dipoles in place. A voltage of up to 1000 V may then appear between the opposite surfaces of the electret. With proper encapsulation, this stored charge may be stable over periods of a year or more, even in the presence of high humidity.

An electret placed with one of its surfaces in contact with the conducting walls of an ion chamber will create an electric field throughout the chamber volume because of the voltage difference on its opposite surface. Any ion pairs created by radiation within the chamber will be separated and collected by this field, serving to partially neutralize the charge carried by the electret. Measurements of the electret voltage before and after the exposure can then be calibrated in terms of the total ionization charge produced in the chamber.

IV. RADIATION DOSE MEASUREMENT WITH ION CHAMBERS

A. Gamma-Ray Exposure

One of the most important applications of ion chambers is in the measurement of gamma-ray exposure. An air-filled ion chamber is particularly well-suited for this application because exposure is defined in terms of the amount of ionization charge created in air.† Under the proper conditions a determination of the ionization charge in an air-filled ionization chamber can give an accurate measure of the exposure, and a measurement of the ionization current will indicate the exposure rate.

The task of measuring exposure is somewhat more complicated than it might first appear, because it is defined in terms of the ionization created by all the secondary electrons generated at the point at which the dose is to be measured. Strictly speaking, one would need to follow each of these secondary electrons over its entire range and measure all the ionization created along its track. Because the range in air of secondary electrons created by typical gamma-ray energies can be several meters, it is impractical to design an instrument that would carry out such a measurement directly. Instead, the principle of compensation is used.

If the test volume of air is surrounded by an infinite sea of equivalent air that is also subject to the same exposure over the course of the measurement, an exact compensation will occur. That is, all the ionization charge created outside the test volume from secondary electrons that were formed within the volume is exactly balanced by charge created within the test volume from secondary electrons formed in the surrounding air. This situation is illustrated in Fig. 5.9.

One popular design based on this compensation is diagrammed in Fig. 5.10 and is called the *free–air ionization chamber*. Each end of the chamber is rendered insensitive by grounding the guard electrodes shown in the diagram. The parallel plate geometry creates electric field lines that are perpendicular to the plates in the space between them, but only the volume defined by the central electrode collects ionization current that is registered by the external circuit. The incident gamma-ray beam is collimated so that it is confined to a region that is far from the chamber electrodes, and secondary electrons created in the sensitive volume cannot reach either electrode. Compensation is therefore not required in the vertical dimension but will take place in the horizontal dimension, provided the intensity of the incident radiation beam is not appreciably reduced in its passage through the

---

†The fundamental SI unit of exposure corresponds to that amount of X-ray or gamma-ray radiation whose associated secondary electrons create an ionization charge of 1 coulomb per kilogram of dry air at STP (see Chapter 2).

Figure 5.9 The principle of compensation in the measurement of gamma-ray exposure. If the density of gamma-ray interactions is uniform, the test volume will record an amount of ionization that is just equal to that produced along the extended tracks of all the secondary electrons formed within the test volume (such as tracks ① and ② above). The ionization produced by these electrons outside the volume is compensated by ionization within the volume which is produced by tracks originating elsewhere (such as tracks ③ and ④).

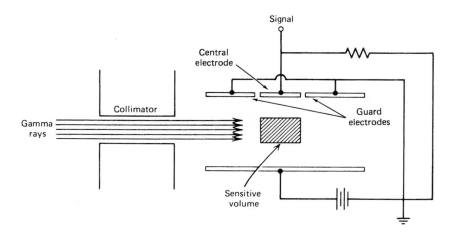

Figure 5.10 The free-air ionization chamber. Because secondary electrons created in the sensitive volume cannot reach the electrodes before stopping, compensation is required only in the dimension parallel to the incident radiation.

chamber. Free-air ionization chambers are widely used for accurate exposure measurements for gamma-ray energies below about 100 keV.

At higher energies, the larger range of the secondary electrons creates some difficulties. In order to prevent ionization loss from secondary electrons reaching the electrodes, the dimensions of the chamber must become impractically large. Therefore, gamma-ray exposure measurements at higher energies are conventionally carried out in *cavity chambers* in which a small volume of air is surrounded by a solid material (chosen because its properties are as similar as possible to air).

To see how compensation can take place in such a situation, first consider the hypothetical arrangement sketched in Fig. 5.11 in which a small volume of air is defined by ideal electrodes that are transparent to both gamma rays and electrons. If this test volume is at the center of a large volume of air that is subject to the same exposure, the compensation previously described will take place. For gamma-ray energies greater than a few hundred keV, however, the extended range of the secondary electrons makes the required surrounding volume of air much too large. The test volume would have to be at the center of a room-sized volume of air throughout which the exposure is the same as at the test volume itself. In practical situations uniform exposure conditions seldom exist over such large volumes.

The situation could be improved by compressing the air surrounding the test volume into a shell, which could then be no more than a centimeter or two in thickness. Because the same number of air molecules would still be present, none of the compensation properties would change and the test volume would still register an accurate measure of gamma-ray exposure. Now the demands of uniformity could be greatly relaxed to requiring only that the exposure be uniform over the much smaller volume defined by the shell of compressed air. Under these conditions, all ionization lost from the test volume will be compensated by ionization from secondary electrons created within the compressed air shell.

It is now only one step further to replace the hypothetical compressed air shell with a more practical wall of solid material (see Fig. 5.11). The wall is said to be *air equivalent* if its compensation properties are similar to those of an air layer. This condition will be met provided the secondary electron yield and rate of electron energy loss per unit mass are similar to those of air. Because both these phenomena depend largely on atomic number, it turns out that virtually any material with atomic number close to that of air (such as aluminum or plastics) is reasonably air equivalent.

If the walls are sufficiently thick compared with secondary electron ranges, a condition of *electronic equilibrium* is established in which the flux of secondary electrons leaving the

Figure 5.11 Compensation in an ideal ion chamber on the left with fully transparent electrodes could be accomplished by secondary electrons created outside the chamber. However, uniform irradiation conditions seldom exist over the required large surrounding volume of air. On the right, the chamber is provided with air-equivalent solid walls that are thicker than the maximum secondary electron range. Now compensation can take place from secondary electrons formed within the much smaller volume of the walls.

inner surface of the wall becomes independent of the wall thickness. Neglecting attenuation of the incident gamma-ray beam in the wall, the measured ion current from a chamber of constant air volume will then be the same, independent of the wall thickness. Table 5.2 lists the minimum air-equivalent wall thickness required to establish electronic equilibrium. For ordinary gamma-ray energies, this thickness is 1 cm or less.

For an air-equivalent ion chamber, the exposure rate R in C/kg · s is simply given by the ratio of the saturated ion current I_s (in amperes) to the mass M (in kg) contained in the active volume:

$$R = \frac{I_s}{M} \tag{5.9}$$

The air mass M is normally calculated from a measurement of the chamber volume and the density at STP,

$$M = 1.293 \, \frac{\text{kg}}{\text{m}^3} \cdot V \cdot \frac{P}{P_0} \cdot \frac{T_0}{T} \tag{5.10}$$

where

V = chamber volume (in m³)

P = air pressure within the chamber

P_0 = standard pressure (760 mm Hg, or 1.013×10^5 Pa)

T = air temperature within the chamber

T_0 = standard temperature (273.15 K)

In routine monitoring, exposure rates of the order of 10^{-3} roentgens/hour (7.167×10^{-11} C/kg · s) are of typical interest. For an ion chamber of 1000 cm³ volume, the saturated ion current at standard temperature and pressure calculated from Eqs. (5.9) and (5.10) is 9.27×10^{-14} A. Because this signal current is very low, sensitive electrometers and careful chamber design are required to minimize leakage currents.

Table 5.2 Thicknesses of Ionization Chamber Walls Required for Establishment of Electronic Equilibrium[a]

Photon Energy (MeV)	Thickness[b] (g/cm²)
0.02	0.0008
0.05	0.0042
0.1	0.014
0.2	0.044
0.5	0.17
1	0.43
2	0.96
5	2.5
10	4.9

[a]From International Commission on Radiation Units and Measurements Report ICRU #20 (1971).

[b]The thickness quoted is based on the range of electrons in water. The values will be substantially correct for tissue-equivalent ionization chamber walls and also for air. Half of the above thickness will give an ionization current within a few percent of its equilibrium value.

From Eq. (5.9), the ion current I_s expected is simply proportional to the mass of the gas within the chamber. Therefore it can only be enhanced by increasing either the chamber volume or pressure. Thus ion chambers designed to measure very low exposure rates are routinely filled to a high pressure to raise their sensitivity. Because of the increased frequency of collisions in the gas at high pressures, argon is often then substituted for air to avoid the greater recombination rate that accompanies negative ion formation.

B. Absorbed Dose

Gas-filled ionization chambers can also be applied indirectly to the measurement of absorbed dose (the energy absorbed per unit mass) in arbitrary materials. The technique is based on application of the *Bragg–Gray principle*, which states that the absorbed dose D_m in a given material can be deduced from the ionization produced in a small gas-filled cavity within that material as follows:

$$D_m = WS_mP \tag{5.11}$$

where

W = average energy loss per ion pair formed in the gas

S_m = relative mass stopping power (energy loss per unit density) of the material to that of the gas

P = number of ion pairs per unit mass formed in the gas

In order for the dose D_m to be measured in grays (J/kg), W must be expressed in J/ion pair and P in ion pairs/kg. Equation (5.11) holds to good approximation for different types of radiation provided several geometric conditions are met. The cavity should be small compared with the range of the primary or secondary charged particles associated with the radiation so that its presence does not greatly affect the particle flux. In the case of gamma rays, the solid medium should also be large compared with the range of the secondary electrons so that electronic equilibrium is established at the inner walls of the cavity.

For an ion chamber, the solid medium is the wall material and the cavity is its internal gas-filled volume. If the gas is air and the wall is air-equivalent material, the factor S_m in Eq. (5.11) is unity. These are the conditions necessary for measurement of absorbed dose in air, which is equivalent to measurement of the gamma-ray exposure discussed in the previous section. If other wall materials or gases are used, the absorbed dose in the wall can be calculated by inserting appropriate values for W and S_m in Eq. (5.11). The dose in biological tissue is of particular interest in radiation protection, so that "tissue-equivalent" ion chambers are widely applied in which the wall is made from a material with an elemental composition similar to that of tissue.

In some applications, there is a problem in keeping the cavity small enough so that the Bragg–Gray principle applies. For example, if the spectrum of gamma rays under measurement extends to very low photon energies, the corresponding secondary electron ranges will be short. An ion chamber with small internal dimensions compared with these ranges would result in very small ion currents, which may cause measurement difficulties. In these applications, use can be made of the "extrapolation chamber," which consists of a pair of electrodes made from the material of interest mounted with variable spacing between them. A series of measurements can be carried out with different electrode spacings and the measured ion current per unit volume extrapolated to the case of zero spacing. With a vanishingly small cavity, the Bragg–Gray principle is once again applicable and the absorbed dose rate can be calculated from this extrapolated ion current.

V. APPLICATIONS OF DC ION CHAMBERS

A. Radiation Survey Instruments

Portable ion chambers of various designs are commonly used as survey instruments for radiation monitoring purposes. These typically consist of a closed air volume of several hundred cm^3 from which the saturated ion current is measured using a battery-powered electrometer circuit. Walls are approximately air equivalent and may be fabricated from plastic or aluminum. These instruments give relatively accurate measurements of the exposure for gamma-ray energies high enough to avoid significant attenuation in the walls or entrance window of the chamber, but low enough so that electronic equilibrium is established in the walls. Figure 5.12 shows calibration curves for two such instruments and illustrates the drop-off in sensitivity for gamma-ray energies that are less than 50–100 keV due to window attenuation.

Other types of portable ion chambers used for dose measurements are based on the charge integration principle. As described previously, the chamber is initially charged, placed in the radiation field for a period of time, and the resulting drop in chamber voltage is used as a measure of the total integrated ionization charge. The "Condenser R-meter" consists of a charger–reader mechanism and several interchangeable ion chambers of different sizes and wall thicknesses. Each is designed to cover a different range of gamma-ray energy and maximum dose. When used in the proper range, these devices can give results that are accurate to within a few percent. Another related instrument is a small-size ion chamber known as a "pocket chamber." These are often provided with an integral quartz fiber electroscope that can be read on an internal scale by holding the pocket chamber up to the light. An initial charging of the chamber zeros the scale of the electroscope. The pocket chamber is then worn during the period of monitoring. The total integrated dose can be checked periodically simply by noting the degree of discharge of the chamber as indicated on the electroscope. The ultimate sensitivity of such devices is normally limited by leakage current, which inevitably occurs across the insulator surface over long periods of time.

Figure 5.12 Energy calibration of two different gamma-ray survey meters. (From Storm et al.[15])

B. Radiation Source Calibrators

Many of the common applications of ion chambers take advantage of their excellent long-term stability. When operated in the saturated region, the ion current depends only on the geometry of the source and detector and can remain stable over very long periods of time. For example, the standardization of gamma-ray emitters is often carried out by comparing the ion current from an unknown source with that generated by a standard source under identical geometry. Typical operating characteristics can remain stable to within ± 0.1% over several years,[16] eliminating the need for frequent recalibration.

One specific design of such a chamber is shown in Fig. 5.13. Typical chamber volumes are several thousand cm^3 and the walls are usually made of brass or steel. The inner collecting electrode is made of a thin foil of aluminum or copper to avoid unnecessary radiation attenuation. The geometry is chosen to avoid low-field regions that could lead to changes in the effective active volume with applied voltage. In such a chamber with a 10^4 cm^3 active volume, the saturation current produced by 1 μCi (3.7×10^4 Bq) of ^{60}Co is on the order of 10^{-13} A, about five times the background current.[18] If higher sensitivity is needed, the gas within the ion chamber may be pressurized. Raising the pressure to 20 atm will increase the ion current by a factor of 20, but the total background arising from all sources will increase by a lesser factor. For large chambers, the component of background arising from alpha activity of the chamber walls should be independent of pressure, provided the alpha particles are already fully stopped within the gas. Pressurizing the chamber, however, requires the use of relatively thick entrance windows, which, although not usually a problem for gamma rays, can interfere with the extension of the applications to include pure beta-emitting isotopes.

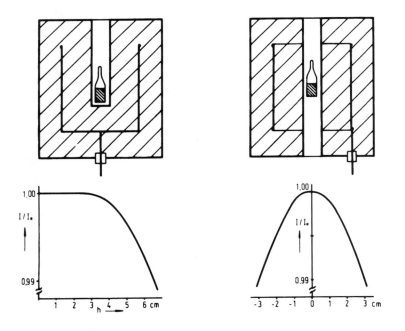

Figure 5.13 Ionization chambers used for calibration of gamma-ray sources. The curve below each configuration shows the behavior of the ionization current for small displacements of the source position. (From Weiss.[17])

C. Measurement of Radioactive Gases

Radioactive gases can conveniently be measured by incorporating them as a constituent of the fill gas of an ionization chamber. Entrance and exit ports are provided to the chamber volume to allow introduction of the gas to be sampled on a continuous flow basis. Examples of specific ion chamber designs of this type are given in Refs. 8, 18, and 19.

The ionization current expected from a given quantity of radioactive gas within the chamber is given by

$$I = \frac{\overline{E}\alpha e}{W} \tag{5.12}$$

where

I = ionization current (in A)

$\overline{E}$ = average energy deposited in the gas per disintegration (in eV)

α = total activity (in Bq)

e = electronic charge (in C)

W = average energy deposited per ion pair in the gas (in eV)

The quantity $\overline{E}$ is simple to predict only in the case of very small radiation energies that are nearly always fully absorbed within the chamber gas. For example, the soft beta particles emitted in the decay of tritium have an average energy of 5.65 keV, and $\overline{E}$ can be taken as equal to this value for chambers that are large compared with the corresponding beta particle range in the gas. Equation (5.12) then predicts a sensitivity of about 1 pA/μCi, or 2.7×10^{-17} A/Bq. Once a beta particle energy is large enough so that typical ranges are comparable to chamber dimensions, more complex procedures are required to estimate the average energy deposited (e.g., see Ref. 20).

A common application is one in which samples of air that contain trace quantities of a radioactive gas are to be continuously monitored. When the gas that flows through the chamber is subject to atmospheric changes, a number of difficulties can arise. These perturbing influences can include effects due to moisture, aerosols, ions, and smoke,[†] which may be found in the sampled air. Prior treatment of the incoming air by filtration or electrostatic precipitation can help to control many of these influences.[21]

The chamber will also be sensitive to the ambient gamma-ray background radiation. If constant, this background can be eliminated by simple subtraction of the signal recorded when the chamber is filled with pure air. In other situations in which the gamma-ray background may change during the course of the measurement, a twin chamber filled with pure air can be used to generate a compensating signal that is subtracted from that produced by the chamber through which the sample gas is circulated.[18]

D. Remote Sensing of Ionization

A novel application of ionization chambers has been demonstrated[22–24] that exploits the fact that positive and negative ions formed in air can survive for periods of seconds or minutes before losing their charge to recombination. These ions can be entrained in a flow of air and, if kept away from surfaces that enhance recombination, may be transported a substantial distance from their point of formation. If a flow of air containing such ions is introduced into an ion chamber, then a current will be observed resulting from the motion of these ions in its

---

[†]The decrease in ionization current from an internal alpha particle source due to smoke particles is the operational basis of the familiar ionization-type home smoke detector.

electric field even if no ions are created within the volume of the ion chamber itself. No radioactivity needs to be transported into the chamber, but the presence of the ions themselves in the air stream introduced into a chamber is sufficient to create the signal.

This principle is the basis of an instrument called the *long-range alpha detector* (*LRAD*). Samples of materials or equipment that might be contaminated by alpha particle emitters are placed into a container through which there is a flow of air. Since the range of the alpha particles is only a few centimeters in air, their direct detection would require bringing a suitable detector very near to the surface, a process that may be impractical because of inaccessibility of the surfaces. Instead, the air stream flowing past these surfaces will pick up some of the positive and negative ions and transport them downstream through a conventional ion chamber. The drift of the ions in the chamber electric field takes place over milliseconds, so the slower motion of the air flow does not play an important role in the detected signal. This application illustrates that the ion current results from the motion of the ions within the field of the ion chamber, regardless of where these ions were originally formed.

VI. PULSE MODE OPERATION

A. General Considerations

Most applications of ionization chambers involve their use in current mode in which the average rate of ion formation within the chamber is measured. Like many other radiation detectors, ion chambers can also be operated in pulse mode in which each separate radiation quantum gives rise to a distinguishable signal pulse. As outlined in Chapter 4, pulse mode operation can offer significant advantages in sensitivity or ability to measure the energy of the incident radiation. Pulse mode ion chambers are used to some extent in radiation spectroscopy, although they have largely been replaced by semiconductor diode detectors (see Chapter 11) for many such applications. However, pulse ion chambers remain important instruments in certain specialized applications such as large-area alpha spectrometers or in neutron detectors of the type described in Chapter 14.

The equivalent circuit of an ion chamber operated in pulse mode is shown in Fig. 5.14. The voltage across the load resistance R is the basic electrical signal. In the absence of any ionization charge within the ion chamber, this signal voltage is zero, and all the applied voltage V_0 appears across the ionization chamber itself. When an ionizing particle passes through the chamber, the ion pairs that are created begin to drift under the influence of the electric field. As will be shown in the analysis below, these drifting charges give rise to *induced charges* on the electrodes of the ion chamber that reduce the ion chamber voltage from its equilibrium value V_0. A voltage then appears across the load resistance, which is equal to the amount by which the chamber voltage has dropped. When all charges within

Figure 5.14 Equivalent circuit of an ion chamber operated in pulse mode. Here C represents the capacitance of the chamber plus any parallel capacitance. V_R is the output pulse waveform.

the chamber have been collected at the opposite electrodes, this voltage reaches its maximum value. There then follows a slow return to equilibrium conditions on a time scale determined by the time constant RC of the external circuit. During this period, the voltage across the load resistance gradually drops to zero and the chamber voltage returns to its original value V_0. If the time constant of the external circuit is long compared with the time required to collect the charges within the chamber, a signal pulse is produced whose amplitude indicates the magnitude of the original charge generated within the ion chamber.

As indicated in the earlier discussion of electron and ion mobilities in gases, typical times required to collect free electrons over several centimeters are a few microseconds. On the other hand, ions (either positive or negative) drift much more slowly and typically require collection times of the order of milliseconds. Therefore, if a signal that accurately reflects the total contribution of both the ions and electrons is to be generated, the collection circuit time constant and subsequent pulse-shaping time constants must be long compared with a millisecond. Under these conditions, the ion chamber must be restricted to very low pulse rates to avoid excessive pulse pile up. Furthermore, the sensitivity of the output of the shaping circuits to low frequencies makes the system susceptible to interference from microphonic signals generated by mechanical vibrations within the ion chamber.

For these reasons, pulse-type ion chambers are more often operated in the electron-sensitive mode. Here a time constant is chosen that is intermediate between the electron collection time and the ion collection time. The amplitude of the pulse that is produced then reflects only the drift of the electrons and will have much faster rise and fall times. Shorter shaping time constants and much higher rates can therefore be tolerated. A significant sacrifice has been made, however, in that the amplitude of the output pulse now becomes sensitive to the position of the original radiation interaction within the chamber and no longer reflects only the total number of ion pairs formed. The use of more complex gridded chambers, described in a later section, can overcome this disadvantage to a large degree. The fill gas of any electron-sensitive ion chamber must of course be chosen from those gases in which the electrons remain as free electrons and do not form negative ions.

B. Derivation of the Pulse Shape

The shape of the pulse produced from the flow of charges within detectors of arbitrary geometry is best addressed using the *Shockley-Ramo theorem* outlined in Appendix D. However, in the simple case of an ion chamber with parallel plate electrodes that are separated by a small spacing compared with the length and width of the plates, edge effects can be neglected and the analysis can be assumed to involve only one variable: the position of the charges in the dimension perpendicular to the plates. It is instructive to then follow a simplified derivation of the pulse shape that invokes arguments based only on the conservation of energy. This derivation gives the correct results and avoids the mathematical complexity that might obscure some of the physical behavior that is important in determining the characteristics of the output pulse expected under these conditions.

In the derivation that follows, we assume that a sufficient electric field is applied so that electron–ion recombination is insignificant and also that the negative charges remain as free electrons. We first derive an expression for the pulse shape for the case in which the collecting circuit time constant is much longer than both the ion and electron collection times.

The pulse shape depends on the configuration of the electric field and the position at which the ion pairs are formed with respect to the equipotential surfaces that characterize the field geometry. To simplify the following analysis, we assume that the chamber electrodes are parallel plates, for which the equipotential surfaces are uniformly spaced planes parallel to the electrode surfaces, and the constant electric field intensity is given by

$$\mathscr{E} = \frac{V}{d} \tag{5.13}$$

Here V is the voltage across the chamber electrodes and d is their spacing. As a further simplification, we assume that all ion pairs are formed at an equal distance x from the positive electrode where the electric potential is equal to $\mathscr{E}x$. This situation is sketched in Fig. 5.15.

The pulse shape is most easily derived based on arguments involving the conservation of energy. Because the time constant of the external circuit is assumed to be large, no appreciable current can flow during the relatively short time required to collect the charges within the ion chamber. Therefore, the energy required to move the charges from their place of origin must come from the energy originally stored across the capacitance C, represented by the ion chamber and associated stray capacitance. This energy is $\frac{1}{2}CV_0^2$, where V_0 is the applied voltage.

After a time t, the ions will have drifted a distance v^+t toward the cathode, where v^+ is the ion drift velocity. Similarly, the electrons will have moved a distance v^-t toward the anode. Both of these motions represent the movement of charge to a region of lower electric potential, and the difference in potential energy is absorbed in the gas through the multiple collisions the charge carriers undergo with gas molecules during their motion. This energy is equal to $Q\Delta\varphi$ for both the ions and electrons, where Q is the total charge and $\Delta\varphi$ is the change in electric potential. The charge $Q = n_0 e$, where n_0 is the number of original ion pairs and e is the electronic charge. The potential difference $\Delta\varphi$ is the product of the electric field $\mathscr{E}$ and the distance traveled toward the electrode. Conservation of energy can therefore be written

original stored energy		energy absorbed by ions		energy absorbed by electrons		remaining stored energy
$\frac{1}{2}CV_0^2$	$=$	$n_0 e \mathscr{E} v^+ t$	$+$	$n_0 e \mathscr{E} v^- t$	$+$	$\frac{1}{2}CV_{ch}^2$

$$\tfrac{1}{2}C(V_0^2 - V_{ch}^2) = n_0 e \mathscr{E}(v^+ + v^-)t$$

$$\tfrac{1}{2}C(V_0 + V_{ch})(V_0 - V_{ch}) = n_0 e\left(\frac{V_{ch}}{d}\right)(v^+ + v^-)t \qquad (5.14)$$

The signal voltage is measured across R in Fig. 5.15 and is denoted as V_R. Its magnitude is almost always small compared with V_0 and is given by $V_0 - V_{ch}$. We can therefore make the approximations

$$V_0 + V_{ch} \cong 2V_0 \quad \text{and} \quad \frac{V_{ch}}{d} \cong \frac{V_0}{d}$$

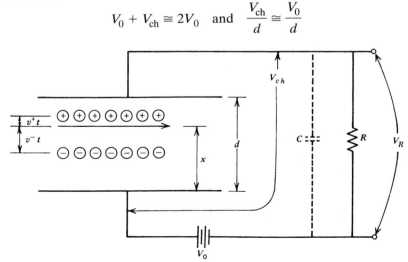

Figure 5.15 Diagram for the derivation of the pulse shape $V_R(t)$ for the signal from an ion chamber. See text for identification.

Putting these substitutions in Eq. (5.14), we obtain

$$\tfrac{1}{2}C(2V_0)V_R = n_0 e \left(\frac{V_0}{d}\right)(v^+ + v^-)t$$

$$V_R = \frac{n_0 e}{dC}(v^+ + v^-)t \tag{5.15}$$

This result describes the initial portion of the signal pulse and predicts a linear rise with time. It is valid only for the period that both the ions and electrons are drifting within the chamber.

The concept of induced charge is sometimes used to describe the changes caused by the drifting charge carriers. By drifting a distance $v^+ t$, the ions cause the chamber voltage to drop by an amount equal to $n_0 e v^+ t / dC$. The same effect would be caused by the reduction of the charge stored across the capacitance C by an amount $n_0 e v^+ t / d$. Therefore, the ion motion can be thought of as inducing a charge of this magnitude. A similar induced charge is created by the electron motion. It should be emphasized that the induced charge results only from the motion of the charge carriers within the chamber volume and does not require their collection at either electrode.

After a time $t^- \equiv x/v^-$, the electrons reach the anode. Their drift has then contributed the maximum possible to the signal voltage, and the second term in Eq. (5.15) becomes a constant equal to its value at t^-. This constant value is $n_0 e v^- t^- / dC$ or $n_0 e x / dC$. For the next period of time, only the ions are still drifting, and Eq. (5.15) takes the form

$$V_R = \frac{n_0 e}{dC}(v^+ t + x) \tag{5.16}$$

The ions reach the cathode after a time $t^+ \equiv (d - x)/v^+$. At this point, the signal voltage no longer increases, and Eq. (5.15) becomes

$$V_R = \frac{n_0 e}{dC}[(d - x) + x]$$

or

$$V_R = \frac{n_0 e}{C} \tag{5.17}$$

The shape of the signal pulse predicted by Eqs. (5.15), (5.16), and (5.17) is shown in Fig. 5.16. When the collection circuit time constant is very large, or $RC \gg t^+$, the maximum amplitude of the signal pulse is given by

$$V_{\max} = \frac{n_0 e}{C} \tag{5.18}$$

and is independent of the position at which the ion pairs were formed within the chamber. Under these conditions, a measurement of the pulse amplitude $V_{\max}$ gives a direct indication of the original number of ion pairs n_0 that contributed to the pulse.

In electron-sensitive operation, however, the portion of the pulse derived above that corresponds to drift of the ions is almost entirely lost by choosing a collection time constant that is much shorter than the ion collection time. The pulse that remains then reflects only the drift of the electrons and will have an amplitude given by Eq. (5.16) (neglecting ion drift)

$$V\big|_{\text{elec}} = \frac{n_0 e}{C} \cdot \frac{x}{d} \tag{5.19}$$

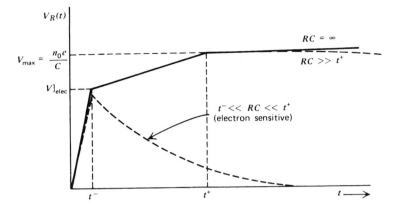

Figure 5.16 Output pulse shape $V_R(t)$ for various time constants RC in the schematic diagram of Fig. 5.15.

The shape of this pulse is also sketched in Fig. 5.16. Only the fast rising portion of the pulse is preserved, and the amplitude is now dependent on the position x at which the electrons were originally formed within the chamber.

In any real situation, the incident radiation creates ion pairs over a range of positions within the chamber. The sharp discontinuities shown in Fig. 5.16 are then somewhat "washed out" in the resulting pulse shape. Electron-sensitive operation will also lead to a situation in which a range of pulse amplitudes will be produced for monoenergetic incident radiation, often a decided disadvantage.

C. The Gridded Ion Chamber

The dependence of the pulse amplitude on position of interaction in electron-sensitive ion chambers can be removed through the use of an arrangement sketched in Fig. 5.17. Here the volume of the ion chamber is divided into two parts by a *Frisch grid*, named after the originator of the design.[25] Through the use of external collimation or preferential location of the radiation source, all the radiation interactions are confined to the volume between the grid and the cathode of the chamber. Positive ions simply drift from this volume to the cathode. The grid is maintained at an intermediate potential between the two electrodes and is made to be as transparent as possible to electrons. Electrons are therefore drawn initially from the interaction volume toward the grid. Because of the location of the load resistor in the circuit, neither the downward drift of the ions nor the upward drift of the electrons as far as the grid produces any measured signal voltage. However, once the electrons pass through the grid on their way to the anode, the grid-anode voltage begins to drop and a signal voltage begins to develop across the resistor. The same type of argument that led to Eq. (5.15) predicts that, for a circuit time constant large compared with the electron collection time, the time-dependent signal voltage across the resistor is

$$V_R = \frac{n_0 e}{dC} v^- t \qquad (5.20)$$

where d is now the grid–anode spacing. This linear rise continues until the electrons reach the anode (see Fig. 5.17b). The maximum signal voltage is therefore

$$V_{max} = \frac{n_0 e}{C} \qquad (5.21)$$

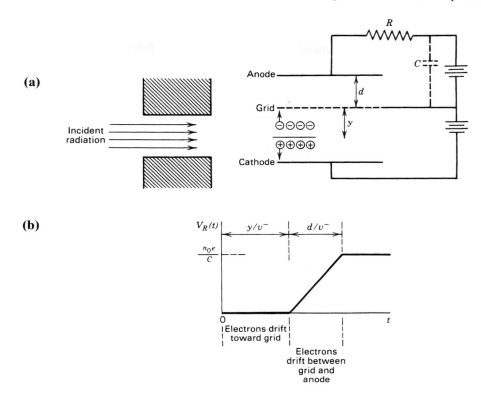

Figure 5.17 (*a*) Operational principle of the gridded ion chamber. All ion pairs must be formed in the lower region of the chamber between the cathode and grid. (*b*) The pulse shape that results from the formation of n_0 ion pairs at a distance y from the grid, where d is the grid–anode spacing [see part (*a*)]. The rise of the pulse results from the drift of the electrons across the grid–anode region. The pulse will decay back to zero with a time constant given by RC.

which is identical to Eq. (5.18). However, now the signal is derived only from the drift of electrons rather than from the motion of both electrons and positive ions. The slow rise corresponding to the drift of ions is eliminated, and the circuit time constant can therefore be set at a much shorter value typical of the electron-sensitive mode of operation described in the previous section. Since each electron passes through the same potential difference and contributes equally to the signal pulse, the pulse amplitude is now independent of the position of formation of the original ion pairs and is simply proportional to the total number of ion pairs formed along the track of the incident particle.

D. Pulse Amplitude

The typical amplitude of an ion chamber pulse is relatively small. From the analysis given in the previous sections, the maximum pulse amplitude to be expected from the creation of n_0 ion pairs in either a standard or gridded ion chamber is given by

$$V_{max} = \frac{n_0 e}{C} \tag{5.22}$$

If a 1 MeV charged particle loses all its energy within the chamber, n_0 can be estimated as

$$n_0 = \frac{E_0}{W} \cong \frac{10^6 \text{ eV}}{35 \text{ eV/ion pair}} = 2.86 \times 10^4$$

For typical ion chambers and associated wiring, the capacitance C will be of the order of 10^{-10} farads. We then calculate a pulse amplitude of

$$V_{max} = \frac{(2.86 \times 10^4)(1.60 \times 10^{-19}\text{ C})}{(10^{-10}\text{ F})} = 4.58 \times 10^{-5}\text{ V}$$

Signals of this size can be processed successfully but require the use of relatively sophisticated preamplifiers and pulse-processing electronics to avoid loss of their inherent resolution.

In ion chambers of ordinary size and pressure, pulses from fast electron or gamma-ray interactions will be much smaller. The value of n_0 in the above example will be smaller by as much as a factor of 100 because of the low dE/dx along the primary or secondary electron tracks through the gas. It then becomes difficult or impossible to amplify these very small pulses successfully without severe deterioration that results from various sources of noise in the signal chain. For these reasons, the detectors described in the following two chapters, which take advantage of internal gas multiplication of the charge, are usually preferred because of their resulting larger signal pulses. Exceptions in which pulse-mode ion chambers can be applied to gamma-ray measurements are given in Chapter 19 in which either a liquid or high-pressure fill gas are of sufficient density to fully stop the secondary electrons and thus restore the value of n_0 to nearer that given in the example.

E. Statistical Limit to Energy Resolution

The theoretical limit to the energy resolution set by the statistics of charge carrier formation in an ion chamber can be predicted from the discussion given in Chapter 4. As an example, let us assume that alpha particles with energy of 5.5 MeV are fully stopped in a gas with W-value of 30 eV/ion pair and a Fano factor of 0.15. As a first step, we can calculate the expected average number of ion pairs n_0 produced in the gas to be

$$n_0 = \frac{E_d}{W} = \frac{5.5 \times 10^6\text{ eV}}{30\text{ eV/ion pair}} = 1.83 \times 10^5\text{ ion pairs}$$

The variance in this number is given by

$$\sigma_{n_0}^2 = Fn_0 = 0.15(1.83 \times 10^5) = 2.75 \times 10^4$$

The standard deviation, or the square root of the variance, is then

$$\sigma_{n_0} = \sqrt{Fn_0} = \sqrt{2.75 \times 10^4} = 166$$

For the expected Gaussian distribution describing n_0, the full-width-at-half-maximum (FWHM) is just the shape factor of 2.35 multiplied by σ, leading to

$$\text{FWHM}(n_0) = 2.35\sigma_{n_0} = 390$$

Finally, since W units of energy are required to create one ion pair, the FWHM in units of particle energy is

$$\text{FWHM(E)} = 2.35\sigma_{n_0} \cdot W = 390(30\text{ eV}) = 11.7\text{ keV}$$

For ion chambers, this figure would conventionally be quoted as the energy resolution. Alternatively, the energy resolution could be quoted as a percentage by dividing the above figure by the deposited energy

$$R = \frac{2.35\sigma_{n_0}W}{E_d} = \frac{11.7 \times 10^3\text{ eV}}{5.5 \times 10^6\text{ eV}} = 0.213\%$$

This excellent energy resolution is almost never achieved in practice (see below) because sources of electronic noise dominate over the statistical contributions.

F. Charged Particle Spectroscopy

There are applications in which pulse-type ion chambers can be used to good advantage for the measurement of charged particle energies. Although it is much more common to use semiconductor detectors for this purpose (see Chapter 11), ion chambers offer features that may be attractive in some circumstances. They can be constructed with almost arbitrary size and geometry, and the gas pressure can be chosen to tailor the stopping power or effective thickness of the active volume. Ion chambers are also far less subject to performance degradation due to radiation damage than are semiconductor detectors. Their design is generally quite simple, and they can be fabricated by many users with standard workshop facilities.

Pulse-type ion chambers have found application in a number of circumstances where their unrestricted size and other characteristics[25] have been exploited. They have proved useful in low-level alpha particle measurements[27, 28] where parallel-plate-type gridded ion chambers have been constructed with cross-sectional areas up to 500 cm^2. Even larger chambers can conveniently be constructed by using cylindrical geometry. In order to achieve good energy resolution, a Frisch grid is normally incorporated into the chamber design. Because of the small amplitude of typical pulses, special care must be taken in choosing a low noise preamplifier and in minimizing mechanical vibrations, which, by modulating the detector capacitance, can give rise to interfering microphonic noise.[29] Using standard techniques, one can achieve energy resolution for 5 MeV alpha particles of 35–45 keV for such chambers.[27] By using a cooled preamplifier to reduce the electronic noise, Bertolini[30] demonstrated an energy resolution of 11.5 keV for alpha particles, essentially the ultimate limit set by charge carrier statistics. This energy resolution is comparable to the best that can currently be achieved using silicon semiconductor detectors.

In the conventional configuration, an entrance window for the chamber is provided so that the charged particles travel perpendicular to the electric field lines as shown in Fig. 5.17a. In that case, all the electrons formed along a given track are approximately equidistant from the grid, and each will have approximately the same drift time. The voltage pulse therefore has the general characteristics shown in Fig. 5.17b and all particle tracks give signal pulses of similar shape. It has been demonstrated[31] that there can be some advantage in designing the chamber so that the particle tracks are parallel rather than perpendicular to the field lines. Then the drift time of electrons to the grid will be different for those formed at the beginning of the track compared with those near the end. The detailed shape of the rise of the output pulse will therefore reflect the spatial distribution of ion pairs as they were formed along the track of the incident particle. Because this spatial distribution is conventionally called a *Bragg curve* (see p. 32), an analysis of the pulse shape to yield information on this spatial distribution is conventionally called *Bragg curve spectroscopy*. This approach has proved to be very useful in extending the information obtained from gridded ion chambers beyond a simple measurement of the energy of the particle. Appropriate analysis of the pulse shape[32–34] can distinguish between particles of different type (atomic number and/or charge state) through differences in the shape of their Bragg curve.

PROBLEMS

5.1 Calculate the charge represented by the positive ions (or free electrons) created when a 5.5 MeV alpha particle is stopped in helium. Find the corresponding saturated current if 300 alpha particles per second enter a helium-filled ion chamber.

5.2 The following data were taken from an ion chamber under constant irradiation conditions:

Voltage (V)	Current (pA)
10	18.72
20	19.41
50	19.93
100	20.12

Using the extrapolation procedure outlined on p. 136, find the saturated ionization current. What minimum voltage must be applied in order to reach a current within 0.5% of saturation?

5.3 An air-equivalent pocket chamber having a capacitance of 75 pF is initially charged to a voltage of 25 V. If the active volume contains 50 cm^3 of air at STP, what value of gamma-ray exposure will reduce the chamber voltage to 20 V?

5.4 An ion chamber is constructed using parallel plate electrodes with a spacing of 5.0 cm. It is filled with pure methane gas at a pressure of 1 atm and operated at an applied voltage of 1000 V. From the data given in Fig. 5.2, calculate the maximum electron collection time.

5.5 An ion chamber of the type shown in Fig. 5.8 has an associated capacitance of 250 pF. It is charged initially to a voltage of 1000V, exposed to a gamma-ray flux for a period of 30 min, at which time a second voltage measurement indicates a value of 850 V. Calculate the average current that would have been measured over the exposure period under these conditions.

5.6 With the aid of Fig. 2.14, estimate the maximum distance of penetration or "range" of an electron with 0.5 MeV initial energy in atmospheric air.

5.7 A free-air ionization chamber of the type sketched in Fig. 5.10 has a sensitive volume of very small dimensions. Estimate the minimum electrode spacing if true compensation is to be maintained up to a maximum gamma-ray energy of 5 MeV.

5.8 The average beta particle energy emitted by ^{14}C is 49 keV. Calculate the saturated ion current if 150 kBq of the isotope in the form of CO_2 gas is introduced into a large-volume ion chamber filled with pressurized argon.

5.9 Calculate the minimum current that must be measured if a 1-liter ion chamber is to be used as a gamma-ray survey meter down to dose rates of 0.5 mR/h (35.8 pC/kg · s).

5.10 A parallel plate ion chamber with 150 pF capacitance is operated in electron-sensitive mode. Calculate the pulse amplitude expected from 1000 ion pairs formed 2 cm from the anode, if the total spacing between the plates is 5 cm.

5.11 A small air-filled ionization chamber of fixed interior dimensions is located 10 m from a 1 MeV gamma-ray source. Imagine that the walls of the chamber are made of aluminum and are of variable thickness. Neglecting background radiation, plot the relative ionization current from the chamber as a function of wall thickness for the two conditions listed below:

(a) The experiment is performed in a normal air-filled laboratory.

(b) The experiment is performed in an earth satellite and the surrounding space is essentially a vacuum.

5.12 An air-equivalent ion chamber is constructed using aluminum walls. What is the minimum thickness of these walls if compensation is to be maintained up to a maximum gamma-ray energy of 10 MeV?

5.13 An air-filled ion chamber is operated at a pressure of 3 atm and a temperature of 100°C. If its active volume is 2500 cm^3, find the saturated ion current corresponding to a gamma-ray exposure rate of 100 pC/kg · s.

5.14 A pocket chamber is constructed with an internal volume of 10 cm^3 and a capacitance of 20 pF. If the smallest voltage change that can be sensed is 50 mV, calculate the minimum gamma-ray exposure to which the pocket chamber will be sensitive.

5.15 Using estimates for the mobility of ions and electrons in air, estimate the ratio of the slopes of the two segments of the rise of the pulse shown in Fig. 5.16.

REFERENCES

1. B. B. Rossi and H. H. Staub, *Ionization Chambers and Counters,* McGraw-Hill, New York, 1949.

2. D. H. Wilkinson*, Ionization Chambers and Counters,* Cambridge University Press, Cambridge, 1950.

3. J. Sharpe, *Nuclear Radiation Detectors*, 2nd ed., Methuen and Co., London, 1964.

4. W. J. Price, *Nuclear Radiation Detection*, 2nd ed., McGraw-Hill, New York, 1964.

5. J. W. Boag, "Ionization Chambers," in *Radiation Dosimetry,* Vol. II (F. H. Attix and W. C. Roesch, eds.), Academic Press, New York, 1966.

6. V. Palladino and B. Sadoulet, *Nucl. Instrum. Meth.* **128**, 323 (1975).

7. T. E. Bortner, G. S. Hurst, and W. G. Stone, *Rev. Sci. Instrum.* **28**, 103 (1957).

8. C. A. Colmenares, *Nucl. Instrum. Meth.* **114**, 269 (1974).

9. N. Takata, *Phys. Med. Biol.* **39**, 1047 (1994).

10. S. M. Mustafa and K. Mahesh, *Nucl. Instrum. Meth.* **150**, 549 (1978).

11. F. Hajnal and J. Pane, *IEEE Trans. Nucl. Sci.* **NS-25**(1), 550 (1978).

12. G. Pretzsch, *Nucl. Instrum. Meth.* **A251**, 386 (1986).

13. P. Kotrappa, J. C. Dempsey, J. R. Hickey, and L. R. Stieff, *Health Phys.* **54**(**1**), 47 (1988).

14. P. E. Cruvinel, S. Mascarenhas and J. Cameron, *Nucl. Instrum. Meth.* **A287**, 580 (1990).

15. E. Storm, D. W. Lier, and H. I. Israel, *Health Phys.* **26**, 179 (1974).

16. G. C. Lowenthal, *Int. J. Appl. Radiat. Isotopes* **20**, 559 (1969).

17. H. M. Weiss, *Nucl. Instrum. Meth.* **112**, 291 (1973).

18. R. A. Jalbert and R. D. Hiebert, *Nucl. Instrum. Meth.* **96**, 61 (1971).

19. M. J. Wood et al., *Health Phys.* **72**(**3**), 423 (1997).

20. T. Torii, *Nucl. Instrum. Meth.* **A356**, 255 (1995).

21. J. R. Waters, *Nucl. Instrum. Meth.* **117**, 39 (1974).

22. M. A. Wolf, J. L. McAtee, and W. P. Unruh, *IEEE Trans. Nucl. Sci.* **37**(**2**), 883 (1990).

23. D. W. MacArthur, K. S. Allander, J. A. Bounds, and J. L. McAtee, *Nucl. Tech.* **102**, 270 (1993).

24. D. W. MacArthur et al., *IEEE Trans. Nucl. Sci.* **40**(**4**), 840 (1993).

25. O. Frisch, *British Atomic Energy Report* **BR-49** (1944).

26. H. W. Fulbright, *Nucl. Instru. Meth.* **162**, 21 (1979)

27. H. Hoetzl and R. Winkler, *Nucl. Instrum. Meth.* **223**, 290 (1984).

28. C. W. Sill, *Health Phys.* **69**(**1**), 21 (1995).

29. G. F. Nowack, *Nucl. Instrum. Meth.* **A255**, 217 (1987).

30. G. Bertolini, *Nucl. Instrum. Meth.* **223**, 285 (1984).

31. C. R. Gruhn et al., *Nucl. Instrum. Meth.* **196**, 33 (1982).

32. N. J. Shenhav and H. Stelzer, *Nucl. Instrum. Meth.* **228**, 359 (1985).

33. R. Kotte, H.-J. Keller, H.-G. Ortlepp, and F. Stary, *Nucl. Instrum. Meth.* **A257**, 244 (1987).

34. A. D Frawley et al., *Nucl. Instrum. Meth.* **A306**, 512 (1991).

Chapter 6

Proportional Counters

$\mathbf{T}$he proportional counter is a type of gas-filled detector that was introduced in the late 1940s. In common with the Geiger–Mueller tubes described in Chapter 7, proportional tubes are almost always operated in pulse mode and rely on the phenomenon of *gas multiplication* to amplify the charge represented by the original ion pairs created within the gas. Pulses are therefore considerably larger than those from ion chambers used under the same conditions, and proportional counters can be applied to situations in which the number of ion pairs generated by the radiation is too small to permit satisfactory operation in pulse-type ion chambers. One important application of proportional counters has therefore been in the detection and spectroscopy of low-energy X-radiation (where their principal competitor is the lithium-drifted silicon detector described in Chapter 13). Proportional counters are also widely applied in the detection of neutrons, and specific examples are given in Chapters 14 and 15.

The general reference texts listed at the beginning of Chapter 5 also contain detailed discussions of many topics dealing with the design and operation of proportional counters. References 1 and 2 at the end of this chapter are journal articles that contain extensive reviews of the early literature.

I. GAS MULTIPLICATION

A. Avalanche Formation

Gas multiplication is a consequence of increasing the electric field within the gas to a sufficiently high value. At low values of the field, the electrons and ions created by the incident radiation simply drift to their respective collecting electrodes. During the migration of these charges, many collisions normally occur with neutral gas molecules. Because of their low mobility, positive or negative ions achieve very little average energy between collisions. Free electrons, on the other hand, are easily accelerated by the applied field and may have significant kinetic energy when undergoing such a collision. If this energy is greater than the ionization energy of the neutral gas molecule, it is possible for an additional ion pair to be created in the collision. Because the average energy of the electron between collisions increases with increasing electric field, there is a threshold value of the field above which this secondary ionization will occur. In typical gases, at atmospheric pressure, the threshold field is of the order of 10^6 V/m.

The electron liberated by this secondary ionization process will also be accelerated by the electric field. During its subsequent drift, it undergoes collisions with other neutral gas

Figure 6.1 A plot of the first Townsend coefficient as a function of electric field for a typical gas.

molecules and thus can create additional ionization. The gas multiplication process therefore takes the form of a cascade, known as a *Townsend avalanche,* in which each free electron created in such a collision can potentially create more free electrons by the same process. The fractional increase in the number of electrons per unit path length is governed by the Townsend equation:

$$\frac{dn}{n} = \alpha \, dx \tag{6.1}$$

Here α is a called the first Townsend coefficient for the gas. Its value is zero for electric field values below the threshold and generally increases with increasing field strength above this minimum (see Fig. 6.1). For a spatially constant field (as in parallel plate geometry), α is a constant in the Townsend equation. Its solution then predicts that the density of electrons grows exponentially with distance as the avalanche progresses:

$$n(x) = n(0)e^{\alpha x} \tag{6.2}$$

For the cylindrical geometry used in most proportional counters, the electric field increases in the direction that the avalanche progresses, and the growth with distance is even steeper. In the proportional counter, the avalanche terminates when all free electrons have been collected at the anode. Under proper conditions, the number of secondary ionization events can be kept proportional to the number of primary ion pairs formed, but the total number of ions can be multiplied by a factor of many thousands. This charge amplification within the detector itself reduces the demands on external amplifiers and can result in significantly improved signal-to-noise characteristics compared with pulse-type ion chambers. The formation of an avalanche involves many energetic electron–atom collisions in which a variety of excited atomic or molecular states may be formed. The performance of proportional counters is therefore much more sensitive to the composition of trace impurities in the fill gas than is the case for ion chambers.

B. Regions of Detector Operation

The differences between various types of gas counters operated in pulse mode are illustrated in Fig. 6.2. The amplitude of the observed pulse from the detector is plotted versus the applied voltage or electric field within the detector. At very low values of the voltage, the field is insufficient to prevent recombination of the original ion pairs, and the collected charge is less than that represented by the original ion pairs. As the voltage is raised, recombination is suppressed and the region of ion saturation discussed in Chapter 5 is achieved. This is the normal mode of operation for ionization chambers. As the voltage is increased still further, the threshold field at which gas multiplication begins is reached. The

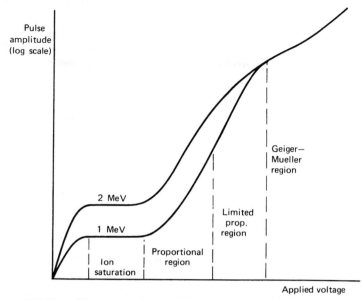

Figure 6.2 The different regions of operation of gas-filled detectors. The observed pulse amplitude is plotted for events depositing two different amounts of energy within the gas.

collected charge then begins to multiply, and the observed pulse amplitude will increase. Over some region of the electric field, the gas multiplication will be linear, and the collected charge will be proportional to the number of original ion pairs created by the incident radiation. This is the region of *true proportionality* and represents the mode of operation of conventional proportional counters, which are the topic of this chapter. Under constant operating conditions, the observed pulse amplitude still indicates the number of ion pairs created within the counter, although their charge has been greatly amplified.

Increasing the applied voltage or electric field still further can introduce nonlinear effects. The most important of these is related to the positive ions, which are also created in each secondary ionization process. Although the free electrons are quickly collected, the positive ions move much more slowly and, during the time it takes to collect the electrons, they barely move at all. Therefore, each pulse within the counter creates a cloud of positive ions, which is slow to disperse as it drifts toward the cathode. If the concentration of these ions is sufficiently high, they represent a space charge that can significantly alter the shape of the electric field within the detector. Because further gas multiplication is dependent on the magnitude of the electric field, some nonlinearities will begin to be observed. These effects mark the onset of the region of *limited proportionality* in which the pulse amplitude still increases with increasing number of initial ion pairs, but not in a linear fashion.

If the applied voltage is made sufficiently high, the space charge created by the positive ions can become completely dominant in determining the subsequent history of the pulse. Under these conditions, the avalanche proceeds until a sufficient number of positive ions have been created to reduce the electric field below the point at which additional gas multiplication can take place. The process is then self-limiting and will terminate when the same total number of positive ions have been formed regardless of the number of initial ion pairs created by the incident radiation. Then each output pulse from the detector is of the same amplitude and no longer reflects any properties of the incident radiation. This is the *Geiger–Mueller* region of operation, and detectors operated under these conditions are discussed in Chapter 7. A related mode of operation, known as *self-quenched streamer* mode, is of interest in position-sensitive detectors and is examined in Chapter 19.

C. Choice of Geometry

Typical proportional counters are constructed with the cylindrical geometry illustrated in Fig. 6.3. The anode consists of a fine wire that is positioned along the axis of a large hollow tube that serves as the cathode. The polarity of the applied voltage in this configuration is important, because the electrons must be attracted toward the center axial wire. This polarity is necessary from two standpoints:

1. Gas multiplication requires large values of the electric field. In cylindrical geometry, the electric field at a radius r is given by

$$\mathscr{E}(r) = \frac{V}{r \ln (b/a)} \tag{6.3}$$

where
V = voltage applied between anode and cathode

a = anode wire radius

b = cathode inner radius.

Large values of the electric field therefore occur in the immediate vicinity of the anode wire where r is small. Because electrons are attracted to the anode, they will be drawn toward the high-field region.

To illustrate, suppose a voltage V of 2000 V is applied to a cylindrical counter with a = 0.008 cm and b = 1.0 cm. The electric field at the anode surface is then 5.18×10^6 V/m. In parallel plate geometry, the field would be uniform between the planar electrodes. With a spacing of 1.0 cm, an applied voltage of 51,800 V would be required to achieve the same electric field. Such a high voltage is practically unworkable and therefore small-diameter anode wires are used in proportional counters to create the high field required.

2. If uniform multiplication is to be achieved for all ion pairs formed by the original radiation interaction, the region of gas multiplication must be confined to a very small volume compared with the total volume of the gas. Under these conditions, almost all primary ion pairs are formed outside the multiplying region, and the primary electron simply drifts to that region before multiplication takes place. Therefore, each electron undergoes the same multiplication process regardless of its original position of formation, and the multiplication factor will be same for all original ion pairs.

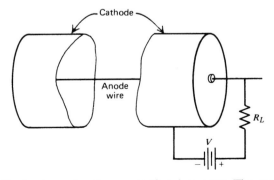

Figure 6.3 Basic elements of a proportional counter. The outer cathode must also provide a vacuum-tight enclosure for the fill gas. The output pulse is developed across the load resistance R_L.

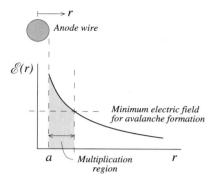

Figure 6.4 The rapid decrease in the electric field strength with distance from the anode wire surface limits the multiplication region to a small volume.

In Fig. 6.4, the electric field is plotted as a function of distance from the anode wire center. The multiplication region begins only when this field becomes larger than the minimum required to support avalanche formation and extends to the anode surface. In the previous example, suppose that the threshold electric field for multiplication in the fill gas is 10^6 V/m. From Eq. (6.3), the field exceeds this value only for radii less than 0.041 cm, or about five times the anode radius. The volume contained within this radius is only about 0.17% of the total counter gas volume.

To help visualize the avalanche formation in the neighborhood of the wire surface, Fig. 6.5 shows results obtained from a Monte Carlo modeling of the electron multiplication and diffusion processes that are important in a typical avalanche. It was assumed that a single free electron drifted into the vicinity of the wire. The resulting avalanche is confined to a small distance along the length of the wire equivalent to only several times its diameter. As a consequence, methods for sensing its position along the wire (discussed later in this chapter) can accurately measure the axial position of the incident electron. The avalanche

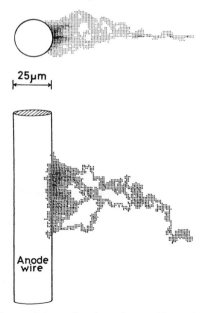

Figure 6.5 Orthogonal views of an avalanche triggered by a single electron as simulated by a Monte Carlo calculation. The density of the shading indicates the concentration of electrons formed in the avalanche. (From Matoba et al.[3])

also covers only a limited range of the wire circumference, oriented generally toward the incident electron direction. Because the avalanche is restricted to a small region around the anode wire, the counter is able to respond to other events that take place elsewhere along the wire before the ions have cleared from the first event. For this reason, the dead time effects that are discussed in Chapter 7 for Geiger–Mueller tubes are much less limiting in proportional counters.

II. DESIGN FEATURES OF PROPORTIONAL COUNTERS

A. Sealed Tubes

A sketch of a proportional counter incorporating many common design features is shown in Fig. 6.6. The thin axial wire anode is supported at either end by insulators that provide a vacuum-tight electrical feedthrough for connection to the high voltage. The outer cathode is conventionally grounded, so that positive high voltage must be applied to ensure that electrons are attracted toward the high-field region in the vicinity of the anode wire. For applications involving neutrons or high-energy gamma rays, the cathode wall can be several millimeters thick to provide adequate structural rigidity. For low-energy gamma rays, X-rays, or particulate radiation, a thin entrance "window" can be provided either in one end of the tube or at some point along the cathode wall.

Good energy resolution in a proportional counter is critically dependent on ensuring that each electron formed in an original ionization event is multiplied by the same factor in the gas multiplication process. The most important mechanical effect that can upset this proportionality is a distortion of the axially uniform electric field predicted by Eq. (6.3). One potential source is any variation in the diameter of the anode wire along its length. Then the value of a in Eq. (6.3) will no longer be constant everywhere along the length of the tube and the degree of gas multiplication will vary from point to point. Because it is easier to make a large diameter wire uniform to within a given fractional tolerance, this effect is minimized by avoiding the use of extremely fine anode wires. However, there is some experimental evidence[5] that the surface roughness of commonly used gold-plated tungsten wire does not detract significantly from the energy resolution even with diameters as small as 12.5 μm. Using an anode wire with as small a diameter as permitted by the surface quality minimizes the high voltage required for a particular gas multiplication value and also tends to promote good energy resolution by minimizing avalanche fluctuations (see p. 177).

ALL DIMENSIONS IN CENTIMETERS

Figure 6.6 Cross-sectional view of a specific proportional tube design used in fast neutron detection. The anode is a 0.025 mm diameter stainless steel wire. The field tubes consist of 0.25 mm diameter hypodermic needles fitted around the anode at either end of the tube. (From Bennett and Yule.[4])

Other common causes of field distortion are effects that occur near the ends of the tube. The electric field can become extremely distorted near the point at which the anode wire enters the insulator, due to the presence of the end wall of the tube or other conducting structures in the vicinity. Unless precautions are taken to avoid these difficulties, ion pairs created in these end regions will undergo a degree of gas multiplication different from those created more generally throughout the volume of the tube.

The most common solution to the end effect problem is to design the counter so that events occurring near the ends do not undergo any gas multiplication and so that an abrupt transition takes place between these "dead" regions and the remainder of the active volume of the proportional tube. One method of achieving this condition is through the use of *field tubes*. As illustrated in Fig. 6.6, field tubes consist of short lengths of conducting tubing (with a diameter many times greater than that of the anode), which are positioned around the anode wire at each end of the counter. The voltage applied to the field tube can be maintained at the anode voltage, but because its diameter is large, no gas multiplication will take place in the region between the field tube and the cathode. The end of the field tube then marks the beginning of the active volume of the proportional counter, which can be some distance away from either end of the overall counter. Alternatively, the field tube may be operated at an intermediate potential between that of the anode and cathode. By setting this potential to that which would exist in its absence at the same radial position, the field within the active volume of the counter is less disturbed by the presence of the field tube, and Eq. (6.3) will describe the radial dependence of the electric field to a very good approximation everywhere within the active volume. Detailed analyses of the electric field in the vicinity of the field tubes can be found in Refs. 4 and 6. Other solutions to the end effect problem involve correction of the field through the use of a semiconducting end plate[7] or by conducting rings on an insulating end plate that are maintained at the proper potential to avoid distortion of the field near the tube end.[8]

B. Windowless Flow Counters

Another common configuration for the proportional counter is sketched in Fig. 6.7. Here it is assumed that the source of radiation is a small sample of a radioisotope, which can then be introduced directly into the hemispherical counting volume of the detector. The great advantage of counting the source internally is the fact that no entrance window need come between the radiation source and active volume of the counter. Because window materials can seriously attenuate soft radiations such as X-rays or alpha particles, the internal source configuration is most widely used for these applications.

The counting geometry can also be made very favorable for this type of detector. In the system shown in Fig. 6.7, virtually any quantum emerging from the surface of the source finds its way into the active volume of the counter and can generate an output pulse. The effective

Figure 6.7 Diagram of a 2π gas flow proportional counter with a loop anode wire and hemispherical volume. The sample can often be inserted into the chamber by sliding a tray to minimize the amount of air introduced.

solid angle is therefore very close to 2π and the detector can have an efficiency that is close to the maximum possible for sources in which the radiation emerges from one surface only.

Some means must be provided to introduce the source into the active volume of internal source counters. In some designs, the base of the counter consists of a rotating table in which several depressions or source wells are provided at regular angular intervals. The table can be rotated so that one well at a time is placed directly under the center of the tube active volume. At the same time, at least one other well is accessible from outside the chamber and can be loaded with a new source to be counted. By rotating the table, the new source can be brought into counting position and the prior sample removed. In other designs, the chamber may simply be opened to allow insertion of the source. In either case, some air will enter the chamber and must be eliminated. The counter is therefore purged by allowing a supply of proportional gas to flow through the chamber for some time and sweep away the residual air. Most proportional counters of this type are also operated as flow counters in which the gas continues to flow at a slower rate through the chamber during its normal use.

Another useful geometry for proportional counters is the "*pancake*" *detector*, illustrated in Fig. 6.8. Often used in systems intended to count alpha and beta activity, the sample is placed near one of the flat surfaces of the detector to provide close to 2π counting geometry. This design has some advantages over the hemispherical type in that it generally has smaller volume and therefore lower background, and also limits the length of beta particle tracks in the gas to minimize their pulse amplitude and thus maximize the separation in amplitude from alpha particles. Pancake detectors also lend themselves to systems in which multiple tubes are mounted to count a number of samples simultaneously. They can be operated with a conventional thin entrance window as a sealed tube, or as a gas flow counter in which the window is removed and a sample holding tray is sealed to the bottom of the chamber. It has been demonstrated that there are beneficial effects of adding a third electrode to this type of proportional counter design. A helical wire, much like a spring, is placed around the anode wire and operated at an intermediate voltage between that of the anode and cathode. As illustrated in Fig. 6.8, its function is to make the electric field around the anode wire more uniform in all azimuthal directions. Without this element, the field strength will vary as a function of azimuthal angle, and therefore the energy resolution of the chamber will suffer. The better energy resolution obtainable with the helical element again aids in the separation of alpha- and beta-induced events (see following discussion on p. 185).

Figure 6.8 In part (*a*), the shape of the "pancake" proportional tube is shown. The effect of the grid on improving the uniformity of the electric field around the anode wire is illustrated in part (*b*).

Figure 6.9 A 4π gas flow proportional counter used to detect radiations that emerge from both surfaces of the sample. The top and bottom halves are provided with separate anode wires and can be separated to introduce the sample that is mounted between them.

If the source is prepared on a backing that is thin compared with the range of the radiation of interest, particles (or photons) may emerge from either surface. In that case, the 4π geometry sketched in Fig. 6.9 can take advantage of the added counting efficiency possible from such sources. The two halves of the chamber can be operated independently or in coincidence to select preferentially only certain types of events.

C. Fill Gases

Because gas multiplication is critically dependent on the migration of free electrons rather than much slower negative ions, the fill gas in proportional counters must be chosen from those species that do not exhibit an appreciable electron attachment coefficient. Because air is not one of these, proportional counters must be designed with provision to maintain the purity of the gas. The gas can be either permanently sealed within the counter or circulated slowly through the chamber volume in designs of the continuous flow type. Sealed counters are more convenient to use, but their lifetime is sometimes limited by microscopic leaks that lead to gradual contamination of the fill gas. Continuous flow counters require gas supply systems that can be cumbersome but bypass many potential problems involving gas purity. The continuous flow design also permits the flexibility of choosing a different fill gas for the counter when desired. These systems may be of the "once through" type, in which the exit gas is vented to the atmosphere, or the gas may be recycled after passage through a purifier. The purifier must remove traces of oxygen or other electronegative impurities, and in its most common form[9] consists of a heated porcelain tube filled with calcium turnings maintained at a temperature of 350°C. The influence of electronegative impurities is most pronounced in large-volume counters and for small values of the electron drift velocity. In carbon dioxide, a gas with relatively slow electron drift, it has been shown[10] that an oxygen concentration of 0.1% results in the loss of approximately 10% of the free electrons per centimeter of travel.

Gas multiplication in the proportional counter is based on the secondary ionization created in collisions between electrons and neutral gas molecules. In addition to ionization, these collisions may also produce simple excitation of the gas molecule without creation of a secondary electron. These excited molecules do not contribute directly to the avalanche but decay to their ground state through the emission of a visible or ultraviolet photon. Under the proper circumstances, these de-excitation photons could create additional ionization elsewhere in the fill gas through photoelectric interactions with less tightly bound electron shells or could produce electrons through interactions at the wall of the counter. Although such photon-induced events are important in the Geiger–Mueller region of operation, they are generally undesirable in proportional counters because they can lead to a loss of proportionality and/or spurious pulses. Furthermore, they cause the avalanches to spread along the anode wire to some extent, increasing possible dead time effects and reducing the spatial resolution in position-sensing detectors. It has been found that the addition of a small amount of polyatomic gas, such as methane, to many of the common fill gases will suppress the photon-induced effects by preferentially absorbing the photons in a mode that does not lead to further ionization. Most monatomic counter gases operated at high values of gas multiplication require the use of such a polyatomic stabilizing additive. This component is often called the *quench gas.*

The noble gases, either pure or in binary mixtures, can be useful proportional gases provided the gas multiplication factor is kept below about 100 (Ref. 11). Beyond this point, adding a quench gas is helpful[12, 13] in reducing instabilities and proportionality loss caused by propagation of ultraviolet photons. Because of cost factors, argon is the most widely used of the inert gases, and a mixture of 90% argon and 10% methane, known as *P−10 gas,* is probably the most common general-purpose proportional gas. When applications require high efficiency for the detection of gamma-ray photons by absorption within the gas, the heavier inert gases (krypton or xenon) are sometimes substituted. Many hydrocarbon gases such as methane, ethylene, and so on are also suitable proportional gases and are widely applied where stopping power is not a major consideration. In applications where the signal is used for coincidence or fast timing purposes, gases with high electron drift velocities are preferred (see Fig. 6.15 later in this chapter). Proportional counters used for thermal neutron detection are operated with BF_3 or ^3He as the proportional gas (see Chapter 14), whereas proportional counters applied in fast neutron spectroscopy are filled with hydrogen, methane, helium, or some other low-Z gas (see Chapter 15). In dosimetry studies, it is often convenient to choose a fill gas that has the approximate composition of biological tissue. For such purposes, a mixture consisting of 64.4% methane, 32.4% carbon dioxide, and 3.2% nitrogen is recommended.[14]

Even though oxygen is electronegative, it has been shown[15,16] that air can serve as an acceptable proportional gas under special circumstances. If the distance that electrons must travel is only 1–2 mm and the electric field is kept high in the drift region, enough electrons escape attachment to form small but detectable avalanches. Alternatively, electrons may attach to oxygen molecules in the air and drift slowly to near the anode, where some are detached through collisions and multiplied by initiating avalanches near the wire.[17,18] For stable operation, ambient air must generally be purged of water vapor and organic vapors[19] before introduction into the counter volume.

The basic properties of a fill gas can be changed significantly by small concentrations of a second gas whose ionization potential is less than that of the principal component. One mechanism, known as the *Penning effect,* is related to the existence of long-lived or metastable excited states in the principal gas. If the excitation energy is larger than the ionization energy of the added component, then a collision between the metastable excited atom and a neutral additive atom can ionize the additive. Because the excitation energy would otherwise be lost without the additive, a greater number of ion pairs will be formed per unit energy lost by the incident radiation. For example the W-value for argon can be reduced from 26.2 to 20.3 eV through the addition of a small concentration of ethylene.[20] Furthermore, because a greater fraction of the incident radiation energy is converted into ions, the relative fluctuation in the total number of ions is decreased by as much as a factor of two.[20] Because of the corresponding improvement in energy resolution, fill gases that consist of Penning mixtures are commonly chosen for proportional counters applied in radiation spectroscopy.[21-26] Some values for W and the Fano factor F for common proportional gases are given in Table 6.2 later in this chapter. Other values measured for noble gases and their mixtures are reviewed and reported in Refs. 27–30. It should be emphasized that these parameters are not to be regarded as universal constants but that some mild dependence of W and F on particle type and/or energy are experimentally observed. The energy dependence is often most evident for X-rays and gamma rays in the energy region near the atomic shell absorption edges (see p. 49) of the fill gas.

The diffusion and drift characteristics of electrons can vary considerably depending on the specific choice of the fill gas. Extensive experimental data are available on these parameters (see Ref. 31), and an example of the differences in electron mobility for different gases is illustrated later in this chapter in Fig. 6.15. Some considerations in choosing fill gas include atomic number and density if gamma rays or X-rays are to be detected in the gas, high drift velocities if fast-rising output pulses are needed, and small values of the electron diffusion coefficient to minimize charge spreading for gas-filled devices in which the position of the primary interaction is to be registered. Many of these parameters can be calculated with reasonable accuracy,[32] and a number of computer codes have been developed based on numerical solutions of the transport equations that describe the electron motion in the gas.[33]

III. PROPORTIONAL COUNTER PERFORMANCE

A. Gas Multiplication Factor

A study of the multiplication process in gases is normally divided into two parts. The single-electron response of the counter is defined as the total charge that is developed by gas multiplication if the avalanche is initiated by a single electron originating outside the region of gas multiplication. This process can be studied experimentally by creating irradiation conditions in which only one electron is liberated per interaction (e.g., by the photoelectric interaction of incident ultraviolet photons). Multiplication factors of the order of 10^5 are adequate to allow direct detection of the resulting pulses, and their amplitude distribution can supply information about the gas multiplication mechanisms within the counter. Studies of this type are described in Refs. 34–37. Results of these experiments are used later in this chapter under the discussion of energy resolution of proportional counters.

If the single-electron response is known, the amplitude properties of pulses produced by many original ion pairs can be deduced. Provided that space charge effects (discussed later in this chapter) are not large enough to distort the electric field, each avalanche is independent, and the total charge Q generated by n_0 original ion pairs is

$$Q = n_0 eM \tag{6.4}$$

where M is the average *gas multiplication factor* that characterizes the counter operation.

Analyses have been carried out[38-42] that attempt to derive a general expression for the expected factor M in terms of the tube parameters and applied voltage. Physical assumptions that are usually made for simplification are that the only multiplication process is through electron collision (any photoelectric effects are neglected), that no electrons are lost to negative ion formation, and that space charge effects are negligible. The solution to the Townsend equation [Eq. (6.1)] in cylindrical geometry must take into account the radial dependence of the Townsend coefficient α caused by the radial variation of the electric field strength. In general, the mean gas amplification factor M can then be written

$$\ln M = \int_a^{r_c} \alpha(r)\, dr \tag{6.5}$$

where r represents the radius from the center of the anode wire. The integration is carried out over the entire range of radii over which gas multiplication is possible, or from the anode radius a to the critical radius r_c beyond which the field is too low to support further gas multiplication. The coefficient α is a function of the gas type and the magnitude of the electric field, $\mathscr{E}(r)$, and data on its behavior can be found in Ref. 43. Equation (6.5) is normally rewritten explicitly in terms of the electric field as

$$\ln M = \int_{\mathscr{E}(a)}^{\mathscr{E}(r_c)} \alpha(\mathscr{E}) \, \frac{\partial r}{\partial \mathscr{E}} \, d\mathscr{E} \tag{6.6}$$

Now by introducing Eq. (6.3) for the shape of the electric field in cylindrical proportional tubes, we can write

$$\ln M = \frac{V}{\ln (b/a)} \int_{\mathscr{E}(a)}^{\mathscr{E}(r_c)} \frac{\alpha(\mathscr{E})}{\mathscr{E}} \, \frac{d\mathscr{E}}{\mathscr{E}} \tag{6.7}$$

Various analytic expressions that appear in the literature to relate the gas multiplication factor to experimental parameters differ from each other in the form of the expression used for $\alpha(\mathscr{E})$. A review of possible options is presented in Refs. 44–47.

By assuming linearity between α and $\mathscr{E}$, Diethorn[48] derived a widely used expression for M:

$$\ln M = \frac{V}{\ln (b/a)} \cdot \frac{\ln 2}{\Delta V} \left(\ln \frac{V}{pa \ln (b/a)} - \ln K \right) \tag{6.8}$$

where
M = gas multiplication factor

V = applied voltage

a = anode radius

b = cathode radius

p = gas pressure

According to the model used in the derivation, ΔV corresponds to the potential difference through which an electron moves between successive ionizing events, and K represents the minimum value of $\mathscr{E}/p$ below which multiplication cannot occur. Both ΔV and K should be constants for any given fill gas.

Hendricks[49] studied the applicability of this expression to a variety of different counters in which the fill gas was either xenon–methane or xenon–CO_2 mixture. The gain versus voltage characteristics are shown in Fig. 6.10, and the same data are reduced to the form of a *Diethorn plot* shown in Fig. 6.11. Wolff[50] demonstrated the satisfactory applicability of Eq. (6.8) to a number of other fill gases listed in Table 6.1. The corresponding values for ΔV and K are shown in the table. The Diethorn approach treats the two constants ΔV and K as empirically fitted parameters. This procedure is normally sufficient to represent experimental data over limited ranges of voltage, pressure, and counter dimensions. Introducing a third fitted parameter[44, 45] can be justified on theoretical grounds and leads to generalized models that can represent proportional counter behavior over much wider ranges of operation.

For a given proportional tube at constant gas pressure, Eq. (6.8) shows that the gas multiplication increases rapidly with applied voltage V. Neglecting the slowly varying logarithmic term, the multiplication M varies primarily as an exponential function of V. Proportional counters must therefore be operated with extremely stable voltage supplies to prevent changes in M over the course of the measurement.

Figure 6.10 Variation of the gas multiplication factor M with voltage applied to various proportional counters. The tubes differ in their physical characteristics, but only the two indicated gases were used. (From Hendricks.[49])

Table 6.1 Diethorn Parameters for Proportional Gases[a]

Gas Mixture	K (10^4 $V/cm \cdot atm$)	ΔV (V)	Reference
90% Ar, 10% CH_4 (P-10)	4.8	23.6	50
95% Ar, 5% CH_4 (P-5)	4.5	21.8	50
100% CH_4 (methane)	6.9	36.5	50
100% C_3H_8 (propane)	10.0	29.5	50
96% He, 4% isobutane	1.48	27.6	50
75% Ar, 15% Xe, 10% CO_2	5.1	20.2	50
69.4% Ar, 19.9% Xe, 10.7% CH_4	5.45	20.3	50
64.6% Ar, 24.7% Xe, 10.7% CO_2	6.0	18.3	50
90% Xe, 10% CH_4	3.62	33.9	49
95% Xe, 5% CO_2	3.66	31.4	49

[a] See Eq. (6.8).

Figure 6.11 A Diethorn plot of the same data shown in Fig. 6.10. As predicted by Eq. (6.8), a graph of $[\ln M \cdot \ln (b/a)/V]$ versus $\ln[V/pa \ln (b/a)]$ for all counters using the same gas closely matches a straight line with slope of $(\ln 2/\Delta V)$ and intercept $(-\ln 2)(\ln K)/\Delta V$. (From Hendricks.[49])

B. Space Charge Effects

In the avalanche process on which proportional counters depend, both electrons and positive ions are created. The electrons are collected relatively quickly (within a few nanoseconds) at the anode, leaving behind the positive ions that move much more slowly and gradually diffuse by drifting outward toward the tube cathode wall. The space charge represented by these net positive charges can, under some circumstances, appreciably distort the electric field from its value if the space charge were absent. Because the ions are formed preferentially near the anode wire where most gas multiplication takes place, the effect of the space charge will be to reduce the electric field at small radii below its normal value. If the magnitude of the effect is sufficiently large, it can have a measurable effect in reducing the size of the output pulse below that which one would ordinarily expect. Furthermore, because the amount of the reduction is likely to vary depending on the detailed geometry of each original ionizing event, the energy resolution of the proportional counter may also be adversely affected by the space charge nonlinearities.

There are two different categories of space charge effects. *Self-induced* effects arise when the gas gain is sufficiently high so that positive ions formed during a given avalanche can alter the field and reduce the number of electrons produced in further stages of the same avalanche. This effect depends on the magnitude of the gas multiplication and the geometry of the tube[51] but does not depend on the pulse rate. The *general* space charge effect includes the cumulative effect of positive ions created from many different avalanches. This effect can be important at lower values of the gas multiplication and becomes more serious as the rate of events within the tube is increased. The general problem of space charge in proportional tubes has been reviewed by Hendricks,[52] who presents a general analytic formulation to allow estimation of the resulting nonlinear effects under a variety of conditions. In general, most authors advocate using gas multiplication factors that are as low as possible consistent with signal-to-noise requirements to avoid potential loss of energy resolution caused by space charge effects within the tube. This recommendation is particularly significant for applications that involve high counting rates.

If the gas multiplication is perfectly linear, one should be able to record exactly the same pulse height spectrum from the detector by decreasing counter voltage and to offset the corresponding drop in gas multiplication by increasing the gain of the amplifier in the subsequent electronics. If space-charge-induced nonlinearities are significant, the large pulse amplitudes will be more severely limited at the higher voltage, and the recorded shape of the spectrum will change. This procedure can serve as a simple check to detect the presence of nonlinearities at the chosen operating voltage of the chamber.

C. Energy Resolution

1. STATISTICAL CONSIDERATIONS

The charge Q, which is developed in a pulse from a proportional counter in the absence of nonlinear effects, can be assumed to be the sum of the charges created in each individual avalanche. There will be n_0 of these avalanches, each triggered by a separate electron from the n_0 ion pairs created by the incident radiation. In the following discussion, we let A represent the electron multiplication factor for any one avalanche triggered by a single electron and M represent the average multiplication factor from all the avalanches which contribute to a given pulse:

$$M = \frac{1}{n_0} \sum_{i=1}^{n_0} A_i \equiv \overline{A} \tag{6.9}$$

Also, because eA_i is the charge contributed by the ith avalanche,

$$M = \frac{Q}{en_0}$$

or

$$Q = n_0 e M \tag{6.10}$$

From the discussions given in Chapter 4, the pulse amplitude from the detector normally is proportional to Q. This amplitude is subject to fluctuation from pulse to pulse even in the case of equal energy deposition by the incident radiation because both n_0 and M in Eq. (6.10) will show some inherent variation. Because these factors are assumed to be independent, we can use the error propagation formula [Eq. (3.41)] to predict the expected relative variance in Q

$$\left(\frac{\sigma_Q}{Q}\right)^2 = \left(\frac{\sigma_{n_0}}{n_0}\right)^2 + \left(\frac{\sigma_M}{M}\right)^2 \tag{6.11}$$

It is convenient to rewrite the second term in the above sum in terms of the variance of the *single-electron* multiplication factor A. Again, because each avalanche is assumed to be independent, we can apply the error propagation formula to Eq. (6.9) to obtain

$$\sigma_M^2 = \left(\frac{1}{n_0}\right)^2 \sum_{i=1}^{n_0} \sigma_A^2$$

$$\sigma_M^2 = \frac{1}{n_0}\sigma_A^2 \tag{6.12}$$

where A represents the typical avalanche magnitude. Combining Eqs. (6.11) and (6.12), we obtain

$$\left(\frac{\sigma_Q}{Q}\right)^2 = \left(\frac{\sigma_{n_0}}{n_0}\right)^2 + \frac{1}{n_0}\left(\frac{\sigma_A}{\overline{A}}\right)^2 \tag{6.13}$$

This expression now allows an analysis of the expected fractional variance of the pulse amplitude $(\sigma_Q/Q)^2$ in terms of the separate contributions of ion pair fluctuations $(\sigma_{n_0}/n_0)^2$ and single-electron multiplication variations $(\sigma_A/\overline{A})^2$.

a. Variations in the Number of Ion Pairs

The first term in Eq. (6.13) represents the relative variance in the original number n_0 of ion pairs. These fluctuations were discussed at the beginning of Chapter 5 and can be expressed in terms of the Fano factor F:

$$\sigma_{n_0}^2 = F n_0$$

or

$$\left(\frac{\sigma_{n_0}}{n_0}\right)^2 = \frac{F}{n_0} \tag{6.14}$$

Estimates of the size of the Fano factor for proportional gases range from about 0.05 to 0.20, and some specific values are given later in Table 6.2. The lowest values of F are generally observed for binary gas mixtures in which the Penning effect is important (see prior discussion of fill gas properties).

b. Variations in Single-Electron Avalanches

The second term in Eq. (6.13) represents the contribution of the fluctuations in single electron avalanche magnitude, a topic that has been the subject of considerable experimental and theoretical investigation. Some examples can be found in Refs. 37, and 53–55. Reference 37 also contains a fairly complete citation of the earlier literature on this topic.

A simple theoretical prediction for the single-electron avalanche distribution can be carried out with the assumption that the probability of ionization by an electron is dependent only on the electric field strength and is independent of its previous history. It can then be shown that the expected distribution in the number of electrons produced in a given avalanche should be predicted by the Furry distribution,

$$P(A) = \frac{(1 - 1/\overline{A})^{A-1}}{\overline{A}} \tag{6.15}$$

where
$\quad A = $ avalanche multiplication, or number of electrons in the avalanche

$\quad \overline{A} = $ mean value of $A\ (= M)$

If $\overline{A}$ is reasonably large, say greater than 50 or 100 as is almost always the case, then the Furry distribution reduces to a simple exponential form,

$$P(A) \cong \frac{e^{-A/\overline{A}}}{\overline{A}} \tag{6.16}$$

which predicts a relative variance

$$\left(\frac{\sigma_A}{\overline{A}}\right)^2 = 1$$

Experiments carried out at low values of the electric field tend to confirm the predicted exponential shape of the avalanche distribution. However, at higher fields, which are more typical of proportional counter operation, a somewhat more complex model must be used to represent experimental results adequately.

In strong electric fields, the probability of ionization by an electron can no longer be considered totally independent of its past history, and the conditions leading to the exponential distribution for single electrons are violated. Single-electron spectra measured under these conditions tend to show a peaked distribution in which the number of low-amplitude events is suppressed below that predicted by the exponential distribution. In a model proposed by Byrne,[56] the simple exponential distribution of Eq. (6.16) is replaced by a Polya distribution,

$$P(A) = \left(\frac{A(1 + \theta)}{\overline{A}}\right)^\theta \exp\left(\frac{-A(1 + \theta)}{\overline{A}}\right) \tag{6.17}$$

where θ is a parameter related to the fraction of electrons whose energy exceeds a threshold energy for ionization[53] and is in the range $0 < \theta < 1$. It can be shown that the relative variance predicted by the Polya distribution is

$$\left(\frac{\sigma_A}{\overline{A}}\right)^2 = \frac{1}{\overline{A}} + b \tag{6.18}$$

where $b \equiv (1 + \theta)^{-1}$ and is observed to have a value of about 0.5. For large values of the multiplication factor $\overline{A}$,

$$\left(\frac{\sigma_A}{\overline{A}}\right)^2 \cong b \tag{6.19}$$

c. Overall Statistical Limit

Regardless of the form assumed for the single-electron avalanche distribution, the distribution in pulse amplitude Q for large values of n_0 (say > 20) approaches a Gaussian distribution in shape. Because this is true in most applications, a symmetric Gaussian-shaped peak in the pulse amplitude distribution from monoenergetic radiation should be expected from fluctuations in n_0 and A.

In order to predict the relative variance of this distribution, we return to Eq. (6.13):

$$\left(\frac{\sigma_Q}{Q}\right)^2 = \left(\frac{\sigma_{n_0}}{n_0}\right)^2 + \frac{1}{n_0}\left(\frac{\sigma_A}{\overline{A}}\right)^2$$

Evaluating $(\sigma_{n_0}/n_0)^2$ from Eq. (6.14) and $(\sigma_A/\overline{A})^2$ from Eq. (6.19), we obtain

$$\left(\frac{\sigma_Q}{Q}\right)^2 = \frac{F}{n_0} + \frac{b}{n_0}$$

$$\left(\frac{\sigma_Q}{Q}\right)^2 = \frac{1}{n_0}(F + b) \tag{6.20}$$

where F is the Fano factor (typical value of 0.05–0.20) and b is the parameter from the Polya distribution that characterizes the avalanche statistics (typical value of 0.4–0.7). From the relative magnitudes of F and b, we see that the pulse amplitude variance is dominated by the fluctuations in avalanche size and that the fluctuations in the original number of ion pairs are typically a small contributing factor.

The relative standard deviation of the pulse amplitude distribution is obtained by taking the square root of Eq. (6.20):

$$\frac{\sigma_Q}{Q} = \left(\frac{F + b}{n_0}\right)^{1/2} \tag{6.21}$$

Because $n_0 = E/W$, where E is the energy deposited by the incident radiation and W is the energy required to form one ion pair

$$\frac{\sigma_Q}{Q} = \left(\frac{W(F + b)}{E}\right)^{1/2} \tag{6.22}$$

$$\frac{\sigma_Q}{Q} = \left(\frac{C}{E}\right)^{1/2} \tag{6.23}$$

where $C = W(F + b)$ and is constant for a given fill gas. The statistical limit of the energy resolution of a proportional counter is thus expected to vary inversely with the square root of the energy deposited by the incident radiation. Using values of $W = 35$ eV/ion pair, $F = 0.20$, and $b = 0.61$ (the lower limit estimated by Byrne), we calculate a value for the constant in Eq. (6.23) of 0.0283 keV. Because the conventional definition of energy resolution (see Fig. 4.5) is given by 2.35 times the relative standard deviation, the expected energy resolution should be about 12.5% at 10 keV and 3.9% at 100 keV for our example.

Various proportional gases will have somewhat different values for W, F, and b, and some specific examples are listed in Table 6.2. The limiting resolution is proportional to $\sqrt{W(F + b)}$, so this parameter can serve as a guide when comparing the potential resolution obtainable from different gases. Later measurements[57] indicate that the parameter b may be somewhat smaller than originally estimated by Byrne, and these values are included in the table.

There is some indication[54,57] that the relative variance of the gas multiplication is smallest at low values of the multiplication factor. This implies that the parameter b may be mildly dependent on the electric field, which is not an unreasonable assumption. Several authors therefore advocate the use of the smallest possible voltage and gas multiplication consistent with keeping electronic noise to a negligible level if the ultimate in proportional counter energy resolution is to be achieved. Low values of the multiplication also will minimize potentially harmful effects that can arise due to nonlinearities caused by space charge

Table 6.2 Resolution-Related Constants for Proportional Gases

Gas	W (eV/ion pair)	Fano Factor F Calculated[a]	Measured	Multiplication Variance b	Energy Resolution at 5.9 keV Calculated[b]	Measured
Ne	36.2	0.17		0.45	14.5%	
Ar	26.2	0.17		0.50	12.8%	
Xe	21.5		≤ 0.17			
Ne + 0.5% Ar	25.3	0.05		0.38	10.1%	11.6%
Ar + 0.5% C_2H_2	20.3	0.075	≤ 0.09	0.43	9.8%	12.2%
Ar + 0.8% CH_4	26.0	0.17	≤ 0.19			
Ar + 10% CH_4	26[c]			0.50	12.8%	13.2%

[a]From Alkhazov et al.[20]

[b]Given by $2.35[W(F + b)/5900 \text{ eV}]^{1/2}$ [see Eq. (6.22)].

[c]From Wolff.[50]

Source: Adapted from Sipila.[57]

effects (see previous discussion). Measurements carried out[5, 58] using common proportional counter gases confirm that the best energy resolutions are observed using low values of applied voltage and also by choosing the smallest diameter anode wire that is consistent with good surface quality and uniformity. Small anode wires result in a very steep gradient in the value of the electric field, and both analytical models and experimental evidence[58] show that avalanche fluctuations will be minimized under these conditions.

2. OTHER FACTORS AFFECTING ENERGY RESOLUTION

Under well-controlled circumstances, the energy resolution actually observed for proportional counters comes very close to the statistical limit predicted by Eq. (6.22). Some data gathered by Charles and Cooke[59] on the observed energy resolution of a proportional counter for low-energy photons are shown in Fig. 6.12. The data points are very well described by the square root energy dependence given by Eq. (6.23). Other data[60] tend to confirm this energy dependence at gas pressures of 1 atm or less, but the resolution was observed to worsen somewhat at higher pressures.

In order to achieve energy resolutions that approach this statistical limit, care must be taken to minimize potentially harmful effects of electronic noise, geometric nonuniformities in the chamber, and variations in the operating parameters of the detector. The noise levels of modern solid-state preamplifiers are generally sufficiently low so that electronic noise is a negligible contributor to energy resolution, even for applications involving low-energy radiation.

The most critical of the geometric factors is the uniformity and smoothness of the anode wire used in the proportional counter. Variations in the wire diameter as small as 0.5% can be significant.[38] Other factors include possible eccentricity of the wire or nonuniformity of the cylindrical cathode, but these are substantially less critical factors. As discussed earlier in the section on design, the ends of the chamber present special problems. Unless care is taken in shaping the electric field through the use of field tubes or by other methods, some interactions may take place in regions of reduced field. The subsequent avalanches will be smaller than normal and can increase the width of the observed pulse amplitude distribution.

Significant operating parameters that can affect energy resolution are gas purity, gas pressure, and stability of the high voltage applied to the chamber. Trace quantities of electronegative gases can significantly reduce the gas multiplication and introduce

Figure 6.12 Fractional standard deviation of the peak (FWHM energy resolution/2.35) for a proportional counter filled with P-10 gas at 1 atm. Two sets of measurements are shown, one for a "standard" tungsten anode wire of 0.05 mm diameter (○) and the other for a similar wire of improved uniformity (●). The corresponding values for $\sqrt{C}$ (in $keV^{1/2}$) in Eq. (6.23) are indicated. (From Charles and Cooke.[59])

Table 6.3 Gain (Pulse Height) Variations in a Proportional Counter

Predictions of the Diethorn model for a P-10 filled proportional tube, a = 0.03 cm, b = 1 cm, V = 1793 V, M = 1000:

1% increase in	Change in M:
Applied voltage V	+17.4%
Gas pressure p	−8.6%
Anode radius a	−6.1%
Cathode radius b	−2.7%

additional fluctuations in the pulse amplitude. Changes in gas pressure of the order of a few tenths of a percent can be significant,[38,61] as reflected in the corresponding change in gas multiplication. This is seldom a problem for sealed tubes (except for window flexing caused by changes in barometric pressure) but can be an important cause of resolution loss for gas flow proportional counters in which the gas pressure is not well regulated. Finally, the extreme dependence of the gas multiplication on applied voltage is indicated by Eq. (6.8). Variations in the applied voltage of 0.1 or 0.2% can significantly change the gas multiplication factor under typical conditions, and thus voltage supplies must be well regulated and free from long-term drifts if the ultimate energy resolution of a proportional counter is to be preserved. The dependence of the gain on these parameters can be predicted using one of the physical models for the gas multiplication process, such as that leading to Eq. (6.8). Some typical results are given in Table 6.3.

It is also commonly observed[62] that changes in the counting rate from proportional counters can influence the gas multiplication factor. If the irradiation rate varies during the course of a measurement, the associated gain change can adversely affect the energy resolution. While a full physical explanation for rate-dependent effects cannot always be identified, some potential factors[63] include the following: space charge buildup in the multiplication region due to slowly moving positive ions, enhanced volume recombination near the anode, possible accumulation on the anode wire of a loosely bound layer of polarizable molecules, and changes in the ballistic deficit (see p. 628) resulting from differences in the mean pulse rise time. Whatever the specific cause(s) may be, the *count rate effect* has proved to be a significant source of resolution loss in many applications of proportional counters.

When used over long periods of time, proportional counters can show "aging" effects[64] that lead to a drop in multiplication factor and deterioration in energy resolution. Leakage of air into sealed tubes is one possible cause, but such changes can occur even in gas flow counters. One potential cause is the gradual buildup of solid deposits or "whiskers" on the anode wire. In gases that contain methane, the conditions created in an avalanche can cause the formation of polymers that collect on the wire surface. In argon–methane and xenon–methane mixtures, significant performance degradation[65] sets in after the accumulation of enough avalanches to represent a total collected charge of about 10^{-6}–10^{-5} C/mm of wire length. Trace contaminants in fill gases can also lead to the development of corrosion products on the wire that may cause spurious pulses from the counter.[66]

D. Time Characteristics of the Signal Pulse

The same type of analysis carried out in Chapter 5 for pulse-type ion chambers can be applied to derive the shape of the output pulse from proportional counters.[67] However, several major differences arising from the following physical considerations change the character of the output pulse rather dramatically.

1. Virtually all the charge generated within the proportional tube originates within the avalanche region, regardless of where the original ion pairs are formed. The time history of the output pulse is therefore conveniently divided into two stages: the *drift time* required for the free electrons created by the radiation to travel from their original position to the region near the anode wire where multiplication can take place, and the *multiplication time* required from the onset of the avalanche to its completion.

   For any given original ion pair, the contribution to the output pulse during the drift time is negligible compared with the contribution of the much larger number of charge carriers formed in the subsequent avalanche. The effect of the drift time therefore is to introduce a delay between the time of formation of the ion pair and the start of the corresponding output pulse. The drift time (perhaps a microsecond or so) is normally much greater than the multiplication time and varies depending on the radial position of the original ion pair within the tube.

2. Because most of the ions and electrons are created very close to the anode wire, the bulk of the output pulse is attributable to drift of the positive ions rather than the movement of the electrons. At first, the positive ions are in a high-field region and move rapidly, leading to a fast-rising initial component of the pulse. Eventually, however, the ions reach regions of the tube at greater radii where the field is smaller, and their drift velocity decreases. The latter part of the pulse therefore rises very slowly and is often not observed in practice due to the finite shaping times of the subsequent electronic circuits.

A simplified analysis of the pulse shape caused by the collection of electrons and positive ions formed in an avalanche follows, using arguments that extend the derivation beginning on p. 149 for parallel plate ion chambers. One major difference is that cylindrical geometry must now be used to represent the shape of the electric field around the anode wire. In general, the energy absorbed dE by the motion of a positive charge Q through a difference in electric potential $d\varphi$ is given by

$$dE = -Q\,d\varphi \tag{6.24}$$

In terms of the electric field $\mathscr{E}(r) = -d\varphi(r)/dr$,

$$\frac{dE}{dr} = Q\mathscr{E}(r) = Q\,\frac{V_0}{r \ln (b/a)} \tag{6.25}$$

where Eq. (6.3) has been used for the value of $\mathscr{E}(r)$.

Let us first assume that n_0 electrons and positive ions are formed in the avalanche at a fixed distance ρ from the surface of the anode wire (see Fig. 6.13). Setting $Q = n_0 e$, the energy absorbed by the motion of the positive ions to the cathode is then

$$E^+ = \int_{a+\rho}^{b} \frac{dE}{dr}\,dr = \frac{QV_0}{\ln (b/a)} \int_{a+\rho}^{b} \frac{dr}{r}$$

$$= \frac{QV_0}{\ln (b/a)} \ln \frac{b}{a + \rho} \tag{6.26}$$

The energy absorbed by the motion of the negatively charged electrons inward to the anode is

$$E^- = -\frac{QV_0}{\ln (b/a)} \int_{a+\rho}^{a} \frac{dr}{r} = \frac{QV_0}{\ln (b/a)} \ln \frac{a + \rho}{a} \tag{6.27}$$

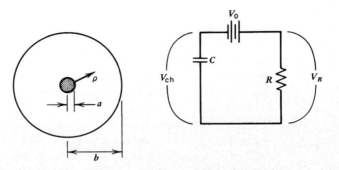

Figure 6.13 On the left is shown a cross-section of the cylindrical geometry used in deriving the pulse shape induced by a single avalanche. The equivalent circuit across which the signal is developed is at the right. Here C represents the capacitance of the detector and associated wiring, V_R is the signal voltage developed across the load resistance R, and V_{ch} is the voltage remaining across the detector.

The sum of the energy absorbed after both species have been collected is, from Eqs. (6.26) and (6.27),

$$\Delta E = E^+ + E^- = \frac{QV_0}{\ln (b/a)} \ln\left(\frac{b}{a + \rho} \cdot \frac{a + \rho}{a}\right)$$

$$= QV_0 \tag{6.28}$$

As in the case of the parallel plate geometry, this energy must come at the expense of the energy stored on the detector capacitance:

$$\tfrac{1}{2}\, CV_{\text{ch}}^2 = \tfrac{1}{2}\, CV_0^2 - \Delta E$$
$$\tfrac{1}{2}C(V_{\text{ch}} + V_0)(V_{\text{ch}} - V_0) = -\Delta E$$

Assuming $V_{\text{ch}} + V_0 \cong 2V_0$ and substituting $V_R = V_0 - V_{\text{ch}}$, we obtain

$$V_R = \frac{\Delta E}{CV_0} = \frac{QV_0}{CV_0} = \frac{Q}{C} \tag{6.29}$$

which is the same result shown for parallel plate geometry. This value is the maximum pulse amplitude that would be developed if the time constant RC is long compared with the ion collection time. In practice, this condition almost never holds for proportional counters. The maximum pulse amplitude then depends on the shape of the voltage–time profile.

   Most of the electrons and ions created in an avalanche are formed close to the anode wire surface. The exponential growth characterized by Eq. (6.2) predicts that half will be formed within one mean free path of the anode, typically only a few micrometers from the surface. From Eqs. (6.26) and (6.27), the ratio of the maximum signal amplitude from electron drift to that from ion drift is given by

$$\frac{E^-}{E^+} = \frac{\ln\,[(a + \rho)/a]}{\ln\,[b/(a + \rho)]} \tag{6.30}$$

Choosing values of $a = 25$ μm and $b = 1$ cm for the tube dimensions, and assuming that $\rho = 3$ μm, we find

$$\frac{E^-}{E^+} = 0.019$$

For this example, less than 2% of the maximum signal results from the motion of electrons, and it is the positive ion drift that dominates the pulse formation. We therefore proceed by neglecting the electron contribution and assuming that the entire signal pulse develops from drift of the ions that are created essentially at the anode wire surface.

From Eqs. (5.3) and (6.3), the drift velocity of the ions varies with radial position as

$$v^+(r) = \frac{\mu \mathscr{E}(r)}{p} = \frac{\mu}{p} \frac{V_0}{\ln(b/a)} \cdot \frac{1}{r}$$ (6.31)

By putting this expression into the law of motion,

$$\int_a^{r(t)} \frac{dr}{v^+(r)} = \int_0^t dt$$

and carrying out the integration, we obtain the following expression for the time-dependent position of the ions:

$$r(t) = \left(2 \frac{\mu}{p} \frac{V_0}{\ln(b/a)} t + a^2\right)^{1/2}$$ (6.32)

The time required to collect the ions can be found by substituting $r(t) = b$ in the above expression:

$$t^+ = \frac{(b^2 - a^2)\, p \ln(b/a)}{2\mu V_0}$$ (6.33)

Using typical values for the parameters, this collection time is very long, with a representative value of several hundred microseconds. However, a large fraction of the signal is developed during the very early phase of the ion drift. The energy absorbed by the motion of the ions as a function of time is

$$E^+(t) = \frac{QV_0}{\ln(b/a)} \int_a^{r(t)} \frac{dr}{r} = \frac{QV_0}{\ln(b/a)} \ln \frac{r(t)}{a}$$ (6.34)

Using Eq. (6.32) for $r(t)$ in the above equation and setting $V_R(t) = E^+(t)/CV_0$, we find the time profile of the signal pulse to be

$$V_R(t) = \frac{Q}{C} \frac{1}{\ln(b/a)} \ln\left(\frac{2\mu V_0}{a^2 p \ln(b/a)} t + 1\right)^{1/2}$$ (6.35)

This equation predicts that the pulse will reach half its maximum amplitude within a time given by

$$t\Big|_{\text{half amplitude}} = \frac{a}{a+b} t^+$$ (6.36)

where t^+ is the full ion collection time given by Eq. (6.33). At this point, the radial position of the ions is given by $\sqrt{ab}$, where the value of the electric field has dropped to a fraction given by $\sqrt{a/b}$ of its value at the anode wire surface.

Again evaluating for $a = 25$ μm and $b = 1$ cm, Eq. (6.36) predicts that the half-amplitude point is reached after only 0.25% of the full ion drift time, typically a fraction of a microsecond. At that point the ions have moved 475 μm from the wire surface, where the electric field is down to 5% of its surface value. This fast leading edge of the pulse is followed by a much slower rise corresponding to the drift of the ions through the lower-field regions found at larger radial distance.

If all original ion pairs are formed at a fixed radius, the electron drift times will all be identical and all avalanches will be synchronized. Then Eq. (6.35) will also describe the

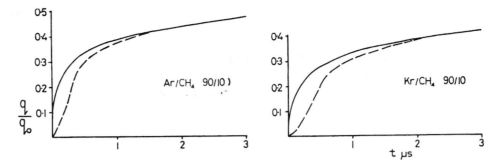

Figure 6.14 Shape of the output pulse leading edge calculated for a typical tube for two different proportional counter gases. In each case, the solid curve represents initial ionization formed at a single radius (constant drift time), whereas the dashed line assumes uniform ionization along a diameter. (From Gott and Charles.[68])

shape of the output pulse for these events. Most situations, however, involve the formation of ion pairs along the track of the incident radiation and thus cover a range of radii. The spread in electron drift times will introduce additional spread in the rise time of the output pulse. Figure 6.14 shows the shape of the expected leading edge of the output pulse under two conditions: ion pairs formed at a constant radius, and ion pairs uniformly distributed throughout the volume of the counter. The rise time of the output pulse is seen to be considerably greater for the latter case. To minimize this rise time and the associated timing uncertainties, a short electron drift time is helpful. This objective is served by keeping the electric field values as high as possible in the drift region and by choosing a gas with high electron drift velocities.[69] Some examples of both "slow" and "fast" fill gases are shown in Fig. 6.15.

When pulses from proportional counters are shaped using time constants of several microseconds, the slow component of the drift of the ions no longer contributes to the pulse amplitude. The shaped pulse therefore has an amplitude that is less than that corresponding to an infinite time constant by an amount known as the *ballistic deficit*. If all ion pairs were

Figure 6.15 The electron drift velocities in various gases, as compiled by Jeavons et al.[70]

formed at a constant radius, the shape of the pulse and the ballistic deficit would always be the same, and the net effect would simply be to reduce all pulse amplitudes by a constant factor. When the interactions are randomly distributed over a variety of radii, the pulse shapes will vary depending on the radial distribution of the original ion pairs. Unless shaping times of many microseconds are used, the ballistic deficit will also vary depending on this radial distribution, and the energy resolution of the detector may therefore suffer. The ballistic deficit caused by various pulse-shaping networks for pulses from proportional counters has been the subject of a number of investigations[68,71–73] that allow estimation of its magnitude. In general, shaping times that are large compared with the variation in the rise time of the pulses (usually several microseconds) should be used to minimize the effect of variations in the ballistic deficit on energy resolution.

The variations in pulse rise time can sometimes be used to good advantage. Radiations with a short range (heavy charged particles or secondary particles from neutron interactions) will often create ions with a limited range of radii within the tube. Background radiation or undesirable events (such as fast electrons or secondary electrons created by gamma-ray interactions) may have much longer ranges and therefore tend to create ions with a greater spread of initial radii. These background pulses will then have a longer rise time than the desired signal pulses and can be eliminated using methods of rise time discrimination outlined in Chapter 17. This method can be applied advantageously for background reduction in the counting of soft X-ray or low-energy beta particles,[74–76] in enhancing the separation between radiation types in the proportional counters used for mixed alpha/beta activity counting described later in this chapter, as well as in the suppression of gamma-ray-induced pulses for the neutron detectors described in Chapters 14 and 15.

E. Spurious Pulses

In some circumstances, satellite pulses may be generated following the primary pulse from a proportional counter. These secondary pulses have nothing to do with the incident radiation but are generated from secondary processes that arise from effects within the primary avalanches. These spurious pulses can lead to multiple counting where only one pulse should be recorded and are a potential cause of counter instability.

Spurious afterpulses are often very small, corresponding to the amplitude of an avalanche triggered by a single electron. Under many circumstances, they may be eliminated by simple amplitude discrimination. At moderate values of the electric field, their rate of occurrence might not exceed a few hundredths of a percent of the primary rate in common counter gases. Needlessly high values of the gas multiplication should be avoided in those situations that are sensitive to these satellite pulses, because their probability of generation increases rapidly with the multiplication factor.

The various physical mechanisms leading to the production of spurious pulses have been reviewed by Campion.[14] One important mechanism originates with the optical photons emitted by excited atoms produced in the avalanche. These photons can reach the cathode wall with virtually no time delay and may produce a low-energy electron at the cathode surface through the photoelectric effect. This single electron will then drift toward the anode and create another avalanche, which will be delayed from the primary pulse by the time required for the electron to drift from the cathode to the anode. In counters of ordinary dimensions, this time will be of the order of several microseconds. In cylindrical counters, all drift times between the cathode and anode will be the same, and consequently, if more than one photoelectron is created, the resulting spurious pulse may have an amplitude considerably greater than that caused by a single primary electron.

Other mechanisms, which are not fully understood, can lead to spurious pulses that are delayed by as much as several hundred microseconds from the primary pulse.[77] Possible causes may involve interaction of the positive ion upon collection at the cathode, field emission of electrons at the cathode wall, metastable states of the counter gas, or electrons delayed by space charge effects from a previous avalanche. Spurious pulses with such long delays almost always have an amplitude corresponding to a single primary electron and therefore are detected only when the gas multiplication factor is rather large.

   The end of the counting plateau discussed in Section IV is reached (under some circumstances) when the applied voltage is sufficiently high to permit counting of these afterpulses. The apparent count rate then abruptly increases beyond the rate at which true interactions are occurring within the counter. A sensitive method for determining the end of the plateau has been suggested by Campion[78] in which the number of pulses that occur within a fixed period of time (several tens of microseconds) is monitored following a primary pulse from the counter. Under normal circumstances, the number of these pulses will be very low and can accurately be predicted from counting statistics based on the observed counting rate. If spurious pulses begin to be detected, however, the rate of afterpulses will abruptly increase and will clearly signal the end of the counting plateau.

IV. DETECTION EFFICIENCY AND COUNTING CURVES

A. Selection of the Operating Voltage

For charged radiations such as alpha or beta particles, a signal pulse will be produced for every particle that deposits a significant amount of energy in the fill gas. At high values of the gas multiplication, a single ion pair can trigger an avalanche with enough secondary ionization to be detectable using preamplifiers of reasonable noise behavior. Unless required by the application, proportional counters are seldom operated in a mode that is sensitive to single avalanches because the measurement is then prone to nonlinearities for larger pulses due to space charge effects. Furthermore, many background and spurious pulses may also be counted, which often correspond to very few original ion pairs.

   Instead, lower values of the gas multiplication are typically used, which require that the pulse originate from a finite number of ion pairs in order to have an amplitude large enough to exceed the discrimination level of the counting system. Because the gas multiplication factor varies with the voltage applied to the detector (see Fig. 6.10), a "counting curve" of the type described in Chapter 4 can be recorded to select an operating voltage appropriate to the specific application. In this measurement, the counting rate is recorded under constant source conditions as the detector voltage is varied. An operating point is then selected that normally corresponds to a flat region or "plateau" on the resulting rate versus voltage curve.

B. Alpha Counting

If almost all pulses from the detector are of the same size, the differential pulse height spectrum has a single isolated peak as shown at the left in the sketch below.

The corresponding counting curve shown at the right then has a simple plateau. An operating point on the counting plateau then ensures that all the pulses of interest from the detector are counted. This simple behavior is characteristic of the G-M counter discussed in Chapter 7 but is observed in proportional counters only in special circumstances. One such case is for monoenergetic charged particles whose range in the counter gas is less than the dimensions of the chamber. Alpha particle sources often fall into this category, and proportional counters can therefore easily record each particle that enters the active volume with virtually 100% efficiency. To avoid energy loss in the entrance windows of sealed tubes, windowless flow counters are often used for alpha counting. Absolute measurements of alpha source activity are therefore relatively straightforward and involve evaluation of the effective solid angle subtended by the counter active volume (which can be close to 2π for internal source flow counters) and small corrections for alpha self-absorption and scattering in the source itself or in the source backing.

C. Beta Counting

For beta particles of typical energies, the particle range greatly exceeds the chamber dimensions. The number of ion pairs formed in the gas is then proportional to only that small fraction of the particle energy lost in the gas before reaching the opposite wall. Beta particle pulses are therefore normally much smaller than those induced by alpha particles of equivalent kinetic energy and will also cover a broader range of amplitude because of the spread in beta particle energies and the variations in possible paths through the gas. A differential pulse height spectrum, which might represent a typical source of mixed beta and alpha activity, is sketched at the left portion of the diagram below. The corresponding counting curve now shows two plateaus: the first at the point at which only alpha particles are counted, and the second where both alpha and beta particles are counted. Because the beta particle pulse height distribution is broader and less well separated from the low-amplitude noise, the "beta plateau" is generally shorter and shows a greater slope than the "alpha plateau."

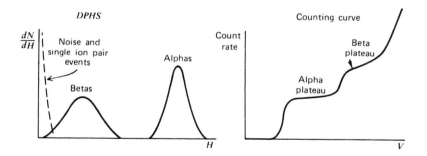

Absolute beta activity measurements are often carried out in a 4π flow counter of the type sketched in Fig. 6.9. The greater range of beta particles allows preparation of sources on backings that are sufficiently thin to allow most of the particles to emerge from either side of the backing, and only small corrections are required to account for those particles that do not reach either half of the chamber active volume.

D. Mixed Sources

In many applications, the objective will be to separately count alpha and beta activity that may be mixed together in a common sample. The inherent separation between the expected amplitudes of alpha- and beta-induced events, illustrated in the left side of the sketch above, can be exploited to provide two separate readout channels, one corresponding to the alpha count and the other to the beta count. This selection is accomplished by electronically setting pulse height windows in parallel branches of the electronic counting circuit, one window set to encompass the alpha peak and the other to include the beta particles. If the gain uniformity and energy resolution of the proportional chamber are good enough and its geometry is such that the two radiations contribute pulses that are widely different in amplitude, then a clean separation will result and there will be very little "cross-talk" between the two counting channels. Poor geometries that allow the beta particles to travel long distances in the gas or that introduce variability in the alpha track lengths will reduce the event separation and may induce some appreciable cross-talk if the amplitudes overlap. This cross-talk can be reduced or eliminated by narrowing the pulse height acceptance windows, but only at the expense of reduced counting efficiencies.

One advantage of proportional counters is that the inherent alpha/beta separation in amplitude results in a very low background counting rate in the alpha channel. Most sources of background (see Chapter 20), such as cosmic rays or ambient gamma rays, produce low density ionization typical of beta particles and therefore have too low an amplitude to fall into the alpha channel. The residual background is largely due to the natural alpha activity of the counter construction materials and can be as low as 1 or 2 counts per hour.

E. X-Ray and Gamma-Ray Sources

Proportional counters can be used for the detection and spectroscopy of soft X-rays or gamma rays whose energy is low enough to interact with reasonable efficiency in the counter gas. Figure 6.16 shows the interaction probability in a 5.08-cm thickness of the common high-Z proportional gases at STP. The discontinuities caused by gas absorption edges (discussed in Chapter 2) are very evident. For xenon, the useful response can be extended appreciably above 100 keV photon energy. The spectroscopy of low-energy X-rays is one of the most important applications of proportional counters and is based on fully absorbing the photoelectrons formed by photon interactions within the gas. Because the photoelectron energy is directly related to the X-ray energy (see Chapter 2), photon energies can be identified from the position of corresponding "full-energy" peaks in the measured pulse height spectrum.

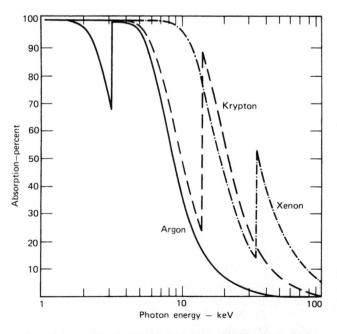

Figure 6.16 Fraction of incident photons absorbed in a 5.08-cm thick layer of several proportional gases at 1 atm pressure.

Low energy X-rays produce photoelectrons with energies of a few tens of keV or less. With typical W-values of 25–35 eV/ion pair, only a few hundred ion pairs are created in the gas along the track of a typical electron to serve as the basis of the signal pulse. The corresponding electrical charge is not much larger than the equivalent noise charge (see Chapter 17) at the input of typical low-noise preamplifiers. Thus additional "internal" gains through the gas amplification process that is inherent in proportional counters is vital to enhance the signal-to-noise ratio of the measured pulses.

The response function for low-energy X-rays can be complicated by several effects related to characteristic X-rays generated by interaction of the primary radiation within the detector. The most significant involves the characteristic K X-rays that usually follow the photoelectric absorption of the primary radiation in the fill gas. Because the corresponding energy can be relatively large (K_α X-ray energies are 2.97, 12.6, and 29.7 keV in argon, krypton, and xenon, respectively), this X-ray may escape from the gas without further interaction. As shown in Fig. 6.17, a corresponding "escape peak" will then appear in the response function which lies below the full-energy peak by an amount equal to the characteristic X-ray energy. This escape peak may be suppressed by choosing a gas whose K-shell binding energy lies above the incident X-ray energy.[†] Other peaks in the response function may also arise from the absorption of characteristic X-rays generated by interaction of the primary radiation in the entrance window or walls of the counter. A low-Z window material such as beryllium is often chosen to minimize this contribution.

Because the entrance window will absorb some incoming radiation no matter how thin it is made, it can play an important role in determining the efficiency for detection of weakly penetrating radiations. At the same time, the window must be gas-tight and often will need to support a considerable pressure difference between the gas in the tube and the external environment. The window may also need to be optically opaque to prevent the liberation of free electrons in the gas by incident visible or ultraviolet light. Commercial proportional tubes intended for X-ray applications are available with either beryllium or aluminum windows with thicknesses ranging from 50–250 μm. The thinnest of the beryllium windows will extend the useful response for incident X-rays down to about 2 keV. For lower attenuation, specialized windows are available that consist of thin films supported by metallic grids.[79,80]

A technique known as *fluorescence gating*[81,82] has been introduced that actually takes advantage of the emission of characteristic X-rays following photoelectric absorption. The technique applies to proportional counters that are constructed as a series of independent

Figure 6.17 Part (*a*) shows the process that gives rise to the X-ray escape peak in the spectrum sketched in part (*b*).

[†]In principle, characteristic X-rays escaping after photoelectric absorption involving the less-tightly bound L or M shells could also give rise to escape peaks, but these are seldom observed for several reasons. The lower X-ray energies from these shells make them less penetrating and thus less likely to escape from the gas volume. Also, the energy of the predicted escape peak is much closer to the full energy peak, and the finite energy resolution of many common detectors is not adequate to separate them.

cells or as a position sensitive chamber[83] in which the photoelectric absorption of the incident photon and the absorption of the resulting characteristic X-ray can be separated electronically. Since the characteristic X-ray will travel some distance through the gas, it likely will be absorbed in a different cell from that of the original photoabsorption. If these two events can be sensed separately as pulses that occur in time coincidence (the flight time of the X-ray will generally be negligible), then the measurement can be restricted to recording only this subset of all possible interactions within the tube. Thus by adding the two pulses together, accepted events will be full energy absorptions in which the fluorescent X-ray is reabsorbed, and the escape peak mentioned in the previous paragraph will be eliminated from the recorded pulse height spectrum. This selection also has the effect of reducing contributions of Compton scattering or background events that are restricted to energy deposition in a single cell only.

At higher photon energies more typical of gamma rays, proportional counters are no longer attractive as a detector. The counting efficiency becomes very small because the direct interaction probability of the photon in the gas drops rapidly with energy (see Fig. 6.16). The remaining gamma sensitivity then arises from photon interactions in the counter wall for which the resulting secondary electron reaches the counter gas. This process results in an intrinsic efficiency of only a percent or so (see Fig. 7.9) and cannot be used as the basis of energy measurement because of the variable loss of electron energy before reaching the gas. When gas counters are applied to gamma-ray detection, they are more commonly used in the Geiger–Mueller region to take advantage of the resulting simplicity of operation.

V. VARIANTS OF THE PROPORTIONAL COUNTER DESIGN

A. Tissue Equivalent Proportional Counter

A specialized type of proportional counter known as the *tissue equivalent proportional counter (TEPC)* plays a useful role in the dosimetry of gamma rays and neutrons.[84] It is based on constructing a chamber whose walls and fill gas mixture mimic the elemental composition of biological tissue (see p. 168). When irradiated by external gamma rays or neutrons, secondary charged particles (primarily fast electrons for gamma rays and recoil protons for fast neutrons) are created in the wall. For walls that are thick compared with the secondary particle ranges, a condition of equilibrium exists in which the flux of secondary particles leaving the inner surface of the wall no longer depends on the wall thickness but is characteristic of the particle flux produced within the bulk of the wall material. One of the most important quantities of interest in dosimetry is the distribution in linear energy transfer (LET) of the secondary particles. As discussed in Chapter 2, the LET value determines the quality factor Q that must be applied to convert measurements of absorbed dose into dose equivalent. The TEPC allows direct measurement of this LET distribution if the fill gas pressure is low enough so that particles entering the gas lose a negligible amount of their energy in passing through to the opposite wall. The pulse amplitude that results will depend on the product of the LET of the particle and the path length that it travels through the gas. In a low pressure gas, the approximation can be made that these paths are roughly straight lines, and if the chamber is spherical, the chord length distribution can be calculated analytically. With knowledge of this distribution and the measured pulse height distribution, the LET distribution of the particles can be unfolded. This process is particularly useful in the analysis of neutron fluxes whose energy distribution is unknown or in mixed neutron/gamma ray fields.[85,86]

B. Parallel Plate Avalanche Counters

A variant of the traditional proportional counter has gained some attention[87–90] for applications in which the need for fast timing information is more important than good energy resolution. The parallel plate avalanche detector is particularly useful in applications involving heavy charged particles where the associated radiation damage in solid-state detectors may prohibit their use. The counter consists of two parallel plate electrodes separated by a gap that is normally kept as small as possible for best timing information. Alternatively, fine mesh grids can be used as electrodes[91] instead of solid plates to reduce their thickness. The electrodes are enclosed in a container in which a proportional gas is introduced, and a homogeneous electric field is produced between the plates which under conditions of relatively low gas pressure, can reach values as high as 4×10^6 V/m · atm (Ref. 88). A charged particle that traverses the gap between the plates leaves a trail of ions and electrons that are multiplied through the usual gas multiplication process. Electrons formed nearest the cathode are obviously subjected to more multiplication than those formed near the anode. Maximum gas gains of the order of 10^4 are possible.

Because typical applications involve particles that fully traverse the gap between the electrodes, the number of ions that are formed reflects the amount of energy lost by the particle in its transit. This type of detector is commonly called a ΔE *detector*, and other examples are given later in this text. Because any typical output pulse contains a mix of gas amplification factors, the energy resolution is seldom better than about 20%. Nonetheless, this performance is often adequate to separate different types of particles with widely different specific energy loss.

The output shape consists of a fast-rising component generated by the electron collection and a slow component resulting from the motion of the positive ions. For timing purposes, only the fast component is used and the slow portion is eliminated by the shorter time constant of the collection circuit. For gaps of the order of 1 mm, the fast component rise time is about 2 ns under high field conditions.[78] If the energy deposited by the incident particle is relatively uniform, the time resolution (as defined in Chapter 17) can be as short as 160 ps.[88,92] The parallel plate avalanche detector can also be made to yield two-dimensional position information[93,94] by using appropriately patterned electrodes.

C. Position-Sensitive Proportional Counters

Because the position of the avalanche in a proportional counter is limited to a small portion of the anode wire length, some schemes have been developed that are capable of sensing the position of an event taking place within the tube. In the common cylindrical geometry, electrons drift inwardly from their place of formation along radial field lines, so that the position of the avalanche is a good indicator of the axial position at which the original ion pair was formed. If the track of incident radiation extends for some distance along the length of the tube, then many avalanches will also be distributed along the anode and only an average position can be deduced.

The most common method of position sensing in proportional tubes is based on the principle of charge division, illustrated in Fig. 6.18. The anode wire is fabricated to have

Figure 6.18 Position localization in proportional counters using the charge-division method. (From Fischer.[96])

significant resistance per unit length, so that the collected charge is divided between the amplifiers placed at either end of the wire in a proportion that is simply related to the position of the interaction. By summing the output of the two amplifiers, a conventional output pulse is produced with an amplitude proportional to the total charge. A position signal is generated by dividing the output of a single amplifier by the summed signal to give a pulse that indicates the relative position along the anode wire length. This division can be carried out using either analog methods[95] or digital techniques.[96]

An alternative to the charge division method, which is capable of excellent spatial resolution, has been developed by Borkowski and Kopp.[97] The method is based on observing the relative rise time from preamplifiers placed at either end of a resistive anode wire. Events occurring far from one of the preamplifiers will exhibit a longer rise time than those near that preamplifier. A position signal can then be derived from the rise time difference between the pulses produced by the two preamplifiers. The spatial resolution reported for this type of detector for well-collimated alpha particles is 0.15 mm FWHM for a tube of 200 mm total length.[98]

Extensive reviews of position-sensitive proportional counters have been compiled describing their application as X-ray and neutron detectors[99] and as focal plane detectors for magnetic spectroscopy of charged particles.[100]

D. Multiwire Proportional Counters

In some situations, it is advantageous to provide more than one anode wire within a proportional counter. For example, detectors with very large surface area can be constructed economically by placing a grid of anode wires between two large flat plates that serve as cathodes on either side of the counter. Figure 6.19 shows a sketch of this arrangement

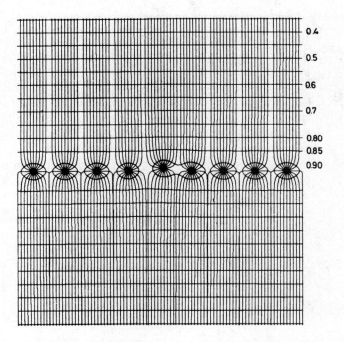

Figure 6.19 A plot of the electric field lines created by a grid of anode wires (running perpendicular to the page) that are placed equidistant between two parallel cathode plates at the top and bottom of the figure. Over much of the volume away from the grid, the field is nearly uniform. A high-field region is created in the immediate vicinity of each grid wire. (From Charpak and Sauli.[101])

together with the configuration of the electric field that results. Electrons formed by ionization of the gas drift inward toward the plane of the anode wires, initially in a nearly uniform field. As they approach, they are accelerated toward the nearest wire into its surrounding high-field region where avalanches are formed. A large negative-polarity-induced pulse appears on the anode wire on which the avalanche is collected, while the neighboring anodes show smaller positive amplitude pulses. The signals from preamplifiers connected to each wire can thus be used to localize the event to the nearest wire, with the finest wire spacing limited to about 1 or 2 mm. To reduce the number of preamplifiers, the anodes can be interconnected using resistors to form a charge-division chain, and the signals from preamplifiers at both ends can then be processed in the same type of scheme shown in Fig. 6.18 to identify centroid of the charge along the chain. The cathode planes also can be fabricated in the form of isolated strips or groups of wires, as shown, for example, in Figure 6.20. Positive signals are induced by the avalanches on the cathode strips, but since they are located some distance away, the induced charge is spread over a wider area. By using electronic center-of-gravity techniques, the centroid of the discharge can be located with reasonable precision. The strips on one cathode plane can be in the x direction while the strips on the opposite cathode are often made to be orthogonal to give an independent y coordinate. Devices of similar design, often extending to several square meters in cross-sectional area, serve as a very common position-sensing detector used in high-energy particle research (see Ref. 103).

While many multiwire chambers are operated in the proportional mode, the relatively low signal amplitude can be a limiting factor in position-sensing applications. It has been demonstrated that the self-quenched streamer mode of operation (see p. 714), a hybrid between proportional and Geiger modes, may be preferable in some applications because of the larger signal size that is produced.

The use of multiple anode wires can also help in the design of fast response proportional counters. Because the speed of response is largely related to the drift time of the initial electron to the multiplying region, the response time can be minimized by providing many multiplying regions throughout the volume of the detector. For example, Breskin[104] describes a 40 mm × 40 mm planar detector in which anode wires are spaced at 1-mm intervals. When operated with ethylene fill gas at a pressure of 5 torr (666 Pa), a time resolution of 0.8-ns FWHM was achieved using 5.5 MeV alpha particle irradiation.

Multiple wires can also be oriented in a single gas enclosure to form layers or cells that function as independent proportional counters. For example, Fig. 6.21 illustrates a cylindrical arrangement in which a center cell is surrounded by an annulus of cells, each made up of an anode wire at the center of a hexagonal array of wires that act as the cell cathode.

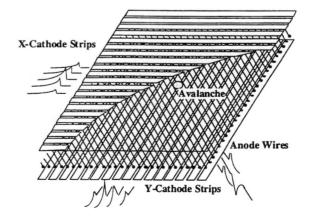

Figure 6.20 Sketch of a two-dimensional position-sensing multiwire proportional counter. (From Sauli.[102])

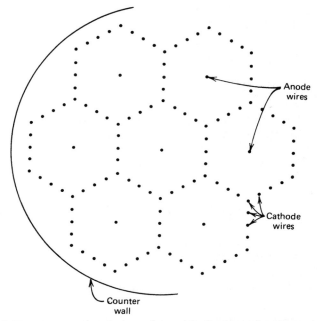

Figure 6.21 The cross-sectional view of a multicell proportional counter. A single anode wire is at the center of each cell, surrounded by a grid of cathode wires.

Independent signals are derived from each anode. This type of design provides multiple detector units that are in direct contact without intervening walls or dead layers. Through the use of coincidence or anticoincidence selection of events that involve more than one cell, those of a specific type can be preferentially accepted. For example, the background in the inner cell when used to count soft beta particles can be reduced[105] by rejecting longer-range particles that create a coincident signal in one of the outer cells. In another application, the complicating effects of particles that strike the wall of gas-filled detectors can be eliminated from measured energy spectra by restricting events to those taking place in the center cell alone.

E. Microstrip Gas Chamber

To address some limitations of multiwire proportional counters, a new type of multi-anode gas-filled detector was first proposed by Oed in 1988.[106] It is based on an insulating substrate on which metallic electrodes are formed by etching techniques. Two examples of the resulting pattern of anode strips are illustrated in Fig. 6.22. An essential feature is that the anode structures are kept quite narrow (typically 10 microns) so that the same type of concentration of the electric field that occurs around a wire also is realized near the surface of the anode strip. Thus avalanches will be formed as electrons are drawn in to the anode strip surface in much the same way as they are produced around wires in standard proportional counters.

An example of a completed *microstrip gas chamber (MSGC)* is illustrated in Fig. 6.23. Because of the requirement for narrow anode strips, the small dimensions shown on the figure are typical for these devices. The region between the drift plane and the anodes contains the fill gas, and the original ionization from the incident radiation is formed in this volume. Electrons from these ion pairs are drawn toward the anodes by an electric field established between the anodes and the drift plane. Once they reach the immediate vicinity of the anode surface, avalanches are generated, and it is here that nearly all the charge is created. As in the conventional proportional counter, it is the motion of the positive ions formed in the avalanches away from the anode surface that generates most of the observed signal.

(a)

(b)

Figure 6.22 Cross sections of the electrode patterns for two configurations of microstrip gas chambers. Part (*a*) is a common pattern of alternating anode and cathode strips, and part (*b*) is a pattern of narrow anodes separated from a continuous cathode by insulating supports (commonly called a *microgap* configuration.)

Replacing wires with etched metallic strips on a substrate has some obvious advantages. Through the photolithography process, the strips can be made with a very narrow pitch so that the spatial resolution can be much finer than that of a multiwire proportional counter. Another advantage is that most of the positive ions formed in the avalanche are quickly drawn to nearby cathodes rather than having to drift from the wire to a much more distant cathode surface. The corresponding positive space charge is therefore cleared away more rapidly, permitting much higher rate operation[108] of the MSGC compared with wire chambers.

In the most common configuration shown in Fig. 6.22*a*, the alternating anode and cathode strips are supported on an insulating substrate. One potential problem is that the substrate will also collect some of these positive ions leading to possible buildup of surface charges causing voltage instabilities and distortions of the electric field. To avoid this surface charging, the substrate must have some finite electrical conductivity, and a number of different recipes for slightly conducting glass and/or coatings on other materials have emerged in the development of these devices. Another potential disadvantage is the fact that electrical sparking can easily occur between the anodes and cathodes because of their small spacing. If sufficiently severe, these discharges can physically damage the structure of the electrodes[109] and ultimately cause permanent damage to the detector. In contrast, wires are reasonably robust and can sustain modest sparking without serious damage.

An alternative structure known as the microgap chamber[110, 111] is also shown in Fig. 6.22*b*. Here the cathode is a continuous conductor and the anodes are supported above

Figure 6.23 Structure of a two-dimensional position-sensing microstrip gas chamber. (From Tanimori et al.[107])

its surface by intermediate insulating strips. Exposing as little as possible of the insulating area helps diminish the accumulation of charges. Structures of this type also have been successfully operated with minimal problems from sparking damage.

These types of detectors lend themselves naturally to two-dimensional position sensing. The individual anode strip where the avalanche occurs will show the largest induced signal and can be identified electronically. The second coordinate is often obtained from a conducting surface called the *backplane*, placed on the bottom side of the substrate. When an event occurs near an anode, a signal is induced on the backplane if the top surface of the substrate has reasonable gaps between the conducting structures. The backplane can be fabricated as strips running orthogonal to the anodes to obtain a second coordinate measurement from the centriod of the induced charges.

Although their primary use has been as position-sensitive detectors for particle tracking, MSGCs can also provide energy information for spectroscopic measurement. The energy resolution attainable is at least as good as that from a conventional proportional counter[112] and is enhanced because the avalanche fluctuations are minimized as a result of the sharp gradient in the electric field strength near the anode surface.[113]

F. Gas Proportional Scintillation Counters

An interesting hybrid detector has been developed that combines some of the properties of a proportional counter with those of the scintillation detector discussed in Chapter 8. The *gas proportional scintillation (GPS) counter* (also known in the literature as the *gas scintillation proportional counter* or the *gas scintillation drift chamber*) is based on generating a signal pulse from the visible and ultraviolet photons emitted by excited gas atoms or molecules. In a conventional gas scintillator, these excited gas molecules are created by direct interaction of incident radiation as it passes through the gas. The lifetimes of the excited states are normally quite short, so that the light appears within a few nanoseconds of the time of passage of the incident particle.

If an electric field is applied to the gas, the free electrons from ion pairs created along the particle track will drift as they do in an ionization chamber or proportional counter. If the field is sufficiently strong, inelastic collisions between these electrons and neutral gas molecules can elevate some of the molecules to excited states, which may then de-excite through the emission of a photon. This process also occurs in a conventional proportional counter, but the photons are treated as something of a nuisance because their subsequent interaction in other parts of the counter can lead to spurious pulses. In the gas proportional scintillation counter, however, the photons emitted by excited gas molecules are detected by a photomultiplier tube, which produces an electrical pulse proportional to the number of photons incident on the tube. The mechanisms involved in the generation of this "secondary scintillation" light in gases have been reviewed in Ref. 114, which also contains an extensive list of citations to early work involving the development of GPS detectors. Other designs and their performance are reviewed in Ref. 115 .

The light output from a GPS detector consists of prompt and delayed responses. The prompt light component is relatively weak and consists of the conventional scintillation light created along the track of the incident radiation. A delay on the order of a microsecond is then typically required for the electrons to drift from the particle track to the high-field region near the anode wire. No light is created during this drift time. The electron then enters the high-field region where excited molecules are created over times that are again of the order of a microsecond, giving rise to a relatively slow but strong secondary signal. Although the intensity of the prompt scintillation in typical gases will be only a few percent of the equivalent light output from the standard scintillator NaI(Tl), the secondary light yield can be several orders of magnitude greater. A single primary electron could, in principle, produce an unlimited number of secondary photons if the high-field region were to extend indefinitely. For designs based on anode wires, the high electric field is created in the vicinity of the wire so that the light-producing region ends at the wire surface. Alternatively, the design of the type shown in Fig. 6.24 creates a uniform high-field region by applying a sufficiently high voltage between two layers of grids that are essentially transparent to the motion of the electrons.[116–118] Through this process, one electron can produce hundreds of photons as it passes between the grid layers.

Gas proportional scintillation counters of this geometry can also be used to provide information on the position of the particle interaction in the gas by using a position-sensitive photomultiplier tube described in Chapter 9. This approach can have significant advantages over using a conventional scintillator for the same purpose because of the much greater light yield characteristic of the GPS detector.

Another version of the GPS chamber eliminates the drift region shown in Fig. 6.24, and a uniform high electric field exists throughout the entire gas volume.[119] This type of design avoids possible loss of charges to recombination in the low-field drift region or any losses due to absorption of electrons that do not pass through the grid. In this case, the light is generated starting when the original ion pairs are formed and persists until they are collected, so a measurement of the time duration can be used to measure the depth of the interaction for position sensing. By dividing the measured total light yield by the time duration, a ratio is obtained that cancels the depth dependence and is then proportional to the particle energy deposited in the gas.

It has been estimated[120,121] that, in pure noble gases, as much as 75–97% of the energy gained by the electron from the electric field is converted into light. Noble gases have no vibrational or rotational states, so excitation requires the elevation of the atom to the lowest electronic excited state at around 10 eV. The electron typically undergoes thousands of nonradiative elastic scattering collisions before gaining enough kinetic energy from the field to create an excited atom. The presence of molecular impurities with low-lying vibrational or rotational states will therefore strongly inhibit this excitation process, and purity of the gas is essential to good secondary scintillation yield. De-excitation of the excited atom R* generally proceeds as a two-step process: collisions with neutral atoms first produce a molecular rare gas excimer, R_2^*. This species then de-excites by breaking up into two ground-state atoms plus an emitted photon ($R_2^* \rightarrow 2R + h\nu$). This light lies in the

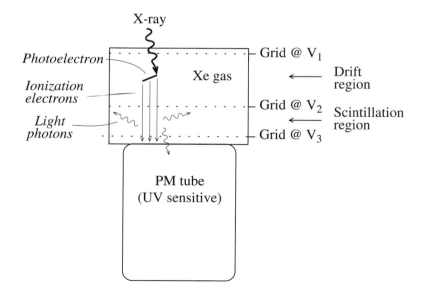

Figure 6.24 Schematic of a gas proportional scintillation counter with a low-field drift region and a higher-field scintillation region. All the light is generated in the scintillation region.

Table 6.4 Spectral Properties of Light Emitted in Gas Proportional Scintillation Counters

Gas	Peak Wavelength of Emission Spectrum (nm)	FWHM of Emission Spectrum (nm)
Argon	128	10
Krypton	147	12
Xenon	173	14

Data from Suzuki and Kubota.[122]

vacuum ultraviolet region of the spectrum for the commonly used noble gases (see Table 6.4). It can be detected directly using special UV-sensitive photomultiplier tubes or, alternatively, with standard tubes after shifting the wavelength to the visible range by absorption and reemission in an organic optical layer on the exit window. (Optical wavelength shifting is discussed more fully beginning on p. 258.) Another option is the use of

Figure 6.25 Curves showing the relative charge gain (1), light gain (2), and secondary light per collected electron (3), for a gas proportional scintillation counter. A numerical scale is shown for curve (2) only. Charge multiplication begins at about 3 kV, but secondary light production is observed at much lower anode voltages. (From Policarpo et al.[127])

photodiodes[123,124] and photoionization devices (see p. 302) that can detect light into the ultraviolet region. In particular, position-sensitive photoionization detectors have proved to be extremely useful in recording both the intensity and the spatial position of the secondary scintillation light in large-area GPS devices applied in radiation imaging.[125,126]

The voltage characteristics of the typical GPS counter are plotted in Fig. 6.25. Increasing the chamber voltage increases the secondary signal amplitude, but an operating point that is short of the gas multiplication region is preferred from energy resolution considerations. By avoiding the formation of an avalanche, the fluctuations in electron multiplication that limit proportional counter resolution are not a factor and the fluctuations in secondary scintillation yield are generally somewhat smaller. The energy resolution for low-energy radiations can be as much as a factor of 2 better than that for proportional counters[127,128] and this major advantage has been the basis of most successful applications of gas proportional scintillation detectors.

One disadvantage of GPS counters is that they are more susceptible to the deleterious effects of gas impurities than are conventional proportional counters. As a result, most of these detectors have been operated using a continuous flow gas purification system to control the impurities. Published efforts,[129] however, have shown that the use of ultrahigh-vacuum techniques and close attention to gas purity can allow the use of sealed detectors over time periods of at least 6 months.

PROBLEMS

6.1 Assuming that $W = 26.2$ eV/ion pair and the Fano factor $F = 0.17$ for argon, find the mean and expected standard deviation in the number of ion pairs formed by the absorption of 1 MeV of radiation energy.

6.2 Explain why air is widely used as a fill gas for ionization chambers but is seldom used in proportional counters.

6.3 **(a)** A proportional counter with anode radius of 0.003 cm and cathode radius of 1.0 cm is filled with P-10 gas at 1 atm. Using the Diethorn parameters from Table 6.1, find the voltage required to achieve a gas multiplication factor of 1000.

(b) At the same voltage, by what factor would the multiplication change if the anode were twice as large?

(c) With the original anode radius, by what factor would the multiplication change if the cathode radius were doubled?

6.4 A windowless flow proportional counter is used to detect 5 MeV alpha particles that are totally stopped in the fill gas. The tube has an anode radius of 0.005 cm, a cathode radius of 5.0 cm and is filled with P-10 gas at 1 atm. Using the data from Table 6.1, estimate the amplitude of the voltage pulses from the counter for an applied voltage of 2000 V and a collection capacitance of 500 pF.

6.5 Describe a measurement that can determine whether a proportional tube is operating in the true proportional region. Be explicit about which quantities are measured and

how the data are used to make the determination. Assume that you have available any necessary type of radioisotope source.

6.6 Explain the following statement: In a windowless proportional counter, the output pulse height from an alpha particle source will increase with increasing alpha energy, whereas for beta particles the opposite is usually true.

6.7 A given voltage-sensitive preamplifier requires a minimum input pulse amplitude of 10 mV for good signal/noise performance. What gas multiplication factor is required in an argon-filled proportional counter with 200 pF capacitance if 50 keV X-rays are to be measured?

6.8 A cylindrical proportional tube has an anode wire radius of 0.003 cm and a cathode radius of 2.0 cm. It is operated with an applied voltage of 2000 V. If a minimum electric field of 1.0 MV/m is required to initiate gas multiplication, what fraction of the internal volume of the tube corresponds to the multiplication region?

6.9 Explain the function of the "field tube" found in some proportional tube designs.

6.10 Explain the sudden discontinuities in the curves for the absorption properties of various gases shown in Fig. 6.16.

6.11 Why are gas scintillation proportional detectors normally operated with a voltage below that required for charge multiplication?

REFERENCES

1. P. J. Campion, *Int. J. Appl. Radiat. Isotopes* **19**, 219 (1968).
2. W. Bambynek, *Nucl. Instrum. Meth.* **112**, 103 (1973).
3. M. Matoba et al., *IEEE Trans. Nucl. Sci.* **NS-32**(1), 541 (1985).
4. E.F. Bennet and T.J. Yule. ANL-7763, (1971).
5. H. Sakurai and B. D. Ramsey, *Nucl. Instrum. Meth.* **A313**, 155 (1992).
6. A. Spernol and B. Denecke, *Int. J. Appl. Radiat. Isotopes* **15**, 195 (1963).
7. F. B. Riggs, Jr., *Rev. Sci. Instrum.* **34**, 392 (1963).
8. H. M. Horstman, G. Ventura, and G. R. Vespignani, *Nucl. Instrum. Meth.* **112**, 619 (1973).
9. J. Byrne and F. Shaikh, *Nucl. Instrum. Meth.* **79**, 286 (1970).
10. P. Povinec, *Nucl. Instrum. Meth.* **163**, 363 (1979).
11. H. Sipila, *Nucl. Instrum. Meth.* **140**, 389 (1977).
12. B. D. Ramsey and P. C. Agrawal, *Nucl. Instrum. Meth.* **A273**, 326 (1988).
13. P. C. Agrawal and B. D. Ramsey, *Nucl. Instrum. Meth.* **A273**, 331 (1988).
14. P. J. Campion, *Nucl. Instrum. Meth.* **112**, 75 (1973).
15. T. Aoyama, H. Miyai, and T. Watanabe, *Nucl. Instrum. Meth.* **221**, 644 (1984).
16. T. Aoyama and T. Watanabe, *Health Phys.* **48**, 773 (1985).
17. T. Aoyama, T. Sato, H. Muraki, and C. Mori, *Nucl. Instrum. Meth.* **A322**, 170 (1992).
18. T. Noguchi, S. Nagashima, and M. Uda, *Nucl. Instrum. Meth.* **A342**, 521 (1994).
19. T. Aoyama, M. Totogawa, and T. Watanabe, *Nucl. Instrum. Meth.* **A255**, 524 (1987).
20. G. D. Alkhazov, A. P. Komar, and A. A. Vorob'ev, *Nucl. Instrum. Meth.* **48**, 1(1967).
21. M.-L. Järvinen and H. Sipilä, *IEEE Trans. Nucl. Sci.* **NS-31**(1), 356 (1984).
22. T. Z. Kowalski and J. Zajac, *Nucl. Instrum. Meth.* **A249**, 426 (1986).
23. J. P. Sephton, M. J. L. Turner, and J. W. Leake, *Nucl. Instrum. Meth.* **A256**, 561 (1987).
24. P. C. Agrawal, B. D. Ramsey, and M. C. Weisskopf, *Nucl. Instrum. Meth.* **A277**, 557 (1989).
25. B. D. Ramsey and P. C. Agrawal, *Nucl. Instrum. Meth.* **A278**, 576 (1989).
26. I. K. Bronic and B. Grosswendt, *Nucl. Instrum. Meth.* **B117**, 5 (1996).
27. F. P. Santos, T. H. V .T. Dias, A. D. Stauffer, and C. A. N. Conde, *Nucl. Instrum. Meth.* **A307**, 347 (1991).
28. T. H. V. T. Dias, F. P. Santos, A. D. Stauffer, and C. A. N. Conde, *Nucl. Instrum. Meth.* **A307**, 341 (1991).
29. T. Doke, N. Ishida and M. Kase, *Nucl. Instrum. Meth.* **B63**, 373 (1992).
30. I. K. Bronic, M. Kimura, M. Inokuti, and M. A. Dillon, *Nucl. Instrum. Meth.* **B71**, 366 (1992).
31. A. Peisert and F. Sauli, CERN 84-08, (1984).
32. J. Va'vra, *Nucl. Instrum. Meth.* **A323**, 34 (1992).
33. R. E. Robson, M. Hildebrandt, and B. Schmidt, *Nucl. Instrum. Meth.* **A394**, 74 (1997).
34. S. C. Curran, A. L. Cockroft, and J. Angus. *Philos. Mag.* **40**, 929 (1949).
35. H. Schlumbohm, *Z. Pys.* **182**, 306 (1965).
36. R. Gold and E. F. Bennett, *Phys. Rev.* **147** (1), 201 (1966).
37. H. Genz, *Nucl. Instrum. Meth.* **112**, 83 (1973).
38. A. Williams and R. I. Sara, *Int. J. Appl. Radiat. Isotopes* **13**, 229 (1962).
39. A. Zastawny, *J. Sci. Instrum.* **43**, 179 (1966).
40. G. D. Alkhazov, *Nucl. Instrum. Meth.* **75**, 161 (1969).
41. P. J. Campion, *Phys. Med. Biol.* **16**, 611 (1971).
42. P. Hopstone and S. Shalev, *Trans. Am. Nuc. Soc.* **21**, 132 (1975).
43. A. Von Engel, *in Handbuch der Physik*, Vol. 21 (S. Flugge, ed.), Springer-Verlag, Berlin, 1956, p. 504.
44. T. Aoyama, *Nucl. Instrum. Meth.* **A234**, 125 (1985).
45. T. Z Kowalski, *Nucl. Instrum. Meth.* **A243**, 501 (1986).
46. Y. Uozumi et al., *Nucl. Instrum. Meth.* **A324**, 558 (1993).
47. A. Zastawny, *Nucl. Instrum. Meth.* **A385**, 239 (1997).
48. W. Diethorn, NYO-6628 (1956).
49. R. W. Hendricks, *Nucl. Instrum. Meth.* **102**, 309 (1972).
50. R. S. Wolff, *Nucl. Instrum. Meth.* **115**, 461 (1974).
51. C. Mori, M. Uno and T. Watanabe, *Nucl. Instrum. Meth.* **196**, 49 (1982).
52. R. W. Hendricks, *Rev. Sci. Instrum.* **40**(9), 1216 (1969).
53. J. Byrne, *Nucl. Instrum. Meth.* **74**, 291 (1969).
54. G.D. Alkhazov, *Nucl. Instrum. Meth.* **89**, 155 (1970).
55. B. Breyer, *Nucl. Instrum. Meth.* **112**, 91 (1973).
56. J. Byrne, *Proc. R. Soc. Edinburg Sect.* **A66**, 33 (1962).
57. H. Sipila, *Nucl. Instrum. Meth.* **133**, 251 (1976).
58. F. Tokanai et al., *IEEE Trans. Nucl. Sci.* **41**(4), 1042 (1994).
59. M. W. Charles and B. A. Cooke, *Nucl. Instrum. Meth.* **61**, 31 (1968).
60. F. Shaikh, J. Byrne, and J. Kyles, *Nucl. Instrum. Meth.* **88**, 317 (1970).
61. W. Hink, A. N. Scheit, and A. Ziegler, *Nucl. Instrum. Meth.* **87**, 137 (1970).
62. B. Bednarek, *Nucl. Instrum. Meth.* **175**, 431 (1980).
63. T. Z. Kowalski, *Nucl. Instrum. Meth.* **A239**, 551 (1985).
64. J. A. Kadyk, *Nucl. Instrum. Meth.* **A300**, 436 (1991).
65. A. Smith and M. J. L. Turner, *Nucl. Instrum. Meth.* **192**, 475 (1982).
66. D. J. Grady and J. C. Robertson, *Nucl. Instrum. Meth.* **179**, 317 (1981).
67. D. H. Wilkinson, *Ionization Chambers and Counters*, Cambridge University Press, Cambridge, 1950.
68. R. Gott and M. W. Charles, *Nucl. Instrum. Meth.* **72**, 157 (1969).
69. L. G. Christophorov, D. L. McCorkle, D. V. Maxey, and J. G. Carter, *Nucl. Instrum. Meth.* **163**, 141 (1979).
70. A. Jeavons et al., *Nucl. Instrum. Meth.* **176**, 89 (1980).
71. E. Mathieson and M. W. Charles, *Nucl. Instrum. Meth.* **72**, 155 (1969).
72. E. Mathieson, *Nucl. Instrum. Meth.* **72**, 355 (1969).
73. B. Breyer and M. Cimerman, *Nucl. Instrum. Meth.* **92**, 19 (1971).
74. P. Gorenstein and S. Mickiewicz, *Rev. Sci. Instrum.* **39**(6), 816 (1968).
75. S. Sudar, L. Vas, and T. Biro, *Nucl. Instrum. Meth.* **112**, 399 (1973).
76. R. C. Hochel and D. W. Hayes, *Nucl. Instrum. Meth.* **130**, 183 (1975).
77. H. Genz, D. S. Harmer, and R. W. Fink, *Nucl. Instrum. Meth.* **60**, 195 (1968).
78. P. J. Campion, *J. Phys. E* **3**, 920 (1970).
79. R. Mutikainen, V.-P. Vittanen, and S. Nenonen, *Jour. X-ray Sci. and Tech.* **4**, 82 (1994).
80. M. Bavdaz et al., *Nucl. Instrum. Meth.* **A345**, 549 (1994).
81. G. Manzo et al., *Nucl. Instrum. Meth.* **177**, 595 (1980).

82. G. Bibbo and P. W. Sanford, *Nucl. Instrum. Meth.* **179**, 189 (1981).

83. B. D. Ramsey et al., *Nucl. Instrum. Meth.* **A383**, 424 (1996).

84. H. H. Rossi and W. Rosenwieg, *Radiology* **64**, 404 (1955).

85. L. Brackenbush, G. W. R. Endres, and D. E. Hadlock, *IEEE Trans. Nucl. Sci.* **33(1)**, 610 (1986).

86. P. Colautti et al., *Nucl. Instrum. Meth.* **A321**, 392 (1992).

87. G. Hempel, F. Hopkins, and G. Schatz, *Nucl. Instrum. Meth.* **131**, 445 (1975).

88. H. Stelzer, *Nucl. Instrum. Meth.* **133**, 409 (1976).

89. G. Charpak et al., *Nucl. Instrum. Meth.* **A274**, 275 (1989).

90. A. Arefiev et al., *Nucl. Instrum. Meth.* **A348**, 318 (1994).

91. D. Fabris et al., *Nucl. Instrum. Meth.* **216**, 167 (1983).

92. A. Breskin and N. Zwang, *Nucl. Instrum. Meth.* **144**, 609 (1977).

93. S. Beghini, L. Corradi, J. H. He, and A. Dal Bellow, *Nucl. Instrum. Meth.* **A362**, 526 (1995).

94. K. D. Joensen et al., *Nucl. Instrum. Meth.* **A356**, 234 (1995).

95. G. P. Westphal, *Nucl. Instrum. Meth.* **134**, 387 (1976).

96. B. E. Fischer, *Nucl. Instrum. Meth.* **141**, 173 (1977).

97. C. J. Borkowski and M. K. Kopp, *Rev. Sci. Instrum.* **39**, 1515 (1968).

98. C. J. Borkowski and M. K. Kopp, *IEEE Trans. Nucl. Sci.* **NS-17(3)**, 340 (1970).

99. R. W. Hendricks, *Trans. Am. Crystallogr. Assoc.* **12**, 103 (1976).

100. J. L. C Ford, Jr., *Nucl. Instrum. Meth.* **162**, 277 (1979).

101. G. Charpak and F. Sauli, *Nucl. Instrum. Meth.* **162**, 405 (1979).

102. F. Sauli, CERN-PPE/94-195 (1994).

103. R. Fourme, *Nucl. Instrum. Meth.* **A392**, 1 (1997).

104. A. Breskin, *Nucl. Instrum. Meth.* **141**, 505 (1977).

105. P. Povinec, J. Szarka, and S. Usacev, *Nucl. Instrum. Meth.* **163**, 369 (1979).

106. A. Oed, *Nucl. Instrum. Meth.* **A263**, 351 (1988).

107. T. Tanimori, A. Ochi, S. Minami, and T. Nagae, *Nucl. Instrum. Meth.* **A381**, 280 (1996).

108. R. Bouclier et al., *Nucl. Instrum. Meth.* **A367**, 168 (1995).

109. H. S. Cho et al., *IEEE Trans. Nucl. Sci.* **44(3)**, 635 (1997).

110. F. Angelini et al., *Nucl. Instrum. Meth.* **A335**, 69 (1993).

111. R. Bellazzini and G. Spandre, *Nucl. Instrum. Meth.* **A368**, 259 (1995).

112. J. van der Marel et al., *Nucl. Instrum. Meth.* **A367**, 181 (1995).

113. J. Miyamoto and G. F. Knoll, *Nucl. Instrum. Meth.* **A399**, 85 (1997).

114. P. E. Thiess and G. H. Miley, *IEEE Trans. Nucl. Sci.* **NS-21**(1), 125 (1974).

115. A. J. P. L. Policarpo, *Space Sci. Instrum.* **3**, 77 (1977).

116. C. A. N. Conde, J. M. F. Dos Santos, and A. C. S. S. M. Bento, *IEEE Trans. Nucl. Sci.* **40(4)**, 452 (1993).

117. D. A. Goganov and A. A. Schultz, *Nucl. Instrum. Meth.* **A394**, 151 (1997).

118. J. M. F. dos Santos et al., *IEEE Trans. Nucl. Sci.* **45(3)**, 229 (1998).

119. D. G. Simons, P. A. J. de Korte, A. Peacock, and J. A. M. Bleeker, SPIE **597**, 190 (1985).

120. R. D. Anderson, E. A. Leimann, and A. Peacock, *Nucl. Instrum. Meth.* **140**, 371 (1977).

121. M. Alegria Feio et al., *Nucl. Instrum. Meth.* **176**, 473 (1980).

122. M. Suzuki and S. Kubota, *Nucl. Instrum. Meth.* **164**, 197 (1979).

123. J. C. Van Staden et al., *Nucl. Instrum. Meth.* **157**, 301 (1978).

124. A. J. de Campos, *IEEE Trans. Nucl. Sci.* **NS-31**(1), 133 (1984).

125. A. J. P. L. Policarpo, *Nucl. Instrum. Meth.* **196**, 53 (1982).

126. D. F. Anderson, *IEEE Trans. Nucl. Sci.* **NS-32**(1), 495 (1985).

127. A. J. P. L. Policarpo, M. A. F. Alves, M. C. M. Dos Santos, and M. J. T. Carvalho, *Nucl. Instrum. Meth.* **102**, 337 (1972).

128. M. Alice, F. Alves, A. J. P. L. Policarpo, and M. Salete, *IEEE Trans. Nucl. Sci.* **NS-22**, 109 (1975).

129. R. D. Andresen, L. Karlsson, and B. G. Taylor, *IEEE Trans. Nucl. Sci.* **NS-23**(1), 473 (1976).

Chapter 7

Geiger–Mueller Counters

The Geiger–Mueller counter (commonly referred to as the G-M counter, or simply Geiger tube) is one of the oldest radiation detector types in existence, having been introduced by Geiger and Mueller in 1928. However, the simplicity, low cost, and ease of operation of these detectors have led to their continued use to the present time. General textbook references in the field of G-M counters include the reviews by Wilkinson,[1] Price,[2] Sharpe,[3] and Emery.[4] Construction details and specific tube designs beyond those given in the following discussion can be found in all these references. Detailed analyses of the physical processes involved in the Geiger discharge are elegantly presented in a series of three articles by Wilkinson.[5–7]

As a complement to the ion chambers and proportional counters discussed in the two previous chapters, G-M counters comprise the third general category of gas-filled detectors based on ionization. In common with proportional counters, they employ gas multiplication to greatly increase the charge represented by the original ion pairs formed along the radiation track, but in a fundamentally different manner. In the proportional counter, each original electron leads to an avalanche that is basically independent of all other avalanches formed from other electrons associated with the original ionizing event. Because all avalanches are nearly identical, the collected charge remains proportional to the number of original electrons.

In the G-M tube, substantially higher electric fields are created that enhance the intensity of each avalanche. Under proper conditions, a situation is created in which one avalanche can itself trigger a second avalanche at a different position within the tube. At a critical value of the electric field, each avalanche can create, on the average, at least one more avalanche, and a self-propagating chain reaction results. At still greater values of the electric field, the process becomes rapidly divergent and, in principle, an exponentially growing number of avalanches could be created within a very short time. Once this *Geiger discharge* reaches a certain size, however, collective effects of all the individual avalanches come into play and ultimately terminate the chain reaction. Because this limiting point is always reached after about the same number of avalanches have been created, *all pulses from a Geiger tube are of the same amplitude regardless of the number of original ion pairs that initiated the process.* A Geiger tube can therefore function only as a simple counter of radiation-induced events and cannot be applied in direct radiation spectroscopy because all information on the amount of energy deposited by the incident radiation is lost.

A typical pulse from a Geiger tube represents an unusually large amount of collected charge, about 10^9–10^{10} ion pairs being formed in the discharge. Therefore, the output pulse amplitude is also large (typically of the order of volts). This high-level signal allows considerable simplification to be made in the associated electronics, often completely eliminating the need for external amplification. Because the tubes themselves are relatively inexpensive, a G-M counter is often the best choice when a simple and economical counting system is needed.

Besides the lack of energy information, a major disadvantage of G-M counters is their unusually large dead time, which greatly exceeds that of any other commonly used radiation detector. These detectors are therefore limited to relatively low counting rates, and a dead time correction must be applied to situations in which the counting rate would otherwise be regarded as moderate (a few hundred pulses per second). Some types of Geiger tube also have a limited lifetime and will begin to fail after a fixed number of total pulses have been recorded (see following discussion of the quench gas).

I. THE GEIGER DISCHARGE

In a typical Townsend avalanche created by a single original electron, many excited gas molecules are formed by electron collisions in addition to the secondary ions. Within usually no more than a few nanoseconds, these excited molecules return to their ground state through the emission of photons whose wavelength may be in the visible or ultraviolet region. These photons are the key element in the propagation of the chain reaction that makes up the Geiger discharge. An energetic optical photon emitted in the filling of an inner electron shell vacancy may be reabsorbed elsewhere in the gas by photoelectric absorption involving a less tightly bound electron, creating a new free electron. Alternatively, the photon may reach the cathode wall where it could release a free electron upon absorption. In either case, the newly created free electron will migrate toward the anode and trigger another avalanche (see Fig. 7.1). If the gas multiplication factor M as defined in Chapter 6 is relatively low (say, 10^2–10^4) as in a proportional tube, the number of excited molecules formed in a typical avalanche (n_0') is not very large. Also, because most gases are relatively transparent in the visible and UV wavelength regions, the probability p of photoelectric absorption of any given photon is also relatively low. Under these conditions, $n_0'p \ll 1$ and the situation is "subcritical" in that relatively few avalanches are formed in addition to those created by the original free electrons. Many proportional gases also contain an additive to absorb the photons preferentially without creation of free electrons, further suppressing the possibility of new avalanches.

In a Geiger discharge, however, the multiplication represented by a single avalanche is much higher (10^6–10^8) and therefore n_0' is also much larger. Now the conditions of "criticality" can be achieved

$$n_0'p \geq 1$$

and an ever-increasing number of avalanches may potentially be created throughout the tube. The time required for the spread of these avalanches is relatively short. Because each avalanche begins only when the free electron has drifted to within a few mean free paths of

Figure 7.1 The mechanism by which additional avalanches are triggered in a Geiger discharge.

the anode wire, the time required to produce all the ion pairs and excited molecules in a given avalanche is a small fraction of a microsecond. The photons preferentially create free electrons that are close to the original avalanche, and these electrons must drift into the high-field region near the anode to generate further avalanches. The net effect is that the Geiger discharge grows (with some of the characteristics of a chemical flame) in both directions along the wire with a propagation velocity[8] that is typically 2–4 cm/μs, until the entire length of the wire is involved. For tubes of usual size, all the charge is generated within a few microseconds of the initiating event and is integrated into the measured pulse amplitude.

In a proportional counter, each avalanche is formed at an axial position within the tube which corresponds to that of the original free electron created by the incident radiation. Furthermore, the free electron approaches the anode wire from one radial direction, and the secondary ions are preferentially formed on that side of the anode wire.[9, 10] In a Geiger discharge, however, the rapid propagation of the chain reaction leads to many avalanches, which initiate at random radial positions along the wire. Secondary ions are therefore formed throughout the cylindrical multiplying region that surrounds the anode. The Geiger discharge therefore grows to envelope the entire anode wire, regardless of the position at which the primary initiating event occurred.

The process that leads to the termination of a Geiger discharge has as its origin the positive ions that are created along with each electron in an avalanche. The mobility of these positive ions is much less than that of the free electrons, and they remain essentially motionless during the time necessary to collect all the free electrons from the multiplying region. When the concentration of these positive ions is sufficiently high, their presence begins to reduce the magnitude of the electric field in the vicinity of the anode wire. Because the ions represent a positive space charge, the region between the ions and the positive anode will have an electric field below that predicted by Eq. (6.3) in the absence of the space charge. Because further gas multiplication requires that an electric field above some minimum value be maintained, the buildup of positive ion space charge eventually terminates the Geiger discharge. Another way to picture the termination mechanism is to recognize that the buildup of the dense cloud of positive ions has the same effect as temporarily increasing the anode wire diameter. The electric field at the outer surface of the cloud then drops to below the critical value for further avalanche formation.

For a fixed applied voltage to the tube, the point at which the Geiger discharge is terminated will always be the same, in the sense that a given density of positive ions will be needed to reduce the electric field below the minimum value required for further multiplication. Thus, each Geiger discharge is terminated after developing about the same total charge, regardless of the number of original ion pairs created by the incident radiation. All output pulses are therefore about the same size, and their amplitude can provide no information about the properties of the incident radiation.

As the high voltage is raised in a Geiger tube, the magnitude of the Geiger discharge increases and the amplitude of the output pulse also increases in size as shown in the sketch below. A higher voltage means a larger initial electric field before the discharge, which then requires a greater buildup of space charge before the field is reduced below the critical value. The pulse amplitude increases roughly in proportion to the *overvoltage*, defined as the difference between the applied voltage and the minimum voltage required to initiate a Geiger discharge.

II. FILL GASES

Gases used in a Geiger tube must meet some of the same requirements discussed in Chapter 6 for proportional counters. Because both types of detectors are based on gas multiplication, even trace amounts of gases that form negative ions (such as oxygen) must be avoided in either case. The noble gases are widely used for the principal component of the fill gas in G-M tubes, with helium and argon the most popular choices. A second component is normally added to most Geiger gases for purposes of *quenching*, discussed in the next section.

The average energy attained by a free electron between collisions with gas molecules depends on the ratio of the electric field magnitude to the gas pressure, $\mathscr{E}/p$ [see Eq. (5.3)]. Because the number of excited molecules n_0' should also increase with this parameter, the development of a full Geiger discharge requires a certain minimum value of $\mathscr{E}/p$, which depends on the specific gas mixture used in the G-M tube. It is common practice to design sealed tubes with a fill gas kept at less than atmospheric pressure in order to minimize the value of $\mathscr{E}$ needed to achieve the necessary $\mathscr{E}/p$ ratio. For normal gases at pressures of several tenths of an atmosphere, operating voltages of about 500–2000 V are required to reach the necessary field strength using anode wires of about 0.1 mm diameter [see Eq. (6.3)].

Geiger tubes can also be designed to operate at atmospheric pressure. In such cases, the mechanical design can be simplified (especially for large tubes) because no pressure differential need be supported across the walls or window of the tube. Just as in proportional counters, either the gas can be permanently sealed within the tube or arrangements can be made to have the gas flow through the counter volume to replenish the gas continuously and to flush out any impurities.

III. QUENCHING

If a Geiger tube is filled with a single gas such as argon, then all the positive ions formed are ions of that same gas species. After the primary Geiger discharge is terminated, the positive ions slowly drift away from the anode wire and ultimately arrive at the cathode or outer wall of the counter. Here they are neutralized by combining with an electron from the cathode surface. In this process, an amount of energy equal to the ionization energy of the gas minus the energy required to extract the electron from the cathode surface (the *work function*) is liberated. If this liberated energy also exceeds the cathode work function, it is energetically possible for another free electron to emerge from the cathode surface. This will be the case if the gas ionization energy exceeds twice the value of the work function. The probability is always small that any given ion will liberate an electron in its neutralization, but if the total number of ions is large enough, there will likely be at least one such free electron generated. This electron will then drift toward the anode and will trigger another avalanche, leading to a second full Geiger discharge. The entire cycle will now be repeated, and under these circumstances the G-M counter, once initially triggered, would produce a continuous output of multiple pulses.

The problem of multiple pulsing is potentially much more severe in Geiger tubes than in proportional counters. In the latter case, the number of positive ions per event is small enough so that only an occasional spurious pulse may result. The amplitude of the spurious pulse will correspond only to a single avalanche and will generally be considerably smaller than that produced by the initial event. With a Geiger tube, the increased number of positive ions reaching the cathode greatly enhances the free electron emission probability. Furthermore, the secondary discharge, even though caused by only a single electron, builds to an amplitude equal to that of the primary discharge.

For these reasons, special precautions must be taken in Geiger counters to prevent the possibility of excessive multiple pulsing. *External quenching* consists of some method for reducing the high voltage applied to the tube, for a fixed time after each pulse, to a value that is too low to support further gas multiplication. Then, secondary avalanches cannot be formed and even if a free electron is liberated at the cathode, it cannot cause another Geiger discharge. The voltage must be suppressed for a time that is greater than the transit time of the positive ion from its place of formation to the cathode (generally a few hundred microseconds), plus the transit time of the free electron (generally about a microsecond).

One method of external quenching is simply to choose R in Fig. 7.2 to be a large enough value (typically 10^8 ohms) so that the time constant of the charge collection circuit is of the order of a millisecond. This "external resistor" method of quenching has the disadvantage of requiring several milliseconds for the anode to return to near its normal voltage, and therefore full Geiger discharges for each event are produced only at very low counting rates.

It is much more common to prevent the possibility of multiple pulsing through *internal quenching*, which is accomplished by adding a second component called the *quench gas* to the primary fill gas. It is chosen to have a lower ionization potential and a more complex molecular structure than the primary gas component and is present with a typical concentration of 5–10%. Although it is given the same name as the component added to proportional counter gases to absorb UV photons, the quench gas in a Geiger counter serves a different function. It prevents multiple pulsing through the mechanism of charge transfer collisions. The positive ions formed by the incident radiation are mostly of the primary component, and they subsequently make many collisions with neutral gas molecules as they drift toward the cathode. Some of these collisions will be with molecules of the quench gas and, because of the difference in ionization energies, there will be a tendency to transfer the positive charge to the quench gas molecule. The original positive ion is thus neutralized by transfer of an electron and a positive ion of the quench gas begins to drift in its place. If the concentration of the quench gas is sufficiently high, these charge-transfer collisions ensure that all the ions that eventually arrive at the cathode will be those of the quench gas. When they are neutralized, the excess energy may now go into disassociation of the more complex molecules in preference to liberating a free electron from the cathode surface. With the proper choice of quench gas, the probability of disassociation can be made much larger than that of electron emission, and therefore no additional avalanches are formed within the tube.

Many organic molecules possess the proper characteristics to serve as a quench gas. Of these, ethyl alcohol and ethyl formate have proved to be the most popular and are widely used in many commercial G-M tubes. Because the quench gas is disassociated in carrying out its function, it is gradually consumed during the lifetime of the tube. An organic

Figure 7.2 Equivalent counting circuit for a G-M tube. The product of resistance R and capacitance C (usually only inherent capacitance of the tube and electronics) determines the time constant of the restoration of the high voltage following a Geiger discharge.

quenched tube may typically have a limit of about 10^9 counts in its useful lifetime before multiple pulsing can no longer be prevented because of depletion of the quench gas.

To avoid the problem of limited lifetime, some tubes use halogens (chlorine or bromine) as the quench gas. Although the halogen molecules also disassociate when carrying out their quenching function, they may be replenished by spontaneous recombination at a later time. In principle, halogen-quenched tubes therefore have an infinite lifetime and are preferred in situations where the application involves extended use at high counting rates. However, there are other mechanisms besides the consumption of the quench gas that limit the lifetime of Geiger tubes. Two of the more important appear to be contamination of the gas by reaction products produced in the discharge and changes on the anode surface caused by the deposition of polymerized reaction products.[11, 12]

IV. TIME BEHAVIOR

A. Pulse Profile

In Chapter 6, the development of the pulse from a single avalanche was shown to derive almost entirely from the motion of the positive ions away from the anode wire. An initial fast component (with rise time less than a microsecond) results from the motion of the ions through the high-field region near the anode wire. A much slower rise follows that corresponds to the slower drift of the ions through the lower-field region that exists in most of the tube volume. In a Geiger discharge, the pulse corresponds to the cumulative effect of many avalanches that have been formed along the entire length of the anode wire. The time required for secondary electrons to reach the multiplying region to trigger these multiple avalanches will be variable, but differences will be limited by the relatively rapid motion of the electrons. The initial pulse rise will be somewhat slower than that for a single avalanche but still will develop over a time that is typically less than a few microseconds. The situation is also complicated by the electric field distortions that take place from the buildup of space charge near the termination of the discharge.

The net effect is that the output pulse from a G-M tube still exhibits a fast rise on the order of a microsecond or less, followed by a much slower rise due to the slow drift of the ions. If the time constant of the collection circuit were infinite, it would take the full ion collection time for the output pulse to reach its theoretical maximum. From considerations of maximum counting rate, time constants are often chosen that are much less than 100 µs, which largely eliminate the slow-rising portion of the pulse and leave only the fast leading edge, as illustrated in Fig. 7.3. Even though a significant fraction of the potential output

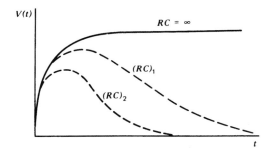

Figure 7.3 Shape of the G-M tube output pulse for different assumed time constants RC of the counting circuit. Here $(RC)_2 < (RC)_1 <$ infinite time constant. The signal voltage $V(t)$ is assumed to be measured as indicated in Fig. 7.2, giving a positive polarity. More conventionally, the pulse is measured with respect to ground within the amplifier, leading to inverted shape or negative polarity.

pulse amplitude may be lost, the large amount of charge generated in the Geiger discharge produces so large a pulse that some amplitude loss can easily be tolerated. Furthermore, because all Geiger discharges are approximately uniform in size and time profile, all pulses will be attenuated by the same fraction in the shaping process and the output pulses will remain almost of one amplitude.

As in proportional counters, there is a delay between the time of formation of the first ion pair within the gas and the initiation of the first avalanche. This delay corresponds to the drift time of the electron from its point of origin to the multiplying region and can be as much as a microsecond in large tubes. When G-M tubes are used for timing purposes, one should therefore expect a time uncertainty of this magnitude if radiation interactions occur randomly throughout the tube volume.

B. Dead Time

The build-up of the positive ion space charge that terminates the Geiger discharge also ensures that a considerable amount of time must pass before a second Geiger discharge can be generated in the tube. As the positive ions drift radially outward, the space charge becomes more diffuse and the electric field in the multiplying region begins to return to its original value. After the positive ions have traveled some of the distance, the field will have recovered sufficiently to permit another Geiger discharge. If the field has not been fully restored, however, the discharge will be less intense than the original because fewer positive ions will be required to shut down the discharge by again reducing the electric field below the critical point. The first pulses that appear are therefore reduced in amplitude compared with the original and may or may not be registered by the counting system, depending on its sensitivity. When the positive ions have eventually drifted all the way to the cathode, the electric field is fully restored and another ionizing event will trigger a second Geiger discharge of full amplitude. This behavior is sketched in Fig. 7.4, which is similar to an oscilloscope trace that can be experimentally observed from Geiger tubes at high counting rates.

Immediately following the Geiger discharge, the electric field has been reduced below the critical point by the positive space charge. If another ionizing event occurs under these conditions, a second pulse will not be observed because gas multiplication is prevented. During this time the tube is therefore "dead" and any radiation interactions that occur in the tube during this time will be lost. Technically, the *dead time* of the Geiger tube is defined as the period between the initial pulse and the time at which a second Geiger

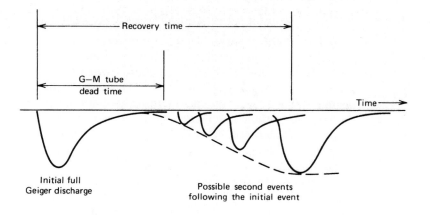

Figure 7.4 Illustration of the dead time of a G-M tube. Pulses of negative polarity conventionally observed from the detector are shown.

discharge, regardless of its size, can be developed. In most Geiger tubes, this time is of the order of 50–100 μs. In any practical counting system, some finite pulse amplitude must be achieved before the second pulse is recorded, and the elapsed time required to develop a second discharge that exceeds this amplitude is sometimes called the *resolving time* of the system. In practice, these two terms are often used interchangeably and the term *dead time* may also be used to describe the combined behavior of the detector–counting system. The *recovery time* is the time interval required for the tube to return to its original state and become capable of producing a second pulse of full amplitude.

We have delayed introducing a discussion of specific detector dead time behavior until now because the G-M tube is a rare example of a radiation detector type that exhibits a long inherent dead time arising from the mechanisms that occur within the detector itself. In most other cases, processes internal to the detector do not inhibit the production of a detector output signal even for two events that may occur with an infinitesimally small time spacing. For example, if two particles interact closely spaced in time in a pulse-type ion chamber, charges from both tracks will drift independently and the induced signal that is observed will be a superposition of the signals expected separately from each. If these two components overlap sufficiently, then only one pulse may be recorded and a loss will occur that is also generally attributed to the dead time of the system. The details of these losses now depend on how the pulses from the detector are shaped, processed, or counted, so the dead time behavior from most measurement systems is critically dependent on the pulse processing electronics as well as the time characteristics of the detector signals. The same general comments apply to the scintillation and semiconductor detectors covered later in this text. Since most of the applications of these detectors involve spectroscopy rather than the simple counting operation to which G-M tubes are restricted, further discussion of these losses is postponed until Chapter 17 under the closely-related topic of *pulse pile-up*.

V. THE GEIGER COUNTING PLATEAU

Because the Geiger tube functions as a simple counter, its application requires only that operating conditions be established in which each pulse is registered by the counting system. In practice, this operating point is normally chosen by recording a *plateau curve* from the system under conditions in which a radiation source generates events at a constant rate within the tube. The counting rate is recorded as the high voltage applied to the tube is raised from an initially low value. The general properties of the resulting counting curve have been discussed in Chapter 4. The results are particularly simple for the case of the Geiger tube in which pulses with amplitudes centered about a single value are involved. The differential pulse height distribution then has the simple appearance sketched in Fig. 7.5a for a fixed value of the voltage applied to the tube. If the voltage is increased, the average pulse amplitude also increases, and the peak in the differential pulse height distribution shifts to the right. Distributions are shown for two hypothetical values of applied voltage in the sketch.

If the minimum amplitude required by the counting system is H_d as indicated on the figure, then no pulses should be counted when the applied voltage is 1000 V, but all the pulses should be counted at 1200 V. The corresponding counting rate therefore abruptly changes from zero to the maximum value between these two voltage settings. Increasing the voltage further only moves the peak more to the right in the differential distribution but does not increase the number of recorded events. Therefore, the counting curve shows a flat plateau at voltages above 1200 V, and in principle any choice of operating voltage above this value would ensure stable operation. The minimum voltage at which pulses are first registered by the counting system is often called the *starting voltage*, whereas the transition between the rapid rise of the curve and the plateau is its *knee*.

If the voltage is raised sufficiently high, the plateau abruptly ends because of the onset of continuous discharge mechanisms within the tube. These can be corona discharges that

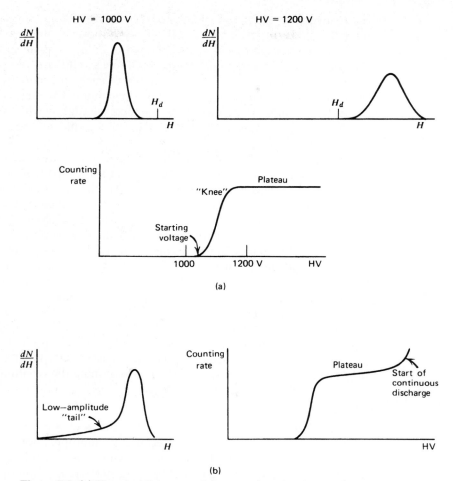

Figure 7.5 (*a*) The establishment of the counting plateau for a G-M tube. As the high voltage is varied in this example from 1000–1200 V, the output pulses change from falling below the counter threshold H_d to a situation in which all pulses are larger than H_d. (*b*) The low-amplitude tail on the pulse height spectrum at the left causes a finite slope of the plateau in the counting curve.

initiate from any sharp irregularities on the anode wire, or multiple pulsing caused by failure of the quenching mechanism. The continuous discharge process is potentially harmful if sustained for any length of time, and one should therefore immediately decrease the applied voltage when the end of the plateau is observed. In the interest of long operating life, an operating point is normally chosen that is only sufficiently far up the plateau to ensure that the flat region has been reached. The Geiger discharges are thus kept near the minimum size compatible with good counting stability and the quench gas is consumed at the minimum rate.

In real cases, the counting plateau always shows some finite slope, as shown in Fig. 7.5*b*. Any effect that adds a low-amplitude tail to the differential pulse height distribution can be a contributing cause of the slope. For example, some regions near the ends of the tube may have a lower than normal electric field strength and the discharges originating in these regions may be smaller than normal. Also, any pulses that occur during the recovery time will also be abnormally small. In order to exclude this latter effect when measuring the plateau characteristics, the counting rate should not exceed a few percent of the inverse of the recovery time. For typical G-M tubes, this restriction means that the plateau measurement should be made at a rate below a few hundred counts per second.

Another cause of slope in the plateau of many G-M tubes is the occasional failure of the quenching mechanism which may lead to a satellite or spurious pulse in addition to the primary Geiger discharge. The probability of such an occurrence is normally low but will increase as the number of positive ions that ultimately reach the cathode also increases. The active volume of the tube may also increase slightly with voltage as the very low electric field near corners of the tube is increased and ion pair recombination is thereby prevented.

The plateau curve may also show a slight hysteresis effect, in that the counting curve recorded by increasing the voltage may differ somewhat from that obtained while decreasing the voltage. This hysteresis arises because the electrical charges on the insulators are slow to equilibrate and can influence the electric field configuration inside the tube.

Geiger tubes are often rated based on the slope shown by the linear portion of the plateau. Organic quench tubes tend to show the flattest plateau, with slopes of the order of 2–3% per 100 V change in the applied voltage. Halogen-quenched tubes have a plateau of greater typical slope, but their longer useful lifetime can sometimes offset this disadvantage. Halogen tubes can also be operated at lower applied voltage values. For example, argon–alcohol counters usually require a starting voltage of at least 1000 V, but neon-filled tubes quenched with bromine can show a starting voltage as low as 200–300 V (Ref. 13). The mechanisms that lead to this unusually low starting voltage strongly depend on the partial pressure of the halogen and the $\mathscr{E}/p$ value within the tube. The interaction of bromine atoms with metastable and excited neon atoms apparently plays an important role in propagation of the discharge and in determining the resulting tube properties.[14]

VI. DESIGN FEATURES

Geiger tubes share some general design features with proportional counters, but in many respects they are less demanding. Because gas multiplication and avalanche formation are again important, the use of a fine anode wire is nearly universal in order to produce the high local values of the electric field necessary to create a multiplying region. In the Geiger tube, less attention to uniformity of the wire is necessary because the Geiger discharge spreads over the entire length of the anode, and nonuniformities are therefore averaged out. Because the pulse amplitude carries no quantitative information, uniformity of response is less important, and therefore the field tubes used in proportional counters to shape the electric field are seldom found in Geiger tubes. Some designs can be used interchangeably as proportional or Geiger tubes, although a change in the fill gas is normally recommended for optimal performance.

A typical design of Geiger tubes is the *end-window* type illustrated in Fig. 7.6. The anode wire is supported at one end only and is located along the axis of a cylindrical cathode made of metal or glass with a metallized inner coating. Radiation enters the tube through the entrance window, which may be made of mica or other material that can maintain its strength in thin sections. Because most Geiger tubes are operated below atmospheric pressure, the window may have to support a substantial differential pressure. The

Figure 7.6 A cross section of a typical end-window Geiger tube.

window should be as thin as possible when counting short-range particles, such as alphas, but may be made more robust for applications that involve beta particles or gamma rays.

Geiger tubes can also be manufactured in many other shapes and configurations. A simple wire loop anode inserted into an arbitrary volume enclosed by a conducting cathode will normally work as a Geiger counter, but some care must be taken to avoid low-field regions in corners where events may be lost. Geiger counters can also take the form of "needle counters," in which the anode consists of a sharply pointed needle supported within a gas enclosed by the cathode. The field in the vicinity of the point varies as $1/r^2$, rather than $1/r$ as in the common cylindrical geometry. Counters with small active volumes can be conveniently manufactured using the needle geometry, and representative designs are described in Refs. 15 and 16.

For applications involving very soft radiations such as low-energy heavy charged particles or soft beta particles, it may be preferable to introduce the source directly into the counting volume. Continuous flow Geiger counters, which are similar in design to the gas flow proportional counters discussed in Chapter 6, are often used for this purpose. By continuously flowing the counter gas through the chamber, the gas purity is maintained despite small amounts of air introduced when samples are changed.

Very high counting efficiencies for some soft radiations can be obtained by introducing the source in the form of a gas mixed directly with the counter gas. For example, low concentrations of ^{14}C are often counted by conversion of the carbon to CO_2 which can then be mixed with a conventional argon–alcohol fill gas. Reasonable properties as a Geiger gas are maintained provided the concentration of CO_2 is not too high. The soft beta particles from tritium can also be efficiently counted using a fill gas consisting of hydrogen in which a small amount of tritiated water vapor has been added.

Because no preamplifier is normally required, a counting system involving Geiger tubes can be as simple as that shown in Fig. 7.7. The series resistance R between the high-voltage supply and the anode of the tube is the load resistance across which the signal voltage is developed. The parallel combination of R with the capacitance of the tube and associated wiring (shown as C_s) determines the time constant of the charge collection circuit. As elaborated in the section on time response, this time constant is normally chosen to be a few microseconds so that only the fast-rising components of the pulse are preserved. Also shown is the coupling capacitor C_c, which is needed to block the tube high voltage but transmit the signal pulse at ground potential to the following circuits. In order to carry out this function without attenuating the pulse amplitude, its value must be large enough so that RC_c is large compared with the pulse duration.

Figure 7.7 Block diagram of the counting electronics normally associated with a G-M tube.

VII. COUNTING EFFICIENCY

A. Charged Particles

Because a single ion pair formed within the fill gas of the Geiger tube can trigger a full Geiger discharge, the counting efficiency for any charged particle that enters the active volume[†] of the tube is essentially 100%. In most practical situations, the effective counting efficiency is therefore determined by the probability that the incident radiation penetrates the window of the tube without absorption or backscattering. With alpha particles, absorption within the window is the major concern and windows with thickness as small as 1.5 mg/cm^2 are commercially available. Thicker windows can generally be used for counting beta particles, but some will be backscattered from the window without penetrating if the window thickness is a significant fraction of the electron range.

B. Neutrons

There are several reasons why Geiger tubes are seldom used to count neutrons. For thermal neutrons, the conventional Geiger gases have low capture cross sections and therefore have an unacceptably low counting efficiency. Gases with a high capture cross section (such as ^{3}He) can be substituted, but then the detector is much more sensibly operated in the proportional region than in the Geiger region. In the proportional counter, neutron-induced events are of much larger amplitude than pulses generated by gamma-ray background and therefore are easily distinguished (see Chapter 14). In the Geiger region, all pulses are of the same amplitude and gamma-ray discrimination is not possible.

Fast neutrons can produce recoil nuclei (see Chapter 15) in a Geiger gas that generate ion pairs and a subsequent discharge. Therefore, Geiger tubes will, to some extent, respond to a fast neutron flux, particularly those that are helium filled. However, gas-filled fast neutron detectors are normally operated as proportional counters rather than as Geiger tubes to take advantage of the spectroscopic information provided only in the proportional region (see Chapter 15).

C. Gamma Rays

In any gas-filled counter, the response to gamma rays of normal energy comes about by way of gamma-ray interactions in the solid wall of the counter. If the interaction takes place close enough to the inner wall surface so that the secondary electron created in the interaction can reach the gas and create ions, a pulse will result. Because only a single ion pair is required, the secondary electron need only barely emerge from the wall near the end of its track in order to generate a pulse from a Geiger tube.

The efficiency for counting gamma rays therefore depends on two separate factors: (1) the probability that the incident gamma ray interacts in the wall and produces a secondary electron, and (2) the probability that the secondary electron reaches the fill gas before the end of its track. As shown in Fig. 7.8, only the innermost layer of the wall near the gas may contribute secondary electrons. This region has a thickness equal to the maximum range of the secondary electrons that are formed, or typically a millimeter or two. Making the wall thicker than this value contributes nothing to the efficiency because electrons formed in regions of the wall farther from the gas have no chance to reach the gas before being stopped.

The probability of gamma-ray interaction within this critical layer generally increases with atomic number of the wall material. Therefore, the gamma-ray efficiency of Geiger

---

[†]A small dead space is often encountered just behind the entrance window of end-window designs where the local electric field may be too small to prevent recombination or electron loss.

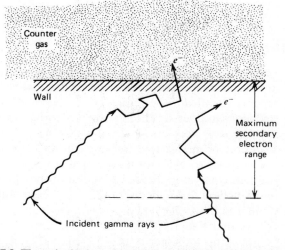

Figure 7.8 The principal mechanism by which gas-filled counters are sensitive to gamma rays involves creation of secondary electrons in the counter wall. Only those interactions that occur within an electron range of the wall surface can result in a pulse.

tubes is highest for those tubes constructed with a cathode wall of high-Z material. G-M tubes with bismuth ($Z = 83$) cathodes have been widely used in the past for gamma-ray detection. As shown in Fig. 7.9, however, the probability of interaction within the sensitive layer remains small even for high-Z materials, and typical gamma-ray counting efficiencies are seldom higher than several percent.

If the photon energy is low enough, direct gamma-ray interaction in the fill gas of the tube may no longer be negligible. Therefore, the counting efficiency for low-energy gamma rays and X-rays can be enhanced by using gases of high atomic number at a pressure as high as possible. Xenon and krypton are common choices in these applications, and

Figure 7.9 The efficiency of G-M tubes for gamma rays normally incident on the cathode. [From W. K. Sinclair, Chap. 5 in *Radiation Dosimetry* (G. J. Hine and G. L. Brownell, eds.). Copyright 1956 by Academic Press. Used with permission.]

counting efficiencies near 100% can be achieved for photon energies below about 10 keV (see Fig. 6.16).

VIII. TIME-TO-FIRST-COUNT METHOD

Because of the long intrinsic dead time of Geiger tubes, they are normally limited to applications in which the counting rate is relatively low. Corrections for dead time losses become significant even at rates of a few hundred counts per second. At rates above several thousand per second, these corrections typically become so large that the true count rate becomes very uncertain and extremely sensitive to small variability in the assumed dead time or model used to represent these losses.

In order to extend the effective counting range of Geiger tubes beyond this normal limit, an innovative mode of operation has been introduced known as the *time-to-first-count method*.[17] As shown in Fig. 7.10, the method is based on switching the high voltage applied to the Geiger tube between two values: the normal operating voltage, and a lower voltage that is set to be less than the minimum voltage needed to develop avalanches in the fill gas. After a Geiger discharge, the voltage is abruptly dropped to the lower value and held there for a fixed "waiting time" (typically 1–2 milliseconds). This waiting time is chosen to be longer than the recovery time of the tube at the reduced applied voltage. During the waiting time, the positive ions generated in the preceding Geiger discharge are swept to the cathode and the tube is restored to its normal or "live" condition. At the end of this period, the voltage is again abruptly switched to the higher operating value. Once this switching has been accomplished, the tube is by definition in a "live" state since all ions have been cleared away. At that time a clock is started, and the elapsed time is then recorded until the next Geiger discharge occurs. The applied voltage is again lowered for the fixed waiting time and the cycle repeated many times.

Through this process, a series of measurements are made of the time between the end of the waiting time and the next occurrence of a true event within the tube. When these time intervals are averaged, one then has a measurement of the average time between an arbitrary point on the time scale and the next true event. In the discussion on p. 98 of Chapter 3, it was shown that such intervals have an average length given by $1/r$, where r is the true event rate. Thus this rate can be determined simply by taking the inverse of the measured average time interval.

Figure 7.10 Illustration of the time-to-first-count method of determining the event rate from a G-M tube.

Note that this technique is totally free of the effects of the detector dead time. Each time measurement period begins with the detector in a live state and ends when the next true event occurs. The precision of the measurement is only limited by the abruptness of the switching of the high voltage that must be used to define time zero and the uncertainty in timing the onset of the Geiger discharge. These uncertainties can be as short as a tenth of a microsecond, so that average time intervals of a few microseconds or more can be measured to reasonable accuracy. This average time interval corresponds to extending the measured counting rate to over 10^5 counts per second, more than two decades higher than allowed by conventional operation of the tube.

There will also be statistical uncertainties associated with this kind of measurement. In general, the number of time intervals measured will play the same role as the number of counts recorded in a conventional counting measurement over a fixed period of time. In this case, a number of counts equal to the number of measured intervals is recorded over a total live time that is equal to the product of the number of intervals multiplied by their average length. In order to achieve a 1% statistical accuracy, therefore, at least 10^4 time intervals need to be measured. However, because the method is most often applied to high counting rate circumstances, recording a large number of intervals will be ensured in a reasonably short total measurement time. In the limit of very high true rates, the rate of measured intervals approaches the reciprocal of the fixed waiting time that is chosen. If this value is 2 ms, then the rate of data acquisition approaches 500/s.

IX. G-M SURVEY METERS

A common type of survey meter used in gamma-ray monitoring consists of a portable Geiger tube, high-voltage supply, and pulse counting rate meter. The pulse rate is then taken as an indication of the intensity of the gamma-ray exposure. The count rate meter scales are often calibrated in terms of exposure rate units, but under some circumstances these readings can be in error by a factor of 2 or 3 or more.

The difficulty arises from the fact that the count rate from a Geiger tube, in contrast to the measured current from an ion chamber, bears no fundamental relation to the gamma-ray exposure rate. For a given gamma-ray energy, the count rate obviously scales linearly with the intensity, but applications are likely to involve gamma rays of many different and variable energies. The exposure rate scale may be calibrated accurately at any fixed gamma-ray energy, but if the survey meter is applied to measurements involving other gamma-ray energies, the variation of the counter efficiency with energy must be considered. Plots of this type are given in Fig. 7.9 for different cathode materials. Ideally, one would like an efficiency versus energy curve that exactly matches a plot of exposure per gamma-ray photon versus energy. If this match were exact, then the G-M survey meter could generally be applied to all energies because the importance of each photon would be weighted correctly by the inherent efficiency of the detector. In an effort to make the match as close as possible, the tube may be covered with a thin external layer of metal such as lead and/or tin,[18] providing what is often called *energy compensation* of the tube. A bare G-M tube will tend to have too high an efficiency at low energies compared with higher energies for a good match to the exposure versus energy curve, and the preferential absorption of the metal filter at lower energies provides something of an empirical compensation. The ratio of the two curves at any given energy is a measure of the correction factor that should be applied to the survey meter readings. A plot of this correction factor versus gamma-ray energy for two commercial survey meters is shown in Fig. 7.11.

Figure 7.11 The sensitivity versus energy of two commercially manufactured Geiger survey meters. Sensitivity is defined as the indicated exposure rate divided by the true exposure rate. (From Storm et al.[19])

PROBLEMS

7.1 Why must the quench gas in a Geiger–Mueller tube have an ionization potential below that of the major fill gas?

7.2 Predict the effect each of the following changes would have on the observed starting voltage for a Geiger–Mueller tube if all other parameters remain the same:

  (a) Doubling the diameter of the anode wire.

  (b) Doubling the fill gas pressure.

  (c) Doubling the trace concentration of the quench gas.

7.3 Why does the pulse height from a Geiger tube continue to increase with applied voltage even after a full Geiger discharge is obtained?

7.4 Estimate the voltage at which the counter of Problem 6.3 enters the Geiger region of operation. Assume that three excited atoms are formed for every ion pair in a typical avalanche and that the subsequent de-excitation photons have a probability of 10^{-5} of creating an additional avalanche.

7.5 Both the proportional counter and Geiger tube are based on internal gas multiplication. Comment on each separately and contrast their behavior with regard to:

  (a) Variation of pulse height with applied voltage.

  (b) The need for a quench gas and its function.

  (c) Ability to differentiate heavy charged particle and electron radiations.

  (d) Ability to register high counting rates.

  (e) Typical counting efficiency for 1 MeV gamma rays.

7.6 The intrinsic detection efficiency for a gas-filled counter (proportional counter or G-M tube) when used with medium-energy gamma rays (say 1 MeV) will depend on the counter wall thickness in the following way:

  (a) Explain the general shape of this curve.

  (b) Give an order-of-magnitude estimate for the optimum wall thickness t_m and relate it to basic physical properties.

7.7 A G-M tube is operated with a counting system whose threshold requires that a full Geiger discharge must take place to register a count. Is its dead time behavior likely to be better described by the paralyzable or nonparalyzable model?

7.8 In a given counter gas operated at a pressure of 0.5 atm, the mobility of a free electron is 1.5×10^{-4} (m/s) $\cdot$ (m/V) $\cdot$ atm. The threshold electric field for the onset of avalanche formation is 2×10^6 V/m. If this gas is used in a cylindrical tube with anode radius of 0.005 cm and cathode radius of 2 cm, calculate the drift time of an electron from the cathode to the multiplying region for an applied voltage of 1500 V.

7.9 A Geiger–Mueller counting system has a resolving time of 350 μs. Find the true rate at which the losses due to this dead time amount to half of all the true events (one could argue that this rate is roughly the maximum rate at which the tube can give reasonably reliable results). Use both the paralyzable and nonparalyzable models to obtain two different predictions that probably bracket the actual value.

7.10 The G-M tube of Problem 7.9 is operated in "time-to-first-count" mode to extend its upper counting rate limit. The voltage to the tube is suppressed for a period of 900 μs following each recorded count to allow its full recovery. Assume that the time required to switch the high voltage on is 0.3 μs, and that the onset of the Geiger discharge can be measured to a much smaller timing uncertainty.

(a) The limit on counting rate now is reached when the probability that a true event occurs during the switching time is no longer small. Calculate the true rate for which this probability reaches 5%.

(b) At this limiting rate, what is the mean time to first count following switching of the voltage?

(c) A survey instrument is based on operating in this mode over a 2-s period to measure the true rate. At the same limiting rate, what is the percent uncertainty in this measurement due to counting statistics?

REFERENCES

1. D. H. Wilkinson, *Ionization Chambers and Counters*, Cambridge University Press, London, 1950.
2. W. J. Price, *Nuclear Radiation Detection*, 2nd ed., Chap. 5, McGraw-Hill, New York, 1964.
3. J. Sharpe, *Nuclear Radiation Detectors*, 2nd ed., Methuen and Co., London, 1964.
4. E. W. Emery, "Geiger-Mueller and Proportional Counters," in *Radiation Dosimetry*, Vol. II (F. H. Attix and W. C. Roesch, eds.), Academic Press, New York, 1966.
5. D. H. Wilkinson, *Nucl. Instrum. Meth.* **A321,** 195 (1992).
6. D. H. Wilkinson, *Nucl. Instrum. Meth.* **A383,** 516 (1996).
7. D. H. Wilkinson, *Nucl. Instrum. Meth.* **A383,** 523 (1996).
8. K. Matsuda and J. Sanada, *Nucl. Instrum. Meth.* **A294,** 268 (1990).
9. J. Fischer, H. Okuno, and A. H. Walenta, *IEEE Trans. Nucl. Sci.* **NS-25**(1), 794 (1978).
10. E. Mathieson, T. J. Harris, and G. C. Smith, *Nature* **272**, 709 (1978).
11. F. E. Lox and Z. Eeckhaut, *Rev. Sci. Instrum.* **40**(9), 1206 (1969).
12. A. Peeva and T. Karatoteva, *Nucl. Instrum. Meth.* **118**, 49 (1974).
13. S. Usacev and M. Seman, *Nucl. Instrum. Meth.* **73**, 41 (1969).
14. D. Srdoc, *Nucl. Instrum. Meth.* **99**, 321 (1972).
15. Y. Fujita, Y. Taguchi, M. Imamura, T. Inoue, and S. Tanaka, *Nucl. Instrum. Meth.* **128**, 523 (1975).
16. T. Aoyama and T. Watanabe, *Nucl. Instrum. Meth.* **197**, 357 (1982).
17. E. J. DiIanni et al., "Radiation Measuring Apparatus," U.S. Patent 4,605,859, issued Aug. 12, 1986.
18. D. Barclay, *IEEE Trans. Nucl. Sci.* **33(1),** 613 (1986).
19. E. Storm, D. W. Lier, and H. Israel, *Health Phys.* **26**, 179 (1974).

Chapter **8**

Scintillation Detector Principles

The detection of ionizing radiation by the scintillation light produced in certain materials is one of the oldest techniques on record. The scintillation process remains one of the most useful methods available for the detection and spectroscopy of a wide assortment of radiations. In this chapter we discuss the various types of scintillators available and the important considerations in the efficient collection of the scintillation light. The following chapters cover modern light sensors—photomultiplier tubes and photodiodes—required to convert the light into an electrical pulse, and the application of scintillation detectors in radiation spectroscopy.

The ideal scintillation material should possess the following properties:

1. It should convert the kinetic energy of charged particles into detectable light with a high scintillation efficiency.
2. This conversion should be linear—the light yield should be proportional to deposited energy over as wide a range as possible.
3. The medium should be transparent to the wavelength of its own emission for good light collection.
4. The decay time of the induced luminescence should be short so that fast signal pulses can be generated.
5. The material should be of good optical quality and subject to manufacture in sizes large enough to be of interest as a practical detector.
6. Its index of refraction should be near that of glass (~1.5) to permit efficient coupling of the scintillation light to a photomultiplier tube or other light sensor.

No material simultaneously meets all these criteria, and the choice of a particular scintillator is always a compromise among these and other factors. The most widely applied scintillators include the inorganic alkali halide crystals, of which sodium iodide is the favorite, and organic-based liquids and plastics. The inorganics tend to have the best light output and linearity, but with several exceptions are relatively slow in their response time. Organic scintillators are generally faster but yield less light. The intended application also has a major influence on scintillator choice. The high Z-value of the constituents and high density of inorganic crystals favor their choice for gamma-ray spectroscopy, whereas organics are often preferred for beta spectroscopy and fast neutron detection (because of their hydrogen content).

The process of *fluorescence* is the prompt emission of visible radiation from a substance following its excitation by some means. It is conventional to distinguish several other processes that can also lead to the emission of visible light. *Phosphorescence* corresponds to the emission of longer wavelength light than fluorescence, and with a characteristic time that is generally much slower. *Delayed fluorescence* results in the same emission spectrum as prompt fluorescence but again is characterized by a much longer emission time following excitation. To be a good scintillator, a material should convert as large a fraction as possible of the incident radiation energy to prompt fluorescence, while minimizing the generally undesirable contributions for phosphorescence and delayed fluorescence. Most of the emphasis in this chapter is on the pulse mode operation of scintillators. Here the light that can contribute to an output pulse is generally limited to the prompt fluorescence because the time constants of the measurement circuit are set much smaller than typical phosphorescence and delayed fluorescence decay times. This long-lived light is then spread out more-or-less randomly in time between signal pulses and arrives at the light sensor as individual photons that often may not be distinguished from random noise. In contrast, scintillators that are operated in current mode under constant illumination will produce a steady-state signal current that is proportional to the total light yield, and all the decay components will contribute in proportion to their absolute intensity. For this reason, the light yield measured from a scintillator operated in pulse mode may appear to be lower than that deduced from the steady-state current recorded from the same scintillator. Current mode scintillation detectors operated under conditions in which the radiation intensity changes rapidly will suffer from memory or "afterglow" effects if long-lived decay components are significant.

A number of general reviews of the basic theory and application of scintillators have been published, with the comprehensive book by Birks[1] as an outstanding example. A useful collection of fundamental papers on scintillators is available[2] that covers the period from the early investigation of many scintillation materials through recent reviews of scintillation mechanisms. In this chapter, we limit the discussions to those processes that are necessary to understand the differences in behavior of various types of scintillators, together with some of their important properties as practical radiation detectors.

I. ORGANIC SCINTILLATORS

A. Scintillation Mechanism in Organics

The fluorescence process in organics arises from transitions in the energy level structure of a single molecule and therefore can be observed from a given molecular species independent of its physical state. For example, anthracene is observed to fluoresce as either a solid polycrystalline material, as a vapor, or as part of a multicomponent solution. This behavior is in marked contrast to crystalline inorganic scintillators such as sodium iodide, which require a regular crystalline lattice as a basis for the scintillation process.

A large category of practical organic scintillators is based on organic molecules with certain symmetry properties which give rise to what is known as a π-electron structure. The π-electronic energy levels of such a molecule are illustrated in Fig. 8.1. Energy can be absorbed by exciting the electron configuration into any one of a number of excited states. A series of singlet states (spin 0) are labeled as $S_0, S_1, S_2, \ldots$ in the figure. A similar set of triplet (spin 1) electronic levels are also shown as $T_1, T_2, T_3, \ldots$ For molecules of interest as organic scintillators, the energy spacing between S_0 and S_1 is 3 or 4 eV, whereas spacing between higher-lying states is usually somewhat smaller. Each of these electronic configurations is further subdivided into a series of levels with much finer spacing that correspond to various vibrational states of the molecule. Typical spacing of these levels is of the order of 0.15 eV. A second subscript is often added to distinguish these vibrational states, and the symbol S_{00} represents the lowest vibrational state of the ground electronic state.

Because the spacing between vibrational states is large compared with average thermal energies (0.025 eV), nearly all molecules at room temperature are in the S_{00} state. In Fig. 8.1 the absorption of energy by the molecule is represented by the arrows pointing upward. In the case of a scintillator, these processes represent the absorption of kinetic

Figure 8.1 Energy levels of an organic molecule with π-electron structure. (From J. B. Birks, *The Theory and Practice of Scintillation Counting.* Copyright 1964 by Pergamon Press, Ltd. Used with permission.)

energy from a charged particle passing nearby. The higher singlet electronic states that are excited are quickly (on the order of picoseconds) de-excited to the S_1 electron state through radiationless internal conversion. Furthermore, any state with excess vibrational energy (such as S_{11} or S_{12}) is not in thermal equilibrium with its neighbors and again quickly loses that vibrational energy. Therefore, the net effect of the excitation process in a simple organic crystal is to produce, after a negligibly short time period, a population of excited molecules in the S_{10} state.

The principal scintillation light (or prompt fluorescence) is emitted in transitions between this S_{10} state and one of the vibrational states of the ground electronic state. These transitions are indicated by the downward arrows in Fig. 8.1. If τ represents the fluorescence decay time for the S_{10} level, then the prompt fluorescence intensity at a time t following excitation should simply be

$$I = I_0 e^{-t/\tau} \tag{8.1}$$

In most organic scintillators, τ is a few nanoseconds, and the prompt scintillation component is therefore relatively fast.

The lifetime for the first triplet state T_1 is characteristically much longer than that of the singlet state S_1. Through a transition called *intersystem crossing*, some excited singlet states may be converted into triplet states. The lifetime of T_1 may be as much as 10^{-3} s and the radiation emitted in a de-excitation from T_1 to S_0 is therefore a delayed light emission characterized as phosphorescence. Because T_1 lies below S_1, the wavelength of this phosphorescence spectrum will be longer than that for the fluorescence spectrum. While in the T_1 state, some molecules may be thermally excited back to the S_1 state and subsequently decay through normal fluorescence. This process represents the origin of the delayed fluorescence sometimes observed for organics.

Figure 8.1 can also be used to explain why organic scintillators can be transparent to their own fluorescence emission. The length of the upward arrows corresponds to those photon energies that will be strongly absorbed in the material. Because all the fluorescence transitions represented by the downward arrows (with the exception of $S_{10} - S_{00}$) have a lower energy than the minimum required for excitation, there is very little overlap between the optical absorption and emission spectra (often called the *Stokes shift*), and consequently little self-absorption of the fluorescence. An example of these spectra for a typical organic scintillator is given in Fig. 8.2.

The *scintillation efficiency* of any scintillator is defined as the fraction of all incident particle energy which is converted into visible light. One would always prefer this efficiency to be as large as possible, but unfortunately there are alternate de-excitation modes available to the excited molecules that do not involve the emission of light and in which the excitation is degraded mainly to heat. All such radiationless de-excitation processes are grouped together under the term *quenching*. In the fabrication and use of organic scintillators, it is often important to eliminate impurities (such as dissolved oxygen in liquid scintillators), which degrade the light output by providing alternate quenching mechanisms for the excitation energy.

In almost all organic materials, the excitation energy undergoes substantial transfer from molecule to molecule before de-excitation occurs. This energy transfer process is especially important for the large category of organic scintillators that involves more than one species of molecules. If a small concentration of an efficient scintillator is added to a bulk solvent, the energy that is absorbed, primarily by the solvent, can eventually find its way to one of the efficient scintillation molecules and cause light emission at that point. These "binary" organic scintillators are widely used both as liquid and plastic solutions incorporating a variety of solvents and dissolved organic scintillants.

A third component is sometimes added to these mixtures to serve as a "waveshifter." Its function is to absorb the light produced by the primary scintillant and reradiate it at a longer wavelength. This shift in the emission spectrum can be useful for closer matching to the spectral sensitivity of a photomultiplier tube or to minimize bulk self-absorption in large liquid or plastic scintillators. Birks and Pringle[3] have reviewed the energy transfer mechanisms in binary and tertiary organic mixtures together with their influence on scintillation efficiency and pulse timing characteristics.

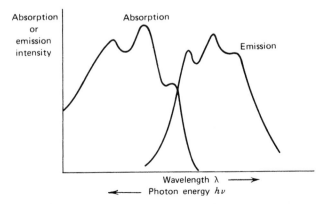

Figure 8.2 The optical absorption and emission spectra for a typical organic scintillator with the level structure shown in Fig. 8.1.

B. Types of Organic Scintillators

1. PURE ORGANIC CRYSTALS

Only two materials have achieved widespread popularity as pure organic crystalline scintillators. Anthracene is one of the oldest organic materials used for scintillation purposes and holds the distinction of having the highest scintillation efficiency (or greatest light output per unit energy) of any organic scintillator. Stilbene has a lower scintillation efficiency but is preferred in those situations in which pulse shape discrimination is to be used to distinguish between scintillations induced by charged particles and electrons (see the later discussion of this technique). Both materials are relatively fragile and difficult to obtain in large sizes. Also, the scintillation efficiency is known to depend on the orientation of an ionizing particle with respect to the crystal axis.[4] This directional variation, which can be as much as 20–30%, spoils the energy resolution obtainable with these crystals if the incident radiation will produce tracks in a variety of directions within the crystal.

2. LIQUID ORGANIC SOLUTIONS

A category of useful scintillators is produced by dissolving an organic scintillator in an appropriate solvent. Liquid scintillators can consist simply of these two components, or a third constituent is sometimes added as a wavelength shifter to tailor the emission spectrum to better match the spectral response of common photomultiplier tubes.

Liquid scintillators are often sold commercially in sealed glass containers and are handled in the same manner as solid scintillators. In certain applications, large-volume detectors with dimensions of several meters may be required. In these cases the liquid scintillator is often the only practical choice from a cost standpoint. In many liquids, the presence of dissolved oxygen can serve as a strong quenching agent and can lead to substantially reduced fluorescence efficiency. It is then necessary for the solution to be sealed in a closed volume from which most of the oxygen has been purged.

Because of their lack of a solid structure that could be damaged by exposure to intense radiation, liquid scintillators are expected to be more resistant to radiation damage effects than crystalline or plastic scintillators. This expectation is borne out by measurements,[5] and reasonable resistance to change up to exposures as high as 10^5 Gy has been demonstrated for some liquids.

Liquid scintillators are also widely applied to count radioactive material that can be dissolved as part of the scintillator solution. In this case, all radiations emitted by the source immediately pass through some portion of the scintillator and the counting efficiency can be almost 100%. The technique is widely used for counting low-level beta activity such as that from carbon-14 or tritium. This large field of *liquid scintillation counting* is described further in Chapter 10.

3. PLASTIC SCINTILLATORS

If an organic scintillator is dissolved in a solvent that can then be subsequently polymerized, the equivalent of a solid solution can be produced. A common example is a solvent consisting of styrene monomer in which an appropriate organic scintillator is dissolved. The styrene is then polymerized to form a solid plastic. Other plastic matrices can consist of polyvinyltoluene or polymethylmethacrylate. Because of the ease with which they can be shaped and fabricated, plastics have become an extremely useful form of organic scintillator.

Plastic scintillators are available commercially with a good selection of standard sizes of rods, cylinders, and flat sheets. Because the material is relatively inexpensive, plastics are often the only practical choice if large-volume solid scintillators are needed. In these cases the self-absorption of the scintillator light may no longer be negligible, and some attention should be given to the attenuation properties of the material. The distance in which the light intensity will be attenuated by a factor of 2 can be as much as several meters, although much smaller attenuation lengths are observed for some plastics.[6]

There is also a wide selection of plastic scintillators available as small-diameter fibers. Used either as single fibers or grouped together as bundles or ribbons, these scintillators lend themselves to applications in which the position of particle interactions must be sensed with good spatial resolution. A detailed discussion of the principles and properties of fiber scintillators is given beginning on p. 253 later in this chapter.

Because of the widespread application of plastic scintillators in particle physics measurements, where they may be exposed to sustained high levels of radiation, considerable attention[7–10] has been paid to the degradation in the scintillation output of plastics due to radiation damage. This process is a complicated one, and many variables such as the dose rate, the presence or absence of oxygen, and the nature of the radiation play important roles. There is also a tendency to observe some recovery or annealing of the damage over periods of time that may be hours or days following an exposure. In typical plastic scintillators, significant degradation in light yield is observed for cumulative gamma-ray exposures in the range of 10^3 or 10^4 Gy, whereas other radiation-resistant formulations[11] show little decrease in light output with doses as high as 10^5 Gy. The changes in measured light can consist either of a decreased light output caused by damage to the fluorescent component or a decrease in the light transmission caused by the creation of optical absorption centers.

4. THIN FILM SCINTILLATORS

Very thin films of plastic scintillator play a unique role in the field of radiation detectors. Because ultrathin films with a thickness as low as 20 μg/cm^2 can be fabricated, it is easy to provide a detector that is thin compared with the range of even weakly penetrating particles such as heavy ions. These films thus serve as transmission detectors, which respond to only the fraction of energy lost by the particle as it passes through the detector. The thickness can be as much as one or two orders of magnitude smaller than the minimum possible with other detector configurations, such as totally depleted silicon surface barriers. The films are available commercially[12] with thicknesses down to approximately 10 μm. Even thinner films can be produced by the user through techniques such as evaporation from a solution of plastic scintillator[13–16] or through the spin coating process.[17,18] The film can be deposited directly on the face of a photomultiplier tube,[14,19] or the light can be collected indirectly through a transparent light pipe in contact with the edges of the film.[20,21] Alternatively, the film can be placed within a reflecting cavity.[20,22–25] The response of these films does not follow directly from the expected energy loss of ions in the detector and is a more complex function of the ion velocity and atomic number.[16,19,26–29] The light yield per unit energy loss increases with decreasing atomic number of the ion, so that thin films can be useful transmission detectors for protons or alpha particles[15,22] even when the energy deposited is relatively small. In common with other organic scintillators, thin film detectors show scintillation decay times of only several ns, and they have proved very useful in fast timing measurements.[30]

5. LOADED ORGANIC SCINTILLATORS

Organic scintillators as a category are generally useful for the direct detection of beta particles (fast electrons) or alpha particles (positive ions). They also are readily adaptable to the detection of fast neutrons through the proton recoil process (see Chapter 15). Because of the low Z-value of their constituents (hydrogen, carbon, and oxygen), however, there is virtually no photoelectric cross section for gamma rays of typical energies. As a result, typical organic scintillators show no photopeak and will give rise only to a Compton continuum in their gamma-ray pulse height spectrum.

To provide some degree of photoelectric conversion of gamma rays, attempts have been made to add high-Z elements to organic scintillators. The most common form is the addition of lead or tin to common plastic scintillators up to a concentration of 10% by weight. It has also been demonstrated[31] that tin can be added to liquid organic scintillator solutions in concentrations of up to 54% by weight while retaining a weak scintillation light output. At low gamma-ray energies, the photopeak efficiency of these materials can be made relatively high,[32] and they have the additional advantages of fast response and low cost compared with more conventional gamma-ray scintillators.[33,34] Unfortunately, the addition of these high-Z elements inevitably leads to a decreased light output, and the energy resolution that can be achieved is therefore considerably inferior to that of inorganic scintillators.

Other examples of loading organic scintillators arise in connection with neutron detection. Liquid or plastic scintillators can be seeded with one of the elements with a high cross section for neutrons such as boron, lithium, or gadolinium. The secondary charged particles and/or gamma rays produced by netron-induced reactions may then be detected directly in the scintillator to provide an output signal. Scintillators of this type are discussed in further detail in Chapter 14.

C. Response of Organic Scintillators

An overall compilation of the properties of organic scintillators is given in Table 8.1.

1. LIGHT OUTPUT

A small fraction of the kinetic energy lost by a charged particle in a scintillator is converted into fluorescent energy. The remainder is dissipated nonradiatively, primarily in the form of lattice vibrations or heat. The fraction of the particle energy that is converted (the scintillation efficiency) depends on both the particle type and its energy. In some cases, the scintillation efficiency may be independent of energy, leading to a linear dependence of light yield on initial energy.

For organic scintillators such as anthracene, stilbene, and many of the commercially available liquid and plastic scintillators, the response to electrons is linear for particle energies above about 125 keV.[35] The response to heavy charged particles such as protons or alpha particles is always less for equivalent energies and is nonlinear to much higher initial energies. As an example, Fig. 8.3 shows the scintillation response of a typical plastic

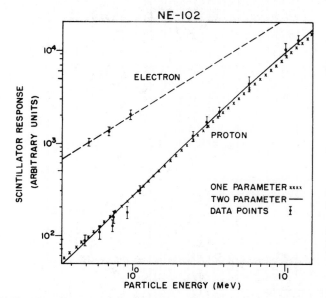

Figure 8.3 The scintillation light yield for a common plastic scintillator (NE 102) when excited by electrons and protons. The data are fit by curves from Eq. (8.3) (one parameter) and Eq. (8.9) (two parameter). (From Craun and Smith.[36])

Table 8.1 Properties of Some Commercially Available Organic Scintillators

Eljen	Bicron	Light Output %Anthracene*	Wavelength of Max Emission (nm)	Decay Constant (ns)	Attenuation Length (cm)	Refractive Index	H/C Ratio	Density	Loading Element % by weight or dist. feature	Softening or Flash Point (°C)	Uses
Crystal											
Anthracene		100	447	30		1.62	0.715	1.25		217	
Stilbene		50	410	4.5		1.626	0.858	1.16		125	
Plastic											
EJ-212	BC-400	65	423	2.4	250	1.581	1.103	1.032		70	General purpose
EJ-204	BC-404	68	408	1.8	160	1.58	1.107	1.032	1.8 ns time constant	70	Fast counting
EJ-200	BC-408	64	425	2.1	380	1.58	1.104	1.032		70	TOF counters, large area
EJ-208	BC-412	60	434	3.3	400	1.58	1.104	1.032	Longest attn. length	70	General purpose, large area, long strips
	BC-420	64	391	1.5	110	1.58	1.100	1.032	1.5 ns time constant	70	Ultrafast timing, sheet areas
EJ-232	BC-422	55	370	1.4	8	1.58	1.102	1.032	1.4 ns time constant	70	Very fast timing, small sizes
	BC-422Q	11	370	0.7	< 8	1.58	1.102	1.032	Benzophenone, 1%	70	Ultrafast timing, ultrafast counting
	BC-428	36	480	12.5	150	1.58	1.103	1.032	Green emitter	70	Photodiodes and CCDs; phoswich detectors
	BC-430	45	580	16.8	NA	1.58	1.108	1.032	Red emitter	70	Silicon photodiodes and red-enhanced PMTs
EJ-248	BC-434	60	425	2.2	350	1.59	0.995	1.049	High temp	100	General purpose
	BC-436	52	425	2.2	NA	1.61	0.960 D:C	1.130	Deuterium, 13.8%	90	Thin disks
EJ-240	BC-444	41	428	285	180	1.58	1.109	1.032		70	Phoswich detectors for dE/dx studies
EJ-256	BC-452	32	424	2.1	150	1.58	1.134	1.080	Lead, 5%	60	X-ray dosimetry (< 100 keV)
	BC-454	48	425	2.2	120	1.58	1.169	1.026	Boron, 5%	60	Neutron spectrometry, thermal neutrons
EJ-252	BC-470	46	423	2.4	200	1.58	1.098	1.037	Air equivalent	65	Dosimetry
	BC-490	55	425	2.3		1.58	1.107	1.030	Casting resin	70	General purpose
	BC-498	65	423	2.4		1.58	1.103	1.032	Applied like paint	70	β, γ detection
Liquid											
EJ-301	BC-501A	78	425	3.2			1.212	0.874	Pulse shape discrim.	26	γ > 100 keV, fast n spectroscopy
EJ-305	BC-505	80	425	2.5			1.331	0.877	High light output	47	γ, fast n, large volume
EJ-313	BC-509	20	425	3.1		1.61	0.0035	1.61	F	10	γ, fast n
EJ-321H	BC-517H	52	425	2.0			1.89	0.86	Mineral oil-based	81	γ fast n, cosmic, charged particles
	BC-517P	28	425	2.2			2.05	0.85	Mineral oil-based	115	γ, fast n, cosmic, charged particles
EJ-325	BC-519	60	425	4.0			1.73	0.875	Pulse shape discrim.	74	γ, fast n, n-γ discrimination
EJ-331	BC-521	60	425	4.0			1.31	0.89	Gd (to 1%)	44	Neutron spectroscopy, neutrino research
EJ-339	BC-523A	65	425	3.7			1.67	0.93	Enriched ^{10}B	1	Total absorption neutron spectrometry
EJ-335	BC-525	56	425	3.8			1.57	0.88	Gd (to 1%)	64	Neutron spectrometry, neutrino research
	BC-533	51	425	3.0			1.96	0.8	Low temp operation	65	γ, fast n, cosmic
	BC-537	61	425	2.8			.99 (D:C)	0.954	^{2}H	−11	Fast n, pulse shape discrimination
	BC-551	40	425	2.2			1.31	0.902	Pb (5% w/w)	44	γ, X-rays < 200 keV
	BC-553	34	425	3.8			1.47	0.951	Sn (10% w/w)	42	γ, X-rays

*NaI(Tl) is 230% on this scale

scintillator. At energies of a few hundred keV, the response to protons is smaller by a factor of 10 compared with the light yield of equivalent energy electrons. At higher energies the discrepancy is less, but the proton response is always below the electron response.

Because of the dependence of the light yield of organics on the type of particle, a special nomenclature is sometimes used to describe the absolute light yield. The term *MeV electron equivalent (MeVee)* is introduced to place the light yield on an absolute basis. The particle energy required to generate 1 MeVee of light by definition is 1 MeV for fast electrons but is several MeV for heavy charged particles because of their reduced light yield per unit energy.

The response of organic scintillators to charged particles can best be described by a relation between dL/dx, the fluorescent energy emitted per unit path length, and dE/dx, the specific energy loss for the charged particle. A widely used relation first suggested by Birks[1] is based on the assumption that a high ionization density along the track of the particle leads to quenching from damaged molecules and a lowering of the scintillation efficiency. If we assume that the density of damaged molecules along the wake of the particle is directly proportional to the ionization density, we can represent their density by $B(dE/dx)$, where B is a proportionality constant. Birks assumes that some fraction k of these will lead to quenching. A further assumption is that, in the absence of quenching, the light yield is proportional to energy loss:

$$\frac{dL}{dx} = S\frac{dE}{dx} \tag{8.2}$$

where S is the normal scintillation efficiency. To account for the probability of quenching, Birks then writes

$$\frac{dL}{dx} = \frac{S\dfrac{dE}{dx}}{1 + kB\dfrac{dE}{dx}} \tag{8.3}$$

Equation (8.3) is commonly referred to as *Birks' formula.* As a practical matter the product kB is treated as an adjustable parameter to fit experimental data for a specific scintillator.[37] In many cases, this single adjustable parameter can give very good fits to the shape of experimental data, with the value of S providing the absolute normalization.

When excited by fast electrons (either directly or from gamma-ray irradiation), dE/dx is small for sufficiently large values of E and Birks' formula then predicts

$$\left.\frac{dL}{dx}\right|_e = S\frac{dE}{dx} \tag{8.4}$$

or the incremental light output per unit energy loss is a constant

$$\left.\frac{dL}{dE}\right|_e = S \tag{8.5}$$

This is the regime in which the light output

$$L \equiv \int_0^E \frac{dL}{dE}\,dE = SE \tag{8.6}$$

is linearly related to the initial particle energy E.

On the other hand, for an alpha particle, dE/dx is very large so that saturation occurs along the track and Birks' formula becomes

$$\left.\frac{dL}{dx}\right|_\alpha = \frac{S}{kB} \tag{8.7}$$

The appropriate value of kB can therefore be determined by taking the ratio of these two responses:

$$kB = \left.\frac{dL}{dE}\right|_e \left/ \left.\frac{dL}{dx}\right|_\alpha \right. \tag{8.8}$$

In order to match experimental data more closely, other formulas for dL/dx have been proposed by a number of authors. These are in effect semiempirical formulas that introduce one or more additional fitting parameters. An extensive analysis of the response of a number of organic scintillators has been carried out by Craun and Smith.[36] Their analysis is based largely on the data of Smith et al.[38] and uses an extended version of Birks' formula

$$\frac{dL}{dx} = \frac{S\dfrac{dE}{dx}}{1 + kB\dfrac{dE}{dx} + C\left(\dfrac{dE}{dx}\right)^2} \tag{8.9}$$

where C is again treated as an empirically fitted parameter. This expression approaches the simple Birks' formula Eq. (8.3) for small values of dE/dx. Parameters for Eq. (8.9) are derived and listed in Ref. 36 for a variety of organic scintillators and exciting particles. Also given is a useful table of dE/dx values for organic scintillators of different composition for protons and deuterons.

The *alpha-to-beta ratio* is a widely used parameter to describe the difference of light output for an organic scintillator for electrons and charged particles of the same energy. The light yield for electrons is always higher than that for a charged particle of the same kinetic energy, and therefore the alpha-to-beta ratio is always less than 1. This ratio will depend on the energy at which the comparison is made, and no fixed value is applicable over the entire energy range. Measurements have been reported by Czirr[39] for a variety of organic scintillators.

Figure 8.4 The light output (expressed in equivalent electron energy deposition) versus proton energy for liquid scintillator NE 213. (From Maier and Nitschke.[40])

The light output of NE 213 liquid scintillator as measured by Maier and Nitschke[40] is shown in Fig. 8.4. Typical of many organic scintillators, the light output can be represented as proportional to $E^{3/2}$, for energies below about 5 MeV, and becomes approximately linear for higher energies.

Clark[41] has compiled a set of data on the absolute scintillation efficiency [S in Eq. (8.5)] of plastic scintillators for gamma-ray excitation. These data are a useful supplement to older tabulations which can be found in Ref. 1. Clark points out the variability of different measurements attempting to determine absolute efficiencies, which in some cases can be as much as a factor of 2 discrepant. This variability can be traced in part to differences in the purity and past history of the scintillation material, and absolute values should always be used with caution. In some organics, the partial overlap of the emission and absorption spectra leads to a size dependence of the apparent efficiency for scintillation. Also, the anisotropic response of anthracene and stilbene further complicates efficiency measurements.

Prolonged exposure to ionizing radiation leads to a general deterioration of the properties of most organic scintillators (for example, see p. 224). Plastic scintillators exposed to light and oxygen have also been shown to undergo a long-term deterioration due to polymer degradation.[42] In addition to internal effects, the surface of plastics can often become crazed on exposure to extreme environments. The surface crazing leads to a substantial drop in observed light output from large scintillators because of the decreased efficiency of internal light reflection.

2. TIME RESPONSE

If it can be assumed that the luminescent states in an organic molecule are formed instantaneously and only prompt fluorescence is observed, then the time profile of the light pulse should be a very fast leading edge followed by a simple exponential decay [Eq. (8.1)]. Much of the published literature quotes a *decay time* that characterizes the prompt scintillation yield from various organic materials. Although this simple representation is often adequate for many descriptions of the scintillator behavior, a more detailed model of the time dependence of the scintillation yield must take into account two other effects: the finite time required to populate the luminescent states, and the slower components of the scintillation corresponding to delayed fluorescence and phosphorescence.

Times of approximately half a nanosecond are required to populate the levels from which the prompt fluorescence light arises. For the very fast scintillators, the decay time from these levels is only three or four times greater, and a full description of the expected pulse shape must take into account the finite rise time as well. One approach[43, 44] assumes that the population of the optical levels is also exponential and that the overall shape of the light pulse is given by

$$I = I_0(e^{-t/\tau} - e^{-t/\tau_1}) \tag{8.10}$$

where τ_1 is the time constant describing the population of the optical levels and τ is the time constant describing their decay. Values for these parameters for several plastic scintillators are given in Table 8.2. Other observations[45] have concluded that the population step is better represented by a Gaussian function $f(t)$ characterized by a standard deviation σ_{ET}. The overall light versus time profile is then described by

$$\frac{I}{I_0} = f(t)e^{-t/\tau} \tag{8.11}$$

Best fit values for σ_{ET} are also shown in Table 8.2. Experimentally, the rise and fall of the light output can be characterized by the full width at half maximum (FWHM) of the resulting light versus time profile, which can be measured using very fast timing procedures. It

Table 8.2 Some Timing Properties of Fast Plastic Scintillators

	Parameters for Eq. (8.10)		Parameters for Eq. (8.11)		Measured FWHM
	τ_1 (rise)	τ (decay)	σ_{ET}	τ	
NE 111	0.2 ns	1.7 ns	0.2 ns	1.7 ns	1.54 ns
Naton 136	0.4 ns	1.6 ns	0.5 ns	1.87 ns	2.3 ns
NE 102A	0.6 ns	2.4 ns	0.7 ns	2.4 ns	3.3 ns

Data from Bengtson and Moszynski.[45]

has become common to specify the performance of ultrafast organic scintillators by their FWHM time rather than the decay time alone.

Studies of the time response of organic scintillators have been reported in Refs. 46–49. In these measurements the emphasis is on conventional plastic scintillators, with the smallest FWHM reported to be 1.3 ns for NE 111. Lynch[44] has also reported results on light output and time response of a variety of liquid scintillators. Some attention[50–52] has focused on plastics to which a quenching agent has been deliberately added. Although the light output is reduced by an order of magnitude or more, the FWHM can be as small as a few hundred picoseconds. For these very fast scintillators, effects other than the scintillation mechanism can sometimes affect the observed time response. Among these is the finite flight time of the photons from the point of scintillation to the photomultiplier tube. Particularly in large scintillators, transit time fluctuation due to multiple light reflections at scintillator surfaces can amount to a sizable spread in the arrival time of the light at the photomultiplier tube photocathode.[53–55] Also, there is evidence[56,57] that self-absorption and reemission of the fluorescence plays an important role in causing an apparent worsening of the time resolution as the dimensions of a scintillator are increased.

Although fast decay time is an advantage in nearly all applications of scintillators, there is at least one application in which a slow decay is needed. In the phoswich detector (see p. 344) two different scintillation materials are employed, one of which must have an appreciably slower decay time than the other. For such applications, a plastic scintillator has been developed[58] with an unusually long decay of 225 ns, more typical of slower inorganic materials.

3. PULSE SHAPE DISCRIMINATION

For the vast majority of organic scintillators, the prompt fluorescence represents most of the observed scintillation light. A longer-lived component is also observed in many cases, however, corresponding to delayed fluorescence. The composite yield curve can often be represented adequately by the sum of two exponential decays—called the fast and slow components of the scintillation. Compared with the prompt decay time of a few nanoseconds, the slow component will typically have a characteristic decay time of several hundred nanoseconds. Because the majority of the light yield occurs in the prompt component, the long-lived tail would not be of great consequence except for one very useful property: *The fraction of light that appears in the slow component often depends on the nature of the exciting particle.* One can therefore make use of this dependence to differentiate between particles of different kinds that deposit the same energy in the detector. This process is often called *pulse shape discrimination* and is widely applied to eliminate gamma-ray-induced events when organic scintillators are used as neutron detectors.

There is strong evidence that the slow scintillation component originates with the excitation of long-lived triplet states (labeled T_1 in Fig. 8.1) along the track of the ionizing

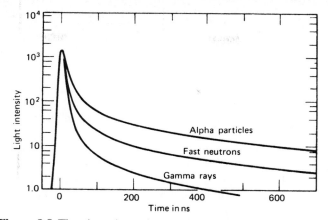

Figure 8.5 The time dependence of scintillation pulses in stilbene (equal intensity at time zero) when excited by radiations of different types. (From Bollinger and Thomas.[60])

particle. Bimolecular interactions between two such excited molecules can lead to product molecules, one in the lowest singlet state (S_1) and the other in the ground state. The singlet state molecule can then de-excite in the normal way, leading to delayed fluorescence. The variation in the yield of the slow component can then be partially explained by the differences expected in the density of triplet states along the track of the particle, because the bimolecular reaction yield should depend on the square of the triplet concentration. Therefore, the slow component fraction should depend primarily on the rate of energy loss dE/dx of the exciting particle and should be greatest for particles with large dE/dx. These predictions are generally confirmed by measurements of the scintillation pulse shape from a wide variety of organics.

Certain organic scintillators, including stilbene crystals[59] and a number of commercial liquid scintillators, are particularly favored for pulse shape discrimination because of the large differences in the relative slow component induced by different radiations. Figure 8.5 shows the differences observed in stilbene for alpha particles, fast neutrons (recoil protons), and gamma rays (fast electrons). In such scintillators, it is not only possible to differentiate radiations with large dE/dx differences (such as neutrons and gamma rays) but also to separate events arising from various species of heavy charged particles as well. Reviews of the pulse shape discrimination properties of different types of organic scintillators and examples of applications are given in Refs. 61–69. Electronic circuits designed to carry out this pulse shape discrimination are described in Chapter 17.

II. INORGANIC SCINTILLATORS

A. Scintillation Mechanism in Inorganic Crystals with Activators

The scintillation mechanism in inorganic materials depends on the energy states determined by the crystal lattice of the material. As shown in Fig. 8.6, electrons have available only discrete bands of energy in materials classified as insulators or semiconductors. The lower band, called the *valence band*, represents those electrons that are essentially bound at lattice sites, whereas the *conduction band* represents those electrons that have sufficient energy to be free to migrate throughout the crystal. There exists an intermediate band of energies, called the *forbidden band*, in which electrons can never be found in the pure

Figure 8.6 Energy band structure of an activated crystalline scintillator.

crystal. Absorption of energy can result in the elevation of an electron from its normal position in the valence band across the gap into the conduction band, leaving a hole in the normally filled valence band. In the pure crystal, the return of the electron to the valence band with the emission of a photon is an inefficient process. Furthermore, typical gap widths are such that the resulting photon would be of too high an energy to lie in the visible range.

To enhance the probability of visible photon emission during the de-excitation process, small amounts of an impurity are commonly added to inorganic scintillators. Such deliberately added impurities, called *activators*, create special sites in the lattice at which the normal energy band structure is modified from that of the pure crystal. As a result, there will be energy states created within the forbidden gap through which the electron can de-excite back to the valence band. Because the energy is less than that of the full forbidden gap, this transition can now give rise to a visible photon and therefore serve as the basis of the scintillation process. These de-excitation sites are called *luminescence centers* or *recombination centers*. Their energy structure in the host crystalline lattice determines the emission spectrum of the scintillator.

A charged particle passing through the detection medium will form a large number of electron–hole pairs created by the elevation of electrons from the valence to the conduction band. The positive hole will quickly drift to the location of an activator site and ionize it, because the ionization energy of the impurity will be less than that of a typical lattice site. Meanwhile, the electron is free to migrate through the crystal and will do so until it encounters such an ionized activator. At this point the electron can drop into the activator site, creating a neutral configuration that can have its own set of excited energy states. These states are illustrated in Fig. 8.6 as horizontal lines within the forbidden gap. If the activator state that is formed is an excited configuration with an allowed transition to the ground state, its de-excitation will occur very quickly and with high probability for the emission of a corresponding photon. If the activator is properly chosen, this transition can be in the visible energy range. Typical half-lives for such excited states are on the order of 50–500 ns. Because the migration time for the electron is much shorter, all the excited impurity configurations are formed essentially at once and will subsequently de-excite with the half-life characteristic of the excited state. It is the decay time of these states that therefore determines the time characteristics of the emitted scintillation light. Some inorganic scintillators can adequately be characterized by a single decay time or a simple exponential, although more complex time behavior is often observed.

There are processes that compete with the one just described. For example, the electron upon arriving at the impurity site can create an excited configuration whose transition to the ground state is forbidden. Such states then require an additional increment of energy to raise them to a higher-lying state from which de-excitation to the ground state is possible. One source of this energy is thermal excitation and the resulting slow component of light is called *phosphorescence*. It can often be a significant source of background light or "afterglow" in scintillators.

A third possibility exists when an electron is captured at an activator site. Certain radiationless transitions are possible between some excited states formed by electron capture and the ground state, in which case no visible photon results. Such processes are called *quenching* and represent loss mechanisms in the conversion of the particle energy to scintillation light.

As an alternative to the independent migration of the electron and hole described above, the pair may instead migrate together in a loosely associated configuration known as an *exciton*. In this case the electron and hole remain associated with each other but are free to drift through the crystal until reaching the site of an activator atom. Similar excited activator configurations can again be formed and give rise to scintillation light in their de-excitation to the ground configuration.

A measure of the efficiency of the scintillation process follows from a simple energy calculation. For a wide category of materials, it takes on the average about three times the bandgap energy to create one electron-hole pair. In sodium iodide, this means about 20 eV of charged particle energy must be lost to create one electron–hole pair. For 1 MeV of particle energy deposited in the scintillator, about 5×10^4 electron–hole pairs are thus created. Various experimental determinations have shown that the absolute scintillation efficiency of thallium-activated sodium iodide is about 12%. Absorption of 1 MeV of energy should therefore yield about 1.2×10^5 eV in total light energy, or 4×10^4 photons with an average energy of 3 eV. The yield is thus very close to 1 photon per electron-hole pair originally formed, and the energy transfer to activator sites must be extremely efficient. The processes of energy transfer in alkali halide scintillators have been reviewed by Murray[70] and by Kaufman et al.[71]

One important consequence of luminescence through activator sites is the fact that the crystal can be transparent to the scintillation light. In the pure crystal, roughly the same energy would be required to excite an electron–hole pair as that liberated when that pair recombines. As a result the emission and absorption spectra will overlap and there will be substantial self-absorption. As we have seen, however, the emission from an activated crystal occurs at an activator site where the energy transition is less than that represented by the creation of the electron–hole pair. As a result the emission spectrum is shifted to longer wavelengths and will not be influenced by the optical absorption band of the bulk of the crystal.

The emission spectrum of the light produced by a number of inorganic scintillators is shown in Fig. 8.7. To make full use of the scintillation light, the spectrum should fall near

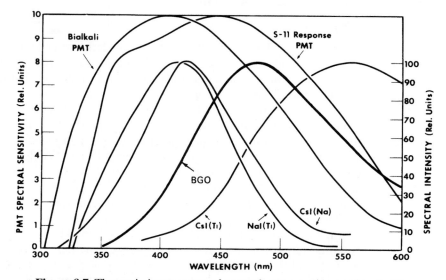

Figure 8.7 The emission spectra of several common inorganic scintillators. Also shown are the response curves for two widely used photocathodes. (Primarily from *Scintillation Phosphor Catalog*, The Harshaw Chemical Company. The emission spectrum for BGO is from Ref. 72.)

the wavelength region of maximum sensitivity for the device used to detect the light. For reference, the responses of several common photocathodes are also plotted in Fig. 8.7 (other photocathode responses are plotted in Fig. 9.2, and that for a typical silicon photodiode in Fig. 9.15).

The scintillation properties at room temperature of a collection of inorganic scintillators are compiled in Table 8.3. The decay times shown in the fourth column are only approximate in that in most cases they represent the dominant decay component only. Many (if not most) inorganic scintillators show more than one decay component, and examples in which a second component is important are also indicated. Where percentages are listed, they represent the relative yields of the components shown. The fifth column is an estimate of the total number of scintillation photons produced over the entire emission spectrum from the deposition of 1 MeV of energy by fast electrons. These values are generally obtained from measurements in which the quantum efficiency of the photomultiplier tube or photodiode used to measure the light (see Chapter 9) is known as a function of wavelength. There is considerable variability in published values for a given material, most likely because of differences in impurity levels or variations in the optical quality of the samples. The sixth column compares the relative pulse amplitude when the scintillator is excited by fast electrons (or gamma rays) and coupled to a glass end-window photomultiplier tube (UV insensitive) with a bialkali photocathode. The variability of photocathode spectral response may also introduce additional variations in these values.

For common inorganic scintillators, the light yield is more nearly proportional to deposited radiation energy than is typically observed in organic scintillators. Quenching processes that are present still lead to some nonlinearity, but often to a much lesser extent than in organics. Variance in the light yield for different types of particles of equal energy is also observed. As in organic scintillators, heavy charged particles produce less light per unit energy. The *alpha-to-beta ratio* (see p. 228) can be much closer to unity than in typical organics (for example, it is in the range of 0.66 to 0.67 for both NaI(Tl) and CsI(Tl) scintillators), but is as low as 0.20 for oxide-based materials such as BGO and GSO described later in this chapter.[73]

B. Characteristics of Alkali Halide Scintillators

1. NaI(Tl)

In 1948, Robert Hofstadter[79] first demonstrated that crystalline sodium iodide, in which a trace of thallium iodide had been added in the melt, produced an exceptionally large scintillation light output compared with the organic materials that had previously received primary attention. This discovery, more than any other single event, ushered in the era of modern scintillation spectrometry of gamma radiation. It is remarkable that the same material remains preeminent in the field despite decades of subsequent research into other scintillation materials. Large ingots can be grown from high-purity sodium iodide to which about 10^{-3} mole fraction of thallium has been added as an activator. Scintillators of unusual size or shape can also be fabricated by pressing small crystallites together. NaI(Tl) is hygroscopic and will deteriorate due to water absorption if exposed to the atmosphere for any length of time. Crystals must therefore be "canned" in an air-tight container for normal use.

The most notable property of NaI(Tl) is its excellent light yield. In common with other typical inorganic scintillators, NaI(Tl) shows a small but measurable nonproportionality of its scintillation response with deposited electron energy (see Fig. 8.8). The departure from proportionality is most pronounced at low energies, and some of the practical effects on scintillator performance are discussed later in Chapter 10. It has come to be accepted as the standard scintillation material for routine gamma-ray spectroscopy and can be

Table 8.3 Properties of Common Inorganic Scintillators

	Specific Gravity	Wavelength of Max. Emission	Refractive Index	Decay Time (μs)	Abs. Light Yield in Photons/MeV	Relative Pulse Height Using Bialk. PM tube	References
Alkali Halides							
NaI(Tl)	3.67	415	1.85	0.23	38 000	1.00	
CsI(Tl)	4.51	540	1.80	0.68 (64%), 3.34 (36%)	65 000	0.49	78, 90, 91
CsI(Na)	4.51	420	1.84	0.46, 4.18	39 000	1.10	92
LiI(Eu)	4.08	470	1.96	1.4	11 000	0.23	
Other Slow Inorganics							
BGO	7.13	480	2.15	0.30	8200	0.13	
CdWO$_4$	7.90	470	2.3	1.1 (40%), 14.5 (60%)	15 000	0.4	98–100
ZnS(Ag) (polycrystalline)	4.09	450	2.36	0.2		1.3[a]	
CaF$_2$ (Eu)	3.19	435	1.47	0.9	24 000	0.5	
Unactivated Fast Inorganics							
BaF$_2$ (fast component)	4.89	220		0.0006	1400	na	107–109
BaF$_2$ (slow component)	4.89	310	1.56	0.63	9500	0.2	107–109
CsI (fast component)	4.51	305	1.80	0.002 (35%), 0.02 (65%)	2000	0.05	113–115
CsI (slow component)	4.51	450	1.80	multiple, up to several μs	varies	varies	114, 115
CeF$_3$	6.16	310, 340	1.68	0.005, 0.027	4400	0.04 to 0.05	76, 116, 117
Cerium-Activated Fast Inorganics							
GSO	6.71	440	1.85	0.056 (90%), 0.4 (10%)	9000	0.2	119–121
YAP	5.37	370	1.95	0.027	18 000	0.45	78, 125
YAG	4.56	550	1.82	0.088 (72%), 0.302 (28%)	17 000	0.5	78, 127
LSO	7.4	420	1.82	0.047	25 000	0.75	130, 131
LuAP	8.4	365	1.94	0.017	17 000	0.3	134, 136, 138
Glass Scintillators							
Ce activated Li glass[b]	2.64	400	1.59	0.05 to 0.1	3500	0.09	77, 145
Tb activated glass[b]	3.03	550	1.5	~3000 to 5000	~50 000	na	145
For comparison, a typical organic (plastic) scintillator:							
NE102A	1.03	423	1.58	0.002	10 000	0.25	

[a] for alpha particles

[b] Properties vary with exact formulation. Also see Table 15.1.

Source: Data primarily from Refs. 74 and 75, except where noted.

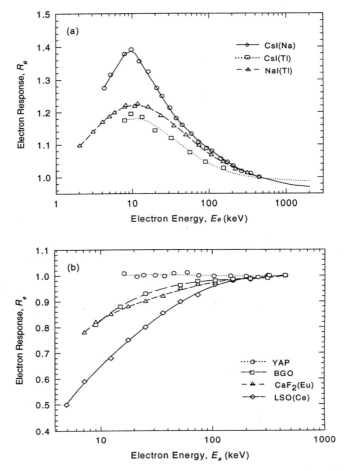

Figure 8.8 The relative scintillation response per unit energy deposited for fast electrons plotted as a function of energy for the scintillation materials shown. The curves are normalized to unity at 445 keV. Perfectly linear response would correspond to a horizontal line on this plot. (From Mengesha et al.[80])

machined into a wide assortment of sizes and shapes. The crystal is somewhat fragile and can easily be damaged by mechanical or thermal shock.

The dominant decay time of the scintillation pulse is 230 ns, uncomfortably long for some fast timing or high counting rate applications. In addition to this prompt yield, a phosphorescence with characteristic 0.15 s decay time has also been measured,[81] which contributes about 9% to the overall light yield. Other longer-lived phosphorescence components have also been measured.[82] Because the anode time constant of photomultiplier tubes is usually set much shorter than these decay times, each photoelectron associated with the phosphorescence is normally resolved individually. At low counting rates, the result is then a series of single-electron pulses that follow the main scintillation pulse and usually are well below the amplitude of interest in the measurement. There are applications, however, in which single-electron sensitivity is needed and the influence of these phosphorescence pulses can be significant. At high counting rates, the phosphorescence will tend to build up due to the multiple overlap from many preceding pulses. This afterglow is often an undesirable characteristic of sodium iodide used in high-rate applications.

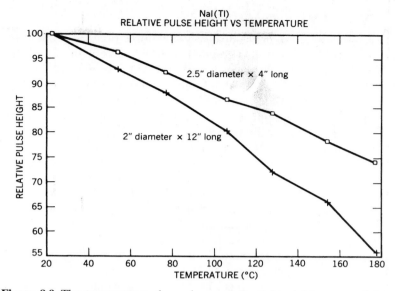

Figure 8.9 The temperature dependence of the light yield measured from two NaI(Tl) crystals. The measurements were made using an oven equipped with a light pipe, and the temperature of the photomultiplier tube was held constant. The difference in behavior between the two crystals is probably due to changes in surface reflectivity. (Data courtesy R. Dayton, Bicron Corporation, Newbury, Ohio.)

The properties summarized above are for thallium-activated sodium iodide operated at room temperature. In some applications, the scintillator must be operated at either lower or higher ambient temperatures. Figure 8.9 shows the dependence of the total light yield in thallium-activated sodium iodide as a function of temperature. The dropoff in scintillation yield with increasing temperature, typical of most scintillation materials, results in generally poorer energy resolution when the scintillator must be used at elevated temperatures. The scintillation decay time in NaI(Tl) is also a function of temperature (see Fig. 8.10), with somewhat faster response at higher temperatures.

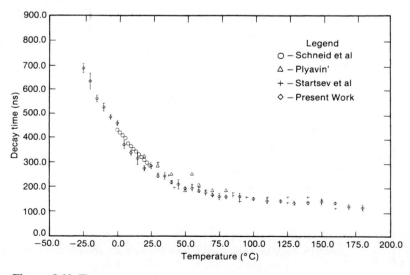

Figure 8.10 Temperature dependence of the scintillation decay time in NaI(Tl). (From Schweitzer and Ziehl.[83])

2. CsI(Tl) and CsI(Na)

Cesium iodide is another alkali halide that has gained substantial popularity as a scintillation material. It is commercially available with either thallium or sodium as the activator element, and very different scintillator properties are produced in the two cases. Cesium iodide has a somewhat larger gamma-ray absorption coefficient per unit size compared to sodium iodide. This advantage is of primary importance for applications such as space instrumentation where size and weight are at a premium. Because it is less brittle than sodium iodide, it can be subjected to more severe conditions of shock and vibration. When cut into thin sheets, cesium iodide may be bent into various shapes without fracturing, and it is reasonably soft and malleable. An extensive bibliography of the properties of cesium iodide (both sodium and thallium activated) can be found in Ref. 84.

A most useful property of CsI(Tl) is its variable decay time for various exciting particles. Pulse shape discrimination techniques can therefore be used to differentiate among various types of radiation.[85–87] Particularly clean separations can be achieved between charged particles such as protons or alpha particles on one hand and electron events on the other hand. The material is less hygroscopic than NaI(Tl) but will deteriorate if exposed to water or high humidity.

CsI(Tl) also has a useful property when grown in thin layers on specially prepared patterned substrates. The material can be produced with a microstructure[88] consisting of columns oriented perpendicular to the surface of the layer. Each column may have a diameter as small as 5 μm, and to some extent behaves as an optically isolated scintillator. Layers with thickness of hundreds of microns are often needed for good detection efficiency in the imaging of X-rays. The columnar microstructure inhibits the lateral spread of the scintillation light that would take place if the layer were grown as a single crystal, and better spatial resolution can be achieved for equivalent thickness. [89]

The emission spectrum of CsI(Tl) is peaked at a much longer wavelength than that for NaI(Tl) (see Fig. 8.7) and is poorly matched to the response of PM tubes with S-11 or bialkali photocathodes. For that reason, the light output is often quoted as being substantially lower in CsI(Tl). However, when measurements are made using photodiodes with extended response into the red region of the spectrum, the scintillation yield is actually higher[75] than that of any other scintillator. The absolute light yield at room temperature is measured[90] to be about 65,000 photons/MeV, with a maximum value about 6% higher at −35°C. The yield falls off with temperature change on either side of this maximum, to 64% of the room temperature value at −100°C and to 95% at +50°C.

The luminescent states in CsI(Tl) are populated through an exponential process that results in an unusually long rise time of 20 ns for the initial appearance of the light. The subsequent decay of these states is among the slowest for the commonly used scintillation materials. The light emission for gamma-ray excitation shows two primary components[91] with decay times and relative intensities at room temperature of 0.68 μs (64%) and 3.34 μs (36%).

CsI(Na) has an emission spectrum that is similar to that of NaI(Tl) with a light yield that is comparable, but its relatively slow decay is a disadvantage. The decay is reported to consist of two components with mean lives of 0.46 and 4.18 μs, and a dependence of the slow to fast component ratio has been shown for various exciting particles.[92] Long-lived phosphorescence components in cesium iodide have also been reported.[93] CsI(Na) is hygroscopic and must be sealed against exposure to ambient atmospheres.

3. LiI(Eu)

Lithium iodide (europium activated) is an alkali halide of special interest in neutron detection. As discussed in Chapter 14, crystals prepared using lithium enriched in ^{6}Li provide for the efficient detection of low-energy neutrons through the ^{6}Li(n, α) reaction. Scintillation properties of this material are included in Table 8.3.

C. Other Slow (> 200 ns) Inorganic Crystals

1. Bismuth Germanate (or BGO)

An alternative scintillation material, $Bi_4Ge_3O_{12}$ (commonly abbreviated as BGO) is commercially available as crystals of reasonable size. A major advantage over many other scintillators is its high density (7.13 g/cm^3) and the large atomic number (83) of the bismuth component. These properties result in the largest probability per unit volume of any commonly available scintillation material for the photoelectric absorption of gamma rays. Its mechanical and chemical properties make it easy to handle and use, and detectors using BGO can be made more rugged than those employing the more fragile and hygroscopic sodium iodide. Unfortunately, the light yield from BGO is relatively low, being variously reported at 10–20% of that of NaI(Tl). Furthermore, its relatively high refractive index (2.15) makes efficient collection of the light more difficult than for scintillators with lower index values. It is therefore of primary interest when the need for high gamma-ray counting efficiency outweighs considerations of energy resolution. Some comparative gamma-ray spectra for BGO and NaI(Tl) are shown in Chapter 10.

Figure 8.11 shows the time profile of the light emitted in a scintillation event in both BGO and NaI(Tl). In BGO, the principal decay time of 300 ns is preceded by a faster component of the light with 60 ns decay time that represents about 10% of the total light yield. On the other hand, the initial decay of the NaI(Tl) light pulse is slightly slower than the

Figure 8.11 Measurements of the light pulse shapes from BGO and NaI(Tl). The abscissa represents time, the ordinate the relative light output. The BGO yield is represented as the sum of separate decay components with 60 and 300 ns decay times. (From Moszynski et al.[94])

Figure 8.12 The dependence of the light output of some common scintillators as a function of temperature. Only the fast (220 nm) component from BaF_2 is included (From Melcher.[97])

230 ns principal decay. These differences, coupled with the much lower light yield from BGO, result in an overall timing resolution for BGO that is about a factor of 2 worse than that for NaI(Tl).[94] In BGO there are almost none of the long decay components that lead to afterglow in sodium iodide and some other scintillators. BGO has therefore found widespread application in X-ray computed tomography scanners where scintillators operated in current mode must accurately follow rapid changes in X-ray intensity.

BGO is an example of a "pure" inorganic scintillator that does not require the presence of a trace activator element to promote the scintillation process. Instead, the luminescence is associated[95] with an optical transition of the Bi^{3+} ion that is a major constituent of the crystal. There is a relatively large shift between the optical absorption and emission spectra of the Bi^{3+} states. Therefore, relatively little self-absorption of the scintillation light occurs, and the crystal remains transparent to its own emission over dimensions of many centimeters. The scintillation efficiency depends strongly on the purity of the crystal, and some of the variability in the light yield reported from BGO in the past can be attributed to using crystals with different residual levels of impurities.[96] The crystals are typically grown by the Czochralski method in which a crystal boule is pulled from a molten mixture of bismuth oxide and germanium oxide at a rate of a few millimeters per hour. The boule can then be cut and polished using conventional methods. BGO remains two to three times more costly than NaI(Tl) and is currently available only in limited sizes.

In common with many other scintillators, the light output from BGO decreases with increasing temperature. Figure 8.12 shows its relative light output together with that of several other scintillators as a function of temperature. Since the light yield is already low at room temperature, its rapid dropoff severely limits the usefulness of BGO in high-temperature applications.

2. Cadmium Tungstate CdWO₄

Cadmium tungstate has been known as a scintillator since about 1950, but it has only been in the 1990s that crystals of interesting size and good optical quality have been widely investigated. The light yield is about 40% of that of NaI(Tl),[98,99] with an emission spectrum peaked in the visible near 470 nm. It has a high density and effective atomic number so that it is an excellent candidate for gamma-ray measurements. Its major drawback is its relatively long decay time, a mixture of two components for gamma ray excitation with 1.1 μs

(40%) and 14.5 µs (60%).[99] The relative intensity of these components changes for different types of particles[100] allowing their separate identification using pulse shape discrimination techniques.

Because of the relatively long decay times, pulse mode applications are limited to those in which the counting rate is not high. Excellent gamma-ray energy resolution of 6.8% at 662 keV has been demonstrated for a small (1 cm³) crystal.[99] Resolutions closer to 10–12% are more typical of larger size crystals.[101] The light collection is somewhat hindered by the high index of refraction (2.3) that makes efficient coupling to photomultiplier tubes or photodiodes more difficult.

The principal application of this material has been as an X-ray detector operated in current mode where the long decay time is not an issue. It is remarkably free of long-lived phosphorescence, making it especially suitable for X-ray computed tomography applications where "afterglow" is highly undesirable in rapidly changing X-ray fluxes.

3. ZnS(Ag)

Silver-activated zinc sulfide is one of the older inorganic scintillators. It has a very high scintillation efficiency comparable to that of NaI(Tl) but is available only as a polycrystalline powder. As a result, its use is limited to thin screens used primarily for alpha particle or other heavy ion detection. Thicknesses greater than about 25 mg/cm² become unusable because of the opacity of the multicrystalline layer to its own luminescence. These scintillation screens played a key role in the early experiments of Rutherford, in which alpha particle interactions were visually observed through a low-power microscope.

4. CaF₂(Eu)

Europium-activated calcium fluoride is notable as a nonhygroscopic and inert inorganic scintillator that can often be used where severe environmental conditions preclude other choices. It is relatively fracture resistant and nonreactive, with a vapor pressure low enough to be usable under vacuum conditions. Its scintillation properties are included in Table 8.3. Even though the light yield is 50% of that of NaI(Tl), the relatively long decay time of 900 ns has inhibited its widespread use as a general scintillator.

D. Unactivated Fast Inorganics with Low Light Yield

1. Barium Fluoride BaF₂

Barium fluoride has the distinction of being the first inorganic crystal discovered to have a very fast component in its scintillation decay.[102-105] It is the only presently known scintillator with high atomic number components that has a decay time of less than 1 ns. This combination of properties therefore makes the material attractive for scintillation detectors in which both high detection efficiency per unit volume and a fast response are required.

Unactivated BaF₂ has been known[106] as a scintillation material since the early 1970s. However, it was not until 1983 that it was shown[102] that the scintillation light actually consists of two components: a fast component with decay time of 0.6 ns emitted in the short wavelength region of the spectrum, and a slower component with 630 ns decay time at somewhat longer wavelengths. The two components are identified in the emission spectrum shown in Fig. 8.13. The fast component went unobserved for many years because most photomultiplier tubes are not sensitive to this short wavelength region of the spectrum. However, if quartz end-window tubes or other light sensors are used that are sensitive in the ultraviolet, about 20% of the total scintillation yield at room temperature is measured in the fast component. It results from the creation of a hole in the outer core band of the ionic crystal, followed by the filling of this hole by an electron from the valence band. This process is characterized by very short transition times and the resulting emission is usually

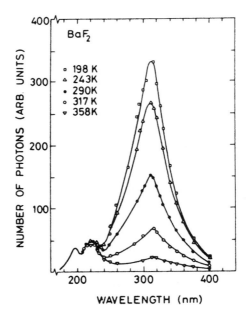

Figure 8.13 The scintillation emission spectra from BaF_2 measured at various temperatures. The fast component (corresponding to the two small peaks at the left) does not display the strong temperature dependence of the slow component. (From Schotanus et al.[107, 108])

in the ultraviolet region of the spectrum. If the principal band gap of the crystal is larger than the energy of the UV photon, then the scintillation light can escape reabsorption and be collected by the photomultiplier tube.

The total light yield in BaF_2 is only about 20% of that observed in NaI(Tl), so the attainable energy resolution is considerably poorer. The light yield in the fast component is quite small, only about 1400 photons per MeV.[109] However, a number of applications have been demonstrated[110–112] in which the fast timing capability coupled with the high density and atomic number of the material have made it the scintillation material of choice.

2. Pure Cesium Iodide CsI

When activated with thallium, cesium iodide has the highest scintillation yield of any known material. In its pure state, it is also a much weaker scintillator (with output between 5 and 8% of that of NaI(Tl) when measured with a photomultiplier tube). However, much of this light shows a mixture of fast components with an effective decay time of about 10 ns that appears in a peak around 305 nm in the ultraviolet region of the spectrum. A broader emission band in the visible range of 350 to 600 nm that has a much longer decay time of up to several microseconds is also often observed. The relative intensities of these components tends to vary among the measurements reported in the literature[113,114] for different samples of the material, but they are roughly comparable in yield in one study of commercially supplied crystals.[115] The slow component is absent in other measurements of highly purified material,[114] indicating that its origin may be related to residual impurities in some crystals.

3. Cerium Fluoride CeF₃

Because of the widespread use of the element cerium as an activator in other crystal matrices, where it acts as a luminescence center, it is not surprising that there has been substantial interest in examining the scintillation potential of compounds of cerium as well. The most widely examined of these is cerium fluoride (CeF_3) that displays weak scintillation light equivalent to about 5% of that of NaI(Tl). The main interest in this material comes about because of its relatively fast decay time. When measured in the visible band, the decay appears to be a single exponential with a decay time of about 27 ns. When the observations are extended in the ultraviolet range,[116] a second component with decay time of about 5 ns is also observable. The total light yield from small volume crystals is reported[117] to be 2100 photons per MeV.

E. Cerium-Activated "Fast" Inorganics

Beginning in the late 1980s, it was realized that the cerium activator could be incorporated into new categories of crystals, resulting in scintillators with reasonably good light yield. The principal decay time of the cerium luminescence ranges from about 20 to 80 ns, depending on the host crystal.[118] Thus the timing characteristics of these scintillators occupy an intermediate position between the organics with a few nanosecond decay time and the older inorganics in which it is several hundred nanoseconds.

1. Gadolinium Silicate Gd_2SiO_5 (or GSO)

This material can be grown in reasonably large sizes as single crystals[119] and the atomic number of 64 for gadolinium offers attractive properties in gamma-ray spectroscopy. Maximum light yield is observed at a cerium doping concentration of about 0.5 mole percent. The scintillation decay time also depends on the cerium doping level, but is dominated by a fast component of about 56 ns at this cerium doping concentration. A longer decay time component of 400 ns is present to 10% intensity ratio as well.[120] The light output also shows an unusually long *rise* time of 10 to 20 ns, depending on the cerium concentration, caused by a slow population process for the fluorescent states. The total light yield is about 20% of that of NaI(Tl). An energy resolution of about 9% for 662 keV gamma rays has been reported as typical performance.[121] Additional properties are given in Table 8.3. It also appears[122] that GSO shows excellent radiation stability, with no appreciable radiation damage observed for gamma-ray exposures up to 10^7 Gy.

2. Yttrium Aluminum Perovskite $YAlO_3$ (or YAP)

This crystal has been known for some time in its role as a laser material when doped with Nd. When doped with cerium, however, it displays some attractive properties as a scintillator. It has a total light output variously reported as 40 to 50% of that of NaI(Tl) and an unusually fast principal decay time of 27 ns. It displays excellent physical properties of hardness, strength, and inertness that readily allow easy fabrication and handling. Its scintillation properties[123-125] show the best performance for small-size crystals, and there is evidence[126] that the light yield and energy resolution suffer for crystals that are larger than a few centimeters in dimension. Small-size crystals show excellent energy resolution, for example, 5.7% at 662 keV,[125] while at the same time offering outstanding timing resolution as low as 160 ps. A slower component of the decay with about 10 μs time constant constitutes less than 10% of the total intensity. YAP demonstrates excellent proportionality of its light output versus energy (see Fig. 8.8), allowing preservation of good energy resolution even when gamma rays undergo multiple interactions in the detector.[124] (See Chapter 10.)

3. Yttrium Aluminum Garnet $Y_3Al_5O_{12}$ (or YAG)

This material when doped with Nd also has a long history as a laser material, but it also has interesting scintillation properties when doped with cerium. The light yield per unit energy is about half that of NaI(Tl), but its spectrum is somewhat unusual in that it is shifted toward longer wavelengths relative to many other scintillators. The peak in the emission spectrum occurs at 550 nm, and it is no longer a good match to many photomultiplier tube spectral responses. Instead, these longer wavelengths are better suited for readout using semiconductor photodiodes. The decay shows several components[127, 128] with principal decay times of 88 ns and 302 ns. For gamma-ray excitation, the relative intensities of these two components are 72% and 28%, respectively. When irradiated with alpha particles, the ratio reverses to 34% and 66%, allowing applications in which pulse shape discrimination can be used to separate different particle types.[127]

4. Lutetium Oxyorthosilicate $Lu_2(SiO_4)O$ (or LSO)

This material was first described as a scintillator in 1991.[129] It has some very interesting properties including a light yield approximately 75% of that of NaI(Tl), a fast decay time of 47 ns,[130] and an emission spectrum conveniently peaked at 420 nm. Lutetium is a rare earth whose high atomic number of 71 makes it an attractive component in scintillators intended for gamma-ray spectroscopy. However, its cost and availability for fabrication into large-volume crystals are still open to some question. The natural element contains the isotope ^{176}Lu which is radioactive and adds an inherent background of about 300 counts/s per cm^3 of LSO. The best energy resolution reported for 662 keV gamma rays of 7.5% is obtainable only in small crystals, and more typical energy resolutions of 10% or more are observed in larger crystals. The scintillation response is known[131] to be rather nonlinear (see Fig. 8.8), so some of the degradation in energy resolution may be due to this influence. The light emission appears to consist strictly of a single 47-ns component and is free of the slower components that sometimes plague other inorganic scintillators.[132] An unusual degree of variability in the light output has been observed from various samples of LSO crystals,[133] possibly caused by different levels of trace impurities that can serve as traps for either electrons or holes before they diffuse to a cerium fluorescence site.

5. Lutetium Orthoaluminate $LuAlO_3$ (or LuAP)

Motivated by the favorable properties of YAP, described above, there has been interest in exploring its chemical analogue LuAP as a host for the cerium activator. YAP is limited somewhat in its application to gamma ray spectroscopy because of the relatively low atomic number ($Z = 39$) of yttrium. Substituting lutetium with $Z = 71$ provides a much more attractive possibility. In fact LuAP has the highest linear attenuation coefficient for common gamma rays of any scintillator, even exceeding that of BGO. Furthermore, most of the scintillation light appears with a very short decay constant of 17 ns[134] and a relatively high light output that may be more than 50% of that of NaI(Tl). This combination of properties would make the material unparalleled for many gamma-ray applications. Thus there has been a great deal of interest[134–139] in developing practical scintillators from this material. Measurements show that there is very little departure from proportionality in its light yield as a function of deposited energy,[140] minimizing the potential loss of energy resolution due to nonproportionality discussed in Chapter 10. An excellent gamma-ray energy resolution of 4.38% at 662 keV was obtained in the same study for a 3 mm thick crystal coupled to a photomultiplier tube. At the present stage of development, however, useful crystals are limited to thicknesses of less than a centimeter because of strong self-absorption[133] of the scintillation light. This limitation offsets some of the inherent advantages of the high atomic number and density of the material in gamma-ray applications. LuAP also shares with LSO the potentially troublesome natural background caused by the radioactivity of the ^{176}Lu component.

F. Glass Scintillators

Silicate glasses containing lithium and activated with cerium are widely used as neutron detectors, and a collection of their physical and scintillation properties may be found in Table 15.1. Bollinger et al.[141] investigated the properties of a variety of lithium and boron glasses as scintillation materials. The light output observed for the lithium formulations averaged over a factor of 10 larger than that for the boron-containing glasses. The lithium formulations have therefore predominated. In silicate glasses, cerium is the only activator that produces fast scintillation light useful for pulse mode operation. The emission is peaked in the blue region of the spectrum and is associated with the Ce^{3+} sites within the glass. Modern scintillation glasses[142] are made from various mixtures of SiO_2, LiO_2, Al_2O_3, MgO, and Ce_2O_3, with BaO sometimes added to increase the glass density.

The absolute scintillation efficiency for converting fast electron energy to light has been measured[143] in the range from 1 to 2%, generally varying inversely with lithium concentration of the glass formulation. The response to charged particles is nonlinear and always considerably below that to electrons of the same energy. For example, the light output for 1 MeV protons, deuterons, and alpha particles is lower than that for 1 MeV electrons by factors of 2.1, 2.8, and 9.5, respectively.[144]

Despite the fact that their relative light output is quite low (typically a maximum of 3500 photons/MeV), glass scintillators are sometimes applied in beta or gamma-ray counting when severe environmental conditions prevent the use of more conventional scintillators. Examples include conditions in which the scintillator must be exposed to corrosive chemical environments or operated at high temperatures. The decay time of glass scintillators (typically 50–75 ns) is intermediate between faster organics and slower crystalline inorganics. Because the material can be fabricated in much the same manner as other glasses, it is widely available in physical forms other than the more common cylindrical shapes. For example, glass scintillators are commercially available as small-diameter fibers whose properties are described later in this chapter. The material is also available as a powder or small-diameter spheres for use in flow cells in which the radioactive material is allowed to pass through a chamber filled with the porous scintillator.

Because glasses may contain naturally radioactive thorium or potassium, precautions must be taken if scintillators are to be applied in low-level counting systems where a minimum background is required. Ordinary glass scintillators will show a spontaneous background rate of about 100–200 disintegrations per minute for every 100 g of material. Glasses are commercially available that are manufactured from low-thorium materials and that are processed to minimize contamination by any other alpha-active materials. These low-background glasses will have a background activity of less than 20 disintegrations per minute per 100 g.

When used as a neutron counter, the lithium content of the glasses is normally enriched to 95% or more in ^6Li. If applied as a beta or gamma counter, this enrichment is not required and can actually be a hindrance if the detector is operated with any appreciable neutron background. Therefore, one of the commercially available forms fabricated using either natural lithium or lithium depleted in ^6Li is normally chosen for these applications.

Another form of glass scintillator, activated with terbium rather than cerium, has found some use in imaging applications. The scintillation decay time is now very much longer (3–5 ms) so that these formulations are no longer useful in pulse mode applications. However, the total light yield is many times larger than from the cerium-activated glass, reaching as high as 50,000 photons per MeV.[145] The light emission is in the green part of the spectrum, peaking at about 550 nm. The specific gravity of these glasses can also be made as high as 3.5, enhancing the interaction probability for X- or gamma rays. These terbium-activated glasses offer some attraction for the imaging of X-rays, where the emitted light is recorded using a position sensitive detector that integrates the light yield over the exposure period.

Some ceramic formulations[146] can also show good light yield, but most have decay times of hundreds of microseconds that are too long to be of interest in pulse mode spectroscopy. They have found current mode application in X-ray computed tomography where their brightness and lack of afterglow on the millisecond time scale are valuable properties.

G. Scintillator Gases

Certain high-purity gases can serve as useful scintillation detection media. Considerable experience has been gained for the noble gases, with xenon and helium receiving the most attention. The mechanism giving rise to the scintillation photons is relatively simple: The incident radiation or charged particle leaves a population of excited gas molecules in its wake as it passes through the scintillator. As these excited molecules return to their ground state, through a variety of different mechanisms, photons will be emitted. The emission spectra correspond to the *second continuum* spectra observed in rare gas discharges. The photons are emitted during the transition from the two lowest molecular excited states to the ground state and much of the scintillation emission lies in the ultraviolet region of the spectrum.[†] As a result, photomultiplier tubes or photodiodes that are sensitive in the near-ultraviolet must be employed if the scintillations are to be detected directly. Alternatively, a small concentration of a second gas such as nitrogen may be added to shift the emission spectrum by absorbing the ultraviolet and reradiating the energy at a longer wavelength. Because of a variety of competing parallel modes of de-excitation, such as intermolecular collisions or internal quenching, the overall scintillation efficiency of gases is characteristically quite low. However, the transitions take place in a very short time, typically a few nanoseconds or less, and gas scintillators can be among the fastest of all radiation detectors.

In addition to their fast response, gas scintillators have the relative advantage of easily variable size, shape, and stopping power for incident radiations. They also tend to be unusually linear over wide ranges of particle energy and dE/dx. Their major disadvantage is the low light yield, which is, at best, over an order of magnitude below that of NaI(Tl) for equivalent particle energy loss. They have been widely applied in the spectroscopy of heavy ions, often with a pressurized active volume. Some scintillation properties of several noble gases are given in Table 8.4. Extensive reviews of gas scintillators can be found in Refs. 1 and 149, and some measurements of specific properties in Refs. 150–153.

Some noble gases when condensed as a cryogenic liquid or solid have also been observed to be quite efficient scintillators. Results have been reported[154–159] on liquid and solid argon, krypton, and xenon and also on liquid helium.[160] The emission spectra are peaked in the UV region around 150 nm. Experimental measurements of the absolute light yield in both liquid argon or liquid xenon show considerable variance, but it is evident that both liquids have relatively high scintillation efficiency. Typical results[161] show a yield of about 40,000 photons/MeV, comparable to that of NaI(Tl) at room temperature. Attenuation lengths for the scintillation light are observed[162] to be between 0.5 and 1.0 m in both liquids. There are obvious difficulties in efficiently collecting the scintillation light while the condensed gas is maintained at cryogenic temperature, and the collection efficiency can be inhibited by optical scattering from density variations in the liquid caused by temperature gradients near the light sensor or walls of the chamber.[163] Xenon in liquid or solid form has the most appeal as a potential gamma-ray detector because of its high atomic number of 54. Kubota et al.[158] present measurements of the decay time characteristics of these condensed gases, with all showing both fast and slow components with relative intensities that depend on the type of exciting particle. Values for the principal decay component range from 6 ns in condensed argon to about 30 ns in condensed xenon.

[†]There is some evidence[147] that Ne, Ar, Kr, and Xe also emit a significant scintillation component in the near-infrared region of the spectrum. By using a PM tube with extended long-wavelength response, this component in argon can be converted into pulses that are large enough to carry out alpha particle spectroscopy with a 5.1% energy resolution.[148]

Table 8.4 Properties of Gas Scintillators at Atmospheric Pressure

Gas	Mean Wavelength of Emission	Number of Photons with $\lambda > 200$ nm per 4.7 MeV Alpha Particle
Xenon	325 nm	3,700
Krypton	318 nm	2,100
Argon	250 nm	1,100
Helium	390 nm	1,100
Nitrogen (for comparison)	390 nm	800
NaI(Tl)	415 nm	41,000

Source: J. B. Birks, *The Theory and Practice of Scintillation Counting.* Copyright 1964 by Pergamon Press, Ltd. Used with permission.

H. Radiation Damage Effects in Inorganic Scintillators

All scintillation materials are subject to radiation damage effects when exposed over prolonged periods to high fluxes of radiation. The damage is most likely evidenced as a reduction in the transparency of the scintillator caused by the creation of color centers that absorb the scintillation light. In addition, there may also be interference with the processes that give rise to the emission of the scintillation light itself. Radiation exposures can also induce long-lived light emission in the form of phosphorescence that can be troublesome in some measurements. The damage effects are often observed to be rate dependent and will vary greatly with the type of radiation involved in the exposure. The effects are often at least partially reversible, with annealing taking place even at room temperature over hours or days following the exposure.

A comprehensive review has been published[164] of the effects of radiation damage in a number of common inorganic scintillators. It was found that the creation of color centers dominates over the damage to the scintillation mechanism in all of the scintillators studied (NaI(Tl), CsI(Tl), CsI, BaF_2, BGO, and $PbWO_4$). The damage in the alkali halides was observed to be closely coupled with the presence of oxygen contamination leading to the formation of hydroxyl species. On the other hand, in the oxide scintillators, the damage was found to be related to structural defects, such as oxygen vacancies in the crystalline lattice.

Because of the many variables such as dose rate, types of particle, energy of particle, and the presence or absence of oxygen or other impurities, it is difficult to quantify the exact radiation dose expected to produce measurable damage in specific circumstances. However, some rough numbers and relative sensitivities have been quoted.[165] Of the scintillation materials discussed in this chapter, the most sensitive appear to be the thallium-activated alkali halides for which exposures of 10 Gy can be significant. At the other extreme, evidence exists that GSO begins to show effects only at doses as high as 10^6 Gy. Between these extremes, roughly in order of increasing radiation resistance, are CsF, BGO, YAP, CeF_3, and BaF_2. Radiation damage in plastic scintillators was discussed earlier in this chapter.

III. LIGHT COLLECTION AND SCINTILLATOR MOUNTING

A. Uniformity of Light Collection

In any scintillation detector, one would like to collect the largest possible fraction of the light emitted isotropically from the track of the ionizing particle. Two effects arise in practical cases that lead to less than perfect light collection: optical self-absorption within the scintillator and losses at the scintillator surfaces. With the exception of very large scintillators (many centimeters in dimension) or rarely used scintillation materials (e.g., ZnS), self-absorption is usually not a significant loss mechanism. Therefore, the uniformity of light collection normally depends primarily on the conditions that exist at the interface between the scintillator and the container in which it is mounted.

The light collection conditions affect the energy resolution of a scintillator in two distinct ways. First, the statistical broadening of the response function discussed in Chapter 10 will worsen as the number of scintillation photons that contribute to the measured pulse is reduced. The best resolution can therefore be achieved only by collecting the maximum possible fraction of all photons emitted in the scintillation event. Second, the *uniformity* of the light collection will determine the variation in signal pulse amplitude as the position of the radiation interaction is varied throughout the scintillator. Perfect uniformity would ensure that all events depositing the same energy, regardless of where they occur in the scintillator, would give rise to the same mean pulse amplitude. With ordinary scintillators of a few centimeters in dimension, uniformity of light collection is seldom a significant contributor to the overall energy resolution. In larger scintillators, particularly those that are viewed along a thin edge, variations in light collection efficiency can often dominate the energy resolution.

Because the scintillation light is emitted in all directions, only a limited fraction can travel directly to the surface at which the photomultiplier tube or other sensor is located. The remainder, if it is to be collected, must be reflected one or more times at the scintillator surfaces. Two situations may prevail when the light photon reaches the surface, as illustrated in Fig. 8.14. If the angle of incidence θ is greater than the critical angle θ_c, total internal reflection will occur. If θ is less than θ_c, partial reflection (called *Fresnel* reflection) and partial transmission through the surface will occur. The fraction of reflected light drops to only a few percent when the angle of incidence is near zero. The critical angle θ_c is determined by the indices of refraction for the scintillation medium n_0 and the surrounding medium (often air) n_1:

$$\theta_c = \sin^{-1} \frac{n_1}{n_0} \tag{8.12}$$

To recapture the light that does escape from the surface, the scintillator is normally surrounded by a reflector at all surfaces except that at which the photomultiplier tube is mounted. Reflectors can be either *specular* or *diffuse*. A polished metallic surface will act as a specular reflector for which the angle of reflection equals the angle of incidence. Better results are usually obtained, however, with a diffuse reflector such as magnesium oxide or aluminum oxide. Here the angle of reflection is approximately independent of the angle of incidence and follows a distribution given by Lambert's law:

$$\frac{dI(\psi)}{dI_0} = \cos \psi \tag{8.13}$$

where ψ is the reflection angle with respect to the direction perpendicular to the surface. Commercially prepared sodium iodide crystals are normally reflected with dry MgO

Figure 8.14 Conditions at the interface of dissimilar optical media ($n_0 > n_1$). Ray ① may escape, but ray ② will be internally reflected at the surface.

powder packed around surfaces of the crystal which have been slightly roughened with an abrasive. Plastic scintillators are usually left with a polished surface, especially when the light must be collected over large dimensions involving multiple reflections. White reflecting paint is easy to apply to scintillators but is an inferior reflector to dry powder. In custom-made scintillators, better results are usually obtained by spraying the surface with a slurry of magnesium oxide in water and allowing the layer to dry. In other circumstances, wrapping with white teflon tape and/or millipore filter paper[166] can also be effective.

Although total internal reflection is desirable at reflecting surfaces, it must be minimized at the surface from which the scintillator is viewed to prevent internal trapping of the light. Ideally, one would like to optically couple the scintillator to the photocathode of the PM tube through a transparent medium of the same index of refraction as the scintillator. Then Eq. (8.12) would predict no internal reflection and all light incident on the surface would be conducted into the glass end window of the PM tube on which the photocathode is deposited. The refractive indices of plastic, liquid, and glass scintillators are fairly close to that of the glass end window (about 1.5). In those cases, near-perfect coupling will be achieved if the interface between the scintillator and PM tube is filled with an optical coupling fluid of the same refractive index. Some internal reflection will inevitably occur for scintillators with a higher index (such as NaI or BGO), and a light photon will be reflected back into the scintillator if its angle of incidence is too large. The usefulness of scintillators with a very high refractive index is seriously hampered because of excessive internal trapping of the scintillation light. High-viscosity silicone oil is commonly used as the coupling agent between the scintillator and PM tube. Design of the scintillation detector must provide a means to prevent the long-term flow of the fluid away from the interface. If the detector will be subjected to severe temperature cycling or vibration, it is sometimes preferable to couple the scintillator with a transparent epoxy cement.

A simple test can be made of light-collecting uniformity by allowing a narrowly collimated gamma-ray beam to strike selected portions of a scintillation crystal. With a uniform surface treatment it is often found that those areas closest to the photocathode give rise to a larger pulse height because of more favorable light collection. In order to compensate for this natural tendency, the surface is often preferentially treated to enhance light collection from the points in the crystal farthest from the photocathode.

It is also possible to model the optical behavior and light collection efficiency in scintillators using Monte Carlo calculations.[167-170] These codes are based on computer modeling of the geometry and optical characteristics of the scintillator under study. Simulations are based on the generation of individual scintillation photons within the scintillator and tracking their fate as they propagate through the scintillator and interact with surfaces. Assumptions must be made about the optical absorption and scattering properties of the scintillator, the reflection and/or absorption characteristics of the surfaces, and the indices of refraction of all optical media through which the scintillation photons pass. These codes prove to be quite accurate in the prediction not only of the total fraction of light collected but also in studying the uniformity and time characteristics of the collected light.

Good light collection in the scintillator also requires that either a single crystal or single piece of plastic be used for the scintillating medium, or that special precautionary techniques be used in the event that separate pieces are cemented together. If separate pieces are used, the additional surfaces will often introduce added reflection and can seriously affect the overall uniformity of light collection. For materials that are difficult to grow in large single crystals, an alternative approach is to suspend granules of the material in a transparent matrix. Examples include the mixture of zinc sulfide powder in clear epoxy[171] or the suspension of barium fluoride powder in liquids.[172] In the ideal case, the indices of refraction of the two materials would be perfectly matched so that the mixture would remain optically clear to allow the propagation of the scintillation light over large distances. As a practical matter, it is very difficult to obtain a perfect match and the mixtures are normally limited to relatively thin layers. The index of refraction for any material is a function of the wavelength of the light, and even if a good match is made at a specific wavelength, it is unlikely that it will remain perfect over the entire emission spectrum of the scintillation material. Care must also be taken to avoid the inclusion of microscopic air bubbles that represent large local changes in refractive index.

Any scintillation counter must be shielded from ambient room light. For temporary arrangements not involving hygroscopic crystals, a simple wrapping of the reflected scintillator and photomultiplier tube with black paper, followed by a layer of the ubiquitous black tape, will often suffice. Commercial sodium iodide scintillation crystals are usually canned in a metallic container and must be hermetically sealed. The surface through which the light is to be collected is provided with a glass or quartz window. The opposite surface is covered with an opaque but thin metallic sheet to provide an entrance window for soft radiations. In cases in which the scintillation counter must be used in low background counting, special care must be given to the proper choice of materials used in contact with the crystal. Some types of glass, for example, are very high in potassium and as a result will generate a large background contribution from the natural ^{40}K activity.

Light collection in large scintillators can often be enhanced by the use of more than one photomultiplier tube. Although this is not usually an attractive option for routine use because of the added complexity, the gains that can be achieved are, in some cases, substantial. The average number of reflections required for a typical event to reach a photomultiplier tube will obviously be less if more than one escape surface is provided. The fewer the reflections, the greater the light collection efficiency, and consequently the greater uniformity of pulse height response. Because of the importance of self-absorption in large scintillators, there often is no substitute for multiple PM tubes in these cases.

More complete discussions of scintillator mounting and reflection are given by Birks[1] and Bell.[173] A useful review of light collection methods and surface properties has also been published by Keil. [174]

B. Light Pipes

It is often unadvisable or even impossible to couple a photomultiplier tube directly to one face of a scintillator. For example, the size or shape of the scintillator may not conveniently match the circular photocathode area of commercially available PM tubes. One solution is to place the unreflected scintillator near the center of a large box, whose interior surfaces are coated with a diffuse reflector. One or more PM tubes can then view the interior of the box to record some fraction of the light that eventually escapes from the scintillator. If the fraction of the box surface replaced by PM tubes is small and the reflectivity of the coating is high,[175] the light is thoroughly randomized before being detected. Very uniform light collection can therefore be achieved even for scintillators with complex or unusual shape. Because of inevitable light losses, however, the total fraction of light collected is typically small. Better light collection efficiency usually can be achieved by using a transparent solid, known as a *light pipe*, to physically couple the scintillator to the PM tube and to act as a guide for the scintillation light.

Light pipes also serve a useful purpose in other situations. If scintillation measurements are to be made in a strong magnetic field, the PM tube must be shielded from the field and this often implies its removal to a location some distance away from the scintillator. Very thin scintillators should not be mounted directly on the PM tube end window to avoid the pulse height variations that can arise due to photocathode nonuniformities. A light pipe between the thin scintillator and the PM tube will spread the light from each scintillation event over the entire photocathode to average out these nonuniformities and improve the pulse height resolution.

Light pipes operate on the principle of total internal reflection outlined in the previous section. They are generally optically transparent solids with a relatively high index of

Figure 8.15 Variation of pulse height with length of light guide for various reflective wrappings. (*a*) Total internal reflection only, (*b*) total internal reflection with reflective covering, (*c*) surface of light guide painted with NE 560 reflector paint, (*d*) specular reflector without light guide, (*e*) diffuse reflector without light guide. (From Kilvington et al.[176])

refraction to minimize the critical angle for total internal reflection. Surfaces are hi polished and are often surrounded by a reflective wrapping to direct back some of the that escapes at angles less than the critical angle. Lucite, with an index of refracti 1.49–1.51, is the most widely used material and can easily be formed into complex sh A study of the effects of various surface and outer reflector arrangements with Lucite pipes has been reported by Kilvington et al.,[176] and the data are summarized in Fig.

For an isotropic light source located near the axis of a transparent cylindrical rod, the following approach can be used to estimate the fraction of light that will be conducted in one direction along the length of the rod by successive internal reflections. Each reflected ray will also pass near the cylinder axis, and these are called *meridional rays* in the discussion of scintillation fibers later in this chapter. In the sketch below, only the light emitted within the cone angle ϕ_c will be incident on the rod surface at the critical angle θ_c or greater and therefore undergoes total internal reflection. Because the angle of reflection equals the angle of incidence, subsequent arrivals of the reflected light at the rod surface will also be above the critical angle, and this light is therefore "piped" along the rod length as in an optical fiber. The fractional solid angle subtended by this angle ϕ_c is calculated as

$$F = \frac{\Omega}{4\pi} = \frac{1}{4\pi} \int_{\phi=0}^{\phi=\phi_c} d\Omega = \frac{1}{4\pi} \int_0^{\phi_c} 2\pi \sin \phi \, d\phi$$

$$= \tfrac{1}{2}(1 - \cos \phi_c) = \tfrac{1}{2}(1 - \sin \theta_c)$$

$$= \tfrac{1}{2}(1 - n_1/n_0) \tag{8.14}$$

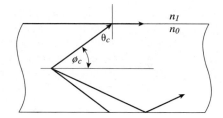

In a cylindrical rod of refractive index $n_0 = 1.5$ surrounded by air ($n_1 \cong 1$), Eq. (8.14) predicts that 16.7% of the isotropically generated light is piped in one direction. Another 16.7% is piped in the opposite direction and may be reflected back depending on the surface conditions at the end of the cylinder.

In slab geometry, the light that reaches a surface with incident angle θ greater than the critical angle θ_c will be trapped and conducted to the slab edges by total internal reflection. For rays with $\theta < \theta_c$, there is some possibility that Fresnel reflection will prevent escape at the first surface. However, then the ray is reflected back into the slab at an angle equal to its angle of incidence, and it arrives at the opposite surface again with $\theta < \theta_c$. Even if several Fresnel reflections happen to occur, eventually the ray will escape. Thus all rays that escape the slab are confined to a double ended cone with apex at the point of origin of the light, axis perpendicular to the slab surfaces, and a vertex angle of θ_c. The total escaping fraction of light is thus given by

$$E = 2 \cdot \frac{\Omega}{4\pi} = \frac{1}{2\pi} \int_{\theta=0}^{\theta=\theta_c} d\Omega = \frac{1}{2\pi} \int_0^{\theta_c} 2\pi \sin\theta \, d\theta$$

$$= 1 - \cos\theta_c = 1 - \sqrt{1 - \left(\frac{n_1}{n_0}\right)^2}$$

The fraction of light trapped in the slab is therefore

$$F = 1 - E = \sqrt{1 - \left(\frac{n_1}{n_0}\right)^2} \tag{8.15}$$

For a slab with refractive index of 1.5 surrounded by air, this result predicts that 75% of the light is piped to the edges.

To maximize the fraction of light that is piped in either geometry, one wants the refractive index of the light pipe n_0 to be as large as possible. In reality, however, the light is generated inside the scintillator, not the light pipe, and it is usually the scintillator refractive index that determines the fraction of light collected. This is particularly true when the scintillator is long in the direction perpendicular to the viewing surface, and a typical scintillation photon is multiply reflected before collection. The scintillator then acts as its own light pipe between the point of scintillation and the exit surface.

For the simple case of a cylindrical scintillation crystal and tube of equal diameter, the light pipe can be a simple cylinder of the same diameter. More often, however, the light pipe cross section shape must vary along its length in order to serve as a smooth transition between the scintillator exit surface and the PM tube end window. No matter how complex the shape of a conventional light pipe may be, the flux of light photons per unit area per unit solid angle can never be greater at any point inside the pipe than at its input. Any light pipe whose cross-sectional area decreases from scintillator to PM tube will therefore result in some light loss. If the cross-sectional area is maintained constant and sharp bends are avoided, however, the pipe can theoretically transmit all the light that enters within the acceptance angle at the input end. Light pipes with this property are called *adiabatic*[177] and obviously require the use of a PM tube with a photocathode area at least as large as the scintillator exit surface.

In cases in which the edge of a thin but large-area scintillator is to be viewed, a unique arrangement known as the strip light guide has found widespread application. As shown in Fig. 8.16, the coupling is accomplished through a number of twisted strips that are aligned at the scintillator edge but converge to a more compact pattern at the photomultiplier end. The unit can easily be made from flat plastic strips, which are then bent following heating and formed to the required shape. Practical guides for these procedures are given by Dougan et al.[178] and Piroue.[179]

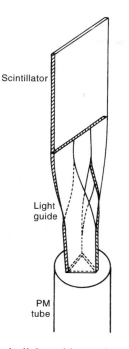

Figure 8.16 A strip light guide can be used to couple the edge of a large, flat scintillator to a PM tube.

C. Fiber Scintillators

1. GENERAL PROPERTIES

Some scintillation materials can also be fabricated as small diameter fibers in which a fraction of the scintillation light is conducted over substantial distance by total internal reflection. General reviews of scintillating fibers of this type are given in Refs. 180–182. As shown in Fig. 8.17, a common configuration consists of a core, in which the scintillation light is generated through interaction of the incident radiation, surrounded by a thin layer of cladding material. Both the core and the cladding are transparent materials, and the index of refraction of the core is higher than that of the cladding. Light rays that arrive at the core–cladding interface with an angle of incidence that is greater than the critical angle for total internal reflection are "piped" down the length of the fiber. In some cases, a light absorbing "extramural absorber" may be applied to the outer surface of the cladding to provide optical isolation.

For a typical core material with index of refraction of 1.58 surrounded by air, the index change would be large enough to pipe over a third of the light toward one end or the other. However, when a cladding material is added to protect the core surface from abrasion or accumulation of foreign material that would inhibit the light piping effect, then the index of refraction change is less pronounced and only 5 to 10% of the light is trapped. Some of

Figure 8.17 Cross section of a typical fiber scintillator. Some fraction of the emitted light is trapped by total internal reflection at the core-cladding interface.

the light escaping into the cladding can be trapped at the cladding–air interface, but this component generally represents a minor contribution because of the inferior optical quality of the outer surface due to inevitable scratching or contamination picked up during handling. Multiclad fibers consist of a core and two layers of cladding, with refractive indices that are sequentially lower for the inner and outer cladding materials. They provide two protected reflection surfaces, and tend to trap about 40% more light than comparable single-clad fibers.[183] Photomultiplier tubes can be used at both ends of the fiber to produce nearly coincident pulses from an interaction of a particle or photon. Alternatively, the light can be measured only at one end of the fiber while the opposite end is mirrored to reflect back as much as possible of the light that initially was trapped in the opposite direction.

2. PLASTIC AND LIQUID CORE FIBERS

Plastic scintillators are readily fabricated (e.g., Ref. 184) into fibers with round, square, or other cross-sectional shapes. The most common formulations are based on a core of polystyrene (refractive index $n = 1.58$) with a few percent of organic fluor. Common cladding materials are polymethylmethacrylate ($n = 1.49$) or fluorinated polymethacrylate ($n = 1.42$).

The typical diameters of common plastic fibers range from a few tenths of a millimeter to several millimeters. For plastic fibers that are smaller in diameter, there is a decrease in the observed light yield due to inhibition of the energy transfer that must occur from the plastic matrix to the organic fluor molecules that are the actual source of the scintillation light. Typical of other plastic scintillators, the emission spectrum is generally peaked in the blue region, with decay times of 2 to 4 ns.

Many plastic fibers are fabricated from this type of two-component solution in which the organic fluor is distributed in a solid matrix. In order to achieve maximum light yield in small dimensions, a relatively high fluor concentration of several percent is normally used. Other plastic fibers consist of three-component systems in which a wavelength shifter is provided as an additional constituent. The wavelength shifter, typically another organic molecule, has the function of absorbing the primary scintillation fluorescence and re-emitting it at a longer wavelength. This spectral shift can reduce the amount of self-absorption that occurs as the light propagates along the length of the fiber.

A glass capillary tube filled with liquid organic scintillator can also function in much the same way as a fiber scintillator if there is a significant reduction in index of refraction from the liquid to the glass.[185–187] This approach exploits the good light yield[188] and fast decay time of liquid scintillators (similar to that of plastic scintillators), while also offering less susceptibility to radiation damage effects compared with plastics.[189] The excellent optical quality of the liquid–glass interface can result in a light loss that is as low as 10^{-6} per reflection,[190] and capillaries with inner diameter as small as 20 μm can have attenuation length (defined below) of up to 3 meters.[191]

Scintillating fibers are sometimes grouped together to form ribbons or bundles to suit various applications. Figure 8.18 shows a ribbon array in which two staggered layers of fibers are arranged so that an incident charged particle will nearly always pass through a substantial chord length in one of the fibers. Glass capillaries can also be easily fused together into bundles as illustrated in Fig. 8.19.

Figure 8.18 A double-layer ribbon made up of fiber scintillators. (From Baumbaugh et al.[192])

Figure 8.19 Glass capillaries fused into bundles (or "multies") and operated filled with liquid scintillator. (From Ferroni and Martellotti.[193])

3. GLASS FIBERS

The glass materials commonly used as scintillators can be easily drawn into small diameter fibers. A typical cerium activated glass scintillator will have an index of refraction of about 1.59. A cladding glass is then chosen with a lower index of refraction, normally around 1.51. If an extramural absorber is required by the application, it can be a coating of black glass or polymer material.

Glasses can be drawn down to fiber diameters as small as 10 μm without appreciable loss in the scintillation efficiency. The energy transfer required from the glass matrix to the activator sites occurs over much smaller distances than in plastics, so that the drop in scintillation efficiency for small diameter plastic fibers is not observed in glass. Compared with plastics, the absolute scintillation efficiency is lower in glasses, so the light yield is less, and the decay time of 50–80 ns is substantially longer. Glass fibers incorporating ^{6}Li as a constituent have proved to be useful in some specialized neutron detection applications.

Glass fibers can also be fused into bundles or large-diameter plates that have very different optical properties than bulk scintillators of the same outer dimension. Because the scintillation light is confined to the individual fiber in which it originated, the loss of spatial resolution caused by the spread of the light in a bulk scintillator is avoided. This property has been exploited in glass fiber plates for X-ray imaging,[194–196] particle tracking,[197] and neutron[198] imaging.

4. LIGHT CAPTURE FRACTION

For the most common type of fiber with a round cross section, the sketch in Fig. 8.20 shows two extremes of the geometric conditions that can hold for the emitted light. At one extreme, if the light is emitted at the exact axis of the fiber, the light consists of "meridional rays" that always pass through the fiber center axis even after many reflections. From

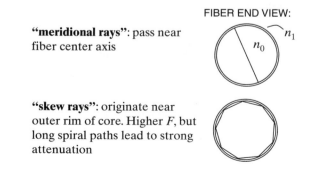

Figure 8.20 Sketch showing the distinction between "meridional rays" that originate near the axis of the fiber, and "skew rays" that represent a much longer transmission path through the fiber.

simple laws of optics [see Eq. (8.14)] one can then write an exact expression for the fraction of these rays that are captured in one direction by total internal reflection as

$$F = \frac{1}{2}\left(1 - \frac{n_1}{n_0}\right)$$

where n_1 and n_0 are the refractive indices of the cladding and core, respectively. For isotropic light emission, equal amounts are captured in both directions, so the total light capture fraction is double that given by the above expression. At the other extreme, light emitted near the outer diameter of the core will often travel a spiral path as it is propagated along the length of the fiber. These "skew rays" theoretically have a higher capture fraction, but in practice must undergo many more reflections at the core–cladding interface that tends to highly attenuate their propagation. The net result is that the actual capture fraction for uniformly distributed points of origin within a circular cross-section fiber is higher by perhaps 10 to 30% than that given by the expression above for meridional rays.

The limited trapping angle within the fiber also influences the timing properties of the signal derived from the light detected at the fiber end. In general, the spread in path lengths will be smaller than that observed for a bulk scintillator of the same length, by as much as a factor of 1.8 over a 2 m distance.[199] Because the variation of optical path lengths can be the limiting factor in the timing resolution of large scintillators, fibers can show superior timing properties under some conditions.

5. LIGHT YIELD AND PROPAGATION

The factors affecting scintillation light yield from plastic, liquid, and glass scintillators are discussed elsewhere in this chapter. Table 8.5 lists some typical values for the type of scintillation materials commonly used for fiber fabrication. These quoted yields are for low dE/dx particles such as fast electrons and will be lower for heavily ionizing particles such as recoil nuclei. As in normal scintillators, this light is emitted in all directions and only the fraction discussed earlier will be captured through total internal reflection.

The intensity of the scintillation light propagated along the length of the fiber is attenuated due to several effects:

1. Any imperfections at the core–cladding interface may disturb the total internal reflection.

2. Some of the scintillation light may be reabsorbed in the fiber due to overlap of the emission and absorption bands of the fluorescent species.

3. Rayleigh scattering from small density fluctuations in the core can deflect an optical ray so that it is no longer totally internally reflected.

The cumulative effect of these attenuation processes is often expressed as a quoted *attenuation length* for the fiber. If the attenuation probability per path length is constant,

Table 8.5 Typical Light Yield for Fiber Scintillators (for low dE/dx particles, in all directions, will be reduced by light capture fraction)

Core material	photons/keV	λ_{peak} (nm)
Glass scintillator	3–5	400
Plastic scintillator	8–10	420
Liquid scintillator	11–13	420
For comparison:		
NaI(Tl)	38	415

Figure 8.21 The spectrum of scintillation light after transmission through various lengths of fiber scintillator, showing the preferential absorption of the shorter wavelengths. (From Davis et al.[200])

one would expect the intensity of light I at a distance x from the original scintillation site to fall off exponentially as

$$\frac{I}{I_0} = e^{-x/L}$$

where I_0 is the intensity close to the site and L is the attenuation length. Typical values of L for fibers range from a few tens of centimeters to several meters.

The behavior actually observed in fibers often departs from this simple exponential behavior. One of the important causes of this deviation is the fact that the short wavelengths in the emission spectrum tend to be more readily reabsorbed than the long wavelengths. This effect is illustrated in Fig. 8.21 where the measured optical spectrum of the transmitted light is shown as a function of distance along the fiber. The average of the spectrum at large distances is shifted substantially toward the longer wavelengths due to the preferential absorption that occurs for the short wavelength components. The measured intensity of the light therefore is more likely to behave as shown in Fig. 8.22. Over short distances, the stronger absorption of the short wavelengths results in a smaller attenuation length, whereas the light that persists over longer distances is attenuated less and is characterized by a larger attenuation length.

Because of the limited energy loss of charged particles in small diameter fibers and the relatively low scintillation efficiency of organic and glass scintillators, the scintillation light generated under typical conditions in fibers is rather small. Furthermore, only a small fraction of this light is captured and piped successfully to the end of the fiber. Therefore, typical conditions may involve scintillation pulses consisting of no more than tens of photons or less. Thus, the pulses are often near the noise level of the measurement system and show relatively large event-to-event fluctuations because of the fundamental statistical fluctuations in the number of detected photons. Therefore, many applications involve using these pulses as simple indicators of the passage of a charged particle rather than in an attempt to

Figure 8.22 The total light transmitted by a fiber as a function of its length. The attenuation length is related to the slope of this plot, and the preferential absorption of the short wavelengths leads to smaller attenuation length for short fibers compared with longer ones.

accurately measure the deposited energy. Light sensors with the lowest possible noise level are important to be able to clearly distinguish the signal pulses under these demanding circumstances. Examples of successful fiber readout sensors include photomultiplier tubes (often the position-sensitive type[201]), avalanche photodiodes (see Chapter 9), and image intensifier tubes viewed by CCDs (see Chapter 11).

6. RADIATION DAMAGE

Fiber scintillators are subject to damage by high doses of radiation because of several effects. The material damage caused by radiation can lower the scintillation light yield and also affect the light absorption along the length of the fiber. Radiation-induced absorption centers tend to attenuate the blue end of the spectrum most severely. It has been observed[202,203] that exposures to gamma rays of about 10^3–10^4 Gy or fast neutron fluences of 10^{12} n/cm^2 can result in a significant lowering of the attenuation length of plastic fibers. There is some indication that liquid core glass capillaries may tolerate an order of magnitude or higher dose before the same effects are observed.[204]

As with the damage in conventional plastic scintillators discussed earlier, the magnitude of the effects in plastic fibers is strongly influenced by the nature of the radiation, the irradiation rate,[202] and the presence or absence of oxygen during the exposure.[205] A partial recovery of the scintillation yield following the exposure is usally observed that results from a slow annealing of the damage over times that may be a month or more at room temperature.

D. Wavelength Shifters

Light collection from large scintillators or complex geometries can sometimes be aided through the use of optical elements that employ *wavelength shifting* techniques. Many liquid or plastic scintillators routinely incorporate an organic additive whose function is to absorb the primary scintillation light and reradiate the energy at a longer wavelength. In that case, the objective is to better match the emission spectrum to the response peak of a photomultiplier tube. The same process can be used to help light collection by exploiting the fact that the reradiated light is emitted isotropically so that its direction bears no relation to the direction of the absorbed light. This feature allows the light to "turn corners" that would otherwise be impossible.

To illustrate, suppose that the primary scintillation light is generated within a large slab scintillator shown in Fig. 8.23. Rather than collect light from the edges of the slab, there are circumstances in which it may be preferable instead to couple the light to one or more cylindrical rods or optical fibers that run perpendicular to the slab surface. (Such could be the case, for example, if a large number of slabs were involved, and a few PM tubes were

Figure 8.23 Sections of two slab scintillators are shown coupled to a common light pipe loaded with wavelength shifter. Light passing across the air gap and entering the pipe can be absorbed, reradiated, and guided to the ends of the rod. (From Eckardt et al.[206])

used at the ends of the rods to record light from all the slabs.) The rods would then be made of optically transparent material to act as light pipes. If the scintillator and rods have similar indices of refraction, it is very difficult to couple light efficiently from the slab to a perpendicular rod in optical contact. Light that is conducted by total internal reflection along the slab arrives at the rod at angles that are not favorable for subsequent internal reflection within the rod. Also, if other slabs are in optical contact with the rod, the light piping properties of the rod are no longer preserved.

A solution to both difficulties follows if the rod is doped with a wavelength shifter[207–210] and passed through the slab in an air-filled hole of slightly larger diameter (see Fig. 8.23). Now the index change is preserved over the entire surface of the rod, and it will therefore act as a near-ideal light pipe. Some fraction of the light from the slab that arrives at a hole may pass across the air gap and enter the rod. The doping level of the wavelength shifter is adjusted so that there is a good probability of absorption of the primary light within the rod. The reradiated light is now isotropic, and one-third or more will typically be piped along the length of the rod. The shifted wavelength is now away from the strong absorption bands of the dopant, so that further loss along the rod length can largely be avoided.

Since there are several inefficient steps in this example, the overall light collection efficiency of such a scheme cannot compare with that obtainable with a more direct coupling of the photomultiplier tube to the scintillator. In some applications, however, the scintillation light yield per event may be large enough to allow one to tolerate considerable loss in the collection process. In such cases, the added flexibility afforded by wavelength shifting techniques has led to successful applications in complex detector geometries[211] or in the compact readout of large-area scintillators.[212]

The same light collection principle can be applied using plastic fibers whose core contains a wavelength-shifting fluor. These can be single- or double-clad fibers that are very similar in structure to the scintillating fibers discussed earlier. For best light propagation along the fiber, one wants a large shift between the optical absorption and emission bands for the fluor so that minimal self-absorption takes place. The most common circumstance is for the fluor to absorb strongly in the blue portion of the spectrum (where most scintillation materials have prominent emission) and to re-emit green or yellow light. Under good conditions, the shifted light can be conducted over a meter or more of the fiber without excessive loss. For circumstances in which the light must be transmitted larger distances from the detector to the light sensor, a clear fiber with better light conduction can be matched to the primary fiber.[213] In some designs,[214,215] the wavelength-shifting fibers are positioned in grooves machined into the surface of the primary scintillation detector. Wavelength-shifting fibers can also be useful to collect the light emitted in other types of detectors, for example, for high-pressure gas proportional scintillation chambers[216,217] where direct viewing with a light sensor may not be convenient.

One of the important properties is the quantum efficiency of the wavelength shifter, or the probability that a wavelength-shifted photon will be emitted per absorbed photon. This efficiency can be as high as 90% for some organic fluors, but it is more typically[218–220] in the range of 70–80%. High quantum efficiencies are obviously needed to preserve the statistical information carried by the number of original photons.

Another important property is the decay time of the wavelength-shifting molecules. The absorption and re-radiation of the primary scintillation light can substantially slow the light emission if the lifetime of the wavelength shifting excited states is comparable with or longer than that of the primary fluor. For example, in the wavelength shifter known as *K27* (a green-emitting fluor), the decay time is 12 ns. Because this time is much longer than the 2–4 ns decay time of typical organic fluors, the time characteristics of the collected light are now dominated by the wavelength shifter rather than the primary scintillator. For critical timing situations, faster wavelength shifters can be chosen.

PROBLEMS

8.1 Calculate the scintillation efficiency of anthracene if 1 MeV of particle energy loss creates 20,300 photons with average wavelength of 447 nm.

8.2 Assuming a decay constant of 230 ns, how much time is required for a NaI(Tl) scintillation event to emit 99% of the total light yield?

8.3 Assuming that the scintillation light pulse in each case is a pure exponential, find the ratio of the maximum brightness (rate of photon emission) of pulses generated by equal electron energy deposition in NaI(Tl) and anthracene.

8.4 Make a selection between a typical inorganic scintillator [say, NaI(Tl)] and a typical organic (say, a plastic scintillator) on the basis of the following properties:

  (a) Speed of response.

  (b) Light output.

  (c) Linearity of light with deposited energy.

  (d) Detection efficiency for high-energy gamma rays.

  (e) Cost.

8.5 Explain the function of the activator added in trace quantities to many inorganic scintillators. Why are they not needed in organic scintillators?

8.6 Which scintillation material is most efficient at converting the energy of a 2 MeV electron into light?

8.7 Explain the following statement: Organic crystalline scintillators remain good energy-to-light converters when dissolved in a solvent, while inorganics no longer function as scintillators if dissolved.

8.8 Scintillation light is emitted isotropically within a slab of plastic scintillator (see Table 8.1). If the other dimensions of the slab are assumed to be infinite, calculate the fraction of the light that escapes from either slab surface.

8.9 The dark-adapted human eye may be able to detect as few as 10 visible photons as a single flash. Will an observer with pupil diameter of 3 mm be able to see individual scintillation events caused by a 1 MeV beta particle in NaI(Tl) while viewing the surface of the scintillator at a distance of 10 cm?

8.10 (a) A 1 MeV fast electron passes across the 0.3 mm diameter of a plastic fiber scintillator. From the data given in Chapter 2, estimate the deposited energy.

  (b) Assuming a reasonable scintillation efficiency, calculate the corresponding number of scintillation photons created along the track.

  (c) The refractive index for the core and the cladding are 1.58 and 1.49, respectively, and the fiber has an attenuation length of 2 m. Estimate the number of scintillation photons arriving at one end of the fiber that is 1 m from the point of interaction.

REFERENCES

1. J. B. Birks, *The Theory and Practice of Scintillation Counting*, Pergamon Press, Oxford, 1964.
2. *Selected Papers on Phosphors, LEDs, and Scintillators*, M. J. Weber, ed., SPIE Optical Engineering Press, Bellingham, Washington, 1998.
3. J. B. Birks and R. W. Pringle, *Proc. R. Soc. Edinburgh Sect. A* **70**, 233 (1972).
4. D. B. Oliver and G. F. Knoll, *IEEE Trans. Nucl. Sci.* **NS-15**(3), 122 (1968).
5. C. Zorn et al., *IEEE Trans. Nucl. Sci.* **37**(2), 487 (1990).
6. J. K. Walker, *Nucl. Instrum. Meth.* **68**, 131 (1969).
7. A. D. Bross and A. Pla-Dalmau, *IEEE Trans. Nucl. Sci.* **39**(5), 1199 (1992).
8. E. Biagtan, E. Goldberg, J. Harmon, and R. Stephens, *Nucl. Instrum. Meth.* **B93**, 296 (1994).
9. V. G. Vasil'chanko et al., *Nucl. Instrum. Meth.* **A369**, 55 (1996).
10. E. Biagtan et al., *Nucl. Instrum. Meth.* **B108**, 125 (1996).
11. V. G. Senchishin et al., *Nucl. Instrum. Meth.* **A364**, 253 (1995).
12. G. Bendiscioli et al., *Nucl. Instrum. Meth.* **206**, 471 (1983).
13. M. L. Muga, D. J. Burnsed, and W. E. Steeger, *Nucl. Instrum. Meth.* **104**, 605 (1972).
14. N. N. Ajitanand and K. N. Iyengar, *Nucl. Instrum. Meth.* **133**, 71 (1976).
15. H. Geissel et al., *Nucl. Instrum. Meth.* **144**, 465 (1977).
16. C. Manduchi, M. T. Russo-Manduchi, and G. F. Segato, *Nucl. Instrum. Meth.* **A243**, 453 (1986).
17. E. Norbeck, T. P. Dubbs, and L. G. Sobotka, *Nucl. Instrum. Meth.* **A262**, 546 (1987).
18. L. W. Weathers and M. B. Tsang, *Nucl. Instrum. Meth.* **A381**, 567 (1996).
19. F. D. Brooks, W. A. Cilliers, and M. S. Allie, *Nucl. Instrum. Meth.* **A240**, 338 (1985).
20. T. Batsch and M. Moszynski, *Nucl. Instrum. Meth.* **125**, 231 (1975).
21. J. W. Kohl, *Nucl. Instrum. Meth.* **125**, 413 (1975).
22. K. Ettling and W. Von Witsch, *Nucl. Instrum. Meth.* **148**, 299 (1978).
23. G. Bendiscioli, V. Filippini, C. Marciano, A. Rotondi, and A. Zenoni, *Nucl. Instrum. Meth..* **227**, 478 (1984).
24. A. Galindo-Uribarri et al., *Nucl. Instrum. Meth.* **A301**, 457 (1991).
25. Z. Youxiong et al., *Nucl. Instrum. Meth.* **A355**, 464 (1995).
26. L. Muga, *Nucl. Instrum. Meth.* **124**, 541 (1975).
27. M. L. Muga and J. D. Bridges, *Nucl. Instrum. Meth.* **134**, 143 (1976).
28. I. Kanno and T. Nakagome, *Nucl. Instrum. Meth.* **A251**, 108 (1986).

29. I. Kanno and T. Nakagome, *Nucl. Instrum. Meth.* **A244**, 551 (1986).

30. T. Batsch and M. Moszynski, *Nucl. Instrum. Meth.* **123**, 341 (1975).

31. C. B. Ashford, I. B. Berlman, J. M. Flournoy, L. A. Franks, S. G. Iversen, and S. S. Lutz, *Nucl. Instrum. Meth. Phys. Res.* **A243**, 131 (1986).

32. Z. H. Cho, I. Ahn, and C. M. Tsai, *IEEE Trans. Nucl. Sci.* **NS-21**(1), 218 (1974).

33. L. A. Eriksson, C. M. Tsai, Z. H. Cho, and C. R. Hurlbut, *Nucl. Instrum. Meth.* **122**, 373 (1974).

34. Z. H. Cho, C. M. Tsai, and L. A. Eriksson, *IEEE Trans. Nucl. Sci.* **NS-22**(1), 72 (1975).

35. E. Brannon and G. L. Olde, *Radiat. Res.* **16**, 1 (1962).

36. R. L. Craun and D. L. Smith, *Nucl. Instrum. Meth.* **80**, 239 (1970).

37. M. Hirschberg et al., *IEEE Trans. Nucl. Sci.* **39(4)**, 511 (1992).

38. D. L. Smith, R. G. Polk, and T. G. Miller, *Nucl. Instrum. Meth.* **64**, 157 (1968).

39. J. B. Czirr, *Nucl. Instrum. Meth.* **25**, 106 (1963).

40. K. H. Maier and J. Nitschke, *Nucl. Instrum. Meth.* **59**, 227 (1968).

41. D. Clark, *Nucl. Instrum. Meth.* **117**, 295 (1974).

42. N. A. Weir, *Int. J. Appl. Radiat. Isotopes* **23**, 371 (1972).

43. A. Raviart and V. Koechlin, *Nucl. Instrum. Meth.* **29**, 45 (1964).

44. F. J. Lynch, *IEEE Trans. Nucl. Sci.* **NS-15**(3), 102 (1968).

45. B. Bengtson and M. Moszynski, *Nucl. Instrum. Meth.* **117**, 227 (1974).

46. T. M. Kelly, J. A. Merrigan, and R. M. Lambrecht, *Nucl. Instrum. Meth.* **109**, 233 (1973).

47. P. B. Lyons and J. Stevens, *Nucl. Instrum. Meth.* **114**, 313 (1974).

48. F. J. Lynch, *IEEE Trans. Nucl. Sci.* **NS-22**(1), 58 (1975).

49. S. Sanyal, S. C. Pancholi, and S. L. Gupta, *Nucl. Instrum. Meth.* **136**, 157 (1976).

50. P. B. Lyons, C. R. Hurlbut, and L. P. Hocker, *Nucl. Instrum. Meth.* **133**, 175 (1976).

51. P. B. Lyons et al., *IEEE Trans. Nucl. Sci.* **NS-24**(1), 177 (1977).

52. K. G. Tirsell et al., *IEEE Trans. Nucl. Sci.* **NS-24**(1), 250 (1977).

53. M. Kurata et al., *Nucl. Instrum. Meth.* **A349**, 447 (1994).

54. S. Albergo et al., *Nucl. Instrum. Meth.* **A362**, 423 (1995).

55. X. H. Yang et al., *Nucl. Instrum. Meth.* **A354**, 270 (1995).

56. M. Moszynski and B. Bengtson, *Nucl. Instrum. Meth.* **142**, 417 (1977).

57. B. Sipp and J. A. Miehe, *Nucl. Instrum. Meth.* **114**, 255 (1974).

58. M. Bantel et al., *Nucl. Instrum. Meth.* **226**, 394 (1984).

59. P. Harihar et al., *Nucl. Instrum. Meth.* **A336**, 176 (1993).

60. L. M. Bollinger and G. E. Thomas, *Rev. Sci. Instrum.* **32**, 1044 (1961).

61. T. G. Miller, *Nucl. Instrum. Meth.* **63**, 121 (1968).

62. J. B. Czirr, *Nucl. Instrum. Meth.* **88**, 321 (1970).

63. R. A. Winyard and G. W. McBeth, *Nucl. Instrum. Meth.* **98**, 525 (1972).

64. D. B. C. B. Syme and G. I. Crawford, *Nucl. Instrum. Meth.* **104**, 245 (1972).

65. I. B. Berlman and O. J. Steingraber, *Nucl. Instrum. Meth.* **108**, 587 (1973).

66. M. Ahmed, *Nucl. Instrum. Meth.* **143**, 255 (1977).

67. M. Moszynski et al., *Nucl. Instrum. Meth.* **A317**, 262 (1992).

68. M. Moszynski et al., *Nucl. Instrum. Meth.* **A350**, 226 (1994).

69. G. Ranucci, A. Goretti, and P. Lombardi, *Nucl. Instrum. Meth.* **A412**, 374 (1998).

70. R. B. Murray, *IEEE Trans. Nucl. Sci.* **NS-22**(1), 54 (1975).

71. R. G. Kaufman, W. B. Hadley, and H. N. Hersh, *IEEE Trans. Nucl. Sci.* **NS-17** (3), 82 (1970).

72. C. L. Melcher et al., *IEEE Trans. Nucl. Sci.* **NS-32**(1), 529 (1985).

73. E. V. Sysoeva, V. A. Tarasov, O. V. Zelenskaya, and V. A. Sulga, *Nucl. Instrum. Meth.* **A414**, 274 (1998).

74. E. Sakai, *IEEE Trans. Nucl. Sci.* **NS-34**(1), 418 (1987).

75. I. Holl, E. Lorenz, and G. Mageras, *IEEE Trans. Nucl. Sci.* **35**(1), 105 (1988).

76. W. W. Moses and S. E. Derenzo, *IEEE Trans. Nucl. Sci.* **36(1)**, 173 (1989).

77. S. Brollo, G. Zanella, and R. Zannoni, *Nucl. Instrum. Meth.* **A293**, 601 (1990).

78. M. Moszynski et al., *IEEE Trans. Nucl. Sci.* **44(3)**, 1052 (1997).

79. R. Hofstadter, *Phys. Rev.* **74**, 100 (1948).

80. W. Mengesha, T. D. Taulbee, B. D. Rooney, and J. D. Valentine, *IEEE Trans. Nucl. Sci.* **45(3)**, 456 (1998).

81. S. Koicki, A. Koicki, and V. Ajdacic, *Nucl. Instrum. Meth.* **108**, 297 (1973).

82. C. F. G. Delaney and A. M. Lamki, *Int. J. Appl. Radiat. Isotopes* **19**, 169 (1968).

83. J. S. Schweitzer and W. Ziehl, *IEEE Trans. Nucl. Sci.* **NS-30**(1), 380 (1983).

84. C. J. Crannell, R. J. Kurz, and W. Viehmann, *Nucl. Instrum. Meth.* **115**, 253 (1974).

85. S. Usuda, A. Mihara, and H. Abe, *Nucl. Instrum. Meth.* **A321**, 247 (1992).

86. M. Moszynski et al., *Nucl. Instrum. Meth.* **A336**, 587 (1993).

87. B. Ye et al., *Nucl. Instrum. Meth.* **A345**, 115 (1994).

88. V. V. Nagarkar et al., *IEEE Trans. Nucl. Sci.* **45(3)**, 492 (1998).

89. T. Jing et al., *IEEE Trans. Nucl. Sci.* **39(5)**, 1195 (1992).

90. J. D. Valentine, D. K. Wehe, G. F. Knoll, and C. E. Moss, *IEEE Trans. Nucl. Sci.* **40(4)**, 1267 (1993).

91. J. D. Valentine et al., *Nucl. Instrum. Meth.* **A325**, 147 (1993).

92. S. Keszthelyi-Landori and G. Hrehuss, *Nucl. Instrum. Meth.* **68**, 9 (1969).

93. P. E. Francois and D. T. Martin, *Int. J. Appl. Radiat. Isotopes* **21**, 687 (1970).

94. M. Moszynski et al., *Nucl. Instrum. Meth.* **188**, 403 (1981).

95. M. J. Weber and R. R. Monchamp, *J. Appl. Phys.* **44**, 5495 (1973).

96. R. G. L. Barnes et al., *IEEE Trans. Nucl. Sci.* **NS-31**(1), 249 (1984).

97. C. L. Melcher, *Nucl. Instrum. Meth.* **B40/41**, 1214 (1989).

98. M. Kobayashi, M. Ishii, Y. Usuki, and H. Yahagi, *Nucl. Instrum. Meth.* **A349**, 407 (1994).

99. D. R. Kinloch, W. Novak, P. Raby, and I. Toepke, *IEEE Trans. Nucl. Sci.* **41(4)**, 752 (1994).

100. T. Fazzini et al., *Nucl. Instrum. Meth.* **A410**, 213 (1998).

101. S. Ph. Burachas et al., *Nucl. Instrum. Meth.* **A369**, 164 (1996).

102. M. Laval et al., *Nucl. Instrum. Meth.* **206**, 169 (1983).

103. K. Wisshak and F. Kaeppeler, *Nucl. Instrum. Meth.* **227**, 91 (1984).

104. Y. C. Zhu et al., *Nucl. Instrum. Meth.* **A244**, 577 (1986).

105. H. J. Karwowski et al., *Nucl. Instrum. Meth.* **A245**, 207 (1986).

106. M. R. Farukhi and C. F. Swinehart, *IEEE Trans. Nucl. Sci.* **NS-18**(1), 200 (1971).

107. P. Schotanus et al., *Nucl. Instrum. Meth.* **A238**, 564 (1985).

108. P. Schotanus et al., *Nucl. Instrum. Meth.* **A259**, 586 (1987).

109. P. Dorenbos et al., *IEEE Trans. Nucl. Sci.* **40(4)**, 424 (1993).

110. M. Moszynski et al., *Nucl. Instrum. Meth.* **226**, 534 (1984).

111. H. Mach, R. L. Gill and M. Moszynski, *Nucl. Instrum. Meth.* **A280**, 49 (1989).

112. R. Novotny, *IEEE Trans. Nucl. Sci.* **38(2)**, 379 (1991).

113. S. Kubota, S. Sakuragi, S. Hashimoto, and J. Ruan (Gen), *Nucl. Instrum. Meth.* **A268**, 275 (1988).

114. S. Keszthelyi-Landori et al., *Nucl. Instrum. Meth.* **A303**, 374 (1991).

115. P. Schotanus, R. Kamermans, and P. Dorenbos, *IEEE Trans. Nucl. Sci.* **37(2)**, 177 (1990).

116. D. F. Anderson, *Nucl. Instrum. Meth.* **A287**, 606 (1990).

117. E. Auffray et al., *Nucl. Instrum. Meth.* **A383**, 367 (1996).

118. C. W. E. van Eijk, *Nucl. Instrum. Meth.* **A392**, 285 (1997).

119. H. Ishibashi et al., *IEEE Trans. Nucl. Sci.* **45(3)**, 518 (1998).

120. M. Tanaka et al., *Nucl. Instrum. Meth.* **A404**, 283 (1998).

121. S. Nakayama et al., *Nucl. Instrum. Meth.* **A404**, 34 (1998).

122. M. Kobayashi and M. Ishii, *Nucl. Instrum. Meth.* **B61**, 491 (1991).

123. S. Baccaro et al., *Nucl. Instrum. Meth.* **A361**, 209 (1995).

124. M. Kapusta, J. Pawelke, and M. Moszynski, *Nucl. Instrum. Meth.* **A404**, 413 (1998).

125. M. Moszynski et al., *Nucl. Instrum. Meth.* **A404**, 157 (1998).

126. A. Del Guerra, G. Di Domenico, R. Pani, and G. Zavattini, *IEEE Trans. Nucl. Sci.* **44(6)**, 2415 (1997).

127. M. Moszynski et al., *Nucl. Instrum. Meth.* **A345**, 461 (1994).

128. T. Ludziejewski et al., *Nucl. Instrum. Meth.* **A398**, 287 (1997).

129. C. L. Melcher and J. S. Schweitzer, *Nucl. Instrum. Meth.* **A314**, 212 (1992).

130. T. Ludziejewski et al., *IEEE Trans. Nucl. Sci.* **42(4)**, 328 (1995).

131. P. Dorenbos et al., *IEEE Trans. Nucl. Sci.* **41(4)**, 735 (1994).

132. H. Suzuki, T. A. Tombrello, C. L. Melcher, and J. S. Schweitzer, *IEEE Trans. Nucl. Sci.* **40(4)**, 380 (1993).

133. A. Lempicki and J. Glodo, *Nucl. Instrum. Meth.* **A416**, 333 (1998).

134. M. Moszynski et al., *Nucl. Instrum. Meth.* **A385**, 123 (1997).

135. W. W. Moses et al., *IEEE Trans. Nucl. Sci.* **42(4)**, 275 (1995).

136. A. Lempicki et al., *IEEE Trans. Nucl. Sci.* **42(4)**, 280 (1995).

137. A. Lampicki et al., *IEEE Trans. Nucl. Sci.* **43(3)**, 1316 (1996).

138. K. S. Shah, P. Bennett, and M. R. Squillante, *IEEE Trans. Nucl. Sci.* **43(3)**, 1267 (1996).

139. C. Dujardin et al., *IEEE Trans. Nucl. Sci.* **45(3)**, 467 (1998).

140. M. Kapusta, M. Balcerzyk, M. Moszynski, and J. Pawelke, *Nucl. Instrum. Meth.* **A421**, 610 (1999).

141. L. M. Bollinger, G. E. Thomas, and R. J. Ginther, *Nucl. Instrum. Meth.* **17**, 97 (1962).

142. A. D. Bross, *Nucl. Instrum. Meth.* **A247**, 319 (1986).

143. A. W. Dalton, *Nucl. Instrum. Meth.* **A259**, 545 (1987).

144. A. W. Dalton, *Nucl. Instrum. Meth.* **A254**, 361 (1987).

145. P. Pavan, G. Zanella, R. Zannoni, and P. Polato, *Nucl. Instrum. Meth.* **B61**, 487 (1991).

146. C. Greskhovich and S. Duclos, *Annu. Rev. Mater. Sci.* **27**, 69 (1997).

147. P. Lindblom and O. Solin, *Nucl. Instrum. Meth.* **A268**, 204 (1988).

148. P. Lindblom and O. Solin, *Nucl. Instrum. Meth.* **A268**, 212 (1988).

149. M. Mutterer, *Nucl. Instrum. Meth.* **196**, 73 (1982).

150. M. Mutterer, J. Pannicke, K. Scheele, W. Spreng, J. P. Theobald, and P. Wastyn, *IEEE Trans. Nucl. Sci.* **NS-27(1)**, 184 (1980).

151. M. Suzuki, J. Ruangen, and S. Kubota, *Nucl. Instrum. Meth.* **192**, 565 (1982).

152. M. Suzuki, *Nucl. Instrum. Meth.* **215**, 345 (1983).

153. P. Grimm, F.-J. Hambsch, M. Mutterer, J. P. Theobald, and S. Kubota, *Nucl. Instrum. Meth.* **A262**, 394 (1987).

154. P. Belli et al., *Nucl. Instrum. Meth.* **A310**, 150 (1991).

155. M. Miyajima, S. Sasaki, H. Tawara, and E. Shibamura, *IEEE Trans. Nucl. Sci.* **39(4)**, 536 (1992).

156. R. van Sonsbeck, C. W. E. van Eijk, and R. W. Hollander, *Nucl. Instrum. Meth.* **A367**, 362 (1995).

157. J. Seguinot, J. Tischhauser, and T. Ypsilantis, *Nucl. Instrum. Meth.* **A354**, 280 (1995).

158. S. Kubota et al., *Nucl. Instrum. Meth.* **196**, 101 (1982).

159. W. Baum, S. Götz, H. Heckwolf, P. Heeg, M. Mutterer, and J. P. Theobald, *IEEE Trans. Nucl. Sci.* **35(1)**, 102 (1988).

160. A. Helaly et al., *Nucl. Instrum. Meth.* **A241**, 169 (1985).

161. T. Doke, K. Masuda, and E. Shibamura, *Nucl. Instrum. Meth.* **A291**, 617 (1990).

162. N. Ishida et al., *Nucl. Instrum. Meth.* **A384**, 380 (1997).

163. N. Ishida et al., *Nucl. Instrum. Meth.* **A327**, 152 (1993).

164. R. Zhu, *Nucl. Instrum. Meth.* **A413**, 297 (1998).

165. M. Ishii and M. Kobayashi, *Prog. in Crystal Growth and Char. of Mat.* **23(1-4)**, 245 (1992).

166. F. Tonetto el al., *Nucl. Instrum. Meth.* **A420**, 181 (1999).

167. J. Bea et al., *Nucl. Instrum. Meth.* **A350**, 184 (1994).

168. V. A. Baranov, V. V. Filchenkov, A. D. Konin, and V. V. Zhuk, *Nucl. Instrum. Meth.* **A374**, 335 (1996).

169. C. E. Ordonez, W. Chang, J. Liu, and D. Gunter, *IEEE Trans. Nucl. Sci.* **44(3)**, 1237 (1997).

170. D. Vozza, C. Moisan, and S. Paquet, *IEEE Trans. Nucl. Sci.* **44(2)**, 179 (1997).

171. S. A. McElhaney, J. A. Ramsey, M. L. Bauer, and M. M. Chiles, *Nucl. Instrum. Meth.* **A299**, 111 (1990).

172. D. R. Winn and M. Whitmore, *IEEE Trans. Nucl. Sci.* **36(1)**, 256 (1989).

173. P. R. Bell, "The Scintillation Method," in *Beta- and Gamma-Ray Spectroscopy* (K. Siegbahn, ed.), Elsevier-North Holland, Amsterdam, 1955.

174. G. Keil, *Nucl. Instrum. Meth.* **87**, 111 (1970).

175. S. P. Ahlen, B. G. Cartwright, and G. Tarle, *Nucl. Instrum. Meth.* **143**, 513 (1977).

176. A. I. Kilvington, C. A. Baker, and P. Illinesi, *Nucl. Instrum. Meth.* **80**, 177 (1970).

177. R. L. Garwin, *Rev. Sci. Instrum.* **23**, 755 (1952).

178. P. Dougan, T. Kivikas, K. Lugner, W. Ramsay, and W. Stiefler, *Nucl. Instrum. Meth.* **78**, 317 (1970).

179. P. A. Piroue, Conference on Instrumentation Techniques in Nuclear Pulse Analysis, National Academy of Sciences, National Research Council Publication 1184 (Nuclear Science Series Report #40) 1964.

180. T. O. White, *Nucl. Instrum. Meth.* **A273**, 820 (1988).

181. R. Ruchti, Nuclear Physics B. (Proc. Suppl.) **44**, 308 (1995).

182. H. Leutz, *Nucl. Instrum. Meth.* **A364**, 422 (1995).

183. A. P. Ivashkin, Yu. G. Kudenko, O. V. Mineev, and J. Imazato, *Nucl. Instrum. Meth.* **A394**, 321 (1997).

184. Ph. Rebourgeard et al., *Nucl. Instrum. Meth.* **A427**, 543 (1999).

185. S. V. Golovkin et al., *Nucl. Instrum. Meth.* **A305**, 385 (1991).

186. M. Adinolfi et al., *Nucl. Instrum. Meth.* **A315**, 177 (1992).

187. P. Annis et al., *Nucl. Instrum. Meth.* **A367**, 377 (1995).

188. A. Cardini et al., *Nucl. Instrum. Meth.* **A361**, 129 (1995).

189. S. V. Golovkin et al., *Nucl. Instrum. Meth.* **A362**, 283 (1995).

190. P. Annis et al., *Nucl. Instrum. Meth.* **A386**, 72 (1997).

191. S. Buontempo et al., *Nucl. Instrum. Meth.* **A360**, 7 (1995).

192. B. Baumbaugh et al., *Nucl. Instrum. Meth.* **A345**, 271 (1994).

193. F. Ferroni and G. Martellotti, *Nucl. Instrum. Meth.* **A368**, 224 (1995).

194. H. Shao, D. W. Miller and C. R. Pearsall, *IEEE Trans. Nucl. Sci.* **38(2)**, 845 (1991).

195. P. Ottonello, G. A. Rottigni, G. Zanella, and R. Zannoni, *Nucl. Instrum. Meth.* **A323**, 485 (1992).

196. G. Gennaro, M. Malvestio, G. Zanella, and R. Zannoni, *Nucl. Instrum. Meth.* **A382**, 567 (1996).

197. C. D'Ambrosio, H. Leutz, T. Shimizu, and O. Shinji, *Nucl. Instrum. Meth.* **A325**, 161 (1993).

198. P. L. Reeder et al., *Nucl. Instrum. Meth.* **A402**, 155 (1998).

199. M. Kuhlen et al., *Nucl. Instrum. Meth.* **A301**, 223 (1991).

200. A. J. Davis et al., *Nucl. Instrum. Meth.* **A276**, 347 (1989).

201. V. Agoritsas et al., *Nucl. Instrum. Meth.* **A357**, 78 (1995).

202. K. Wick et al., *Nucl. Instrum. Meth.* **B61**, 472 (1991).

203. A. Murakami, H. Yoshinaka, and M. Goto, *IEEE Trans. Nucl. Sci.* **40(4)**, 495 (1993).

204. A. Cardini et al., *Nucl. Instrum. Meth.* **A346**, 163 (1994).

205. K. G. Young et al., *IEEE Trans. Nucl. Sci.* **40(4)**, 461 (1993).

206. V. Eckardt et al., *Nucl. Instrum. Meth.* **155**, 389 (1978).

207. W. Selove, W. Kononenko, and B. Wilsker, *Nucl. Instrum. Meth.* **161**, 233 (1979).

208. W. Viehmann and R. L. Frost, *Nucl. Instrum. Meth.* **167**, 405 (1979).

209. W. Kononenko, W. Selove, and G. E. Theodosiou, *Nucl. Instrum. Meth.* **206**, 91 (1983).

210. C. Aurouet et al., *Nucl. Instrum. Meth.* **211**, 309 (1983).

211. V. I. Kryshkin and A. T. Ronzhin, *Nucl. Instrum. Meth.* **A247**, 583 (1986).

212. G. Keil, *Nucl. Instrum. Meth.* **83**, 145 (1970).

213. T. Asakawa et al., *Nucl. Instrum. Meth.* **A340**, 458 (1994).

214. K. Hara et al., *Nucl. Instrum. Meth.* **A365**, 370 (1995).

215. D. V. H. Bengis and R. M. Sealock, *IEEE Trans. Nucl. Sci.* **42(4)**, 374 (1995).

216. A. Parsons et al., *IEEE Trans. Nucl. Sci.* **36(1)**, 931 (1989).

217. T. K. Edberg et al., *Nucl. Instrum. Meth.* **A316**, 38 (1992).

218. P. L. Hink et al., SPIE Proceedings **1549**, 193 (1991).

219. A. Pla-Dalmau, G. W. Foster, and G. Zhang, *Nucl. Instrum. Meth.* **A361**, 192 (1995).

220. C. H. Lally, G. J. Davies, W. G. Jones, and N. J. T. Smith, *Nucl. Instrum. Meth.* **B117**, 421 (1996).

Chapter 9

Photomultiplier Tubes and Photodiodes

I. INTRODUCTION

The widespread use of scintillation counting in radiation detection and spectroscopy would be impossible without the availability of devices to convert the extremely weak light output of a scintillation pulse into a corresponding electrical signal. The photomultiplier (PM) tube accomplishes this task remarkably well, converting light signals that typically consist of no more than a few hundred photons into a usable current pulse without adding a large amount of random noise to the signal. Although there has been some progress (described later in this chapter) in the development of semiconductor photodiodes for use with scintillators, the PM tube remains the most widely used device for this purpose. A great variety of commercially available PM tubes are sensitive to radiant energy in the ultraviolet, visible, and near-infrared regions of the electromagnetic spectrum. They have many applications in optical spectroscopy, laser measurements, and astronomy. Useful reviews of PM tube properties and design characteristics can be found in Ref. 1. In this chapter the discussion is limited to those designs of primary interest for scintillation counting. Morton has published[2] a very readable historical account of the development of tubes for this purpose. Useful guides and standards for the testing of PM tubes for scintillation counting have been developed as part of a series of such standards published[3] by the IEEE.

The simplified structure of a typical photomultiplier tube is illustrated in Fig. 9.1. An outer (usually glass) envelope serves as a pressure boundary to sustain vacuum conditions inside the tube that are required so that low-energy electrons can be accelerated efficiently by internal electric fields. The two major components inside the tube are a photosensitive layer, called the *photocathode*, coupled to an electron multiplier structure. The photocathode serves to convert as many of the incident light photons as possible into low-energy electrons. If the light consists of a pulse from a scintillation crystal, the photoelectrons produced will also be a pulse of similar time duration. Because only a few hundred photoelectrons may be involved in a typical pulse, their charge is too small at this point to serve as a convenient electrical signal. The electron multiplier section in a PM tube provides an efficient collection geometry for the photoelectrons as well as serving as a near-ideal amplifier to greatly increase their number. After amplification through the multiplier structure, a typical scintillation pulse will give rise to $10^7 - 10^{10}$ electrons, sufficient to serve as the charge signal for the original scintillation event. This charge is conventionally collected at the anode or output stage of the multiplier structure.

Figure 9.1 Basic elements of a PM tube. (From Ref. 1.)

Most photomultipliers perform this charge amplification in a very linear manner, producing an output pulse at the anode that remains proportional to the number of original photoelectrons over a wide range of amplitude. Much of the timing information of the original light pulse is also retained. Typical tubes, when illuminated by a very short duration light pulse, will produce an electron pulse with a time width of a few nanoseconds after a delay time of 20–50 ns.

PM tubes are commercially available in a wide variety of sizes and properties. We begin our discussion with the important elements of PM tube design and their influence on overall performance.

II. THE PHOTOCATHODE

A. The Photoemission Process

The first step to be performed by the PM tube is the conversion of incident light photons into electrons. This process of photoemission can be thought of as occurring in three sequential stages: (1) the absorption of the incident photon and transfer of energy to an electron within the photoemissive material, (2) the migration of that electron to the surface, and (3) the escape of the electron from the surface of the photocathode.

The energy that can be transferred from the photon to an electron in the first step is given by the quantum energy of the photon $h\nu$. For blue light typical of that emitted by many scintillators, the quantum energy is about 3 eV. In step 2, some of that energy will be lost through electron–electron collisions in the migration process. Finally, in step 3, there must be sufficient energy left for the electron to overcome the inherent potential barrier that always exists at any interface between material and vacuum. This potential barrier (often called the *work function*) is normally greater than 3 or 4 eV for most metals but can be as low as 1.5–2 eV for suitably prepared semiconductors.

From these energy considerations, some general comments can be made regarding photocathodes. First, the finite potential barrier in step 3 imposes a minimum energy on the incoming light photon even if all other energy losses are zero. All photocathodes therefore have a long-wavelength (small v) cutoff that is usually in the red or near-infrared portion of the spectrum.[†] Even for higher-energy light photons, the surface barrier should be as low as possible to maximize the number of escaping electrons. The rate of energy loss as the electron migrates to the surface should be kept small in order to maximize the depth in the material (called the *escape depth*) at which electrons may originate and still reach the surface with sufficient energy to overcome the potential barrier. The rate of energy loss in metals is relatively high, and an electron can travel no more than a few nanometers before its energy drops below the potential barrier. Therefore, only the very thin layer of material lying within a few nanometers of the surface will contribute any photoelectrons from common metals. In semiconductors, the rate of energy loss is much lower and the escape depth can extend to about 25 nm. This, however, is still a very small thickness even with respect to stopping visible light. Photocathodes of this thickness are semitransparent and will cause less than half the visible light to interact within the photosensitive layer. Therefore, such photocathodes cannot come close to converting all the visible light photons into electrons, no matter how low the potential barrier may be.

In order for an incident light photon to be absorbed in a semiconductor, its energy must exceed the bandgap energy E_g. (For a discussion of the band structure in semiconductors, see Chapter 11.) The absorption process consists simply of elevating an electron from the normally populated valence band to the conduction band. Within about a picosecond, these electrons rapidly lose energy through phonon interactions with the crystal until their energy is at the bottom of the conduction band. In normal semiconductors, the electron potential outside the surface is higher than the bottom of the conduction band by an amount called the *electron affinity*. If an electron is to escape, it must reach the surface in the short time before phonon interactions have reduced its energy to the bottom of the conduction band. The electron, however, will remain at the bottom of the conduction band for perhaps another 100 ps before recombining with a hole and dropping to the valence band. The use of *negative electron affinity* materials, discussed more fully later in this chapter, leads to a much greater escape depth by allowing electrons that have dropped to the bottom of the conduction band to also escape if they reach the surface (see Fig. 9.4).

B. Spontaneous Electron Emission

The surface potential barrier influences another important property of photocathodes: *thermionic noise.* Normal conduction electrons within the photocathode material will always have some thermal kinetic energy that, at room temperature, will average about 0.025 eV. There is a spread in this distribution, however, and those electrons at the extreme upper end of the distribution can occasionally have an energy that exceeds the potential barrier. If that electron is close enough to the surface, it may escape and give rise to a spontaneous thermally induced signal. In metals, the thermal emission rate is low ($\sim 100/\text{m}^2 \cdot \text{s}$)

[†]The presence of low concentrations of impurity states in some photocathode materials can result in a small but measurable sensitivity beyond the normal cutoff wavelength into the infrared region of the spectrum.

because of the relatively high potential barrier. In semiconductors, the lower potential barrier leads to thermal emission rates as high as 10^6–$10^8/\mathrm{m}^2 \cdot \mathrm{s}$. Their superior photosensitivity is therefore achieved only at the price of a higher noise rate from thermally stimulated electron emission.

On theoretical grounds, the rate of thermionic emission should rise exponentially with photocathode temperature. The observed rate of spontaneous electron emission does increase with temperature, but the dependence is generally much milder than this prediction, indicating the influence of nonthermal effects in determining the overall emission rate.

C. Fabrication of Photocathodes

Photocathodes can be constructed as either opaque or semitransparent layers. Each type is used in a somewhat different geometric arrangement. An opaque photocathode is normally fabricated with a thickness somewhat greater than the maximum escape depth and is supported by a thick backing material. Photoelectrons are collected from the same surface on which the light is incident. Semitransparent photocathodes generally are no thicker than the escape depth and are deposited on a transparent backing (often the glass end window of the PM tube). Light first passes through the transparent backing and subsequently into the photocathode layer, and photoelectrons are collected from the opposite surface. Because they are more readily adaptable to tube designs that use a flat end window, semitransparent photocathodes are more common in PM tubes designed for scintillation counters.

An important practical property of photocathodes is the uniformity to which their thickness can be held over the entire area of the photocathode. Variations in thickness give rise to corresponding changes in the sensitivity of the photocathode and can be one source of resolution loss in scintillation counters. This problem is especially serious for large-diameter PM tubes.

D. Quantum Efficiency and Spectral Response

The sensitivity of photocathodes can be quoted in several ways. When applied to dc light measurements, it is traditional to quote an overall photocathode efficiency in terms of current produced per unit light flux on its surface (amperes per lumen). A unit of greater significance in scintillation counting is the *quantum efficiency* (*QE*) of the photocathode. The quantum efficiency is simply defined as

$$QE = \frac{\text{number of photoelectrons emitted}}{\text{number of incident photons}} \tag{9.1}$$

The quantum efficiency would be 100% for an ideal photocathode. Because of the limitations mentioned earlier, practical photocathodes show maximum quantum efficiencies of 20–30%.

The quantum efficiency of any photocathode will be a strong function of wavelength or quantum energy of the incident light, as shown in Fig. 9.2. To estimate the effective quantum efficiency when used with a particular scintillator, these curves must be averaged over the emission spectrum of the scintillator. One consideration in selecting a photocathode is to seek a high quantum efficiency over the wavelength range in which the emission spectrum of the scintillator is concentrated.

The sensitivity at the long-wavelength or low photon energy end of the scale is largely limited by the reduced absorption of light in the photocathode and the low energy imparted to the photoelectron. At a sufficiently high λ this electron no longer has sufficient energy to escape the surface of the photocathode and the response drops to zero. The response at the opposite end of the scale is normally not a function of the photocathode

Figure 9.2 The spectral sensitivity of a number of photocathode materials of interest in PM tubes. The use of silica or quartz windows is necessary to extend the response into the ultraviolet region. (Courtesy of EMI GENCOM Inc., Plainview, NY.)

itself but rather of the window through which the light must enter to reach the photoemissive layer. For normal glass, the cutoff will be at a wavelength in the region of 350 nm, which is usually adequate for most scintillation materials. For some scintillators (e.g., noble gases), however, a significant part of the emission spectrum can be in the ultraviolet region with shorter wavelength. For such applications, special PM tubes with entrance windows made from fused silica or quartz can be used to extend the sensitivity to wavelengths as short as about 160 nm.

An alternative measure of quantum efficiency is sometimes quoted for PM tubes used in scintillation counting. The number of photoelectrons produced per unit energy lost in a scintillator mounted with close optical coupling to the photocathode is proportional to the quantum efficiency averaged over the emission spectrum of the scintillator. Because of the widespread use of thallium-activated sodium iodide as a scintillation crystal, the standard for quotation is the number of photoelectrons produced from a given photocathode per keV of energy loss by fast electrons in a NaI(Tl) crystal from which nearly all the light is collected. For photocathodes with peak quantum efficiency of 25–30%, measurements give about 8–10 photoelectrons per keV energy loss.[4,5] The reciprocal of this value, or the average energy loss required to create one photoelectron, is therefore approximately 100–120 eV. As emphasized elsewhere in this text, this value for the energy loss required to produce one basic information carrier in a typical scintillation detector is much larger than the equivalent value in gas-filled or semiconductor detectors.

Presently available materials for photocathodes include a *multialkali* material based on the compound Na_2KSb. Prepared by activation with a small amount of cesium, this material was the first to show a relatively high quantum efficiency of up to 30% in the blue region of the spectrum. A later formulation based on K_2CsSb activated with oxygen and cesium is given the name *bialkali* and can show an even higher efficiency in the blue. Furthermore, thermionic emission from bialkali photocathodes tends to be significantly lower than that from the multialkali materials, leading to lower spontaneous noise rates from tubes with this photocathode.

III. ELECTRON MULTIPLICATION

A. Secondary Electron Emission

The multiplier portion of a PM tube is based on the phenomenon of secondary electron emission (see p. 42). Electrons from the photocathode are accelerated and caused to strike the surface of an electrode, called a *dynode*. If the dynode material is properly chosen, the energy deposited by the incident electron can result in the reemission of more than one electron from the same surface. The process of secondary electron emission is similar to that of photoemission discussed in the previous section. In this case, however, electrons within the dynode material are excited by the passage of the energetic electron originally incident on the surface rather than by an optical photon.

Electrons leaving the photocathode have a kinetic energy on the order of 1 eV or less. Therefore, if the first dynode is held at a positive potential of several hundred volts, the kinetic energy of electrons on arrival at the dynode is determined almost entirely by the magnitude of the accelerating voltage. The creation of an excited electron within the dynode material requires an energy at least equal to the bandgap, which typically may be of the order of 2–3 eV. Therefore, it is theoretically possible for one incident electron to create on the order of 30 excited electrons per 100 V of accelerating voltage. Because the direction of motion of these electrons is essentially random, many will not reach the surface before their de-excitation. Others that do arrive at the surface will have lost sufficient energy so that they cannot overcome the potential barrier at the surface and are therefore incapable of escaping. Therefore, only a small fraction of the excited electrons ultimately contribute to the secondary electron yield from the dynode surface.

The secondary electron yield is a sensitive function of incident electron energy. If a relatively low-energy electron strikes the dynode surface, little energy is available for transfer to electrons in the dynode material, and relatively few electrons will be excited across the gap between the valence and conduction bands. At the same time, because the distance of penetration is not large, most of these excited electrons will be formed near the surface. For incident electrons of higher energy, more excited electrons will be created within the dynode but at greater average depth. Because the probability of escape will diminish with

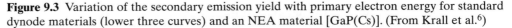

Figure 9.3 Variation of the secondary emission yield with primary electron energy for standard dynode materials (lower three curves) and an NEA material [GaP(Cs)]. (From Krall et al.[6])

increasing depth, the observed electron yield will be maximized at an optimum incident electron energy.

The overall multiplication factor for a single dynode is given by

$$\delta = \frac{\text{number of secondary electrons emitted}}{\text{primary incident electron}} \tag{9.2}$$

and should be as large as possible for maximum amplification per stage in the photomultiplier tube. A plot of δ versus incident electron energy is given in Fig. 9.3 for several dynode materials. For the conventional dynode materials of BeO, MgO, and Cs_3Sb, the maximum multiplication factor reaches about 10 for an optimum incident energy near 1 keV, although values of 4–6 are more typical at conventional interdynode voltages of a few hundred volts.

B. Negative Electron Affinity Materials

The secondary emission yield of dynodes can be increased significantly through the use of *negative electron affinity* (NEA) materials.[6-10] The most successful of these materials has been gallium phosphide (GaP), heavily doped to a concentration of about 10^{19} atoms/cm^3 with a p-type material such as zinc. The zinc creates acceptor sites within the bulk of the gallium phosphide. A thin, nearly monatomic layer of an electropositive material such as cesium is then applied to one surface. The acceptors at the surface attract an electron from the electropositive cesium, and each cesium atom becomes ionized and is held to the surface by electrostatic forces.

The effect of this surface treatment can best be illustrated through the band structure diagram shown in Fig. 9.4. At the left is shown a conventional band diagram for undoped gallium phosphide, which is also representative of conventional dynode materials. The series of arrows on the left shows a typical history of an electron that does not escape. The original excitation causes a number of electrons to be elevated from the valence band to some point well up into the conduction band. As these electrons diffuse, they lose energy primarily through phonon interactions, such that within about a picosecond, these "hot" electrons have come into thermal equilibrium with their local environment and their energy has relaxed to near the bottom of the conduction band. If this electron is to escape, it must reach the surface with an energy greater than the potential that exists on the vacuum side of the surface. Once the electron drops to the bottom of the conduction band, its energy is normally below the vacuum potential and is too low to permit escape. In this case only a short time is available for the electron to escape and it cannot travel large distances from its point of origin. Therefore, relatively small escape depths of only a few nanometers are possible. However, the electron will tend to diffuse for a substantially longer time (typically 100 ps) before dropping across the gap to rejoin the valence band.

On the right is shown the band bending created by the filling of acceptor sites at the surface by the thin cesium layer. The effect of the bending is to bring the vacuum potential below that of the bottom of the conduction band in the interior of the material. Therefore, electrons

Figure 9.4 Band structure near the surface for conventional semiconductors (left) and NEA materials (right). (Adapted from Krall et al.[6])

that have already dropped to the bottom of the conduction band still have sufficient energy to escape if they happen to diffuse to the surface. Because the thickness of the bent-band region is very small, it can be less than a mean free path and the electron may escape without further energy loss in the surface region. The net effect is that the electrons that have already reached the bottom of the conduction band are still candidates for escape and remain so for a period of time that is about 100 times greater than in the previous case. The average escape depth will therefore tend to be much greater and can reach tens or hundreds of nanometers.

The effect of this change on the secondary electron yield is profound. The increased time over which electrons may escape enhances the escape probability for any typical electron. Furthermore, excited electrons created deep within the dynode material remain candidates for escape. Therefore, the secondary electron yield will continue to increase with increasing primary electron energy until the distance of penetration of the primary is very large. These effects are reflected in the yield curve for a GaP(Cs) dynode shown in Fig. 9.3. Secondary electron yields of 50 or 60 are readily achieved with an interstage voltage of 1000 V, and much higher values are possible in principle if even larger voltages are permitted by the PM tube design.

A secondary advantage of NEA materials is evident in PM tubes used for ultrafast timing applications. Because almost all escaping electrons have previously dropped to the bottom of the conduction band, their average energy is lower and much more uniform than secondary electron energies from conventional materials. Because variations in initial energy contribute to the time spread in the multiplier section, a narrower distribution leads to less time broadening. Furthermore, the higher gain per stage permits a reduction in the number of stages required for a given total gain, also reducing the overall time spread.

C. Multiple Stage Multiplication

To achieve electron gains on the order of 10^6, all PM tubes employ multiple stages. Electrons leaving the photocathode are attracted to the first dynode and produce δ electrons for each incident photoelectron. The secondaries that are produced at the surface of the first dynode again have very low energies, typically a few eV. Thus, they are quite easily guided by another electrostatic field established between the first dynode and a second similar dynode. This process can be repeated many times, with low-energy secondary electrons from each dynode accelerated toward the following dynode. If N stages are provided in the multiplier section, the overall gain for the PM tube should be given simply by

$$\text{overall gain} = \alpha\delta^N \qquad (9.3)$$

where α is the fraction of all photoelectrons collected by the multiplier structure. Conventional dynode materials are characterized by a typical value of $\delta = 5$, and α is near unity for well-designed tubes. Ten stages will therefore result in an overall tube gain of 5^{10}, or about 10^7. If high-yield NEA dynodes are used with $\delta = 55$, the same gain can be achieved with only four stages.

The overall gain of a PM tube is a sensitive function of applied voltage V. If δ were a linear function of interdynode voltage, then the overall gain of a 10-stage tube would vary as V^{10}. As shown in Fig. 9.3, however, δ for conventional dynodes varies as some fractional power of interdynode voltage so that the overall gain is more typically proportional to $V^6 - V^9$.

D. Statistics of Electron Multiplication

If δ were strictly a constant, each photoelectron would be subject to exactly the same multiplication factor. Under fixed operating conditions, all output pulses that originate from a single photoelectron would then have the same amplitude. In practice, the emission of secondary electrons is a statistical process, and therefore the specific value of δ at a given dynode will fluctuate from event to event about its mean value. The shape of the single photoelectron pulse height spectrum observed from a real PM tube is an indirect measure of the degree of fluctuation in δ and has thus been the subject of extensive investigation.

In the most simple model, the production of secondary electrons at a dynode can be assumed to follow a Poisson distribution about the average yield. For a single photoelectron incident on the first dynode, the number of secondaries produced has a mean value of δ and standard deviation σ of $\sqrt{\delta}$ (see Chapter 3). The relative variance, defined as $(\sigma/\delta)^2$, is thus equal to $1/\delta$. When this process is now compounded over N identical stages of the PM tube, the mean number of electrons collected at the anode (and hence the pulse amplitude) is given by δ^N. It can be demonstrated from the properties of Poisson statistics that the relative variance in this number is now

$$\frac{1}{\delta} + \frac{1}{\delta^2} + \frac{1}{\delta^3} + \ldots + \frac{1}{\delta^N} \cong \frac{1}{\delta - 1} \tag{9.4}$$

Thus, if $\delta \gg 1$, the relative variance or spread in the output pulse amplitude is dominated by fluctuations in the yield from the first dynode where the absolute number of electrons is smallest.

In many applications of scintillators, hundreds or thousands of photoelectrons contribute to each pulse, and they are therefore much larger than single photoelectron pulses. When poor light collection or low-energy radiations are involved, however, signal pulses corresponding to only a few photoelectrons may be involved. Then the fluctuations in electron multiplication may interfere with the ability to discriminate against noise events, many of which correspond to single photoelectrons. Figure 9.5 shows the expected distribution in the number of secondaries produced by the first dynode when struck by different numbers of photoelectrons. If the value of δ is small, it is impossible to separate cleanly the events caused by one photoelectron from those in which more photoelectrons are involved. If the dynodes are characterized by a larger value of δ, however, the separation is much more distinct and it is possible to distinguish peaks in the distribution corresponding to discrete numbers of photoelectrons up to about 4 or 5. This behavior is demonstrated in Fig. 9.6 in the pulse height spectrum observed from a PM tube with a high-δ first dynode made from a NEA material.

Experimental measurements of the single photoelectron pulse height spectra from PM tubes generally show a peaked distribution,[12,13] but with a larger relative variance than that predicted by the Poisson model. In fact, observations made under some conditions show no

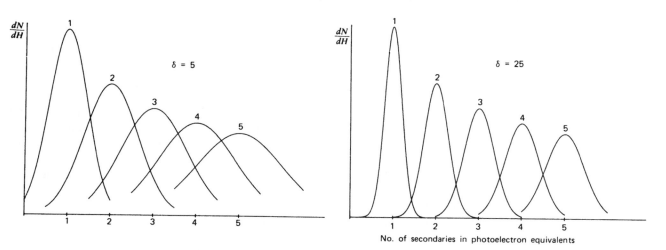

Figure 9.5 Statistical broadening of the secondary electron yield from the first dynode of a PM tube. Numbers identify the number of incident photoelectrons. Two cases are shown representing conventional dynodes (δ = 5) and NEA materials (δ = 25).

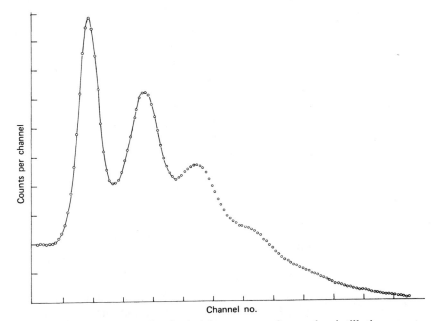

Figure 9.6 The measured pulse height spectrum for weak scintillation events obtained from a RCA 8850 photomultiplier tube. The high-gain first dynode results in distinguishable peaks in the spectrum corresponding to 1, 2, and 3 photoelectrons per pulse. (From Houtermans.[11])

peak at all, but rather an exponential-like distribution.[14] These discrepancies have led to alternate models of the multiplication statistics in which a Polya distribution[15] or compound Poisson[16,17] is substituted for the simple Poisson description of electron multiplication. No universal descriptions have as yet emerged that can accommodate all experimental measurements, and it is possible that differences in specific electron trajectories and dynode properties may preclude a general model applicable to all PM tubes.

IV. PHOTOMULTIPLIER TUBE CHARACTERISTICS

A. Structural Differences

Figure 9.7 shows some representative construction details of PM tubes of various designs. All consist of a semitransparent photocathode, a photoelectron collection region between the photocathode and the first dynode, a multistage electron multiplier section, and an anode for collection of the amplified charge. These structures are enclosed in a glass vacuum envelope, through which electrical leads are conducted at the base. Tubes with flat end plate windows are the only type in general use for scintillation counting, so cylindrical scintillation crystals can easily be mounted directly on the end window adjacent to the photocathode. Tubes are available commercially with circular photocathodes ranging in diameter from a few millimeters to over 20 cm. A nominal 2-in. (5 cm) diameter is one of the common choices in many scintillation applications, and the widest selection is generally available in this size. PM tubes with square and hexagonal photocathodes have also been introduced for specialized applications in which an array of tubes must be closely packed.

The "venetian blind" type of construction is one of the oldest used for photomultiplier tubes and is now virtually obsolete. It is readily adaptable to tubes of varying numbers of stages but suffers from a relatively slow response time due to low electric fields at the surface of the dynodes. The "box and grid" structure is also fairly old and slow, but it is still

Figure 9.7 Configurations of some common types of PM tubes. (*a*) Focused linear structure. (*b*) Circular grid. (*c*) Venetian blind. (*d*) Box-and-grid. (Courtesy of EMI GENCOM Inc., Plainview, NY.)

standard in many tubes of large diameter. The circular grid and linear multiplier structures were introduced to speed up the electron transit time through the multiplier structure and are used in those PM tubes with the fastest response time.

Photomultiplier tubes should be protected from excessive mechanical shock or vibration to avoid physical damage to internal components. Also, some modulation of the output signal can be induced by vibration of the multiplier structures. Most tube manufacturers offer a rugged version of some of their photomultiplier tubes (often originally designed for space applications). These are provided with electron multiplier structures and other internal components mechanically arranged to resist vibration and shock. Standard tubes are quite adequate in this respect for most scintillation counting applications, but rugged versions can be helpful in counters designed for field use where they may be subjected to rough handling.

Another type of specialized electron multiplier is the continuous channel, illustrated in Fig. 9.8. This device is extremely simple, consisting of a hollow glass tube whose inner surface acts as a secondary electron emitter. A potential difference is applied across the length of the tube, thereby attracting electrons from the entrance end to the exit end. Electrons entering the tube will eventually strike the wall giving rise to a pulse of secondary electrons. These will be further accelerated along the length of the tube until they in turn also strike the wall, giving rise to further secondaries. The device acts much like a photomultiplier tube with continuous dynodes available along its entire length. The number of times an entering electron and its subsequent secondaries strike the wall is an accident of past orientation and individual trajectories and will vary considerably for different entering electrons. When operated at high voltages, the electron gain may be large enough so that

Detailed view of typical electron trajectories

Figure 9.8 Continuous channel electron multiplier.

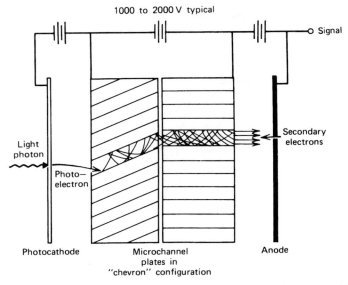

Figure 9.9 Elements of a PM tube based on microchannel plate electron multiplication.

the resulting space charge near the exit of the channel limits the total charge per pulse to about 10^6 or 10^7 electrons at saturation for channels of small diameter.

Channel multipliers must be shaped to prevent feedback problems that can arise when positive ions occasionally formed within the channel are accelerated in the reverse direction from the electrons. By forming the channel as a curve or chevron, these ions can be made to strike a wall before their energy is high enough to create secondary electrons.

Clusters of many thousands of tubes can be fabricated to form a *microchannel plate*. Each channel is of very small diameter (typically 15–50 μm) and acts as an independent electron multiplier. Their application as the multiplier element in PM tubes is reviewed in Refs. 18–21. In the configuration shown in Fig. 9.9, each photoelectron generally enters a separate channel. If all channels are operated in charge saturation and are identical, then the total collected charge is proportional to the number of original photoelectrons.

The primary advantage of PM tubes based on microchannel plates lies in their excellent timing properties.[22–27] The total electron transit time through a channel is a few nanoseconds, compared with 20–50 ns for conventional multiplier structures (see the following section). The spread in transit time, which determines timing performance, is about 100 ps, a factor of 2 or 3 better than the fastest conventional PM tube currently available.

B. Pulse Timing Properties

Because the time required for photoemission in the photocathode or secondary emission from dynodes is very short (0.1 ns or less), the time characteristics of the PM tube are determined exclusively by the electron trajectories. The *electron transit time* of a PM tube is defined as the average time difference between the arrival of a photon at the photocathode and the collection of the subsequent electron burst at the anode. In PM tubes of various designs, electron transit times range from 20–80 ns. In most timing applications, however, the transit time itself is not of primary importance because if it were always a constant, it would introduce only a fixed delay in the derived signal. Instead, the *spread in transit time* is a more important quantity because it determines the time width of the pulse of electrons arriving at the anode of the tube. The timing response of a typical PM tube is illustrated in Fig. 9.10.

Figure 9.10 The response of a PM tube to a short pulse of light on the photocathode.

The region between the photocathode and first dynode is critical in determining the timing properties. To allow uniform collection over large photocathodes, this distance is kept fairly large compared with interdynode distances (see Fig. 9.11). The difference in paths between a photoelectron leaving the center of the photocathode and one at the edge is often a dominant factor in the observed spread in transit time. The photocathode is often curved to minimize the transit time spread across its diameter. It is convenient to have a flat outer surface for scintillator mounting, so an end window with a plano-concave shape is frequently used with the photocathode deposited on the inner curved surface. A second source of transit time spread arises from the distribution in initial velocities of photoelectrons leaving the photocathode. This effect can be minimized by using a large voltage difference between the photocathode and first dynode.

Figure 9.11 Computer-generated trajectories of electrons accelerated from the photocathode to the first dynode in a PM tube. In this design, the photocathode is curved to minimize the spread in transit time. Structures labeled F_1 and F_2 are electrodes carrying adjustable voltage that electrostatically focus the electrons for optimum performance. (From Kume et al. [28])

Figure 9.12 Measurements of the transit time spread observed from the PM tube shown in Fig. 9.11 as a function of the average number of photoelectrons generated per pulse. (From Kume et al. [28])

The amount of transit time spread observed for a specific pulse also depends on the number of initial photoelectrons per pulse. To simplify the analysis and comparison of different photomultipliers, many of the measurements reported in the literature concentrate on the transit time spread due to a *single* photoelectron. This parameter does not include the important contribution of the cathode to first dynode space,[29,30] and is only a measure of the time spreads introduced subsequent to the first dynode. As the number of photoelectrons increases, a larger number of samplings of various possible electron trajectories through the tube are made. If the distribution in the various possible transit times is assumed to be Gaussian, then statistical theory predicts that the relative spread in transit times should vary inversely with the square root of the number of photoelectrons. In Fig. 9.12, this behavior is verified for a typical PM tube. Thus, high light yield from a scintillator is important in timing applications as well as in pulse height measurements. The time spread attributable to the multiplier section also decreases with increasing interdynode voltage, and the best timing performance is normally obtained by operating the tube at maximum voltage permitted by the ratings.

The timing performance of PM tubes is a continuing area of development, and some trends are reviewed in Ref. 31. Table 9.1 lists data for a number of representative tubes on the electron transit time and anode rise time (related to the spread in transit time). When used with slow inorganic scintillators, PM tubes are fast enough so that their contribution to the overall time response usually is not an important factor. It is only when scintillators with shorter decay times are employed to derive a fast timing signal that the PM tube can become a significant element in determining the resultant timing properties.

C. Maximum Ratings

All commercial photomultipliers are supplied with a set of maximum voltage and current ratings that are not to be exceeded during routine use. Detailed specifications will often give individual values for maximum photocathode to first dynode, dynode to dynode, last dynode to anode, and cathode to anode maximum voltages as well. These limits are set by a variety of conditions including ohmic leakage and dark current considerations. The user is often more interested in a single figure for the maximum applied voltage when using the

Table 9.1 Properties of Some Commercially Available Photomultiplier Tubes

A	B	C	D	E	F	G	H	I	J	K	L	M	N
Ham	1635	10	8	L8	BA	1250	1500	1.1	95	76	1	0.8	8.5
Ham	1450	19	15	L10	BA	1500	1800	1.7	115	88	3	1.8	19
Ham	380	38	34	L10	BA	1250	1750	1.1	95	88	3	2.7	37
Ham	1306	51	46	B8	BA	1000	1500	0.27	110	95	2	7.0	60
Ham	3318	51 sq	45	BM10	BA	1000	1500	0.27	110	95	2	4.8	45
Ham	3336	60 h	55	BM10	BA	1000	1500	0.27	110	95	2	6.0	47
Burle	4516	19	13	L10	BA	1500	1800	0.52	66	—	0.2	1.8	20
Burle	S83010E	38	32	C10	RbCsSb	1000	1000	2.4	100	92	1	2.8	32
Burle	S83054F	51	47	B8	BA	800	1200	0.10	10.5[a]	103	3	11	63
Burle	S83020F	60 h	56	L10	BA	1100	1700	0.10	71	100	1	10	69
Burle	S83079F	76 sq	—	B8	BA	800	1200	0.21	11.3[a]	100	3	14	73
Burle	S83006F	130	111	T10	BA	1100	1650	0.07	92	105	1	22	105
ADIT	B29B02H	29	24	B11	BA	1100	1500	1.0	70	80	10	11	35
ADIT	B51B01	51	46	B10	BA	1100	1500	1.0	70	80	10	17	45
ADIT	B76B01	76	70	B10	BA	1100	1500	1.0	70	80	10	17	50
ADIT	B133D01	127	119	B10	BA	1100	1500	1.0	120	97	10	17	50
ETL	9078	19	15	L10	BA	900	1200	0.71	70	90	0.1	1.8	20
ETL	9924	30	23	B11	RbCs	870	1050	2.1	95	100	0.1	15	80
ETL	9266	52	45	L10	BA	900	1100	0.6	80	100	0.2	4	37
ETL	9350	200	190	L14	BA	1550	2300	67	75	100	15	5	80

A = manufacturer: Ham = Hamamatsu, ETL = Electron Tubes Limited.

B = model number.

C = diameter or dimension of tube outline (sq = square, h = hex) in mm.

D = minimum usable photocathode dimension.

E = dynode structure: L = linear focused, B = box and grid, BM = box and mesh, C = circular.

F = photocathode material: BA = bialkali.

G = recommended operating voltage.

H = maximum tube voltage.

I = gain $\times 10^6$ at voltage in G.

J = cathode luminous sensitivity (μA/lm) measured with 2854 K tungsten source.

K = cathode radiant sensitivity (mA/W) measured at or near the wavelength of photocathode peak sensitivity.

L = dark current (an approximate number due to large variations in the method of measurement between different manufacturers) (nA).

M = anode rise time at voltage in G (ns).

N = transit time at voltage in G (ns).

[a]cathode luminous sensitivity is measured using a blue Corning C.S. No. 5-58 filter.

manufacturer's recommended voltage divider string. Because virtually every PM tube will show an increase in gain as the voltage is increased, the maximum value for applied voltage practically determines the maximum gain obtainable from the tube.

Maximum values are also often given for photocathode current and anode current. The first limit is usually set by consideration of photocathode resistivity, which can give rise to distorted electrostatic fields at high currents, whereas the anode limit is set by heat dissipation considerations. In normal pulse-mode scintillation counting, current limits seldom are of concern because the maximum counting rate permitted by pile up and pulse resolution considerations nearly always corresponds to an average current far less than the maximum ratings.

D. Photomultiplier Tube Specifications

Manufacturers will conventionally quote performance of PM tubes in terms of certain characteristics, which are defined here.

1. *Overall Luminous Sensitivity.* Defined as the ratio of the measured anode current at operating voltage to the luminous flux from a tungsten light source of specified temperature incident on the photocathode. This quantity is an overall measure of the expected current from the PM tube per unit incident light from a broad-band source. The units are amperes per lumen (A/lm).

2. *Cathode Luminous Sensitivity.* Defined as above, except that the current of photoelectrons leaving the photocathode is substituted in the numerator for the anode current. This quantity is again measured in amperes per lumen, is a characteristic only of the photocathode, and is independent of the electron multiplier structure.

3. *Overall Radiant Sensitivity.* This parameter is defined as the ratio of anode current to radiant power at a given wavelength incident on the photocathode. Units are amperes per watt (A/W).

4. *Cathode Radiant Sensitivity.* Defined as above, except that the photocathode current is substituted for the anode current.

5. *Dark Current.* Normally specified in terms of anode current measured without photocathode illumination when the tube is operated to provide a given overall luminous sensitivity.

6. *Anode Pulse Rise Time.* Quoted as the time taken for the output pulse to rise from 10 to 90% of the peak when the photocathode is illuminated by a flash of light of very short duration.

7. *Anode Pulse Width.* Normally quoted as the time width of the output pulse measured at half maximum amplitude, again for short-duration illumination of the photocathode.

Table 9.1 lists specifications for a number of commercial photomultiplier tubes of significant current interest in scintillation counting. Because PM tube design remains a rapidly developing area, potential users should always seek up-to-date information directly from the tube manufacturers.

E. Linearity

The electron multiplication factor in nearly all PM tubes remains constant for pulses that range in size from a single photoelectron to many thousands. Under these conditions the amplitude of the pulse collected at the anode is linearly related to the number of photoelectrons, and consequently to the intensity of the scintillation light flash. Nonlinearities can arise for very large pulses caused by space charge effects between the last dynode and anode where the number of electrons is greatest. The buildup of space charge affects the trajectories of electrons in this region and causes some to be lost that would otherwise be collected. Another factor that can cause nonlinearities at high pulse amplitudes is any deviation of dynode voltages from their equilibrium value during the course of the pulse. Under normal circumstances in scintillation pulse counting, these effects are seldom important with an adequately designed tube base (see later discussion) and the photomultiplier tube remains in the linear range.

F. Noise and Spurious Pulses

Usually the most significant source of random noise from a photomultiplier tube results from thermionic electrons that are spontaneously emitted by the photocathode. The pulses that result from this process correspond to a single photoelectron, so their amplitude is limited to the lowest end of the scale. Because most scintillation counting is done under conditions in which a scintillation pulse corresponds to many photoelectrons, amplitude discrimination is usually sufficient to eliminate all contributions of this thermal noise. In applications in which very-low-energy radiation is measured or in which single photons are detected, valid signal events may also correspond to a single photoelectron and therefore may be indistinguishable from thermionic noise. In that event there is no alternative but to try to reduce the noise contribution as much as possible. The rate at which these pulses are observed is proportional to the area of the photocathode, and therefore one should select a tube of the smallest diameter required for a specific application in order to minimize these dark pulses. The rate at which thermionic electrons are emitted per unit area varies greatly between photocathode materials, and bialkali photocathodes are among the most quiet. Typical spontaneous emission rates at room temperature are in the range of 10^2–10^4 electrons/cm$^2 \cdot$ s. The dark current specifications given in Table 9.1 largely reflect the contribution of thermionic electrons, and the tubes with lowest noise rates are those with the lowest dark currents for equivalent luminous sensitivity.

The significance of this spontaneous electron emission from the photocathode can be quite different depending on whether the scintillator–PM tube combination is operated in pulse or current mode. Choosing for illustration a value of 10^3 electrons/s randomly emitted from the entire photocathode, a tube operated in pulse mode will produce small noise pulses with amplitude that corresponds to a single photoelectron with an average time spacing of 1 ms. Typical shaping times used for scintillation detectors are in the microsecond range, so the probability is very small that even one such photoelectron will contribute to a given scintillation pulse. If the single photoelectron pulses are much smaller than the signal pulses so that they can be amplitude discriminated, then there is no appreciable contribution to the noise characteristics of the measurement. In current mode, however, all the dark current adds to the signal current, and the fluctuation in its value can be a significant contributor to the uncertainty in measuring the signal. To estimate the steady-state dark current associated with the same example as above for the spontaneous emission rate, we assume a gain of 10^6 in the electron multiplier section of the PM tube:

$$I_{dark} = (10^3 \text{ elec/s})(10^6)(1.6 \times 10^{-19} \text{ C/elec}) = 160 \text{ pA}$$

The measurement of currents of this order of magnitude or less from the PM tube anode will be subject to noise introduced by the dark current fluctuations.

For some photocathodes the rate at which thermionic electrons are emitted can be drastically reduced by cooling the tube. Reductions by a factor of 100 or more may be observed with proper temperature reduction. Dry ice or liquid nitrogen are often used for cooling, but self-contained refrigerators are also commercially available for this purpose. Problems that can arise in connection with PM tube cooling include water vapor condensation on exposed cold surfaces and increased photocathode electrical resistance at lower temperatures. Large photocathode resistance can distort the electrostatic field between the photocathode and first dynode and may lead to a loss in photoelectron collection efficiency.

Photomultiplier tubes should be stored in the dark when not in use. Exposure to room light is disastrous while voltage is supplied to the tube because very high illumination levels lead to anode currents that greatly exceed the maximum ratings and that can quickly damage the multiplier structures. Incident light, especially from fluorescent tubes, is to be avoided even when no voltage is applied to the tube because of the temporary increase in dark pulses that will result. It is not unusual to observe an increase of 100 or more in the rate of dark pulses immediately after exposure to intense room light, and a measurable increase can persist for several hours. It appears that the major cause of this increased noise rate is the emission of light from phosphorescence states in the glass envelope of the PM tube.

Another source of dark pulses originates with natural radioactivity in the structure of the tube itself. The most important components are usually ^{40}K and thorium contained in the glass envelope. A beta particle produced in radioactive decay will give rise to a flash of Cherenkov radiation, which can liberate photoelectrons from the photocathode in much the same way as normal scintillation events. For applications in which the ultimate in low background is a necessity, special tubes with low-activity glass to minimize these events are available. Scintillation or Cherenkov light produced in the glass by external radiation can also be a significant source of dark pulses.[32,33] One such source is cosmic radiation (see p. 761), which generally results in dark pulses of small amplitude from Cherenkov light produced in the thin end window of the tube. Because of the low specific energy loss of the secondary cosmic radiations and the low light yield from the Cherenkov process, the corresponding dark pulses correspond to only a few photoelectrons and are therefore usually discarded by amplitude discrimination in typical scintillation applications. However, if very weak scintillation events must be recorded, these dark pulses can be of the same size as the signal. In such cases, the rate from cosmic radiation can be minimized by operating the tube with its major axis oriented horizontally so that the end window presents minimum cross section to the cosmic secondaries that are directed preferentially in a downward vertical direction.

Afterpulses are another type of noise sometimes observed in PM tubes.[34] These are satellite pulses that will sometimes follow a true signal pulse after a short delay period. One mechanism that can give rise to afterpulsing is the emission of light from the latter stages of the multiplier structure, which finds its way back to the photocathode. Such afterpulses will be delayed by a time characteristic of the electron transit time through the tube, or roughly 20–50 ns. Because these pulses often correspond to a single photon, their amplitude is usually quite small. Another cause of afterpulsing can be an imperfect vacuum within the tube. Traces of residual gas can be ionized by the passage of electrons through the multiplier structure. The positive ions that are formed will drift in the reverse direction and some may find a path back to the photocathode. Typical ions will liberate tens or hundreds of photoelectrons when they strike the photocathode, and the resulting pulse will be of rather large size. Because the velocity of the positive ion is relatively low, the time it takes to drift back to the photocathode can range from hundreds of nanoseconds to a microsecond or more. The time spacing between the primary pulse and the afterpulse from this mechanism is therefore relatively large. Because the amount of residual gas can vary considerably between tubes of identical design, problems caused by afterpulsing in some applications may often be eliminated by simple substitution of another tube. The probability of production of an afterpulse increases linearly with primary pulse amplitude in at least one tube design but is usually no more than a few percent. Afterpulsing is therefore of little consequence in ordinary scintillation spectroscopy but can become a serious perturbing effect in timing measurements where the pulses of interest follow a preceding high-intensity burst of radiation. Discussions of the afterpulsing characteristics of some PM tubes, together with electronic means of afterpulse suppression, may be found in Refs. 35–39.

G. Photocathode Nonuniformities

Direct measurement (e.g., see Ref. 9, 40, or 41) has shown that the sensitivity of photo-cathodes, especially those with large diameters, is far from uniform across the entire photocathode area. This problem is further compounded by the difficulties of achieving uniform photoelectron collection to the first dynode from the entire photocathode area. The combination of these two effects can lead to situations in which the anode pulse observed for a given flash of light may vary by as much as 30–40% as the position of the illumination is moved across the photocathode area. This nonuniformity is a potentially serious problem in scintillation counting because response variations will tend to spoil the energy resolution of the system. One means of reducing the problem is to place a light pipe between the scintillator and end window of the PM tube. This is especially important for thin scintillators and tends to spread light from any scintillation event over the entire photocathode, thereby averaging out much of the nonuniformity.

H. Gain Variations with Counting Rate

Another nonideality of PM tubes of which the user should be aware is the possibility of gain changes during the course of a measurement. The most common situation is one in which the counting rate changes by a large factor. If the divider string current is too low (see p. 284), changes in count rate can lead to gain changes due to resulting variations in the dynode potentials.[42] Even if this effect is totally eliminated, the tube may require several hours to stabilize at a constant count rate as thermal and space charge effects created by the electron current through the multiplier structure of the tube reach equilibrium.[43] The gain changes may not be fully reversible, and hysteresis effects have been observed experimentally.[44] Photomultiplier structures can be designed to minimize these drifts, and a good tube will not change its gain by more than 1% as the counting rate is varied from 10^3 to 10^4 per second.[3] Specifications of this type can often be found for many PM tube designs. The gradual drift in tube gain that often follows a large change in tube current or counting rate is called *fatigue*[45, 46] and can be a serious problem if the tube current changes by orders of magnitude during the course of the measurement.

V. ANCILLARY EQUIPMENT REQUIRED WITH PHOTOMULTIPLIER TUBES

A. High-Voltage Supply and Voltage Divider

An external voltage source must be connected to the PM tubes in such a way that the photocathode and each succeeding multiplier stage are correctly biased with respect to one another. Because electrons must be attracted, the first dynode must be held at a voltage that is positive with respect to the photocathode, and each succeeding dynode must be held at a positive voltage with respect to the preceding dynode. For efficient photoelectron collection, the voltage between photocathode and first dynode is often several times as great as the dynode-to-dynode voltage differences.

The interstage voltage requirements of a PM tube may, in principle, be supplied by individual voltage sources, such as a multicell battery. Battery supplies are practical in some applications where counting rates are low, but often are unattractive because of the rapid discharge rates due to the current demands of the latter stages of the PM tube. The internal current in the PM tube is at a maximum between the last dynode and the anode, so the battery cell that supplies the voltage to the last dynode must be capable of a current drain equal to the average dc current leaving this dynode.

In the vast majority of cases, the voltage differences are provided instead by a resistive voltage divider and a single source of high voltage. Figure 9.13*a* shows a typical wiring diagram for the base of a PM tube in which a positive polarity high voltage is used. In this case

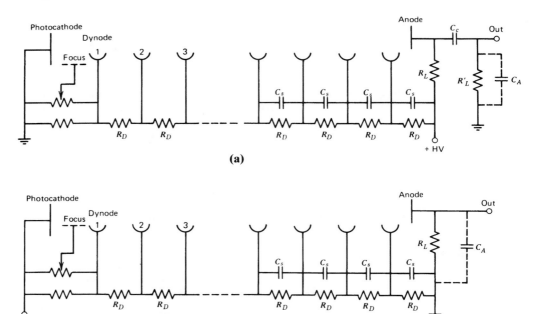

Figure 9.13 Typical wiring diagrams for the base of a PM tube. Scheme (*a*) utilizes positive high voltage and a grounded photocathode. Scheme (*b*) uses negative high voltage, and the photocathode must be isolated from ground. Values of the divider string resistors R_D are chosen using criteria given in the text. The equivalent anode load resistance is R_L in (*b*), and the parallel combination of R_L and R'_L in (*a*). Other identifications are given in the text.

the cathode of the PM tube is grounded, and the divider string supplies successively increasing positive voltages to each dynode down the multiplying string. The anode is at a dc potential equal to the supply voltage, and signal pulses must therefore be capacitively coupled from the anode to allow the pulse component to be passed on at ground potential to succeeding electronic devices. The load resistor R_L can be chosen by the experimenter so that the resulting anode circuit time constant is of proper magnitude. The anode capacitance C_A usually is not a physical capacitor, but only the stray capacitance associated with the anode structure and connecting cables. An analysis of the pulse shape expected from this anode circuit is given later in this chapter.

The direct current through the divider string is determined simply by the ratio of the applied high voltage and the summed resistance of all the resistors in the divider string. In order to use the simplest and least expensive voltage supply, one would like to keep this current at a minimum. Small currents also minimize problems caused by heat dissipation in the divider resistors. However, the divider string current should in principle be kept large compared with the internal PM tube current represented by the pulse of electrons flowing from dynode to dynode. If the internal current at the peak of a pulse becomes comparable with the divider current, the voltages of the dynodes normally will begin to deviate from their equilibrium values, leading to drift of the PM tube gain. This problem is especially serious for the last few stages of the PM tube where peak currents are at a maximum. To suppress this effect, it is commonplace to provide *stabilizing capacitors* (labeled as C_s in Fig. 9.13) to the stages of the divider string near the anode to help hold these latter dynode voltages at a constant value throughout the pulse. The stabilizing capacitor momentarily supplies the current lost from the dynode during the pulse and is recharged by the divider string current during the period between pulses. To prevent a more than 1% interdynode voltage change, the charge stored on the stabilizing capacitor (given by the product of capacitance value and the interdynode voltage) must be 100 times greater than the charge emitted by that dynode during the pulse.

A numerical example is instructive at this point. Assume we start with a scintillation event that liberates 1000 photoelectrons from the photocathode of the PM tube. Further assume that the PM tube provides an overall gain of 10^6, so that 10^9 electrons per pulse leave the last dynode and are collected by the anode. If these scintillation pulses are occurring at a rate of 10^5 per second, then the average dc anode current can easily be calculated as

$$I_{\text{avg}} = 10^9 \, \frac{\text{electrons}}{\text{pulse}} \times 1.6 \times 10^{-19} \, \frac{\text{coulomb}}{\text{electron}} \times 10^5 \, \frac{\text{pulses}}{\text{second}}$$

$$= 1.6 \times 10^{-5} \, \text{A} = 0.016 \, \text{mA} \tag{9.5}$$

Because this current appears in discrete pulses, however, the peak current during a pulse is substantially higher. As an extreme case, assume we have a very fast organic scintillator whose decay time combined with the transit time spread in the PM tube produces a narrow electron pulse of 5-ns width. The peak pulse current is then approximately

$$I_{\text{peak}} = 10^9 \, \text{electrons} \times 1.6 \times 10^{-19} \, \frac{\text{coulomb}}{\text{electron}} \times \frac{1}{5 \times 10^{-9} \, \text{s}}$$

$$= 0.032 \, \text{A} = 32 \, \text{mA} \tag{9.6}$$

For photomultipliers operated in either current or pulse mode, the current through the voltage divider string should be kept large compared with the average dc signal current. In the example shown, this criterion can easily be met with a voltage supply capable of a few tenths of a milliampere. In pulse mode, however, the instantaneous demands in the last few stages of the PM tube can be many times greater. In the case of the example above, it would be impractical to attempt to design the divider string with a dc current much larger than the maximum pulse current as well. In that case, a divider string current of 10 or 20 mA might be used and stabilizing capacitors provided for the last few stages in which the peak pulse current is greater than 1–2 mA. Alternate schemes which employ zener diodes or transistors to stabilize the dynode voltages are described in Refs. 47 and 48.

The polarity (+ or −) of the high voltage used with PM tubes is in some sense an arbitrary choice. In the example given previously, the photocathode end of the divider string was grounded, which then required the application of positive high voltage to the anode end of the string. Exactly the same interdynode voltages can be achieved by grounding the opposite (or anode) end of the divider string and applying negative high voltage to the photocathode end. This latter arrangement is shown in Fig. 9.13b. The designer of the tube base circuit must choose one of these alternatives, and both are in common use. It is therefore important that users be aware of which convention has been chosen by the manufacturer of their own tube base before initial use of the equipment with a PM tube. Mistakenly applying the wrong polarity to a PM tube is not usually fatal to the tube, but electrons will refuse to swim uphill and the PM tube will not work. Designing the tube base for positive high voltage is in some ways simpler because it allows the photocathode of the PM tube to be grounded. Because this end is often in contact with a scintillation crystal whose cover must be exposed, it is convenient to keep the crystal at ground potential as well. Operating the tube with negative high voltage means that the photocathode will be at the full high voltage supplied to the tube, and care must be taken to prevent spurious pulses due to high-voltage leakage through the glass tube envelope to nearby grounded structures. A high electric field across the tube end window can also induce spurious pulses in some tubes due to electroluminescence in the glass. An advantage that stems from use of negative high voltage is the elimination of the coupling capacitor (C_c in Fig. 9.13a) required if positive polarity is used. The anode is now at ground potential, and signal pulses can be direct-coupled into subsequent measuring circuits. This advantage is particularly important for fast pulse applications in which it is often desirable to couple the anode directly into a 50-ohm transmission line structure. It is also the necessary choice when the tube is to be dc-coupled to allow its operation in current mode.

Because the gain of a PM tube is extremely sensitive to changes in voltage, it is very important that sources of high voltage be well regulated and free of ripple. Drifts in the high voltage will show up as corresponding changes in tube gain and can, if sufficiently large, deteriorate the energy resolution of a scintillation counter. Any ripple is likely to be superimposed on the output signal, especially if the tube is operated with the photocathode grounded. In that case, any fluctuations in the supply voltage can be capacitively coupled directly onto the signal output. If the tube is operated with negative high voltage, the anode end of the tube is at ground potential, and ripples in the high voltage supply are important only to the extent that they modulate the overall gain of the tube.

The very large changes in gain that accompany changes in voltage with a PM tube are often a great convenience in setting up a counting system. Although the timing properties of the tube are optimized only when operated near the recommended voltage, other general properties such as linearity and relative signal-to-noise are not seriously changed over wide ranges of voltage. Therefore, it is often possible to operate a PM tube well below its recommended voltage without appreciably hurting its performance. The experimenter can therefore conveniently change the gain of the PM tube by orders of magnitude to suit the needs of the remainder of the signal chain simply by changing the applied voltage. This procedure is often followed in simple scintillation counters to determine counting plateaus and optimum operating conditions.

In those cases in which the tube voltage is likely to be varied over wide regions, the performance sometimes can suffer at the lower voltages due to reduced gain at the critical first dynode. It is here that the statistical properties of the PM tube are largely determined. In these cases a zener diode is sometimes used in conjunction with the divider string to hold the voltage between the photocathode and first dynode at a constant value, whereas the remainder of the divider string is varied.

In the usual case, the electrical signal from a PM tube is derived from the anode. The burst of electrons in the last stage of the PM tube is simply collected, resulting in a charge Q of negative polarity. As shown in Section VII of this chapter, a corresponding voltage pulse is then developed across the load resistance in the anode circuit. In some specialized applications, there may be some advantage in deriving an alternative signal from a preceding dynode. At each dynode stage, a net positive charge is induced during a pulse since more electrons leave the stage than were originally collected. This positive charge becomes progressively larger for each succeeding dynode stage due to the electron multiplication produced by each dynode. If a load resistor is placed between a given dynode and the voltage divider chain shown in Fig. 9.13, then a positive voltage pulse of smaller amplitude than the negative anode pulse can be observed. There has been some demonstration[22, 49, 50] that a dynode signal taken several stages before the anode has some advantage in fast timing measurements. However, the vast majority of scintillation pulses are taken from the PM tube anode, and the analysis of pulse shape beginning on p. 292 applies equally well to dynode or anode pulses. For simplicity, the analysis assumes that the anode pulse also corresponds to a positive charge Q, even though in practice the polarity of the anode pulse will be negative.

B. Magnetic Shielding

The electron optics within a PM tube are particularly sensitive to stray magnetic fields because of the low average energy (on the order of 100 eV) of the electrons traveling from stage to stage. Even the influence of the earth's magnetic field is sufficient to have an appreciable effect on the trajectories of these electrons. In situations in which the tube is likely to be physically moved or brought near equipment with stray magnetic fields, it is essential that a magnetic shield be provided to prevent gain shifts of the PM tube. The most common form consists of a thin cylinder of mu-metal that fits closely around the outside glass envelope of the PM tube. For most tube designs, this shield must be held at photo-cathode potential in order to avoid noise due to electroluminescence in the glass envelope.

VI. PHOTODIODES AS SUBSTITUTES FOR PHOTOMULTIPLIER TUBES

A. Potential Advantages

Photomultiplier tubes are the most common light amplifiers used with scintillators, both in pulse and current mode operation. However, advances in the development of semiconductor photodiodes have led to the substitution of solid-state devices for PM tubes in some applications. In general, photodiodes offer the advantages of higher quantum efficiency (and therefore the potential for better energy resolution), lower power consumption, more compact size, and improved ruggedness compared with PM tubes used in scintillation counting. Photodiodes are also virtually insensitive to magnetic fields and therefore can sometimes be substituted in experiments where magnetic fields prevent the use of PM tubes. Because of the relatively small dimensions over which charges must move in these devices, their time response is comparable to that of conventional PM tubes, and they can be used to good advantage in coincidence and other timing applications.

There are two general designs that have received attention as possible substitutes for PM tubes. *Conventional photodiodes* have no internal gain and operate by directly converting the optical photons from the scintillation detector to electron–hole pairs that are simply collected. *Avalanche photodiodes* incorporate internal gain through the use of higher electric fields that increase the number of charge carriers that are collected.

B. Conventional Photodiodes

Reference 51 presents a useful review of the application of conventional photodiodes for scintillator readout. When light is incident on a semiconductor, electron–hole pairs are generated in a manner similar to that detailed in Chapter 11 for incident ionizing radiation. Photons corresponding to typical scintillation light carry about 3–4 eV of energy, sufficient to create electron–hole pairs in a semiconductor with a bandgap of approximately 1–2 eV. (This process of conversion of visible light to electrical carriers is the basis of the common solar cell.) The conversion is not limited by the need for charge carriers to escape from a surface as in a conventional photocathode, so the maximum quantum efficiency of the process can be as high as 60–80%, several times larger than in a PM tube. This high quantum efficiency also spans a much wider wavelength range than is typical for photocathodes in PM tubes, so a much higher primary charge usually is created by the light from the scintillator. However, there is no subsequent amplification of this charge as in a PM tube, so the output signal is smaller by orders of magnitude.

A common configuration for a silicon photodiode is shown in Fig. 9.14. Light is incident on a *p*-layer entrance window that is kept as thin as possible to enhance transmission

Figure 9.14 Basic configuration of a conventional photodiode.

of the light to the active volume of the silicon. Electrons and holes (see Chapter 11) produced by the light are collected at the boundaries of the central *i*-region driven by the electric field resulting from the applied voltage. The corresponding induced charge is processed in an attached preamplifier to produce the output signal pulse.

In a typical scintillation event, only a few thousand visible photons are produced, so the size of the charge pulse that can be developed is limited at best to no more than the same number of electronic charges. Because of the small signal amplitude, electronic noise is a major problem in pulse mode operation, especially for large-area detectors and low-energy radiations. In an extensive study of the performance of many different scintillators with a silicon photodiode, Sakai[5] measured energy resolution values at 662 keV that were always much poorer (often by a factor of 2) than in equivalent measurements with a PM tube. Differences were smaller for higher-energy gamma rays, but the PM tube performance was always superior. Successful applications to date have largely been limited to high-energy radiations[52–55] and/or small-diameter diodes[56–60] for which the associated dark current and capacitance are also small. In current mode, the cumulative effect of many scintillation events at high rates can override the inherent noise, resulting in excellent operational characteristics. For example, photodiodes have become the light detector of choice for current mode scintillators used in X-ray computed tomography (CT) scanners for medical imaging.[61, 62]

Because photodiodes are similar to the silicon particle detectors discussed in Chapter 11, they share many of the same properties. Photodiodes are generally designed as fully depleted detectors, consisting of high purity *p*- or *n*-type silicon with highly doped *p*- and *n*-type contacts on opposite surfaces. This so-called PIN configuration is discussed in further detail in Chapter 11. The noise behavior also follows that of other semiconductor detectors and a detailed description is given in Chapter 17. Briefly, the noise arises largely from two separate sources. The first, generally called *series noise*, primarily originates from sources within the preamplifier input stage. Its relative importance increases with detector capacitance. For that reason, photodiodes with area beyond about 1 cm^2 show high noise levels and therefore are not used widely in the readout of scintillators. A second noise source, generally called *parallel noise*, is due largely to fluctuations in the leakage current in the photodiode. Since the leakage current will also tend to increase with the size of the photodiode, this second source of noise also inhibits the use of large areas. The capacitance of the photodiode will decrease as its thickness is increased, but the leakage current will tend to increase. Photodiodes common for scintillator applications are fabricated using silicon wafers of 300–500 μm thickness as something of a compromise between these conflicting trends. This thickness is much greater than would be required simply to be opaque to the incident light.

The relative contributions of series and parallel noise also depend on the shaping time constant chosen in the pulse processing electronics. Longer shaping times will tend to reduce the relative importance of series noise but will increase the importance of parallel noise (see Fig. 17.16). There will be an optimum in the choice of shaping time that will minimize the noise contribution. There are other considerations that enter this choice, however, such as the need to keep the shaping time longer than the scintillator decay time to avoid signal loss or the preference for short shaping times to minimize pulse pileup at high rates.

The spectral response of a typical silicon photodiode is plotted in Fig. 9.15. The quantum efficiency reaches higher values and extends much farther into the long wavelength region than that of typical photocathodes. This extended spectral response is particularly important for scintillators (such as CsI(Tl) or BGO) with emission spectra that have significant yield at longer wavelengths (see Fig. 8.7). The relative light yields for various scintillation materials given previously in Table 8.3 were measured using a typical PM tube photocathode response, and the values change significantly when measured with a photodiode. For example, the pulse amplitude from CsI(Tl) is smaller than the equivalent pulse from NaI(Tl) when measured with PM tubes but becomes over twice as large if the measurement is made with a silicon photodiode.[5]

Figure 9.15 A comparison of the quantum efficiency of a silicon photodiode (labeled #458) with representative bialkali and S-20 photocathode quantum efficiencies. The emission spectrum from a BGO scintillator is shown for reference. (From Groom.[63])

For typical photodiodes fabricated from silicon, the noise levels are much larger than in an equivalent photomultiplier tube. Since the dark current is one of the contributors to the noise, its contribution and the overall noise figure can be reduced by cooling the photodiode (see Fig. 9.16). The rapid rise in dark current above room temperature has generally prevented the use of silicon photodiodes in applications requiring operation at elevated temperatures.

Another way to reduce dark current is to choose semiconductor materials with a wider band gap than silicon. For example, successful photodiodes have been fabricated using mercuric iodide crystals that have several orders of magnitude lower dark current. As a result, excellent energy resolution has been demonstrated[65–67] in scintillators coupled to small mercuric iodide photodiodes. The attainable energy resolution, because of the high quantum efficiency of the semiconductor material, is actually superior to that obtainable even with the best photomultiplier tubes. One example,[66] a 1.27-cm diameter mercuric iodide photodiode coupled to a CsI(Tl) scintillator, achieved an energy resolution of 4.58% at 662 keV.

An alternative approach to improve the noise figure in photodiodes is to reduce their capacitance. One interesting possibility[68, 69] is to choose the silicon drift configuration described in Chapter 13. These "drift photodiodes" have the combined property of relatively large area (up to 1 cm²) and exceptionally low capacitance. To further reduce their noise, the leakage current can be suppressed by modest cooling to around 0°C. While still limited to small diameters, silicon drift photodiodes coupled to CsI(Tl) have shown the best energy resolution (4.34% at 662 keV) observed to date from any scintillation

Figure 9.16 The temperature dependence of the leakage current for some typical silicon photodiodes. (From Bian et al.[64])

detector. Again, much of the improved energy resolution arises because of the improved statistics resulting from the high-quantum efficiency of the photodiode.

The rise time of the pulse from a conventional photodiode is ultimately limited by the charge collection time from the active volume, typically only a few nanoseconds. However, timing resolution is normally determined by the uncertainties introduced by the relatively high noise level. Because series noise tends to be dominant, best spectroscopic behavior is obtained by using long shaping times (see Chapter 17). Techniques for generating timing pulses generally dictate a somewhat shorter shaping time and hence introduce a larger degree of noise into the signal.

Because noise considerations limit the size of photodiodes to relatively small sizes (typically 1-cm diameter or less) it is difficult to obtain good light collection from larger size scintillators that are more typical of gamma-ray spectroscopy applications. It is tempting to think about the alternative of using multiple photodiodes to cover a greater fraction of the exit window of the scintillator, and to then add together the signals from all the photodiodes. It turns out that this approach is not very satisfactory, since not only the signal but also the noise will be additive from the multiple diodes. First stated as *Groom's theorem*,[63] the light collection and hence the signal will increase roughly in proportion to the total area of photodiodes, but the noise will also increase by the same factor. The net result is very little change in the signal-to-noise ratio.

Both conventional and avalanche photodiodes respond directly to ionizing radiation in a manner similar to that described in Chapter 11 for silicon junction detectors. When used to read out the light from scintillators, there is the possibility that, in addition to registering the pulses of light, the photodiode may also respond directly if the incident radiation can penetrate the scintillator and reach the photodiode active volume. This so-called *nuclear counter effect* can be particularly troublesome because the size of the pulse produced by these direct interactions in the silicon will always be many times greater than the pulses produced from the scintillator light. Direct interactions in the silicon produce an electron–hole pair for every 3.6 eV deposited, while 15 to 20 times that energy deposition is needed in a good scintillator to produce the same charge. The difference comes about because scintillators typically convert only about 5–10% of the particle energy into light, not all that light enters the photodiode, and there is less than 100% quantum efficiency in converting that light into electron–hole pairs. In order to minimize this nuclear counter effect, the photodiode should be kept thin. However, the capacitance of the diode increases as its thickness is decreased, adding to the noise level. Thus some compromise must be struck between avoidance of these direct interactions and the preservation of reasonable energy resolution.

C. Avalanche Photodiodes

The small amount of charge that is produced in a conventional photodiode by a typical scintillation event can be increased through an *avalanche* process that occurs in a semiconductor at high values of the applied voltage. The charge carriers are accelerated sufficiently between collisions to create additional electron–hole pairs along the collection path, in much the same way that gas multiplication occurs in a proportional counter. (This same process is described in Chapter 13 in connection with *avalanche diode* detectors for ionizing radiations.) The internal gain helps pull the signal up from the electronic noise level and permits good energy resolution in pulse mode at lower radiation energy than possible using conventional photodiodes. Because the gain factor is very sensitive to temperature and applied voltage, avalanche photodiodes require well-regulated high-voltage supplies for stable operation. For current mode applications, the inherent stability of conventional photodiodes (without gain) is usually preferred.

Avalanche photodiodes can be fabricated in a number of different ways, but a common choice known as the *reach-through configuration* is shown in Fig. 9.17. Light enters through the thin p^+ layer on the left of the diagram and interacts somewhere within the π region that constitutes most of the diode thickness. The results of interactions are electron–hole pairs, and the electron is drawn to the right through the drift portion and into the multiplying

Figure 9.17 The reach-through configuration for an avalanche photodiode is sketched at the top of the figure. Below is a plot of the resulting electric field when a bias voltage is applied.

region where a high electric field exists. Here additional electron–hole pairs are created, increasing the measured signal. Gain factors of a few hundred are typical under normal circumstances. This enhancement of the signal is sufficient to allow much lower light levels to be sensed, or lower energies measured in their use with scintillators. Through the use of antireflective coatings on the external surface of the photodiode, quantum efficiencies as high as 80% are achievable at the peak wavelength of the response, typically in the 500–600 nm range. Enhanced shorter wavelength response is important for scintillators, many of which have their peak emission in the blue region of the spectrum. Measures can be taken[70] to enhance the blue sensitivity of photodiodes by proper choice of antireflective coating and the minimization of surface dead layer thickness.

In the multiplication process, electrons are drawn through the high field region and create additional electron–hole pairs. The electrons continue on in the same direction, but the holes will be attracted in the opposite direction. At sufficiently high field values, the holes can also multiply and, since hole multiplication also produces additional free electrons, this process leads to a runaway. That occurs at the breakdown voltage, and the applied voltage under normal circumstances is kept somewhat below this level. In this region, the overall gain will be an exponential function of applied voltage, accounting for the extreme sensitivity to applied voltage. The gain factor is also a strong function of temperature,[71] decreasing by a few percent per degree as the temperature is increased.

The timing properties of avalanche photodiodes can be quite favorable. In ideal cases, the limiting factor will be the variations in the electron drift time with position of interaction of the incident photon within the photodiode. These variations lead to a typical variability on the order of a nanosecond in the time at which the avalanche is triggered. In addition, the noise present with the signal also leads to timing uncertainty that often dominates the achievable timing performance. In favorable circumstances, a timing resolution of well under a nanosecond can be obtained,[72] but it is more typically a few nanoseconds or more.[73] Because of the higher signal/noise provided by the gain, shorter shaping times can be used with avalanche photodiodes than would be possible with an equivalent conventional photodiode, permitting higher rate operation and better timing resolution.

The multiplication process in an avalanche photodiode involves electrons undergoing collisions at random positions throughout the avalanche. When an ionizing collision occurs, one free electron is increased to two. The pulse-to-pulse fluctuations in gain are somewhat more significant than those observed from a typical PM tube. We have seen from Eq. (9.4) that the relative fluctuations in the gain of a PM tube decrease as the gain per dynode δ increases. Although the analogy is not exact, avalanche multiplication can be thought of as having a δ value of 2, much lower than the values typical for PM tube dynodes, and therefore the overall fluctuations might be expected to be larger. These fluctuations can be a significant contributor to the overall energy resolution observed when avalanche photodiodes are used with scintillators. The same categories of electronic noise previously mentioned for standard photodiodes are also still present, potentially degrading the energy resolution. On the other hand, the high quantum efficiency of avalanche photodiodes compared with PM tubes helps to offset these disadvantages, and excellent energy resolution has been demonstrated[74–76] in their use with small-volume scintillators.

VII. SCINTILLATION PULSE SHAPE ANALYSIS

The shape of the voltage pulse produced at the anode of a PM tube following a scintillation event depends on the time constant of the anode circuit. As discussed in Chapter 4, two extremes can be identified, both of which are commonly used in connection with scintillation counting. The first corresponds to those situations in which the time constant is chosen to be large compared with the decay time of the scintillator. This is the situation usually chosen if good pulse height resolution is a major objective and pulse rates are not excessively high. Then each pulse of electrons is integrated by the anode circuit to produce a voltage pulse whose amplitude is equal to Q/C, the ratio of the collected electron charge to the anode circuit capacitance. The second extreme is obtained by setting the anode circuit time constant to be much smaller than the scintillator decay time. As the following analysis will show, a much faster pulse results, which can often be an advantage in fast timing applications or when high pulse rates are encountered. At the same time, a sacrifice is then made in pulse amplitude and resolution.

The anode circuit can be idealized as shown in Fig. 9.18. C represents the capacitance of the anode itself, plus capacitance of the connecting cable and input capacitance of the circuit to which the anode is connected. The load resistance R may be a physical resistor wired into the tube base (see Fig. 9.13) or, if none is provided, the input impedance of the connected circuit. The current flowing into the anode $i(t)$ is simply the current of electrons from a single pulse, assumed to begin at $t = 0$. The shape of $i(t)$ will obviously influence the shape of the anode voltage pulse, and we choose for analysis a simplified representation of a typical electron pulse following a scintillation event. The principal component of emitted light from most scintillators can be adequately represented as a simple exponential decay. If the spread in transit time of the PM tube is small compared with this decay time, then a realistic model of the electron current arriving at the PM tube anode is simply

$$i(t) = i_0 e^{-\lambda t} \tag{9.7}$$

where λ is the scintillator decay constant. The initial current i_0 can be expressed in terms of the total charge Q collected over the entire pulse by noting

$$Q = \int_0^\infty i(t)\, dt = i_0 \int_0^\infty e^{-\lambda t}\, dt = \frac{i_0}{\lambda} \tag{9.8}$$

Therefore

$$i_0 = \lambda Q \tag{9.9}$$

and

$$i(t) = \lambda Q e^{-\lambda t} \tag{9.10}$$

To derive the voltage pulse $V(t)$ expected at the anode, we first note that the current flowing into the parallel RC circuit must be the sum of the current flowing into the capacitance i_C and the current through the resistance i_R

$$i(t) = i_C + i_R \tag{9.11}$$

$$i(t) = C\frac{dV(t)}{dt} + \frac{V(t)}{R} \tag{9.12}$$

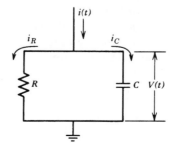

Figure 9.18 Simple parallel RC circuit representing a PM tube anode circuit.

Inserting Eq. (9.10) for $i(t)$ and dividing by C, we obtain

$$\frac{dV(t)}{dt} + \frac{1}{RC}V(t) = \frac{\lambda Q}{C}e^{-\lambda t} \tag{9.13}$$

The solution to this first-order inhomogeneous differential equation with initial condition $V(0) = 0$ can be shown to be

$$V(t) = \frac{1}{\lambda - \theta} \cdot \frac{\lambda Q}{C}(e^{-\theta t} - e^{-\lambda t}) \tag{9.14}$$

where $\theta \equiv \dfrac{1}{RC}$ is the reciprocal of the anode time constant.

Case 1. Large Time Constant

If the anode time constant is made large compared with the scintillator decay time, then $\theta \ll \lambda$ and Eq. (9.14) can be approximated by

$$V(t) \cong \frac{Q}{C}(e^{-\theta t} - e^{-\lambda t}) \tag{9.15}$$

A plot of this pulse form is shown in Fig. 9.19. Because $\theta \ll \lambda$, the first exponential in Eq. (9.15) decays slowly and the short time behavior is approximately

$$V(t) \cong \frac{Q}{C}(1 - e^{-\lambda t}) \qquad \left(t \ll \frac{1}{\theta}\right) \tag{9.16}$$

After a sufficiently long time, the second exponential decays to zero, and the long-time behavior is determined by the first exponential:

$$V(t) \cong \frac{Q}{C}e^{-\theta t} \qquad \left(t \gg \frac{1}{\lambda}\right) \tag{9.17}$$

The following important observations can now be made:

1. The leading edge of the pulse has the time behavior $(1 - e^{-\lambda t})$ and its rise time therefore is determined by the scintillator decay constant λ. Fast scintillators have large λ values that lead to fast-rising pulses.
2. The tail of the pulse has the time behavior $e^{-\theta t}$ and therefore decays away at a rate determined by the anode circuit time constant $RC \equiv 1/\theta$.
3. The amplitude of the pulse is given simply by Q/C, but this value is reached only if $\theta \ll \lambda$. Restated, the anode circuit time constant must be large compared with the scintillator decay time.

Most scintillation counting is carried out in this mode because the pulse height is maximized and subsequent sources of noise will have minimum degrading effect on pulse height resolution. Furthermore, the pulse amplitude achieved is not sensitive to changes in load resistance or to small changes in time characteristics of the electron pulse.

The experimenter must then choose a time constant that is at least 5–10 times greater than the scintillator decay time but that is not excessively long to prevent needless pulse pileup with the tail from a preceding pulse at high rates. The time constant is determined by the product RC, and either the anode capacitance or load resistance can be varied to change its value. In most applications, however, it is the resistance that must be tailored to achieve the desired time constant because the capacitance is intentionally held at its minimum value to maximize the pulse amplitude (Q/C).

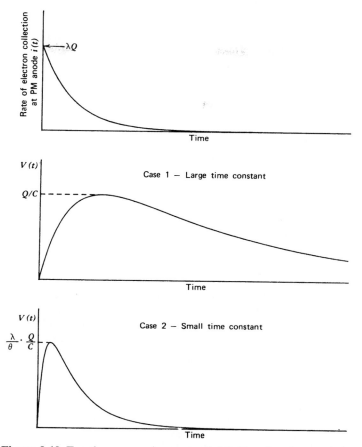

Figure 9.19 For the assumed exponential light pulse shown at the top, plots are given of the anode pulse $V(t)$ for the two extremes of large and small anode time constant. The duration of the pulse is shorter for Case 2, but the maximum amplitude is much smaller.

Case 2. Small Time Constant
In the opposite extreme, the anode time constant is set at a small value compared with the scintillator decay time, or $\theta \gg \lambda$. Now Eq. (9.14) becomes

$$V(t) = \frac{\lambda}{\theta} \cdot \frac{Q}{C}(e^{-\lambda t} - e^{-\theta t}) \tag{9.15a}$$

This pulse shape is also graphed in Fig. 9.19. The behavior at small values of t is now

$$V(t) = \frac{\lambda}{\theta} \cdot \frac{Q}{C}(1 - e^{-\theta t}) \qquad \left(t \ll \frac{1}{\lambda}\right) \tag{9.16a}$$

whereas for large t

$$V(t) = \frac{\lambda}{\theta} \cdot \frac{Q}{C}e^{-\lambda t} \qquad \left(t \gg \frac{1}{\theta}\right) \tag{9.17a}$$

The following general conclusions now apply:

1. The leading edge of the pulse has the time behavior $(1 - e^{-\theta t})$, which is determined by the anode time constant $RC \equiv 1/\theta$.
2. The tail of the pulse has the time behavior $e^{-\lambda t}$, which is identical to that of the scintillator light.
3. The maximum amplitude of the pulse is now $(\lambda Q/\theta C)$, a great deal smaller than the Case 1 maximum (Q/C) because, by definition of Case 2, $\lambda \ll \theta$.

The voltage pulse is now of much shorter duration than in Case 1, and its shape approaches that of the scintillator-produced PM tube current as the time constant is made smaller and smaller. This short duration comes at the price of a much reduced pulse amplitude, which varies linearly with the time constant and inversely with the scintillator decay time. Under fixed conditions, however, the pulse amplitude still is a linear measure of the charge Q collected by the anode, although it is more likely to be subject to fluctuations due to noise and component instabilities.

The simplified model we have used assumes a continuous and smooth current $i(t)$, which does not fully represent the discreteness or "clumped" nature of the anode current that ultimately arises from discrete photoelectrons. In Case 1, effects of the discreteness are largely smoothed out by the current integration process that takes place. In Case 2, however, no integration is carried out and the pulse form is much more sensitive to fluctuations that originate from the statistical nature of the photoelectron production. These fluctuations in pulse shape and amplitude are most serious for weak scintillation events, which produce only a small number of photoelectrons. The effect is illustrated in Fig. 9.20. When the anode time constant is very small, no significant integration of the current occurs, and the anode voltage signal reproduces the time dependence of the current shown at the middle of the figure. Only those photoelectrons contribute to the amplitude of the output pulse that, when broadened by the transit time spread of the PM tube, pile up during the early part of the scintillation event. Thus the light generated later in the decay of the scintillator does not contribute to the pulse amplitude. In contrast, the pulse shown at the bottom of the figure is taken with a large enough anode time constant to fully integrate all the light from the scintillator, and the statistical fluctuations in its amplitude will be minimized.

Even if a long anode time constant is chosen, the effective number of photoelectrons contributing to a typical pulse may be reduced in the pulse shaping step carried out in the subsequent pulse processing electronics. To avoid compromising the measured energy reso-

Figure 9.20 The sketch at the top represents the emission of electrons from the photocathode as a function of time following a scintillation event. The resulting current arriving at the anode is shown in the middle plot, and the bottom curve is the leading edge of the voltage pulse produced by integrating this current on an anode circuit with a long time constant.

lution, shaping times must be chosen that are several times longer than the scintillator decay time to ensure that most of the amplitude information survives through the shaping process. In order to have 95% of the scintillation light contribute to the pulse amplitude, an effective integration time that is three times the decay constant is required. Even using pulse shaping methods such as delay line clipping or gated integration (see Chapter 16) that lead to integrated pulses of minimum width, this requirement sets a lower limit on the pulse width, which in turn determines the maximum counting rate that can be tolerated. At a fundamental level, it is the overlap of the light itself from successive events within the scintillator that sets this limit. As an example, NaI(Tl) with a decay time of 230 ns thus requires a minimum pulse width of 690 ns for near-complete integration of the light. In order to avoid losing more than 10% of these pulses to pile-up (see Chapter 17), the counting rate must be held to below 74,000/s. To extend scintillation spectroscopy to higher rates, either complex pulse shape analysis schemes[77] must be employed to separate the pulses that fall near each other (not yet a common practice), or a scintillation material with faster decay time must be substituted.

VIII. HYBRID PHOTOMULTIPLIER TUBES

An interesting variant of the traditional photomultiplier tube (PMT) design is most frequently called the *hybrid photomultiplier tube* (*HPMT*) or, alternatively, the *hybrid photodiode* (*HPD*). As shown in Fig. 9.21, the basic principle involves a fundamentally different way of multiplying the charges created in a photocathode by incident light. As in a conventional PMT, the light is converted to electrons with a wavelength-dependent quantum efficiency in a photocathode. The conventional electron multiplier structure is now replaced by a silicon detector (of the type described in Chapter 11) placed in the same vacuum housing. A large voltage difference, typically between 10 and 15 kV, is applied between the photocathode and the silicon detector to accelerate electrons through the vacuum between the two elements of the tube. Photoelectrons emerge from the photocathode with very little energy—typically 1 eV or less. However, as they are drawn toward the silicon detector, they undergo continuous acceleration through the vacuum and strike the silicon detector front surface as high energy electrons. For example, if a voltage of 10 kV is applied, then the electrons arrive with a kinetic energy of 10 keV. As detailed in Chapter 11, an energetic electron of this type will lose its energy in the silicon detector through the creation of multiple electron–hole pairs. A 10-keV electron will create about 2800 such pairs if its entire energy is deposited in the active volume of the detector. Since each pair carries an electronic charge, the process just described effectively results in the multiplication of the unit charge of each photoelectron by a factor of 2800. Although this level of amplification is far less than that typical of PMTs (10^6–10^7), it nonetheless can produce signals of sufficient size to be successfully amplified in the succeeding electronic components.

Figure 9.21 Elements of hybrid photomultiplier tubes. Part (*a*) shows the electrostatic focusing arrangement, while part (*b*) shows a proximity-focused configuration. (From Basa, et al.[78])

A. HPMT Designs

The fabrication of HPMTs involves overcoming potential difficulties of compatibility of the components in the same high vacuum environment. The photocathode is easily "poisoned" or made inefficient by trace residual gas impurities, and the silicon detector can develop high noise levels if surface contaminants produce excess leakage current. Overcoming these difficulties has led to the commercial availability of HPMT beginning in the 1990s.

As shown in Fig. 9.21, there are two general types of HPMTs: those that employ electrostatic focusing, and those that are proximity-focused. The first type uses focusing electrodes to create electron optics that guide the electrons from the large-area photocathode to a small-area silicon detector. This design has the advantage that the small silicon diode has a low capacitance and leakage current, and therefore operates at a relatively low noise level. The second type (proximity-focused) uses a larger silicon detector of about the same diameter as the photocathode and provides a much smaller gap between the two. In this case, the electrons are drawn from the photocathode to the detector along electric field lines that are essentially parallel to the tube axis. Proximity-focused tubes may incorporate a gap of only a few millimeters and therefore can be very compact with overall length of 3 cm or less. This type, with its short and parallel electron paths, is perturbed by external magnetic fields much less than electrostatically focused designs, particularly when they are oriented with the magnetic field parallel to the tube axis.

B. Operational Characteristics

HPMTs have been successfully demonstrated to read out the light from typical scintillation gamma spectrometers[79] and are often applied in situations in which the detected light levels are very low. The most striking advantage of the HPMT over conventional tubes is the much lower statistical spread in the amplitude of the output pulse. As previously illustrated in Fig. 9.5, this statistical spread is determined almost exclusively by the number of charge carriers produced per incident electron in the first stage of amplification. The larger this number, the smaller will be the relative statistical broadening. In ordinary PMTs, a typical number for the first stage multiplication might be 5, while in high gain NEA dynode materials it could be as large as 25. The corresponding number in the HPMT is typically 2500–3500. Thus if repeated events occur in which exactly a fixed number of photoelectrons are released from the photocathode, the pulse height spectrum recorded from these events will be much broader from a PMT than from a HPMT. One way of illustrating the statistical effects is to record the pulse height spectrum while the tube is illuminated with a weak light source that creates a series of pulses whose average number of photoelectrons is relatively small. One example of this type of spectrum was given previously for a NEA dynode PM tube in Fig. 9.6. With a perfect light sensor, the recorded spectrum would consist of a series of separately resolved peaks corresponding to $1, 2, 3, \ldots$ photoelectrons per pulse. Because of statistical spreading, the spectrum resolves only the first three of these peaks. Contrast that circumstance with the spectrum shown in Fig. 9.22, taken from a HPMT. Here the peaks are separately resolvable all the way to a value of 14 photoelectrons. These peaks are seen to be superimposed on a broad continuum that arises largely because some of the incident electrons are backscattered from the silicon detector before they deposit all their energy. For incident 13 keV electrons, about 18% of the electrons are backscattered.[80]

One of the major benefits of the superior statistical behavior of the amplification is the ability to better separate those events that correspond to a single photoelectron from those that arise from multiple photoelectrons. For example, thermionic noise arises from single photoelectrons emitted randomly in time from the photocathode, and therefore such noise pulses will appear in the single photoelectron peak. If that peak is cleanly separated from those corresponding to multiple photoelectrons, then a simple pulse height discrimination

Figure 9.22 Pulse height spectrum taken from a hybrid photomultiplier tube when illuminated by light pulses of low intensity. Separate peaks can be distinguished for events that originate from the discrete number of photoelectrons shown along the top of the plot. (From D'Ambrosio et al.[80])

can eliminate the thermionic noise without sacrificing more than absolutely necessary of the signal events. The clean resolution of single photon events from electronic noise has also been exploited in many applications in which the signal itself consists of only single photoelectrons released from the photocathode. If the only broadening effects were statistical, the individual photoelectron peaks shown in Fig. 9.22 should have a predicted pulse height resolution of under 1% (see discussion of energy resolution in Chapter 11). The widths actually observed are much broader than this because of the added contribution of electronic noise in the amplifying elements necessary to process these low-level signals.

HPMTs also have a number of other operational advantages. Because the gain factor depends only on the energy deposited in the silicon detector, it should vary linearly with the value of the applied high voltage. This behavior is observed experimentally if allowance is made for the energy loss of the electrons in the entrance window of the detector (typically a few hundred to a thousand eV). In contrast, for a conventional PMT the gain typically varies as the voltage raised to the seventh or eighth power [see Eq. (9.3)]. Thus there is much less sensitivity to instability or drift in the voltage supply. Furthermore, the only current drawn from the high voltage supply is the internal current represented by the flow of photoelectrons. This current is many orders of magnitude smaller than the current typically required in the resistive divider string of a PMT to stabilize the interdynode voltages. For these reasons, much simpler and more compact high voltage supplies can be used with HPMTs, despite the somewhat higher values of the voltages required. The simplicity of the single-stage multiplication also generally results in much less temperature sensitivity of the gain than in a conventional tube.

The timing properties of HPMTs are also quite favorable. In either the electrostatic-focused or proximity-focused tubes, there is very little spread in the transit time compared with that observed from conventional PMTs. The accelerated photoelectrons are stopped near the entrance of the silicon detector, forming electron–hole pairs. The silicon detector, which typically might be 300 μm thick, is normally biased to collect positive charges (holes) at the front surface and negative charges (electrons) at the rear. Thus the holes are quickly collected, and it is the flow of electrons through most of the thickness of the depletion region that gives rise to the observed signal current (see discussions in Chapter 11). The corresponding current pulse from small-diameter (low capacitance) silicon detectors has a duration of a few nanoseconds with rise and fall times of less than 1 ns. Timing precision of 100 ps has been demonstrated,[81] but best performance is only achieved in applications for which large light pulses are produced because of the influence of the electronic noise in the external electronics on the timing precision.

C. Variants of the HPMT

Several other versions of the HPMT have been investigated that employ different types of silicon detectors. Because the gain achievable using a conventional silicon detector is only a few thousand, designs have been produced that substitute an avalanche silicon detector that provides an additional internal gain factor. Small diameter avalanche photodiodes are capable of providing an internal gain of several hundred, thereby boosting the overall gain of the tube to over 10^5. This type of amplification is now much more typical of conventional PMTs, and overcomes many of the problems of low amplitude pulses. However, the statistics of the gain multiplication caused by avalanching adds a large broadening to the output pulses, and the individual photoelectron peaks illustrated in Fig. 9.22 can no longer be observed. Another variation is to substitute a position-sensitive silicon detector. This implementation converts the HPMT into a position-sensitive PMT of the type discussed in the next section. In one such design,[82] 1024 separate pixels were provided on the silicon detector, each bonded to its own read-out electronics. Further developments of this type should be expected in this rapidly developing area of technology.

IX. POSITION-SENSING PHOTOMULTIPLIER TUBES

In ordinary scintillation counting, the light from a typical scintillation event is spread over the entire PM tube photocathode, and therefore the point of interaction within the photocathode is of no particular interest. However, there are some specialized applications, for example, in particle physics or in nuclear medicine, in which the position of the arriving photon on the photocathode carries some information. For these applications, there has been interest in developing PM tubes that can provide some position information regarding the incident light.

A first requirement in any such design is to use an electron-multiplying structure that maintains spatial separation between the multiplied electron clouds originating from photoelectrons generated at separate locations on the photocathode. This requirement rules out the common multiplier structures such as those illustrated in Fig. 9.7 in which electrons from all areas of the photocathode are focused onto a common dynode. However, several of the electron multiplier structures we have already discussed meet this requirement of confining the multiplied electrons to a localized position. These include the microchannel plate multiplier in which each multiplied electron cloud is confined to a single small channel, and the hybrid photomultiplier tube in which electrons from the photocathode are accelerated onto a corresponding position on a silicon detector. PMTs based on these multiplier configurations are commercially available and can be used in position sensing or imaging applications.

Other specialized dynode multiplier structures have also been developed specifically to produce position-sensitive photomultiplier tubes. The first of these consists of a fine mesh structure in which as many as 16–20 stages of multiplication are produced by layering metal dynodes that are perforated with many holes, as illustrated in Fig. 9.23. When an appropriate surface treatment is given to the dynode structures and they are operated at optimum potential between layers, secondary electron yields of about three per stage are achievable. Thus an overall gain of 10^6–10^9 can be realized,[83] using up to 20 stages. With this configuration, there is a natural tendency for the charge to spread as the multiplication takes place from stage to stage. This spreading is clearly undesirable if the ultimate in spatial resolution is needed. Thus other dynode designs have evolved using fine meshes[84, 85] or a metal channel dynode structure (see Fig. 9.24) that do a better job of confining the multiplied charge to a narrow spot. A concern in the design of this type of position-sensitive PMT is the uniformity of response to light across the entrance area of the photocathode. Factors of two deviation in the gain of the tube can be observed[87] in some types of tubes.

Figure 9.23 Cross section of a focused mesh electron multiplier. Each cusp-shaped dynode layer has an associated guard plate with its holes aligned over the points of the cusps. The guard plate helps focus the emitted secondary electrons onto the next dynode layer and also stops any secondary ions (the upward arrow shown) from causing ion feedback. (From Vallerga et al.[83])

These variations involve not only the nonuniformity of the photocathode but also variations in the probability of collecting a photoelectron depending on its emission position.

At the anode end or output of the tube, several schemes can be used to retain the position information. The most direct method is to simply divide the anode into separate discrete anodes or pixels, each with its own individual readout connection. As long as the cross-talk caused by electron spreading is kept small, an array of 8×8 anodes is equivalent to packaging 64 separate photomultiplier tubes in a common vacuum enclosure. Such tubes have been made available commercially that have sufficient pulse height resolution to distinguish the peak in the spectrum from the detection of single photons. An alternative method is to provide two layers of anode wires, each layer consisting of multiple parallel wires. The two layers are oriented with their wires perpendicular to those in the other layer to produce separate x and y position coding. This "crossed anode wire" readout is widely used for position-sensitive PM tubes because the charge division scheme illustrated in Fig. 9.25 can be used to reduce the number of readout channels to four. Alternatively, each separate anode wire can be read out with its own preamplifier[87] to allow greater flexibility for correction of inevitable nonlinearities in the spatial signals using the charge-sharing technique.

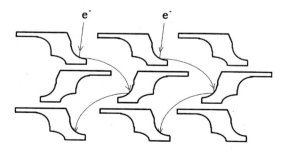

Figure 9.24 Cross section of a metal channel dynode structure. (From Yoshizawa and Takeuchi.[86])

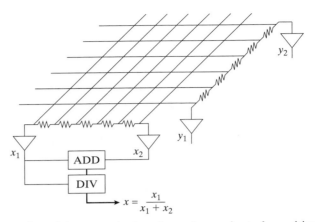

Figure 9.25 Illustration of the crossed-wire scheme for readout of a position-sensing photomultiplier tube. A resistor chain interconnecting all the wires running in one dimension is provided with preamplifiers on each end. Carrying out the operations shown produces a pulse that ideally is proportional only to the event coordinate in that dimension.

X. PHOTOIONIZATION DETECTORS

There is another alternative to PM tubes that has been exploited for applications in which the light to be detected is in the ultraviolet portion of the spectrum. Certain organic compounds in the gaseous phase can be ionized by UV photons to form ion pairs. If the organic vapor is incorporated as a component of the fill gas of a conventional detector that is sensitive to ionization (such as a proportional counter), then the signal pulse amplitude will reflect the number of incoming photons that have undergone conversion to ions. Furthermore, if position-sensitive detectors such as multiwire proportional counters are used, the spatial position of the conversion point can also be determined.

Some organic compounds that are of interest as photoionization agents are listed in Table 9.2. Of these, TMAE and TEA have received the most attention. It has been shown that each can be successfully incorporated into standard proportional gases with typical concentration of up to 3%. In the case of TMAE, its vapor pressure of 0.35 torr at 20°C (Ref. 89) is a limit on its room temperature absolute concentration in the gas phase. At that partial pressure, its quantum efficiency is approximately 2% per mm of path length for the fast component of the emission spectrum (see Fig. 8.13) from BaF_2 (Ref. 90). The vapor reacts with oxygen and organic materials such as vacuum grease and O-rings, so that some care is needed in chamber design and gas handling procedures. TEA is somewhat easier to handle and has a higher vapor pressure, but its higher photoionization potential limits its response to shorter wavelengths.

Table 9.2 Some Organics Potentially Useful in Photoionization Detectors

Acronym	Chemical Name	Vapor Phase Ionization Potential (eV)	Long-Wavelength Cutoff (nm)
TMAE	Tetrakis(dimethylamino)ethylene	5.36	231
TMBI	1,1′,3,3′-Tetramethyl-2,2′-biimidazolidinylidene	5.41	229
TMAB	1,1,4,4-Tetrakis(dimethylamino)butadiene	5.60	221
TMPD	*N,N,N′,N′*-tetramethyl-*p*-phenylenediamine	6.20	200
TEA	Triethylamine	7.5	165

Data as tabulated by Policarpo.[88]

Photoionization detectors are not commercially available, but they can offer interesting properties for specialized applications. They can be made in almost unlimited size with excellent uniformity of response over the large area entrance window. Good spatial resolution can also be achieved (of interest in imaging applications) by incorporating the position-sensing techniques discussed in Chapter 5 for large-area proportional detectors. They have found useful application in the detection of UV emissions from scintillators,[90–92] Cherenkov detectors,[93–95] and gas proportional scintillation counters.[88, 93]

PROBLEMS

9.1. Calculate the long-wavelength limit of the sensitivity of a photocathode layer with work function of 1.5 eV.

9.2. Find the transit time for an electron between typical dynodes in a PM tube if the interdynode spacing is 12 mm and the potential difference is 150 V per stage. For simplicity, assume a uniform electric field.

9.3. Using the data plotted in Fig. 9.3, find the total applied voltage necessary for a PM tube with a six-stage multiplier using GaP(Cs) dynodes to achieve an electron gain factor of 10^6.

9.4. The dark current from a PM tube with electron gain of 10^6 is measured to be 2 nA. What is the corresponding electron emission rate from the photocathode?

9.5. The gain per dynode δ of a 10-stage PM tube varies as $V^{0.6}$ where V is the interdynode voltage. If the tube is operated at an overall voltage of 1000 V, how much voltage fluctuation can be tolerated if the gain is not to change by more than 1%?

9.6. The decay time (inverse of the decay constant) for scintillations in NaI(Tl) is 230 ns. Neglecting any time spread introduced by the PM tube, find the maximum value of the voltage pulse amplitude for anode circuit time constants of 10, 100, and 1000 ns. Express your answer as a ratio to the amplitude that would be observed for an infinite time constant. What is the minimum value of the time constant if this ratio is to be at least 0.9?

9.7. A current pulse of the following shape flows into a parallel RC circuit:

$$i(t) = I \qquad 0 \le t < T$$
$$i(t) = 0 \qquad T \le t$$

Find the general solution $V(t)$ for the voltage appearing across the circuit, assuming $V(0) = 0$. Sketch the solution in the two limits:

(a) $RC \gg T$.

(b) $RC \ll T$.

9.8. Calculate the amplitude of the signal pulse expected from a NaI(Tl)–PM tube combination under the following circumstances:

Radiation energy loss: 1.2 MeV.

Light collection efficiency: 70%.

Photocathode quantum efficiency: 20%.

PM tube electron gain: 100,000.

Anode capacitance: 100 pF.

Anode load resistance: 10^5 ohms.

Any other physical parameters you may need can be found in the text.

9.9. What is the principal advantage of microchannel plate PM tubes compared with more conventional designs?

9.10. The bandgap energy in silicon at room temperature is 1.11 eV. Calculate the longest wavelength of light that is energetically capable of exciting an electron across this gap to create an electron–hole pair in a photodiode.

9.11. A scintillator absorbs an incident flux of 5 MeV alpha particles that totals 10^6 particles/s. The scintillation efficiency for these particles is 3%, and the average wavelength of the emitted light is 420 nm. If the scintillator is coupled to a photodiode with an average quantum efficiency of 75% for the scintillation light and the light collection efficiency is 80%, estimate the expected signal from the photodiode when operated in current mode.

9.12. Estimate the acceleration voltage required for a hybrid photomultiplier tube using a silicon diode to have a charge gain of 5000.

REFERENCES

1. *RCA Photomultiplier Manual*. Technical Series PT-61, RCA Solid State Division, Electro Optics and Devices, Lancaster, PA, 1970. (See also other manufacturer's publications such as *Photomultiplier Tubes, Principles and Applications*, Philips Photonics, Philips Export B. V., 1994; *Photomultiplier Tube, Principle to Application*, Hamamatsu Photonics K. K., 1994.)

2. G. A. Morton, *IEEE Trans. Nucl. Sci.* **NS-22** (1), 26 (1975).

3. "Test Procedures for Photomultipliers for Scintillation Counting and Glossary for Scintillation Counting Field," IEEE Standard 398-1990, (1990).

4. M. Miyajima, S. Sasaki, and E. Shibamura, *Nucl. Instrum. Meth.* **224**, 331 (1984).

5. E. Sakai, *IEEE Trans. Nucl. Sci.* **NS-34** (1), 418 (1987).

6. H. R. Krall, F. A. Helvy, and D. E. Persyk, *IEEE Trans. Nucl. Sci.* **NS-17**(3), 71 (1970).

7. R. E. Simon and B. F. Williams, *IEEE Trans. Nucl. Sci.* **NS-15**(3), 167 (1968).

8. G. A. Morton, H. M. Smith, Jr., and H. R. Krall, *IEEE Trans. Nucl. Sci.* **NS-16** (1), 92 (1969).

9. B. Leskovar and C. C. Lo, *IEEE Trans. Nucl. Sci.* **NS-19**(3), 50 (1972).

10. H. R. Krall and D. E. Persyk, *IEEE Trans. Nucl. Sci.* **NS-19**(3), 45 (1972).

11. H. Houtermans, *Nucl. Instrum. Meth.* **112**, 121 (1973).

12. T. Sandor, *Nucl. Instrum. Meth.* **78**, 8 (1970).

13. R. Bosshard et al., *IEEE Trans. Nucl. Sci.* **NS-19**(3), 107 (1972).

14. F. W. Inman and J. J. Muray, *IEEE Trans. Nucl. Sci.* **NS-16**(2), 62 (1969).

15. J. R. Prescott, *Nucl. Instrum. Meth.* **39**, 173 (1966).

16. J. P. Ballini, *Nucl. Instrum. Meth.* **116**, 109 (1974).

17. J. Ballini, P. Cazes, and P. Turpin, *Nucl. Instrum. Meth.* **134**, 319 (1976).

18. B. Leskovar, *Phys. Today* **30**(11), 42 (1977).

19. G. Pietri, *IEEE Trans. Nucl. Sci.* **NS-22**(5), 2084 (1975).

20. S. Dhawan and R. Majka, *IEEE Trans. Nucl. Sci.* **NS-24**(1), 270 (1977).

21. C. C. Lo, P. Lecomte, and B. Leskovar, *IEEE Trans. Nucl. Sci.* **NS-24**(1), 302 (1977).

22. B. Leskovar and C. C. Lo, *IEEE Trans. Nucl. Sci.* **NS-25**(1), 582 (1978).

23. M. Moszynski, J. Vacher, and R. Odru, *Nucl. Instrum. Meth.* **204**, 141 (1982).

24. M. Moszynski, J. Vacher, and R. Odru, *Nucl. Instrum. Meth.* **204**, 471 (1983).

25. J. De Vries and C. W. E. Van Eijk, *Nucl. Instrum. Meth.* **A239**, 243 (1985).

26. B. Leskovar and T. T. Shimizu, *IEEE Trans. Nucl. Sci.* **NS-34**, 427 (1987).

27. V. V. Anashin et al., *Nucl. Instrum. Meth.* **A357**, 103 (1995).

28. H. Kume et al., *IEEE Trans. Nucl. Sci.* **NS-33**(1), 364 (1986).

29. F. de la Barre, *Nucl. Instrum. Meth.* **102**, 77 (1972).

30. B. Sipp and J. A. Miehe, *Nucl. Instrum. Meth.* **114**, 249 (1974).

31. M. Moszynski, *Nucl. Instrum. Meth.* **A337**, 154 (1993).

32. F. A. Johnson, *Nucl. Instrum. Meth.* **87**, 215 (1970).

33. G. Gangopadhyay, D. Banerjee, and R. Bhattacharya, *Nucl. Instrum. Meth.* **A336**, 186 (1993).

34. G. A. Morton, H. M. Smith, and R. Wasserman, *IEEE Trans. Nucl. Sci.* **NS-14**(1), 443 (1967).

35. S. J. Hall and J. McKeown, *Nucl. Instrum. Meth.* **112**, 545 (1973).

36. R. Staubert et al., *Nucl. Instrum. Meth.* **84**, 297 (1970).

37. S. S. Stevens and J. W. Longworth, *IEEE Trans. Nucl. Sci.* **NS-19**(1), 356 (1972).

38. G. P. Lamaze, J. K. Whittaker, R. A. Schrack, and O. A. Wasson, *Nucl. Instrum. Meth.* **123**, 403 (1975).

39. R. Mirzoyan, E. Lorenz, D. Petry, and C. Prosch, *Nucl. Instrum. Meth.* **A387**, 74 (1997).

40. J. M. Paul, *Nucl. Instrum. Meth.* **89**, 285 (1970).

41. H. J. Kellermann et al., *Nucl. Instrum. Meth.* **115**, 301 (1974).

42. A. K. Gupta and N. Nath, *Nucl. Instrum. Meth.* **53**, 352 (1967).

43. C. Weitkamp, G. G. Slaughter, W. Michaelis, and H. Schmidt, *Nucl. Instrum. Meth.* **61**, 122 (1968).

44. M. Yamashita, *Nucl. Instrum. Meth.* **142**, 435 (1977).

45. L. Cathy, "Control of Fatigue in Photomultipliers," DP-642 (1961).

46. R. D. Conner and M. K. Husain, *Nucl. Instrum. Meth.* **6**, 337 (1960).

47. R. D. Hiebert, H. A. Thiessen, and A. W. Obst, *Nucl. Instrum. Meth.* **142**, 467 (1977).

48. C. R. Kerns, *IEEE Trans. Nucl. Sci.* **NS-24**(1), 353 (1977).

49. J. De Vries and F. E. T. Kelling, *Nucl. Instrum. Meth.* **A262**, 385 (1987).

50. B. Bengtson and M. Moszynski, *Nucl. Instrum. Meth.* **204**, 129 (1982).

51. M. Suffert, *Nucl. Instrum. Meth.* **A322**, 523 (1992).

52. R. Glasow et al., *Nucl. Instrum. Meth.* **228**, 354 (1985).

53. G. Hall et al., *IEEE Trans. Nucl. Sci.* **NS-33**(1), 310 (1986).

54. W. G. Gong et al., *Nucl. Instrum. Meth.* **A268**, 190 (1988).

55. V. Prat et al., *Nucl. Instrum. Meth.* **A380**, 209 (1996).

56. S. E. Derenzo, *Nucl. Instrum. Meth.* **219**, 117 (1984).

57. H. Grassmann et al., *Nucl. Instrum. Meth.* **A234**, 122 (1985).

58. E. Gramsch et al., *Nucl. Instrum. Meth.* **A311**, 529 (1992).

59. A. J. Bird et al., *IEEE Trans. Nucl. Sci.* **40**(4), 395 (1993).

60. V. Prat, H. Simon, and A. Kazandjian, *Nucl. Instrum. Meth.* **A369**, 617 (1996).

61. D. P. Boyd, *IEEE Trans. Nucl. Sci.* **NS-26**(2), 2836 (1979).

62. T. Takahashi, H. Itoh, T. Shimada, and H. Takeuchi, *IEEE Trans. Nucl. Sci.* **37**(3), 1478 (1990).

63. D. E. Groom, *Nucl. Instrum. Meth.* **219**, 141 (1984).

64. Z. Bian, J. Dobbins, and N. Mistry, *Nucl. Instrum. Meth.* **A239**, 518 (1985).

65. J. Markakis, *IEEE Trans. Nucl. Sci.* **NS-35**(1), 356 (1988).

66. Y. J. Wang et al., *IEEE Trans. Nucl. Sci.* **43**(3), 1277 (1996).

67. M. P. Tornai et al., *IEEE Trans. Nucl. Sci.* **44**(3), 1127 (1997).

68. C. Fiorini et al., *Nucl. Instrum. Meth.* **A409**, 395 (1998).

69. C. Fiorini, F. Perotti, and C. Labanti, *IEEE Trans. Nucl. Sci.* **45**(3), 483 (1998).

70. E. Gramsch and R. E. Avila, *Nucl. Instrum. Meth.* **A414**, 299 (1998).

71. S. Kobayashi, K. Yamaoka, M. Amami, and M. Kobayashi, *Nucl. Instrum. Meth.* **A364**, 95 (1995).

72. M. Moszynski et al., *IEEE Trans. Nucl. Sci.* **43**(3), 1298 (1996).

73. C. Carrier and R. Lecomte, *Nucl. Instrum. Meth.* **A299**, 115 (1990).

74. K. M. James and M. J. Masterson, *Nucl. Instrum. Meth.* **A313**, 196 (1992).

75. A. Ochi, Y. Nishi, and T. Tanimori, *Nucl. Instrum. Meth.* **A378**, 267 (1996).

76. M. Moszynski et al., *IEEE Trans. Nucl. Sci.* **45**(3), 472 (1998).

77. W. -H. Wong and H. Li, *IEEE Trans. Nucl. Sci.* **45**(3), 838 (1998).

78. S. Basa et al., *Nucl. Instrum. Meth.* **A330**, 93 (1993).
79. G. Bellia, L. Cosentino, P. Finocchiaro, and K. Loukachine, *Nucl. Instrum. Meth.* **A385**, 116 (1997).
80. C. D'Ambrosio et al., *Nucl. Instrum. Meth.* **A345**, 279 (1994).
81. P. Benetti et al., *Nucl. Instrum. Meth.* **A367**, 384 (1995).
82. T. Gys et al., *Nucl. Instrum. Meth.* **A387**, 131 (1997).
83. J. Vallerga, J. Hull, and M. Lampton, *IEEE Trans. Nucl. Sci.* **35(1)**, 539 (1988).
84. S. Suzuki et al., *IEEE Trans. Nucl. Sci.* **40(4)**, 431 (1993).
85. D. Grigoriev, O. Johnson, W. Worstell, and V. Zavarzin, *IEEE Trans. Nucl. Sci.* **44(3)**, 990 (1997).
86. Y. Yoshizawa and J. Takeuchi, *Nucl. Instrum. Meth.* **A387**, 33 (1997).
87. R. A. Kroeger et al., *IEEE Trans. Nucl. Sci.* **44(3)**, 881 (1997).
88. A. J. P. L. Policarpo, *Nucl. Instrum. Meth.* **196**, 53 (1982).
89. D. F. Anderson, *IEEE Trans. Nucl. Sci.* **NS-28**(1), 842 (1981).
90. C. L. Woody, C. I. Petridou, and G. C. Smith, *IEEE Trans. Nucl. Sci.* **NS-33**(1), 136 (1986).
91. D. F. Anderson et al., *Nucl. Instrum. Meth.* **225**, 8 (1984).
92. D. Visvikis et al., *Nucl. Instrum. Meth.* **A392**, 414 (1997).
93. D. F. Anderson, *IEEE Trans. Nucl. Sci.* **NS-32** (1), 495 (1985).
94. P. Krizan et al., *Nucl. Instrum. Meth.* **A367**, 257 (1995).
95. J. Va'vra, J. Kadyk, J. Wise, and P. Coyle, *Nucl. Instrum. Meth.* **A370**, 352 (1996).

Chapter **10**

Radiation Spectroscopy with Scintillators

The widespread availability in the early 1950s of thallium-activated sodium iodide scintillation detectors allowed the spectroscopy of gamma rays to be carried out for the first time with small, portable instruments. With this development, a practical detector was available that could provide a high efficiency for the detection of gamma rays and, at the same time, was capable of sufficiently good energy resolution to be useful in separating the contributions of polyenergetic gamma-ray sources. Gamma-ray spectroscopy using scintillators has since developed into a mature science with applications in an impressive array of technical fields.

Despite the fact that it was virtually the first practical solid detection medium used for gamma-ray spectroscopy, NaI(Tl) remains the most popular scintillation material for this application. This extraordinary success stems from its extremely good light yield, excellent linearity, and the high atomic number of its iodine constituent. Other scintillators mentioned in Chapter 8, notably cesium iodide and some of the newer inorganic scintillators, have also achieved some success in gamma-ray spectroscopy, but the combined use of all other materials extends to only a small fraction of the cases in which sodium iodide is found. In the sections that follow, we concentrate on NaI(Tl), with the understanding that most of the discussion and general conclusions can be extended to other scintillation materials by taking into account the differences in their gamma-ray interaction probabilities and scintillation properties.

Useful textbook reviews of scintillation spectroscopy of gamma radiation have been published by Birks,[1] Shafroth,[2] and Siegbahn.[3]

I. GENERAL CONSIDERATIONS IN GAMMA-RAY SPECTROSCOPY

An X-ray or gamma-ray photon is uncharged and creates no direct ionization or excitation of the material through which it passes. The detection of gamma rays is therefore critically dependent on causing the gamma-ray photon to undergo an interaction that transfers all or part of the photon energy to an electron in the absorbing material. These interaction processes are detailed in Chapter 2 and represent sudden and major alterations of the photon properties, as opposed to the continuous slowing down of heavy charged particles or electrons through many simultaneous interactions.

Because the primary gamma-ray photons are "invisible" to the detector, it is only the fast electrons created in gamma-ray interactions that provide any clue to the nature of the incident gamma rays. These electrons have a maximum energy equal to the energy of the incident gamma-ray photon and will slow down and lose their energy in the same manner as any other fast electron, such as a beta particle. Energy loss is therefore through ionization and excitation of atoms within the absorber material and through bremsstrahlung emission (see Chapter 2).

In order for a detector to serve as a gamma-ray spectrometer, it must carry out two distinct functions. First, it must act as a conversion medium in which incident gamma rays have a reasonable probability of interacting to yield one or more fast electrons; second, it must function as a conventional detector for these secondary electrons. In the discussion that follows, we first assume that the detector is sufficiently large so that the escape of secondary electrons (and any bremsstrahlung created along their track) is not significant. For incident gamma rays of a few MeV, the most penetrating secondary electrons will also be created with a few MeV kinetic energy. The corresponding range in typical solid detector media is a few millimeters (most bremsstrahlung photons generated along the electron track will be considerably less penetrating). The assumption of complete electron absorption therefore implies a detector whose minimum dimension is at least about a centimeter. Then only a small fraction of the secondary electrons, which are created more or less randomly throughout the volume of the detector, lie within one range value of the surface and could possibly escape. Later in this chapter we discuss the complicating effects of electron and bremsstrahlung escape in small detectors as a perturbation on the simpler model that follows.

The following discussions are kept relatively general so that they apply not only to other scintillation materials but also to other solid or liquid detection media used in gamma-ray spectroscopy. Chapters 12 and 13 discuss semiconductor detectors, which also have been widely applied to gamma-ray spectroscopy. The following section serves as a general introduction to these chapters as well, because the basic modes of gamma-ray interactions are identical for all detector types.

Because of the low stopping power of gases, the requirement of full energy absorption for the secondary electrons generally rules out gas-filled detectors for the spectroscopy of gamma rays. Exceptions to this statement are possible only if the gas is at very high pressure (see the discussion of high pressure xenon chambers in Chapter 19) or if the incident photon has very low energy. The penetration distance of a 1 MeV electron in STP gases is several meters, so normal gas-filled detectors of practical size can never come close to absorbing all the secondary electron energy. To complicate the situation further, most gamma-ray-induced pulses from a gas-filled counter arise from gamma-ray interactions taking place in the solid counter wall, following which the secondary electron finds its way to the gas. Under these conditions, the electron loses a variable and indeterminate amount of energy in the wall, which does not contribute to the detector output pulse, and virtually all hope of relating the electron to incident gamma-ray energy is lost.

II. GAMMA-RAY INTERACTIONS

Of the various ways gamma rays can interact in matter, only three interaction mechanisms have any real significance in gamma-ray spectroscopy: photoelectric absorption, Compton scattering, and pair production. As detailed in Chapter 2, photoelectric absorption predominates for low-energy gamma rays (up to several hundred keV), pair production predominates for high-energy gamma rays (above 5–10 MeV), and Compton scattering is the most probable process over the range of energies between these extremes. The atomic number of the interaction medium has a strong influence on the relative probabilities of these three interactions, as can be seen from the formulas and plots given in Chapter 2. The most striking of these variations involves the cross section for photoelectric absorption, which varies approximately as $Z^{4.5}$. As we shall see from the following discussion, because photoelectric absorption is the preferred mode of interaction, there is a premium on choosing detectors for gamma-ray spectroscopy from materials that incorporate elements with high atomic number.

A. Photoelectric Absorption

Photoelectric absorption is an interaction in which the incident gamma-ray photon disappears. In its place, a photoelectron is produced from one of the electron shells of the absorber atom with a kinetic energy given by the incident photon energy $h\nu$ minus the binding energy of the electron in its original shell (E_b). This process is shown in the diagram below. For typical gamma-ray energies, the photoelectron is most likely to emerge from the K shell, for which typical binding energies range from a few keV for low-Z materials to tens of keV for materials with higher atomic number. Conservation of momentum requires that the atom recoils in this process, but its recoil energy is very small and usually can be neglected.

$$h\nu \qquad\qquad E_{e^-} = h\nu - E_b$$

The vacancy that is created in the electron shell as a result of the photoelectron emission is quickly filled by electron rearrangement. In the process, the binding energy is liberated either in the form of a characteristic X-ray or Auger electron. In iodine, a characteristic X-ray is emitted in about 88% of the cases.[4] The Auger electrons have extremely short range because of their low energy. The characteristic X-rays may travel some distance (typically a millimeter or less) before being reabsorbed through photoelectric interactions with less tightly bound electron shells of the absorber atoms. Although escape of these X-rays can at times be significant, for now we assume that they are also fully absorbed in keeping with our simplified model.

Thus, the effect of photoelectric absorption is the liberation of a photoelectron, which carries off most of the gamma-ray energy, together with one or more low-energy electrons corresponding to absorption of the original binding energy of the photoelectron. If nothing escapes from the detector, then the sum of the kinetic energies of the electrons that are created must equal the original energy of the gamma-ray photon.

Photoelectric absorption is therefore an ideal process if one is interested in measuring the energy of the original gamma ray. The total electron kinetic energy equals the incident gamma-ray energy and will always be the same if monoenergetic gamma rays are involved. Under these conditions, the differential distribution of electron kinetic energy for a series of photoelectric absorption events would be a simple delta function as shown below. The single peak appears at a total electron energy corresponding to the energy of the incident gamma rays.

B. Compton Scattering

The result of a Compton scattering interaction is the creation of a recoil electron and scattered gamma-ray photon, with the division of energy between the two dependent on the scattering angle. A sketch of the interaction is given below.

Before   After

The energy of the scattered gamma ray hv' in terms of its scattering angle θ is given by

$$hv' = \frac{hv}{1 + (hv/m_0c^2)(1 - \cos\theta)} \tag{10.1}$$

where m_0c^2 is the rest mass energy of the electron (0.511 MeV). The kinetic energy of the recoil electron is therefore

$$E_{e-} = hv - hv' = hv\left(\frac{(hv/m_0c^2)(1 - \cos\theta)}{1 + (hv/m_0c^2)(1 - \cos\theta)}\right) \tag{10.2}$$

Two extreme cases can be identified:

1. A grazing angle scattering, or one in which $\theta \cong 0$. In this case, Eqs. (10.1) and (10.2) predict that $hv' \cong hv$ and $E_{e-} \cong 0$. In this extreme, the recoil Compton electron has very little energy and the scattered gamma ray has nearly the same energy as the incident gamma ray.

2. A head-on collision in which $\theta = \pi$. In this extreme, the incident gamma ray is backscattered toward its direction of origin, whereas the electron recoils along the direction of incidence. This extreme represents the maximum energy that can be transferred to an electron in a single Compton interaction. Equations (10.1) and (10.2) yield for this case

$$hv'\big|_{\theta=\pi} = \frac{hv}{1 + 2hv/m_0c^2} \tag{10.3}$$

$$E_{e-}\big|_{\theta=\pi} = hv\left(\frac{2hv/m_0c^2}{1 + 2hv/m_0c^2}\right) \tag{10.4}$$

In normal circumstances, all scattering angles will occur in the detector. Therefore, a continuum of energies can be transferred to the electron, ranging from zero up to the maximum predicted by Eq. (10.4). Figure 10.1 shows the shape of the distribution of Compton recoil electrons predicted by the Klein–Nishina cross section (Chapter 2) for several different values of the incident gamma-ray energy. For any one specific gamma-ray energy, the electron energy distribution has the general shape shown in the sketch below.

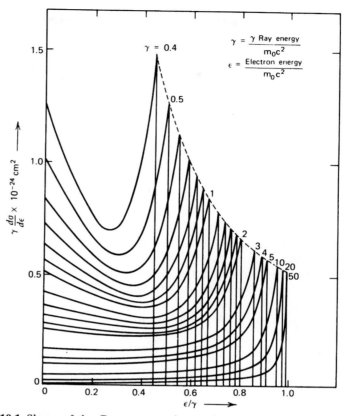

Figure 10.1 Shape of the Compton continuum for various gamma-ray energies. (From S. M. Shafroth (ed.), *Scintillation Spectroscopy of Gamma Radiation.* Copyright 1964 by Gordon & Breach, Inc. By permission of the publisher.)

The gap between the maximum Compton recoil electron energy and the incident gamma-ray energy is given by

$$E_C \equiv h\nu - E_{e^-}\big|_{\theta = \pi} = \frac{h\nu}{1 + 2h\nu/m_0 c^2} \tag{10.5}$$

In the limit that the incident gamma-ray energy is large, or $h\nu \gg m_0 c^2/2$, this energy difference tends toward a constant value given by

$$E_C \cong \frac{m_0 c^2}{2} \, (= 0.256 \text{ MeV}) \tag{10.6}$$

The preceding analysis is based on the assumption that Compton scattering involves electrons that are initially free or unbound. In actual detector materials, the binding energy of the electron prior to the scattering process can have a measurable effect on the shape of the Compton continuum. These effects will be particularly noticeable for low incident gamma-ray energy. They involve a rounding-off of the rise in the continuum near its upper extreme and the introduction of a finite slope to the abrupt drop of the Compton edge. These effects are often masked by the finite energy resolution of the detector but can be evident in the spectra from detectors with high inherent resolution (see Fig. 13.9). The finite momentum of orbital electrons also causes gamma-ray photons that are scattered at a fixed angle from a monoenergetic source to have a narrow distribution in their energy (the "Doppler spread"), as contrasted with a single energy predicted by Eq. (10.1).

C. Pair Production

The third significant gamma-ray interaction is pair production. The process occurs in the intense electric field near the protons in the nuclei of the absorbing material and corresponds to the creation of an electron–positron pair at the point of complete disappearance of the incident gamma-ray photon. Because an energy of $2m_0c^2$ is required to create the electron–positron pair, a minimum gamma-ray energy of 1.02 MeV is required to make the process energetically possible. If the incident gamma-ray energy exceeds this value, the excess energy appears in the form of kinetic energy shared by the electron–positron pair. Therefore, the process consists of converting the incident gamma-ray photon into electron and positron kinetic energies, which total

$$E_{e^-} + E_{e^+} = hv - 2m_0c^2 \tag{10.7}$$

For typical energies, both the electron and positron travel a few millimeters at most before losing all their kinetic energy to the absorbing medium. A plot of the total (electron + positron) charged particle kinetic energy created by the incident gamma ray is again a simple delta function, but it is now located $2m_0c^2$ below the incident gamma-ray energy, as illustrated in the sketch below. In our simple model, this amount of energy will be deposited each time a pair production interaction occurs within the detector. As introduced in the next section, this energy corresponds to the position of the *double escape peak* in actual gamma-ray pulse height spectra.

The pair production process is complicated by the fact that the positron is not a stable particle. Once its kinetic energy becomes very low (comparable to the thermal energy of normal electrons in the absorbing material), the positron will annihilate or combine with a normal electron in the absorbing medium. At this point both disappear, and they are replaced by two annihilation photons of energy m_0c^2 (0.511 MeV) each. The time required for the positron to slow down and annihilate is small, so that the annihilation radiation appears in virtual coincidence with the original pair production interaction.

III. PREDICTED RESPONSE FUNCTIONS

A. "Small" Detectors

As an example of one extreme in gamma-ray detector behavior, we first examine the expected response of detectors whose size is small compared with the mean free path of the *secondary gamma radiations* produced in interactions of the original gamma rays. These secondary radiations consist of Compton scattered gamma rays, together with annihilation photons formed at the end of the tracks of positrons created in pair production. Because the mean free path of the secondary gamma rays is typically of the order of several centimeters, the condition of "smallness" is met if the detector dimensions do not exceed 1 or 2 cm. At the same time, we retain our original simplifying assumption that all charged particle energy (photoelectron, Compton electron, pair electron, and positron) is completely absorbed within the detector volume.

The predicted electron energy deposition spectra under these conditions are illustrated in Fig. 10.2. If the incident gamma-ray energy is below the value at which pair production is significant, the spectrum results only from the combined effect of Compton scattering

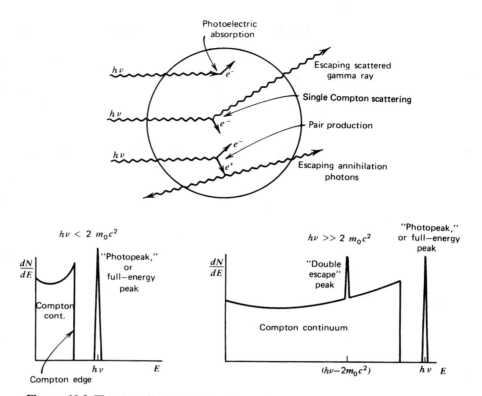

Figure 10.2 The "small detector" extreme in gamma-ray spectroscopy. The processes of photoelectric absorption and single Compton scattering give rise to the low-energy spectrum at the left. At higher energies, the pair production process adds a double escape peak shown in the spectrum at the right.

and photoelectric absorption. The continuum of energies corresponding to Compton scattered electrons is called the *Compton continuum*, whereas the narrow peak corresponding to photoelectrons is designated as the *photopeak*. For the "small" detector, only single interactions take place, and the ratio of the area under the photopeak to the area under the Compton continuum is the same as the ratio of the photoelectric cross section to the Compton cross section in the detector material.

If the incident gamma-ray energy is sufficiently high (several MeV), the results of pair production are also evident in the electron energy spectrum. For a small detector, only the electron and positron kinetic energies are deposited, and the annihilation radiation escapes. The net effect is to add a *double escape peak* to the spectrum located at an energy of $2m_0c^2$ (~1.02 MeV) below the photopeak. The term *double escape* refers to the fact that both annihilation photons escape from the detector without further interaction.

B. Very Large Detectors

As an opposite extreme case, imagine that gamma rays could be introduced near the center of a very large detector, perhaps in an arrangement resembling that of Fig. 10.3. The detector dimensions are now assumed to be sufficiently large so that all secondary radiations, including Compton scattered gamma rays and annihilation photons, also interact within the detector active volume and none escape from the surface. For typical gamma-ray energies, this condition would translate into requiring detector dimensions on the order of many tens of centimeters, unrealistically large for most practical cases.

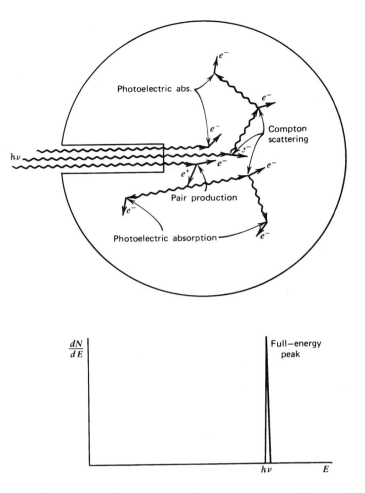

Figure 10.3 The "large detector" extreme in gamma-ray spectroscopy. All gamma-ray photons, no matter how complex their mode of interaction, ultimately deposit all their energy in the detector. Some representative histories are shown at the top.

Nonetheless, it is helpful to see how increasing the detector size greatly simplifies its response function. Some typical histories, obtained by following a particular source gamma ray and all subsequent secondary radiation, are sketched in Fig. 10.3. If the initial interaction is a Compton scattering event, the scattered gamma ray will subsequently interact at some other location within the detector. This second interaction may also be a Compton scattering event, in which case a scattered photon of still lower energy is produced. Eventually, a photoelectric absorption will occur and the history is terminated at that point.

It is important to appreciate the small amount of time required for the entire history to take place. The primary and secondary gamma rays travel at the speed of light. If the average migration distance of the secondary gamma rays is of the order of 10 cm, the total elapsed time from start to finish of the history will be less than a nanosecond. This time is substantially less than the inherent response time of virtually all practical detectors used in gamma-ray spectroscopy. Therefore, the net effect is to create the Compton electrons at each scattering point and the final photoelectron in time coincidence. The pulse produced by the detector will therefore be the sum of the responses due to each individual electron. If the detector responds linearly to electron energy, then a pulse is produced which is proportional to the *total* energy of all the electrons produced along the history. Because nothing escapes from the detector, this total electron energy must simply be the original energy of the gamma-ray photon, no matter how complex any specific history may be. *The detector response is the same as if the original gamma-ray photon had undergone a simple photoelectric absorption in a single step.*

The same type of argument can be used if the history involves a pair production event. The annihilation photons formed when the positron is stopped are now assumed to interact through Compton scattering or photoelectric absorption elsewhere in the detector. Again, if the detector is large enough to prevent any secondary radiation from escaping, the sum of the kinetic energies of the electron–positron pair and subsequent Compton and photoelectrons produced by interaction of the annihilation radiation must equal the original gamma-ray photon energy. Therefore, the detector response is again simply proportional to the original gamma-ray energy.

The conclusion to be reached is therefore very simple: If the detector is sufficiently large and its response linearly dependent on electron kinetic energy, then the signal pulse is identical for all gamma-ray photons of the same energy, regardless of the details of each individual history. This circumstance is very fortunate because the detector response function now consists of the single peak shown in Fig. 10.3 rather than the more complex function shown in Fig. 10.2. The ability to interpret complex gamma-ray spectra involving many different energies is obviously enhanced when the response function consists of a single peak.

By common usage, the corresponding peak in the response function is often called the *photopeak*, just as in the case of the small detector. It should be realized, however, that in addition to simple photoelectric events, much more complex histories involving multiple Compton scattering or pair production also contribute pulses that fall within this peak. A better name is the *full-energy peak* because it represents all histories in which all of the original gamma-ray energy is fully converted to electron kinetic energy.

C. Intermediate Size Detectors

Real detectors of the sizes in common use for gamma-ray spectroscopy are neither small nor large by the standards given above. For usual geometries in which the gamma rays are incident externally on the surface of the detector, even large-volume detectors appear finite because some interactions will take place near the entrance surface. Normal detector response functions therefore combine some of the properties discussed for the two previous cases, as well as additional features related to *partial* recovery of the secondary gamma-ray energy. Some representative histories that illustrate these added possibilities are shown in Fig. 10.4, together with corresponding features in the response function.

The spectrum for low to medium gamma-ray energies (where pair production is not significant) again consists of a Compton continuum and photopeak. Now, however, the ratio of the area under the photopeak to that under the Compton continuum is significantly enhanced over that for the very small detector due to the added contribution of multiple events to the photopeak. The lower the incident gamma-ray energy, the lower will be the average energy of a Compton scattered photon and the corresponding average distance of migration. Thus, even detectors of moderate size will appear to be large, and the relative area under the photopeak increases with decreasing incident photon energy. At very low energies (say, < 100 keV) the Compton continuum may effectively disappear.

At medium energies, the possibility of multiple Compton scattering followed by escape of the final scattered photon can lead to a total energy deposition that is greater than the maximum predicted by Eq. (10.4) for single scattering. These multiple events can thus partially fill in the gap between the Compton edge and the photopeak, as well as distort the shape of the continuum predicted for single scattering.

If the gamma-ray energy is high enough to make pair production significant, a more complicated situation prevails. The annihilation photons now may either escape or under-

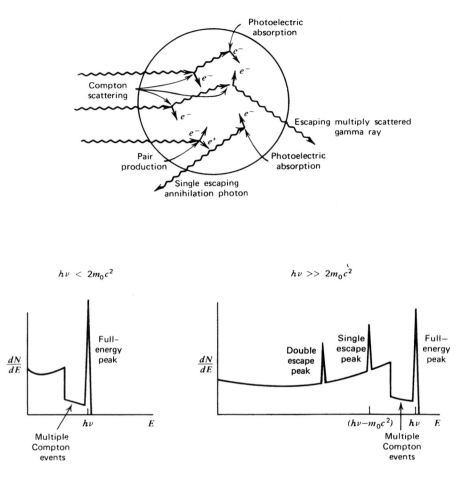

Figure 10.4 The case of intermediate detector size in gamma-ray spectroscopy. In addition to the continuum from single Compton scattering and the full-energy peak, the spectrum at the left shows the influence of multiple Compton events followed by photon escape. The full-energy peak also contains some histories that began with Compton scattering. At the right, the single escape peak corresponds to initial pair production interactions in which only one annihilation photon leaves the detector without further interaction. A double escape peak as illustrated in Fig. 10.2 will also be present due to those pair production events in which both annihilation photons escape.

go further interaction within the detector. These additional interactions may lead to either partial or full-energy absorption of either one or both of the annihilation photons.

If both annihilation photons escape without interaction, events occur that contribute to the double escape peak discussed previously. Another relatively frequent occurrence is a history in which one annihilation photon escapes but the other is totally absorbed. These events contribute to a *single escape peak*, which now appears in the spectrum at an energy of m_0c^2 (0.511 MeV) below the photopeak. A continuous range of other possibilities exists in which one or both of the annihilation photons are partially converted to electron energy through Compton scattering and subsequent escape of the scattered photon. Such events accumulate in a broad continuum in the pulse height spectrum lying between the double escape peak and the photopeak.

The response function to be expected for a real gamma-ray detector will depend on the size, shape, and composition of the detector, and also the geometric details of the irradiation conditions. For example, the response function will change somewhat if a point gamma-ray source is moved from a position close to the detector to one that is far away. The variation is related to the differences in the spatial distribution of the primary interactions that occur within the detector as the source geometry is changed. In general, the response function is too complicated to predict in detail other than through the use of Monte Carlo calculations, which simulate the histories actually taking place in a detector of the same size and composition.

Some properties of the response function are of general interest in gamma-ray spectroscopy. The *photofraction* is defined as the ratio of the area under the photopeak (or full-energy peak) to that under the entire response function. It is a direct measure of the probability that a gamma ray that undergoes interaction of any kind within the detector ultimately deposits its full energy. Large values of the photofraction are obviously desirable to minimize the complicating effects of Compton continua and escape peaks in the spectrum.

At high gamma-ray energies, the single and double escape peaks are quite prominent parts of the response function and can, under some circumstances, become larger than the photopeak. The ratio of the area under the single or double escape peak to the area under the photopeak is also a widely quoted property of the response function that can help in the interpretation of complex spectra.

D. Complications in the Response Function

1. SECONDARY ELECTRON ESCAPE

If the detector is not large compared with typical secondary electron ranges, a significant fraction of the electrons may leak from the detector surface and their energy will not be fully collected. This effect is enhanced for high gamma-ray energies for which the average secondary electron energy is also high. Electron leakage will tend to distort the response function by moving some events to a lower amplitude from that which would be observed if the entire electron energy were collected. The shape of the Compton continuum will therefore be altered somewhat to favor lower amplitudes. Because some events will be lost from the photopeak, the photofraction will be reduced as compared with the situation in which electron leakage is not important.

2. BREMSSTRAHLUNG ESCAPE

One of the mechanisms by which secondary electrons lose energy is by the radiation of bremsstrahlung photons. The fraction lost by this process increases sharply with electron energy and becomes the dominant process for electrons with energy over a few MeV. Bremsstrahlung production scales approximately as Z^2 of the absorber [see Eq. (2.11)] so its importance is greatest in detectors with high atomic number. Even though the electron itself may be fully stopped within the detector, there is a possibility that some fraction of the bremsstrahlung photons may escape without being reabsorbed. The effects on the response function are similar to those described in the previous paragraph for electron escape and are again most important when the incident gamma-ray energy is large. For both secondary electron or bremsstrahlung escape, the effects are to change the shape of the response function somewhat, but additional peaks or sharp features are not introduced.

3. CHARACTERISTIC X-RAY ESCAPE

In the photoelectric absorption process, a characteristic X-ray often is emitted by the absorber atom. In the majority of cases this X-ray energy is reabsorbed near the original interaction site. If the photoelectric absorption occurs near a surface of the detector, however, the X-ray photon may escape as shown in the sketch below.

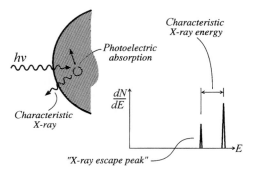

In this event, the energy deposited in the detector is decreased by an amount equal to the X-ray photon energy. Without the X-ray escape, the original gamma ray would have been fully absorbed and the resulting pulse would have contributed to the photopeak. With escape, a new category of events is created in which an amount of energy equal to the original gamma-ray energy minus the characteristic X-ray energy is repeatedly deposited in the detector. Therefore, a new peak will appear in the response function and will be located at a distance equal to the energy of the characteristic X-ray below the photopeak. These peaks are generally labeled "X-ray escape peaks" and tend to be most prominent at low incident gamma-ray energies and for detectors whose surface-to-volume ratio is large. Examples are shown in the spectra of Figs. 10.10 and 13.7.

For gamma rays whose energies are above the K-shell binding energy of the absorber, most photoelectric absorptions involve these most tightly-bound electrons in the atom. The main characteristic X-ray escape peak is therefore located below the full-energy peak by an amount given by the K-shell binding energy. More subtle effects are also present in principle because of interactions in and transitions to the more weakly-bound electron shells, but corresponding escape peaks are usually difficult to resolve from the full-energy peak because of the much lower X-ray energies that are involved.

4. SECONDARY RADIATIONS CREATED NEAR THE SOURCE

a. Annihilation Radiation

If the gamma-ray source consists of an isotope that decays by positron emission, an additional peak in the spectrum at 0.511 MeV is to be expected from the annihilation photons created when the positron is stopped. Most standard gamma-ray sources are encapsulated in a covering sufficiently thick to fully stop all the positrons, and thus they undergo annihilation in the region immediately surrounding the source. This region therefore acts as a source of 0.511 MeV annihilation radiation, which is superimposed on the gamma-ray spectrum expected from decay of the source itself. For detector geometries in which it is possible to detect both annihilation photons from a single decay simultaneously (as in a well counter), then a peak at 1.022 MeV may also be observed in the recorded spectrum.

b. Bremsstrahlung

Most commonly-available gamma-ray sources decay by beta-minus emission, and the source encapsulation is usually also thick enough to stop these beta particles. In other cases, an external absorber may be used to prevent the beta particles from reaching the detector where their energy deposition would needlessly complicate the gamma-ray spectrum. In the absorption process, however, some secondary radiation in the form of bremsstrahlung will be generated and may reach the detector and contribute to the measured spectrum. In principle, the bremsstrahlung spectrum may extend to an energy equal to the maximum beta particle energy, but significant yields are confined to energies that are

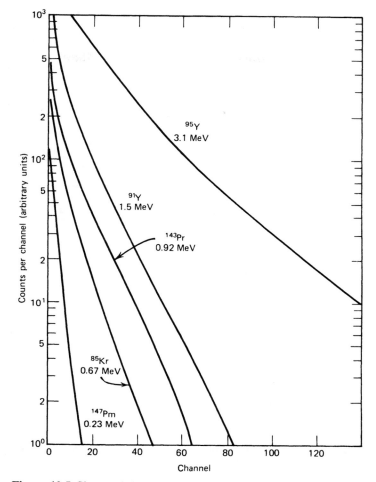

Figure 10.5 Shape of the bremsstrahlung spectra produced by beta particles with the indicated endpoint energies. (From Heath.[5])

much lower than this value. Some examples of bremsstrahlung energy spectra are given in Fig. 10.5, which illustrates the shape of the spectrum favoring low-energy bremsstrahlung photon emission. Because these spectra are continua, they do not lead to peaks in the recorded spectra but rather can add a significant continuum on which all other features of the gamma-ray spectra are superimposed. Because the bremsstrahlung contribution cannot simply be subtracted as a background, its inclusion can lead to errors in quantitative measurements of areas under peaks in the gamma-ray spectrum. To minimize the generation of bremsstrahlung, the use of beta absorbers made from low atomic number materials, such as beryllium, is often preferred.

5. EFFECTS OF SURROUNDING MATERIALS

In any practical application, a detector used for gamma-ray spectroscopy is surrounded by other materials that can have a measurable influence on its response. At a very minimum, the detector is encapsulated to provide a barrier against moisture and light or is mounted within a vacuum enclosure. To reduce natural background, most gamma-ray detectors are also operated within a shielded enclosure. The gamma-ray source itself is often part of a larger sample of material or is contained within some type of encapsulation. All these

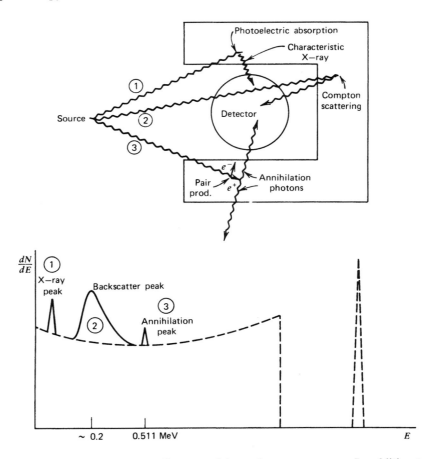

Figure 10.6 Influence of surrounding materials on detector response. In addition to the expected spectrum (shown as a dashed line), the representative histories shown at the top lead to the indicated corresponding features in the response function.

materials are potential sources of secondary radiations that can be produced by interactions of the primary gamma rays emitted by the source. If the secondary radiations reach the detector, they can influence the shape of the recorded spectrum to a noticeable extent. Some possibilities are illustrated in Fig. 10.6.

a. Backscattered Gamma Rays

Pulse height spectra from gamma-ray detectors often show a peak in the vicinity of 0.2–0.25 MeV, called the *backscatter peak*. The peak is caused by gamma rays from the source that have first interacted by Compton scattering in one of the materials surrounding the detector. Figure 10.7 shows the energy dependence of these scattered gamma rays as a function of the scattering angle. From the shape of these curves, it can be seen that any scattering angle greater than about 110–120° results in scattered photons of nearly identical energy. Therefore, a monoenergetic source will give rise to many scattered gamma rays whose energy is near this minimum value, and a peak will appear in the recorded spectrum. The energy of the backscatter peak will correspond to Eq. (10.3):

$$hv'\big|_{\theta = \pi} = \frac{hv}{1 + 2hv/m_0c^2}$$

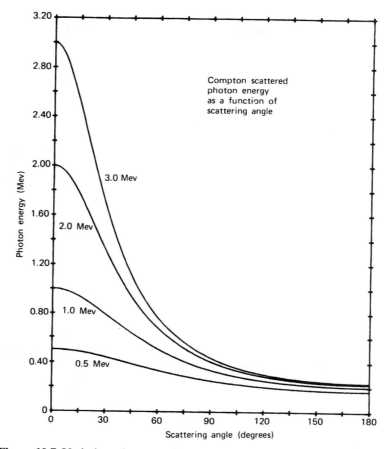

Figure 10.7 Variation of scattered gamma-ray energy with scattering angle.

In the limit that the primary gamma-ray energy is large ($hv \gg m_0 c^2/2$), this expression reduces to

$$hv'\big|_{\theta = \pi} \cong \frac{m_0 c^2}{2} \tag{10.8}$$

Thus, the backscatter peak always occurs at an energy of 0.25 MeV or less.

b. Other Secondary Radiations

In addition to Compton scattering, other interactions of the primary gamma rays in the surrounding materials can give noticeable peaks in the recorded spectrum. For example, photoelectric absorption in the materials immediately surrounding the detector can lead to generation of a characteristic X-ray that may reach the detector. If the atomic number of the material is high, the X-ray photon will be relatively energetic and can penetrate significant thicknesses of intervening material. Therefore, high-Z materials should be avoided in the immediate vicinity of the detector. On the other hand, the most effective shielding materials are those with high atomic numbers such as lead. A *graded shield* is one in which the bulk of the shield is made from high-Z materials, but the inner surface is lined with a material with lower atomic number. This inner lining serves to absorb the characteristic X-ray emitted by the bulk of the shield, at the same time emitting only low-energy or weakly penetrating X-rays of its own.

If the energy of the primary gamma rays is high, pair production within high-Z surrounding materials can give a significant yield of annihilation radiation. A peak can therefore appear at 0.511 MeV in the spectrum from the detection of these secondary photons. There is a danger of confusing this peak with that expected from annihilation radiation produced by radioactive sources that are positron emitters, and care must therefore be exercised in identifying the source of these annihilation photons.

E. Summation Effects

Additional peaks caused by the coincident detection of two (or more) gamma-ray photons may also appear in the recorded pulse height spectrum. The most common situation occurs in applications involving an isotope that emits multiple cascade gamma rays in its decay, as illustrated in Fig. 10.8. If we assume that no isomeric states are involved, the lifetime of the intermediate state is generally so short that the two gamma rays are, in effect, emitted in coincidence. It is then quite possible for both gamma-ray photons from a single decay to interact and deposit all their energy within a time that is short compared with the response time of the detector or the resolving time of the following electronics. If enough of these events occur, a *sum coincidence peak* will be observable in the spectrum that occurs at a pulse height that corresponds to the sum of the two individual gamma-ray energies. A continuum of sum events will also occur at lower amplitudes due to the summation of partial energy loss interactions.

The relative number of events expected in the sum peak depends on the branching ratio of the two gamma rays, the angular correlation that may exist between them, and the solid angle subtended by the detector. A complete analysis is often quite complex, but the following simplified derivation illustrates the general approach that can be applied.

Let ϵ_1 be the intrinsic peak efficiency of the detector for gamma ray ①, and let Ω be the fractional solid angle (steradians/4π) subtended by the detector. Then the full-energy peak area for gamma ray ① in the absence of summing effects is

$$N_1 = \epsilon_1 \Omega S y_1 \tag{10.9}$$

where S is the number of source decays over the observation period and y_1 is the yield of gamma ray ① per disintegration. Applying the same definitions to gamma ray ②, we obtain

$$N_2 = \epsilon_2 \Omega S y_2 \tag{10.10}$$

The probability of simultaneous detection of both gamma rays is the product of both individual detection probabilities, multiplied by a factor $W(0°)$ to account for any angular correlation between the gamma-ray photons. $W(0°)$ is defined as the relative yield of γ_2 per

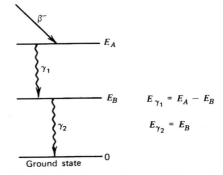

Figure 10.8 Simplified nuclear decay scheme which can lead to summation effects in gamma spectra. Provided the intermediate state (at E_B) is short-lived, γ_1 and γ_2 are emitted in virtual coincidence.

unit solid angle about the 0° direction defined by the detector position, given that γ_1 is emitted in the same direction. Then the sum peak area should be

$$N_{12} = S\,(\epsilon_1\Omega y_1)(\epsilon_2\Omega y_2)W(0°)$$
$$= S\epsilon_1\epsilon_2 y_1 y_2 \Omega^2 W(0°) \tag{10.11}$$

The summation process not only creates the sum peak but also removes events that would otherwise fall within individual gamma-ray full-energy peaks. The remaining number of full-energy events for γ_1 is [from Eqs. (10.9) and (10.11)]

$$N_1\Big|_{\substack{\text{with}\\\text{summation}}} = N_1 - N_{12}$$
$$= \epsilon_1\Omega S y_1[1 - \epsilon_2\Omega y_2 W(0°)] \tag{10.12}$$

Because a coincident event of any kind from γ_2 (not just a photopeak event) will remove a count from N_1, the detection efficiency ϵ_2 should now be interpreted as the intrinsic *total* efficiency. For these losses to remain small, the fractional solid angle Ω is often restricted to small values to keep the second term in the above equation much smaller than the first. If the solid angle is too large, quantitative measurements based on determination of the area under full-energy peaks can be in error unless an accounting of the second term is provided.[6]

The summation process described above involves multiple radiations from the same nuclear decay event and therefore is classified as a *true coincidence* by the definitions given in Chapter 17. Another process can also lead to summed pulses due to the accidental combination of two separate events from independent decays that occur closely spaced in time. Because the time intervals separating adjacent events are randomly distributed, some will be less than the inherent resolving time of the detector or pulse-processing system. These *chance coincidences* increase rapidly with increasing counting rate and will occur even in the absence of true coincidences. A corresponding sum peak can therefore appear in spectra from isotopes that emit only a single radiation per decay. For both true coincidence and chance summing, there will also be many cases in which a second interaction involves partial rather than full energy absorption in the detector. For this category of events, the first event (assumed here to be a full-energy absorption) still is lost from the peak where it normally would be expected in the absence of summing. Now, however, the summed pulse is stored into a continuum that lies above the expected peak and it does not contribute to a sum peak in the spectrum.

Chance coincidences will occur if a second pulse arrives within the resolving time t_r following a typical signal pulse. For a random pulse rate of r_s and $r_s t_r \ll 1$, the rate at which coincidences occur should be the fraction of all time that lies within t_r of a preceding pulse (given by $r_s t_r$) multiplied by the rate of pulse arrival (r_s), or

$$r_{ch} = r_s^2 t_r \tag{10.13}$$

Therefore, the accidental sum peak will have an intensity that is proportional to the square of the counting rate, whereas both the true sum peak or normal photopeaks will be linearly related to the counting rate. When multiple radiations are involved, accidental sum peaks may potentially occur at all possible combinations of any two single energies. At normal rates and typical detector solid angles, however, sum peaks are usually lost in fluctuations in the continua and background present from other energies, except at the upper energy extremes of the spectrum where such backgrounds are low.

As a practical matter, the resolving time t_r is normally set by the shaping time constants of the linear amplifier used in the pulse-processing chain from the detector. The chance coincidences therefore take the form of *pulse pile-up* in the amplifier, which is further detailed in the discussions of Chapter 17.

F. Coincidence Methods in Gamma-Ray Spectrometers

1. CONTINUUM REDUCTION

For an ideal gamma-ray detector, the response function would simply be a single well-resolved peak with no associated continuum. Then the pulse height spectrum from a complex gamma-ray source could be most easily interpreted, and the presence of high-energy gamma rays would not hinder the detection of weak radiations at lower energies.

At the price of added complexity, some steps can be taken to approach this ideal more closely, even for gamma-ray detectors with response functions that are inherently more complicated. These methods involve placing other detectors around the primary detector and employ coincidence or anticoincidence techniques to select preferentially those events that are most likely to correspond to full-energy absorption. For the case of sodium iodide spectrometers, the most common methods involve the use of an annular detector surrounding the primary crystal for Compton suppression by anticoincidence, or the use of two or more adjacent crystals in the *sum-coincidence* mode. Representative descriptions of sodium iodide spectrometers in which one or both of these methods of continuum suppression have been applied are given in Refs. 7–11.

An explanation of these techniques is postponed until Chapter 12, where their use with germanium detectors is detailed. Although significant improvements in peak-to-continuum ratios can be achieved by applying these methods to NaI(Tl) spectrometers, current attention has focused on their application to germanium systems where continua are much more prominent and greater gains can be achieved through their suppression.

2. THE COMPTON SPECTROMETER

The combination of two separated gamma-ray detectors operated in coincidence, as shown in Fig. 10.9, is another configuration that can simplify the response function at the expense of detection efficiency. A collimated beam of gamma rays is allowed to strike the first

$$E_c = h\nu \left[\frac{\frac{h\nu}{m_0 c^2}(1 - \cos\theta)}{1 + \frac{h\nu}{m_0 c^2}(1 - \cos\theta)} \right]$$

Figure 10.9 The geometry of the Compton spectrometer is shown at the top. The spectrum of those events from detector ① that are in coincidence with pulses from detector ② is shown as the solid peak at the bottom. The normal spectrum from detector ① is shown as the dashed curve.

detector in which the desired mode of interaction is now Compton scattering. Some fraction of the scattered gamma rays will travel to the second detector where they may also interact to give a second pulse. Because the separation distance is normally no greater than a few tens of centimeters, the pulses are essentially in time coincidence. By selectively recording only those pulses from the first crystal that are in coincidence with a pulse from the second crystal, the recorded spectrum largely reflects only single Compton scattering events. Because the angle of scattering is fixed, a constant amount of energy is deposited for each scattering interaction involving monoenergetic incident gamma rays. Photoelectric absorption and all other events that do not lead to coincidence between the two detectors are excluded. The response function is thus reduced to a single peak, which appears at a position within the original Compton continuum determined by the scattering angle. For typical gamma ray energies, the Doppler spread in the scattered photon energy (see p. 311) is usually small compared with the broadening effects of the variation in scattering angles introduced by the finite sizes of the detectors.

Figure 10.10 Pulse height spectrum from a NaI(Tl) scintillator for gamma rays emitted by ^{57}Co at 122 and 136 keV. The iodine X-ray escape peak lies 28 keV below the corresponding full energy peak and is evident only for the more intense 122 keV gamma ray.

Figure 10.11 Response functions for a 7.62 cm $\times$ 7.62 cm cylindrical NaI(Tl) scintillator for gamma rays from 0.335 to 2.75 MeV. (From Heath.[5])

IV. PROPERTIES OF SCINTILLATION GAMMA-RAY SPECTROMETERS

A. Response Function

Sodium iodide gained much of its early popularity because the relatively high atomic number ($Z = 53$) of its iodine constituent ensures that photoelectric absorption will be a relatively important process. The corresponding high intrinsic detection efficiency and large photofraction have contributed to the success of sodium iodide scintillation spectrometers. Other materials, such as cesium iodide or BGO, have even higher density or effective atomic number, and therefore the response function for these materials shows an even greater detection efficiency and photofraction. However, the relatively high light output and smaller decay time have led to the dominance of NaI(Tl) in spectroscopy with scintillators.

The importance of many of the factors that influence the shape of the response function for NaI(Tl) scintillators is detailed by Mueller and Maeder.[12] An extensive catalog of experimentally measured gamma-ray spectra for nearly 300 radionuclides as recorded by a 3 in. × 3 in. NaI(Tl) spectrometer has been published by Heath.[5] A later compilation by Adams and Dams[6] contains spectra for both 3 in. × 3 in. and 4 in. × 4 in. cylindrical sodium iodide crystals. These published data can be of considerable help in predicting the response function to be expected from a scintillation spectrometer when applied to gamma-ray-emitting isotopes. Examples of gamma-ray spectra are given in Figs. 10.10 through 10.13. Many of the features described in the previous discussion can be observed in these spectra.

A direct comparison of gamma-ray spectra observed under similar conditions from equal size crystals of sodium iodide and BGO is shown in Fig. 10.14. The considerably better energy resolution of sodium iodide is evidenced by the smaller widths of the peaks in the bottom spectrum. The higher efficiency of BGO is apparent, especially at the higher energy, from the larger area under the photopeaks in the upper spectrum. The larger photofraction of BGO also results in less prominent Compton continua and annihilation quanta escape peaks. A collection of spectra from gamma rays covering the range from 124 keV to 11.67 MeV taken with 2.54 × 2.54 and 7.62 × 7.62 cm BGO detectors can be found in Ref. 14.

Because of the lack of suitable sources, it is often difficult experimentally to measure the response function for all gamma-ray energies of potential interest. It is then necessary

Figure 10.12 Pulse height spectra recorded from NaI(Tl) scintillation detectors. (*a*) A spectrum for a ^{86}Rb source (1.08 MeV gamma rays) showing the contribution at the lower end of the scale from bremsstrahlung generated in stopping the beta particles emitted by the source. (*b*) Spectrum from a ^{60}Co source (1.17 and 1.33 MeV gamma rays emitted in coincidence) taken under conditions in which the solid angle subtended by the detector is relatively large, enhancing the intensity of the sum peak at 2.50 MeV. (From F. Adams and R. Dams, *Applied Gamma-Ray Spectrometry,* 2nd ed. Copyright 1970 by Pergamon Press, Ltd. Used with permission.)

Figure 10.13 Pulse height spectrum from a NaI(Tl) scintillator for gamma rays emitted by [24]Na at 1369 and 2754 keV. The single and double escape peaks corresponding to pair production interactions of the higher energy gamma rays are very apparent, as is the annihilation radiation peak at 511 keV due to pair production interactions in surrounding materials.

Figure 10.14 Comparative pulse height spectra measured for BGO (top) and NaI(Tl) (bottom) scintillators of equal 7.62 cm × 7.62 cm size for gamma rays from [24]Na. (From Moss et al.[13])

Figure 10.15 A comparison of a measured pulse height spectrum (points) with a theoretical spectrum calculated by Monte Carlo methods. (From Berger and Seltzer.[15])

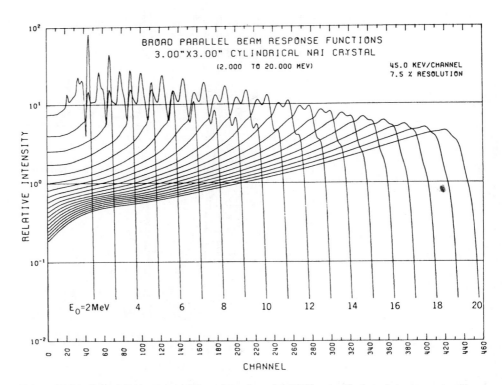

Figure 10.16 Calculated response functions for a NaI(Tl) scintillation detector extending to 20 MeV. (From Berger and Seltzer.[15])

to resort to calculations to derive the response function. Because of the complexity of the situation in which multiple interactions play an important role, virtually all practical calculations are done using the Monte Carlo method. The work of Berger and Seltzer[15] is a good example of such a calculation for sodium iodide cylindrical detectors. This publication also contains an extensive list of references to prior experimental and calculational efforts. Figure 10.15 shows the good match to experimental data that can be obtained when the computational model contains sufficient detail to provide adequate representation of all important interactions taking place within the detector. Figure 10.16 illustrates a set of calculated response functions for a sodium iodide scintillator extending to relatively high gamma-ray energies. The gradual disappearance of the photopeak and the broadening of the escape peaks as the gamma-ray energy increases is clearly illustrated by this series of plots. The photopeak disappears because pair production becomes the dominant mode of interaction, and the high-energy pair that is produced is difficult to absorb fully because of losses from bremsstrahlung emission and leakage from the surface of the crystal. The broadening of the escape peak is due to the additional statistical fluctuations in the number of photoelectrons produced as their number increases (see Section B.2 below).

B. Energy Resolution

In contrast to the theoretical energy deposition spectra shown in Figs. 10.2 to 10.4, the measured response functions shown above contain the "blurring" effects due to the finite energy resolution of the detector. The most striking difference is the fact that all peaks now have some finite width rather than appearing as narrow, sharp lines.

As introduced in Chapter 4, the *energy resolution R* is defined as

$$R = \frac{\text{FWHM}}{H_0} \tag{10.14}$$

where   FWHM = full width at half maximum of the full-energy peak
H_0 = mean pulse height corresponding to the same peak

The energy resolution of scintillators is the poorest of any commonly used detector, and therefore the spectra have relatively broad peaks.

1. ORIGINS OF RESOLUTION LOSS

From the arguments of Chapter 4, the finite energy resolution of any detector may contain contributions resulting from the separate effects of charge collection statistics, electronic noise, variations in the detector response over its active volume, and drifts in operating parameters over the course of the measurement. For scintillation detectors, the fluctuations in PM tube gain from event to event (see p. 273) can also add to the measured resolution. Also in some scintillators, the departure from exact proportionality between absorbed particle energy and light yield can be a significant contributor to the energy resolution for measurements on gamma rays (see p. 332). For the majority of applications of scintillators to gamma-ray spectroscopy, the first of these sources is normally the most significant. Contributions of electronic noise are usually negligible, so that preamplifiers and other electronic components used with scintillators need not include elaborate schemes for noise reduction. Variations in the light collected from scintillation events over the volume of the crystal can be a significant problem, so the techniques discussed in Chapter 8 for promoting uniform light collection are quite important. In commercially prepared crystals of typical sizes, these nonuniformities normally are a small part of the total peak width. Drifts in the operating parameters are usually associated with the PM tube and can be severe if the detector is subject to large changes in counting rate or temperature. Some methods of spectrum stabilization discussed in Chapter 18 can be applied in these cases but usually are necessary only under extreme conditions.

2. PHOTOELECTRON STATISTICS

Statistical spreads are therefore left as the single most important cause of peak broadening in scintillators. The statistical fluctuations will be most significant at the point in the signal chain at which the numbers of information carriers are at a minimum. For scintillation counters, this point is reached after conversion of the scintillation light to electrons by the photocathode of the PM tube.

A representative numerical example is helpful in illustrating this point. In the signal chain illustrated below, it is assumed that exactly 0.5 MeV of electron energy is deposited by the gamma-ray photon in the scintillation crystal. In thallium-activated sodium iodide, the scintillation efficiency is about 12%. In the example, 60 keV is thus converted by the scintillator into visible light with an average photon energy of about 3 eV. Therefore, about 20,000 scintillation photons are produced per pulse. With allowance for some light loss at the crystal surface and at the crystal–phototube interface, perhaps 15,000 of these light photons ultimately reach the photocathode. If we assume that the average quantum efficiency of the photocathode over the scintillation spectrum is 20%, then 3000 photoelectrons are produced. This number is now the minimum in the signal chain because subsequent stages in the PM tube multiply the number of electrons.

The amplitude of the signal produced by the PM tube is proportional to this number of photoelectrons. For repeated events in which exactly the same energy is deposited in the crystal, the pulse amplitude will fluctuate as a result of the inherent statistical variation in the number of photoelectrons collected for each event. If we assume that Poisson statistics hold, then the standard deviation of the number of photoelectrons should be the square root of the mean number produced. Therefore, in our example we would expect a standard deviation of $\sqrt{3000}$, or about 1.8% of the mean value. Energy resolution is formally defined in terms of the full width at half maximum of the peak rather than its standard deviation. For an assumed Gaussian shape, the FWHM is 2.35 values of the standard deviation, and therefore the statistical contribution to the energy resolution for the example should be 4.3%.

In many practical cases, the statistical broadening of the peak predominates over other potential sources of resolution loss. In that event, the variation of the resolution with gamma-ray energy can be predicted simply by noting that the FWHM of the peak is proportional to the square root of the gamma-ray photon energy. The average pulse height produced is directly proportional to the gamma-ray energy. Therefore, from the definition of energy resolution,

$$R \equiv \frac{\text{FWHM}}{H_0} = K \frac{\sqrt{E}}{E} = \frac{K}{\sqrt{E}} \tag{10.15}$$

where K is a constant of proportionality. The energy resolution should thus be inversely proportional to the square root of the gamma-ray energy. If we take the logarithm of both sides of Eq. (10.15), we derive

$$\ln R = \ln K - \tfrac{1}{2} \ln E \tag{10.16}$$

Therefore, a plot of $\ln R$ versus $\ln E$ should be a straight line with slope of $-\tfrac{1}{2}$.

Figure 10.17 shows a plot of experimentally determined resolution values for a sodium iodide scintillator as a function of the gamma-ray energy at which they were measured. The data adhere fairly closely to a straight line, but the slope is not as steep as predicted,

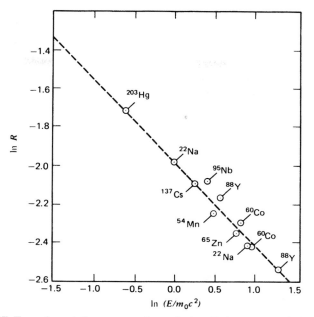

Figure 10.17 Experimentally measured resolution R from a NaI(Tl) scintillation detector for various gamma-ray energies E. (From Beattie and Byrne.[16])

indicating the influence of nonstatistical sources of the peak broadening. A more adequate representation of measured data can take the form

$$R = \frac{(\alpha + \beta E)^{1/2}}{E} \tag{10.17}$$

where α and β are constants particular to any specific scintillator–PM combination.

3. OTHER FACTORS IN ENERGY RESOLUTION

Other sources of resolution loss in scintillation spectrometers are conveniently categorized into three groups: those that are characteristic of the crystal itself (the *intrinsic crystal resolution*), those effects that are characteristic of the PM tube, and the variable probability that a visible photon generated by a scintillation event in the crystal produces a photoelectron that is collected by the first dynode of the PM tube (the *transfer variance*).

The intrinsic crystal resolution includes any variation caused by local fluctuations in the scintillation efficiency of the crystal. With modern fabrication techniques, the uniformity of sodium iodide crystals is generally sufficiently good so that the line broadening caused by this effect alone would be very small (less than 2%). More significant fluctuations arise because of the less-than-perfect reflection conditions that exist at the surface of the crystal. The consequent nonuniform light collection efficiency can introduce significant line broadening, especially in crystals of large size.

The variance introduced by the photomultiplier tube can be a significant contribution. Uniformity of photoelectron collection from the photocathode is an important factor, as is the statistical fluctuation in the electron multiplication. There is considerable variation in the performance of different photomultipliers in this regard, even among different samples of the same design. For example, in a study of several hundred PM tubes sampled from a few standard types, Persyk and Moi[17] observed an average NaI scintillator energy resolution of 10–11% for [57]Co radiation (122 keV). The best PM tube included in the sample, however, gave a corresponding value of 8.5%.

A more subtle component of the intrinsic crystal resolution arises from the nonlinearity in the response of most scintillation materials (see Fig. 8.8). If all incident gamma rays underwent an interaction in which their entire energy were converted to a single electron, this nonlinearity would not be a source of resolution loss. However, the incident gamma-ray energy may be subdivided among two or more secondary electrons through single or multiple Compton scattering followed by photoelectric absorption. Furthermore, even if simple photoelectric absorption occurs, the excited atom that remains may convert its excitation energy in a number of ways which lead to varying electron energy spectra, primarily in the form of Auger electrons. Consequently, even a monoenergetic flux of incident gamma rays will lead to a wide distribution of electron energies within the crystal. If the response of the crystal is not linear with electron energy, the total light yield will be different from event to event, depending on details of the energy subdivision between the various electrons that are produced. A number of studies[18, 19] conclude that the effects of nonlinearity can have a significant impact on the observed energy resolution from scintillators. As one example, Fig. 10.18 shows that the calculated effect of nonproportionality in sodium iodide can be as large a contributor to the overall energy resolution as the contribution of photoelectron statistics at energies above about 300 keV. This source of peak broadening would clearly be more serious in scintillators with a larger degree of nonlinearity than seen in NaI(Tl). In discussing the effects of nonlinearity of scintillators, it is important to distinguish between the response to fast electrons as opposed to the response to X-rays or gamma rays. The first is expected to be an intrinsic property of the scintillator, whereas the second depends on the particular mix of secondary electrons produced by photon interactions. Since this mix can change with the size of the crystal because of the increased importance of multiple scattering, the two behaviors will not be expected to be identical.[20] One should also be careful to distinguish between the terms *nonlinearity* and *nonproportionality*.[20] A linear behavior between light output and energy deposited merely requires that a straight-line fit can be made to the data, whereas the term *proportionality* requires that, in addition, the best-fit line pass through the origin. The response of a given scintillator may be perfectly linear over a range of energies, but there can still be some contribution to gamma-ray peak broadening unless the response is also proportional to energy.

Because the energy resolution varies with energy, values are usually specified at a fixed gamma-ray energy for comparison purposes. It is conventional to quote the energy resolution for gamma rays from ^{137}Cs (0.662 MeV) as a standard. The energy resolution for other gamma-ray energies can then be estimated through the use of Eq. (10.15).

Good quality solid sodium iodide scintillators coupled to modern PM tubes can achieve an energy resolution of about 6–7% at 0.662 MeV. If the shape of the crystal is more complicated than a simple right cylinder, the added difficulty of attaining uniformity in light collec-

Figure 10.18 A comparison of the predicted contributions of scintillator nonproportionality and Poisson photoelectron statistics to the energy resolution of a 2.54 cm × 2.54 cm NaI(Tl) scintillator. The points labeled "Total" are measured resolution values. (From Valentine et al.[19])

tion will often make the resolution somewhat worse. For example, the energy resolution for crystals with a center well is generally 1–3% larger than the equivalent solid cylinder.

4. PREVENTION OF RESOLUTION LOSS CAUSED BY LONG-TERM DRIFT

In scintillator measurements that must extend over many hours or days, some resolution loss can be experienced as a result of drifts in the gain of the PM tube and other circuit components. Some electronic methods that can be used to minimize or completely eliminate these effects are outlined in Chapter 18. These techniques work best when there is a single isolated peak in the spectrum from which an error signal can be derived to adjust a variable gain component in the signal chain. If a strong isolated peak does not exist in the measured spectrum, or if the counting rates are low, these methods may be impractical.

For scintillation counters, an alternative method can be used, which is based on providing a reference light source within the scintillation package to produce an artificial peak in the spectrum. If the light pulses are of constant intensity, a feedback signal can be generated to adjust the system gain to hold the resulting peak at a constant position in the measured spectrum. Light sources used for this purpose fall into two general categories: those that consist of a combination of a radioactive isotope with a suitable phosphor, and those that are basically electronic in design. A common requirement is that their yield be extremely stable over long periods of time because any change in light output will be interpreted as a drift in the gain of the counting system. Discussions of the design of light sources used for this purpose are given in Refs. 21–25.

C. Linearity

For all scintillators, the *scintillation efficiency* or amount of light generated per unit energy loss (dL/dE) depends both on the particle type and its kinetic energy. For an ideal spectrometer, dL/dE would be a constant independent of particle energy. The total light yield would then be directly proportional to the incident particle energy, and the response of the scintillator would be perfectly linear.

For electrons in NaI(Tl), the scintillation efficiency does vary mildly with electron energy (see Fig. 8.8). For direct electron spectroscopy, some significant nonlinearity should thus be expected. In gamma-ray measurements, monoenergetic gamma rays create varied combinations of secondary electrons with widely different energies. Figure 8.8 is therefore sampled at many points over the energy scale even for a single photon energy, and some of the nonlinearity is averaged out. The average pulse height observed for gamma rays is therefore closer to being linear with photon energy, but measurable nonlinearities remain.

Figure 10.19 shows some experimental results for linearity measurements with gamma rays on NaI(Tl). This nonlinearity must be taken into account when relating the pulse

Figure 10.19 The differential linearity measured for a NaI(Tl) scintillator. (From Heath. [5])

Figure 10.20 Measured light output per unit deposited energy for NaI(Tl), normalized to unity at 88 keV. (From Wayne et al.[26])

height scale to gamma-ray energy. In practice, a calibration is usually carried out in which peak positions are plotted versus known gamma-ray energies. Because Fig. 10.19 represents the slope of this calibration line, a small degree of curvature or nonlinearity should be expected. For interpolation between narrowly spaced peaks of known energy, however, the assumption of linearity normally leads to a negligible error.

Figure 10.20 shows the measured relative light output from sodium iodide over the low-energy range. The dip that is evident in this plot occurs near the K-shell absorption edge of iodine. The response of the scintillator really depends on the energy that is deposited by secondary electrons produced by the incident photon, and a complex mix of photoelectrons and/or Auger electrons will result from various types of photon interactions as illustrated in Fig. 10.21. The detailed yields of these secondary electrons will change abruptly at the

Figure 10.21 Representation of the possible origins of electrons and photons following the photoelectric absorption of an incident X-ray or gamma ray with energy E that is above the K-shell binding energy of 33.17 keV. (From Rooney and Valentine.[27])

K-shell absorption edge, so it is not surprising that the discontinuities in the observed light output also occur near this energy.

D. Detection Efficiency

One of the marked advantages in using scintillation crystals for gamma-ray measurements is the fact that many standard size detectors can be mass produced with virtually identical properties. Because the physical dimensions of these crystals can be controlled to within very small tolerances, the interaction probability for gamma rays will be identical for all crystals of the same size and shape.

One of the common applications of sodium iodide scintillators is to measure the absolute intensity of a given source of gamma rays. From the discussion in Chapter 4, such a measurement based on simple counting data requires a prior knowledge of the efficiency of the radiation detector. Undoubtedly, there are more published data available on the detection efficiency of sodium iodide scintillators for gamma rays than for any other detector type or application. The number of different sizes and shapes of NaI(Tl) crystals in routine use is relatively limited so that reasonably complete data can be compiled on each of the common configurations.

1. CRYSTAL SHAPES

Two general crystal shapes are in widespread use for applications in gamma-ray detection. The solid right circular cylinder is simple to manufacture and encapsulate and can be mounted directly to the circular face plate of most PM tubes. If the height-to-diameter ratio of the cylinder does not greatly exceed unity, the light collection properties are quite favorable in this geometry. If the height-to-diameter ratio is much less than 1, the pulse height resolution can often be improved by interposing a light pipe between the crystal and photomultiplier tube to spread the light more uniformly from each scintillation event over the entire photocathode, averaging out spatial variations of its quantum efficiency.

A *well crystal* is a right circular cylinder into which a cylindrical well has been machined, usually along the cylindrical axis. A significant advantage of this geometry is the very high counting efficiency that can be achieved by placing the samples to be counted at the bottom of the well. In this position, almost all the gamma rays that are emitted isotropically from the source are intercepted by at least a portion of the crystal. For low-energy gamma rays, the counting efficiency in this geometry can therefore approach 100%. At higher energies, some of the advantage is lost because the average path length through the crystal is somewhat less than if the gamma rays were externally incident on a solid crystal. Because the efficiency for sources near the bottom of the well is not a sensitive function of position, well counters can also simplify the counting of multiple samples with different physical properties while providing nearly identical counting efficiency.

2. EFFICIENCY DATA

Data on detector efficiency are commonly presented in the form of a graph of the efficiency value as a function of gamma-ray energy. Some examples are shown in Figs. 10.22 through 10.25. It is important to point out those parameters that must be specified before using this type of data:

1. The specific category of efficiency which is being tabulated must be clearly identified. As defined in Chapter 4, detector efficiencies are classified as either absolute or intrinsic, with the latter being the more common choice. An additional specification must be made that deals with the type of event accepted by the counting system. Here the most common choices are either peak or total efficiencies, with the distinction hinging on whether only full-energy events or all events are accepted (see Chapter 4).

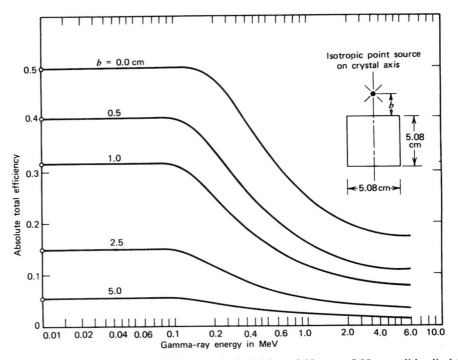

Figure 10.22 The absolute total efficiency calculated for a 5.08 cm × 5.08 cm solid cylindrical NaI(Tl) scintillator. Different values of the source location are shown. (From Snyder.[28])

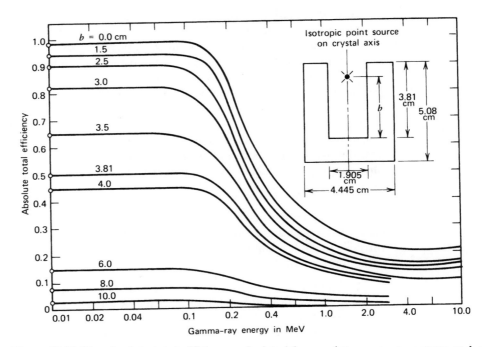

Figure 10.23 The absolute total efficiency calculated for a point gamma-ray source and a NaI(Tl) well-type scintillator with the dimensions shown. The parameter b is the source height above the well bottom. (From Snyder.[28])

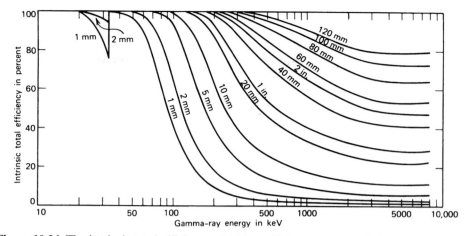

Figure 10.24 The intrinsic total efficiency of various thicknesses of NaI(Tl) for gamma rays perpendicular to its surface.

2. The size and shape of the scintillation crystal have a strong influence on the counting efficiency. Although the major influence on the intrinsic efficiency is the thickness of the crystal in the direction of the incident gamma radiation, mild variation with other detector dimensions should also be expected.

3. The size and physical nature of the source also influence the counting efficiency. Data are widely available for the relatively simple case of an isotropic point gamma-ray source located a specified distance from the detector face along its axis. Although absolute efficiencies are quite sensitive to the source–detector spacing, this dependence is much milder for intrinsic efficiencies and vanishes entirely if the source is sufficiently far from the detector. Other common source conditions for which data can be found include the case of a parallel beam of gamma rays uniformly irradiating one surface of the detector (equivalent to a point source at an infinite distance) and a narrowly collimated beam (a "pencil beam") incident at a specified point on the detector surface. Limited data also are available for distributed sources consisting of disks or volumetric sources under specified source–detector geometries.

4. Any absorption taking place between the point of gamma-ray emission and the scintillation crystal will also influence the detection efficiency. Published data normally neglect the effects of the crystal housing and other material between the source and detector, but some data can be found that account for self-absorption within large-volume gamma-ray sources.

There are two general methods by which efficiency data may be generated. The most straightforward is simply to measure the appropriate counting rate induced by a gamma-ray source of known activity. Sets of "standard" gamma-ray sources can be purchased whose absolute activity can be quoted to about 1% precision. If the experimenter has a set of such sources available covering the energy range of interest, then an efficiency curve can be determined experimentally for the specific detector in use. In many cases, however, a set of absolutely calibrated sources is not available or they do not adequately represent the geometric irradiation conditions of the actual experiment. Then the experimenter must turn to published efficiency data, which can include a greater variety of energies and experimental conditions.

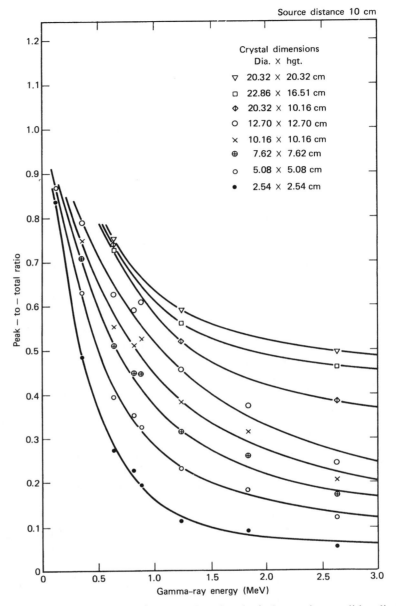

Figure 10.25 Peak-to-total ratio (or the photofraction) for various solid cylinders of NaI(Tl) for a point gamma-ray source 10 cm from the scintillator surface. (Courtesy of Harshaw Chemical Company.)

A second means of obtaining efficiency data is through calculation based on an assumed knowledge of the various gamma-ray interaction probabilities. The simplest case is the total efficiency, which is completely determined by the total linear absorption coefficient μ that characterizes the detector material. The intrinsic total efficiency is just the value of the gamma-ray interaction probability $(1 - e^{-\mu l})$ integrated over all path lengths l taken by those gamma rays that strike the detector. For simple geometries, this integration can be carried out analytically. In most cases, however, the peak efficiency is of greater interest. Because more complex processes involving multiple interactions contribute to full-energy events, the peak efficiency, in general, no longer can be calculated analytically. Instead, recourse must be made to Monte Carlo calculations, which attempt to simulate the behavior of gamma-ray photons based on knowledge of the individual probabilities for photoelectric absorption, Compton scattering, and pair production. Because each Monte Carlo calculation is, in effect, a computer experiment, the results obtained are subject to statistical uncertainties determined by the number of histories that have been calculated. Furthermore, the results are specific to the detector geometry and gamma-ray energy assumed and cannot be generalized further.

Extensive tables and graphs of sodium iodide detector efficiencies can be found in Refs. 5, 6, and 29. A good review of both experimentally determined and calculated efficiencies for solid cylindrical sodium iodide crystals has been published by Grosswendt and Waibel.[30] Extensive references are also tabulated in Refs. 31 and 32 for calculations and measurements of efficiencies for well-type crystals, and in Ref. 33 for other crystal shapes.

When the gamma-ray-emitting sample is not negligibly small, self-absorption effects can substantially reduce the efficiency that would be obtained for point sources. The usual procedure is to apply a multiplicative correction factor to standard efficiencies to account for absorption within the sample itself.[34] This approach gives adequate representation for a wide range of physical situations.[35] Reviews of scintillator efficiencies for absorbing disk sources are given in Refs. 36 and 37.

Most of the available efficiency data concern sources that are located along the axis of symmetry of the detector, and only limited data are available regarding the directional response of scintillation crystals to off-axis sources. One such analysis for cylindrical scintillators is given in Ref. 38, which also contains references to other previously published data on directional characteristics.

Efficiency data can also be found for scintillation materials other than sodium iodide. Some measurements comparing cesium iodide with sodium iodide are shown in Figs. 10.26 and 10.27. CsI(Tl) exhibits both a higher efficiency and a greater photofraction, but its somewhat lower light output (as measured using standard PM tubes) leads to a poorer energy resolution. Corresponding differences are even more extreme in the case of BGO. Figure 10.28 shows that the intrinsic peak efficiency for an equivalent size BGO crystal can be greater by almost a factor of 10 than for sodium iodide at high gamma-ray energies, but

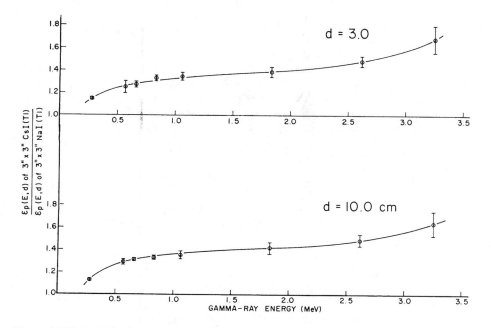

Figure 10.26 Relative intrinsic peak efficiencies of 7.62 cm × 7.62 cm cylindrical crystals of CsI(Tl) and NaI(Tl); d is the crystal–source distance. (From Irfan and Prasad.[39])

Figure 10.27 Experimental photofractions of 7.62 cm × 7.62 cm cylindrical crystals of CsI(Tl) and NaI(Tl). (From Irfan and Prasad.[39])

Figure 10.28 Intrinsic peak efficiencies for BGO and NaI(Tl) scintillators of equal 38 mm × 38 mm size. Radioisotope sources used for various photon energies are indicated. (From Evans and Orndoff.[40])

the energy resolution is even poorer than in cesium iodide. Gamma-ray detection efficiencies for NE-213 organic liquid scintillators are given in Refs. 41 and 42, for plastic scintillators in Ref. 43, and for lead-doped plastic scintillators in Ref. 44.

3. PEAK AREA DETERMINATION

To apply the peak efficiency data for any detector, the area under the full-energy peaks that appear in its spectrum must be determined. Even after subtraction of the normal background, nearly all such peaks will be superimposed on a continuum caused by many of the complicating effects described earlier in this chapter. It is therefore not always a simple task to determine the number of events that contribute to a given full-energy peak.

If the peak were a simple isolated one without any superimposed continuum, as shown in Fig. 10.29a, its area could be determined by simple integration between the limits shown. When the spectrum is recorded in a multichannel analyzer, the equivalent process is a simple addition of the content of each channel between the indicated limits. If a continuum is also present, as in Fig. 10.29b, some additional unwanted counts are included in this process and must be subtracted. Some shape must therefore be assumed for the continuum within the region under the peak, and a number of fitting procedures of varying degrees of complexity can be applied. A linear interpolation between the continuum values on either side of the peak is the easiest approach and will give sufficient accuracy for many purposes.

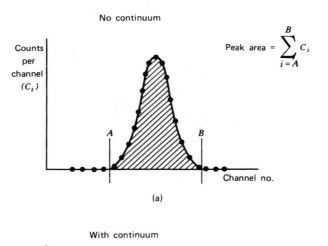

$$\text{Peak area} = \sum_{i=A}^{B} C_i$$

(a)

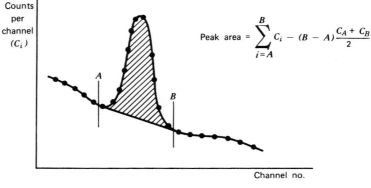

$$\text{Peak area} = \sum_{i=A}^{B} C_i - (B - A)\frac{C_A + C_B}{2}$$

(b)

Figure 10.29 Methods of obtaining peak areas from multichannel spectra. In part (a) the continuum under the peak present in part (b) is negligible.

At times, closely spaced or overlapping peaks do not allow the straightforward summation method to be applied. More complex methods must then be used to separate the individual contributions of each of the closely lying peaks. These methods normally involve fitting an analytic shape to that portion of the peak that can be clearly resolved, and assuming that the remainder of the peak is described by the fitted function. A Gaussian curve fitted to the points that lie within one standard deviation on either side of the peak value adequately represents the shape of the measured photopeak from a NaI(Tl) scintillator over an assortment of source geometry and counting conditions.[45] More complex shapes are sometimes necessary for spectra recorded at high rates or under nonideal circumstances.

V. RESPONSE OF SCINTILLATION DETECTORS TO NEUTRONS

Some types of scintillators specifically designed to be used as neutron detectors are described in Chapters 14 and 15. However, virtually any scintillation material will respond to some extent if exposed to a neutron flux. Fast neutrons are sometimes unavoidably present in gamma-ray measurements made around accelerators or reactors, and the pulses they produce can be an undesirable background. These pulses fall into two general categories: prompt pulses that are produced within a few nanoseconds of the time the neutron enters the scintillator, and pulses that occur after some delay.

In NaI(Tl) and BGO, the prompt pulses are principally caused by the detection of gamma rays produced in inelastic scattering interactions of the neutron with the scintillator.[46] Figure 10.30 shows the detection efficiency corresponding to the prompt pulses in these two materials. BGO has a better gamma ray to neutron sensitivity ratio compared with NaI(Tl).

The delayed pulses can be triggered by two categories of events. A neutron first may be moderated (requiring perhaps 100 μs) and then captured in the detector as a thermal neutron. The resulting capture gamma rays may lead to a detected pulse. Pulses that occur after a much longer delay may result if the neutron capture produces a radioactive species that subsequently decays. Examples with their corresponding half-lives are ^{24}Na (15 h) and ^{128}I (25 min) in sodium iodide, and ^{75}Ge (83 min) and ^{77}Ge (11.3 h) in BGO.

Figure 10.30 The prompt pulse response to fast neutrons in 7.6 cm × 7.6 cm scintillation detectors of NaI(Tl) and BGO. (From Hausser et al.[46])

VI. ELECTRON SPECTROSCOPY WITH SCINTILLATORS

Scintillators can also be applied to the measurement of fast electrons (such as beta particles) that are incident on one surface of the crystal. Although it has become more common to use lithium-drifted silicon detectors for this purpose (see Chapter 13), applications sometimes arise in which the size limitation of silicon detectors or other considerations dictate the use of scintillators.

The nature of the electron response function depends on the scintillation material, its physical thickness, and the angle of incidence of the electrons. Electrons from an external source normally must pass through some protective covering and/or light reflector before reaching the surface of the scintillator itself. In the discussion that follows, the energy loss that may occur in these intervening materials is not explicitly considered but may be important if the electron energy is small. We also assume that the scintillators under consideration are thicker than the maximum range of the incident electrons. Even so, the detector may not be totally opaque to the secondary bremsstrahlung photons that will be generated along the path of the electron.

In general, the response functions show a pronounced full-energy peak corresponding to the total absorption of the incident electron energy, together with a tail extending to lower energies. The major cause of such partial energy absorption is backscattering, in which the electron reemerges from the surface through which it entered after having undergone only partial energy loss. Other events that contribute to the tail are those electrons that are fully stopped within the scintillator but that generate bremsstrahlung photons that escape from the front or back surface of the detector.

Both the probability of backscattering and the fraction of the electron energy loss due to bremsstrahlung increase markedly with the atomic number of the scintillator. Because both processes detract from the full-energy peak and add the unwanted tail to the response function, scintillators with low atomic number are generally preferred for electron spectroscopy (just the opposite criterion than that desired for gamma-ray spectroscopy). Therefore, organic scintillators such as anthracene or plastics are most commonly applied in electron measurements. Table 10.1 lists the probability for backscattering of normally incident electrons for some common scintillator materials. The much lower backscattering probabilities for the low-Z materials are evident.

Figure 10.31 shows measured pulse height spectra in both cesium iodide and a plastic scintillator for normally incident 1 MeV electrons. The low-energy tail is more pronounced for cesium iodide because of its higher atomic number. Additional data for other scintillation materials, electron energies, and varying angles of incidence are given in Ref. 47.

Table 10.1 Fraction of Normally Incident Electrons Backscattered from Various Detector Surfaces

Scintillator	Electron Energy (MeV)				
	0.25	0.50	0.75	1.0	1.25
Plastic	0.08 ± 0.02	0.053 ± 0.010	0.040 ± 0.007	0.032 ± 0.003	0.030 ± 0.005
Anthracene	0.09 ± 0.02	0.051 ± 0.010	0.038 ± 0.004	0.029 ± 0.003	0.026 ± 0.004
NaI(Tl)	0.450 ± 0.045	0.410 ± 0.010	0.391 ± 0.014	0.375 ± 0.008	0.364 ± 0.007
CsI(Tl)	0.49 ± 0.06	0.455 ± 0.023	0.430 ± 0.013	0.419 ± 0.018	0.404 ± 0.016

Source: Titus.[47]

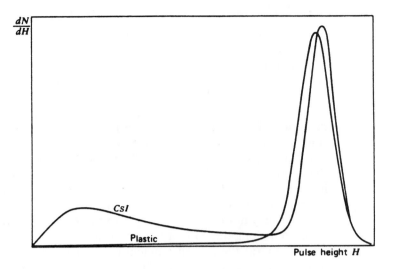

Figure 10.31 Experimental pulse height spectra from CsI(Tl) and plastic scintillators for 1.0 MeV electrons at normal incidence. The spectra are normalized to the same maximum pulse height. (From Titus. [47])

VII. SPECIALIZED DETECTOR CONFIGURATIONS BASED ON SCINTILLATION

A. The Phoswich Detector

The combination of two dissimilar scintillators optically coupled to a single PM tube is often called a *phoswich* (or *phosphor sandwich*) detector. The scintillators are chosen to have different decay times so that the shape of the output pulse from the PM tube is dependent on the relative contribution of scintillation light from the two scintillators. Most applications involve the use of this pulse shape difference to distinguish events that have occurred in only one scintillator from those that occur in both. For example, lightly penetrating radiations can be made to stop fully in the first scintillator, but more penetrating particles may generate light in both. Sodium iodide and cesium iodide are often chosen as the two materials because their decay times are quite different (0.23 versus 0.68 + 3.34 μs), and pulses arising from only one decay are easily distinguished from those with both components, using the pulse shape discrimination methods of Chapter 17. Other common scintillator combinations include BGO and cesium iodide[48,49] or two different plastic scintillators,[50,51] one with a fast decay time and the other with a slow decay. Examples of additional combinations that have been used include CsI(Na) with GSO,[52] BGO with GSO,[53] plastic scintillator with BaF_2,[54–56] and YSO with LSO.[57] Phoswich detectors employing pulse shape discrimination have proved to be useful in suppressing background in the counting of X-rays and beta particles.[58,59]

Alternatively, separate electronic pulses may be derived from the fast and slow components of the PM tube signal.[51,60] Independent measurements of the energy deposited in each scintillator can then be obtained without the need for a second PM tube. Using a thin fast scintillator in front of a thick slow scintillator allows simultaneous measurements of dE/dx and E for particles that penetrate the thin detector. The particle identification techniques described on p. 396 can then be applied to distinguish one type of heavy charged particle from another.

A somewhat simpler version of a phoswich detector has been commercially implemented as a combined alpha/beta particle probe. It consists of a thin ZnS(Ag) screen mounted behind a thin aluminum entrance window to serve as the alpha particle detector. This screen is thick enough to stop the alphas, but thin enough so that incident beta particles deposit a very small amount of energy while passing through. A plastic scintillator that is a few millimeters thick is mounted behind this screen and provides the useful response to beta particles. The two scintillators are in optical contact and are viewed by a single photomultiplier tube. The contributions of the alpha and beta particles can be separated by simple amplitude selection because the energy of the alpha particles is generally much larger than the betas, and the scintillation efficiency of the ZnS(Ag) scintillator is much higher than that of the plastic scintillator.

B. The Moxon–Rae Detector

Another type of gamma-ray detector can be built based on the detection of secondary electrons produced by gamma-ray interactions in a converter that is external to the detector. Although the resulting detection efficiency is typically quite low, the efficiency versus energy dependence can be tailored in certain ways to suit the needs of specialized applications. For example, Moxon and Rae[61] first described the properties of a thick low-Z converter used to produce secondary electrons that were then detected by a thin plastic scintillator. By choosing a converter that consisted of graphite with a small component of bismuth, they found that the detection efficiency could be made nearly proportional to the incident gamma-ray energy. This proportionality allows a simplified analysis to be carried out for a class of experiments involving neutron capture because the detection probability of any given radiative capture event becomes independent of the specific de-excitation cascade mechanisms following the capture.[61] The proportionality of efficiency with gamma-ray energy results from the combined probabilities of electron production within the converter and of their escape from the surface facing the electron detector. Detailed analyses of the theory and performance of Moxon–Rae detectors can be found in Refs. 61–63.

C. Liquid Scintillation Counters

The liquid scintillation media discussed in Chapter 8 can be applied to avoid some of the difficulties that arise when measuring low-energy beta particles or alpha particles using conventional methods. The approach, sometimes called *internal source liquid scintillation counting*, involves dissolving the sample to be counted directly into the liquid scintillator. Under these conditions, problems relating to sample self-absorption, attenuation of particles by detector windows, and beta backscattering from the detector are completely avoided. These advantages are particularly important for low-energy radiations such as the beta particles emitted by tritium and ^{14}C. The endpoint energies for these beta spectra are 18 and 160 keV, respectively, and average beta energies are about one-third these values. Because these isotopes are particularly important in chemical and biomedical applications, much of the development of the liquid scintillation technique has taken place in connection with these sciences. Several texts have been published[64–66] that thoroughly review the fundamental principles of liquid scintillation counting.

The first step in the technique involves incorporation of the sample within a suitable liquid scintillation solution. Problems can often arise in this step because most liquid scintillation solutions are based on toluene or other organic solvents, whereas many samples are often more conveniently prepared in an aqueous solution. Detailed discussions of various methods for obtaining stable solutions through the use of solubilizing or complexing agents are given in Refs. 67 and 68. A common problem is that the introduction of the sample tends to reduce the scintillation light output compared with the pure scintillator. This phenomenon, commonly called *quenching*, often limits the amount of sample material that can effectively be incorporated within the scintillator solution. The quenching can be due to either alteration of the optical properties of the solution by the sample (*color quenching*) or interference with the energy transfer process within the scintillator itself. Insoluble materials can sometimes be introduced as a suspension of fine particles. In those cases where rapid settling of the suspension is a problem, some methods have been developed that involve converting the solution to a gel immediately after preparation of these suspensions.

After the sample has been prepared, the solution is normally loaded into a glass vial and placed in a light-tight enclosure viewed by one or more PM tubes. Because of the effects of quenching and the fact that typical applications involve radiations of low energy, pulses produced in the PM tube often correspond to no more than a few photoelectrons. Under these circumstances, the measurements are potentially very sensitive to sources of noise that may interfere with accurate and reproducible counting of the sample. Significant sources of noise are thermally generated electrons from the photocathode of the PM tube, long-lived phosphorescence in the scintillator, and chemiluminescence (light generated by chemical reactions within the sample–scintillator solution). Virtually all these noise sources correspond to the generation of only a single electron per pulse, so they can normally be rejected by placing a discrimination level in the signal chain to eliminate those pulses whose amplitude corresponds to a single photoelectron. However, because the signal also consists of only a few photoelectrons, there is a risk that this discrimination process will also eliminate some of the signal.

Because all beta particles emitted by the sample pass through some portion of the scintillator, and the great majority are fully stopped within the solution, the counting efficiency can potentially be close to 100%. The degree to which the few-photoelectron signal can be distinguished from single-electron noise determines the practical counting efficiency. As a gauge of the development of the technique, it is interesting to note the improvement in the counting efficiency for tritium as improvements in PM tube design and other techniques have been implemented. A counting efficiency of about 20% in 1960 was improved to about 60% in 1970 through the use of low-noise bialkali PM tubes, and to a value of about 90% through the application of PM tubes with gallium phosphide high-gain dynodes.[69]

One method of eliminating the PM tube noise is to use two PM tubes to view the scintillator from opposite sides. Only those pulses that are observed in coincidence between the two tubes are counted. Because the noise generated in each tube will be uncorrelated, a true coincidence will not be observed for these events, and the recorded counts will correspond only to events generated within the liquid scintillator. The summed output from both PM tubes can then be used to record the pulse height spectrum for the sample.

Although most commonly applied to samples emitting beta particles or conversion electrons, liquid scintillation counting has also been used to count samples that are alpha active.[70–72] Although the best reported energy resolution[71] of 5–8% is much inferior to that attainable with semiconductor diode detectors, the advantages of high counting efficiency and uniform counting geometry offer some attraction for applications such as counting low-level environmental samples.[72] Because typical alpha energies are several MeV, the light output is much greater than for low-energy beta particle counting, and therefore a counting efficiency that approaches 100% is relatively easy to achieve. Pulse shape discrimination can be applied to eliminate backgrounds due to beta particles or gamma-ray-induced events.[73] By using these techniques, an exceptionally low background level of 0.01 count/min has been reported[71] while maintaining essentially 100% counting efficiency for alpha particles within the sample. To provide an even cleaner separation between the alpha particle events and those from the background, the use of a system with a two-parameter multichannel analyzer[74–76] has been shown to be very effective. Simultaneous measurements are made of the pulse amplitude and rise time that allow a cleaner separation by defining areas of interest in a two-dimensional memory. The background can then be as low as 0.001 counts/min. An illustration of the principle of this technique is given later in Fig. 18.13.

D. Position-Sensitive Scintillators

1. ONE-DIMENSIONAL POSITION SENSING

Because the light from a scintillator is generated along the track of the ionizing particle, it is possible to sense the position of interaction by localizing the source of the scintillation light. For sensing position in one dimension, a long rod or bar of scintillation material can be used with PM tubes or photodiodes positioned at either end as in the sketch below:

In this type of geometry, it is generally observed that the intensity of the light measured at one end of the rod drops off exponentially with the distance at which the scintillation light is generated. Thus, we can write for the signal from PM tube ①

$$E_1 = \frac{E_\gamma P}{E_0} \exp\left[-\alpha(L/2 + x)\right] \tag{10.18}$$

where E_γ = energy deposited by gamma ray

P = probability that light quantum produced at one end will generate a photoelectron in adjacent tube

E_0 = energy deposited per light photon created in scintillator

α = light attenuation coefficient

Similarly, for PM tube ②,

$$E_2 = \frac{E_\gamma P}{E_0} \exp\left[-\alpha(L/2 - x)\right] \tag{10.19}$$

By dividing one signal by the other, we obtain

$$\frac{E_2}{E_1} = \frac{\exp\left[-\alpha(L/2 - x)\right]}{\exp\left[-\alpha(L/2 + x)\right]} = \exp(+2\alpha x)$$

$$\ln \frac{E_2}{E_1} = 2\alpha x$$

$$x = \frac{1}{2\alpha} \ln \frac{E_2}{E_1} \tag{10.20}$$

Therefore, by electronically deriving the logarithm of the ratio of the two PM tube signals, we obtain a linear indication of the position at which the scintillation occurs.

By multiplying Eq. (10.18) and (10.19), we obtain

$$E_\gamma^2 = E_1 E_2 \left(\frac{E_0}{P}\right)^2 e^{\alpha L}$$

Now taking the square root of both sides, we obtain

$$E_\gamma = \sqrt{E_1 E_2}\, \frac{E_0}{P} e^{\alpha L/2} \tag{10.21}$$

Thus, the square root of the product of the two PM tube signals serves as a measure of the total scintillation light, independent of the position within the bar.

For best spatial resolution,[77] the light attenuation coefficient α in the expressions above should have an optimal value of $2.9/L$. This light attenuation can be accomplished either by adjusting the reflection conditions at the surface of the scintillator or by allowing some absorption of the light to occur in the bulk of the scintillator. In a typical application of this technique,[77] a spatial resolution of 10 mm was obtained in a 500 mm long sodium iodide rod, using gamma rays of 662 keV energy.

An alternative method of sensing the scintillation position in one dimension is to use the same geometry described above but to exploit the fact that there will be a slight difference in the time of arrival of the scintillation light at the two ends of the bar. This time difference is maximum for an event that occurs at either end of the bar and decreases linearly to zero for an event at the center. The time differences are small, since the velocity of propagation of the light traveling directly from the scintillation site is c/n, where c is the velocity of light and n is the index of refraction of the scintillation medium. For typical scintillators, this corresponds to a flight time of about 5 ps/mm. However, most of the light is reflected many times from the surface of the scintillator as it travels along the length of the bar and therefore the actual flight path and propagation time are substantially extended. In one application of this method using a 250-mm long plastic scintillator rod,[78] a spatial resolution of between 17 and 23 mm was obtained for 511 keV gamma rays.

2. TWO-DIMENSIONAL POSITION SENSING (IMAGING DETECTORS)

In nuclear medicine, it is often necessary to form the image of the distribution of gamma-ray-emitting isotopes distributed throughout the patient. The *gamma-ray camera* is a device that senses the two-dimensional coordinates of a gamma-ray photon as it interacts in a large-area detector and forms an image through the accumulation of many such events over the exposure time. A lead pinhole or parallel hole collimator is used to restrict the gamma rays that strike the detector so that the image can be directly interpreted as the spatial distribution of the emitting isotope.

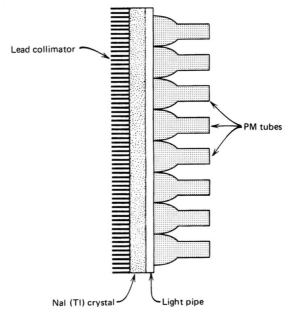

Figure 10.32 Elements of a two-dimensional position-sensitive scintillation detector, commonly called a *gamma camera*.

The most common type of gamma-ray camera is based on an original design by Anger.[79] Its basic elements are diagrammed in Fig. 10.32. The detection medium consists of a flat single scintillation crystal (generally sodium iodide) with length and width up to 50 cm and thickness of about 1 cm. The light generated by gamma-ray interactions in this crystal is sensed by an array of PM tubes that completely cover one of its flat faces. The two-dimensional position of each event across the area of the crystal is deduced from the relative size of the signals produced from these tubes.

Each scintillation event will generate output pulses of significant amplitude from all the PM tubes that are near the location of the interaction. The largest signal will generally be from the tube nearest the position, with smaller pulses from tubes at a greater distance. The "center of gravity" of the light is interpolated from these signals using either a charge division process in which a portion of each signal is coupled to X and Y output lines, or through the use of readout schemes based on delay line encoding of the position.[80] The resulting analog X and Y position signals are typically digitized using fast analog-to-digital

Figure 10.33 Example of a human bone scan produced by using a gamma camera to image the distribution of methylene diphosphonate labeled with ^{99m}Tc. The upper and lower halves of the image each were produced by recording approximately 10^6 counts over a 3-min period. (Courtesy of W. L. Rogers, University of Michigan Medical Center.)

converters and accumulated in a two-dimensional digital memory to form the image. The summed output from all the tubes is a good measure of the total energy deposited in the crystal and is normally passed through a single-channel analyzer to record selectively only those events that correspond to the full source energy. This selection eliminates loss of contrast in the image caused by gamma rays that have initially scattered in the patient or elsewhere before reaching the camera and would therefore give a false position indication.

An example of an image from a typical gamma-ray camera is given in Fig. 10.33. The intrinsic spatial resolution of the camera when recording the most common type of image generated by ^{99m}Tc (140 keV) is limited to about 3-mm FWHM. In practice, additional spatial resolution loss normally occurs due to the geometric uncertainties in the source position when it is some distance from the face of the collimator.

PROBLEMS

10.1 A gamma-ray photon after Compton scattering through an angle of 90° has an energy of 0.5 MeV. Find its energy before the scattering.

10.2 A 2 MeV gamma-ray photon is incident on a detector, undergoes two sequential Compton scatterings, and then escapes. If the angles of scattering are 30° and 60°, respectively, how much total recoil electron energy is deposited in the detector? Does the answer change if the sequence of the scattering angles is reversed?

10.3 Find the maximum energy that can be deposited by a 1 MeV gamma-ray photon if it undergoes two successive Compton scattering events and then escapes the detector.

10.4 Estimate the time that separates two successive gamma-ray scattering interactions that are 3 cm apart in sodium iodide. Compare with the characteristic decay time for the light that is generated in the same material.

10.5 The cross sections for photoelectric, Compton, and pair production interactions in sodium iodide at 2 MeV are in the ratio 1:20:2, respectively. Will the pulse height spectrum from 2 MeV gamma rays incident on a sodium iodide scintillator give a peak-to-total ratio of less than, more than, or about equal to $1/23$?

10.6 If the energy resolution of a particular NaI(Tl) scintillation detector is 7% for ^{137}Cs gamma rays (0.662 MeV), estimate its energy resolution for the 1.28 MeV gamma rays from ^{22}Na.

10.7 The mass attenuation coefficient of NaI at 0.5 MeV is 0.955 cm²/g. Find the intrinsic total efficiency of a slab detector 0.50 cm thick at this energy. If the photofraction is 40% at the same energy, what is the intrinsic peak efficiency?

10.8. (a) Find the energy of the Compton edge for the 1.17 MeV gamma rays from ^{60}Co.

(b) Calculate the backscatter peak energies corresponding to incident gamma rays of 1, 2, and 3 MeV.

10.9 Listed below are a number of parameters of interest in gamma-ray spectroscopy using scintillation detectors:

(a) Density of the detector medium.

(b) Kinetic energy required to create a scintillation photon in the crystal.

(c) Average atomic number (Z-value) of the detector medium.

(d) Geometry of the source–detector system.

(e) Gain of the photomultiplier tube.

(f) Quantum efficiency of the photocathode in the photomultiplier.

(g) Gain of the amplifier used between the detector and pulse analysis system.

(h) Fraction of light generated in the crystal that reaches the photocathode of the PM tube (light collection efficiency).

Identify those parameters from this list that have a major influence on the detector *intrinsic peak efficiency*. Repeat, but now identify those that have a major influence on *energy resolution*.

10.10 Calculate the pulse amplitude from the anode of a PM tube used with a NaI(Tl) scintillator under the following conditions: A 1-MeV electron loses all its energy in the scintillator, the light collection efficiency to the photocathode is 50%, the average quantum efficiency of the photocathode is 20%, and 80% of the photoelectrons are collected at the first dynode. Assume that the PM tube has 10 stages with a multiplication factor $\delta = 2.5$ per stage. The anode load resistance is 100 kΩ, and the anode capacitance is 100 pF.

10.11 A particular radioisotope emits two coincident gamma rays, each with 100% yield per decay, with no angular correlation between the photon directions. A sample is placed 10 cm from the surface of a 5 cm radius cylindrical detector along its axis. The intrinsic peak efficiency of the detector for γ_1 is 50%, and for γ_2 it is 30%.

(a) If the sample activity is low enough so that chance coincidences are negligible, calculate the ratio of the counts under the sum peak in the recorded pulse height spectrum to the counts under the γ_1 full-energy peak.

(b) Calculate the rate at which events are recorded in the sum peak if the source activity is 100 kBq. For a detector resolving time of 3 μs, what additional rate should be expected from chance coincidences between γ_1 and γ_2?

10.12 From Fig. 2.14, estimate the range of a 1-MeV electron in sodium iodide. From your answer, calculate the percentage of the total volume of a 5.08 cm $\times$ 5.08 cm cylindrical crystal that lies near enough to the surface so that electron escape is possible.

10.13 A radioisotope source is known not to emit any gamma-ray photons with energy of 511 keV, but a peak is observed at this position in the recorded gamma-ray spectrum. Give two possible origins for this peak.

10.14 If the energy resolution of a scintillator is 8.5% at 662 keV, find the standard deviation (in energy units) of the Gaussian curve that would be a fit to the photopeak at that energy.

10.15 Why are materials with low atomic number often preferred as scintillators for electron spectroscopy, while the opposite is true for gamma-ray spectroscopy?

10.16 Explain the major advantage of liquid scintillation counting when applied to low-energy beta emitters compared with conventional solid scintillation detectors.

REFERENCES

1. J. B. Birks, *The Theory and Practice of Scintillation Counting,* Pergamon Press, Oxford, 1964.
2. S. M. Shafroth (ed.), *Scintillation Spectroscopy of Gamma Radiation,* Gordon & Breach, London, 1967.
3. K. Siegbahn (ed.), *Alpha-, Beta-, and Gamma-Ray Spectroscopy,* North Holland, Amsterdam, 1968.
4. C. D. Broyles, D. A. Thomas, and S. K. Haynes, *Phys. Rev.* **89**, 715 (1953).
5. R. L. Heath, *Scintillation Spectrometry Gamma-Ray Spectrum Catalogue,* IDO-16880, Vols. 1 and 2, 1964.
6. F. Adams and R. Dams, *Applied Gamma-Ray Spectrometry,* 2nd ed. and revision of original publication by C. E. Crouthamel, Pergamon Press, Oxford, 1970.
7. N. A. Wogman, D. E. Robertson, and R. W. Perkins, *Nucl. Instrum. Meth.* **50**, 1 (1967).
8. N. A. Wogman, R. W. Perkins, and J. H. Kaye, *Nucl. Instrum. Meth.* **74**, 197 (1969).
9. B. A. Euler, D. F. Covell, and S. Yamamoto, *Nucl. Instrum. Meth.* **72**, 143 (1969).
10. B. Bengtson and M. Moszynski, *Nucl. Instrum. Meth.* **85**, 133 (1970).
11. M. D. Hasinoff, S. T. Lim, D. F. Measday, and T. J. Mulligan, *Nucl. Instrum. Meth.* **117**, 375 (1974).
12. R. Mueller and D. Maeder, "Single Crystal Spectroscopy," Chap. VII in *Scintillation Spectroscopy of Gamma Radiation* (S. M. Shafroth, ed.), Gordon & Breach, New York, 1967.
13. C. E. Moss et al., *Nucl. Instrum. Meth.* **219**, 558 (1984).
14. R. R. Kiziah and J. R. Lowell, *Nucl. Instrum. Meth.* **A305**, 129 (1991).
15. M. J. Berger and S. M. Seltzer, *Nucl. Instrum. Meth.* **104**, 317 (1972).
16. R. J. D. Beattie and J. Byrne, *Nucl. Instrum. Meth.* **104**, 163 (1972).
17. D. E. Persyk and T. E. Moi, *IEEE Trans. Nucl. Sci.* **NS-25**(1), 615 (1978).
18. P. Dorenbos, J. T. M. de Haas, and C. W. E. van Eijk, *IEEE Trans. Nucl. Sci.* **42**(6), 2190 (1995).
19. J. D. Valentine, B. D. Rooney, and J. Li, *IEEE Trans. Nucl. Sci.* **45**(3), 512 (1998).
20. J. D. Valentine, B. D. Rooney, and P. Dorenbos, *IEEE Trans. Nucl. Sci.* **45**, 1750 (1998).
21. J. Berger et al., *Nucl. Instrum. Meth.* **A279**, 343 (1989).
22. J. Zabierowski, *Nucl. Instrum. Meth.* **A338**, 577 (1994).
23. M. Kobayashi et al., *Nucl. Instrum. Meth.* **A337**, 355 (1994).
24. D. Autiero et al., *Nucl. Instrum. Meth.* **A372**, 556 (1996).
25. A. Fyodorov, M. Korzhik, A. Lopatik, and O. Missevitch, *Nucl. Instrum. Meth.* **A413**, 352 (1998).
26. L. R. Wayne, W. A. Heindl, P. L. Hink, and R. E. Rothschild, *Nucl. Instrum. Meth.* **A411**, 351 (1998).
27. B. D. Rooney and J. D. Valentine, *IEEE Trans. Nucl. Sci.* **44**(3), 509 (1997).
28. B. J. Snyder, "Calculation of Gamma Ray Scintillation Detector Efficiencies and Photofraction by Monte Carlo Methods," Ph.D. Dissertation, The University of Michigan, 1965.
29. C. C. Grosjean and W. Bossaert, *Table of Absolute Detection Efficiencies of Cylindrical Scintillation Gamma-Ray Detectors,* Computing Laboratory, University of Ghent, 1965.
30. B. Grosswendt and E. Waibel, *Nucl. Instrum. Meth.* **133**, 25 (1976).
31. P. Holmberg and R. Rieppo, *Int. J. Appl. Radiat. Isotopes* **24**, 99 (1973).
32. R. Rieppo, *Int. J. Appl. Radiat. Isotopes* **27**, 453 (1976).
33. R. Rieppo, *Nucl. Instrum. Meth.* **115**, 541 (1974).
34. R. B. Galloway, *Nucl. Instrum. Meth.* **A300**, 367 (1991).
35. Y. S. Horowitz, S. Mordechai, and A. Dubi, *Nucl. Instrum. Meth.* **122**, 399 (1974).
36. T. Nakamura, *Nucl. Instrum. Meth.* **105**, 77 (1972).
37. M. Belluscio, R. De Leo, A. Pantaleo, and A. Vox, *Nucl. Instrum. Meth.* **118**, 553 (1974).
38. I. Petr, A. Adams, and J. B. Birks, *Nucl. Instrum. Meth.* **95**, 253 (1971).
39. M. Irfan and R. D. G. Prasad, *Nucl. Instrum. Meth.* **107**, 583 (1973).
40. A. E. Evans and J. D. Orndoff, NUREG/CR-1398 (LA-8301), 1980.
41. N. A. Lurie, L. Harris, Jr., and J. P. Wondra, *Nucl. Instrum. Meth.* **129**, 619 (1975).
42. R. De Leo, G. D'Erasmo, and A. Pantaleo, *Nucl. Instrum. Meth.* **129**, 501 (1975).
43. E. V. Bonzi and R. T. Mainardi, *Nucl. Instrum. Meth.* **B72**, 477 (1992).
44. A. Schaarschmidt and H. Durner, *Nucl. Instrum. Meth.* **105**, 504 (1972).
45. T. S. Mudhole and N. Umakantha, *Nucl. Instrum. Meth.* **116**, 401 (1974).
46. O. Hausser et al., *Nucl. Instrum. Meth.* **213**, 301 (1983).
47. F. Titus, *Nucl. Instrum. Meth.* **89**, 93 (1970).
48. E. Costa, E. Massaro, and L. Piro, *Nucl. Instrum. Meth.* **A243**, 572 (1986).

49. L. Piro et al., *Nucl. Instrum. Meth.* **A257**, 429 (1987).
50. M. Bantel et al., *Nucl. Instrum. Meth.* **226**, 394 (1984).
51. F. Lidén, J. Nyberg, A. Johnson, and A. Kerek, *Nucl. Instrum. Meth.* **A253**, 305 (1987).
52. Z. He and D. Ramsden, *Nucl. Instrum. Meth.* **A336**, 330 (1993).
53. T. Kamae et al., *IEEE Trans. Nucl. Sci.* **40**(2), 204 (1993).
54. G. Lanzano et al., *Nucl. Instrum. Meth.* **A323**, 694 (1992).
55. Y. Futami et al., *Nucl. Instrum. Meth.* **A326**, 513 (1993).
56. R. Novotny et al., *IEEE Trans. Nucl. Sci.* **43**(3), 1260 (1996).
57. M. Dahlbom et al., *IEEE Trans. Nucl. Sci.* **44**(3), 1114 (1997).
58. M. R. Mayhugh, A. C. Lucas, and B. K. Utts, *IEEE Trans. Nucl. Sci.* **NS-25**(1), 569 (1978).
59. B. H. Erkkila et al., *IEEE Trans. Nucl. Sci.* **NS-32(1)** , 969 (1985).
60. J. Pouliot et al., Lawrence Berkeley Laboratory Report LBL-24396, Dec. 1987.
61. M. C. Moxon and E. R. Rae, *Nucl. Instrum. Meth.* **24**, 445 (1963).
62. S. S. Malik, *Nucl. Instrum. Meth.* **125**, 45 (1975).
63. S. S. Malik and C. F. Majkrzak, *Nucl. Instrum. Meth.* **130**, 443 (1975).
64. Y. Kobayashi and D. V. Maudsley, *Biological Applications of Liquid Scintillation Counting,* Academic Press, New York, 1974.
65. D. L. Horrocks, *Applications of Liquid Scintillation Counting,* Academic Press, New York, 1974.
66. M. Crook and P. Johnson (eds.), *Liquid Scintillation Counting,* Vol. 4, Heyden and Son, London, 1977.
67. J. H. Parmentier and F. E. L. Ten Haaf, *Int. J. Appl. Radiat. Isotopes* **20**, 305 (1969).
68. B. W. Fox, *Techniques of Sample Preparation for Liquid Scintillation Counting,* North-Holland, Amsterdam, 1976.
69. R. Vaninbroukx and I. Stanef, *Nucl. Instrum. Meth.* **112**, 111 (1973).
70. J. W. McKlveen, H. W. Berk, and W. R. Johnson, *Int. J. Appl. Radiat. Isotopes* **23**, 337 (1972).
71. J. W. McKlveen and W. J. McDowell, *Trans. Am. Nuc. Soc.* **22**, 149 (1975).
72. W. J. McDowell, *IEEE Trans. Nucl. Sci.* **NS-22**(1), 649 (1975).
73. P. Cross and G. W. McBeth, *Health Phys.* **30**, 303 (1976).
74. W. J. McDowell, *Radioactivity & Radiochemistry* **3**(2), 26 (1992).
75. J. R. Cadieux et al., *Nucl. Instrum. Meth.* **A353**, 534 (1994).
76. S. Hamanaka et al., *Nucl. Instrum. Meth.* **A410**, 314 (1998).
77. J. N. Carter et al., *Nucl. Instrum. Meth.* **196**, 477 (1982).
78. R. Myllyla, H. Heusala, and M. Karras, *IEEE Trans. Nucl. Sci.* **NS-28**(1), 167 (1981).
79. H. O. Anger, *Rev. Sci. Instrum.* **29**, 27 (1958).
80. T. Hiramoto, E. Tanaka, and N. Nohara, *J. Nucl. Med.* **12**, 160 (1971).

Chapter 11

Semiconductor Diode Detectors

In many radiation detection applications, the use of a solid detection medium is of great advantage. For the measurement of high-energy electrons or gamma rays, detector dimensions can be kept much smaller than the equivalent gas-filled detector because solid densities are some 1000 times greater than that for a gas. Scintillation detectors offer one possibility of providing a solid detection medium, and their application to the detection and measurement of various radiations has been described in Chapter 10.

One of the major limitations of scintillation counters is their relatively poor energy resolution. The chain of events that must take place in converting the incident radiation energy to light and the subsequent generation of an electrical signal involves many inefficient steps. Therefore, the energy required to produce one information carrier (a photoelectron) is of the order of 100 eV or more, and the number of carriers created in a typical radiation interaction is usually no more than a few thousand. The statistical fluctuations in so small a number place an inherent limitation on the energy resolution that can be achieved under the best of circumstances, and nothing can be done about improving the energy resolution beyond this point. As detailed in Chapter 10, the energy resolution for sodium iodide scintillators is limited to about 6% when detecting 0.662 MeV gamma rays and is largely determined by the photoelectron statistical fluctuations.

The only way to reduce the statistical limit on energy resolution is to increase the number of information carriers per pulse. As we show in this chapter, the use of semiconductor materials as radiation detectors can result in a much larger number of carriers for a given incident radiation event than is possible with any other common detector type. Consequently, the best energy resolution from radiation spectrometers in routine use is achieved using semiconductor detectors. The fundamental information carriers are *electron–hole pairs* created along the path taken by the charged particle (primary radiation or secondary particle) through the detector. The electron–hole pair is somewhat analogous to the ion pair created in gas-filled detectors. Their motion in an applied electric field generates the basic electrical signal from the detector.

Devices employing semiconductors as the basic detection medium became practically available in the early 1960s. Early versions were called *crystal counters*, but modern detectors are referred to as *semiconductor diode detectors* or simply *solid-state detectors*. Although the latter term is somewhat ambiguous in the sense that technically scintillation counters can also be thought of as solid-state detectors, it has come into widespread use to characterize only those devices that are based on electron–hole pair collection from semiconductor media.

In addition to superior energy resolution, solid-state detectors can also have a number of other desirable features. Among these are compact size, relatively fast timing characteristics, and an effective thickness that can be varied to match the requirements of the application. Drawbacks may include the limitation to small sizes and the relatively high susceptibility of these devices to performance degradation from radiation-induced damage.

Of the available semiconductor materials, silicon predominates in the diode detectors used primarily for charged particle spectroscopy and discussed in this chapter. Germanium is more widely used in the gamma-ray measurements described in Chapter 12, whereas devices that use other semiconductor materials are covered in Chapter 13.

Several comprehensive books are available on the topic of solid-state detectors, including Refs. 1–6. Each of these contains a rather complete citation of the literature up to the time of publication, and the other references in these chapters are largely limited to those that have appeared more recently.

I. SEMICONDUCTOR PROPERTIES

A. Band Structure in Solids

The periodic lattice of crystalline materials establishes allowed energy bands for electrons that exist within that solid. The energy of any electron within the pure material must be confined to one of these energy bands, which may be separated by gaps or ranges of forbidden energies. A simplified representation of the bands of interest in insulators or semiconductors is shown in Fig. 11.1. The lower band, called the *valence band*, corresponds to those outer-shell electrons that are bound to specific lattice sites within the crystal. In the case of silicon or germanium, they are parts of the covalent bonding that constitute the interatomic forces within the crystal. The next higher-lying band is called the *conduction band* and represents electrons that are free to migrate through the crystal. Electrons in this band contribute to the electrical conductivity of the material. The two bands are separated by the *bandgap*, the size of which determines whether the material is classified as a semiconductor or an insulator. The number of electrons within the crystal is just adequate to fill completely all available sites within the valence band. In the absence of thermal excitation, both insulators and semiconductors would therefore have a configuration in which the valence band is completely full and the conduction band completely empty. Under these circumstances, neither would theoretically show any electrical conductivity.

In a metal, the highest occupied energy band is not completely full. Therefore, electrons can easily migrate throughout the material because they need achieve only small incremental energy to be above the occupied states. Metals are therefore always characterized by very high electrical conductivity. In insulators or semiconductors, on the other hand, the electron must first cross the bandgap to reach the conduction band and the

Figure 11.1 Band structure for electron energies in insulators and semiconductors.

conductivity is therefore many orders of magnitude lower. For insulators, the bandgap is usually 5 eV or more, whereas for semiconductors, the bandgap is considerably less.

B. Charge Carriers

At any nonzero temperature, some thermal energy is shared by the electrons in the crystal. It is possible for a valence electron to gain sufficient thermal energy to be elevated across the bandgap into the conduction band. Physically, this process simply represents the excitation of an electron that is normally part of a covalent bond such that it can leave the specific bonding site and drift throughout the crystal. The excitation process not only creates an electron in the otherwise empty conduction band, but it also leaves a vacancy (called a *hole*) in the otherwise full valence band. The combination of the two is called an *electron–hole pair* and is roughly the solid-state analogue of the ion pair in gases. The electron in the conduction band can be made to move under the influence of an applied electric field. The hole, representing a net positive charge, will also tend to move in an electric field, but in a direction opposite that of the electron. The motion of both of these charges contributes to the observed conductivity of the material.

The probability per unit time that an electron–hole pair is thermally generated is given by

$$p(T) = CT^{3/2} \exp\left(-\frac{E_g}{2kT}\right) \tag{11.1}$$

where

T = absolute temperature

E_g = bandgap energy

k = Boltzmann constant

C = proportionality constant characteristic of the material

As reflected in the exponential term, the probability of thermal excitation is critically dependent on the ratio of the bandgap energy to the absolute temperature. Materials with a large bandgap will have a low probability of thermal excitation and consequently will show the very low electrical conductivity characteristic of insulators. If the bandgap is as low as several electron volts, sufficient thermal excitation will cause a conductivity high enough for the material to be classified as a semiconductor. In the absence of an applied electric field, the thermally created electron–hole pairs ultimately recombine, and an equilibrium is established in which the concentration of electron–hole pairs observed at any given time is proportional to the rate of formation. From Eq. (11.1), this equilibrium concentration is a strong function of temperature and will decrease drastically if the material is cooled.[†]

After their formation, both the electron and the hole take part in a random thermal motion that results in their diffusion away from their point of origin. If all electrons (or holes) were initially created at a single point, this diffusion leads to a broadening distribution of the charges as a function of time. A cross section through this distribution would be approximated by a Gaussian function with a standard deviation σ given by

$$\sigma = \sqrt{2Dt} \tag{11.2}$$

[†]Because the ionization potential for gases is typically 15 eV or more, the probability of a thermally generated ion pair is negligibly small in gas ionization chambers, even at room temperature.

where D is the diffusion coefficient and t is the elapsed time. Values for D can be predicted from the relationship

$$D = \mu \frac{kT}{e} \tag{11.3}$$

where μ is the mobility of the charge carrier, k is the Boltzmann constant, and T is the absolute temperature. At 20°C (293K), the numerical value of kT/e is 0.0253 V.

C. Migration of Charge Carriers in an Electric Field

If an electric field is applied to the semiconductor material, both the electrons and holes will undergo a net migration. The motion will be the combination of a random thermal velocity and a net *drift velocity* parallel to the direction of the applied field. The motion of the conduction electrons is a relatively easy process to visualize, but the fact that holes also contribute to conductivity is less obvious. A hole moves from one position to another if an electron leaves a normal valence site to fill an existing hole. The vacancy left behind by the electron then represents the new position of the hole. Because electrons will always be drawn preferentially in an opposite direction to the electric field vector, holes move in the same direction as the electric field. This behavior is consistent with that expected of a point positive charge, because the hole actually represents the absence of a negatively charged electron.

At low-to-moderate values of the electric field intensity, the drift velocity v is proportional to the applied field. Then a *mobility* μ for both electrons and holes can be defined by

$$v_h = \mu_h \mathscr{E} \tag{11.4}$$
$$v_e = \mu_e \mathscr{E} \tag{11.5}$$

where $\mathscr{E}$ is the electric field magnitude. In gases, the mobility of the free electron is much larger than that of the positive ion, but in semiconductor materials the mobility of the electron and hole are roughly of the same order. Numerical values for common semiconductor materials are given in Table 11.1.

At higher electric field values, the drift velocity increases more slowly with the field. Eventually, a *saturation velocity* is reached which becomes independent of further increases in the electric field. Figure 11.2 shows the dependence of the drift velocity on field magnitude for silicon and germanium.

Many semiconductor detectors are operated with electric field values sufficiently high to result in saturated drift velocity for the charge carriers. Because these saturated velocities are of the order of 10^7 cm/s, the time required to collect the carriers over typical dimensions of 0.1 cm or less will be under 10 ns. Semiconductor detectors can therefore be among the fastest-responding of all radiation detector types.

In addition to their drift, the charge carriers will also undergo the influence of diffusion mentioned in the previous section. Without diffusion, all charge carriers would travel to the collecting electrodes following exactly the electric field lines that connect their point of origin to their collection point. The effect of diffusion is to introduce some spread in the arrival position that can be characterized as a Gaussian distribution whose standard deviation can be predicted by combining Eqs. (11.2), (11.3), and (11.4)

$$\sigma = \sqrt{\frac{2kTx}{e\mathscr{E}}} \tag{11.6}$$

where x represents the drift distance. In small-volume detectors, a typical value for σ would be less than 100 μm. This diffusion broadening of the charge distribution limits the precision to which position measurements can be made using the location at which charges are collected at the electrodes in semiconductor detectors.

Table 11.1 Properties of Intrinsic Silicon and Germanium

	Si	Ge
Atomic number	14	32
Atomic weight	28.09	72.60
Stable isotope mass numbers	28-29-30	70-72-73-74-76
Density (300 K); g/cm^3	2.33	5.32
Atoms/cm^3	4.96×10^{22}	4.41×10^{22}
Dielectric constant (relative to vacuum)	12	16
Forbidden energy gap (300 K); eV	1.115	0.665
Forbidden energy gap (0 K); eV	1.165	0.746
Intrinsic carrier density (300 K); cm^{-3}	1.5×10^{10}	2.4×10^{13}
Intrinsic resistivity (300 K); $\Omega \cdot cm$	2.3×10^5	47
Electron mobility (300 K); $cm^2/V \cdot s$	1350	3900
Hole mobility (300 K); $cm^2/V \cdot s$	480	1900
Electron mobility (77 K); $cm^2/V \cdot s$	2.1×10^4	3.6×10^4
Hole mobility (77 K); $cm^2/V \cdot s$	1.1×10^4	4.2×10^4
Energy per electron–hole pair (300 K); eV	3.62	
Energy per electron–hole pair (77 K); eV	3.76	2.96
Fano factor (77 K)	0.143 (Ref. 7)	0.129 (Ref. 9)
	0.084 (Ref. 8)	0.08 (Ref. 10)
	0.085 to 0.137 }(Ref. 12)	< 0.11 (Ref. 11)
		0.057 0.064 }(Ref. 12)
	0.16 (Ref. 13)	0.058 (Ref. 14)

Source: G. Bertolini and A. Coche (eds.), *Semiconductor Detectors*, Elsevier-North Holland, Amsterdam, 1968, except where noted.

The collection time of the charges is also spread out by diffusion by an amount that can be estimated from the spatial broadening divided by the drift velocity, or generally less than 1 ns for small volumes. In many cases, these diffusion effects are negligible, and the charges can be pictured as moving along the electric field lines, all with the same drift velocity. However, the consequences of diffusion can become significant for large-volume detectors or when position or timing measurements of high precision are involved.

D. Effect of Impurities or Dopants

1. INTRINSIC SEMICONDUCTORS

In a completely pure semiconductor, all the electrons in the conduction band and all the holes in the valence band would be caused by thermal excitation (in the absence of ionizing radiation). Because under these conditions each electron must leave a hole behind, the number of electrons in the conduction band must exactly equal the number of holes in the valence band.

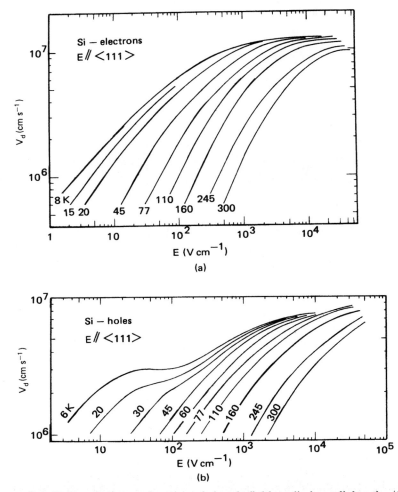

Figure 11.2 Drift velocity as a function of electric field applied parallel to the ⟨111⟩ crystallographic direction. Absolute temperature is the parameter for the different curves. (*a*) Electrons in silicon; (*b*) holes in silicon; (*c*) electrons in germanium; (*d*) holes in germanium. (From Ottaviani et al.[15])

Such material is called an *intrinsic* semiconductor. Its properties can be described theoretically, but in practice it is virtually impossible to achieve. The electrical properties of real materials tend to be dominated by the very small levels of residual impurities; this is true even for silicon and germanium, which are the semiconductors available in the highest practical purities.

In the discussions that follow, we let n represent the concentration (number per unit volume) of electrons in the conduction band. Also, p represents the concentration of holes in the valence band. In the intrinsic material (subscript i), the equilibrium established by the thermal excitation of electrons from the valence to conduction band and their subsequent recombination leads to equal numbers of electrons and holes, or

$$n_i = p_i \tag{11.7}$$

The quantities n_i and p_i are known as the intrinsic carrier densities. From Eq. (11.1), it is clear that these densities will be lowest for materials with large bandgap energy and when the material is used at low temperature. Intrinsic hole or electron densities at room temperature are 1.5×10^{10} cm^{-3} in silicon, and 2.4×10^{13} cm^{-3} in germanium.

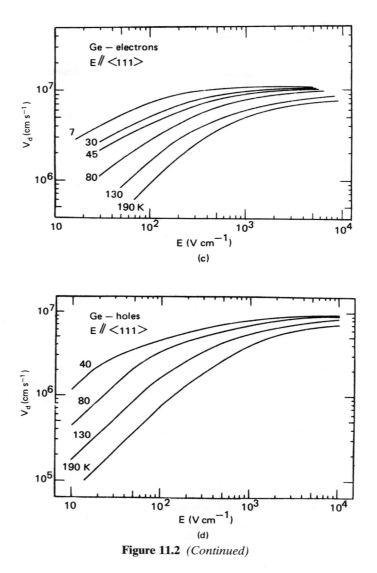

Figure 11.2 *(Continued)*

In a metallic conductor, only the flow of negatively charged electrons contributes to its electrical conductivity. In contrast, the flow of both negatively charged electrons and positively charged holes contribute to the conductivity of an intrinsic semiconductor. The value of the conductivity (or its inverse, the resistivity ρ) is determined by the intrinsic carrier density n_i and the mobilities μ_h and μ_e of the holes and electrons. If we have a slab of a semiconductor with thickness t and surface area A, the current I that will flow when a voltage V is applied across the thickness is

$$I = \frac{AV}{\rho\, t} \qquad \text{or} \qquad \rho = \frac{AV}{I\, t}$$

The current is made up of two separate components: the current due to the flow of holes I_h and that due to the flow of electrons I_e. Note that, although the two types of charge carriers move in opposite directions, the separate currents are additive because of the opposite charges of holes and electrons. Thus an external measurement of the current by itself cannot distinguish between the flow of holes or electrons. The total observed current will be their sum, with each term given by the product of the area, intrinsic carrier density, electronic charge e, and the drift velocity of the charge carrier. Thus

$$I = I_e + I_h = An_i e(v_e + v_h)$$

From Eqs. 11.2 and 11.3

$$I = An_i e \mathscr{E}(\mu_e + \mu_h) = An_i e \frac{V}{t}(\mu_e + \mu_h)$$

Combining

$$\rho = \frac{1}{en_i(\mu_e + \mu_h)} \tag{11.8}$$

Inserting numerical values for intrinsic silicon at room temperature:

$$\rho = \frac{1}{(1.6 \times 10^{-19}C)(1.5 \times 10^{10}/cm^3)(1350 + 480)\ cm^2/V \cdot s}$$

$$\rho = 2.3 \times 10^5 \frac{V \cdot s \cdot cm}{C} = 230{,}000\ \Omega \cdot cm$$

Presently available silicon material of the highest purity falls short of achieving this resistivity value because of the effect (discussed in the following sections) of residual impurities.

2. n-TYPE SEMICONDUCTORS

To illustrate the effect of doping on semiconductor properties, we use crystalline silicon as an example. Germanium and other semiconductor materials behave in a similar way. Silicon is tetravalent and in the normal crystalline structure forms covalent bonds with the four nearest silicon atoms. A sketch of this situation is shown in Fig. 11.3a, where each of the dashes represents a normal valence electron involved in a covalent bond. Thermal excitation in the intrinsic material consists of breaking loose one of these covalent electrons, leaving behind an unsaturated bond or hole.

We now consider the effect of the small concentration of impurity that may be present in the semiconductor either as a residual amount after the best purification processes, or as a small amount intentionally added to the material to tailor its properties. We first assume that the impurity is pentavalent or is found in group V of the periodic table. When present

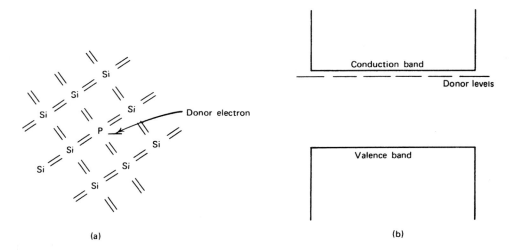

(a)   (b)

Figure 11.3 (a) Representation of a donor impurity (phosphorus) occupying a substitutional site in a silicon crystal. (b) Corresponding donor levels created in the silicon bandgap.

in small concentrations (of the order of a few parts per million or less) the impurity atom will occupy a substitutional site within the lattice, taking the place of a normal silicon atom. Because there are five valence electrons surrounding the impurity atom, there is one left over after all covalent bonds have been formed. This extra electron is somewhat of an orphan and remains only very lightly bound to the original impurity site. It therefore takes very little energy to dislodge it to form a conduction electron without a corresponding hole. Impurities of this type are referred to as *donor impurities* because they readily contribute electrons to the conduction band. Because they are not part of the regular lattice, the extra electrons associated with donor impurities can occupy a position within the normally forbidden gap. These very loosely bound electrons will have an energy near the top of the gap as shown in Fig. 11.3*b*. The energy spacing between these donor levels and the bottom of the conduction band is sufficiently small so that the probability of thermal excitation given by Eq. (11.1) is high enough to ensure that a large fraction of all the donor impurities are ionized. In nearly all cases, the concentration of impurity N_D is large compared with the concentration of electrons expected in the conduction band for the intrinsic material. Therefore, the number of conduction electrons becomes completely dominated by the contribution from the donor impurities, and we can write

$$n \cong N_D \tag{11.9}$$

The added concentration of electrons in the conduction band compared with the intrinsic value increases the rate of recombination, shifting the equilibrium between electrons and holes. As a result, the equilibrium concentration of holes is decreased by an amount such that the equilibrium constant given by the product of n and p is the same as for the intrinsic material:

$$np = n_i p_i \tag{11.10}$$

For example, in room-temperature silicon, the intrinsic carrier densities are about 10^{10} cm^{-3}. If a donor impurity is present at a concentration of 10^{17} atoms/cm^3 (about 2 parts per million), the density of conduction electrons n will be 10^{17} cm^{-3} and the concentration of holes p will be 10^3 cm^{-3}. Because the *total* number of charge carriers of both types is now much greater (10^{17} cm^{-3} versus 2×10^{10} cm^{-3}), the electrical conductivity of a doped semiconductor is always much larger than that of the corresponding pure material.

Even though conduction electrons now greatly outnumber the holes, charge neutrality is maintained because of the presence of ionized donor impurities. These sites represent net positive charges that exactly balance the excess electron charges. They are not, however, to be confused with holes because the ionized donors are fixed in the lattice and cannot migrate.

The net effect in *n*-type material is therefore to create a situation in which the number of conduction electrons is much greater and the number of holes much smaller than in the pure material. The electrical conductivity is then determined almost exclusively by the flow of electrons, and holes play a very small role. In this case, the electrons are called the *majority carriers* and holes the *minority carriers*.

The resistivity of doped material can be calculated from the dopant concentration and the mobility of the majority carrier, since only the flow of majority carriers is important in measured currents. As an example, assume we have silicon with a donor density of $10^{13}/$cm^3, which will also be the concentration of conduction electrons. Then the resistivity will be

$$\rho = \frac{1}{eN_D\mu_e} \tag{11.11}$$

$$\rho = \frac{1}{(1.6 \times 10^{-19}C)(10^{13}/\text{cm}^3)(1350 \text{ cm}^2/V \cdot s)}$$

$$\rho = 463 \ \Omega \cdot \text{cm}$$

3. p-TYPE SEMICONDUCTORS

The addition of a trivalent impurity such as an element from group III of the periodic table to a silicon lattice results in a situation sketched in Fig. 11.4a. If the impurity occupies a substitutional site, it has one fewer valence electron than the surrounding silicon atoms and therefore one covalent bond is left unsaturated. This vacancy represents a hole similar to that left behind when a normal valence electron is excited to the conduction band, but its energy characteristics are slightly different. If an electron is captured to fill this vacancy, it participates in a covalent bond that is not identical to the bulk of the crystal because one of the two participating atoms is a trivalent impurity. An electron filling this hole, although still bound to a specific location, is slightly less firmly attached than a typical valence electron. Therefore, these *acceptor impurities* also create electron sites within the normally forbidden energy gap. In this case, the acceptor levels lie near the bottom of the gap because their properties are quite close to sites occupied by normal valence electrons.

Normal thermal excitation in the crystal ensures that there will always be some electrons available to fill the vacancies created by the acceptor impurities or to occupy the acceptor sites shown in Fig. 11.4b. Because the energy difference between typical acceptor sites and the top of the valence band is small, a large fraction of all the acceptor sites are filled by such thermally excited electrons. These electrons come from other normal covalent bonds throughout the crystal and therefore leave holes behind in the valence band. To a good approximation, an extra hole is created in the valence band for every acceptor impurity that is added. If the concentration N_A of acceptor impurities is made to be large compared with the intrinsic concentration of holes p_i, then the number of holes is completely dominated by the concentration of acceptors, or

$$p \cong N_A \qquad (11.12)$$

The increased availability of holes enhances the recombination probability between conduction electrons and holes and therefore decreases the equilibrium number of conduction electrons. Again, the same equilibrium constant discussed earlier holds, and $np = n_i p_i$. In p-type material, holes are the majority carrier and dominate the electrical conductivity. The filled acceptor sites represent fixed negative charges that balance the positive charge of the majority holes.

The equilibrium that is at work between electrons and holes is illustrated by the nomogram in Fig. 11.5. Two logarithmic scales are shown, on the left the concentration of con-

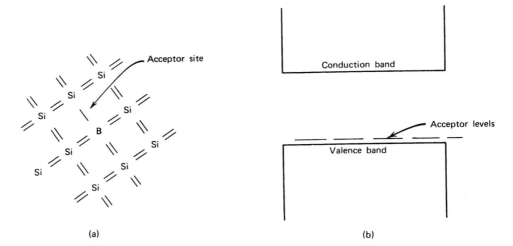

(a) (b)

Figure 11.4 (a) Representation of an acceptor impurity (boron) occupying a substitutional site in a silicon crystal. (b) Corresponding acceptor levels created in the silicon bandgap.

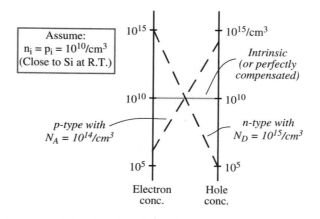

Figure 11.5 Nomogram showing the relationship between electron and hole concentrations in a semiconductor. Lines connecting points on the two logarithmic scales always pass through the center of the diagram for any type or degree of doping.

duction electrons, and on the right the concentration of holes. In intrinsic or perfectly compensated (see following section) semiconductors, the concentrations are equal and have a value given by the intrinsic carrier density. In silicon at room temperature, that value is $1.5 \times 10^{10}/\text{cm}^3$. Intrinsic (or perfectly compensated) material can thus be represented as a horizontal line connecting the points on the scales at about $10^{10}/\text{cm}^3$. Doping the material will tip the equilibrium so that one carrier dominates, but the product of the concentrations of electrons and holes must still be the same as shown in Eq. (11.10). Since the scales are logarithmic on the nomogram, the two concentrations will be linked by a line that passes through the center of the previous line. The same statement can be made for any level of doping, whether with donors or acceptors, so this point at the center of the nomogram acts as a "pivot point" that is common to all lines. Material that is n-type will result in lines with negative slope, while p-type material corresponds to lines with positive slope.

Figure 11.5 also illustrates that the total concentration of charge carriers (electrons plus holes) is at a minimum for the intrinsic material. This condition corresponds to a minimum in electrical conductivity if the mobilities of the two carriers are about the same. Tipping the equilibrium in either direction due to the presence of either donors or acceptors will raise the majority carrier concentration by an absolute amount that is always greater than the amount by which the minority carrier concentration is decreased. The effect is shown in Fig. 11.6, where the minimum conductivity occurs for the pure or intrinsic material, and either excess donors or acceptors result in a higher conductivity. Thus one measure of the impurity level of semiconductor materials is the electrical conductivity, or

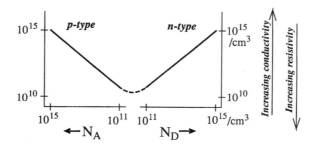

Figure 11.6 Plot using logarithmic scales of the conductivity of a semiconductor as a function of the net concentration of acceptors (N_A) or donors (N_D).

its inverse, the resistivity. For the intrinsic material in which all the charge carriers are produced through thermal excitation, the resistivity value can be calculated from Eq. (11.8) using the intrinsic carrier density and mobilities. The corresponding values for silicon and germanium are listed in Table 11.1. In practice, these theoretical values of resistivity are never observed because of the unavoidable residual impurities. Using the most advanced purification methods available at this writing, silicon resistivity of about 50,000 Ω-cm can be achieved, compared with a theoretical value of over 200,000 Ω-cm.

4. COMPENSATED MATERIAL

If donor and acceptor impurities are present in a semiconductor in equal concentration, the material is said to be *compensated*. Such material has some of the properties of an intrinsic semiconductor because electrons contributed by donor impurities are removed to some extent by their capture at the site of acceptor impurities. Despite the potential confusion with purified intrinsic material, compensated regions in semiconductors are commonly given the designation *i* because of their near intrinsic properties.

In practice, it is impossible to achieve exact compensation at the time of fabrication of the doped material because any small imbalance in the acceptor or donor concentration quickly leads to *n*-type or *p*-type behavior. At present, the only practical means for achieving compensation over large volumes in silicon or germanium is through the lithium ion drifting process after the crystal has been fabricated. This procedure is discussed in Chapter 13.

5. HEAVILY DOPED MATERIAL

Thin layers of semiconductor material that have an unusually high concentration of impurity are often given a special notation. Thus, n^+ and p^+ designate heavily doped *n*- and *p*-type layers that, as a result, have very high conductivity. These layers are often used in making electrical contact with semiconductor devices, because the very low minority carrier density allows their application as "blocking" contacts described later in this chapter.

E. Trapping and Recombination

Once electrons and holes are formed in a semiconductor, they will tend to migrate either spontaneously or under the influence of an applied electric field until they are either collected at an electrode or recombination takes place. There are theoretical predictions[16] that the average lifetime of charge carriers before recombination in perfectly pure semiconductors could be as large as a second. In practice, lifetimes at least three or four orders of magnitude smaller than a second are actually observed that are dominated entirely by the very low level of impurities remaining in the material. Some of these impurities, such as gold, zinc, cadmium, or other metallic atoms occupying substitutional lattice positions, introduce energy levels near the middle of the forbidden gap. They are therefore classified as "deep impurities" (as opposed to acceptor or donor impurities that, because the corresponding energy levels lie near the edges of the forbidden band, are called *shallow impurities*). These deep impurities can act as *traps* for charge carriers in the sense that if a hole or electron is captured, it will be immobilized for a relatively long period of time. Although the trapping center ultimately may release the carrier back to the band from which it came, the time delay is often sufficiently long to prevent that carrier from contributing to the measured pulse.

Other types of deep impurities can act as *recombination centers*. These impurities are capable of capturing both majority and minority carriers, causing them to annihilate. An impurity level near the center of the forbidden gap might, for example, first capture a conduction electron. At a slightly later time, a hole from the valence band might also be captured, with the electron then filling the hole. The impurity site is thus returned to its original state and is capable of causing another recombination event. In most crystals, recombination through such centers is far more common than direct recombination of electrons and holes across the full bandgap.

Both trapping and recombination contribute to the loss of charge carriers and tend to reduce their average lifetime in the crystal. For the material to serve as a good radiation detector, a large fraction (preferably 100%) of all the carriers created by the passage of the incident radiation should be collected. This condition will hold provided the collection time for the carriers is short compared with their mean lifetime. Collection times of the order of 10^{-7} to 10^{-8} s are fairly common, so that carrier lifetimes of the order of 10^{-5} s or longer are usually sufficient.

Another widely quoted specification is the *trapping length* within the material. This quantity is simply the mean distance traveled by a carrier before trapping or recombination and is given by the product of the mean lifetime and the average drift velocity. In order to have an acceptable detector, the trapping length should be long compared with the physical dimensions over which the charge must be collected.

In addition to impurities, structural defects within the crystal lattice can also lead to trapping and charge carrier loss. These imperfections include point defects such as vacancies or interstitials that tend to behave as acceptors or donors, respectively. Carrier loss may also occur at line defects or dislocations that may be produced in stressed crystals. A dislocation represents the slippage of one crystal plane with respect to another, and its intersection with the surface of the crystal leads to a pit upon chemical etching. The density of these etched pits is often quoted as a measure of the crystalline perfection of a semiconductor sample.

II. THE ACTION OF IONIZING RADIATION IN SEMICONDUCTORS

A. The Ionization Energy

When a charged particle passes through a semiconductor with the band structure shown in Fig. 11.1, the overall significant effect is the production of many electron–hole pairs along the track of the particle. The production process may be either direct or indirect, in that the particle produces high-energy electrons (or *delta rays*) that subsequently lose their energy in producing more electron–hole pairs. Regardless of the detailed mechanisms involved, the quantity of practical interest for detector applications is the average energy expended by the primary charged particle to produce one electron–hole pair. This quantity, often loosely called the *ionization energy* and given the symbol ϵ, is experimentally observed to be largely independent of both the energy and type of the incident radiation. This important simplification allows interpretation of the number of electron–hole pairs produced in terms of the incident energy of the radiation, provided the particle is fully stopped within the active volume of the detector.

When radiation interacts in a semiconductor, the energy deposition always leads to the creation of equal numbers of holes and electrons. This statement holds regardless of whether the host semiconductor is pure or intrinsic, or doped as *p*-type or *n*-type. Just as equal numbers of free electrons and positive ions are created in a gas, every conduction electron produced in a semiconductor must also create a hole in the valence band, leading to an exact balance in the initial number of created charges. It should also be emphasized that the doping levels typical in *p*- or *n*-type semiconductors are so low that these atoms play no significant role in determining the nature of the radiation interactions in the material. Thus *p*-type or *n*-type silicon of equal thickness will present identical interaction probabilities for gamma rays, and the range of charge particles in either type will also be the same.

The dominant advantage of semiconductor detectors lies in the smallness of the ionization energy. The value of ϵ for either silicon or germanium is about 3 eV (see Table 11.1), compared with about 30 eV required to create an ion pair in typical gas-filled detectors. Thus, the number of charge carriers is 10 times greater for the semiconductor case, for a given energy deposited in the detector. The increased number of charge carriers has two beneficial effects on the attainable energy resolution. The statistical fluctuation in the number of carriers per pulse becomes a smaller fraction of the total as the number is increased. This factor often is predominant in determining the limiting energy resolution of a detector for medium to high radiation energy. At low energies, the resolution may be limited by electronic noise in the pre-amplifier, and the greater amount of charge per pulse leads to a better signal/noise ratio.

More detailed examination shows that ϵ depends on the nature of the incident radiation. Most detector calibrations are carried out using alpha particles, and the values for ϵ shown in Table 11.1 are based on this mode of excitation. All experimental values obtained using other light ions or fast electrons seem to be fairly close,[17–19] but differences as large as 2.2% have been reported[20] between proton and alpha particle excitation in silicon. These observed differences point up the need to carry out detector calibration using a radiation type that is identical to that involved in the measurement itself if precise energy values are required.

A much larger difference is measured for ϵ when heavy ions or fission fragments are involved. The value of ϵ is significantly higher than for alpha particle excitation, leading to a lower than anticipated number of charge carriers. The physical origins of this *pulse height defect* are discussed later in this chapter.

The ionization energy is also temperature dependent. For the most significant detector materials, the value of ϵ increases with decreasing temperature. As shown in Table 11.1, ϵ in silicon is about 3% greater at liquid nitrogen temperature compared with room temperature.[18]

There is also evidence[21–23] that the ionization energy can be dependent on the energy of the radiation, especially in the soft X-ray energy range. It appears that its value increases with decreasing X-ray energy below about 1 keV, and that the Fano factor (described below) also increases significantly with decreasing energy in this range.

B. The Fano Factor

In addition to the mean number, the fluctuation or variance in the number of charge carriers is also of primary interest because of the close connection of this parameter with energy resolution of the detector. As in gas counters, the observed statistical fluctuations in semiconductors are smaller than expected if the formation of the charge carriers were a Poisson process. The Poisson model would hold if all events along the track of the ionizing particle were independent and would predict that the variance in the total number of electron–hole pairs should be equal to the total number produced, or E/ϵ. The Fano factor F is introduced as an adjustment factor to relate the observed variance to the Poisson predicted variance:

$$F \equiv \frac{\text{observed statistical variance}}{E/\epsilon} \qquad (11.13)$$

For good energy resolution, one would like the Fano factor to be as small as possible. Although a complete understanding of all the factors that lead to a nonunity value for F does not yet exist, rather sophisticated models have been developed[24] that at least qualitatively account for experimental observations. Some numerical values for silicon and germanium are given in Table 11.1.

There is considerable variation in reported experimental values, particularly for silicon. The Fano factor is usually measured by observing the energy resolution from a given detector under conditions in which all other factors that can broaden the full-energy peak (such as electronic noise or drift) can be estimated and taken into account. The assumption is then made that the residual width can be attributed to statistical effects only. If nonstatistical residual factors remain, however, the Fano factor will appear to be larger than it actually is. This may explain the historical trend toward lower values as measurement procedures are refined. It has also been postulated[13] that the value of the Fano factor may depend on the nature of the particle that deposits the energy. Some measurements[25] suggest that its value in silicon may also vary significantly with radiation energy, especially in the energy range typical of X-ray photons.

III. SEMICONDUCTORS AS RADIATION DETECTORS

A. Pulse Formation

When a particle deposits energy in a semiconductor detector, equal numbers of conduction electrons and holes are formed within a few picoseconds along the particle track. The detector configurations that are discussed in the following sections all ensure that an electric field is present throughout the active volume, so that both charge carriers feel electrostatic forces that cause them to drift in opposite directions. The motion of either the electrons or holes constitutes a current that will persist until those carriers are collected at the boundaries of the active volume. With the simplifying assumption that all charge carriers are formed at a single point, the resulting currents can be represented by the plot at the top of Fig. 11.7. Since the charge collection times are not likely to be the same because of differences in drift distance and carrier mobilities, one of the two currents will persist for a longer time than the other. When these currents are integrated on a measuring circuit with long time constant, the measured induced charge has the time characteristics shown in the lower plot in Fig. 11.7. This time profile will also be that of the rise of the pulse produced by a conventional preamplifier used to process the pulses from the detector. This pulse profile is similar to that derived in Chapter 5 for ion chambers (see Fig. 5.16), with the exception of the time scale. In gases, the collection time for the positive charges (ions) is greater by orders of magnitude than that for the negative (free electron) charges, so the ion motion adds a very long component to the pulse rise and, as a practical matter, does not contribute to the output pulse. In silicon or germanium semiconductor detectors, the hole mobility is within a factor of about 2 or 3 of the electron mobility, so the collection times are much closer to being equivalent. As a result, while pulse-type ion chambers almost never include the motion of the ions in the output pulse, standard silicon and germanium semiconductor detectors rely on complete integration of the currents due to both the electrons and holes. Both carrier types must therefore be completely collected for the resulting pulse to be a faithful measure of the energy deposited by the particle.

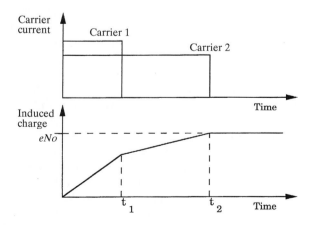

Figure 11.7 The upper plot shows an idealized representation of the electron and hole currents flowing in a semiconductor following the creation of N_0 electron–hole pairs. In the lower plot, t_1 represents the collection time for the carrier type (either electrons or holes) that is collected first, and t_2 is the collection time for the other carrier. If both are fully collected, a charge of eN_0 is induced to form the signal, where e is the electronic charge.

B. Electrical Contacts

In order to construct a practical radiation detector, some means must be provided to collect the electrical charges created by the radiation at either boundary of the semiconductor material. An *ohmic* contact is a nonrectifying electrode through which charges of either sign can flow freely. If two ohmic contacts are fitted on opposite faces of a slab of semiconductor and connected to a detection circuit, the equilibrium charge carrier concentrations in the semiconductor will be maintained. If an electron or hole is collected at one electrode, the same carrier species is injected at the opposite electrode to maintain the equilibrium concentrations in the semiconductor.

The steady-state leakage currents that are observed using ohmic contacts are too high, even with the highest resistivity material available (see following section), to permit their general application to silicon or germanium detectors. Instead, *noninjecting* or *blocking* electrodes are universally employed to reduce the magnitude of the current through the bulk of the semiconductor. If blocking electrodes are used, charge carriers initially removed by the application of an electric field are not replaced at the opposite electrode, and their overall concentration within the semiconductor will drop after application of an electric field. The leakage current can thus be reduced to a sufficiently low value to allow the detection of the added current pulse created by the electron–hole pairs produced along the track of an ionizing particle.

The most appropriate type of blocking contacts are the two sides of a *p-n* semiconductor junction. It is very difficult to inject electrons from the *p* side of this junction because holes are the majority carrier and free electrons are relatively scarce. At the opposite side, electrons are the majority carrier and holes cannot readily be injected. In this chapter, we discuss detectors that are created by placing the *p*- and *n*-type materials in direct contact, forming a *p-n* junction. In Chapters 12 and 13, detectors in which the *p* and *n* regions are separated by an intrinsic or compensated region (the *i* region) are described.

C. Leakage Current

In order to create an electric field large enough to achieve an efficient collection of the charge carriers from any semiconductor detector, an applied voltage of typically hundreds or thousands of volts must be imposed across the active volume. Even in the absence of ionizing radiation, all detectors will show some finite conductivity and therefore a steady-state *leakage current* will be observed. Random fluctuations that inevitably occur in the leakage current will tend to obscure the small signal current that momentarily flows following an ionizing event and will represent a significant source of noise in many situations. Methods of reducing the leakage current are therefore an important consideration in the design of semiconductor detectors.

The resistivity of the highest purity silicon currently available is about 50,000 Ω-cm. If a 1-mm thick slab of this silicon were cut with 1-cm^2 surface area and fitted with ohmic contacts, the electrical resistance between faces would be 5000 Ω. An applied voltage of 500 V would therefore cause a leakage current through the silicon of 0.1 A. In contrast, the peak current generated by a pulse of 10^5 radiation-induced charge carriers would only be about 10^{-6} A. It is therefore essential to reduce this bulk leakage current greatly through the use of blocking contacts. In critical applications, the leakage current must not exceed about 10^{-9} A to avoid significant resolution degradation.

At these levels, leakage across the surface of the semiconductor can often become more significant than bulk leakage. Great care is taken in the fabrication of semiconductor detectors to avoid contamination of the surfaces, which could create leakage paths. Some configurations may also use grooves in the surface or guard rings to help suppress surface leakage (see Chapter 5).

D. The Semiconductor Junction

1. BASIC JUNCTION PROPERTIES

The radiation detectors described in this chapter are based on the favorable properties that are created near the junction between n- and p-type semiconductor materials. Charge carriers are able to migrate across the junction if the regions are brought together in good thermodynamic contact. Simply pressing together two pieces of the material will not suffice because gaps will inevitably be left that will be large compared with the interatomic lattice spacing. In practice, the junction is therefore normally formed in a single crystal by causing a change in the impurity content from one side of the junction to the other.

As an illustration, assume that the process begins with a p-type crystal that has been doped with a uniform concentration of acceptor impurity. In the concentration profile at the top of Fig. 11.8, this original acceptor concentration N_A is shown as a horizontal line. We now assume that the surface of the crystal on the left is exposed to a vapor of an n-type impurity that diffuses some distance into the crystal. The resulting donor impurity profile is labeled N_D on the figure and falls off as a function of distance from the surface. Near the surface, the donor impurities can be made to outnumber the acceptors, converting the left portion of the crystal to n-type material.

The approximate variation of equilibrium charge carrier concentration is also plotted at the top of Fig. 11.8 and labeled as p (hole concentration) and n (conduction electron concentration). These profiles are subsequently altered in the vicinity of the p-n junction because of the effects of charge carrier diffusion. In the n-type region at the left, the density of conduction electrons is much higher than in p-type. The junction between the two regions therefore represents a discontinuity in the conduction electron density. Wherever such a sharp gradient exists for any carrier that is free to migrate, a net diffusion from regions of high concentration to those of low concentration must take place. Thus, there will be some net diffusion of conduction electrons into the p-type material, where they will quickly combine with holes. In effect, this annihilation represents the capture of the conduction electron by one of the vacancies existing in the covalent bonds in the p-type material. The diffusion of conduction electrons out of the n-type material leaves behind immobile positive charges in the form of ionized donor impurities. A similar and symmetric argument leads to the conclusion that holes (the majority in the p-type material) must also diffuse across the junction because they also see an abrupt density gradient. Each hole that is removed from the p side of the junction leaves behind an acceptor site that has picked up an extra electron and therefore represents a fixed and immobile negative charge. The combined effect is to build up a net negative space charge on the p side and a positive space charge on the n side of the junction.

The accumulated space charge creates an electric field that diminishes the tendency for further diffusion. At equilibrium, the field is just adequate to prevent additional net diffusion across the junction, and a steady-state charge distribution is therefore established.

The region over which the charge imbalance exists is called the *depletion region* and extends into both the p and n sides of the junction. If the concentrations of donors on the n side and acceptors on the p side are equal, the diffusion conditions are approximately the same for both holes and electrons, and the depletion region extends equal distances into both sides. Usually, however, there is a marked difference in the doping levels on one side of the junction compared with the other. For example, if the donor concentration in the n-type material is higher than that of acceptor atoms in the p-type, the electrons diffusing across the junction will tend to travel a greater distance into the p-type material before all have recombined with holes. In this case, the depletion region would extend farther into the p side.

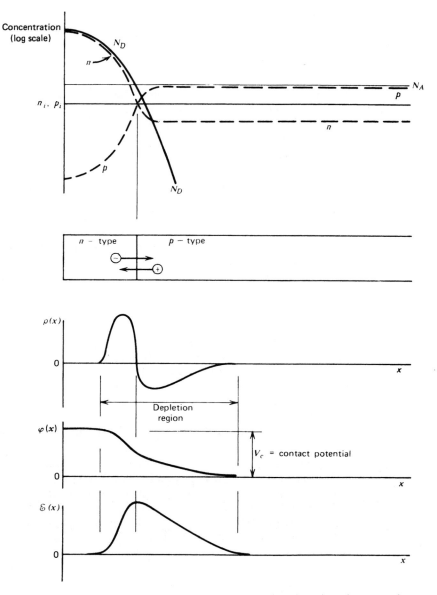

Figure 11.8 The assumed concentration profiles for the *n-p* junction shown at the top are explained in the text. The effects of carrier diffusion across the junction give rise to the illustrated profiles for space charge $\rho(x)$, electric potential $\varphi(x)$, and electric field $\mathscr{E}(x)$.

The buildup of net charge within the region of the junction leads to the establishment of an electric potential difference across the junction. The value of the potential φ at any point can be found by solution of Poisson's equation

$$\nabla^2\varphi = -\frac{\rho}{\epsilon} \tag{11.14}$$

where ϵ is the dielectric constant of the medium, and ρ is the net charge density. In one dimension, Eq. (11.14) takes the form

$$\frac{d^2\varphi}{dx^2} = -\frac{\rho(x)}{\epsilon} \tag{11.15}$$

so that the shape of the potential across the junction can be obtained by twice integrating the charge distribution profile $\rho(x)$. Graphical examples are shown in Fig. 11.8. At equilibrium, the potential difference across the junction (called the *contact potential*) amounts to nearly the full bandgap value of the semiconductor material. The direction of this potential difference is such that it opposes the further diffusion of electrons from left to right and holes from right to left in Fig. 11.8.

Where a difference in electrical potential exists, there must also be an electric field $\mathscr{E}$. Its magnitude is found by taking the gradient of the potential

$$\mathscr{E} = -\text{grad } \varphi \tag{11.16}$$

which, in one dimension, is simply

$$\mathscr{E}(x) = -\frac{d\varphi}{dx} \tag{11.17}$$

The electric field will extend over the width of the depletion region, in which charge imbalance is significant and the potential has some gradient. Its variation is also sketched in Fig. 11.8.

The depletion region exhibits some very attractive properties as a medium for the detection of radiation. The electric field that exists causes any electrons created in or near the junction to be swept back toward the *n*-type material, and any holes are similarly swept toward the *p*-type side. The region is thus "depleted" in that the concentration of holes and electrons is greatly suppressed. The only significant charges remaining in the depletion region are the immobile ionized donor sites and filled acceptor sites. Because these latter charges do not contribute to conductivity, the depletion region exhibits a very high resistivity compared with the *n*- and *p*-type materials on either side of the junction. Electron–hole pairs that are created within the depletion region by the passage of radiation will be swept out of the depletion region by the electric field, and their motion constitutes a basic electrical signal.

The thermal generation of charge carriers will continue to take place in the depletion region, contributing a component sometimes called the *generation current* to the observed leakage current. These charges are swept away typically within a few nanoseconds, however, a time that is many orders of magnitude shorter than the time required to establish thermal equilibrium. Thus, the steady-state concentration of carriers is strongly reduced in the depletion region because the removal of charges is a much faster process than their creation. The small concentration of carriers created by an ionizing particle is therefore easily detected above this highly suppressed, thermally generated concentration.

2. REVERSE BIASING

Thus far, we have discussed a semiconductor diode junction to which no external voltage is applied. Such an unbiased junction will function as a detector, but only with very poor performance. The contact potential of about 1 V that is formed spontaneously across the junction is inadequate to generate a large enough electric field to make the charge carriers move very rapidly. Therefore, charges can be readily lost as a result of trapping and recombination, and incomplete charge collection often results. The thickness of the depletion region is quite small, and the capacitance of an unbiased junction is high. Therefore, the noise properties of an unbiased junction connected to the input stage of a preamplifier are quite poor. For these reasons, unbiased junctions are not used as pulse mode radiation detectors, but instead, an external voltage is applied in the direction to cause the semiconductor diode to be reverse biased.

The *p-n* junction is most familiar in its role as a diode. The properties of the junction are such that it will readily conduct current when voltage is applied in the "forward" direction, but it will conduct very little current when biased in the "reverse" direction. In the configuration of Fig. 11.8, first assume that a positive voltage is applied to the p side of the junction with respect to the n side. The potential will tend to attract conduction electrons from the n side as well as holes from the p side across the junction. Because, in both cases, these are the majority carriers, conductivity through the junction is greatly enhanced. The contact potential shown in Fig. 11.8 is reduced by the amount of the bias voltage that is applied, which tends to lessen the potential difference seen by an electron from one side of the junction to the other. This is the direction of forward biasing, and only small values of the forward bias voltage are needed to cause the junction to conduct large currents.

If the situation is reversed, and the p side of the junction is made negative with respect to the n side, the junction is reverse biased. Now the natural potential difference from one side of the junction to the other is enhanced, as shown in Fig. 11.9c. Under these

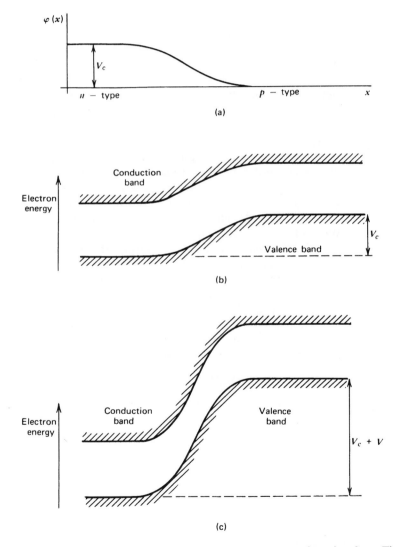

Figure 11.9 (a) The variation of electric potential across an n-p junction from Fig. 11.8. (b) The resulting variation in electron energy bands across the junction. The curvature is reversed because an increase in electron energy corresponds to a decrease in conventional electric potential $\varphi(x)$ defined for a positive charge. (c) The added displacement of the bands caused by application of a reverse bias V across the junction.

circumstances, it is the minority carriers (holes on the n side and electrons on the p side) that are attracted across the junction and, because their concentration is relatively low, the reverse current across the diode is quite small. Therefore, the p-n junction serves as a rectifying element, allowing relatively free flow of current in one direction while presenting a large resistance to its flow in the opposite direction. If the reverse bias is made very large, a sudden breakdown in the diode will occur and the reverse current will abruptly increase, often with destructive effects.

3. PROPERTIES OF THE REVERSE BIAS JUNCTION

When a reverse bias is applied to the junction, virtually all the applied voltage will appear across the depletion region, because its resistivity is much higher than that of the normal n- and p-type material. Because the effect of the reverse bias is to accentuate the potential difference across the junction, Poisson's equation [Eq. (11.14)] demands that space charge must also increase and extend a greater distance on either side of the junction. Thus, the thickness of the depletion region is also increased, extending the volume over which radiation-produced charge carriers will be collected. Practical detectors are almost always operated with a bias voltage that is very large compared with the contact potential, so that the applied voltage completely dominates the magnitude of the potential difference across the junction.

In the analysis that follows, we first assume that the semiconductor wafer in which the junction is formed is sufficiently thick so that the depletion region does not reach either surface and is contained within the interior volume of the wafer. This condition holds for *partially depleted* detectors in which some portion of the wafer thickness remains undepleted. Many semiconductor detectors are operated with sufficient reverse bias voltage so that the depletion region extends through the full wafer thickness, creating a *fully depleted* (or *totally depleted*) detector. These configurations share many of the properties derived below, except that the depletion region is obviously limited by the physical thickness of the wafer.

Some properties of the reverse bias junction can be derived if we represent the charge distribution sketched in Fig. 11.8 by the idealized distribution shown below:

$$\rho(x) = \begin{cases} eN_D & (-a < x \leq 0) \\ -eN_A & (0 < x \leq b) \end{cases}$$

Here the electron diffusion is assumed to result in a uniform positive space charge (the ionized donor sites) over the region $-a < x \leq 0$ on the n side of the junction. A corresponding negative space charge (the filled acceptor sites) resulting from hole diffusion is assumed to extend over the region $0 < x \leq b$ on the p side. Because the net charge must be zero, $N_D a = N_A b$.

Equation (11.15) applied to this case takes the form

$$\frac{d^2\varphi}{dx^2} = \begin{cases} -\dfrac{eN_D}{\epsilon} & (-a < x \leq 0) \\ +\dfrac{eN_A}{\epsilon} & (0 < x \leq b) \end{cases}$$

We now carry out an integration and apply the boundary conditions that the electric field $\mathcal{E} = -d\varphi/dx$ must vanish at both edges of the charge distribution:

$$\frac{d\varphi}{dx}(-a) = 0 \quad \text{and} \quad \frac{d\varphi}{dx}(b) = 0$$

The result is then

$$\frac{d\varphi}{dx} = \begin{cases} -\dfrac{eN_D}{\epsilon}(x + a) & (-a < x \le 0) \\[2mm] +\dfrac{eN_A}{\epsilon}(x - b) & (0 < x \le b) \end{cases}$$

The corresponding shape of the electric field $\mathcal{E} = -d\varphi/dx$ is sketched below:

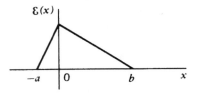

Another integration will now yield the electric potential $\varphi(x)$. The difference in potential from the n side to the p side of the junction, if we neglect the relatively small contact potential, is just the value of the applied reverse bias V. We can therefore apply the boundary conditions

$$\varphi(-a) = V \quad \text{and} \quad \varphi(b) = 0$$

The solution then takes the form

$$\varphi(x) = \begin{cases} -\dfrac{eN_D}{2\epsilon}(x + a)^2 + V & (-a < x \le 0) \\[2mm] +\dfrac{eN_A}{2\epsilon}(x - b)^2 & (0 < x \le b) \end{cases}$$

Since the solutions for either side of the junction must match at $x = 0$, we can write

$$V - \frac{eN_D a^2}{2\epsilon} = \frac{eN_A b^2}{2\epsilon}$$

or

$$N_A b^2 + N_D a^2 = \frac{2\epsilon V}{e}$$

Now since $N_D a = N_A b$, the expression above can be rewritten:

$$(a + b)b = \frac{2\epsilon V}{eN_A}$$

The total width of the depletion region d is the entire distance over which the space charge extends, or $d = a + b$.

For purposes of the present example, we have assumed that the n-side doping level is much higher than on the p side, so that $N_D \gg N_A$. Because $N_D a = N_A b$, it follows that $b \gg a$, and therefore the space charge extends much farther into the p side than the n side. Then $d \cong b$ and we can write

$$d \cong \left(\frac{2\epsilon V}{e N_A} \right)^{1/2}$$

If we had started from the opposite assumption that the p-side doping level was predominant, a similar result would be obtained except that N_A in the above expression would be replaced by N_D. A generalized solution for the thickness of the depletion region is therefore

$$d \cong \left(\frac{2\epsilon V}{e N} \right)^{1/2} \tag{11.18}$$

In this expression, N now represents the dopant concentration (either donors or acceptors) on the side of the junction that has the lower dopant level. (For surface barriers described later in this chapter, N is the dopant concentration in the bulk of the crystal.)

The resistivity ρ_d of the doped semiconductor [see Eq. (11.11)] is given by $1/e\mu N$, where μ is the mobility of the majority carrier. Equation (11.18) may thus be written

$$d \cong (2\epsilon V \mu \rho_d)^{1/2} \tag{11.19}$$

Because one often would like the largest depletion width possible for a given applied voltage, it is advantageous to have the resistivity as high as possible. This resistivity is limited by the purity of the semiconductor material before the doping process, because enough dopant must be added to override the nonuniform effects of the residual impurities. A premium is therefore placed on obtaining detectors fabricated from the highest purity material possible.

Because of the fixed charges that are built up on either side of the junction, the depletion region exhibits some properties of a charged capacitor. If the reverse bias is increased, the depletion region grows thicker and the capacitance represented by the separated charges therefore decreases. The value of the capacitance per unit area is

$$C = \frac{\epsilon}{d} \cong \left(\frac{e\epsilon N}{2V} \right)^{1/2} \tag{11.20}$$

Good energy resolution under conditions in which electronic noise is dominant depends on achieving a small detector capacitance and is thus promoted by using the largest possible applied voltage, up to the point that the detector becomes fully depleted.

The maximum electric field will occur at the point of transition between the n- and p-type material. Its magnitude is given by

$$\mathscr{E}_{max} \cong \frac{2V}{d} = \left(\frac{2VNe}{\epsilon} \right)^{1/2} \tag{11.21}$$

and can easily reach $10^6 - 10^7$ V/m under typical conditions. For partially depleted junctions, the depletion layer thickness d is proportional to $\sqrt{V}$ so that the value of $\mathscr{E}_{max}$ increases with applied voltage as $\sqrt{V}$.

The interrelation between these parameters is illustrated in the nomogram for silicon detectors given in Fig. 11.10. Also shown are scales corresponding to the ranges of various charged particles to allow selection of conditions required to produce a depletion depth that exceeds the range.

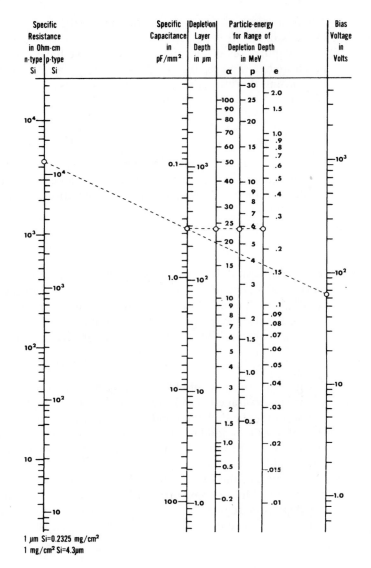

Figure 11.10 Nomogram illustrating interrelation between parameters for silicon junction detectors. (Similar to nomogram originally published by Blankenship.[26])

The maximum operating voltage for any diode detector must be kept below the breakdown voltage to avoid a catastrophic deterioration of detector properties. Commercially manufactured detectors are supplied with a maximum voltage rating that should always be strictly observed. Additional protection can be provided by monitoring the leakage current during application of the voltage (see the discussion later in this chapter).

To summarize, the reverse biased *p-n* junction makes an attractive radiation detector because charge carriers created within the depletion region can be quickly and efficiently collected. The width of the depletion region represents the active volume of the detector and is changed in partially depleted detectors by varying the reverse bias. The variable active volume of semiconductor junctions is unique among radiation detectors and sometimes is used to good advantage. The capacitance of a partially depleted detector also varies with applied voltage, and stable operation therefore requires the use of charge sensitive preamplifiers (see Chapter 17).

IV. SEMICONDUCTOR DETECTOR CONFIGURATIONS

A. Diffused Junction Detectors

One common fabrication method for semiconductor diode detectors starts with a homogeneous crystal of *p*-type material. One surface is treated by exposing it to a vapor of *n*-type impurity (typically phosphorus), which then converts a region of the crystal near the surface from *p*-type to *n*-type material. A junction is therefore formed some distance from the surface at the point at which the *n*- and *p*-type impurities reverse their relative concentration. Typical depths of the diffused *n*-type layer range from 0.1 to 2.0 μm. Because the *n*-type surface layer is heavily doped compared with the *p*-type original crystal, the depletion region extends primarily into the *p* side of the junction. Therefore, much of the surface layer remains outside the depletion region and represents a *dead layer* or window through which the incident radiation must pass before reaching the depletion region. In charged particle spectroscopy, this dead layer can be a real disadvantage because a portion of the particle energy will be lost before the active region of the detector is reached. Methods for experimentally determining its thickness are given later in this chapter.

To avoid the disadvantages of the dead layer, diffused junction detectors have been replaced in many applications by other configurations described in the following sections. Diffused junction detectors are still commercially manufactured, however, and offer some advantage over surface barrier detectors. They are somewhat more rugged and less prone to the problems that can arise due to the accumulation of oil or other foreign matter on the surface of the detector.

B. Surface Barrier Detectors

The role of the *p*-type material in forming the junction can be assumed by a high density of electron traps formed at the surface of an *n*-type crystal. The resulting depletion region behaves in much the same way as discussed earlier for a diffused junction detector. Formation of the surface states is carried out using recipes that have evolved somewhat empirically. One such set of typical procedures is described in Ref. 27. The usual treatment is etching of the surface, followed by evaporation of a thin gold layer for electrical contact. Best results are obtained if the evaporation is carried out under conditions that promote slight oxidation of the surface; the resulting oxide layer between the gold and silicon apparently plays an important role in the resulting properties of the surface barrier. Surface barriers can also be produced by starting with a *p*-type crystal and evaporating aluminum to form an equivalent *n*-type contact. The very thin dead layers that characterize surface barrier detectors are further discussed later in this chapter.

One potential disadvantage of surface barriers is their sensitivity to light. The thin entrance windows are optically transparent, and photons striking the detector surface can reach the active volume. The energy of visible light photons of 2–4 eV is greater than the bandgap energy of most semiconductors, and electron–hole pairs can therefore be produced by photon interactions. A very high noise level is produced by normal room lighting, but the vacuum enclosure required for most charged particle applications usually reduces light-induced noise to insignificant levels. The thin entrance window also makes the detector sensitive to damage from exposure to vapors, and the front surface must never be directly handled.

A cross-sectional diagram of a typical mounting arrangement for a surface barrier detector is shown in Fig. 11.11*a*. The outer housing and front surface are normally grounded, and an electrical lead from the back surface of the semiconductor wafer attaches to the center electrode of the coaxial connector at the rear. Because normal surface barriers are usually created on *n*-type crystals, a positive polarity voltage is required to reverse bias the junction.

Figure 11.11 Construction and mounting of silicon junction detectors shown in cross-sectional view. (*a*) Surface barrier mount with coaxial connector (M) at rear. The silicon wafer (S) is mounted in a ceramic ring (I) with electrical contact made between either side of the junction and opposite metalized surfaces of the ring. The front surface is connected to the outer case (C) and grounded, whereas the back surface is connected to the center conductor of the coaxial connector. (*b*) Cutaway view of a transmission mount, in which both surfaces of the silicon wafer are accessible. The coaxial connector is placed at the edge of the ceramic ring. (Courtesy of EG & G ORTEC, Oak Ridge, TN.)

C. Ion Implanted Layers

An alternative method of introducing doping impurities at the surface of the semiconductor is to expose that surface to a beam of ions produced by an accelerator. This method is known as *ion implantation* and can be used to form n^+ or p^+ layers by accelerating, for example, either phosphorus or boron ions, respectively. At a fixed accelerator voltage (typically about 10 kV) monoenergetic ions are produced that have a well-defined range in the semiconductor material. By changing the energy of the incident ions, the concentration profile of the added impurity can be closely controlled. Following exposure to the ion beam, an annealing step is normally carried out to reduce the effects of radiation damage caused by the incident ions. One of the advantages of ion implantation is that the annealing temperature required (less than 500°C) is much lower than that needed for the thermal diffusion of dopants to form a diffused junction. The structure of the crystal is therefore less disturbed, and carrier lifetimes are not unnecessarily reduced. Compared with surface barriers, ion-implanted detectors tend to be more stable and less subject to ambient conditions. Also, they can be formed with entrance windows as thin as 34 nm silicon equivalent[28,29] and they are available commercially. A review of the use of ion implantation to form radiation detectors can be found in Ref. 30.

D. Fully Depleted Detectors

As shown by Eq. (11.18), the width of the depletion region associated with a *p-n* junction increases as the reverse bias voltage is increased. If the voltage can be increased far enough, the depletion region eventually extends across virtually the entire thickness of the silicon wafer, resulting in a fully depleted (or totally depleted) detector. Because of the several advantages this configuration presents over partially depleted detectors, the fully depleted configuration is the preferred type in most applications.

In the usual case, one side of the junction is made up of a heavily doped n^+ or p^+ layer or, alternatively, a surface barrier. The opposite side of the junction generally consists of high-purity semiconductor material that is only mildly n or p type. (Such material is often designated v or π respectively.) The reason that high-purity material is important is reflected in Eq. (11.18). For a given applied voltage, the depletion depth is maximized by minimizing the concentration of doping impurities on the higher-purity side of the junction. Thick depletion regions can therefore only be obtained by starting from semiconductor material with the lowest possible impurity concentration. Also, with a large difference in the doping levels, the depletion layer essentially extends only into the high-purity side of the junction. The heavily doped layer can then be very thin, providing an *entrance window* for weakly penetrating radiations.

In Fig. 11.12, we assume that we have such a junction formed between a heavily doped p^+ surface layer and a high-purity n-type silicon wafer. As the reverse bias voltage applied to the detector is raised from zero, the depletion region extends further from the p^+ surface into the bulk of the wafer. For low values of the voltage, the wafer is only partially depleted and the electric field goes to zero at the far edge of the depletion region. Between this point and the back surface of the wafer, a region of undepleted silicon exists in which there is no electric field. This region then represents a very thick dead layer from which charge carriers are not collected. For all practical purposes, partially depleted detectors are therefore only sensitive to charged particles incident on the front surface.

If the applied voltage is increased further, the depletion region may be made to extend all the way to the back surface of the wafer. The voltage required to achieve this condition is sometimes called the *depletion voltage*. Its value is found by setting the depletion depth d in Eq. (11.18) equal to the wafer thickness T:

$$V_d = \frac{eNT^2}{2\epsilon}$$

When this stage is reached, a finite electric field exists all the way through the wafer, and the back dead layer thickness is reduced to that of the surface electrical contact that is employed. This condition is represented by the middle plot in Fig. 11.12. Once the wafer is

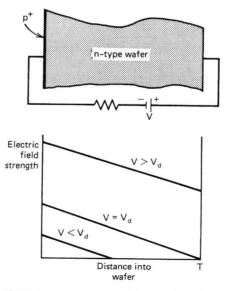

Figure 11.12 The electric field shape in a reverse bias semiconductor detector. Three plots are shown for bias voltages that are below, equal to, and above the depletion voltage V_d.

fully depleted, raising the applied voltage further simply results in a constant increase in the electric field everywhere in the wafer. At voltages much larger than the depletion voltage, the electric field profile therefore tends to become more nearly uniform across the entire wafer thickness. Under these conditions, the detector is sometimes said to be *over-depleted*. Because of the advantages of having a high electric field everywhere within the detector active volume, virtually all totally depleted detectors are operated at sufficient voltage to achieve this condition.

Figure 11.13 shows several configurations that are typical of fully depleted detectors, together with the corresponding electric field profile through the wafer. In order to deplete the wafer fully at as low a voltage as possible, one normally starts with material with the highest available purity, either *n* or *p* type. The junction is then formed by providing a heavily doped surface layer of the opposite type. This is often called the *rectifying contact*. Because of its high doping level, it also serves as an excellent blocking contact in which the minority carrier concentration is very low. In the nearly pure bulk of the wafer, however, the minority carriers are not highly suppressed and an additional blocking contact is normally provided at the opposite face of the wafer. If the high-purity silicon is mildly *n* type, then a thin n^+ layer is provided at this back surface. Since both materials are *n* type, no semiconductor junction exists at this surface. Instead, the n^+ layer provides the noninjecting conditions necessary to suppress leakage current due to minority carrier motion across the junction.

As shown on the right in Fig. 11.13, the roles of the n^+ and p^+ surface layers are reversed if one starts with high-purity material that is mildly *p* type. In both cases, the electric field is a maximum at the rectifying contact and decreases linearly to a minimum at the blocking contact. Shown in the center in Fig. 11.13 is the case in which intrinsic or perfectly compensated material is used for the wafer. In this case, the distinction between the two contacts disappears and the electric field is uniform throughout the entire wafer. The detector is fully depleted even for very low values of applied voltage. This *p-i-n* configuration is discussed in connection with lithium-drifted detectors in the next chapter.

Fully depleted silicon detectors are very useful as *transmission detectors* for incident particles that have sufficient energy to pass completely through the wafer. The pulse amplitude then indicates the energy lost by the incident radiation during its transit through the device. Totally depleted silicon detectors are commercially available in thicknesses from about 50 to 2000 μm. Transmission mounts of the type shown in Fig. 11.11*b* allow access to both surfaces of the wafer.

Several properties of totally depleted detectors are of primary importance. The dead layers must be as small as possible at both the front and rear surfaces of the detector if the

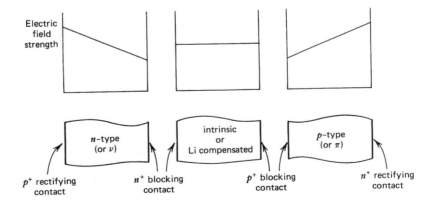

Figure 11.13 The electric field shapes for fully depleted planar semiconductor detectors of different configurations.

pulse is to indicate accurately the energy loss of the particle during its transit. An empirical test is often carried out to determine the minimum bias voltage at which these detectors are totally depleted. The pulse height from a monoenergetic source of charged particles is recorded for the particles incident on both the front and back face of the detector. When the detector is totally depleted, the pulse height should be approximately the same for either orientation. In interpreting such measurements, allowance must be made for the fact that the inherent window thicknesses of the front and back contacts of these detectors are often not the same.

In partially depleted detectors, the thickness uniformity of the crystal from which the detector is fabricated is not critical because the active volume of the detector is determined by the limited depletion depth. In fully depleted detectors, however, the wafer thickness must be kept quite uniform to avoid energy loss variations across the surface of the detector. Consequently, considerable effort is taken to provide uniform crystal wafers when totally depleted configurations are produced.

Fully depleted detectors have other advantages over partially depleted configurations in which there is an undepleted back dead layer. The finite electrical resistance of this dead layer is a source of Johnson noise that can contribute to the degradation of energy resolution. It is eliminated in a fully depleted detector by extending the depletion region all the way to the back contact. Timing properties also tend to be superior in fully depleted configurations. In a partially depleted detector, the electric field drops to zero at the edge of the depletion region. Charge carrier velocities therefore become very low in these low-field regions, slowing the rise of the signal pulse. In a totally depleted detector, the electric field can be maintained at a high value everywhere within the detector volume. Finally, some added stability results from the fact that the active volume and capacitance of a fully depleted detector are no longer functions of the applied voltage as they are in a partially depleted configuration.

The thickness of wafer that can be fully depleted using voltages short of catastrophic breakdown depends on the purity of the semiconductor. In this respect, there is a significant difference between silicon and germanium. Using ultrapure germanium described in the following chapter, a depletion thickness of several centimeters can be achieved. The impurity levels in currently available silicon are somewhat higher, and depletion thicknesses are generally limited to no more than several millimeters. Greater thicknesses in silicon are currently possible only through the use of material compensated by the lithium-drifting process described in Chapter 13.

E. Passivated Planar Detectors

The newest method of fabricating silicon junction detectors combines the techniques of ion implantation and photolithography to produce detectors with very low leakage currents and excellent operational characteristics.[31–33] Methods that were first developed in the semiconductor industry to produce integrated circuits have now been adapted successfully[34, 35] to the fabrication of detectors. The techniques described below lend themselves to the batch production of multiple detectors simultaneously starting with a large-area silicon wafer, thus providing potential cost savings. These techniques can also accommodate the type of complex electrode geometry required, for example, in the silicon microstrip detectors described in Chapter 13.

The planar fabrication process generally begins with high-purity silicon that is mildly *n* type due to residual donor impurities. The steps in the fabrication process are shown in Fig. 11.14. After the wafer has been polished and cleaned, the surface is "passified" through the creation of an oxide layer at elevated temperature. Next, the techniques of photolithography are used to remove selectively areas of the oxide where the entrance windows of the finished detectors are to be located. The junction is then formed by converting a very

Figure 11.14 Steps in the fabrication of passivated planar silicon diode detectors. (From Kemmer.[34])

thin layer of silicon within the windows into *p*-type material through the implanting of acceptor ions (boron) using an accelerator. To serve as a blocking electrical contact, the rear surface of the wafer is converted into n^+ material through implantation of donor (As) ions. The radiation damage in the implanted layers is next removed through annealing at elevated temperature. Finally, aluminum is evaporated and patterned by photolithography to provide thin ohmic electrical contacts at the front and rear surfaces. The individual detectors are then separated and encapsulated.

One advantage of this planar fabrication process is that the junction edges are defined by the ion implantation pattern and can be kept within the bulk of the wafer. The oxide-passivated surface keeps leakage currents much lower than in surface barrier detectors, where the junction edge extends all the way to the edge of the wafer. Much of the leakage current in these designs then occurs where these edges are encapsulated in epoxy or similar material. Formation of the p^+ layer through ion implantation also provides planar detectors with a very thin and uniform entrance window or dead layer, an important consideration in preserving good energy resolution for the detector. The aluminized front surface is more rugged and less subject to damage compared with the gold front surface used in surface barrier fabrication.

V. OPERATIONAL CHARACTERISTICS

A. Leakage Current

When voltage is applied to a junction detector in the normal fashion, that is, to reverse bias the junction, a small current of the order of a fraction of a microampere is normally observed. The origins of this *leakage current* are related both to the bulk volume and surface of the detector. Bulk leakage currents arising internally within the volume of the detector can be caused by either of two mechanisms.

The direction of the electric field across the depletion region is such that any majority carriers that diffuse from the normal p and n regions of the detector to the edges of the depletion region will be repelled away from the junction. However, the minority carriers in either case are attracted and will therefore be conducted across the junction. Because the minority carriers are generated continuously on both sides of the junction and are free to diffuse, a steady-state current will result that will be roughly proportional to the area of the junction. In most cases, the minority carrier current is small and is seldom an important leakage source.

A second source of bulk leakage is the thermal generation of electron–hole pairs within the depletion region. This rate will obviously increase with the volume of the depletion region and can be reduced only by cooling the material. Silicon detectors of usual dimensions have a sufficiently low thermally generated current to allow their use at room temperature, but germanium detectors, because of the lower gap energy, must always be operated at reduced temperatures.

Surface leakage effects take place at the edges of the junction where relatively large voltage gradients must be supported over small distances. The amount of surface leakage can vary greatly, depending on such factors as the type of detector encapsulation used, humidity, and any contamination of the detector surface by fingerprints, vacuum pump oil, or other condensable vapors. Guard rings analogous to those described in Chapter 5 are sometimes incorporated into the design of semiconductor diode detectors[36,37] to reduce the surface leakage, but in commercial detectors the normal approach is to rely on clean encapsulation techniques to keep the surface leakage within tolerable levels. The introduction of the planar fabrication process has allowed the production of detectors in which the junction edges are buried within the silicon wafer. As a result, leakage current is reduced to a small fraction of that typically observed in either diffused junction or surface barrier devices.

In addition to the effects on energy resolution discussed in the following section, the leakage current has another practical influence on detector operation. The bias voltage to the detector is always supplied through a large-value series resistor (R_L in Fig. 17.5) for signal isolation purposes. Therefore, the true bias voltage applied to the junction is reduced from that of the voltage source by the product of the leakage current and the series resistance. If the leakage current is large enough, the drop across the resistor can appreciably diminish the actual voltage applied to the detector, and the supply voltage must then be raised to compensate for this loss. It is therefore a fairly common practice to monitor the leakage current with an ammeter in series with the voltage supply.

Monitoring the leakage current can also detect the onset of abnormal detector behavior. During steady operation, the leakage should normally maintain a steady value, and any abrupt change or increase in the leakage current can indicate a change in detector performance, which may degrade the energy resolution. Also, it is useful to monitor the leakage current as the bias voltage is first applied to the detector. Normally, the leakage current will increase as the bias voltage is raised. However, any sudden increases can signal the approach of the breakdown of the diode, and the voltage should therefore be reduced to a lower value. Finally, the long-term behavior of the leakage current is often a useful monitor on the degree of radiation damage suffered by a given detector when used under conditions in which such damage is significant.

B. Detector Noise and Energy Resolution

Sources of electronic noise in spectroscopic measurements fall into two main categories: *series noise* and *parallel noise*. A discussion of the distinctions between these categories and their variations with operating parameters such as the choice of electronic shaping time is postponed until Chapter 17. For the type of silicon diode detectors used for charged particle measurements, three main contributors to the electronic noise are most significant:

1. Fluctuations in the bulk generated leakage current, a component of parallel noise.
2. Fluctuations in the surface leakage current, another component of parallel noise.
3. Noise associated with series resistance or poor electrical contacts to the detector, a contributor to series noise.

The relative importance of these sources will depend on the magnitude of the leakage currents for the specific detector, the capacitance of the detector, and whether the diode is partially or fully depleted. The overall peak broadening due to electronic noise is often measured experimentally by injecting the output of a stable pulse generator into the preamplifier input while the detector remains connected. If the pulse generator has negligible spread in the amplitude of its output pulses, then the width of the corresponding "pulser peak" in the recorded spectrum is a direct measure of the electronic noise contribution.

This noise width combines in quadrature with other sources of peak broadening, such as the contributions of charge carrier statistics and fluctuations in particle energy loss in dead layers to determine the peak widths actually observed. Examples of the energy resolution attainable with semiconductor diode detectors are given in the section on alpha particle spectroscopy later in this chapter.

If trapping effects become significant in the detector, it is usually evidenced by the appearance of low-energy tails on the peaks observed from monoenergetic sources of radiation. These tails correspond to pulses in which less than the total amount of charge generated by the radiation has been collected. Because the amount of trapping varies according to the distance traveled by the carriers before reaching the collection electrodes, the amount of energy loss is variable and the full-energy peaks are spread only to the low-energy side.

C. Changes with Detector Bias Voltage

When the bias voltage and electric field are low, the pulse height from radiations that are fully stopped within the depletion layer continues to rise with applied voltage. This variation is caused by the incomplete collection of charge carriers because of trapping or recombination along the track of the incident particle. The fraction that escape collection will decrease as the electric field is increased. Similar losses to recombination are observed in a gas-filled ion chamber at low values of the electric field. Once the electric field is sufficiently high, charge collection becomes complete and the pulse height no longer changes with further increases in the detector bias voltage. This region of operation is called the *saturation region* and corresponds to the region of ion saturation in a gas-filled ion chamber.

If radiations of a single energy and type are involved, it is sometimes possible to operate the detector at a bias voltage that is short of true saturation without significant deterioration in the energy resolution, because the fraction of charge lost for each event is likely to be nearly constant. When measuring radiations of diverse energy and specific ionization, however, it is quite important to ensure that the detector is operating in the region of true saturation to avoid significant deterioration in the energy resolution. To reach the saturation region, somewhat higher electric fields are generally required as the detector undergoes radiation damage.[38]

If the electric field is made sufficiently high, multiplication effects can be induced in a semiconductor detector that are analogous to gas multiplication in proportional counters or Geiger–Mueller tubes. The multiplication arises when electrons liberated in the initial radiation interaction gain sufficient energy from the field to create further electron–hole pairs as they drift toward the collecting electrodes. These multiplication effects are discussed in some detail in Ref. 39 and are the basis of operation for silicon avalanche detectors described in Chapter 13.

D. Pulse Rise Time

Semiconductor diode detectors are generally among the fastest of all commonly used radiation detectors. A general review of theoretical and experimental work on their timing properties is given in Ref. 40. Under normal conditions, the observed pulse rise time is of the order of 10 ns or less. The detector contribution to this rise time is composed of the *charge transit time* and the *plasma time.*

The charge transit time corresponds to the migration of the electrons and holes formed by the incident radiation across the region of high electric field in the depletion region. The rise time of the output pulse is therefore limited by the time required for complete migration of these charges from their point of formation to the opposite extremes of the depletion region. These times are minimized in detectors with high electric fields and small depletion widths. In totally depleted detectors, the depletion width is fixed by the physical thickness of the silicon wafer, and therefore the transit time is decreased as the bias voltage is increased. In partially depleted detectors, however, the depletion width increases with increasing bias [see Eq. (11.18)], and therefore the effect of a larger bias voltage is to increase both the electric field and the distance over which charges must be collected. Furthermore, because the electric field is not uniform, the drift velocity of electrons and holes will vary as they move across the depletion region. The dependence of the charge transit time on bias voltage in these detectors is therefore somewhat more complicated, but it can be shown to be independent of the voltage if certain simplifying assumptions are made.[41] A derivation is given in Chapter 12 (see p. 421) of the time profile of the signal pulse attributable to charge migration in solid-state detectors in which the electric field is uniform.

For the case of a particle range that is much less than the width of the depletion region, all the charge carriers are created near one boundary. The collection time of one type of carrier corresponds to its migration across the entire depletion region and is therefore much longer than that for the other carrier. For a surface barrier on an *n*-type crystal, it is thus the electron collection time that dominates the time response for weakly penetrating particles.

A second component called the *plasma time* is observed when heavy charged particles, such as alpha particles or fission fragments, comprise the incident radiation. For these radiations, the density of electron–hole pairs along the track of the particle is sufficiently high to form a plasma-like cloud of charge that shields the interior from the influence of the electric field. Only those charge carriers at the outer edge of the cloud are subject to the influence of the field, and they begin to migrate immediately. The outer regions are gradually eroded away until the charges at the interior are finally subject to the applied field and also begin to drift. The plasma time is roughly defined as the time required for the charge cloud to disperse to the point where normal charge collection proceeds.

A number of theoretical models have been developed to describe the plasma erosion process;[42–47] it is predicted[42] that the plasma time should vary inversely with the electric field strength at the position of the track and increase as the cube root of the linear carrier density along the track. The effects of the plasma formation are observed to be a fixed delay of several nanoseconds between the time of track formation and the onset of the rise of the output pulse together with a slowing of the rise time of the output pulse. Measurements of the delay time with silicon surface barrier detectors,[48–53] give typical values of 1–3 ns for alpha particles, and 2–5 ns for heavy ions and fission fragments.

The actual rise time observed from a detector–preamplifier combination may also be influenced by the preamplifier properties. The time constant of the equivalent input circuit must be short if the rise time is to be held to that determined by the detector charge collection and plasma time properties only. One contributor to the input time constant is the series resistance of the undepleted region in partially depleted detectors. Therefore, fully depleted detectors in which the series resistance is largely eliminated are often favored in fast-timing situations.

E. Entrance Window or Dead Layer

When heavy charged particles or other weakly penetrating radiations are involved, the energy loss that may take place before the particle reaches the active volume of the detector can be significant. Because the thickness of the dead layer includes not only the metallic electrode but also an indeterminate thickness of silicon immediately beneath the electrode in which charge collection is inefficient, the dead layer can be a function of the applied voltage. Its effective thickness must often be measured directly by the user if accurate compensation is to be made.

The simplest and most frequently used technique is to vary the angle of incidence of a monoenergetic charged particle radiation. When the angle of incidence is zero (i.e., perpendicular to the detector surface), the energy loss in the dead layer is given by

$$\Delta E_0 = \frac{dE_0}{dx} t \tag{11.22}$$

where t is the thickness of the dead layer. The energy loss for an angle of incidence of θ is given by

$$\Delta E(\theta) = \frac{\Delta E_0}{\cos \theta} \tag{11.23}$$

Therefore, the difference between the measured pulse height for angles of incidence of zero and θ is given by

$$E' = (E_0 - \Delta E_0) - (E_0 - \Delta E(\theta))$$

$$E' = \Delta E_0 \left(\frac{1}{\cos \theta} - 1 \right) \tag{11.24}$$

If a series of measurements are made as the angle of incidence is varied, a plot of E' as a function of $(1/\cos \theta - 1)$ should be a straight line whose slope is equal to ΔE_0. Using tabular data for dE_0/dx for the incident radiation, we can calculate the dead layer thickness from Eq. (11.22).

One possible flaw in this method involves the assumption that the energy loss through the dead layer depends only on the total path length traversed and not on the relative orientation of the particle path with respect to the detector axis. There is some evidence that recombination should be more severe for particle paths parallel to the direction of the electric field in the detector compared with paths perpendicular to the field. This recombination would tend to cause a lower than expected response for paths near normal incidence and should be evidenced by a curvature in the plot described above.

The thinnest dead layers are produced in semiconductor detectors of the ion implanted or surface barrier types. Typical values of 100 nm of silicon equivalent correspond to an energy loss of 4 keV for 1 MeV protons, 14 keV for 5 MeV alpha particles, and several hundred keV for fission fragments. Because variations in this energy loss due to straggling or variable angle of incidence will potentially detract from energy resolution, thin dead layers are quite important in high-resolution charged particle spectroscopy. Using special techniques, dead layers of less than 30 nm have been successfully fabricated.[54]

F. Channeling

In crystalline materials, the rate of energy loss of a charged particle can depend on the orientation of its path with respect to the crystal axes.[55] Particles that travel parallel to crystal planes can, on the average, show a rate of energy loss that is lower than that for particles directed in some arbitrary direction. Therefore, these "channeled" particles can penetrate significantly farther through the crystal. The effects are particularly significant for thin totally depleted detectors because the amount of energy deposited is then dependent on the orientation of the crystal planes with respect to the particle direction. To minimize the tendency for incident particles to channel, detectors are normally fabricated from silicon cut so that the ⟨111⟩ crystal orientation is perpendicular to the wafer surface.

Channeling can affect the recorded pulse height even in situations in which the particle is fully stopped within the active volume. Nuclear collisions are less probable for channeled particles and therefore the pulse height defect for heavy ions (discussed later in this chapter) may be reduced.

G. Radiation Damage

The proper operation of any semiconductor detector depends on the near perfection of the crystalline lattice to prevent defects that can trap charge carriers and lead to incomplete charge collection. Any extensive use of these detectors, however, ensures that some damage to the lattice will take place because of the disruptive effects of the radiation being measured as it passes through the crystal. While the energy that goes into the creation of electron–hole pairs leads to fully reversible processes that leave no damage, nonionizing energy transfers to the atoms of the crystal lead to irreversible changes. The effects tend to be relatively minor for lightly ionizing radiations (beta particles or gamma rays) but can become quite significant under typical conditions of use for heavy charged particles. For example, prolonged exposure of silicon surface barrier detectors to heavy ions or fission fragments will lead to a measurable increase in the leakage current and a significant loss in energy resolution of the detector. With extreme radiation damage, multiple peaks may appear in the pulse height spectrum recorded for monoenergetic particles. Furthermore, the time characteristics of the detector may be degraded even at doses that are too low to show measurable spectral effects.[56] In silicon microstrip detectors of the type discussed in Ch. 13, radiation damage can lead to a decrease in the inter-strip resistance and a loss of spatial resolution.[57]

The radiation-induced damage can be classified into the two categories of bulk and surface effects.[58] The most fundamental type of bulk radiation damage is the *Frenkel defect*, produced by the displacement of an atom of the semiconductor material from its normal lattice site. The vacancy left behind, together with the original atom now at an interstitial position, constitutes a trapping site for normal charge carriers. These are sometimes called *point defects* to distinguish them from more complex "clusters" of crystalline damage that are formed along the track of a primary "knock-on" atom if sufficient energy is transferred. Gamma rays and electrons with energy of a few MeV or less create only point defects, whereas heavy charged particles of equivalent energy are generally more damaging because they also form clusters. The number of Frenkel defects produced by a fission fragment is estimated to be about 100–1000 times greater than that produced by an alpha particle.[59] At the other extreme, an incident electron or beta particle requires a minimum of about 145 keV to produce a defect, and very little damage is observed for electrons whose energy is much below 250 keV.[60] The severity of damage to be anticipated is therefore a strong function of the nature of the radiation involved. When enough of these defects have been formed, carrier lifetime and charge collection efficiency are reduced and the energy resolution of the detector is degraded due to fluctuations in the amount of charge lost. Some minor annealing of the radiation damage can occur[61] over long periods of time (see Fig. 11.17), but for all intents and purposes, the damage is permanent.

The increase in leakage current appears to be more directly related to surface effects[59] and also contributes to a loss of detector resolution from the corresponding increase in leakage current fluctuation. In devices such as silicon microstrip detectors that include oxide passivation layers, the surface effects are closely related to the ionization created within the oxide and its trapping at interfaces. For penetrating radiations including gamma rays or neutrons, the damage is generally distributed throughout the detector and the direction of incidence of the radiation has little effect. For electrons or charged particles, however, the orientation with respect to the detector is important. Irradiation of the front (or gold) surface of totally depleted detectors requires exposures that are several orders of magnitude less than those needed to produce the same effects by irradiation of the back (or aluminum) contact.[62]

For silicon surface barriers irradiated on the gold or front surface, various data have been published on the integrated flux of charged particles required to produce a significant deterioration in detector performance. Although subject to a great deal of variability, depending on the specifics of each experiment, serious changes appear to take place for irradiations of about 10^{14} fast electrons/cm^2 (Ref. 60), 10^{12} to 10^{13} protons/cm^2 (Refs. 62 and 63), 10^{11} alpha particles/cm^2 (Ref. 64), and about 3×10^8 fission fragments/cm^2 (Ref. 59). Exposure to fast neutron fluxes of about 3×10^{11} neutrons/cm^2 (Ref. 65) and gamma-ray doses of about 10^6 R (Ref. 66) are also sufficient to lead to significant performance degradation. In general, fully depleted detectors are less sensitive than partially depleted devices because the average electric field throughout the detector is somewhat higher. The effects of radiation-induced charge trapping are minimized when the highest possible electric field is present in the active volume.

An effect known as *type inversion*[67,68] has been observed to occur in high resistivity *n*-type silicon after prolonged exposure to fast neutrons or high-energy particles with integrated fluence of about 10^{13}/cm^2. The effective concentration of donors gradually decreases with exposure, until a transition to *p*-type behavior is observed. Some models for this change[67,69] postulate the radiation-induced formation of deep acceptor levels, close to the center of the bandgap, that become electrically active when voltage is applied to the detector. In detectors with a p$^+$-ν-n$^+$ structure (see Fig. 11.13), an interesting consequence of the type inversion is the effective shift of the rectifying contact from one side of the detector to the other. The device may continue to function after the inversion with the same polarity of applied voltage, but eventually the growing concentration of acceptors raises the voltage level required for full depletion to levels beyond those causing breakdown.

H. Energy Calibration

When applied to the measurement of fast electrons or light ions such as protons or alpha particles, semiconductor diode radiation detectors respond very linearly, and the energy calibration obtained for one particle type is very close to that obtained using a different radiation type. Some observations show that there is a small difference in the pulse height observed for protons and alpha particles of the same energy,[70,71] but such differences are usually on the order of 1% or less. It is not clear whether these differences reflect a fundamental difference in the ionization energy ϵ for various particles or whether other factors related to the modes of energy loss may be responsible. For any application in which an absolute energy calibration of about 1% or less is required, it is always best to calibrate the detector using the same type of particle that is involved in the measurement itself.

The most common calibration source is the alpha-emitting isotope ^{241}Am. This isotope emits alpha particles of 5.486 MeV (85%) and 5.443 MeV (13%), and a representative pulse height spectrum is illustrated in Fig. 11.15. An accurate calibration of the energy scale requires that account be taken of the energy loss of these alpha particles in the source itself, in any intervening material between the source and detector, and in the window or dead layer of the detector.

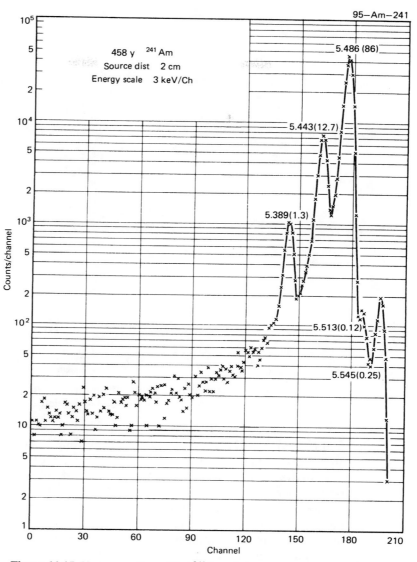

Figure 11.15 Upper portion of the ^{241}Am alpha spectrum as recorded by a high-resolution surface barrier detector. (From Chanda and Deal.[72])

I. Pulse Height Defect

The response of semiconductor detectors to very heavy ions such as fission fragments is less straightforward. There is abundant evidence that the pulse height observed for heavy ions is substantially less than that observed for a light ion of the same energy. The *pulse height defect* is defined in units of energy as the difference between the true energy of the heavy ion and its apparent energy, as determined from an energy calibration of the detector obtained using alpha particles.

A plot of the effects of the pulse height defect is given in Fig. 11.16. Measurements of the pulse height defect for surface barriers when irradiated by fission fragments[74] show that a value as large as 15 MeV for the defect is possible, compared with an average energy

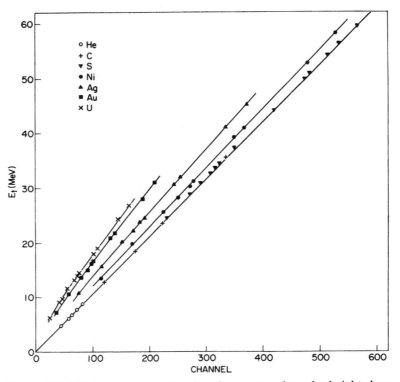

Figure 11.16 The true energy of various ions versus the pulse height channel number from a silicon surface barrier detector. (From Wilkins et al.[73])

of about 80 MeV. Other measurements[75–77] using fragments that were separated in mass and energy obtained smaller defect values of 3–7 MeV.

Analysis has shown[73,78–80] that three separate phenomena contribute to the observed pulse height defect. The first and simplest contributor is the energy loss of the ion in the entrance window and dead layer of the detector. The magnitude of the contribution to the pulse height defect can be calculated from the stopping power of the ion and measurements of the dead layer thickness using methods discussed earlier. Because heavy ions such as fission fragments show maximum dE/dx at the start of their range, whereas light ions such as alpha particles show the reverse behavior, the fractional energy loss in the dead layer will be much more significant for the heavy ions.

A second contributor to the pulse height defect involves the tendency for heavy ions to lose energy by means other than simple electronic collisions. As the velocity of the ion decreases, nuclear collisions become important, and recoil nuclei are the direct result of such interactions. Because of the low velocity of these recoil nuclei, the probability of electronic interactions is reduced and a net decrease in the efficiency with which electron–hole pairs are produced is observed. The contribution of nuclear collisions increases with the effective charge on the ion and is therefore most significant for heavy ions. A calculation of the magnitude of this effect has been carried out by Haines and Whitehead[81] from basic theories of ion slowing down in solids. An experimental study of the effect in surface barriers is described in Ref. 82.

A third factor in the pulse height defect involves the high rate of electron-hole recombination expected in the dense plasma created along the ion track, particularly near its end.[74,83,84] The magnitude of the recombination would be expected to decrease with increasing bias voltage and may also depend on the relative orientation of the particle path

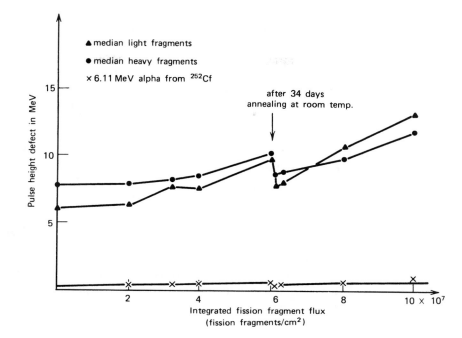

Figure 11.17 Dependence of measured pulse height defect versus fission fragment exposure for a heavy ion silicon detector. (From Bozorgmanesh.[85])

with respect to the electric field within the detector. A practical method of reducing the pulse height defect is to minimize the effect of recombination by creating the largest possible electric field within the detector.

Because the effects of trapping and recombination are influenced by radiation damage within the detector, it should be anticipated that the pulse height defect may increase with normal use. Figure 11.17 shows the measured pulse height defect for a silicon heavy ion detector as a function of integrated fission fragment flux. The general trend toward increasing pulse height defect is evident, together with the effects of partial annealing of the radiation damage over several weeks following removal of the detector from the flux.

VI. APPLICATIONS OF SILICON DIODE DETECTORS

A. General Charged Particle Spectroscopy

Since their development as practical detectors in the early 1960s, silicon diodes have become the detectors of choice for the majority of applications in which heavy charged particles are involved. Some of the more common applications involving the spectroscopy of alpha particles and fission fragments and the measurement of energy loss of charged particles in transmission detectors are discussed in the following section. Most of the conclusions carry over into other charged particle spectroscopy involving protons, deuterons, or other heavy ions.

Silicon diode detectors are sometimes used for the measurement of beta particles and fast electrons, particularly as thin totally depleted transmission detectors. However, because it is more common to use ion-drifted detectors for this purpose, a discussion of the response of semiconductor detectors to fast electrons will be postponed until Chapter 13.

Compared with competing techniques, the use of semiconductor detectors often provides advantages in a number of key areas. These include exceptionally good energy resolution, good stability and freedom from drift, excellent timing characteristics, very thin entrance windows, and simplicity of operation. The relatively small size can be an advantage in some situations but is also a limitation in those applications in which a detector with a large surface area is required. Silicon diodes are commercially available with surface area up to 20 cm^2, but the corresponding large capacitance results in a poorer energy resolution than is attainable with smaller detectors. More usual sizes range from 1 to 5 cm^2. Depletion depths up to 5 mm can be obtained commercially in some special configurations, but the more common detectors are limited to a depletion depth of 1 mm or less.

In the event that the detector depletion depth is greater than the range of the incident heavy ions, the response of the detector is very simple. For monoenergetic incident particles, only a single full-energy peak is observed because there are no competing processes that can significantly scatter out the ion or otherwise lead to partial energy deposition.[†] For fully depleted detectors, the depletion depth is given simply by the thickness of the silicon wafer. In partially depleted detectors, the depletion depth increases with applied bias and is therefore limited by the maximum detector bias that can be applied without risking detector breakdown. The maximum bias and the corresponding depletion depth are normally provided as a specification by the detector manufacturer. If a measurement of the radiation energy is not required, simple counting of charged particle radiation can be carried out with semiconductor detectors whose depletion depth is less than the range, provided the energy deposited within the depletion region is sufficiently high to generate a pulse that lies above the noise level of the instrumentation system.

B. Alpha Particle Spectroscopy

Silicon diodes operated at room temperature are near-ideal detectors for alpha particles and other light ions. Because of the wide availability of convenient monoenergetic sources of alpha particles, the performance of semiconductor detectors conventionally is tested by recording the pulse height spectrum from such sources. The most common of these is ^{241}Am, and the corresponding alpha spectrum is widely used for comparison of the energy resolution of solid-state detectors. A representative spectrum taken with a detector of good resolution is shown in Fig. 11.15.

With alpha particles in this energy range (5.486 MeV), the noise contribution of the preamplifier and other electronic components can be much smaller than the inherent energy resolution of the detector itself. In this case, one might expect that the statistics of charge carrier formation would limit the energy resolution that is achievable. This limiting resolution can be calculated beginning with Eq. (4.15) in terms of the Fano factor F and the average number of charge carriers per pulse N, or the energy of the alpha particle E and the ionization energy ϵ

$$R_{\text{lim}} = 2.35 \sqrt{\frac{F}{N}} = 2.35 \sqrt{\frac{F\epsilon}{E}} \tag{11.25}$$

In this equation, R is expressed as a fractional (or percentage) energy resolution. The convention with silicon semiconductor detectors is to instead quote the FWHM in units of energy, so we multiply both sides of the above equation by the alpha particle energy E

$$\text{FWHM}]_{\text{lim}} = 2.35 \sqrt{FE\epsilon} \tag{11.26}$$

[†]For incident ions with relatively high energies, some complications in the response can be observed due to nuclear reactions induced by the incident particle in the material of the detector. One example occurs in silicon detectors due to the inelastic scattering of light ions from ^{28}Si nuclei in the detector that are followed by the emission of a 1.78 MeV gamma ray. Since the probability is high that this gamma ray escapes from the detector, a satellite peak located 1.78 MeV below the full-energy peak can sometimes be observed.[86] In the case of incident protons, the probability for this process remains lower than 0.1% per incident particle for energies below 10 MeV.

Putting in values for silicon of $F = 0.11$, $\epsilon = 3.62$ eV, and evaluating at $E = 5.486$ MeV, we obtain

$$\text{FWHM}]_{\text{lim}} = 3.47 \text{ keV}$$

for the predicted statistical limit. Under very carefully controlled conditions, an energy resolution for alpha particles of 8 keV has been demonstrated[87] for passivated planar silicon detectors. In commercially available small-size detectors, however, the energy resolution that is achievable in practice tends to be no better than about 10 keV. The discrepancy from the predicted statistical limit represents the additional contributions to peak broadening that are still significant in these detector systems.

One of these contributions is due to the fact that a small portion of the alpha particle energy is transferred to recoil nuclei rather than to electrons. These low-energy recoil nuclei lose their energy in quasielastic collisions with surrounding atoms and form almost no additional electron–hole pairs. If the fraction of energy lost in this manner were constant for each alpha particle, there would be no effect on energy resolution. However, this energy loss is subject to large fluctuation since it is influenced by a few relatively large events. It has been estimated[88] that the FWHM contribution due to these fluctuations in silicon amount to about 3.5 keV for 6 MeV alpha particles.

Other significant contributions to peak broadening are the effects of incomplete charge collection and variations in the energy lost by the particle in dead layers at the detector surface.[89] For lower charged particle energies and detectors of large capacitance, the electronic noise can also be a significant contribution. The noise level from the detector–preamplifier–amplifier combination is dominated by fluctuations in the detector leakage current, the inherent preamplifier noise, and the characteristics of the FET used in the input stage of the preamplifier.

Since all these sources of peak broadening are normally independent, the square of each FWHM that would theoretically be observed for each source alone can be summed together to give the square of the overall FWHM.

C. Heavy Ion and Fission Fragment Spectroscopy

The energy measurement of fission fragments or other ions of large mass involves several special concerns. Most stem from the high density of charge carriers that are created along the track of these ions. Recombination of electron–hole pairs is accentuated, and the detector may require a higher bias voltage than that required to saturate the signal from alpha particles. The pulse height defect discussed previously is also accentuated by this high carrier density, complicating the energy calibration procedures. Also, the prolonged exposure of detectors to heavy ions or fission fragments creates rapid performance deterioration due to radiation damage in the detector.

Detector manufacturers generally offer some silicon detectors especially tailored for heavy ion spectroscopy. They are designed to minimize the problems of slow rise time and pulse height defect caused by the high specific energy loss along the particle track. The most effective step is to ensure that the electric field is as high as possible. One approach is to use thin slices of low-resistivity silicon to prepare totally depleted detectors that require a large value of the reverse bias to become fully depleted. Special methods of preparing contacts can also be used[90] to allow extreme overbiasing (beyond the point of full depletion), which can raise the field magnitude by as much as a factor of 20 over its value at the depletion voltage.[91] The silicon used for heavy ion detectors should also have large carrier lifetimes to help reduce the pulse height defect due to recombination.

Typical measurements[92–94] of the response function of silicon diodes to monoenergetic heavy ions show an asymmetric peak with significant tailing toward the low energy side. The cause of the tailing is probably related to fluctuations in the energy loss due to the entrance window or in the charge lost to recombination along the particle track. Representative energy resolution figures range from about 100 keV FWHM for 35–50 MeV oxygen ions, to about 300 keV for 50 MeV sulfur ions.[93]

Recording the fission fragment spectrum from the spontaneously fissioning isotope ^{252}Cf serves as a standard test of heavy ion detector performance. Thin sources of this isotope are readily available and are widely used to monitor detector properties. Figure 11.18 shows a ^{252}Cf fission fragment spectrum taken with a good quality silicon surface barrier detector. Schmitt et al. have shown that a careful analysis of this spectrum can give an accurate energy calibration of the detector. The spectrum also provides a check on detector properties, such as resolution, low-energy tailing, and any possible effects due to internal multiplication in the detector.

In the first work reported by these investigators,[38,96] a series of surface barrier and diffused junction silicon detectors were carefully calibrated by recording the pulse height spectrum from monoenergetic heavy ions produced in an accelerator. These results show that for a given heavy ion species, the fragment energy versus pulse height relation is of the form $E = ax + b$, where E is the ion energy and x is the measured pulse height. This relation was found to hold over a wide energy range and illustrates the constancy of the pulse height defect for a given particle type. By studying a number of different ions, a general relation was demonstrated for an ion of mass m of the form

$$E(x, m) = (a + a'm)x + b + b'm \qquad (11.27)$$

where the constants a, a', b, and b' assume different values depending on the specific detector.

The californium fission fragment spectrum was also recorded for the same detectors. A scheme was then developed by which the values of the constants in Eq. (11.27) can be extracted from properties of the ^{252}Cf spectrum. A more recent study by Weissenberger et al.[97] has confirmed this general approach to energy calibration of the detector, with the numerical results shown below. Referring to spectrum features defined in Fig. 11.18, the following values were obtained for each of the constants:

$$a = \frac{24.300}{P_L - P_H} \qquad \left(\frac{\text{MeV}}{\text{channel}}\right)$$

$$a' = \frac{0.0283}{P_L - P_H} \qquad \left(\frac{\text{MeV}}{\text{channel} \cdot \text{amu}}\right)$$

$$b = 90.397 - aP_L \qquad (\text{MeV})$$

$$b' = 0.1150 - a'P_L \qquad \left(\frac{\text{MeV}}{\text{amu}}\right)$$

where P_L and P_H are the channel numbers corresponding to L and H on the figure.

Schmitt and Pleasonton[95] have published a set of acceptance criteria based on the ^{252}Cf spectrum that should be displayed by a good quality heavy ion detector. These are shown in Table 11.2 and refer to the same spectrum shape parameters. The physical significance of each of these terms is detailed in Ref. 95. Deviations from these criteria can indicate poor detector resolution or other undesirable performance characteristics. Periodic monitoring of the californium spectrum can, for example, indicate the onset of performance degradation caused by radiation damage over prolonged periods.

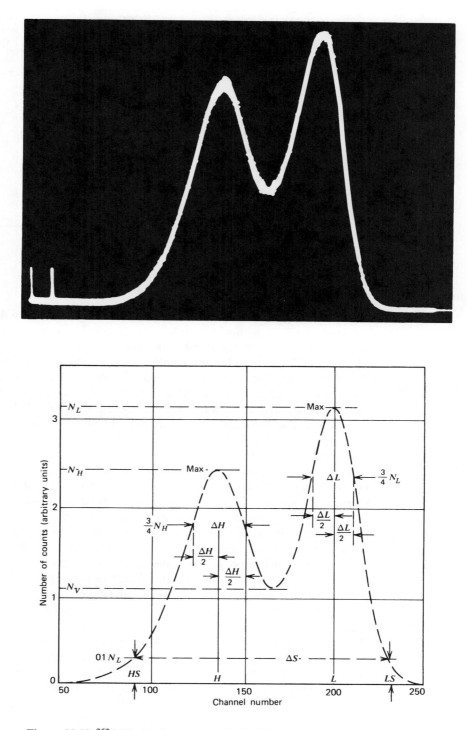

Figure 11.18 ^{252}Cf fission fragment pulse height spectrum. The spectrum parameters defined on the diagram can be used for energy calibration and detector evaluation (see text). (From Bozorgmanesh[85] and Schmitt and Pleasonton.[95])

Table 11.2 Parameters of the ^{252}Cf Fission Fragment Spectrum

Spectrum Parameter[a]	Reasonable Limit	Expected Value
N_L/N_V	> 2.85	~ 2.9
N_H/N_V	~2.2	~ 2.2
N_L/N_H	—	~ 1.30
$\Delta L/(L - H)$	< 0.38	~ 0.36
$\Delta H/(L - H)$	≤ 0.45	≤ 0.44
$(H - HS)/(L - H)$	< 0.70	≤ 0.69
$(LS - L)/(L - H)$	≤ 0.49	≤ 0.48
$(LS - HS)/(L - H)$	≤ 2.18	~ 2.17

[a]Definitions are given in Fig. 11.18.

Source: Schmitt and Pleasonton.[95]

D. Energy Loss Measurements—Particle Identification

Previously, we gave examples of charged particle spectroscopy that involved totally stopping these particles within the depletion region of the semiconductor detector. Neglecting the pulse height defect, the number of charge carriers created is proportional to the total energy of the incident radiation. Applications sometimes arise in which the specific energy loss dE/dx of the incident radiation is of interest, rather than its total energy. For these applications, detectors that are thin compared with the particle range are chosen. The number of charge carriers created within a detector of small thickness Δt will simply be $(dE/dx)\Delta t/\epsilon$. The particle passes completely through the detector, retaining most of its initial energy, and a signal proportional to dE/dx is observed. In such applications, the device is often called a ΔE detector.

A number of detector types are used for such measurements. Thin film scintillators discussed in Chapter 8 can be made in uniform thicknesses that are the smallest of any solid detector, but they do not offer good energy resolution. A totally depleted detector, on the other hand, can be manufactured from a semiconductor wafer as thin as 10 μm and will provide the excellent energy resolution generally observed for all solid-state detectors. The thickness uniformity of transmission detectors is very important if the inherently good energy resolution is to be preserved. For example, a variation of only 1 μm in a wafer of 20-μm thickness will introduce a 5% variation in the signal—considerably larger than a typical energy resolution figure. The fragile nature of such thin semiconductor detectors presents a potential operational problem, and efforts have been made to integrate the thin and thick silicon detectors into a single unit using wafer bonding.[98] Alternatively, monolithic combinations of ΔE and E detectors produced by ion implantation[99] have also been demonstrated.

For low-energy particles, the range may be too small to allow the use of even the thinnest available silicon detector as a ΔE detector. A number of applications[100–104] therefore use a gas-filled ionization or proportional counter for this purpose. Their advantage is that their thickness can be made very uniform and adjustable by changing the gas pressure, but the limiting energy resolution is theoretically poorer. In many applications, however, the ΔE signal fluctuations are dominated by energy straggling of the incident particles, and the energy resolution of gas-filled detectors may be adequate.

Transmission detectors are commonly used in conjunction with a normal surface barrier or other "thick" detector in the *particle identifier telescope* arrangement diagrammed in Fig. 11.19. By accepting only those events that occur in coincidence between the two detectors, a simultaneous measurement of dE/dx and E is carried out for each incident particle.

Figure 11.19 (a) A particle identifier arrangement consisting of tandem ΔE and E detectors operated in coincidence. (b) Experimental spectrum obtained for the $\Delta E \cdot E$ signal product for a mixture of different ions. (From Bromley.[105])

For nonrelativistic charged particles of mass m and charge ze, Bethe's formula [Eq. (2.2)] predicts that

$$\frac{dE}{dx} = C_1 \frac{mz^2}{E} \ln C_2 \frac{E}{m} \tag{11.28}$$

where C_1 and C_2 are constants. If we form the product $E(dE/dx)$, the result is only mildly dependent on the particle energy but is a sensitive indicator of the mz^2 value that characterizes the particle involved. If the incident radiation consists of a mixture of different particles whose energies do not differ by large factors, the product of the pulse amplitudes from both detectors will therefore be a nearly unique parameter for each different particle type. Because the incident energy can be obtained by summing the pulse amplitudes from the ΔE and E detectors, simultaneous determination of both mass and energy of each incident particle is therefore possible. Figure 11.19 shows a distribution of the $E\Delta E$ product for a particle identifier telescope in which a large number of individual particle types can be distinguished.

An alternative approach described by Goulding and Landis[106] is based on the observation that the range R and energy E for a wide variety of charged particles can be related by a power-law approximation (see Fig. 2.7)

$$R(E) = aE^b \tag{11.29}$$

where a and b are constants. If the incident particle deposits an energy ΔE in the transmission detector of thickness Δt, and the remainder of its energy E_r in the thick detector, then

$$\Delta t = R(E_r + \Delta E) - R(E_r)$$

or

$$\frac{\Delta t}{a} = (E_r + \Delta E)^b - E_r^b \tag{11.30}$$

The value of a is a constant for a given particle type roughly proportional to $1/mz^2$. The value of b does not change greatly if ions of similar mass are involved ($b = 1.73$ for protons and $b = 1.65$ for carbon ions). Therefore, choosing a reasonable value for b and using a signal-processing element to carry out the operations on ΔE and E_r indicated in Eq. (11.30) will yield a parameter that is characteristic of the particle type and independent of its energy.

Particle identification schemes based on the above methods and their subsequent modifications are compared in the review article by Goulding and Harvey.[107] Their applicability is always limited by the approximations made to the actual energy loss or range behavior of the particles that are involved. An alternative "brute force" approach,[108] which can avoid any approximations, is to make use of a table of actual energy loss data stored in a computer memory and to carry out on-line digital comparisons with these data. The fundamental limit to the ability of any scheme based on ΔE and E measurements to distinguish particle types generally is set by the fluctuations in the ΔE signal produced by energy straggling.[109]

E. X-ray Spectroscopy with Silicon *p-i-n* Diodes

Even though silicon has the relatively low atomic number of 14, photoelectric absorption of incident photons is still predominant in the soft X-ray region below 20 keV. While the lithium-drifted and other silicon configurations described in Chapter 13 have come to predominate for X-ray spectroscopy, a simpler silicon *p-i-n* diode can be a useful alternative. In this configuration, a high resistivity *i*-region is provided with *p* and *n* noninjecting contacts at either surface to help reduce the leakage current below that which would be observed with a simple diode. A typical thickness of 300 micrometers is sufficient to provide useful detection efficiency up to 20 or 30 keV. The relatively small number of electron–hole pairs created by a low energy X-ray photon requires that the noise level of the detector be reduced to an absolute minimum. For this reason, these diodes are often cooled to further reduce the leakage current to insignificant levels. Cooling to about −30°C, using a compact thermoelectric cooler, is commonplace. These *p-i-n* diodes have found useful application in the measurement of fluorescent X-rays and are commercially available with energy resolution values down to about 250 eV FWHM measured at a 5.9-keV X-ray energy. While this energy resolution is not as good as that achievable with the alternative instruments described in Chapter 13, it is sufficient to be useful in a number of practical applications of X-ray spectroscopy.

F. Current Mode Operation

The overwhelming majority of applications of silicon junction detectors involve pulse mode operation to take advantage of their excellent pulse height resolution. However, it has been demonstrated[32] that these detectors can also be useful for monitoring high radiation fields when operated in current mode. In these applications, the output current is directly measured from the detector and has been demonstrated to have excellent proportionality to the intensity of the incident radiation up to output currents of several amperes. In any such application, one must bear in mind the susceptibility of semiconductor detectors to radiation damage and limit the integrated particle flux below those listed on p. 388 as sufficient to cause appreciable internal detector damage.

It is possible to measure radiation-induced current from a semiconductor detector, even in the absence of an applied voltage.[110,111] This *photovoltaic* mode of operation is similar to that of a common solar cell, except that the charge carriers are created by the ionizing radiation rather than by incident light. As can be seen from Fig. 11.8, the simple junction of *n*- and *p*-type semiconductors creates a contact potential and a depletion region. Electrons and holes formed within the depletion region will drift under the influence of the corresponding electric field, and current will be measured in an external circuit connected across the junction. This photovoltaic mode is not widely used because the contact potential is small, amounting to less than 1 V in silicon. The thickness of the depletion region is also small, usually less than the range of charged particles of interest. Junction detectors are therefore normally operated with a large reverse bias voltage to improve their charge collection efficiency and increase the thickness of their active volume. However, the small depletion region present without applied voltage may be adequate to result in a useful current when the incident radiation intensity is sufficiently high.

G. Silicon Diodes as Personnel Monitors

Silicon diodes, generally of the *p-i-n* configuration, have become commercially available as compact radiation sensors to be worn as personnel monitors. When compared with passive monitors such as radiographic film or thermoluminescent dosimeters discussed in Chapter 19, they offer the advantage of real-time readout. Other active detectors such as G-M tubes have been used for similar monitoring, but silicon diodes offer a more compact and rugged radiation detector.

When exposed to X- rays or gamma rays, the silicon diode will produce pulses from secondary electrons that are created either in the active volume of the diode, or in surrounding materials from which secondary electrons reach the detector active volume. Ideally, one count recorded from the diode should represent one unit of exposure, independent of the energy of the incident X-ray or gamma ray. The actual behavior of silicon diodes is very different from this ideal, since the probability of recording a count is not closely related to the dose delivered by that incident photon. As shown in Fig. 11.20, the sensitivity for a typical diode in units of counts/mR falls off by several orders of magnitude as the incident energy progresses from the keV to the MeV range. Just as in the case of G-M tubes (see Chapter 7), designers attempt to flatten this sensitivity as a function of energy by employing metallic absorbers around the detector to achieve what is normally called "energy compensation." These absorbers have the effect of differentially attenuating the lower energies that are incident on the device, and also can boost the response at high energies because of the greater yield of electrons they produce that can enter the diode from outside its volume. Figure 11.20 also shows the response of the same diode with various thicknesses of copper and tin filters. Although these filters help in flattening the response somewhat, there is still a large deviation from uniform sensitivity over the full energy range. Therefore if the

Figure 11.20 Calculated sensitivities (in counts/mR) of a typical photodiode to X- or gamma-ray photons as a function of their energy. Plots are shown for the bare photodiode as well as for the cases in which filters of the composition and thicknesses shown were included. (From Chen et al.[112])

monitors are employed in circumstances in which the incident spectrum is variable, substantial errors will result if they are used far from the calibration energy.

Rather than operating the silicon diode in simple pulse counting mode, some improvements in its applicability to gamma ray dosimetry can be accomplished by different modes of operation. Because the atomic number of silicon (14) is not greatly different from that of soft tissue, the mix of conversion processes that convert gamma rays photon energy into energetic electrons is similar. In fact, the mass absorption coefficients for silicon are within 10% of soft tissue values from 150 keV to well over 1 MeV. Therefore, a current-mode measurement of the response of a silicon diode will be a reasonable surrogate for the dose rate delivered by the same gamma ray photons in tissue. As an alternative offering higher sensitivity than obtainable with current mode operation, the pulses from the detector can be electronically weighted in proportion to their amplitude and summed to give a signal that has similar properties.

Silicon diodes are also incorporated into commercially-available personnel monitors intended to respond to neutron exposures. Used in conjunction with converters such as ^{6}Li, ^{10}B, or ^{157}Gd that are detailed in Chapter 14, pulses from reaction products produced by slow neutrons can be recorded from the diode. The use of a hydrogen-containing converter such as a thin layer of polyethylene extends the response to fast neutrons through the recoil protons (see Chapter 15) detected by the diode. In some designs, a pair of diodes using converters of both types are used to provide sensitivity over the full range of neutron energies. By taking a weighted sum of the output of both diodes, the combined detection efficiency can be made to approach the fluence-to-dose variation with the neutron energy illustrated in Chapter 2. If this ideal is met, the monitor responds to neutrons in proportion to their importance in delivering effective dose equivalent, and can then be applied in circumstances in which there is little or no knowledge of the incident neutron energy spectrum. Actual instruments depart from this ideal, but can produce results that are within a factor of two of the desired response when applied in a wide assortment of spectra encountered around common nuclear facilities.[113] The sensitivity to gamma ray photons can be suppressed because of the small pulse amplitudes they produce compared with the neutron-induced reaction products when measured using diodes with thin depletion regions.

PROBLEMS

11.1 From the values for intrinsic carrier densities given in Table 11.1, estimate the impurity levels (in parts per billion) that begin to change intrinsic silicon and germanium into doped materials at room temperature.

11.2 Find the ratio of the number of charge carriers created in silicon by a 1-MeV proton to the number created by the same energy deposition in air.

11.3 From the data in Table 11.1, calculate the mean value and variance in the number of electron–hole pairs created by the loss of 100 keV of particle energy in silicon.

11.4. By what factor is the rate of thermal generation of electron–hole pairs in germanium reduced by cooling from room temperature to liquid nitrogen temperature (77 K)?

11.5 In order to reverse bias the junction, should the positive terminal of the voltage supply be connected to the *p* or *n* side of the junction? Justify your answer.

11.6 Indicate the functional dependence of the following properties of a *p-n* junction on the magnitude of the applied reverse bias:

(a) Depletion width.

(b) Capacitance.

(c) Maximum electric field.

11.7 There is often a premium on fabricating detectors with the largest possible depletion width for a given applied voltage. Explain why starting with semiconductor material of the highest available purity enhances this objective.

11.8 Use the nomogram of Fig. 11.10 to find the bias voltage necessary to create a depletion depth of 0.1 mm in a junction detector prepared from *n*-type silicon with 1000 Ω-cm resistivity.

11.9 The primary alpha peak from a ^{241}Am calibration source was centered in channel number 461 of a multichannel analyzer when the alpha particles were collimated to be perpendicular to the surface of a silicon junction detector. The geometry was then changed to cause the alpha particles to be incident at an angle of 35° to the normal, and the observed peak shifted to channel number 455. Assuming no zero offset in the MCA, find the dead layer thickness in units of alpha energy loss.

11.10 Why does the typical energy resolution for surface barrier detectors worsen as the surface area of the detector increases?

11.11 A totally depleted silicon detector with 0.1-mm thickness is operated with large overbias so as to saturate the carrier velocities everywhere within the wafer. Estimate the maximum electron and hole collection times.

11.12 A partially depleted silicon surface barrier is operated with sufficient bias voltage to ensure that the depletion depth greatly exceeds the range of incident 5 MeV alpha particles. If used with a voltage-sensitive preamplifier, how much will the pulse amplitude change if the bias voltage changes by 5%?

11.13 A 10-MBq source of alpha particles is located 10 cm in front of a silicon surface barrier detector. After what length of exposure time is radiation damage to the detector likely to become significant?

11.14 In a given heavy ion detector in which zero energy deposition corresponds to channel zero in an associated multichannel pulse height analyzer, the 5.486 MeV alpha particles from ^{241}Am are recorded in channel number 116. If heavy ions of 21.0 MeV energy are recorded in channel 402, what is their pulse height defect?

11.15 Sketch the expected differential pulse height spectra from a silicon surface barrier detector under the following conditions:

(a) Incident 5 MeV alpha particles, depletion depth of the detector greater than the alpha range.

(b) On the same graph, the corresponding spectrum for a depletion depth of one-half the alpha range.

(c) Again on the same graph, the spectrum with depletion depth as in part (a), but for the same alpha particles after they have passed through an absorber whose thickness is equal to one-half the alpha range in the absorber material.

11.16 A planar germanium detector is operated at a temperature of 77 K with a nearly uniform electric field value of 1000 V/cm throughout its volume. Charges are created at a point within the detector volume by the interaction of a low-energy X-ray photon. Using the mobility values given in Table 11.1, estimate the spatial broadening that will occur due to diffusion of the cloud of electrons as they drift over a distance of 1 cm.

REFERENCES

1. G. Dearnaley and D. C. Northrop, *Semiconductor Counters for Nuclear Radiations*, 2nd ed., Wiley, New York, 1966.
2. G. Bertolini and A. Coche (eds.), *Semiconductor Detectors*, Elsevier-North Holland, Amsterdam, 1968.
3. G. T. Ewan, "Semiconductor Spectrometers," in *Progress in Nuclear Techniques and Instrumentation*, Vol. 3, F. J. M. Farley (ed.), Elsevier-North Holland, New York, 1968.
4. A. Kuhn, *Halbleiter und Kristallzähler*, Akademische Verlagsgesellschaft, Geest and Portig, Leipzig, 1969.
5. W. L. Brown, W. A. Higinbotham, G. L. Miller, and R. L. Chase (eds.), *Semiconductor Nuclear-Particle Detectors and Circuits*, Publication 1593, National Academy of Sciences, Washington, DC, 1969.
6. H. Büker, *Theorie und Praxis der Halbleiterdetektoren für Kernstrahlung*, Springer-Verlag, Berlin, 1971.
7. J. M. Palms, P. V. Rao, and R. E. Wood, *Nucl. Instrum. Meth.* **76**, 59 (1969).
8. J. E. Eberhardt, *Nucl. Instrum. Meth.* **80**, 291 (1970).
9. H. R. Bilger, *Phys. Rev.* **163**, 238 (1967).
10. R. H. Pehl and F. S. Goulding, *UCRL* **19530**, 333 (1969).
11. A. H. Sher and W. J. Keery, *IEEE Trans. Nucl. Sci.* **NS-17**(1), 39 (1970).
12. H. R. Zulliger and D. W. Aikten, *IEEE Trans. Nucl. Sci.* **NS-17**(3), 187 (1970).
13. T. Yamaya, R. Asano, H. Endo, and K. Umeda, *Nucl. Instrum. Meth.* **159**, 181 (1979).
14. N. Stroken, V. Ajdacic, and B. Lalovic, *Nucl. Instrum. Meth.* **94**, 147 (1971).
15. G. Ottaviani, C. Canali, and A. Alberigi Quaranta, *IEEE Trans. Nucl. Sci.* **NS-22**(1), 192 (1975).
16. W. Van Roosbroeck and W. Shockley, *Phys. Rev.* **94**, 1558 (1954).
17. R. H. Pehl, F. S. Goulding, D. A. Landis, and M. Lenzlinger, *Nucl. Instrum. Meth.* **59**, 45 (1968).
18. C. Canali, M. Martini, G. Ottaviani, and A. Alberigi Quaranta, *IEEE Trans. Nucl. Sci.* **NS-19**(4), 9 (1972).
19. R. D. Ryan, *IEEE Trans. Nucl. Sci.* **NS-20**(1), 473 (1973).
20. R. A. Langley, *Nucl. Instrum. Meth.* **113**, 109 (1973).
21. A. Owens et al., *Nucl. Instrum. Meth.* **A382**, 503 (1996).
22. P. Lechner, R. Hartmann, H. Soltau, and L. Strüder, *Nucl. Instrum. Meth.* **A377**, 206 (1996).
23. B. G. Lowe, *Nucl. Instrum. Meth.* **A399**, 354 (1997).
24. C. A. Klein, *IEEE Trans. Nucl. Sci.* **NS-15**(3), 214 (1968).
25. E. Perotti and C. Fiorini, *Nucl. Instrum. Meth.* **A423**, 356 (1999).
26. J. L. Blankenship, *IRE Trans. Nucl. Sci.* **NS-7**(2-3), 190 (1960).
27. N. J. Hansen, R. G. Scott, and D. J. Henderson, *Nucl. Instrum. Meth.* **104**, 333 (1972).
28. H. R. Zulliger, W. E. Drummond, and L. M. Middleman, *IEEE Trans. Nucl. Sci.* **NS-19**(3), 306 (1972).
29. H. Grahmann and S. Kalbitzer, *Nucl. Instrum. Meth.* **136**, 145 (1976).
30. J. W. Mayer, *Nucl. Instrum. Meth.* **63**, 141 (1968).
31. J. Von Borany, G. Mende, and B. Schmidt, *Nucl. Instrum. Meth.* **212**, 489 (1983).
32. P. Burger, M. O. Lampert, R. Henck, and J. Kemmer, *IEEE Trans. Nucl. Sci.* **NS-31**(1), 344 (1984).
33. I. Ahmad, *Nucl. Instrum. Meth.* **A242**, 395 (1986).
34. J. Kemmer, *Nucl. Instrum. Meth.* **226**, 89 (1984).
35. P. Burger and Y. Beroud, *Nucl. Instrum. Meth.* **226**, 45 (1984).
36. J. M. Jaklevic and F. S. Goulding, *IEEE Trans. Nucl. Sci.* **NS-19**(1), 384 (1972).
37. E. Belcarz et al., *Nucl. Instrum. Meth.* **77**, 21 (1970).
38. H. W. Schmitt et al., *Proceedings of the IAEA Conference on The Physics and Chemistry of Fission*, Salzburg, 1965, p. 531.
39. F. J. Walter, *IEEE Trans. Nucl. Sci.* **NS-11**(3), 232 (1964).
40. A. Alberigi Quaranta, M. Martini, and G. Ottaviani, *IEEE Trans. Nucl. Sci.* **NS-16**(2), 35 (1969).
41. H. M. Mann, J. W. Haslett, and G. P. Lietz, *IRE Trans. Nucl. Sci.* **NS-8**(1), 157 (1961).
42. W. Seibt, K. E. Sundstrom, and P. A. Tove, *Nucl. Instrum. Meth.* **113**, 317 (1973).
43. E. C. Finch, *Nucl. Instrum. Meth.* **121**, 431 (1974).
44. E. C. Finch, *Nucl. Instrum. Meth.* **129**, 617 (1975).
45. C. F. G. Delaney and E. C. Finch, *Nucl. Instrum. Meth.* **215**, 219 (1983).
46. I. Kanno, *Rev. Sci. Instrum.* **58**(10), 1926 (1987).
47. I. Kanno, T. Inbe, S. Kanazawa, and I. Kimura, *Rev. Sci. Instrum.* **65**(9), 3040 (1994).
48. R. N. Williams and E. M. Lawson, *Nucl. Instrum. Meth.* **120**, 261 (1974).
49. H. Henschel, H. Hipp, A. Kohnle, and F. Gonnenwein, *Nucl. Instrum. Meth.* **125**, 365 (1975).
50. E. C. Finch, C. F. G. Delaney, and M. Asghar, *IEEE Trans. Nucl. Sci.* **NS-27**(1), 286 (1980).
51. E. C. Finch, A. A. Cafolla, and M. Asghar, *Nucl. Instrum. Meth.* **198**, 547 (1982).
52. H.-O. Neidel, H. Henschel, H. Geissel, and Y. Laichter, *Nucl. Instrum. Meth.* **212**, 299 (1983).
53. R. Butsch, J. Pochodzalla, and B. Heck, *Nucl. Instrum. Meth.* **228**, 586 (1985).
54. C. Inskeep, E. Elad, and R. A. Sareen, *IEEE Trans. Nucl. Sci.* **NS-21**(1), 379 (1974).
55. G. Dearnaley, *IEEE Trans. Nucl. Sci.* **NS-11**(3), 249 (1964).
56. P. Mulas and E. L. Haines, *Rev. Sci. Instrum.* **40**, 507 (1969).
57. K. Livingston, P. J. Woods, T. Davinson, and A.C. Shotter, *Nucl. Instrum. Meth.* **A370**, 445 (1996).
58. R. Wunstorf, *IEEE Trans. Nucl. Sci.* **44**(3), 806 (1997).
59. F. Shiraishi, *Nucl. Instrum. Meth.* **69**, 316 (1969).
60. Y. M. Liu and J. A. Coleman, *IEEE Trans. Nucl. Sci.* **NS-18**(1), 192 (1971).
61. H. J. Ziock et al., *IEEE. Trans. Nucl. Sci.* **40**(4), 344 (1993).
62. J. A. Coleman, D. P. Love, J. H. Trainor, and D. J. Williams, *IEEE Trans. Nucl. Sci.* **NS-15**(3), 363 (1968).
63. K. Ohba, T. Shoji, S. Ito, and J. Hiratate, *IEEE Trans. Nucl. Sci.* **NS-30**(1), 371 (1984).
64. F. A. Hanser and B. Sellers, *Rev. Sci. Instrum.* **41**, 780 (1970).
65. G. Dearnaley and A. B. Whitehead, *Nucl. Instrum. Meth.* **12**, 205 (1961).
66. R. W. Klingensmith, *IRE Trans. Nucl. Sci.* **NS-8**(1), 112 (1961).
67. J. Matheson et al., *Nucl. Instrum. Meth.* **A371**, 575 (1996).
68. R. Wheadon, *Nucl. Instrum. Meth.* **A386**, 143 (1997).
69. G. Hall, *Nucl. Instrum. Meth.* **A388**, 283 (1997).
70. K. W. Kemper and J. D. Fox, *Nucl. Instrum. Meth.* **105**, 333 (1972).
71. W. N. Lennard et al., *Nucl. Instrum. Meth.* **A248**, 454 (1986).
72. R. N. Chanda and R. A. Deal, "Catalogue of Semiconductor Alpha-Particle Spectra," IN-1261 (1970).

73. B. D. Wilkins et al., *Nucl. Instrum. Meth.* **92**, 381 (1971).
74. E. C. Finch and A. L. Rodgers, *Nucl. Instrum. Meth.* **113**, 29 (1973).
75. H. Wohlfarth et al., *Nucl. Instrum. Meth.* **140**, 189 (1977).
76. E. C. Finch et al., *Nucl. Instrum. Meth.* **142**, 539 (1977).
77. E. C. Finch et al., *Nucl. Instrum. Meth.* **228**, 402 (1985).
78. E. P. Steinberg, S. B. Kaufman, B. D. Wilkins, and C. E. Gross, *Nucl. Instrum. Meth.* **99**, 309 (1972).
79. M. Ogihara, J. Nagashima, W. Galster, and T. Mikumo, *Nucl. Instrum. Meth.* **A251**, 313 (1986).
80. E. C. Finch, *Nucl. Instrum. Meth.* **A257**, 381 (1987).
81. E. L. Haines and A. B. Whitehead, *Rev. Sci. Instrum.* **37**, 190 (1966).
82. G. Forcinal, P. Siffert, and A. Coche, *IEEE Trans. Nucl. Sci.* **NS-15**(1), 475 (1968).
83. E. C. Finch, M. Asghar, and M. Forte, *Nucl. Instrum. Meth.* **163**, 467 (1979).
84. I. Kanno, *Jour. Nucl. Sci. Tech.* **29**(7), 690 (1992).
85. H. Bozorgmanesh, Ph.D. dissertation, The University of Michigan, 1976.
86. T. H. Zabel et al., *Nucl. Instrum. Meth.* **174**, 459 (1980).
87. E. Steinbauer et al., *Nucl. Instrum. Meth.* **B85**, 642 (1994).
88. G. D. Alkhazov, A. P. Komar, and A. A. Vorob'ev, *Nucl. Instrum. Meth.* **48**, 1 (1967).
89. E. Steinbauer et al., *Nucl. Instrum. Meth.* **A339**, 102 (1994).
90. J. B. A. England and V. W. Hammer, *Nucl. Instrum. Meth.* **96**, 81 (1971).
91. J. B. A. England, *Nucl. Instrum. Meth.* **102**, 365 (1972).
92. E. D. Klema, J. X. Saladin, J. G. Alessi, and H. W. Schmitt, *Nucl. Instrum. Meth.* **178**, 383 (1980)
93. E. D. Klema, F. J. Camelio, and T. K. Saylor, *Nucl. Instrum. Meth.* **225**, 72 (1984).
94. J. C. Overley and H. W. Lefevre, *Nucl. Instrum. Meth.* **B10/11**, 237 (1985).
95. H. W. Schmitt and F. Pleasonton, *Nucl. Instrum. Meth.* **40**, 204 (1966).
96. H. W. Schmitt, W. E. Kiker, and C. W. Williams, *Phys. Rev.* **137**, B837 (1965).
97. E. Weissenberger et al., *Nucl. Instrum. Meth.* **A248**, 506 (1986).
98. G. Thungström et al., *Nucl. Instrum. Meth.* **A391**, 315 (1997).
99. G. Cardella et al., *Nucl. Instrum. Meth.* **A378**, 262 (1996).
100. J. Barrette, P. Braun-Munzinger, and C. K. Gelbke, *Nucl. Instrum. Meth.* **126**, 181 (1975).
101. D. R. Maxson, D. C. Palmer, and J. P. Bading, *Nucl. Instrum. Meth.* **142**, 479 (1977).
102. B. Sundqvist et al., *IEEE Trans. Nucl. Sci.* **NS-24**(1), 652 (1977).
103. E. Rosario-Garcia and R. E. Benenson, *Nucl. Instrum. Meth.* **143**, 245 (1977).
104. A. Honkanen et al., *Nucl. Instrum. Meth.* **A395**, 217 (1997).
105. D. A. Bromley, *IRE Trans. Nucl. Sci.* **NS-9**(3), 135 (1962).
106. F. S. Goulding and D. A. Landis, "Recent Advances in Particle Identifiers at Berkeley," in *Semiconductor Nuclear-Particle Detectors and Circuits*, Publication 1593, National Academy of Sciences, Washington, DC, 1969, p. 757.
107. F. S. Goulding and B. G. Harvey, *Ann. Rev. Nucl. Sci.* **25**, 167 (1975).
108. D. G. Perry and L. P. Remsberg, *Nucl. Instrum. Meth.* **135**, 103 (1976).
109. A. G. Seamster, R. E. L. Green, and R. G. Korteling, *Nucl. Instrum. Meth.* **145**, 583 (1977).
110. S. C. Klevenhagen, *Phys. Med. Biol.* **22**, 353 (1977).
111. A. Maruhashi, *Nucl. Instrum. Meth.* **141**, 87 (1977).
112. C–R. Chen et al., *IEEE Trans. Nucl. Sci.* **40**(4), 857 (1993).
113. M. Sasaki et al., *Nucl. Instrum. Meth.* **A418**, 465 (1998).

Chapter **12**

Germanium Gamma-Ray Detectors

I. GENERAL CONSIDERATIONS

The simple junction and surface barrier detectors discussed in Chapter 11 find widespread application for the detection of alpha particles and other short-range radiations but are not easily adaptable for applications that involve more penetrating radiations. Their major limitation is the maximum depletion depth or active volume that can be created. Using silicon or germanium of normal semiconductor purity, depletion depths beyond 2 or 3 mm are difficult to achieve despite applying bias voltages that are near the breakdown level. Much greater thicknesses are required for the detectors intended for gamma-ray spectroscopy, which are the topic of this chapter. From Eq. (11.18) the thickness of the depletion region is given by

$$d = \left(\frac{2\epsilon V}{eN} \right)^{1/2} \tag{12.1}$$

where V is the reverse bias voltage and N is the net impurity concentration in the bulk semiconductor material. (ϵ is the dielectric constant and e is the electronic charge.) At a given applied voltage, greater depletion depths can only be achieved by lowering the value of N through further reductions in the net impurity concentration.

There are two general approaches that can be taken to accomplish this goal. The first is to seek further refining techniques capable of reducing the impurity concentration to approximately 10^{10} atoms/cm^3. At this impurity level in germanium, Eq. (12.1) predicts that a depletion depth of 10 mm can be reached using a reverse bias voltage of less than 1000 V. However, such a low impurity concentration corresponds to levels that are less than 1 part in 10^{12}, a virtually unprecedented degree of material purity. Techniques have been developed to achieve this goal in germanium, but not in silicon.[†] Detectors that are manufactured from this ultrapure germanium are usually called *intrinsic germanium* or *high-purity germanium* (HPGe) detectors, and they have become available with depletion depths of several centimeters.

The second approach to reducing net impurity concentration is to create a compensated material in which the residual impurities are balanced by an equal concentration of dopant atoms of the opposite type. This compensation cannot be carried out simply by adding the appropriate amount of dopant to the semiconductor before the crystal is grown, because the balance between acceptors and donors will never be exactly right. Therefore, the material always turns out to be *n* or *p* type depending on which impurity shows a predominance, no matter how slight. Instead, the process of *lithium ion drifting* has been applied in both silicon and germanium crystals to compensate the material after the

---

[†]One reason for this difference is that the higher melting point for silicon (1410°C versus 959°C for germanium) makes the exclusion of impurities in the refining process more difficult.

crystal has been grown. Residual acceptor impurities are exactly balanced over a thickness of up to 2 cm by the addition of interstitial lithium donor atoms. The resulting compensated material has many of the properties of intrinsic or pure material. Even if the compensation is not perfect, the residual net impurity level may be low enough so that the drifted region can easily be depleted over its entire thickness.

Germanium detectors produced by the lithium drifting process are given the designation Ge(Li). They became commercially available in the early 1960s and served as the common type of large-volume germanium detector for two decades. The widespread availability of high-purity germanium in the early 1980s provided an alternative to lithium drifting, and manufacturers have now discontinued production of Ge(Li) detectors in favor of the HPGe type. A major reason for this evolution, discussed later in this chapter, is the much greater operational convenience afforded by HPGe detectors. Whereas Ge(Li) detectors must be continuously maintained at low temperature, HPGe detectors can be allowed to warm to room temperature between uses. In general, the important performance characteristics such as detection efficiency and energy resolution are essentially identical for Ge(Li) and HPGe detectors of the same size, so the application examples shown later in this chapter are generally applicable to both types. A useful collection of test procedures and standard nomenclature for germanium detectors can be found in the ANSI/IEEE Standard #325.[1] Procedures involved in their manufacture are detailed in Ref. 2.

In silicon, large depletion thicknesses can only be achieved by lithium drifting because of the present limit on available purity. A detailed discussion of the lithium-drifting process is therefore postponed until Chapter 13 in which lithium-drifted silicon detectors are described.

II. CONFIGURATIONS OF GERMANIUM DETECTORS

A. High-Purity Germanium (HPGe) Detector Fabrication

Techniques for the production of ultrapure germanium with impurity levels as low as 10^{10} atoms/cm^3 first were developed in the mid 1970s. The starting material is bulk germanium intended for use in the semiconductor industry. This material, already of high purity, is further processed using the technique of zone refining. The impurity levels are progressively reduced by locally heating the material and slowly passing a melted zone from one end of the sample to the other. Since impurities tend to be more soluble in the molten germanium than in the solid, impurities are preferentially transferred to the molten zone and are swept from the sample. After many repetitions of this process, impurity levels as low as 10^9 atoms/cm^3 have been achieved. The resulting germanium is perhaps the most highly purified and completely analyzed material of any kind that has ever been produced in commercial volume. Large single crystals of germanium are then slowly grown from this purified feedstock. If the remaining low-level impurities are acceptors (such as aluminum) the electrical properties of the semiconductor crystal grown from the material is mildly p type. (The designation π type is often used to represent this high-purity p-type material.) Alternatively, if donor impurities remain, high purity n type (designated ν type) is the result. Figure 12.1 shows the electrical conductivity of germanium as a function of the net donor or acceptor concentrations ("dopants") that are present. Practical germanium detectors are always fabricated from material in which the net dopant concentration is very small and the conductivity is correspondingly low. In more widely used nomenclature, this condition is usually identified as being "high-resistivity material."

B. Planar Configuration

A representative configuration for a *planar* HPGe detector fabricated from high-purity p-type (or π-type) germanium is shown in Fig. 12.2. In a planar configuration, the electrical contacts are provided on the two flat surfaces of a germanium disk. The n^+ contact can be

Figure 12.1 The variation of electrical conductivity in germanium as a function of net dopant concentration. Perfectly pure (intrinsic) material is at the zero point along the abscissa.

formed either by lithium evaporation and diffusion onto one surface of the wafer, or by direct implantation of donor atoms using an accelerator. The detector depletion region is formed by reverse biasing this n^+-p junction. The contact at the opposite face of the crystal must be a noninjecting contact for a majority carrier. It may consist of a p^+ contact produced by ion implantation of acceptor atoms, or a metal–semiconductor surface barrier that acts as the electrical equivalent. Ion implantation techniques have the advantage that the contact layers can be made very thin to serve as entrance windows for weakly penetrating radiations such as low-energy X-rays. The formation of p-type contacts is straightforward by implanting boron. Producing n-type contacts is less common since the radiation damage caused by the implantation process produces acceptor sites in germanium. Contacts that are n-type can be made by phosphorus implantation, but a difficult annealing step is required to ensure that these donor atoms become electrically active. Consequently, the majority of n-type contacts are still produced by lithium diffusion, resulting in a somewhat thicker surface dead layer.

High-purity germanium detectors are generally operated as fully depleted detectors (see Chapter 11). Reverse biasing requires that a positive voltage be applied to the n^+ contact with respect to the p^+ surface. The depletion region effectively begins at the n^+ edge of the central region and extends deeper into the π region as the voltage is raised until the detector becomes fully depleted. Further increases in the voltage then serve to increase the electric field everywhere in the detector by a uniform amount (see Fig. 11.12). It is preferable to apply sufficient overvoltage so that the minimum electric field is high enough to impart saturated drift velocities to the charge carriers, minimizing the collection time and the detrimental effects due to carrier recombination and trapping. In germanium at the normal operating temperature of 77 K, saturated electron velocities are reached at a minimum field of about 10^5 V/m, but field strengths three to five times larger are required to fully saturate the hole velocity (see Fig. 11.2). Practical problems related to breakdown and surface leakage often limit the maximum voltage (typically 3–5 kV) to values at which electrons but not holes will reach saturated drift velocity.

Figure 12.2 Configuration of a planar HPGe detector. The Ge semiconductor may be ν type (p^+ contact is rectifying), π type (n^+ contact is rectifying), or lithium drifted. See Fig. 11.13 for sketches of the electric field shape for these three cases.

Germanium detectors may also be fabricated starting with high-purity n-type material (also known as v-type) rather than with p-type material as described above. In this case, n^+ and p^+ contacts are again provided at each surface of the wafer. Reverse biasing still requires the application of a positive voltage to the n^+ contact with respect to the p^+ surface. However, the roles of the two contacts are reversed. The p^+ layer now serves as the rectifying contact and the depletion region extends from this contact as the voltage is raised. The n^+ contact now functions as a *blocking* or noninjecting contact in which the population of holes is very low.

A third type of diode configuration represents the case of lithium-drifted detectors. In this case, the lithium-drifting process (see p. 458) is used to convert the bulk of the wafer to compensated or *intrinsic* material. Here the concentrations of donors and acceptors exactly balance, and the material (often designated *i type*) is neither n nor p type. Again, n^+ and p^+ contacts are provided on each face of the wafer. In this case, there is no distinction between the rectifying contact and the blocking contact (see Fig. 11.13).

When fully depleted and operated with a large overvoltage, the electric field in planar detectors is almost uniform from one contact to the other. Holes or electrons therefore drift under the influence of a nearly constant field throughout the detector active volume. Under these circumstances, both the n^+-π-p^+ configuration and the n^+-v-p^+ configuration described above show very similar properties as planar geometry detectors. In contrast, the electric field is quite nonuniform in the coaxial geometry detectors discussed below, and therefore the configuration choice has a more significant influence on the details of the charge carrier motion.

C. Coaxial Configuration

For the planar detectors outlined above, the diameter of the cylindrical crystal from which the wafer is cut is typically no more than a few centimeters. The maximum depletion depth (or thickness of the lithium-drifted region) is limited to less than 1 or 2 cm. The total active volume available in planar detectors therefore does not exceed about 10–30 cm³. To produce a detector with a larger active volume as needed in gamma-ray spectroscopy, a different approach is used. The detector is then constructed in cylindrical or *coaxial* geometry as illustrated in Fig. 12.3. In this case, one electrode is fabricated at the outer cylindrical surface of a long germanium cylindrical crystal. A second cylindrical contact is provided by removing the core of the crystal and placing a contact over the inner cylindrical surface. Because the crystal can be made long in the axial direction, much larger active volumes can be produced (up to 750 cm³ at the time of this writing). An added advantage of coaxial geometry is that, by using a small inner diameter, large-volume detectors can be fabricated with lower capacitance than would be possible using planar geometry.

A *closed-ended coaxial* configuration is one in which only part of the central core is removed and the outer electrode is extended over one flat end of the cylindrical crystal (see Fig. 12.3). Most commercial fabricators of HPGe detectors choose the closed-ended configuration over a true coaxial geometry to avoid the complications of dealing with leakage currents at the front surface. Also, the closed-ended configuration provides a planar front surface that can serve as an entrance window for weakly penetrating radiations if fabricated with a thin electrical contact. In the closed-ended configuration, the electric field lines are no longer completely radial as they are in the true coaxial case, and there is some tendency to produce regions of reduced field strength near the corners of the crystal where carrier velocities may be lower than normal. Extending the central hole to reach close to the front surface helps keep the field lines as nearly radial as possible, and some manufacturers round the front corners of the crystal and hole (called *bulletizing*) to help eliminate the low-field regions.

Coaxial HPGe detectors are also available in *well* configurations in which the housing is shaped to allow external access to the central hole. Small radioisotope sources can then

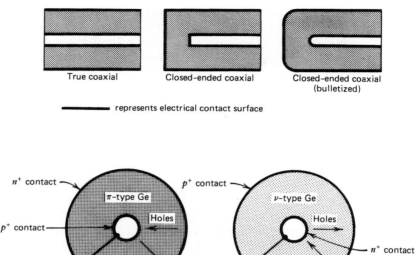

True coaxial   Closed-ended coaxial   Closed-ended coaxial
                                                            (bulletized)

━━━ represents electrical contact surface

p-type coaxial   n-type coaxial

Figure 12.3 At the top are shown the three common shapes of large-volume coaxial detectors. Each represents a cross-sectional view through the axis of a cylindrical crystal. The outer electrode is extended over the flat front (left) surface in both closed-ended cases. Cross sections perpendicular to the cylindrical axis of the crystal are shown at the bottom. The HPGe material may be either high-purity p or n type. The corresponding electrode configurations are shown for each type.

be placed within this well for measurements in which the source is nearly surrounded by germanium and the detection efficiency can be unusually high.

In coaxial geometry, the rectifying contact that forms the semiconductor junction can in principle be placed either at the inner or outer surface of the crystal. The electric field conditions that result are quite different. If the rectifying contact is on the outer surface, the depletion layer grows inward as the voltage is raised until reaching the inner hole surface at the depletion voltage. If the inner surface has the rectifying contact, the depletion region grows outward, and a much larger voltage is required to fully deplete the detector volume. The first choice also maintains a higher electric field value in the outer regions of the cylindrical crystal (where most of the volume lies) as the voltage is raised above full depletion. For these reasons, manufacturers universally choose to locate the rectifying contact at the outer surface. Specifically, the outer contact will be n^+ for p-type HPGe and p^+ for n-type HPGe. The inner surface is provided with a blocking contact of the opposite type. The derivations given in the next section of this chapter assume this configuration for coaxial detectors. To provide the electrical output signal, the gate of a field effect transistor (FET) is connected to the inner contact. Thus the high voltage is applied to the outer surface, and the inner contact is near ground potential.

D. Electric Field and Capacitance

The electric field in germanium detectors determines the drift velocity of the charge carriers, so its configuration is important in considerations of pulse shape, timing behavior, and completeness of the charge collection process. The spatial variation of the field strength across the detector active volume is markedly different in planar and coaxial geometry, and the two cases are considered separately in the following discussion.

In each geometry, we solve Poisson's equation

$$\nabla^2 \varphi = -\frac{\rho}{\epsilon} \tag{12.2}$$

to find the electric potential φ in the presence of the charge density ρ (ϵ is the dielectric constant). The charge density is dependent on the material from which the detector is fabricated. For π-type germanium, $\rho = -eN_A$, where e is the electronic charge and N_A is the density of acceptor impurities. This negative charge represents the filled acceptor sites on the π side of the junction. Similarly, the ionized donor sites in v-type germanium represent a positive charge density given by $\rho = eN_D$. In lithium-drifted germanium, impurities of each type are exactly compensated, and $\rho = 0$. In the derivations that follow, we state the results explicitly for the first case of π-type germanium. The corresponding results for v-type and lithium-drifted detectors can be obtained by substituting the appropriate value for ρ.

1. PLANAR GEOMETRY

From the results derived in Chapter 11 for planar diodes [Eq. (11.18)], the detector depletion depth is

$$d = \left(\frac{2\epsilon V}{\rho}\right)^{1/2} \tag{12.3}$$

Full depletion requires a minimum applied voltage V_d (the *depletion voltage*) at which the depletion depth extends entirely across the slab thickness T:

$$V_d = \frac{\rho T^2}{2\epsilon} \tag{12.4}$$

In one-dimensional slab geometry, Poisson's equation [Eq. (12.2)] becomes

$$\frac{d^2\varphi}{dx^2} = -\frac{\rho}{\epsilon} \tag{12.5}$$

For an applied voltage less than that required for full depletion, the electric field $\mathscr{E} = -\text{grad } \varphi = -d\varphi/dx$ is obtained by solution of Eq. (12.5) with the boundary condition

$$\varphi(d) - \varphi(0) = V$$

The result is

$$-\mathscr{E}(x) = \frac{V}{d} + \frac{\rho}{\epsilon}\left(\frac{d}{2} - x\right) \tag{12.6a}$$

or

$$|\mathscr{E}(x)| = \frac{V}{d} + \frac{eN_A}{\epsilon}\left(x - \frac{d}{2}\right) \tag{12.6b}$$

where x is the distance from the p^+ contact. For $V < V_d$, the portion of this solution corresponding to the undepleted region of the detector is not applicable, and the field is zero. Equation (12.6) also holds for $V > V_d$, because the effect of the overvoltage $(V - V_d)$ is to increase the field by a constant amount $(V - V_d)/T$ everywhere within the detector. Plots of some electric field configurations predicted by Eq. (12.6) are shown in Fig. 12.4.

Figure 12.4 The variation of the electric field strength throughout the active volume of a planar HPGe detector for different values of the reverse bias V. The bias value at which full depletion is achieved is labeled V_d. The uniform field present in a Ge(Li) detector of the same dimension is shown for comparison. (From Llacer.[3])

As in surface barrier or junction detectors, the capacitance varies with the applied bias voltage up to the value at which the detector is fully depleted. In planar geometry, the detector capacitance per unit area prior to the point of full depletion is given by

$$C = \left(\frac{\epsilon\rho}{2V}\right)^{1/2} \qquad (12.7)$$

For $V > V_d$, the detector capacitance is a constant obtained by setting $V = V_d$ in Eq. (12.7). The independence of detector capacitance on applied bias is often taken as an indication of full depletion within the detector.

2. COAXIAL GEOMETRY

Poisson's equation [Eq. (12.2)] in cylindrical coordinates becomes

$$\frac{d^2\varphi}{dr^2} + \frac{1}{r}\frac{d\varphi}{dr} = -\frac{\rho}{\epsilon} \qquad (12.8)$$

We treat the case of a true coaxial detector with inner and outer radii of r_1 and r_2. A boundary condition is that the potential difference between these radii is given by the applied voltage V, or $\varphi(r_2) - \varphi(r_1) = V$. Solving Eq. (12.8) for $\mathscr{E}(r) = -d\varphi/dr$, we find that the resulting electric field configuration is

$$-\mathscr{E}(r) = -\frac{\rho}{2\epsilon}r + \frac{V + (\rho/4\epsilon)(r_2^2 - r_1^2)}{r \ln(r_2/r_1)} \qquad (12.9)$$

Figure 12.5 The variation of the electric field strength with radius in a HPGe coaxial detector. For the example shown, a minimum value of the reverse bias voltage V of 2070 V is required to fully deplete the detector. The $1/r$ field variation in a coaxial Ge(Li) detector is shown for comparison. (From Llacer.[3])

or

$$\left| \mathscr{E}(r) \right| = \frac{eN_A}{2\epsilon} r + \frac{V - (eN_A/4\epsilon)(r_2^2 - r_1^2)}{r \ln (r_2/r_1)} \tag{12.10}$$

provided N_A (the acceptor concentration) is constant over the detector volume. Some plots of this field shape for different values of V are shown in Fig. 12.5. More detailed plots are presented in Ref. 4 for different detector dimensions and acceptor concentrations.

The absolute value of the electric field profile remains the same for both the p-type and n-type coaxial configurations shown in Fig. 12.3, provided the same voltage and dopant concentrations are involved. Only the polarity of the field changes, reversing the drift directions of the holes and electrons. Note that both cases maintain the rectifying contact at the outer surface.

The voltage V_d needed to fully deplete the detector can be found by setting $\mathscr{E}(r_1) = 0$ in Eq. (12.9) above. The result is

$$V_d = \frac{\rho}{2\epsilon} \left[r_1^2 \ln\left(\frac{r_2}{r_1}\right) - \frac{1}{2}(r_2^2 - r_1^2) \right]$$

The depletion voltage decreases linearly with ρ (or dopant level) and is theoretically zero for perfectly compensated germanium.

The capacitance per unit length of a fully depleted true coaxial detector is

$$C = \frac{2\pi\epsilon}{\ln\left(r_2/r_1\right)} \tag{12.11}$$

Because there is generally an advantage in minimizing detector capacitance, the radius of the central core r_1 is normally kept to a minimum consistent with allowing the fabrication of an appropriate electrical contact.

E. Surface Dead Layer

To a first approximation, the active volume of a germanium detector is simply the region between the n^+ and p^+ contacts. However, these contacts may have appreciable thickness and can represent a dead layer on the surface of the crystal through which the incident radiation must pass. For example, some traditional methods of fabricating contacts, such as evaporating and diffusing lithium into the surface to form an n^+ layer, can produce thicknesses of several hundred micrometers. For gamma rays of about 200 keV energy or greater, the attenuation in such layers is generally negligible, and the gamma-ray detection efficiency is not appreciably affected by the presence of the dead layer. However, if gamma rays or X-rays of lower energy are to be measured, the presence of layers of this thickness must be avoided to prevent attenuation.

For such applications, the ion implantation technique has become more popular for the formation of the contacts. For example, the p^+ contact can be produced by implanting boron ions that have been accelerated to about 20 keV into the crystal surface. Such contacts are only a few tenths of a micrometer in thickness and can serve as suitable entrance windows for soft X-rays. A large-volume germanium detector fabricated with a thin ion-implanted surface contact will therefore be useful for the measurement of photons over a wide energy range, from X-rays of a few keV through gamma rays of many MeV energy.

The surface dead layer on germanium detectors may vary slowly over periods of time because of the formation of so-called surface channels[5] in which the electric field and charge collection efficiency are reduced. It is therefore advisable that efficiency measurements be repeated periodically, especially at lower gamma-ray energies where these effects will be most significant.

III. GERMANIUM DETECTOR OPERATIONAL CHARACTERISTICS

A. Detector Cryostat and Dewar

Because of the small bandgap (0.7 eV), room-temperature operation of germanium detectors of any type is impossible because of the large thermally-induced leakage current that would result. Instead, germanium detectors must be cooled to reduce the leakage current to the point that the associated noise does not spoil their excellent energy resolution. Normally, the temperature is reduced to 77 K through the use of an insulated dewar in which a reservoir of liquid nitrogen is kept in thermal contact with the detector.

For Ge(Li) detectors, the low temperature must be maintained continuously to prevent a catastrophic redistribution of the drifted lithium that will rapidly take place at room temperature. Lithium drifting is eliminated in HPGe detectors, and they can be allowed to warm to room temperature between uses. Fabrication techniques have been developed to the point that modern detectors will withstand indefinite cycling of the temperature. This operational advantage is the major reason that HPGe detectors have supplanted Ge(Li) configurations.

The detector must be housed in a vacuum-tight cryostat to inhibit thermal conductivity between the crystal and the surrounding air. The cryostat is normally evacuated and sealed by the manufacturer, although a pumping port may be provided to facilitate unusual operations such as high-temperature annealing (see p. 452). In order to provide some passive pumping inside the vacuum space, a molecular sieve or other absorber is normally provided within the sealed volume. A thin end window is usually located near the crystal to minimize attenuation of gamma rays before they enter the germanium. The cryostat can be mounted on the liquid nitrogen dewar in a number of orientations, including horizontal or vertical extensions to suit specific shielding or other application requirements. An example is shown in Fig. 12.6. To avoid the necessity of filling the dewar more frequently than weekly, its capacity must be about 30 liters. The corresponding size and weight of the dewar are much greater than those of the detector itself, and portable operation is difficult. For such applications, some manufacturers provide a smaller dewar with a more limited holding time. Sealed detector capsules are also available so that the same germanium crystal can easily be attached to many different cryostat–dewar configurations (horizontal, vertical, portable, etc.).

Although it is conventional to operate HPGe detectors at liquid nitrogen temperature, in some applications it may be more convenient to allow the detector temperature to rise above this nominal 77 K value. Several studies[6–8] have shown that the performance of coaxial detectors does not suffer until the absolute temperature rises to about 130 K. There are several alternatives to liquid nitrogen cooling that may be used. Mechanical closed-cycle refrigerators have been demonstrated[9,10] that are capable of cooling detectors to as low as 50–60 K, but care must be exercised to prevent electronic noise caused by microphonics from mechanical vibrations. Commercial systems based on the Solvay cycle are sufficiently free of vibration to allow their use while the detector is in operation. Coolers based on the Joule–Thomson effect have also been used in detector applications.[11] Here a supply of high-pressure gas, typically nitrogen at 100 atm, is expanded in a capillary tube to provide the cooling. These systems provide alternatives in situations where liquid nitrogen is either not available or impractical.

Because of the potential condensation of impurity gases on the detector surface, there is a premium on establishing high vacuum conditions in the cryostat. Leakage of water vapor and other gases through imperfect vacuum seals is the most common cause of failure in germanium detectors operated over long periods of time. The surface contamination leads to increases in the leakage current from the detector, potentially spoiling its energy resolution. Careful attention to vacuum sealing techniques can maintain the leakage current in the picoampere range necessary to retain the original performance specifications of the detector. Because the charcoal trap or molecular sieve, present to help establish high vacuum conditions, is effective only at liquid nitrogen temperatures, the presence of water and other contaminants in the vicinity of the detector will generally be more significant when the detector is warmed to room temperature. For that reason, it may be advisable to maintain germanium detectors constantly at liquid nitrogen temperature as a precaution and to help ensure a longer lifetime.[12]

Germanium detectors also are universally fitted with an interlock that prevents application of high voltage to the detector unless it has reached a low temperature. This interlock is necessary since any inadvertent application of high voltage at room temperature leads to excessively high leakage current that will likely destroy the input FET of the preamplifier. Because the FET is normally mounted inside the cryostat for cooling, its replacement is not routine.

The preamplifier is normally incorporated as part of the cryostat package in modern HPGe systems. It is always advantageous to locate the preamplifier as near to the detector as possible to minimize capacitance. The input stages of the preamplifier normally are also cooled along with the detector to reduce electronic noise. There is also some advantage in keeping the exterior of the cryostat free of electronic "boxes" to facilitate close-fitting shielding and/or Compton suppression systems (see p. 437).

Figure 12.6 Diagram showing the location of a HPGe detector within its vacuum capsule. In this design, the capsule can be connected, without using vacuum pumps, to a variety of cryostats or cryostat–dewar assemblies. The example shown is the common horizontal configuration of the cryostat above the liquid nitrogen dewar (see inset). (Courtesy of M. Martini, EG & G ORTEC, Oak Ridge, TN.)

B. Energy Resolution

The dominant characteristic of germanium detectors is their excellent energy resolution when applied to gamma-ray spectroscopy. In Fig. 12.7, comparative pulse height spectra are shown for a NaI(Tl) scintillator and a germanium detector for identical incident gamma-ray spectra. The great superiority of the germanium system in energy resolution allows the separation of many closely spaced gamma-ray energies, which remain unresolved in the NaI(Tl) spectrum. Consequently, virtually all gamma-ray spectroscopy that involves complex energy spectra is now carried out with germanium detectors.

The overall energy resolution achieved in a germanium system is normally determined by a combination of three factors: the inherent statistical spread in the number of charge carriers, variations in the charge collection efficiency, and contributions of electronic noise. Which of these factors dominate depends on the energy of the radiation and the size and inherent

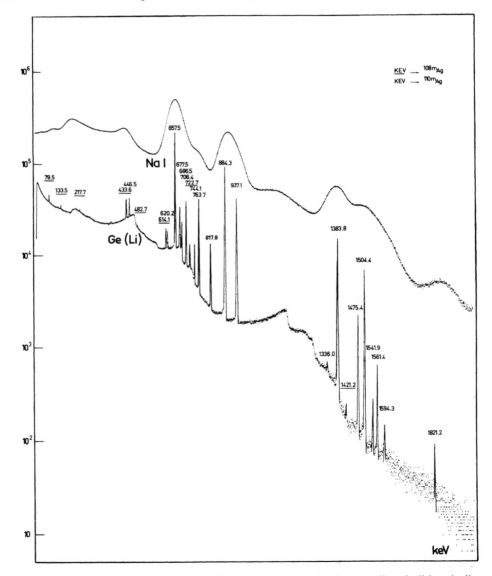

Figure 12.7 Comparative pulse height spectra recorded using a sodium iodide scintillator and a Ge(Li) detector. The source was gamma radiation from the decay of [108m]Ag and [110m]Ag. Energies of peaks are labeled in keV. (From Philippot.[13])

quality of the detector in use. The full width at half maximum, W_T, of a typical peak in the spectrum due to the detection of a monoenergetic gamma ray can be synthesized as follows

$$W_T^2 = W_D^2 + W_X^2 + W_E^2 \tag{12.12}$$

where the W values on the right-hand side are the peak widths that would be observed due only to effects of carrier statistics, charge carrier collection, and electronic noise.

The first of these factors, W_D^2, represents the inherent statistical fluctuation in the number of charge carriers created and is given by

$$W_D^2 = (2.35)^2 F \epsilon E \tag{12.13}$$

where F is the Fano factor, ϵ is the energy necessary to create one electron–hole pair, and E is the gamma-ray energy. If we assume a Fano factor value of 0.08 and use 2.96 eV for ϵ, Eq. (12.13) predicts a statistical broadening of 1.32 keV at the common calibration energy of 1.333 MeV. As pointed out at the end of this section, small coaxial germanium detectors achieve a best energy resolution at that energy of about 1.7 keV, indicating either that the remaining terms in Eq. (12.12) are still significant, or that the assumed Fano factor value is too small. The variation of the statistical contribution to the observed FWHM of a small germanium detector is shown in Fig. 12.8. These data would indicate that the value for the Fano factor might be in the 0.10 to 0.15 range, but more recent evaluations tend to support a value nearer 0.08.

The contribution of the second term, W_X^2, is due to incomplete charge collection and is most significant in detectors of large volume and low average electric field. Its magnitude can often be experimentally estimated by carrying out a series of FWHM measurements as the applied voltage is varied. The necessary assumption is that, if the electric field could be made infinitely large, the effects of incomplete charge collection could be reduced to an

Figure 12.8 The measured statistical contribution W_D for an 8 mm diameter $\times$ 4 mm planar Ge(Li) detector. Lines are shown corresponding to various assumed values for the Fano factor. (From Palms et al.[14])

insignificant level. Therefore, a plot of the observed FWHM versus the reciprocal of applied voltage, such as shown in Fig. 12.9, allows an extrapolation to infinite field conditions. The residual value of FWHM given by this extrapolation is then assumed to arise only from the remaining two factors in Eq. (12.12). One should use this type of analysis with caution, however, since the drift velocities of the carriers eventually saturate at high electric field values. Data taken under these conditions would no longer be expected to show a continuing reduction in the effects of incomplete charge collection with further increases in voltage.

The third factor, W_E^2, represents the broadening effects of all electronic components following the detector. Its magnitude can be conveniently measured by supplying the output of a precision pulser with a highly stable amplitude to the preamplifier and recording the corresponding peak in the pulse height spectrum. These measurements should be made while the detector is normally connected to the preamplifier so that capacitive loading of the preamplifier input is typical of conditions under actual use. A parallel test pulse input to the preamplifier is normally provided for this purpose (see Fig. 17.6).

As an illustration of the relative importance of the three major contributions to the overall energy resolution, Fig. 12.10 presents the variation with gamma-ray energy of each of the three terms in Eq. (12.12) for a typical HPGe detector. At low energies, the contributions from electronic noise and charge collection dominate, while the additional broadening due to carrier statistics becomes significant at higher energies. Small-volume detectors will usually show a better overall energy resolution than larger detectors of the same quality due to two factors. Small detectors will generally have lower capacitance values, and the electronic noise of the system increases with detector capacitance. Also, the effects of carrier loss due to trapping are magnified in large-volume detectors with relatively large charge collection distances. The best energy resolution figures are therefore obtained using detectors that are tailored to the energy range and application of interest. For example, small planar detectors give the best performance for X-rays and low-energy gamma rays where a large active volume may not be necessary.

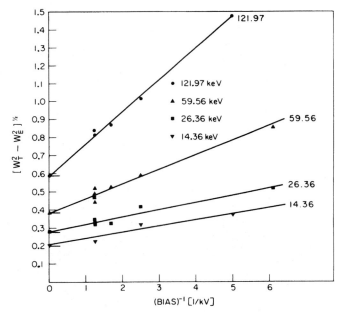

Figure 12.9 Extrapolation of peak width data from an Si(Li) detector to infinite field conditions. The electronic broadening (W_E) has already been subtracted quadratically from the measured FWHM, so that the remaining contribution given by the extrapolation should only be due to statistical effects. (From Zulliger et al.[15])

Figure 12.10 Example of the variation of the FWHM of the full-energy peak of an 86 cm³ HPGe detector with gamma-ray energy. Each of the three terms in Eq. (12.12) is shown separately, together with the total FWHM. (From Owens.[16])

Energy resolution figures for germanium detectors are often specified at 5.9 keV (^{55}Fe), 122 keV (^{57}Co), 662 keV (^{137}Cs), or 1333 keV (^{60}Co). Representative FWHM values for commercially available systems with a small planar germanium detector are about 150–250 eV at 5.9 keV, increasing to 400–600 eV at 122 keV. Larger coaxial detectors will produce FWHM values of 800–1200 eV at 122 keV, rising to 1.7–2.3 keV at 1333 keV.

C. Pulse Shape and Timing Properties

1. THE CHARGE COLLECTION PROCESS

The detailed shape of the signal pulse developed by a germanium detector is important on several counts. In standard pulse height spectroscopy, it is necessary for the shaping times of the pulse-processing electronics to be substantially larger than the longest rise time likely to be encountered from the detector if resolution loss due to ballistic deficit is to be avoided. In applications where timing information must be obtained from the pulse, both the rise time and the detailed shape of the leading edge of the pulse become important when considering various time pick-off methods. The ultimate time resolution obtainable from germanium detectors is critically dependent both on the overall average rise time and the significant variation in the pulse shape from event to event.

Assuming that the equivalent circuit of the measuring electronics presents a large time constant compared with the largest rise time produced by the detector, the leading edge of the signal pulse is almost entirely determined by the details of the charge collection process within the detector. Because it is always advantageous to have the smallest possible rise time, conditions are normally sought in which charge collection occurs in the minimum possible time. In Fig. 11.2, drift velocity plots were shown for electrons and holes in germanium as a function of the applied electric field. For low values of the electric field, the velocity increases linearly with the field, which implies a constant value for the electron or hole mobility. At sufficiently high electric fields, however, the velocity ceases to increase and approaches a constant saturation value. For electrons in germanium at 77 K, this saturation drift velocity is approximately 10^5 m/s and is achieved at a field value of about 10^5 V/m. The saturated velocity for holes is similar but requires a minimum field of approximately 3×10^5 V/m. Pulses of minimum rise time are therefore obtained by operating the detector with sufficient applied voltage so that an electric field of at least this magnitude is present everywhere within the active volume. Of the various detector configurations, the planar and true coaxial geometries are those in which regions of low electric fields are most easily avoided. Therefore, these types have predominated when the detector must be used in critical timing situations.

Two factors limit the ultimate time resolution that can be achieved with germanium detectors. One is that the charge collection process is inherently slow. Even at the saturated drift velocity, the time required for a charge carrier to travel a distance of 1 cm (which may be of the order of the detector thickness) is approximately 100 ns. Typical pulse rise times will therefore be of the same order of magnitude. These times are much longer than the output from fast detectors (such as organic scintillators) and therefore the time performance can never be as good. A second factor, however, makes the timing properties even worse. As will be illustrated below, the detailed shape of the pulse rise from germanium detectors can change substantially from event to event, depending on the position at which the electron–hole pairs are created within the active volume. When the volume of the detector is uniformly irradiated, these positions are more or less randomly distributed, and therefore the output pulses show a great variation in the shape of their leading edge. This variable pulse shape creates difficulties with many methods of deriving timing signals, and special time pick-off techniques are required to minimize the accompanying difficulties (see Chapter 17).

A general review of the pulse shape and timing properties of germanium detectors is given in Ref. 17. More detailed experimental and analytic investigations can be found in Refs. 18–23.

2. MODELS FOR THE PULSE SHAPE

Just as in other detectors in which signal carriers must be collected over appreciable distances, the shape of the leading edge of pulses from germanium detectors depends on the position at which the charge carriers are formed within the active volume. The simplest case is that of a very short-range particle, which, to first approximation, creates all the electron–hole pairs at one location within the detector. If that location lies within the interior of the active volume, there will be unique and separate collection times for holes and electrons because each species must travel a fixed distance before being collected. If the point of interaction is near either edge of the active volume, the observed pulse rise will then be due primarily to the motion of only one type of charge carrier. For charged radiations whose range is not small compared with the active volume, a distribution of collection times will result from the corresponding spatial distribution of the points at which holes and electrons are formed. If the orientation of the particle track can change significantly from event to event, an additional variation in the pulse rise time will be introduced.

The shape of the leading edge of the output pulse is subject to much the same analysis as was given for ion chambers in Chapter 5. One major difference from gas-filled counters is that the mobilities of the positive and the negative charge carriers (holes and electrons) are somewhat similar, whereas in a gas, the positive ion mobility is much less than that of free electrons. If a sufficient number of simplifying assumptions are made, analytic expressions can be derived to describe the shape of the pulse expected in simple germanium detector configurations. The common assumptions are:

1. All charge carriers are created at a fixed position within the detector active volume. This assumption ignores the finite range of secondary electrons produced by gamma-ray interactions, as well as the fact that many gamma-ray pulses arise from multiple interactions (e.g., Compton scattering followed by photoelectric absorption) that occur at widely separated points within the detector.

2. Trapping and detrapping of the charge carriers are ignored. The capture of charges in shallow traps followed by a rapid release can significantly alter the transit time of charge carriers and can be important when the concentration of such shallow traps is not negligible.

3. All charge carriers are assumed to be generated entirely within the active volume of the detector where the electric field has its full expectation value.

4. The electric field within the active volume of the detector is sufficiently high to cause saturation of the drift velocity of both electrons and holes.

With this set of simplifying assumptions, analytic expressions can be obtained for the expected pulse shape for two relatively simple geometries: the planar detector in which the electric field is approximately uniform, and the true coaxial configuration in which the electric field varies inversely with the radial distance from the detector axis.

a. Planar Geometry

In general, the energy dE absorbed by the motion of a positive charge q_0 through a potential difference $d\varphi$ is given by

$$dE = -q_0 \, d\varphi \tag{12.14}$$

In terms of the electric field $\mathscr{E}(x) = -d\varphi(x)/dx$

$$\frac{dE}{dx} = q_0 \mathscr{E}(x) = q_0 \frac{V_0}{d} \tag{12.15}$$

where V_0 is the applied voltage and d is the detector thickness. The energy absorbed by the motion from x_0 to x is

$$\Delta E = \int_{x_0}^{x} dE = \frac{q_0 V_0}{d} \int_{x_0}^{x} dx = \frac{q_0 V_0}{d} (x - x_0) \tag{12.16}$$

Using the same arguments introduced for parallel plate ion chambers beginning on p. 148, we obtain the corresponding signal voltage:

$$\Delta V_R = \frac{\Delta E}{C V_0} = \frac{q_0}{C} \frac{(x - x_0)}{d} \tag{12.17}$$

This voltage corresponds to an *induced charge* of

$$\Delta Q = C \Delta V_R = q_0 \frac{(x - x_0)}{d} \tag{12.18}$$

The development of the signal pulse can thus be represented by the growth of a time-dependent induced charge $Q(t)$. There are two separate components of the induced charge, one corresponding to the motion of the electrons and the other to the motion of the holes. From Eq. (12.18), we can write the combination as

$$Q(t) = \frac{q_0}{d} \left(\begin{array}{c} \text{electron drift} \\ \text{distance} \end{array} + \begin{array}{c} \text{hole drift} \\ \text{distance} \end{array} \right) \tag{12.19}$$

This induced charge starts at zero when the electrons and holes are first formed by the ionizing particle and reaches its maximum of q_0 when both species have been collected. As in the ion chamber case, $q_0 = n_0 e$, where n_0 is now the number of electron–hole pairs and e is the electronic charge.

We now assume that these charges are formed at a distance x from the n^+ contact of the planar detector (see Fig. 12.11). The following definitions are used:

$$t_e \equiv \text{electron collection time}$$

$$= x/v_e, \text{where } v_e \text{ is the saturation velocity}$$

$$t_h \equiv \text{hole collection time}$$

$$= \frac{d - x}{v_h}, \text{where } v_h \text{ is the saturation hole velocity}$$

Equation (12.19) can be divided into four possible time domains as follows. While both holes and electrons are drifting ($t < t_h$ and $t < t_e$):

$$Q(t) = q_0\left(\frac{v_e}{d}t + \frac{v_h}{d}t\right) \tag{12.20a}$$

If electrons have been collected, but holes are still drifting ($t_e < t < t_h$):

$$Q(t) = q_0\left(\frac{x}{d} + \frac{v_h}{d}t\right) \tag{12.20b}$$

If holes have been collected, but electrons are still drifting ($t_h < t < t_e$):

$$Q(t) = q_0\left(\frac{v_e}{d}t + \frac{(d - x)}{d}\right) \tag{12.20c}$$

Figure 12.11 Shape of the leading edge of the output pulse $Q(t)$ for various interaction points within the detector [from Eq. (12.20)]. The initial slope of all pulses is the same, corresponding to the period when both electrons and holes are drifting. For point ①, electrons are collected very quickly and most of the rise corresponds to the drift of holes to the p^+ layer. The opposite is true for point ⑤. t_h and t_e are the assumed drift times for holes and electrons, respectively, across the entire thickness d.

After both holes and electrons have been collected ($t > t_h$ and $t > t_e$):

$$Q(t) = q_0 \qquad (12.20d)$$

Figure 12.11 shows several such pulse shapes for varying values of x. The variation in effective rise time of the pulses amounts to approximately a factor of 2.

b. Coaxial Geometry

The pulse shape generated by the collection of electrons and holes in a coaxial detector shares many of the general features derived above for a planar type, but the linear portions of the pulse leading edge take on a curvature because of the different geometry. The equivalent to Eq. (12.15) for cylindrical geometry is

$$\frac{dE}{dr} = q_0 \mathscr{E}(r) \qquad (12.21)$$

The absolute value of the electric field $\mathscr{E}(r)$ in p-type HPGe true coaxial detectors [Eq. (12.10)] can be written

$$\mathscr{E}(r) = 2\alpha r + \frac{\beta}{r} \qquad (12.22)$$

where

$$\alpha \equiv \frac{eN_A}{4\epsilon} \quad \text{and} \quad \beta \equiv \frac{V_0 - \alpha(r_2^2 - r_1^2)}{\ln (r_2/r_1)}$$

Putting this expression into Eq. (12.21) and integrating, we obtain

$$\Delta E = q_0 \int_{r_0}^{r} \mathscr{E}(r)\, dr$$

$$= q_0\alpha(r^2 - r_0^2) + q_0\beta \ln \frac{r}{r_0} \qquad (12.23)$$

The time-dependent charge induced by the motion of the electrons outward toward the n^+ contact can then be expressed in terms of the radial position of the electrons $r_e(t)$:

$$Q^-(t) = \frac{\Delta E^-}{V_0} = \frac{q_0\alpha}{V_0}[r_e^2(t) - r_0^2] + \frac{q_0\beta}{V_0} \ln \frac{r_e(t)}{r_0} \qquad (12.24)$$

Similarly, the induced charge due to the motion of the holes inward to the p^+ contact is

$$Q^+(t) = \frac{q_0\alpha}{V_0}[r_0^2 - r_h^2(t)] + \frac{q_0\beta}{V_0} \ln \frac{r_0}{r_h(t)} \qquad (12.25)$$

and the total induced charge is

$$Q(t) = Q^-(t) + Q^+(t) \qquad (12.26)$$

The shape of $Q(t)$ depends on assumptions made about the motion of the electrons and holes as reflected in $r_e(t)$ and $r_h(t)$. Because of the radial variation of the electric field strength, the drift velocities may change as the carriers are collected. One approach[24] is to assume an empirical dependence of the drift velocities with electric field strength $\mathscr{E}$ of the form

$$v = \frac{\mu_0 \mathscr{E}}{[1 + (\mathscr{E}/\mathscr{E}_0)^\gamma]^{1/\gamma}}$$

where μ_0 is the mobility at low $\mathscr{E}$ values. Both γ and $\mathscr{E}_0$ are treated as adjustable parameters fitted to experimental drift velocity measurements. The radial position of each species can then be solved numerically. Plots of $Q(t)$ derived in this way are given in Ref. 24 for typical p-type HPGe coaxial detectors.

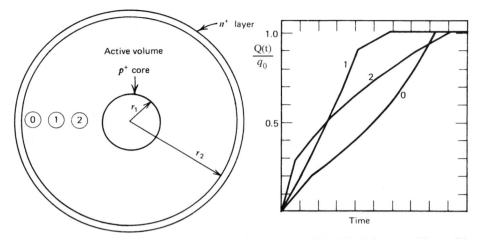

Figure 12.12 Leading edge of the output pulse for a coaxial drifted detector. Three different interaction points are indicated as 0, 1, and 2. (From Gadeken and Robertson.[21])

If the assumption of constant drift velocities is retained, the radial position of each carrier species is

$$r_e(t) = r_0 + v_e t \quad \text{and} \quad r_h(t) = r_0 - v_h t$$

and solutions to Eq. (12.26) are straightforward. For the simple case of Ge(Li) configurations where no net fixed charge is present in the intrinsic region, $\alpha = 0$ and $\beta = V_0 / \ln (r_2 / r_1)$. The result is then given by

$$Q(t) = \frac{q_0}{\ln (r_2/r_1)} \left[\ln \left(1 + \frac{v_e t}{r_0} \right) - \ln \left(1 - \frac{v_h t}{r_0} \right) \right] \tag{12.27}$$

Equation (12.27) holds only for the first time domain, during which both holes and electrons are drifting. As in the previous case for planar geometry, the first and second terms in the square brackets become constants [equal to $\ln (r_2/r_0)$ and $\ln (r_1/r_0)$] when the electrons or holes are collected, and a slope change occurs in the waveform at the corresponding times. After both are collected, $Q(t) = q_0$. Plots of this pulse shape for various assumed interaction radii are given in Fig. 12.12. Here the variation in effective rise time can be even greater than for the planar case.

The changes in the shape of the pulse observed from a coaxial detector as a function of the radial position of the interaction illustrated in Figure 12.12 can be exploited for some purposes. With the availability of digital techniques to analyze pulse shapes, studies have demonstrated the practicality of measuring the radial position of the interaction through changes in these shapes. This information can be used for general position-sensing purposes, or to correct for the radially dependent effects of hole trapping in damaged detectors.[25,26] If the electrodes on the germanium detector are segmented so that separate pulses are obtained from each segment, then the position-sensing can be extended beyond just the radial coordinate by comparing the size of the individual electrode signals. Such two- or three-dimensional position-sensing germanium detectors are under development,[27,28] and potentially they offer the ability to resolve not only the energy of the event but also its position within the detector volume.

Pulses arising from single-site gamma ray interactions can also be distinguished by their shape from those in which the gamma ray undergoes multiple interactions at different radii. In general gamma-ray spectroscopy, most of the full-energy events consist of multiple-site interactions (see Fig. 12.16), while single-site events are most likely to be

Compton scatterings. Thus improvements in the peak-to-Compton ratio can be obtained[29–31] by preferentially selecting multiple-site events only. These differences also can be exploited in background suppression arising from beta particles[32,33] that consist only of single-site interactions, while preserving most of the full-energy gamma-ray events.

3. EFFECTS OF TRAPPING AND DETRAPPING

Martini and McMath[34] have made an analysis of the effects of trapping and detrapping on the expected pulse shape. To simplify the following discussion, we assume that all the electron–hole pairs are created near one boundary of the active volume. The signal pulse then corresponds to the motion of only that carrier species that is collected at the opposite boundary. If we assume that the carriers are formed at the p^+ contact of a planar detector, we can set $x = d$ in Eq. (12.20), which then reduces to

$$Q(t) = \begin{cases} q_0 \dfrac{t}{t_e} & (t \leq t_e) \\[2ex] q_0 & (t > t_e) \end{cases} \tag{12.28}$$

and the simple linear pulse rise corresponds to the drift of electrons across the active volume.

If we now assume that a uniform concentration of electron traps exists in this region, some of the charge may be lost, at least temporarily, from the signal pulse. If no detrapping takes place, the loss is permanent and the resulting pulse takes the form

$$Q(t) = \begin{cases} \dfrac{q_0 \tau_T}{t_e} \left(1 - e^{-t/\tau_T}\right) & (t \leq t_e) \\[3ex] \dfrac{q_0 \tau_T}{t_e} \left(1 - e^{-t_e/\tau_T}\right) & (t > t_e) \end{cases} \tag{12.29}$$

where τ_T is the *mean free drift time* of the electrons before trapping occurs. Permanent trapping adds curvature to the rise of the pulse and attenuates its amplitude, as shown in Fig. 12.13.

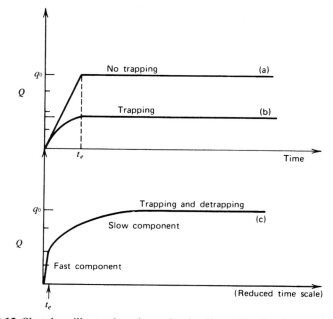

Figure 12.13 Sketches illustrating the pulse leading edge for interactions in a planar detector at one extreme of the active volume: (*a*) no trapping, (*b*) permanent trapping, and (*c*) trapping with slow detrapping. (From Martini and McMath.[34])

Ultimately, the trapped electrons may be released by thermal excitation if the trap is shallow. If τ_D is the mean lifetime of a trapped electron before detrapping occurs, the resulting pulse shape is composed of "fast" and "slow" components. The fast component or leading edge is characterized by

$$Q(t) = q_0 \frac{\tau_e}{t_e} \left(\frac{t}{\tau_D} + \frac{\tau_e}{\tau_T} \left(1 - e^{-t/\tau_e} \right) \right) \qquad (t \leq t_e) \qquad (12.30)$$

where

$$\tau_e \equiv \frac{\tau_T \tau_D}{\tau_T + \tau_D}$$

The slow component does not lend itself to a simple analytic expression but serves to build up $Q(t)$ to an asymptotic value of q_0 with a characteristic time on the order of τ_D. A sketch of this behavior is also illustrated in Fig. 12.13.

Trapping can thus lead to a reduction in expected pulse amplitude or, if detrapping occurs, to a slowing of a portion of the pulse's leading edge. The latter can have a significant effect on the timing properties of germanium detectors in which trapping is important. Furthermore, trapping can cause a deterioration in energy resolution due to the variable amount of charge lost per pulse. Detrapping is helpful in this respect only if it occurs on a short time scale compared with pulse-shaping times in the subsequent electronics. Pulse amplitudes that are expected when trapping is significant can be treated using the Hecht equation [Eq. (13.5)], presented in the following chapter for compound semiconductors where trapping is generally much more significant than in germanium.

In coaxial detectors, the cylindrical geometry adds further complication to the prediction of the expected pulse amplitude in the presence of trapping. In Ref. 35, a detailed analysis is presented that is shown to model accurately the behavior of coaxial detectors in which hole trapping has been made more severe by radiation damage.

4. SLOW OR DEFECTIVE PULSES

A fraction of the pulses observed from some germanium detectors have rise times substantially longer than predicted by the above analysis. Furthermore, these slow pulses are also "defective" in amplitude and can contribute to a loss of energy resolution or low-energy tails on peaks recorded in the pulse height spectrum. There is some evidence[36,37] that these defective pulses correspond to events that originate near the edges of the active volume or in other regions where the local electric field is below normal.[38] Charge collection is then slowed, and losses due to trapping are more severe than for normal pulses. These effects are magnified in older detectors with a high concentration of defects or trapping centers, often the result of radiation damage from prolonged exposure of the detector to fast neutrons (see p. 451).

Because the slow-rising pulses are also deficient in amplitude, gains can sometimes be made in both the energy resolution and peak-to-Compton ratio (see p. 428) in gamma-ray spectra by discarding such pulses based on sensing their long rise times in an external circuit.[39] However, this approach will also reduce the detection efficiency if a substantial fraction of all pulses are involved. A more elaborate pulse height correction technique[40] can instead be used that avoids this efficiency loss. The principle of this technique is based on making a measurement of each pulse rise time and recording a set of independent pulse height spectra, each corresponding to pulses with nearly equal rise times. If a correlation between pulse height and rise time can be found, these multiple spectra can then be reassembled into a single spectrum by applying an appropriate correction factor to each.

IV. GAMMA-RAY SPECTROSCOPY WITH GERMANIUM DETECTORS

In the measurement of gamma-ray energies above several hundred keV, there are, at present, only two detector categories of major importance: inorganic scintillators, of which NaI(Tl) is the most popular, and germanium semiconductor detectors. Although there are many other potential factors, the choice in a given application most often revolves about a trade-off between counting efficiency and energy resolution. Sodium iodide scintillators have the advantage of availability in large sizes, which, together with the high density of the material, can result in very high interaction probabilities for gamma rays. The relatively high atomic number of iodine also ensures that a large fraction of all interactions will result in complete absorption of the gamma-ray energy, and therefore the photofraction (or fraction of events lying under the full-energy peak in the pulse height spectrum) will also be relatively high.

The energy resolution of scintillators is poor. The comparative spectra shown in Fig. 12.7 illustrate the clear superiority of germanium detectors in situations where many closely spaced gamma-ray energies must be separated. Good germanium systems will have a typical energy resolution of a few tenths of a percent compared with 5–10% for sodium iodide. This improved energy resolution does not come without a price, however. The smaller sizes available and lower atomic number of germanium combine to give photopeak efficiencies an order of magnitude lower in typical cases. Furthermore, the photofractions are low and thus the continua in the pulse height spectra are more of a problem.

This latter disadvantage is somewhat offset by the superior energy resolution of germanium. Not only does good resolution help separate closely spaced peaks, but it also aids in the detection of weak sources of discrete energies when superimposed on a broad continuum. Detectors with equal efficiency will result in equal areas under the peak, but those with good energy resolution produce a narrow but tall peak that may then rise above the statistical noise of the continuum. This effect is illustrated in Fig. 12.14 in which the only variation between spectra is the resolution.

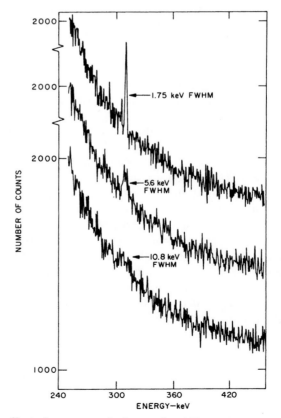

Figure 12.14 The effect of energy resolution on the ability to identify a weak peak superimposed on a statistically uncertain continuum. The area under the peak is the same in all three cases. (From Armantrout et al.[41])

Germanium detectors are clearly preferred for the analysis of complex gamma-ray spectra involving many peaks. The choice is less obvious when only a few widely-spaced gamma-ray energies are involved, particularly if a measurement of their intensity rather than an accurate energy determination is the prime objective. Then the greater efficiency, larger photofraction, and lower cost of scintillators may well tip the balance in their favor.

Because the concentration of dopants is very small, the gamma-ray interaction characteristics are identical for lithium-drifted or high-purity germanium. The detector properties of response function and detection efficiency discussed in the remainder of this chapter are therefore the same for germanium detectors of similar size and shape, regardless of whether they are of the Ge(Li) or HPGe type. The extensive experience gained with Ge(Li) detectors over several decades is therefore still applicable in many ways to the more modern HPGe devices. We show examples of spectra and other detection properties published for both types, with the understanding that they are broadly representative of the behavior of germanium detectors in general. Detailed discussions of the use of germanium detectors in gamma-ray spectroscopy can be found in Ref. 42.

A. Response Function

1. GENERAL CHARACTERISTICS

Examples of pulse height spectra measured with a germanium detector for monoenergetic incident gamma rays are given in Fig. 12.15. The processes of photoelectric absorption, Compton scattering, and pair production all may contribute to the observed response in the manner described in Chapter 10 for the case of sodium iodide scintillators. The lower atomic number of germanium and the smaller typical active volume of germanium detectors compared with NaI scintillators lead to significant differences in the importance of various features of the pulse height spectrum.

The lower atomic number of germanium results in a photoelectric cross section that is smaller by a factor of 10–20 compared with sodium iodide, and an even greater discrepancy exists in comparison with high-Z scintillators such as BGO. As a direct result, the intrinsic peak efficiency of a germanium detector is many factors smaller than that of a sodium iodide scintillator of equivalent active volume. Even though the area under the full-energy peak is much smaller, these peaks still are a prominent and obvious part of the spectrum because their width is so small due to the superior energy resolution of semiconductor detectors. Events contributing to the full-energy peak are now much more likely to consist of multiple interactions, such as Compton scattering followed by photoelectric absorption of the scattered photon, and absorption of the full photon energy in a single photoelectric event is relatively rare. In Fig. 12.16, it can be seen that multiple events are the dominant contributor to the full-energy peak over all but the lowest range of gamma-ray energies.

The Compton continuum, as described in Chapter 10, is also a prominent part of germanium detector spectra. Because the ratio of the Compton to photoelectric cross section is much larger in Ge than in NaI, a much greater fraction of all detected events lie within this continuum rather than under the photopeak. The improved energy resolution leads to a more faithful reproduction of the shape of the distribution, including the rise in the continuum near the Compton edge. Because Compton scattering is an interaction with electrons, the position of the Compton edge is at the same energy for all detector types [see Eq. (10.4)].

The *peak-to-Compton ratio* is sometimes quoted as one feature of germanium detector gamma-ray spectra. This index of detector performance is defined[44] as the ratio of the count in the highest photopeak channel to the count in a typical channel of the Compton continuum associated with that peak. This sample of the continuum is to be taken in the relatively flat portion[†] of the distribution lying just to the left of the rise toward the

---

[†]Officially defined[44] as the interval from 358 to 382 keV for the 662 keV gamma rays from ^{137}Cs, and from 1040 to 1096 keV for the 1332 keV gamma rays from ^{60}Co.

Figure 12.15 Pulse height spectra taken with a *p*-type HPGe detector with 110% relative efficiency, defined on p. 450. (Spectra courtesy R. Keyser, EG&G ORTEC, Oak Ridge, Tennessee.) (*a*) Spectrum recorded from 662 keV gamma rays emitted by a ^{137}Cs source showing the effects of photoelectric absorption and Compton scattering in the detector. In addition to the backscatter peak, Compton continuum, and full energy peak features, a small sum peak is evident from pulse pile-up (see Ch. 17), as well as a small peak from 1460 keV background gamma rays from ^{40}K. (*b*) Spectrum from 1460 keV gamma rays emitted by ^{40}K that now shows the additional effects of pair production taking place in the detector and surrounding materials. Both single escape (SE) and double escape (DE) peaks can be seen, along with the peak at 511 keV due to annihilation radiation produced by pair production interactions in surrounding materials.

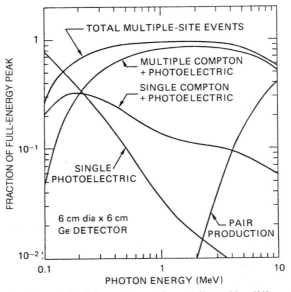

Figure 12.16 Fractions of the full-energy peak contributed by different energy loss mechanisms in a 6 cm × 6 cm coaxial HPGe detector, as predicted by Monte Carlo simulation. Absorption of the gamma-ray photon in a single photoelectric interaction predominates only for energies below about 140 keV. (From Roth et al.[43])

Compton edge (see Fig. 12.15). The ratio is conventionally quoted for the 1332 keV gamma ray from ^{60}Co and is a measure of the combined effects of detector energy resolution and photofraction. The ratio is adversely affected by any scattered gamma rays that enter the detector and it therefore also is influenced by structural materials near the detector or source. Typical values of the ratio range from 30 to 60 for representative coaxial germanium detectors measured under low-scatter conditions. For detectors with equal photofraction, the ratio will vary inversely with the FWHM value of the full-energy peak. For detectors with equal energy resolution, the ratio will be approximately proportional to the photofraction.

Because of the greater transparency of the detector to secondary gamma rays, escape peaks play a much more important role than in NaI scintillators. Escape peaks arise whenever a fixed amount of energy may be lost from the detector with a significant probability. The escape of the characteristic X-rays from germanium following photoelectric absorption can be significant, especially for small detectors with a large surface-to-volume ratio. A small peak will then be found in the spectrum at 11 keV below the photopeak, with the energy difference corresponding to the characteristic K X-ray energy for germanium. These X-ray escape peaks will be most prominent for incident low-energy gamma rays, because photoelectric absorption is then most probable, and interactions will tend to occur near the detector surface.

At high gamma-ray energies, the escape of annihilation radiation following pair production within the detector is very significant. The pair production process, described in Chapter 10, consists of creating an electron–positron pair at the site of the original gamma-ray interaction. The kinetic energy of both particles is expended in creating charge carriers, but the positron near the end of its track will annihilate and create two 0.511 MeV annihilation photons. In germanium detectors, there is a high probability that one or both of these photons will escape. Escape peaks will therefore appear in the spectrum corresponding to events in which one or both of the annihilation photons carry away a portion of the original gamma-ray energy. If both annihilation photons escape, a *double escape peak* appears at an energy of 1.022 MeV less than the full-energy peak. Events in which one annihilation photon escapes but the other is fully absorbed lead to a *single escape peak* appearing at an energy of 0.511 MeV below the full-energy peak. If the incident gamma-ray energy is sufficiently high, the probability of full-energy absorption may be very small, and the escape peaks often are the most prominent peaks that appear in the response function of the detector. Their location is illustrated in the spectra shown in Fig. 12.17. Because of their prominence, the escape peaks are often used in place of the full-energy peak to derive the primary energy of incident gamma rays. The primary energy can be obtained from the position of the single and double escape peaks by adding one or two units, respectively, of the annihilation photon energy (511.003 keV).

Careful measurements have demonstrated that the annihilation single escape peak can display subtle effects arising from the finite momentum of the positron–electron pair immediately before annihilation. This momentum causes one annihilation photon to have a slightly higher energy than its corresponding partner. Therefore, if only one photon is reabsorbed, a variation is introduced in the escaping energy. This variation results in a slight broadening of the single escape peak compared with either the full-energy peak or the double escape peak. It has also been demonstrated[46] that the centroid of the single escape peak in a coaxial detector can be slightly shifted (on the order of 50 eV) to higher energies because of effects due to the positron mobility in the electric field of the detector. One should therefore use single escape peak positions and widths with caution in precision measurements where accuracies of the order of 100 eV or less are to be achieved.

Germanium detectors of a few millimeters thickness are also sometimes applied in the low-energy range typical of X-ray spectroscopy. Although the lithium-drifted silicon detectors described in Chapter 13 tend to dominate for this application, germanium offers some alternative properties that can at times be advantageous. The higher Z-value and density of germanium allow coverage of a wider range of energies for a given thickness. For example, a 3-mm thickness of germanium remains more than 50% efficient for the detection of X-rays

Figure 12.17 The relative intensity of the full-energy, single escape, and double escape peaks for a 100 cm³ Ge(Li) detector for gamma-ray energies from 4.439 to 17.64 MeV. (From Berg et al.[45])

Figure 12.18 Pulse height spectrum taken from a ^{241}Am source with a 4-mm thick planar HPGe detector. A number of the X-ray escape peaks are identified on the figure. (From Rossington et al.[47])

Figure 12.19 Pulse height spectrum from 1173 keV and 1332 keV gamma rays emitted by ^{60}Co measured using a p-type HPGe detector with 110% relative efficiency. Evident in the spectrum are characteristic X-rays from photoelectric absorptions in the lead shield, the backscatter peak, barely perceptible single escape (SE) and double escape (DE) peaks from pair production interactions of the 1332 keV gamma ray (the escape peaks are weak because the 1332 keV gamma ray energy is only slightly above the threshold for pair production), a peak at 511 keV from annihilation radiation produced in the shield, the Compton edges and full energy peaks from the two primary gamma rays, a small peak at 1460 keV from ^{40}K in the background, a small "shoulder" from the summing of two 1173 keV events through pile-up, a pronounced sum peak at 2506 keV due to the full absorption of both primary gamma ray photons that are emitted in coincidence, a small peak at 2614 keV from ^{228}Th in the background, and a very small sum peak at 2665 keV from the pile-up of two 1332 keV pulses. (Spectrum courtesy R. Keyser, EG&G ORTEC, Oak Ridge, Tennessee.)

Figure 12.20 Spectrum from a 35 cm^3 Ge(Li) coaxial detector exposed to a source of ^{56}Mn emitting multiple gamma rays with the energies (in keV) indicated on the plot. Since many of these are relatively high energy, a large number of single escape (SE) and double escape (DE) peaks are evident. (From Helmer et al.[50])

up to 100 keV, while the same thickness of a silicon detector drops below 50% detection efficiency at about 35 keV. Also, the energy required to create an electron–hole pair and the Fano factor are both slightly smaller in germanium than in silicon, leading to a predicted statistical limit to energy resolution that is about 27% better in germanium than in silicon.[47] One drawback of germanium is that it is more difficult to achieve a thin front dead layer to minimize attenuation of incident low-energy X-rays. Another disadvantage is that the higher atomic number of germanium results in more energetic K X-rays being created after photoelectric absorption. These have a much higher probability of escaping from one of the surfaces of the detector leading to the "X-ray escape peaks" evident in Fig. 12.18. These escape peaks complicate interpretation of the resulting spectrum, and are much less prominent in spectra from silicon detectors because of the lower K X-ray energy (see Fig. 13.8).

2. TYPICAL GERMANIUM PULSE HEIGHT SPECTRA

Because germanium detectors tend to be fabricated with nonstandard dimensions, catalogs of gamma-ray spectra are not as widely published as those that can be found for common size sodium iodide scintillators. Some publications that do contain spectra recorded for a collection of radioisotopes using relatively small-volume germanium detectors are given in Refs. 48–50. Examples of gamma ray spectra, including some recorded from larger-volume detectors, are shown in Figs. 12.19 through 12.21 and previously in Fig. 12.15.

Figure 12.21 ^{60}Co gamma-ray spectra from germanium detectors of three different sizes with the relative efficiencies shown. The plots have been normalized to the same area of the upper full-energy peak. Notice the relative decrease in the importance of the Compton continuum as the detector volume (and efficiency) increases. Note also that the vertical scale is logarithmic, so that the escape peaks are actually somewhat less intense for the largest detector, even though they are quite evident because of the lower Compton continuum on which they are superimposed. (From Helmer et al.[50])

Good results also have been achieved[51] in calculating the response function of germanium detectors based on a knowledge of the cross section for the primary gamma rays and assumed behavior of the secondary radiations. As an example, some results from the Monte Carlo program of Meixner[52] are shown in Fig. 12.22. The excellent agreement between calculation and measurement lends credence to the assumptions used in the Monte Carlo code, which can then be applied to a wide range of experimental situations. The plot shown in Fig. 12.23 shows a perspective diagram of pulse height spectra calculated by the same code as the gamma-ray energy is varied. The enhancement of the single and double escape peaks as the gamma-ray energy increases is quite evident from this diagram.

3. PEAK SHAPE

The detailed shape of the peaks observed in germanium spectra is important if the area under the peak is to be accurately measured. Some details of possible features of the peak shape are shown in Fig. 12.24. Most empirical fits to the shape use modifications of a Gaussian distribution allowing for tailing on the low-energy side of the distribution.[54,55] Tailing can arise from several physical effects, including imperfect charge collection in some regions of the detector, or secondary electron and bremsstrahlung escape from the active volume. The distinction between short-term and long-term tails in the figure is that the short-term tailing has a more serious effect on the shape of the peak near its base, whereas long-term tailing often can be treated as an additional portion of the background.

Figure 12.22 A comparison of the measured Ge(Li) spectrum for gamma rays from ^{182}Ta with a Monte Carlo calculated spectrum. (From Meixner.[52])

One method of specifying the severity of tailing for a specific detector is to quote the full width at one-tenth maximum (FW.1M) of the full-energy peak, in addition to the more conventional full width at half maximum (FWHM). For good quality germanium detectors with minimal tailing, the FW.1M should be less than double the FWHM. (The FW.1M/FWHM ratio for a pure Gaussian peak is 1.823.) Another index even more sensitive to tailing is the ratio of the full width at 1/50 maximum (FWFM) to the FWHM, conventionally measured at 1.333 MeV. Good germanium detectors give values for this ratio between 2.5 and 3.0, compared with 2.376 for a pure Gaussian.

B. Methods for Continuum Reduction

As outlined in Chapter 10, the ideal gamma-ray spectrometer should have a response function consisting of a single peak only, with no associated continuum. As can be seen from the examples given earlier in this chapter, germanium spectra are characterized by very prominent continua, which can obscure low-intensity peaks from other gamma-ray energies. The photofraction values are typically much poorer than those of NaI(Tl) scintillators, and therefore the methods of continuum reduction outlined in Chapter 10 can potentially provide even greater benefits when applied to germanium detectors.

Figure 12.23 Perspective diagram of calculated Ge(Li) spectra for energies up to 3 MeV. (From Meixner.[52])

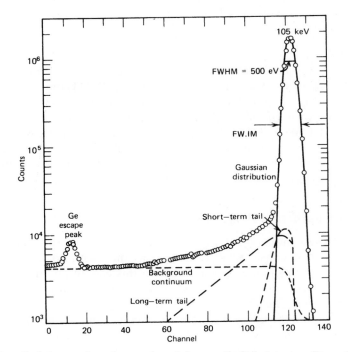

Figure 12.24 Detailed shape of an observed peak from a Ge(Li) detector. (From Meyer et al.[53])

1. COMPTON REJECTION BY ANTICOINCIDENCE

The Compton continuum in gamma-ray spectra from germanium detectors is generated primarily by gamma rays that undergo one or more scatterings in the detector followed by escape of the scattered photon from the active volume. In contrast, the full-energy absorption events result in no escaping photons. Therefore, coincident detection of the escaping photons in a surrounding annular detector can serve as a means to reject preferentially those events that only add to the continuum, without affecting the full-energy events. The rejection is carried out by passing the pulses from the germanium detector through an electronic gate that is closed if a coincident pulse is detected from the surrounding detector (called *anticoincidence mode*).

To be effective, the surrounding detector must be large enough to intercept most of the escaping photons and should have a good efficiency for their detection. Large scintillation detectors most readily meet these requirements, and both NaI(Tl) and BGO have been used for this purpose. BGO has the strong advantage that its high density and atomic number allow a more compact configuration compared with a sodium iodide detector of the same detection efficiency.

Good Compton suppression also requires that there be a minimum amount of absorbing material between the inner and outer detectors that could attenuate the escaping scattered photons. The relatively thick outer contact on *p*-type coaxial detectors is one example of such an absorber, and *n*-type germanium detectors with thinner contacts are generally preferred in this application.

Figure 12.25 shows one example of a Compton suppression system that combines BGO and NaI(Tl) scintillators in an annular geometry surrounding a germanium detector of the HPGe type. The reduction in the Compton continuum from the germanium detector by operating it and the scintillation detectors in anticoincidence mode is illustrated in Fig. 12.26. Other examples of the design and performance of BGO-based Compton suppression systems are described in Refs. 57–61.

Because they also result in some coincident events in both detectors, pair production interactions of the incident gamma rays followed by escape of one or both of the annihilation photons are also suppressed by these systems. Single and double escape peaks in the recorded spectrum are therefore less prominent than in the original spectrum from the germanium detector.

One potential disadvantage of Compton suppression systems becomes apparent if a radioisotope source with a complex decay scheme is being measured. Many gamma rays can then be emitted in coincidence, and it is possible for independent gamma rays from the same disintegration to interact in both detectors. These events are therefore rejected, leading to unwanted suppression of some full-energy peaks.

Figure 12.25 An example of a germanium detector surrounded by a Compton suppression system made up of NaI(Tl) and BGO scintillators. (From Nolan et al.[56])

Figure 12.26 Pulse height spectra from a ^{60}Co gamma-ray source using the system shown in Fig. 12.25. A collimator shielded the scintillators from direct irradiation. (*a*) Normal spectrum recorded from the central germanium detector. (*b*) Spectrum recorded in anticoincidence with the scintillation detectors, showing the suppression of the Compton continuum. The full-energy peak areas were unaffected to within 1%. (From Nolan et al.[56])

2. SUM–COINCIDENCE MODE

Most of the Compton continuum consists of single Compton scatterings followed by escape of the scattered gamma ray, whereas full-energy events at typical gamma-ray energies are primarily comprised of multiple scattering sequences followed by a photoelectric absorption. The peak-to-Compton ratio can therefore be enhanced by requiring a recorded event to correspond to more than one interaction within the detector before its acceptance. In germanium detectors, this selection is usually accomplished by subdividing the detector into several segments (or providing several adjacent independent detectors) and seeking coincident pulses from two or more of the independent segments. When coincidences are found, the output from all detector segments is summed and recorded. The resulting spectrum is made up only of the full-energy peak lying above a featureless continuum that is greatly suppressed and has no abrupt Compton edges.

As an example, Palms et al.[62] describe a sum–coincidence spectrometer consisting of two concentric coaxial germanium detectors. For gamma-ray energies between 300 and 1800 keV, the ratio of the full-energy peak to the average Compton continuum was increased by a factor of 4 or 5 compared with a single germanium detector of the same volume. Compton edges and double escape peaks were essentially eliminated. These improvements are accomplished at the expense of a reduction in the full-energy peak efficiency, however, because events that correspond to full-energy absorption confined within a single detector segment are inevitably lost. The absolute reduction in the full-energy peak efficiency over the same energy range was approximately a factor of 5. Another example of the effectiveness of the sum-coincidence mode of operation is shown in Fig. 12.27.

Larsen and Strauss[64] describe the performance of separately housed germanium detectors in the sum-coincidence mode. A severe loss of efficiency takes place because the two detectors must then be separated significantly, and most success with sum-coincidence operation has been achieved by using segmented or split coaxial detectors mounted in close proximity within the same cryostat.[65]

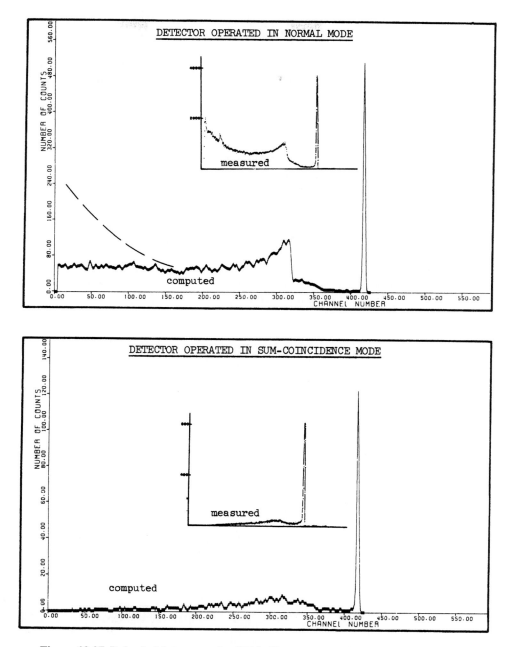

Figure 12.27 Pulse height spectra for 835 keV gamma rays from a Ge(Li) detector in normal operation (top) compared with spectra from the same detector operated in sum–coincidence mode. (From Walker and Palms.[63])

3. THE PAIR SPECTROMETER

A different approach to simplifying the recorded germanium detector spectrum is to attempt to select only the double escape peak. If the gamma-ray energy is sufficiently high, a significant fraction of all interactions will correspond to pair production in which both photons produced by positron annihilation escape from the primary detector. Because these annihilation photons are always emitted in opposite directions, two additional detectors placed on opposite sides of the primary detector can intercept them with a reasonable efficiency. If coincidence is demanded among all three detectors, the selection of double escape events will be highly specific, and most of the continuum (as well as the full-energy peak) will be suppressed. A considerable sacrifice in counting efficiency must always be made, but the isolation of the double escape peak and suppression of other backgrounds can be very effective. Systems consisting of a central germanium detector and two surrounding NaI(Tl) scintillators are a common configuration whose performance has been well documented.[66,67]

Figure 12.28 shows the very simple response function that can be obtained from such a pair spectrometer. Although the majority of the recorded events correspond to the isolated double escape peak, a small continuum still persists due to several nonidealities. Pulses above the peak can be recorded from events in which an annihilation photon undergoes small-angle scattering before leaving the central detector but still falls within the accepted pulse height window from the outer detector. Events below the peak can be generated from gamma-ray interactions that occur near a boundary of the central detector and for which either the positron or pair electron escapes without depositing all its energy. Robertson et al.[66] have shown that this latter continuum can be suppressed significantly by using pulse shape discrimination to discard slow-rising pulses, which are most likely to be generated from interactions near the edges of the germanium central detector (see p. 426).

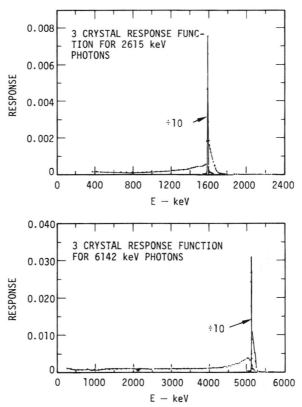

Figure 12.28 Response functions of a Ge(Li) crystal operated as a pair spectrometer in conjunction with two NaI(Tl) detectors. (From Stelts and Browne.[67])

C. Energy Calibration

1. CALIBRATION SOURCES

In gamma-ray spectroscopy with germanium detectors, the pulse height scale must be calibrated in terms of absolute gamma-ray energy if various peaks in the spectrum are to be properly identified. In many routine applications, the gamma rays expected to appear in the spectrum are well known in advance and the corresponding peaks can readily be identified by inspection. In other applications, unknown gamma-ray spectra may be encountered which do not provide an unambiguous calibration of the energy scale. In such cases, a separate calibration gamma-ray source is conventionally used to supply peaks of known energy in the spectrum. Accurate calibration should involve a standard source with gamma-ray energies that are not widely different from those to be measured in the unknown spectrum. Because even the best spectrometer systems often show nonlinearities of a channel or two over a full range of several thousand channels, it is also useful to have multiple calibration peaks at various points along the measured energy range to account for these nonlinearities.

The precision to which the centroid of a peak in a pulse height spectrum can be localized is dependent on the spectrometer system resolution and its stability over the period of the measurement. With high-quality germanium systems, the uncertainty in the peak position can approach one part in 10^5, which is of the same order as the uncertainty in the calibration energy standards. Therefore, an important goal is to define closely the energy of the standards so that their energy uncertainty does not contribute unnecessarily to the overall imprecision of a gamma-ray measurement.

Criteria that enter into the selection of standards to be used for germanium spectrometer calibration have been reviewed by Kern.[68] When the ultimate in precision is required, use of one of two primary standards is recommended: the tungsten K-α_1 X-ray (near 59 keV) or the ^{198}Au gamma ray (near 411 keV). Precise values and estimated uncertainties for these energies are listed in Table 12.1. Some authors have also treated annihilation radiation (near 511 keV) as a primary standard, but its use in high-precision calibrations should be avoided. The observed line is always several keV wide due to finite momentum effects at the annihilation point and will also be shifted below the electron rest-mass energy (511.003 keV) by up to 10 eV.[71,72] This shift reflects the effects of the finite binding energy of the electron involved in the annihilation process and therefore will be dependent on specific materials around the source. A detailed investigation of the annihilation peak[72] also shows that errors of 40–50 eV can easily be made in its centroid location unless account is taken of its slightly asymmetric shape.

Secondary gamma-ray standards more widely separated over the energy range are also listed in Table 12.1. Other energy standards between 26 and 100 keV are discussed in Ref. 73, between 50 and 420 keV in Ref. 69, and between 400 and 1300 keV in Ref. 74. For higher-energy gamma rays, the position of the double escape peak often will lie within the energy range of one of the listed standards, and the energy scale can thus be "bootstrapped" to higher energies by noting the position of the corresponding full-energy peak.

Another method of extrapolation to higher energies (often called the *cascade-crossover* method) can be carried out if gamma rays from a cascade such as that illustrated in the sketch below are involved.

Table 12.1 Gamma Rays Used as Energy Calibration Standards[a]

Source	Energy (keV)	Source	Energy (keV)
^{241}Am	59.536 ± 0.001	^{192}Ir	468.060 ± 0.010
^{109}Cd	88.034 ± 0.010	Annihilation	511.003 ± 0.002
^{182}Ta	100.106 ± 0.001	^{207}Bi	569.690 ± 0.030
^{57}Co	122.046 ± 0.020	^{208}Tl	583.139 ± 0.023
^{144}Ce	133.503 ± 0.020	^{192}Ir	604.378 ± 0.020
^{57}Co	136.465 ± 0.020	^{192}Ir	612.430 ± 0.020
^{141}Ce	145.442 ± 0.010	^{137}Cs	661.615 ± 0.030
^{182}Ta	152.435 ± 0.004	^{54}Mn	834.840 ± 0.050
^{139}Ce	165.852 ± 0.010	^{88}Y	898.023 ± 0.065
^{182}Ta	179.393 ± 0.003	^{207}Bi	1063.655 ± 0.040
^{182}Ta	222.110 ± 0.003	^{60}Co	1173.231 ± 0.030
^{212}Pb	238.624 ± 0.008	^{22}Na	1274.550 ± 0.040
^{203}Hg	279.179 ± 0.010	^{60}Co	1332.508 ± 0.015
^{192}Ir	295.938 ± 0.010	^{140}La	1596.200 ± 0.040
^{192}Ir	308.440 ± 0.010	^{124}Sb	1691.022 ± 0.040
^{192}Ir	316.490 ± 0.010	^{88}Y	1836.127 ± 0.050
^{131}I	364.491 ± 0.015	^{208}Tl	2614.708 ± 0.050
^{198}Au	411.792 ± 0.008	^{24}Na	2754.142 ± 0.060

[a]The primary X-ray standard: Tungsten $K - \alpha_1 = 59.31918 \pm 0.00035$ keV (from Greenwood et al.).[69]

Source: Values from Gunnink et al.[70]

Here, the gamma ray labeled C should have an energy equal to that of the sum of the lower-energy gamma rays labeled A and B, after a small correction is made for the difference in nuclear recoil energy. If the energies of A and B lie within the calibration range, a new calibration point at the energy of C is thereby obtained.

Even if there are no crossover transitions in the radioisotope decay scheme, the coincidence sum peaks discussed earlier sometimes can be useful for extending the energy range of a limited set of calibration sources. For the example of the ^{60}Co spectrum previously shown in Fig. 12.19, the simultaneous detection of the 1.173 MeV and 1.332 MeV coincident gamma ray photons leads to a sum peak at 2.506 MeV that can also be used in the energy calibration of the detector.

2. CALIBRATION CURVE

Once energy calibration points have been established over the entire energy range of interest, a calibration curve relating energy to channel number is normally derived. Common techniques involve the least-square fitting of a polynomial of the form $E_i = \Sigma_{n=0}^{N} a_n C_i^n$, where E_i is the energy corresponding to the channel number C_i. A polynomial of order $N = 4$ or 5 normally is adequate for typical germanium spectrometers, depending on the severity of nonlinearity that is present.

Because the degree of nonlinearity is often quite small, the calibration curve is sometimes presented as a plot of the deviation from perfect linearity versus channel number. A representative plot of this type is given in Fig. 12.29.

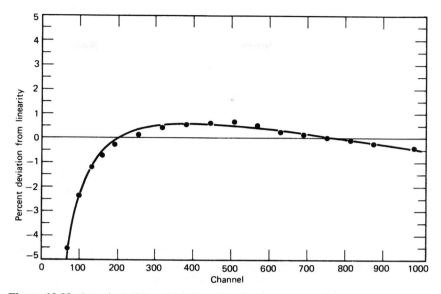

Figure 12.29 A typical differential linearity plot for a germanium detector system. (From Zulliger et al.[75])

3. DIRECTIONAL DEPENDENCE OF PEAK POSITIONS

When carrying out gamma-ray energy measurements of very high precision, some care must be taken that the unknown source and calibration source are placed so that the emitted gamma rays strike the detector from the same direction in either case. Shifts up to 110 eV have been reported in apparent peak position as the source of gamma rays was moved around the detector.[76,77] Such shifts are large enough to be of significance in high-precision measurements.

The peak shift with direction is thought to arise from two effects.[77,78] One involves the tendency for the secondary electrons produced in gamma-ray interactions to gain a slight amount of energy in the electric field of the detector. Because there is some correlation between the secondary electron direction and the incoming gamma-ray direction, the amount of energy gained from the field will differ slightly, depending on orientation. A second potential source of peak shift is any difference in charge collection efficiency in the detector. These differences often are a source of peak broadening when the detector is uniformly irradiated over its entire volume. However, gamma rays incident from a specific direction will tend to interact preferentially in certain regions of the detector volume. If the charge collection efficiency varies appreciably between regions, then the average peak position can also shift with incident gamma-ray direction.

D. Detection Efficiency

1. EFFICIENCY CHARACTERISTICS

A sketch of the typical variation of the intrinsic full-energy peak efficiency for coaxial type germanium detectors is shown in Fig.12.30. Curves are shown for representative n- and p-type detectors of the same volume. The different behavior has nothing to do with the dopant in the bulk material but only with the nature of the contact that must be made on the two types of material. The efficiency difference arises because of the thicker outer contact (generally formed by a diffusion process) on the p-type material contrasted with a

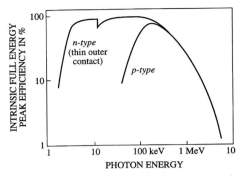

Figure 12.30 General character of the variation of intrinsic peak efficiency for coaxial germanium detectors with incident photon energy. The difference between *n*- and *p*-type detector behavior at low energies is caused by the attenuation due to different contact thicknesses.

much thinner contact typical on *n*-type detectors. Thus the roll-off of the detection efficiency for the *p*-type detectors below about 150 keV is due to the absorption of low-energy photons as they pass through the dead layer of the outer contact.

In the extended response of the *n*-type detector, the intrinsic full-energy peak efficiency remains near 100% between a few keV and about 100 keV. This is the region where photoelectric absorption dominates in germanium, and even a few millimeters of thickness are sufficient to intercept nearly all the incident photons. As the energy increases between 100 keV and 1 MeV, most gamma-ray photons still interact in the detector, but now some will undergo Compton scattering and escape rather than contributing to the full-energy peak. At energies beyond 1 MeV, a significant fraction of all the incident gamma rays can pass through the detector without interaction of any kind, and the peak efficiency drops more rapidly.

Also shown on the curve for the *n*-type detector is the "notch" that occurs at the *K*-shell binding energy of germanium at 11 keV. For energies just above this value, the incident gamma ray is strongly absorbed without deep penetration beyond the detector surface. If the subsequent characteristic *K* X-ray escapes, the event does not contribute to the full energy peak. For slightly lower energies, *K*-shell absorption is no longer possible and *L*-shell interactions predominate. Now the incident gamma ray tends to penetrate somewhat deeper and the energy of the fluorescent *L* X-ray is much lower, both leading to diminished probability of escape.

2. EFFICIENCY CALIBRATION

Any measurement of absolute emission rates of gamma rays (i.e., not relative to a similar source of known activity) requires knowledge of the detector efficiency. The emission rate for a point source can then be calculated from Eq. (4.19) by measuring the full-energy peak area over a fixed time and by determining the detector solid angle from its dimensions and the source–detector spacing. In germanium gamma-ray spectroscopy, an efficiency based on the area under the single or double escape peak is sometimes used in place of that based on the full-energy peak. Although efficiencies of germanium detectors can be estimated from published measurements or calculations for detectors of similar size, the accuracy of results based on these values will not be much better than 10–20%. One major difficulty is that the dimensions of these detectors are not standardized to any degree, and it is very difficult to determine precisely their active volume. Furthermore, long-term changes in charge collection efficiency and/or window thickness can lead to drifts in the detector efficiency over periods of time.

For these reasons, users will normally carry out their own periodic efficiency calibrations of their germanium detectors using sources calibrated by some other means. Any error in assumed detector dimensions will then apply both to the calibration and actual measurements and will not affect the accuracy of activity measurements. The source-detector distance still must be accurately measured and reproduced to avoid errors in the relative solid angle. The calibration is normally carried out for an assortment of gamma-ray energies covering the range of interest to allow construction of an empirical efficiency versus energy curve.

Many of the national standards laboratories will provide isotopes whose gamma-ray emission rates have been calibrated to a precision ranging from 0.5 to 2.0%. Single isotope sources provide a few well-separated gamma-ray peaks whose area can be determined by simple methods to calibrate the detector efficiency. If the energy scale must extend over a relatively wide range, multiple sources must be used, either sequentially or in combination. Table 12.2 lists radionuclides used for efficiency calibrations, together with decay data necessary to compute gamma-ray yields from absolute activity.

Table 12.2 Decay Data for Radionuclides Used as Efficiency Standards

Nuclide	$T_{1/2}$		E (keV)	I^a(%)	$\Delta I/I^b$ (%)
^{22}Na	2.60	y	1274.5	99.95	0.0
^{24}Na	15.0	h	1368.5	100.0	0.0
			2754.0	99.85	0.0
^{46}Sc	83.7	d	889.2	99.98	0.0
			1120.5	99.99	0.0
^{54}Mn	312.5	d	834.8	99.98	0.0
^{57}Co	272	d	14.4	9.6	1.0
			122.1	85.6	0.3
^{60}Co	5.27	y	1173.2	99.88	0.0
			1332.5	99.98	0.0
^{85}Sr	64.8	d	13.4	50.7	1.5
			514.0	99.28	0.0
^{88}Y	106.6	d	14.2	52.5	1.5
			1836.1	99.4	0.2
^{95}Nb	35.15	d	765.8	99.80	0.0
^{113}Sn	115.2	d	24.1	79.5	2.0
^{131}I	8.02	d	364.5	82.4	0.5
^{134}Cs	2.06	y	604.6	97.5	0.2
^{137}Cs	30.0	y	31.8/32.2	5.64	2.0
			661.6	85.3	0.4
^{139}Ce	137.6	d	33.0/33.4	64.1	2.0
			165.8	80.0	0.4
^{141}Ce	32.5	d	35.6/36.0	12.6	2.0
			145.5	48.4	0.9
^{140}La	40.27	h	1596.6	95.6	0.3
^{198}Au	2.696	d	411.8	95.53	0.1
^{203}Hg	46.6	d	70.8/72.9	10.1	1.5
			279.2	81.3	0.2
^{241}Am	432	y	59.5	36.0	1.0

$^a I$: Gamma-ray photon yield per disintegration.

$^b \Delta I/I$: Uncertainty in yield figure.

Note: Only those gamma-ray lines are listed for which the yield uncertainty is 2.0% or less. Some of these nuclides emit other gamma rays in addition to those shown.

Source: Debertin et al.[79]

Standard calibration sources are available as small deposits on thin backing material, so that they may closely approximate nonabsorbing point sources. However, radioactive samples to be measured often have nonnegligible volume and mass, and gamma rays can be attenuated by self-absorption within the sample material itself. In such cases, correction must be made for this attenuation if an accurate determination of the gamma-ray emission rate from the entire sample is needed. To aid in the calibration of large-volume samples, typically of water solutions or soil, a standard sample container called a *Marinelli beaker* has been specified[44] that closely fits over the endcap of the detector cryostat. Several standard sizes of Marinelli beakers are specified to accommodate samples of various volumes. With a standard geometry, the self-absorption in samples of similar composition and volume will be comparable. Some calibrated standard sources are available in Marinelli beaker form to simplify the efficiency determination for such bulk samples.

Recording of separate spectra for each source provides results that are simple to interpret, but the calibration process is then time consuming and tedious. The substitution of a single source that emits many different gamma-ray energies is a tempting alternative because only a single calibration spectrum need be recorded and analyzed. The problems of interference between multiple responses become much more severe, however, and if precise results are to be obtained, more sophisticated methods of determining peak area must be used that take into account these interference effects. Several authors[55,80,81] suggest the use of ^{226}Ra in equilibrium with its decay products and provide tables of gamma-ray intensities

Table 12.3 Gamma Rays Emitted by ^{226}Ra in Equilibrium with its Daughters[a]

Isotope	Gamma-Ray Energy (keV)	Relative Intensity
^{226}Ra	186.211 ± 0.010	9.00 ± 0.10
^{214}Pb	241.981 ± 0.008	16.06 ± 0.19
^{214}Pb	295.213 ± 0.008	42.01 ± 0.53
^{214}Pb	351.921 ± 0.008	80.42 ± 0.81
^{214}Bi	609.312 ± 0.007	100 ± 0.92
^{214}Bi	768.356 ± 0.010	10.90 ± 0.15
^{214}Bi	934.061 ± 0.012	6.93 ± 0.10
^{214}Bi	1120.287 ± 0.010	32.72 ± 0.39
^{214}Bi	1238.110 ± 0.012	12.94 ± 0.17
^{214}Bi	1377.669 ± 0.012	8.87 ± 0.15
^{214}Bi	1509.228 ± 0.015	4.78 ± 0.09
^{214}Bi	1729.595 ± 0.015	6.29 ± 0.10
^{214}Bi	1764.494 ± 0.014	34.23 ± 0.44
^{214}Bi	1847.420 ± 0.025	4.52 ± 0.09
^{214}Bi	2118.551 ± 0.030	2.53 ± 0.05
^{214}Bi	2204.215 ± 0.040	10.77 ± 0.20
^{214}Bi	2447.860 ± 0.100	3.32 ± 0.08

[a]Only the strongest transitions are shown. Energies are measured relative to an assumed 411.794 keV gamma ray from ^{198}Au. Quoted errors do not include any error contribution from this reference standard.

Source: Zobel et al.[55]

Table 12.4 Multiple Gamma Rays Emitted in the Decay of ^{152}Eu

Energy (keV)	Relative Intensity
121.8	141. ± 4.[a]
244.7	36.6 ± 1.1
344.3	127.2 ± 1.3
367.8	4.19 ± 0.04
411.1	10.71 ± 0.11
444.0	15.00 ± 0.15
488.7	1.984 ± 0.023
586.3	2.24 ± 0.05
678.6	2.296 ± 0.028
688.7	4.12 ± 0.04
778.9	62.6 ± 0.6
867.4	20.54 ± 0.21
964.0	70.4 ± 0.7
1005.1	3.57 ± 0.07
1085.8	48.7 ± 0.5
1089.7	8.26 ± 0.09[b]
1112.1	65.0 ± 0.7
1212.9	6.67 ± 0.07
1299.1	7.76 ± 0.08
1408.0	100.0 ± 1.0[c]
1457.6	2.52 ± 0.09

[a] In order to use this line, no ^{154}Eu should be present.

[b] Not intended for use in calibrations because of the proximity to the more intense nearby energy.

[c] The absolute yield of this line is 0.2101 per decay.[83]

Source: Data taken from Gehrke et al.[82]

per disintegration (see Table 12.3). These cover the energy range from 186 to 2448 keV and are particularly useful because a source of ^{226}Ra is often available in radioisotope laboratories. References 53 and 82 provide similar lists of gamma-ray intensities per disintegration for a number of other isotopes proposed for detector efficiency calibration. Of these, ^{152}Eu has gained popularity because of its convenient half-life (13 y) and the wide range of gamma-ray energies produced in its decay (see Table 12.4).

3. SUM–COINCIDENCE EFFECTS

In employing any source that emits more than one radiation in coincidence, some care must be taken to ensure that measured peak intensities are not affected by the sum-coincidence effects previously described beginning on p. 438. If two gamma rays are emitted in coincidence from the source, they may interact simultaneously in the detector, and the resulting pulse will not, in general, lie within the full-energy peak corresponding to either one. The problem is particularly severe for those sources that involve many cascade gamma rays and, in the specific examples given above, can lead to errors when using reported peak intensities from ^{226}Ra (Ref. 84) or ^{152}Eu (Ref. 82). Not only must coincident detection of gamma rays be avoided, but any other coincident radiation, such as characteristic X-rays or bremsstrahlung, can also lead to summing effects that may reduce the apparent peak intensities. Depending on the decay scheme, coincidence summing may also increase the apparent intensity of some peaks whose energy corresponds to the sum of two lower energy peaks. Quantitative assessments of the changes that may occur in the full-energy peak intensity due to summing effects can be found in Refs. 82 and 85, together with suggested calculation methods to correct for such errors. The effects of coincidence summing are most significant for detectors with thin entrance windows (the usual case for *n*-type Ge detectors) that admit low-energy X-rays or bremsstrahlung that may be emitted in the decay of the isotope of interest. Because of the geometry effect discussed in the next paragraph, well-type germanium detectors in which the source is nearly completely surrounded by the germanium detector are also particularly vulnerable to summing effects.[86]

Because the summing effects will depend on the square of the detector solid angle [see Eq. (10.11)], whereas the simple peaks vary linearly, the relative effect of summing can be reduced by reducing the solid angle. Meyer et al.[53] have presented data to show the variation of apparent peak intensities with source-detector spacing and recommend that for typical germanium detector sizes, the source–detector spacing be maintained to be at least 10 cm and preferably 30–40 cm. Of course, there is a limit as to how far away one can practically place a source of low activity, and a compromise between summing inaccuracies and statistical errors in the peak must ultimately be made.

Other potential errors in the peak area measurement can arise from inaccurate treatment of system dead time and from pulse pile-up effects (see Chapter 17). Both phenomena are most important when pulse rates are relatively high, and their effect on absolute detector calibration has been reviewed by Müller.[87]

4. INTERPOLATION USING FITTED EFFICIENCY FUNCTIONS

Once the efficiency of a detector has been measured at several energies using calibrated sources, it is useful to fit a curve to these points that describes the efficiency over the entire energy range of interest. A number of different empirical formulas have been described in the literature and incorporated into software packages used for gamma-ray spectrum analysis. In general, somewhat different functions are needed for planar and coaxial geometries.

For planar detectors, one formula originally suggested by Mowatt[88] has been found to give a very good representation of the full-energy peak efficiency of several different detectors over the energy range from 60 keV to 1863 keV. It takes the form

$$\epsilon = \frac{K[\tau + \sigma Q \exp{(-RE)}]}{\tau + \sigma} \{1 - \exp[-P(\tau + \sigma)]\} \qquad (12.31)$$

where τ = photoelectric absorption coefficient in germanium at energy E

σ = Compton absorption coefficient at energy E

and $K, Q, R,$ and P are parameters fitted to the experimental points.

For coaxial detectors, a variety of published fitting functions have been evaluated and intercompared over the 50 keV to 8500 keV energy range in Ref. 89. Common formulas incorporate 3–9 empirical parameters that are fitted to the experimentally measured points. A small number of parameters can be sufficient if the energy range is limited. Formulas with more parameters are generally required to cover wider energy ranges, but are also more

Figure 12.31 The points shown are measured full-energy peak efficiency values for a 25% relative efficiency n-type germanium detector. The cryostat had a 0.5 mm thick aluminum entrance window. The points at lower energies were obtained using standard radioisotope calibration sources, whereas neutron-induced capture gamma rays provided the upper energy values. The solid line is a 9-parameter analytical fit to all the points using a function of the form given in Eq. (12.32). (From Kis et al.[89])

prone to the possible appearance of nonphysical oscillations in the fitted function if the data points are widely spaced. In some cases, the energy range is subdivided into two or more sections, and separate fits carried out for each range. For covering wide energy ranges, one commonly-used formula is a linear function relating the logarithm of the efficiency to the logarithm of the energy

$$\ln \epsilon = \sum_{i=1}^{N} a_i \left(\ln \frac{E}{E_0} \right)^{i-1} \tag{12.32}$$

where E_0 is a fixed reference energy and the values of a_i are the fitted parameters. An example of this type of fit using $N = 9$ to a large set of experimentally measured efficiency points is given in Fig. 12.31. When fewer data points are available than shown in the example or in smaller regions of the energy are to be covered, the value of N might be limited to as few as 2 or 3 terms to yield a stable solution free of oscillations. A number of alternative fitting functions are discussed in Ref. 42.

5. CONVENTIONS IN QUOTING DETECTION EFFICIENCY

The full-energy or peak efficiency of germanium detectors can be quoted in several different ways. The quantity most often directly measured is the absolute full-energy peak efficiency, defined as the ratio of counts in the full-energy peak divided by the number of gamma rays emitted by the source. A widely accepted standard energy at which this measurement is made is the 1.333 MeV gamma ray emitted by ^{60}Co. The absolute efficiency is obviously a strong function of both gamma-ray energy and the exact source–detector spacing used for the measurement. This dependence on spacing is greatly diminished if we instead use the intrinsic efficiency, defined as the number of full-energy peak counts divided by the number of gamma rays incident on the detector. At gamma-ray energies sufficiently high so that the detector is relatively transparent (interaction probabilities less than 20 or 30%), the intrinsic efficiency is approximately proportional to the total active volume of the detector. Consequently, most commercial detectors are specified by their total active volume measured in cm^3. For detectors with nonsymmetrical shapes, the intrinsic efficiency can vary substantially depending on orientation of the source relative to the detector axis, because the average thickness of the detector along the direction of the incident radiation is most critical in determining the intrinsic efficiency. For coaxial or closed-ended detectors in which the geometry is complex, the intrinsic efficiency is often difficult to calculate exactly from the observed count rate because of uncertainties in the effective detector solid angle.

In an effort to eliminate some of these difficulties in quoting detector efficiency, many commercial manufacturers specify the photopeak efficiency relative to that of a standard 3 in. × 3 in. (7.62 cm × 7.62 cm) cylindrical NaI(Tl) scintillation crystal. A source–detector spacing of 25 cm is assumed in both cases for standardization. The value is normally specified for the 1.333 MeV gamma-ray photopeak from ^{60}Co. The efficiency ratio can be directly measured by simply determining the photopeak areas from both detectors, using an uncalibrated ^{60}Co source. Alternatively, only the germanium photopeak area may be directly measured, and the NaI(Tl) value calculated by using a calibrated source and by assuming an absolute peak efficiency value of 1.2×10^{-3} (Ref. 90). An approximate rule of thumb for germanium coaxial detectors is that the efficiency ratio in percent is given by the detector volume in cm^3 divided by a factor of 5. The efficiency ratio of high-purity germanium detectors when they were first introduced in the 1960s was a few percent but has reached over 200% for the largest detectors currently available. Further increases are likely in the future as crystal fabrication techniques continue to improve.

6. EFFICIENCY DATA

If a set of calibrated sources is not available to the experimenter, an estimate of the detector efficiency may be made by referring to published data. In the case of sodium iodide scintillators discussed in Chapter 10, relatively few standard sizes of crystals have predominated, so it is often possible to find published data for a detector that is identical to that of the user. In the case of germanium detectors, there is much less standardization of sizes, and the variety of geometries (planar, coaxial, closed-ended) adds to the complexity of the situation. It is therefore much less likely that published data can be found for any specific detector geometry, and considerable interpolation between available values is often needed. The variation of efficiency with energy will be quite similar for detectors of roughly the same size and shape, even if the absolute efficiency values may differ.

Another approach is to derive efficiency data from calculations based on knowledge of the probability of each of the primary gamma-ray interactions within the detector. Monte Carlo codes provide the most accurate calculational methods for efficiency determination, and examples of the Monte Carlo approach applied to germanium detectors can be found in Refs. 91–93. Compared with similar calculations of detection efficiency in NaI(Tl), the added importance of the escape of secondary radiations in germanium detectors (secondary electrons, bremsstrahlung, etc.) makes accurate modeling more difficult.

7. COMPARISON WITH OTHER GAMMA-RAY SPECTROMETERS

Figure 12.32 shows the intrinsic peak detection efficiency as a function of energy for some typical examples of X- and gamma-ray spectrometers of different types. These plots should be regarded as only showing general trends, in that there will always be a great deal of variation within any one type depending on detector size and configuration. At energies below 20 keV where fluorescent X-rays from the lighter elements are found, an argon-filled proportional counter can show good efficiency (see Fig. 6.16). Substituting xenon (with its higher atomic number) for the argon fill gas will extend the useable energy range by a significant margin. To achieve better detection efficiency at higher energies, greater stopping power is required for the incident photons than can be provided conveniently in a low-density gas. Lithium-drifted silicon detectors (described in Chapter 13) of a few millimeter thickness provide excellent detection efficiency throughout the full X-ray energy range up to about 100 keV. At higher energies typical of gamma rays, germanium detectors can be highly efficient up to about 1 MeV. At higher gamma-ray energies, their detection

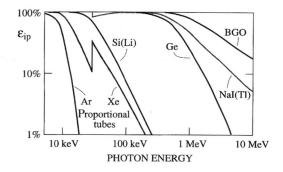

Figure 12.32 The general dependence of intrinsic full-energy peak efficiency on incident photon energy for a number of common detectors used for X- and gamma-ray spectroscopy. The curves are to be interpreted as showing broad trends only, and there will be large differences between specific detectors within each of the categories shown.

efficiency is generally lower than that of sodium iodide scintillators of reasonable size that could be considered for the same application because of their limited total volume and lower atomic number of germanium compared with iodine. To achieve the greatest detection efficiency for high energy gamma rays, large-volume scintillators with high-Z constituents (such as BGO) provide the highest efficiency of common gamma-ray spectrometers.

E. Miscellaneous Effects on Detector Performance

1. RADIATION DAMAGE

As discussed in Chapter 11, semiconductor detectors are relatively sensitive to performance degradation caused by damage created within the detector by incident radiation. The large volume and long charge collection paths in germanium gamma-ray detectors make them more susceptible to such degradation compared with the much thinner silicon diode detectors typically applied to charged particle detection. Because the amount of damage created by fast neutrons of a given fluence is large compared with the damage from an equivalent fluence of gamma rays, the most significant effects often arise in reactor or accelerator laboratories where fast neutrons may be present. An extensive review of the effects of this damage in both Ge(Li) and HPGe detectors is given in Refs. 94 and 95.

The principal consequence of radiation damage is to increase the amount of hole trapping within the active volume of the detector. In a damaged detector, the amount of charge collected is subject to a loss caused by this trapping, which will vary from pulse to pulse depending on the position of the interaction. Measured peaks in the pulse height spectrum will then show a tailing toward the low-energy side. The spectra shown in Fig. 12.33 illustrate a gradual broadening of gamma-ray peaks from a Ge(Li) detector as the fast neutron exposure is increased. Exposure of thick planar germanium detectors to a fast neutron fluence of about 10^9 n/cm^2 is sufficient to risk measurable changes in the detector resolution[97, 98] and many become totally unusable after exposure to a fluence of 10^{10} n/cm^2.

In coaxial detectors, the specific detector configuration can have a strong influence on the measured spectral effects. In HPGe coaxials fabricated from high-purity p-type germanium, holes are the carrier type that are drawn inward to the p^+ contact near the cylindrical axis. For HPGe coaxials made from high-purity n-type germanium, the electrode polarity is reversed, and holes are instead collected at the p^+ contact now fabricated on the outer cylindrical surface (see Fig. 12.3). It is found that these n-type HPGe detectors (sometimes also called *reverse electrode* configurations) show much less performance degradation from radiation damage when compared with the more common p-type detectors. In one study[99] of similar coaxial detectors of both types, the n-type detector was able to withstand 28 times the neutron fluence

Figure 12.33 The deterioration of a Ge(Li) γ-spectrum with increasing neutron fluence. Only the upper part of the spectrum is shown. (From Kraner et al.[96])

before showing the same peak broadening as that observed from the p-type configuration. The explanation for this difference lies in the fact that the damage sites preferentially trap holes rather than electrons. Because of the cylindrical geometry and attenuation of the incident gamma rays, more interactions occur at large values of the detector radius than at small values. For such locations, the rise of the signal pulse is dominated by the species that drifts inward toward the detector center (see Fig. 12.3). If these are holes, the corresponding pulse is much more likely to be affected by hole trapping than if the holes must only travel the shorter distance to the outer electrode. As a result of their superior performance in the presence of damage, n-type HPGe coaxial detectors have become the configuration of choice for applications in which exposure to fast neutrons may occur over time.

One of the advantages of HPGe detectors over the older Ge(Li) types is that, should radiation damage occur, it can be repaired by annealing the detector. It has been shown[100] that raising the temperature to 120°C for 72 h can restore lightly damaged ($< 4 \times 10^9$ n/cm^2) n-type HPGe detectors to their original resolution performance. During the heating process, an external vacuum pump must be connected to preserve the cryostat vacuum. The process is simple enough so that many users are able to carry out this step in their own laboratories.

2. NEUTRON-INDUCED PULSES

When detectors are used under conditions in which fast neutrons may be present as a background, account must be taken of their effect on the recorded pulse height spectrum. The most obvious contribution is the appearance of spurious peaks that arise mainly due to excitation of germanium nuclei by inelastic neutron scattering, followed by the emission of de-excitation gamma rays, internal conversion electrons, or X-rays. The neutron-induced peaks are usually identifiable in the spectrum because their width is normally larger than that for gamma-ray-induced events. The peak broadening takes place because a fraction of the excitation energy goes into the recoiling germanium nucleus, which subsequently contributes a variable yield of electron–hole pairs, adding to those created by the de-excitation radiation.

Chasman et al.[101] investigated the response of a Ge(Li) detector to monoenergetic fast neutrons ranging from 1.2 to 16.3 MeV. Lines were observed that resulted from the inelastic excitation of four germanium isotopes, as well as from inelastic neutron excitation of other materials in the detector vacuum enclosure and cryostat. At neutron energies above 3 MeV, additional peaks were detected due to neutron-induced charged particles created in reactions with germanium atoms. Other peaks below 200 keV have been observed[102] that arise from isomers produced by neutron reactions with several germanium isotopes. These lower-energy peaks tend not to be broadened and can easily be confused with peaks from external gamma rays. A spectrum showing many of the neutron-induced peaks is given in Fig. 12.34.

Figure 12.34 Prompt pulse height spectrum produced by fast neutron irradiation of a 50 cm^3 Ge(Li) detector. (From Bunting and Kraushaar.[102])

The radiation damage that may result from a given exposure to neutrons can be appraised from the previous section if an estimate of the fast neutron flux is obtained. The neutron-induced pulse height spectrum can be used for such an estimate, as suggested by Bell.[103] The neutron fluence is approximately given by

$$\frac{\text{neutrons}}{\text{cm}^2} = \frac{300 \times (\text{counts in 693.4 keV peak})}{\text{detector volume in cm}^3}$$

so that a quick determination of the spectrum can indicate whether further exposure of the detector to the neutron environment is advisable.

PROBLEMS

12.1 Assume that the Fano factor in germanium were half the currently assumed value. What quantitative effect would this change have on the theoretically obtainable energy resolution?

12.2 A planar germanium detector with a 10-mm thick intrinsic region is operated with sufficient applied voltage to saturate the carrier velocities. What is the approximate value of the required voltage? What must be the minimum charge carrier lifetimes if no more than 0.1% of either holes or electrons are to be lost from any pulse?

12.3 The detector described above is operated with a pulse-processing system that produces a peak with 1.2 keV equivalent FWHM from a pulser input. Estimate the energy resolution of the detector–electronics system for incident 140 keV gamma rays.

12.4 The Compton edge in a gamma-ray spectrum recorded with a germanium detector for a given isotope lies at an energy of 1.16 MeV. Find the energy of the incident gamma rays and the equivalent energy of the Compton edge in a sodium iodide detector.

12.5 Why are escape peaks generally more prominent in germanium detector gamma-ray spectra compared with sodium iodide detectors?

12.6 A germanium detector has a photopeak efficiency of 40% relative to that of a standard 7.62 cm × 7.62 cm NaI(Tl) scintillator. Find the counting rate for the 1.333 MeV peak for a ^{60}Co point source of 150 kBq activity at a distance of 40 cm.

12.7 Assuming that charge collection is complete and that electronic noise is negligible, find the expected energy resolution (in percent) of a germanium detector for the 0.662 MeV gamma rays from ^{137}Cs.

12.8 For incident 2.10 MeV gamma rays, at what energy does the peak appear in the spectrum from a pair spectrometer?

REFERENCES

1. American National Standard, IEEE Standard 325-1996 (1996).
2. R. M. Keyser and T. W. Raudorf, *Nucl. Instrum. Meth.* **A286**, 357 (1990).
3. J. Llacer, *Nucl. Instrum. Meth.* **98**, 259 (1972).
4. J. M. Marler and P. V. Hewka, *IEEE Trans. Nucl. Sci.* **NS-21**(1), 287 (1974).
5. E. L. Hull et al., *Nucl. Instrum. Meth.* **A364**, 488 (1995).
6. G. A. Armantrout, *IEEE Trans. Nucl. Sci.* **NS-19**(3), 289 (1972).
7. G. H. Nakano, D. A. Simpson, and W. L. Imhof, *IEEE Trans. Nucl. Sci.* **NS-24**(1), 68 (1977).
8. R. H. Pehl, E. E. Haller, and R. C. Cordi, *IEEE Trans. Nucl. Sci.* **NS-20**(1), 494 (1973).
9. E. Sakai, Y. Murakami, and H. Nakatani, *IEEE Trans. Nucl. Sci.* **NS-29**(1), 760 (1982).
10. R. E. Stone, V. A. Barkley, and J. A. Fleming, *IEEE Trans. Nucl. Sci.* **NS-33**(1), 299 (1986).
11. G. Alberti, R. Clerici, and A. Zambra, *Nucl. Instrum. Meth.* **158**, 425 (1979).
12. F. Taccetti et al., *Nucl. Instrum. Meth.* **A398**, 238 (1997).
13. J. Cl. Philippot, *IEEE Trans. Nucl. Sci.* **NS-17**(3), 446 (1970).
14. J. M. Palms et al., *Nucl. Instrum. Meth.* **64**, 310 (1968).
15. H. R. Zulliger, L. M. Middleman, and D. W. Aitken, *IEEE Trans. Nucl. Sci.* **NS-16**(1), 47 (1969).
16. A. Owens, *Nucl. Instrum. Meth.* **A238**, 473 (1985).
17. A. Alberigi Quaranta, M. Martini, and G. Ottaviani, *IEEE Trans. Nucl. Sci.* **NS-16**(2), 35 (1969).
18. T. W. Raudorf, T. J. Paulus, M. O. Bedwell, and M. Martini, *IEEE Trans. Nucl. Sci.* **NS-24**(1), 78 (1977).
19. B. C. Robertson and H. L. Malm, *Nucl. Instrum. Meth.* **150**, 401 (1978).
20. D. C. S. White and W. J. McDonald, *Nucl. Instrum. Meth.* **115**, 1 (1974).
21. L. L. Gadeken and B. C. Robertson, *Nucl. Instrum. Meth.* **136**, 255 (1976).
22. R. Chun, L. L. Gadeken, and B. C. Robertson, *Nucl. Instrum. Meth.* **137**, 295 (1976).
23. H. Engel, H. Schneider, and R. Spitz, *Nucl. Instrum. Meth.* **142**, 525 (1977).
24. T. W. Raudorf, M. O. Bedwell, and T. J. Paulus, *IEEE Trans. Nucl. Sci.* **NS-29**(1), 764 (1982).
25. F. S. Goulding and D. A. Landis, *IEEE Trans. Nucl. Sci.* **41**(4), 1145 (1994).
26. W. C. G. Ho et al., *Nucl. Instrum. Meth.* **A412**, 507 (1998).
27. I. Y. Lee, *Nucl. Instrum. Meth.* **A422**, 195 (1999).
28. J. Blair, D. Beckedahl, J Kammeraad, and G. Schmid, *Nucl. Instrum. Meth*, **A422**, 331, (1999).

29. B. Aspacher and A. C. Rester, *Nucl. Instrum. Meth.* **A338**, 511 (1994).
30. B. Aspacher and A. C. Rester, *Nucl. Instrum. Meth.* **A338**, 516 (1994).
31. G. J. Schmid et al., *Nucl. Instrum. Meth.* **A422**, 368 (1999).
32. B. Philhour et al., *Nucl. Instrum. Meth.* **A403**, 136 (1998).
33. F. Petry et al., *Nucl. Instrum. Meth.* **A332**, 107 (1993).
34. M. Martini and T. A. McMath, *Nucl. Instrum. Meth.* **79**, 259 (1970).
35. T. W. Raudorf and R. H. Pehl, *Nucl. Instrum. Meth.* **A255**, 538 (1987).
36. E. Sakai, *IEEE Trans. Nucl. Sci.* **NS-18**(1), 208 (1971).
37. L. Karlsson, *Nucl. Instrum. Meth.* **109**, 101 (1973).
38. G. White, F. A. Smith, and C. F. Coleman, *Nucl. Instrum. Meth.* **A234**, 535 (1985).
39. M. Moszynski and B. Bengtson, *Nucl. Instrum. Meth.* **80**, 233 (1970).
40. N. Matsushita, J. Kasagi, and W. C. McHarris, *Nucl. Instrum. Meth.* **201**, 433 (1982).
41. G. A. Armantrout, A. E. Bradley, and P. L. Phelps, *IEEE Trans. Nucl. Sci.* **NS-19**(1), 107 (1972).
42. K. Debertin and R. G. Helmer, *Gamma- and X-Ray Spectrometry with Semiconductor Detectors,* North Holland Publishers, Amsterdam, 1988.
43. J. Roth, J. H. Primbsch, and R. P. Lin, *IEEE Trans. Nucl. Sci.* **NS-31**(1), 367 (1984).
44. ANSI/IEEE Standard 325-1986, *Test Procedures for Germanium Gamma Ray Detectors* (1986).
45. U. E. P. Berg, H. Wolf, B. Schaeffer, and K. Wienhard, *Nucl. Instrum. Meth.* **129**, 155 (1975).
46. I. K. MacKenzie and J. L. Campbell, *Nucl. Instrum. Meth.* **101**, 149 (1972).
47. C. S. Rossington, R. D. Giauque, and J. M. Jaklevic, *IEEE Trans. Nucl. Sci.* **39**(4), 570 (1992).
48. F. Adams and R. Dams, *Applied Gamma-Ray Spectrometry,* 2nd ed. and revision of original publication by C. E. Crouthamel, Pergamon Press, Oxford, 1970.
49. J. P. Balagna and S. B. Helmick, *An Atlas of Gamma-Ray Spectra,* LA-4312 (1970).
50. R. G. Helmer, J. R. Davidson, and R. J. Gehrke, "Gamma-Ray Spectrum Catalog," CD-ROM published by Idaho National Engineering and Environmental Laboratory, 1999.
51. C. E. Moss and J. R. Streetman, *Nucl. Instrum. Meth.* **A299**, 98 (1990).
52. Ch. Meixner, *Nucl. Instrum. Meth.* **119**, 521 (1974).
53. R. A. Meyer, K. G. Tirsell, and G. A. Armantrout, *Proceedings, ERDA X- and Gamma-Ray Symposium,* Ann Arbor, MI (CONF-760539), p. 40 (1976).
54. H. H. Jorch and J. L. Campbell, *Nucl. Instrum. Meth.* **143**, 551 (1977).
55. V. Zobel, J. Eberth, U. Eberth, and E. Eube, *Nucl. Instrum. Meth.* **141**, 329 (1977).
56. P. J. Nolan, D. W. Gifford, and P. J. Twin, *Nucl. Instrum. Meth.* **A236**, 95 (1985).
57. C. W. Beausang et al., *Nucl. Instrum. Meth.* **A313**, 37 (1992).
58. S. L. Micek et al., *Nucl. Instrum. Meth.* **B64**, 282 (1992).
59. M. P. Carpenter et al., *Nucl. Instrum. Meth.* **A353**, 234 (1994).
60. K. Fukuda, J. Okhuma, T. Asano, and Y. Satoh, *Nucl. Instrum. Meth.* **B114**, 379 (1996).
61. E. Mauerhofer et al., *Nucl. Instrum. Meth.* **A371**, 465 (1996).
62. J. M. Palms, R. E. Wood, and O. H. Puckett, *IEEE Trans. Nucl. Sci.* **NS-15**(3), 397 (1968).

63. D. M. Walker and J. M. Palms, *IEEE Trans. Nucl. Sci.* **NS-17**(3), 296 (1970).
64. R. N. Larsen and M. G. Strauss, *IEEE Trans. Nucl. Sci.* **NS-17**(3), 254 (1970).
65. A. R. Sayres and J. A. Baicker, *IEEE Trans. Nucl. Sci.* **NS-15**(3), 393 (1968).
66. A. Robertson, G. C. Cormick, T. J. Kennett, and W. V. Prestwich, *Nucl. Instrum. Meth.* **127**, 373 (1975).
67. M. L. Steits and J. C. Browne, *Nucl. Instrum. Meth.* **133**, 35 (1976).
68. J. Kern, *Gamma Ray Standards*, CONF-7210117, p. 345 (1974).
69. R. C. Greenwood, R. G. Helmer, and R. J. Gehrke, *Nucl. Instrum. Meth.* **77**, 141 (1970).
70. R. Gunnink, J. B. Niday, R. P. Anderson, and R. A. Meyer, *Gamma-Ray Energies and Intensities*, UCID-15439 (1969).
71. K. Shizuma, H. Inoue, and Y. Yoshizawa, *Nucl. Instrum. Meth.* **137**, 599 (1976).
72. K. Fransson, A. Nilsson, J. DeRaedt, and K. G. Rensfelt, *Nucl. Instrum. Meth.* **138**, 479 (1976).
73. G. C. Nelson and B. G. Saunders, *Nucl. Instrum. Meth.* **84**, 90 (1970).
74. R. G. Helmer, R. C. Greenwood, and R. J. Gehrke, *Nucl. Instrum. Meth.* **96**, 173 (1971).
75. H. R. Zulliger, L. M. Middleman, and D. W. Aitken, *IEEE Trans. Nucl. Sci.* **NS-16**(1), 47 (1969).
76. P. C. Lichtenberger and I. K. MacKenzie, *Nucl. Instrum. Meth.* **116**, 177 (1974).
77. R. G. Helmer et al., *Nucl. Instrum. Meth.* **123**, 51 (1975).
78. H. I. Bak, Y. D. Bae, M. S. Kim, and H. D. Choi, *Nucl. Instrum. Meth.* **A366**, 332 (1995).
79. K. Debertin et al., *Proceedings, ERDA X- and Gamma-Ray Symposium,* Ann Arbor, MI (CONF-760539), p. 59 (1976).
80. G. Walford and C. E. Doust, *Nucl. Instrum. Meth.* **62**, 353 (1968).
81. G. Wallace and G. E. Coote, *Nucl. Instrum. Meth.* **74**, 353 (1969).
82. R. J. Gehrke et al., *Nucl. Instrum. Meth.* **147**, **405** (1977).
83. R. B. Firestone, S. Y. F. Chu, and C. M. Baglin, "Table of Isotopes CD-ROM," Eighth Ed., Wiley Interscience, New York, 1998.
84. A. Dolev et al., *Nucl. Instrum. Meth.* **68**, **176** (1969).
85. G. J. McCallum and G. E. Coote, *Nucl. Instrum. Meth.* **130**, 189 (1975).
86. S. J. Gelsema and M. Blaauw, *Nucl. Instrum. Meth.* **A368**, 410 (1996).
87. J. W. Müller, *Nucl. Instrum. Meth.* **112**, 47 (1973).
88. R. S. Mowatt, *Nucl. Instrum. Meth.* **70**, 237 (1969).
89. Z. Kis et al., *Nucl. Instrum. Meth.* **A418**, 374 (1998).
90. R. L. Heath, *Scintillation Spectrometry, Gamma-Ray Spectrum Catalog*, 2nd ed., Vol. 1, IDO-16880-1 (1964).
91. A. Owens et al., *IEEE Trans. Nucl. Sci.* **38**(2), 221 (1991).
92. L. K. Herold and R. T. Kouzes, *IEEE Trans. Nucl. Sci.* **38**(2), 231 (1991).
93. S. Kamboj and B. Kahn, *Health Phys.* **70**(4), 512 (1996).
94. H. W. Kraner, *IEEE Trans. Nucl. Sci.* **NS-27**(1), 218 (1980).
95. L. S. Darken, *Nucl. Instrum. Meth.* **B74**, 523 (1993).
96. H. W. Kraner, C. Chasman, and K. W. Jones, *Nucl. Instrum. Meth.* **62**, 173 (1968).
97. P. H. Stelson, J. K. Dickens, S. Raman, and R. C. Trammell, *Nucl. Instrum. Meth.* **98**, 481 (1972).
98. H. W. Kraner, R. H. Pehl, and E. E. Haller, *IEEE Trans. Nucl. Sci.* **NS-22**(1), 149 (1975).
99. R. H. Pehl et al., *IEEE Trans. Nucl. Sci.* **NS-26**(1), 321 (1979).

100. T. W. Raudorf et al., *IEEE Trans. Nucl. Sci.* **NS-31**(1), 253 (1984).

101. C. Chasman, K. Jones, and R. Ristinen, *Nucl. Instrum. Meth.* **37**, 1 (1965).

102. R. L. Bunting and J. J. Kraushaar, *Nucl. Instrum. Meth.* **118**, 565 (1974).

103. R. A. I. Bell, "Tables for Calibration of Radiation Detectors," Australian National University Report ANU-P/606 (1974).

Chapter 13

Other Solid-State Detectors

I. LITHIUM-DRIFTED SILICON DETECTORS

Using silicon of the highest currently available purity, the depletion depth that can be achieved by reverse biasing a conventional silicon diode detector of the type discussed in Chapter 11 is limited to 1–2 mm. If thicker silicon detectors are required, a different approach must be used to fabricate the device. The process of *lithium drifting* can instead be applied to create a region of compensated or "intrinsic" silicon in which the concentrations of acceptor and donor impurities are exactly balanced over thicknesses up to 5–10 mm. When provided with noninjecting electrodes (often called *blocking contacts*, see p. 368), this region then serves as the active volume of a detector. Such detectors are called *lithium-drifted silicon detectors*, and they are given the designation Si(Li). Their thickness is limited only by the distance over which the lithium drifting can be carried out successfully.

The process of compensating impurities by slowly drifting lithium into a semiconductor crystal was originally developed for producing detectors from either germanium or silicon. For two decades, drifted germanium [or Ge(Li)] detectors represented the only configuration in which a relatively large active volume could be provided in germanium. More recently, detectors of equivalent volume have become available in the high-purity germanium (or HPGe) configuration, and lithium drifting in germanium is no longer popular. However, the process remains important in silicon because its available material purity has not equaled that of germanium.

The lower atomic number of silicon ($Z = 14$) compared with that of germanium ($Z = 32$) means that for typical gamma-ray energies its photoelectric cross section is lower by about a factor of 50 (see Chapter 2). The gamma-ray full-energy peak efficiency for silicon detectors is therefore very low, and consequently they are not widely used in general gamma-ray spectroscopy. However, there are two main application areas in which the lower atomic number of silicon is not a hindrance but a help. One area involves the detection of very-low-energy gamma rays or X-rays, where the probability for photoelectric absorption can be reasonably high, even for silicon detectors of a few millimeters in thickness. Because of some advantages of silicon over germanium in this application, Si(Li) detectors have become the most common choice for low-energy photon spectrometry (LEPS) systems. These advantages include less prominent X-ray escape peaks from silicon, and the fact that the greater transparency of silicon for high-energy gamma rays is actually helpful when low-energy X-rays must be measured in the presence of a gamma ray background. The second common application area is the detection and spectroscopy of beta particles or other externally incident electrons. Here the lower atomic number of silicon is a distinct advantage because fewer electrons will backscatter from the detector without depositing their full energy. Both of these application areas are more fully described later in this chapter.

The semiconductor properties of the two materials favor silicon over germanium to some extent. Its larger bandgap ensures that the thermally generated leakage current at any given temperature will be smaller per unit volume in silicon than in germanium. The energy required to create an electron-hole pair and the Fano factor are not greatly different in the two materials, so the statistical contribution to energy resolution is within about 25% of being the same. If charge trapping is negligible, the energy resolution using equivalent electronic components may be better in silicon because of its lower leakage current.

In thin silicon surface barrier detectors, this bulk-generated leakage current is not normally a significant contributor to the noise. With Si(Li) detectors, however, the compensated region is sufficiently thick so that at room temperature the fluctuations in this leakage current can be a significant noise source. As a result, nearly all low-noise applications are carried out by cooling the detector to liquid nitrogen temperature in much the same manner as described in Chapter 12 for germanium detectors. Manufacturers generally recommend that these systems be continuously maintained at low temperature to prevent unnecessary thermal stresses due to temperature gradients, and to avoid the possibility of gradual redistribution of the drifted lithium at room temperature. In contrast with Ge(Li) detectors, the lithium migration at room temperature in Si(Li) detectors is slow enough that they can usually survive room temperature storage (without applied voltage) for days or weeks without performance degradation.

A. The Ion Drift Process

In both silicon and germanium, the material with highest available purity tends to be *p*-type, in which the best refining processes have left a predominance of acceptor impurities. Donor atoms must therefore be added to the material to accomplish the desired compensation. The alkali metals such as lithium, sodium, and potassium tend to form interstitial donors in crystals of silicon or germanium. The ionized donor atoms that are created when the donated electron is excited into the conduction band are sufficiently mobile at elevated temperatures so that they can be made to drift over periods of days under the influence of a strong electric field. Of the examples mentioned above, only lithium can be introduced into silicon or germanium in sufficient concentration to serve as a practical compensating dopant. The positively charged lithium ion has a helium-like core that gives it a small radius that promotes its diffusion through other solids.

The fabrication process begins by diffusing an excess of lithium through one surface of the *p*-type crystal so that the lithium donors greatly outnumber the original acceptors, creating an *n*-type region near the exposed surface. The resulting *p-n* junction is then reverse biased while the temperature of the crystal is elevated to enhance the mobility of the ionized lithium donors. The lithium ions are slowly drawn by the electric field into the *p*-type region where their concentration will increase and approach that of the original acceptor impurities. At a typical drifting temperature of 40°C, several days or weeks are required for adequate results. A remarkable feature of the drifting process is that a nearly exact compensation must automatically take place because the lithium distribution in the drifted region tends toward a state in which the total space charge is zero at every point. An equilibrium is thereby established in which the lithium ion continues to drift an increasing distance into the *p*-type region, but in which the only net change in lithium concentration occurs at the boundaries. It has been demonstrated[1] that any departure from exact compensation is unstable as the ions are drifting, and any imbalance in lithium concentration is quickly eroded away until a uniform concentration is reached. Through this drifting procedure, it is possible to achieve compensated regions that extend up to 5–10 mm in silicon and 10–15 mm in germanium.

Detailed analyses of the lithium-drifting process[2–4] demonstrate that the presence of thermally excited electron–hole pairs present during the drift contribute to the net space charge and can upset the exact compensation between lithium donor and acceptor impurities. The fabrication therefore normally takes place in two steps, in which a relatively long duration "clean-up" drift follows the primary drifting process. The temperature of the second step is kept much lower than that of the first so that the thermally excited carriers exert less influence, and a gradual redistribution of the lithium can take place to again approach near-perfect compensation. However, because the operating temperature of the detector is always considerably lower than the clean-up temperature, the potential for some residual charge imbalance and imperfect compensation remains. The resulting effects can be deleterious to the energy resolution of the detector, in that the electric field in some parts of the crystal may be weaker than would otherwise be expected.[3] With careful fabrication techniques, the level of uncompensated impurities in the drifted region can be kept below $10^9/\text{cm}^3$.

Lithium ions can be drifted into both silicon and germanium crystals to produce detectors. The lithium ion mobility is much greater in germanium and remains high enough at room temperature to permit an undesirable redistribution of the lithium from the compensated situation achieved during the drift. The lithium profile must therefore be preserved immediately after the drift in germanium by drastically reducing the crystal temperature, typically to that of liquid nitrogen (77 K). In silicon, the ion mobility is low enough at room temperature to permit temporary storage of lithium-drifted silicon detectors without cooling.

B. The *p-i-n* Configuration

Once the drifting process is completed, the resulting detector has the simplified configuration shown in Fig. 13.1. The excess lithium at the surface from which the drift was started

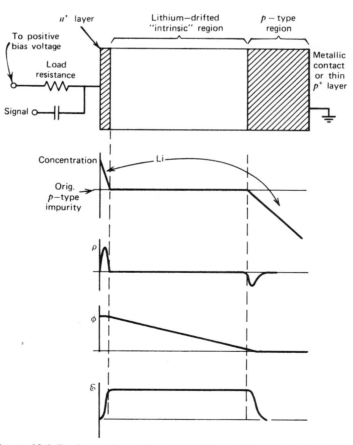

Figure 13.1 Basic configuration of a lithium-drifted *p-i-n* junction detector. Also shown are the corresponding profiles for impurity concentration, charge density ρ, electric potential φ, and electric field $\mathscr{E}$.

serves to convert that surface into an n^+ layer that can be used as an electrical contact. The noncompensated p region at the opposite side is often given a metallic coating that acts as an ohmic contact.

The lifetime of charge carriers created within the compensated region (often simply called the intrinsic or i region) can be substantially greater than the time required to collect them at either boundary, and therefore good charge collection properties can result. It is necessary to collect the charges quickly, however, and substantial voltages are normally applied to ensure that few electrons or holes are lost before collection is complete. Typical bias voltages are 500–4000 V across intrinsic regions of 5–10 mm thickness.

Because ideally, no net charge exists in the i region, the resulting electric potential predicted by Eq. (12.6a) for $\rho = 0$ varies linearly across the simple planar configuration illustrated in Fig. 13.1. The electric field will therefore be uniform across the i region. Because the resistivity of the compensated material is much higher than either the p or n^+ regions, virtually all the applied voltage appears across the i region, and the electric field drops sharply to zero at its boundaries. Therefore, the i-region dimensions determine the active volume of the detector, and the migration of charge carriers to the p-i and i-n boundaries gives rise to the basic signal pulse.

C. Electric Field and Pulse Shape

Most Si(Li) detectors are fabricated in planar geometry in which the lithium is drifted in from the flat surface of a thick silicon wafer. Although some large-volume coaxial configurations have been produced, the common applications of Si(Li) detectors to the measurement of soft X-rays or electrons benefit more from the large flat entrance surface that

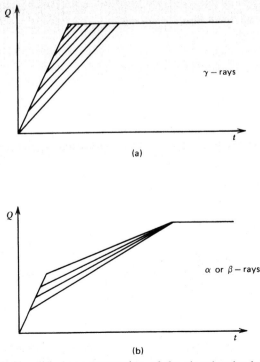

Figure 13.2 Simplified representation of the signal pulse leading edge for a Si(Li) detector. Variations arise from differences in the position of interaction within the detector active volume. (From Moszynski et al.[5])

the planar configuration provides. In planar geometry, the magnitude of the constant electric field is

$$\mathscr{E} = \frac{V}{d} \tag{13.1}$$

where V is the applied voltage and d is the thickness of the i region. This simple result is predicated on the assumption that the lithium compensation is perfect, so that no net charge exists in the i region. In practical Si(Li) detectors, departure from ideal compensation may introduce sufficient space charge to distort significantly the shape of the electric field from that described by this simple model (see below).

As in germanium detectors, the time profile $Q(t)$ of the induced charge at the detector electrodes depends on the position in the i region at which the charge carriers are formed. These variations are directly reflected in the shape of the leading edge of the pulses shown in Fig. 13.2. The variations in pulse shape are most severe for X-ray or gamma-ray irradiation of the detector, because then the interaction positions are randomly distributed throughout the detector volume. When a charged particle is incident on one face of the detector, the variation in pulse shape is usually somewhat less severe. Figure 13.3 shows measured rise times for several different Si(Li) detectors using beta particle excitation. The rise time was measured for the initial portion of the pulse and was defined as the time required for the pulse to change by 10% of the final amplitude. The time resolution (see Chapter 17) achievable with Si(Li) detectors depends both on the average rise time of the pulses and the spread in this rise time caused by the effects mentioned above. In this same study,[5] time resolutions of a few nanoseconds are quoted for beta particle excitation. Somewhat poorer time resolution should be expected for pulses from X-rays or gamma rays.

The output pulse shape can also change if the electric field is not uniform throughout the active volume of the detector. The electric field is expected to be uniform in Si(Li) planar detectors only if the compensation of donors and acceptors within the active volume is perfect. It has been shown,[5,6] however, that inhomogeneities in the compensation of impurities can lead to an electric field that decreases strongly near the n side of the detector.

Figure 13.3 Observed rise time (see text for definition) for several Si(Li) planar detectors of different thicknesses. Bottom curve is for 77 K temperature; others are for 293 K. (From Moszynski et al.[5])

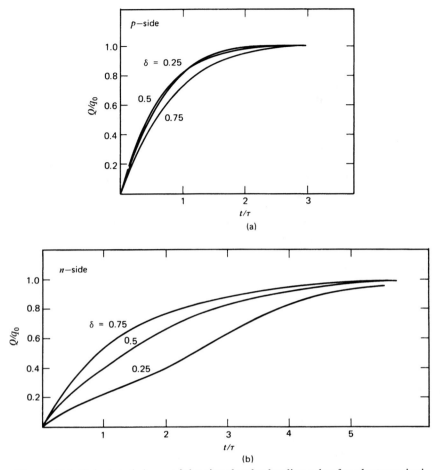

Figure 13.4 Calculated shape of the signal pulse leading edge for electrons incident on a Si(Li) detector. Various assumed electron ranges are shown, defined as $\delta \equiv$ electron range/detector thickness. The time scale is in units of $\tau \equiv d/\mu_e \mathscr{E}_0$, where d = detector thickness, μ_e is the electron mobility, and $\mathscr{E}_0$ is the maximum electric field whose shape is assumed to be given by Eq. (13.2). (a) Electrons incident on the p side (high-field side) of the detector. (b) Electrons incident on the n side (low-field side) of the detector. (From Moroz and Moszynski.[7])

If the residual space charge due to uncompensated impurities varies linearly across the active volume, then Poisson's equation [Eq. (12.2)] predicts a quadratic shape for the electric field. Moroz and Moszynski[7] showed that a quadratic variation of the form

$$\mathscr{E}(x) + \mathscr{E}_0 \left(1 - \frac{x^2}{d^2} \right)$$ (13.2)

where $\mathscr{E}_0$ = electric field strength at the negative electrode

x = distance from the negative electrode

d = thickness of the detector active volume

adequately predicted pulse shapes observed for monoenergetic electrons incident on both sides of the detector. Some of the resulting pulse shapes are shown in Fig. 13.4. Pulse shapes for alpha particle excitation from the p side of another Si(Li) detector are shown in Fig. 13.5.

As in the case of germanium detectors, "defective" pulses are sometimes observed in Si(Li) detectors from regions of poor charge collection (often near the boundaries of the active volume). These defective pulses have both a reduced amplitude and an unusually long rise time, because regions in which the effective charge collection is slow will also exhibit a high trapping probability. Figure 13.6 illustrates that many of these undesirable pulses may be rejected from an amplitude spectrum by excluding all signal pulses whose rise time exceeds a certain upper limit.

Figure 13.5 Observed shape of the signal pulse leading edge for two different Si(Li) detectors. In both cases, short-range alpha particles were incident on the p side of the planar detector. If the electric field were constant, the predicted pulse shape for various applied voltage values would be the linear waveforms shown. The actual data points are shown fitted by a curve of the form $Q/q_0 = tgh(t/\tau)$, where τ is treated as an adjustable parameter. Since this behavior is predicted for a nonuniform electric field with parabolic shape, the data give further evidence for the validity of Eq. (13.2). (From Moszynski and Przyborski.[8])

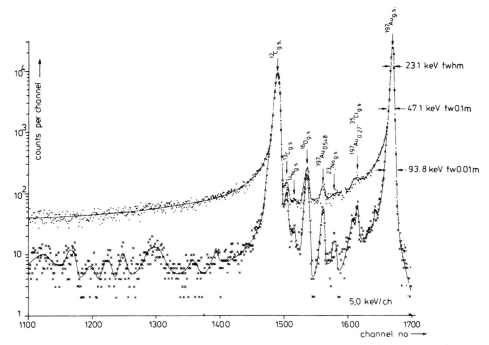

Figure 13.6 Comparison of pulse height spectra from a Si(Li) detector with (×) and without (●) rejection of pulses with rise times greater than 70 ns. The source was a proton beam scattered from several target nuclides. (From Lindstrom and Lisdat.[9])

D. Low-Energy Photon Spectroscopy

1. RESPONSE FUNCTION

In silicon, the photoelectric process is more probable than Compton scattering for photon energies below about 55 keV (see Chapter 2). For energies below about 30 keV, photoelectric absorption is predominant, so that the response function of a Si(Li) detector to gamma rays below this energy is dominated by a single full-energy peak caused by absorption of the resulting photoelectron. If the electric field in the detector is sufficiently high, charge collection is complete and the shape of the full-energy peak can often be represented adequately by a Gaussian, although slight modifications of the Gaussian shape have been shown to be beneficial in exacting cases.[10]

Some photoelectric absorptions cause the prompt emission of a characteristic X-ray as the resulting vacancy in the electron shell is filled. At low incident gamma-ray energies, most of the absorptions occur near the surface of the detector where escape of the resulting X-ray can be significant. The result will be the appearance of a small X-ray escape peak in the response function, which in silicon will be located 1.8 keV below the full-energy peak (see Fig. 13.7). Figure 13.8 shows the intensity of the escape peak for a planar Si(Li) detector. The corresponding escape peak in a germanium detector spectrum will be much more prominent because of the smaller average penetration distance of the incident radiation and the greater escape probability of the Ge characteristic X-rays due to their higher energy (11 keV). The fluorescent yield in Ge (50%) is also much larger than that in Si (5%), so that the probability of emission of a characteristic X-ray is higher.

When the intensity of incident photons is measured from the area under the photopeak, often a key issue is the clean separation of its area from a low energy tail[13,14] that

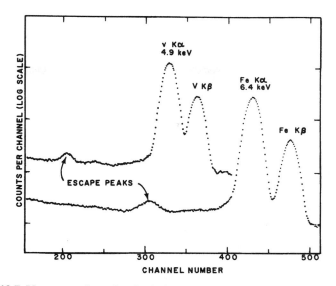

Figure 13.7 Upper portion of pulse height spectra of *K* X-rays from the decay of ^{51}Cr and ^{57}Co. Peaks caused by the escape of the Si X-ray from the detector are evident in both spectra. (From Wood et al.[11])

results from partial energy loss processes. Therefore a number of careful studies have been carried out[13-19] to describe the detailed response functions of Si(Li) spectrometers to monoenergetic X- and gamma-rays. These studies reveal that there can be observable contributions to the tailing as a result of incomplete charge collection in a dead layer at the detector surface, and to secondary radiations such as characteristic X-rays that can be contributed by interactions in the surface electrode or subsurface dead layer.[20]

At higher gamma-ray energies, a significant contribution of Compton scattering adds a continuum to the spectrum. Figure 13.9 shows the shape of the Compton continuum for an incident photon of 140.5 keV energy. Also shown are the computed shapes of the electron energy distributions from single Compton scattering assuming both free electrons and bound electrons in silicon. It is evident that the binding effects must be taken into account

Figure 13.8 Ratio of the X-ray escape peak intensity to the full energy peak intensity for Si(Li) and germanium planar detectors. (From Rossington et al.[12]).

Figure 13.9 Calculated and observed shape of the Compton continuum for a ^{99m}Tc source (140.5 keV). The dashed curve is predicted based on scattering of free electrons, whereas the solid curve is calculated taking into account the finite electron binding energy in silicon. The points are experimental data. (From Felsteiner et al.[21])

if the shape of this continuum is to be described adequately. Because of the low photoelectric cross section in silicon, Si(Li) detectors are seldom used to measure gamma rays with energy above about 150 keV.

2. LINEARITY AND ENERGY RESOLUTION

The response of Si(Li) detectors to low-energy X-rays and gamma rays has been shown to be very linear, provided the applied voltage is high enough (approximately 250 V/mm) to avoid significant loss of charge from trapping and recombination. In Fig. 13.10, the observed channel number for the photopeak is plotted versus X-ray energy for a number of characteristic X-rays produced by proton excitation.[22] The maximum possible nonlinearity is quoted as 1% over this low-energy range, where many potential causes of nonlinearity are most significant.

The resolution and noise behavior of a high-resolution Si(Li) detector have been studied by Hollstein[23] for energies between 5 and 280 keV. The results are shown in Fig. 13.11. As in germanium systems, the observed full width at half maximum of the full-energy peak W_T can be expressed as the quadrature sum of a number of independent components:

W_E—the FWHM-equivalent noise that can be attributed to the electronic components of the signal-processing chain.

W_D—the FWHM-equivalent spread due to the charge generation statistics within the detector. It is assumed that $W_D^2 = 2.35^2 \epsilon F E$, where ϵ is the energy necessary to create one electron–hole pair, F is the Fano factor, and E is the photon energy.

W_x—the FWHM equivalent attributable to detector leakage current and any charge collection problems within the detector.

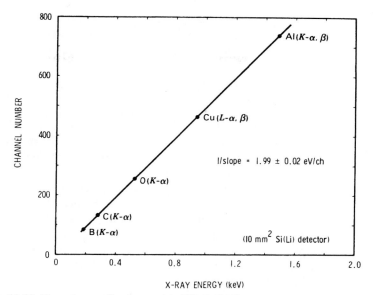

Figure 13.10 Experimentally determined linearity between pulse height and X-ray energy for a Si(Li) detector. (From Musket.[22])

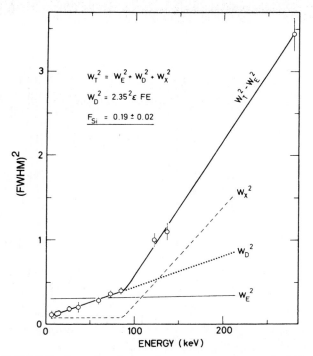

Figure 13.11 Contribution of various noise sources to the total width of the full-energy peak from a Si(Li) photon spectrometer. For definitions of the various terms, see the text. (From Hollstein.[23])

It is the charge generation statistics that set the ultimate limit to the energy resolution, even if all other sources of peak broadening could be eliminated. A calculation from the expression above using a Fano factor of 0.11 and ϵ of 3.76 eV predicts a value of W_D of 116 eV at the standard calibration energy of 5.9 keV from a ^{55}Fe source. Small-area detectors are available commercially that approach an energy resolution of 130 eV, remarkably close to the statistical limit. More typical systems tend to have measured energy resolutions in the 140 to 170 eV range. Large-area detectors involve greater contributions of leakage current and electronic noise. For example, results with a commercially available Si(Li) spectrometer with a surface area as large as 300 mm^2 have been reported[24] with an energy resolution of 255 eV. Improvements in energy resolution have come about through the use of double guard ring structures to reduce the surface leakage currents, and the introduction of active reset low-noise preamplifiers (see Chapter 17). Careful design of the detector-preamplifier first-stage coupling is also required to prevent additional broadening from microphonics (see p. 631) caused by mechanical vibrations and/or acoustical noise. Loss of resolution from this cause has been shown[25] to be significant for some commercial systems.

The relative magnitudes of the serial and parallel electronic noise sources discussed in Chapter 17 for typical Si(Li) detectors generally leads to an optimum shaping time of tens of microseconds[26] for minimum noise contribution. This shaping time is much longer than would be required to ensure complete charge collection from the detector (which takes place in tens of nanoseconds or less), but is necessary to realize the best energy resolution by minimizing the contribution of electronic noise. Maximum counting rates are therefore limited to 10^4 per second or less to avoid excessive losses caused by pulse pile-up (see Chapter 17). Higher rates can be accommodated only by reducing the shaping time and accepting some deterioration in the achievable energy resolution.

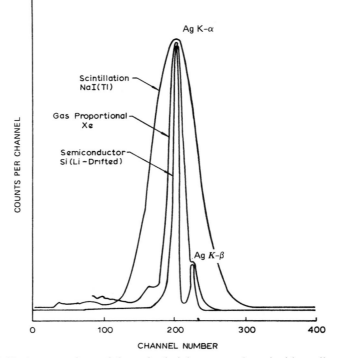

Figure 13.12 A comparison of the pulse height spectra from incident silver K-series X-rays as recorded by three different detectors with varying energy resolution. The K-α and K-β energies are approximately 21 and 25 keV. (From Muggleton.[27])

The comparative energy resolution of three different detector types commonly applied to X-ray spectroscopy is shown graphically in Fig. 13.12. The generally superior energy resolution of Si(Li) detectors in this application allows a much better separation of the X-ray energy components compared with either a proportional counter or scintillation detector.

3. DETECTION EFFICIENCY

In principle, the full-energy peak efficiency of any Si(Li) detector can be determined from knowledge of its size and shape, together with the appropriate gamma-ray interaction cross sections. In practice, however, manufacturers often give only nominal values for detector dimensions, and these may be somewhat uncertain due to the effects of incomplete charge collection near the edges of the active volume. Furthermore, the detector itself is usually mounted inside a vacuum cryostat to permit its cooling to liquid nitrogen temperature. The exact spacing between source and detector is then sometimes difficult to determine. For these reasons, it is almost always necessary to carry out an experimental determination of the detector efficiency using calibrated sources, if reasonably accurate data are required. The task is complicated by the fact that there are relatively few standard sources available in the energy range of 5–50 keV, which is of primary interest in Si(Li) detectors.

One suitable calibration source is the alpha-active nuclide ^{241}Am, with a 458-year half-life. Two relatively intense gamma rays are emitted with energies of 26.35 and 59.54 keV, together with a number of characteristic L X-rays from the Np decay product. Data on relative yields of these photons are listed in Table 13.1, and a Si(Li) spectrum is shown in Fig. 13.13. A calibrated ^{241}Am source will therefore provide efficiency calibration points over the very important energy range of about 10–60 keV.

If absolutely calibrated sources are unavailable, an alternative approach can be used. The *X/gamma method*[30] makes use of radioisotopes that emit a dominant high-energy gamma ray, together with K X-rays from internal conversion or electron capture. The X-ray intensity can be determined absolutely from decay data if the gamma-ray intensity is somehow measured in an absolute manner. The gamma-ray measurement can be carried out to a high degree of accuracy at many laboratories, using standard NaI(Tl) scintillators or germanium detectors. With the data given in Table 13.2, these sources then provide X-ray calibration points with a similar absolute accuracy. The presence of the high-energy gamma ray is seldom a problem with Si(Li) detectors of typical dimensions because their probability of interaction is negligibly small.

As with other solid-state detectors, there is a need to repeat efficiency calibrations periodically because there is good evidence that the efficiency may change substantially over periods of weeks or months. These efficiency variations, although not fully understood, probably arise due to variations in charge collection efficiency and changes in dead layer thicknesses.

Table 13.1 Photon Intensities per Disintegration of ^{241}Am

Line	Energy (keV)	Percentage per Disintegration
L-α	13.9	13.3 ± 0.4
L-$\eta\beta$	17.8	19.4 ± 0.6
L-γ	20.8	4.9 ± 0.2
γ	26.35	2.4 ± 0.1
γ	59.54	35.82 ± 0.12

Data from Campbell and McGhee.[28]

Figure 13.13 Photon spectrum from ^{241}Am obtained using a 30 mm^2 × 3 mm Si(Li) detector. (From Gehrke and Lokken.[29])

The energy-dependent efficiency of a representative Si(Li) detector is plotted in Fig. 13.14. Over a substantial range of incident energy (from about 7 to 20 keV in the example shown) the intrinsic peak efficiency of the detector is very close to 100%. This is the energy interval in which virtually all incident photons are absorbed in the detector by simple photoelectric absorption. The absolute counting efficiency then depends primarily on the surface area of the detector. As a result, essentially all low-energy photon detectors of this type are fabricated in planar geometry to maximize the surface area. At higher energies, the peak efficiency begins to fall off rapidly, reflecting the similar fall-off of the photoelectric cross section in silicon. Thicker detectors will help somewhat, but for all practical purposes, planar Si(Li) detectors are limited to energies below about 100 keV.

The efficiency at low energies is critically dependent on the thickness of window materials or dead layers associated with the detector and cryostat. (For very soft radiations, even a few centimeters of air between the source and detector can lead to significant attenuation.) In Fig. 13.14, the individual transmission efficiencies of typical window and dead layer values are shown separately, together with the resulting composite efficiency for the detector. The discontinuity at about 1.8 keV is a result of the K absorption edge of the silicon dead layer assumed for this detector. In practice, the dead layer thickness can sometimes be determined by measuring the relative intensities of two or more X-rays emitted by a single source for which the emission yields are well known[32] or from separate sources with known absolute yields.[33]

A number of semiempirical relations have been used to serve as a functional fit to interpolate between measured efficiency points. One of these of the form

$$\epsilon = \Omega \exp(\alpha E^\beta)[1 - \exp(\gamma E^\delta)] \tag{13.3}$$

was suggested by Gallagher and Cipolla[31] in which the five parameters α, β, γ, δ and Ω are fit to the experimentally measured efficiency points. The applicability of Eq. (13.3) has been demonstrated for several different irradiation conditions,[31,34] and a least-square fit can normally be made to within the uncertainty of the data points themselves.

The intrinsic peak efficiency for four common types of detectors used in low-energy photon spectroscopy is plotted in Fig. 13.15. The relative radiation transparency of the xenon proportional counter makes its efficiency drop off most rapidly with energy. Furthermore, above

Table 13.2 Nuclides Suitable for Use in the X/γ Calibration Method

Nuclide	Half-Life (days)	X-Ray or Low-Energy Gamma	Energy (keV)	Energy of High-Energy Gammas (keV)	Intensity Ratio X/γ	
[51]Cr	27.710	K-α	4.95	320.11	2.018	$\pm$ 0.021
		K-β	5.43		0.274	$\pm$ 0.0056
[54]Mn	312.14	K-α	5.412	834.83	0.2234	$\pm$ 0.0011
		K-β	5.95		0.0305	$\pm$ 0.00034
[57]Co	271.80	K-α	6.397	122.063	0.5863	$\pm$ 0.016
		K-β	7.06		0.0807	$\pm$ 0.0024
		γ	14.41		0.1086	$\pm$ 0.0023
[65]Zn	244.0	K-α	8.041	1115.55	0.6790	$\pm$ 0.0044
		K-β	8.91		0.0946	$\pm$ 0.00077
[75]Se	119.76	K-α	10.327	264.65	0.8273	$\pm$ 0.013
		K-β	11.72		0.1278	$\pm$ 0.0021
[85]Sr	64.851	K-α	13.375	514.00	0.5035	$\pm$ 0.0035
		K-β	15.86		0.089	$\pm$ 0.00077
[88]Y	106.62	K-α	14.142	898.03	0.5425	$\pm$ 0.0035
		K-β	16.765		0.0986	$\pm$ 0.00081
[109]Cd	462.7	K-α	22.10	88.036	22.47	$\pm$ 0.58
		K-β'_1	24.93		4.022	$\pm$ 0.107
		K-β'_2	25.46		0.706	$\pm$ 0.023
[113]Sn	115.09	K-α	24.14	391.69	1.225	$\pm$ 0.009
		K-β'_1	27.27		0.220	$\pm$ 0.005
		K-β'_2	27.86		0.0414	$\pm$ 0.00087
[137]Cs	10964	K-α	32.06	661.63	0.0663	$\pm$ 0.0008
		K-β'_1	36.4		0.0127	$\pm$ 0.00016
		K-β'_2	37.3		0.00322	$\pm$ 0.00008
[133]Ba	3841	K-α	30.85	356.0	1.604	$\pm$ 0.023
		K-β'_1	35.0		0.374	$\pm$ 0.006
		K-β'_2	35.8			
		γ	53.2		0.0352	$\pm$ 0.006
[139]Ce	139.69	K-α	33.30	165.85	0.792	$\pm$ 0.011
		K-β'_1	37.8		0.151	$\pm$ 0.004
		K-β'_2	38.7		0.0388	$\pm$ 0.0009

Data from Campbell and McGhee.[28]

Figure 13.14 Calculated full-energy peak efficiency for a 3-mm thick Si(Li) detector. (From Gallagher and Cipolla.[31])

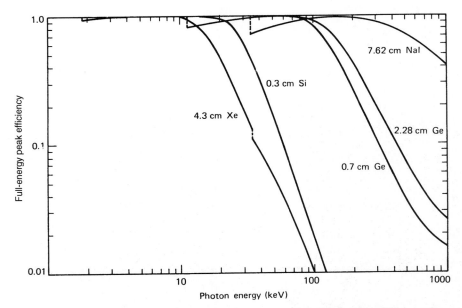

Figure 13.15 Calculated full-energy peak efficiency for five different detectors as a function of incident X- or gamma-ray energy. The thicknesses in the direction of the incident radiation are indicated on the figure. The detectors are a xenon-filled proportional counter, a Si(Li) detector, two different germanium detectors, and a NaI(Tl) scintillator. (From Israel et al.[35])

its *K*-edge energy of 34 keV, photoelectron absorption in the xenon results in a characteristic X-ray with a fluorescent yield of 81%. This X-ray stands a very good chance of escaping from the detector without further interaction, and the pulse does not fall within the full-energy peak in the spectrum. Therefore, the photopeak efficiency of the detector does not follow the abrupt increase in the attenuation coefficient at this energy and actually shows a slight drop.

The efficiency for silicon is comparatively good over most of the X-ray range, and Si(Li) detectors are widely applied to this energy region. Characteristic escape effects in silicon are not serious because of its low *K*-edge energy and small fluorescent yield of 5%. The corresponding slight dip in the photopeak efficiency at 1.8 keV is barely perceptible in Fig. 13.15. Both germanium and sodium iodide, in reasonable thicknesses, are essentially opaque to electromagnetic radiation below 100 keV. The photopeak efficiencies also show the effect of characteristic X-ray escape, and the fluorescent yields in germanium and iodine are 50 and 88%, respectively.

4. APPLICATION IN FLUORESCENCE SPECTROSCOPY

Commercially available low-energy photon spectrometers (LEPS) consisting of a cooled Si(Li) planar detector, preamplifier, and liquid nitrogen dewar have come into widespread application for the measurement of X-ray spectra in the energy region of 1–50 keV. A typical system is shown in Fig. 13.16. For the ultimate in low-noise performance, the input stages of the preamplifier will also be cooled to reduce electronic noise.

These systems are widely applied for the analysis of materials by means of the characteristic X-rays emitted following excitation by one of several means. If the sample is irradiated by X-rays with a photon energy above the *K*-shell binding energy of a specific element, photoelectric absorption will induce characteristic X-ray emission. This process is called *X-ray fluorescence*. Other means of excitation can include alpha, electron, or other charged particle bombardment. Separate elements within the sample then can be identified by observing peaks in the pulse height spectrum corresponding to the characteristic X-rays of that particular element.

Figure 13.16 A typical low-energy photon spectrometer system consisting of a Si(Li) detector within a vacuum cryostat, cooled by a liquid nitrogen dewar. (Courtesy of Princeton Gamma-Tech, Princeton, NJ.)

Figure 13.17 X-ray spectrum produced by electron excitation of the mineral glaucophane. All the elemental constituents (except hydrogen) are identifiable as peaks in the spectrum. (From Jaklevic and Goulding.[36])

As an example, Fig. 13.17 shows the X-ray spectrum of a sample that has been excited by electron bombardment. Because the characteristic X-ray energies of light elements are quite small, a premium is placed on system energy resolution in order to resolve separate contributions from adjacent low-Z elements. Detection of characteristic X-rays for elements as light as boron (183 eV K-α energy) has been reported.[22] Many of the instrumental problems of importance in X-ray fluorescence analysis are discussed in the review by Goulding.[37]

E. Electron Spectroscopy

Lithium-drifted silicon detectors are also well suited to the measurement of electron energies. Thicknesses can be obtained which easily exceed the maximum distance of penetration of beta particles or other electrons of routine interest. The relatively low Z-value of silicon also ensures that backscattering of electrons incident on the face of the detector will be minimized. Because the detector is usually enclosed within a vacuum cryostat, electrons from external sources must penetrate the cryostat entrance window and may lose a significant amount of energy in the process. For low-energy electron spectroscopy, the sample must be introduced into the vacuum envelope to avoid window losses and air absorption (e.g., see Ref. 38).

The response of silicon detectors to normally incident monoenergetic electrons with energies between 0.15 and 5.0 MeV has been investigated both experimentally and through the use of a computational model by Berger et al.[39] A typical set of pulse height spectra is shown in Fig. 13.18. When the thickness of the detector is considerably larger than the maximum penetration distance of the electron, the response function consists of a full-energy peak plus a continuum of lower amplitude events. The full-energy peak corresponds to all those electrons that are fully stopped within the active volume of the detector, and for which bremsstrahlung photons generated along its path are fully absorbed within the detector. The continuum[40] represents partial energy loss events, which can correspond to either the backscattering of electrons that reemerge from the incident surface without having lost all their energy, or the escape of bremsstrahlung photons generated along the track

Figure 13.18 Experimental and calculated response functions for silicon detectors for 0.50-MeV electrons. The experimental data are shown as points, whereas the continuous plot is the result of a Monte Carlo calculation. z is the detector thickness, and z/r_0 is the ratio of this thickness to the mean range of the electrons. (From Berger et al.[39])

Figure 13.19 Conversion electron spectrum of a ^{244}Cm sample determined using an 80 mm^2 × 3 mm Si(Li) detector. (From Ahmad and Wagner.[38])

of an electron that otherwise may be fully stopped. As the detector is made thinner, some electrons begin to penetrate fully through the detector without losing their entire energy. The corresponding pulses shift from the full-energy peak to lower amplitudes in the spectrum. For very thin detectors, few electrons are fully absorbed and the pulse height spectrum corresponds to the energy loss distribution of electrons before they escape from the active volume. Numerical tables of response functions of the type shown in Fig. 13.18 can be found in Ref. 41 for detector thicknesses from 0.05 to 10.0 mm and electron energies from 0.15 to 5 MeV.

A Si(Li) detector whose intrinsic thickness exceeds that of the maximum penetration distance of electrons of interest makes a very suitable electron spectrometer. Figure 13.19 shows a recorded spectrum in which peaks appear due to monoenergetic electrons from the internal conversion process.

For electrons that are incident at oblique angles to the surface, a larger fraction are scattered back out of the detector before losing all their energy. As a result, the continuum in the response function is enhanced relative to the case of normal incidence. In one measurement[42] in which many angles were sampled by placing the detector close to the source to subtend a solid angle of 25% of 2π, the fraction of all events falling under the continuum rose to 65% compared with about 20% for small solid angles. If the source intensity permits, it is obviously of advantage to restrict the incident electrons to near-normal incidence only.

Silicon detectors can also be applied in the measurement of positrons produced at accelerators or from radioactive sources with a thin covering. Since positrons lose their energy in much the same way as fast electrons, there is little difference in the direct production of electrons and holes along the track of the particle, and the ranges of electrons and positrons of equal energy are similar. However, when the positron reaches the end of its track, it will annihilate creating two 511 keV annihilation photons. Most of these photons typically will escape from the detector, but some can undergo additional interactions within its active volume. Because these processes occur on time scales that are short compared with typical pulse shaping times, the measured pulse amplitude represents the sum of the energies deposited directly by the positron plus any additional energy from the photons. This effect is illustrated in the spectrum shown in Fig. 13.20, where the shoulder of pulse amplitudes beyond that of the "full energy" peak represents the contribution of the annihilation photons.

Figure 13.20 Response of a silicon detector with 2-mm depletion depth to monoenergetic positrons with the indicated energies. (From Frommhold et al.[43])

II. SEMICONDUCTOR MATERIALS OTHER THAN SILICON OR GERMANIUM

In contrast to semiconductors consisting of a single chemical element represented by silicon (Si) and germanium (Ge), it was a compound semiconductor (AgCl) that was first used to demonstrate that radiation induced signals could be extracted from a semiconducting device.[44] Yet, compound semiconductors have experienced slow development over the years, largely inhibited by crystal growth issues involving electrically active impurities and native defects. Handling issues have also been formidable, in that some compound semiconductors do not have the robust material properties required to withstand common semiconductor process technology. In many cases, compound semiconductor detectors have shown temporal instability, often referred to as *polarization.*

The large majority of semiconductor radiation detectors presently in use are manufactured from either Si or Ge. The widespread popularity of these materials is attributable to their excellent charge carrier transport properties, which allow for the use of large crystals without excessive charge carrier losses from trapping and recombination. Detectors of practical size can be manufactured in which virtually all electron–hole pairs excited by the incident radiation are collected to make up the basic signal pulse. Si- and Ge-based detectors, however, are less than perfect in several respects. The ideal semiconductor detector would have a large active volume, high charge carrier mobilities, and a high atomic number if applied in gamma-ray spectroscopy, and be operable at room temperature without any cooling apparatus.

Good energy resolution requires a relatively small bandgap, thereby reducing the average ionization energy required to excite an electron–hole pair (see Fig. 13.21). The small ionization energy increases the number of charge carriers excited per unit deposited energy, decreasing statistical fluctuations. Additionally, efficient charge collection from the device requires large values of the carrier mobilities (μ) and long charge carrier lifetimes (τ^*). Large detector volumes are realized through the growth and production of high-purity or properly compensated semiconductor crystals. Both Si and Ge fit these requirements reasonably well, hence their great success.

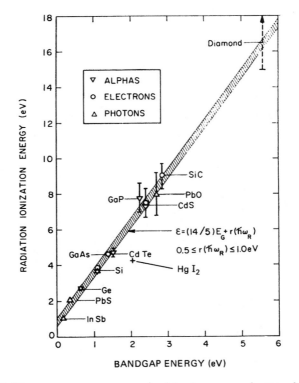

Figure 13.21 The average energy required to form one electron–hole pair (ϵ) versus bandgap energy for a number of semiconductor materials. (From Klein.[45])

Photoelectric absorption is the favored gamma-ray interaction within a spectrometer, so one would prefer that the detector be composed of elements with high atomic numbers. A great deal of attention has therefore been focused on seeking other suitable semiconductor materials that incorporate at least one element of higher atomic number than Si($Z = 14$) or Ge($Z = 32$). Figure 13.22 shows a comparison of gamma-ray linear attenuation coefficients for Si, Ge, CdTe, and HgI$_2$.[46] Si is weakest in this comparison, with its Compton interaction cross section becoming larger than its photoelectric absorption cross section at approximately 60 keV. The much higher effective atomic numbers of CdTe and HgI$_2$ translate into significantly higher X-ray and gamma-ray photoelectric absorption than Si and Ge. The detection efficiency per unit thickness will therefore be higher, and the photofraction for an equivalent volume also will be larger.

Ge detectors must always be operated at low temperatures to reduce thermally generated leakage current. In low-noise applications, such as X-ray spectroscopy, silicon detectors must also be cooled for the same reason. In principle, a different semiconductor material with a wider bandgap (e.g., greater than 1.5 eV) could reduce the bulk-generated leakage current so that use at room temperature would be possible. In many applications, the convenience of room temperature operation would probably outweigh the disadvantages of a wider bandgap, such as the greater energy required to excite an electron–hole pair. A number of semiconductor materials other than Si or Ge have been investigated as radiation detectors, and an extensive review on wide bandgap compound semiconductors and their use as radiation detectors can be found in the literature.[47]

In some compound semiconductor materials, the best purification available still leaves behind trace impurities that act as electrically active dopants. For example, shallow acceptor impurities near the bottom of the bandgap make the material slightly *p*-type and reduce its resistivity. Intentionally including a donor in the material with an energy level deep in

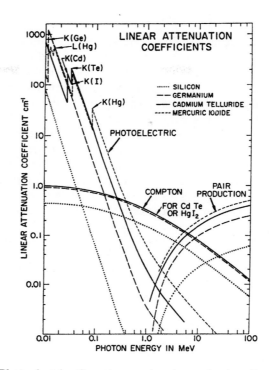

Figure 13.22 Photoelectric, Compton, and pair production linear attenuation coefficients for Si, Ge, CdTe, and HgI$_2$. *K*-shell absorption edges are shown. (From Malm.[46])

the bandgap (a deep donor) can serve to compensate the material and greatly increase the resistivity. The parallel process is to add a deep acceptor to material that is slightly *n*-type. Deep donors and acceptors may be created by adding dopant during material growth or by intentionally disrupting the crystal stoichiometry to produce native defects that act as deep donors or deep acceptors. In either case, the change in resistivity may be beneficial in decreasing the bulk leakage current, yet the donors and acceptors may also interfere with charge carrier transport by acting as trapping sites. The net result is a complicated interplay regarding the role of impurities and native defects in establishing the electric field shape and the charge carrier collection efficiency.

The most widely explored compound semiconductors for radiation measurements are CdTe, Cd$_{1-x}$Zn$_x$Te, and HgI$_2$, and they represent the only compound semiconductor detectors that are presently commercially available. Their bandgap values are sufficiently large to allow for room temperature operation for typical gamma-ray applications. However, many of the most promising applications involve X-ray measurements in which the deposited energy and resulting induced charge signal is relatively small. For such cases, modest cooling can reduce the leakage current and thereby improve the energy resolution of these detectors. HgI$_2$ detectors need only be cooled to 0°C to achieve leakage currents less than 0.1 pA in typical cases.[48] Additional temperature reduction can reduce the noise level of the input FET on the preamplifier, but any more reductions in the already small detector leakage current do not have an appreciable effect on detector energy resolution. In the case of CdTe, the leakage current is high enough so that further cooling to −40°C can produce useful improvements in their energy resolution. Compactness is preserved through the use of miniature electrical Peltier coolers that are built into the detector housing and whose presence is virtually invisible to the user.

A. General Operation of Planar Devices

As a rule, high-quality compound semiconductor single crystals are relatively small by comparison with commercially available Si and Ge single crystal ingots. Because of process difficulties imposed by either the small crystal sizes or other material issues, planar geometry devices are the most common configuration for compound semiconductor radiation detectors. As described on p. 407, this simple configuration generally consists of two conductive electrodes, either ohmic or blocking, on either side of a semiconductor crystal. A voltage is applied across the device, thus creating an internal electric field. Electron–hole pairs excited within the semiconductor by ionizing radiation are separated by the electric field and swept to their respective electrodes (see Fig. 13.23). The motion of the charges induces a corresponding current to flow in the external circuit. By integrating the current pulse, the induced charge can be measured and serves to determine the energy deposited by the incident radiation.

Assuming that the detector is fully active across its width, the general relationship for the induced charge due to a motion dx of the carriers ($dx|_e$ for electrons and $dx|_h$ for holes) can be stated from the Shockley–Ramo Theorem[49,50] (see Appendix D) as

$$dQ^* = \frac{-eN_0}{W}\left(dx|_e + dx|_h\right) \tag{13.4}$$

where Q^* is the induced charge, N_0 is the initial number of electron hole pairs, W is the detector width, e is the electronic charge, and the opposite charge signs and motion of the carriers are taken into account. For a very pure and defect free semiconductor material, full charge collection of the electrons and holes is possible and the integral of Eq. (13.4) reduces to $Q^* = eN_0$. This is essentially the result obtained in Chapter 5 for pulse-type ion chambers in which recombination and loss of charges is negligible. Generally, this circumstance is not realized with compound semiconductor radiation detectors, and the effect of charge carrier trapping losses on the induced charge signal must be considered.

When significant trapping occurs for either charge carrier, the induced charge becomes a function of the distance over which the charges travel, leading to interaction position dependence of the amplitude of the output pulse. Hecht's relation[51] describes the effect of charge transport in a uniform electric field, in which the induced charge as a function of interaction location becomes[52, 53]

$$Q^* = eN_0\left\{\frac{v_h\tau_h^*}{W}\left(1 - \exp\left[\frac{-x_i}{v_h\tau_h^*}\right]\right) + \frac{v_e\tau_e^*}{W}\left(1 - \exp\left[\frac{(x_i - W)}{v_e\tau_e^*}\right]\right)\right\} \tag{13.5}$$

where v is the charge carrier velocity, τ^* is the charge carrier lifetime, x_i represents the radiation interaction location measured from the cathode, W is the detector width, and the

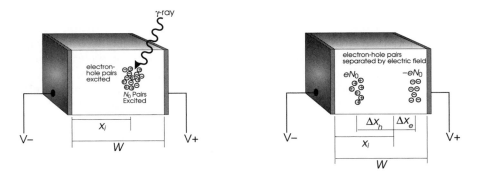

Figure 13.23 Basic operating configuration for a planar semiconductor detector. Electrons and holes are swept to their respective electrodes by an externally applied voltage. (From McGregor and Hermon.[47])

(a)

(b)

Figure 13.24 (*a*) Relative position-dependent charge collection for cases in which hole and electron transport properties are identical. Various values of $v\tau^*/W$ are assumed to show the dependence. (*b*) Same plot as above, but for cases in which the electron carrier extraction factor is always 10 times greater than that for holes.

e and *h* subscripts represent electrons and holes, respectively. The induced charge for a single event (photoelectric absorption, for instance) becomes a function of the interaction location and the ratio $v\tau^*/W$ (referred to as the *carrier extraction factor*[53]). Consequently, deviations in the induced charge from position-dependent charge collection can severely degrade energy resolution. Good energy resolution can be obtained if both carrier extraction factors are greater than about 50. Since charge carrier velocities are usually near saturation and charge carrier lifetimes are limited by properties of the starting bulk material, the energy resolution is usually improved by reducing the device width (W) until both carrier extraction factors reach an acceptable value.

For simplicity, let us assume for the following discussion that the interactions are purely photoelectric and so they occur only at a single site. With the condition that a planar device is either irradiated from the side or is exposed to gamma rays of high enough energy to interact uniformly throughout the device, Fig. 13.24*a* shows the normalized position-dependent induced charge for cases in which the electron and hole transport characteristics are the same. Notice that interactions in the detector center will result in the highest possible induced charge and interactions near the contacts result in the lowest. The induced charge

Figure 13.25 Drift velocity as a function of electric field strength for charge carriers in CdTe and HgI_2. In both materials, holes are much less mobile than electrons. (From Scharager et al.[54])

variation over the detector volume will cause energy resolution degradation. Flat regions on the induced charge map will appear as "peaks" in the pulse height spectrum, whereas sloped regions will appear as "tails." Notice that the deviation is very slight when $v\tau^*/W$ for both electrons and holes is greater than 50.

Usually for compound semiconductors, the mobility for holes is much lower than the mobility of electrons (see Fig. 13.25), resulting in lower hole extraction factors at common operating voltages. Figure 13.24b shows cases in which the carrier extraction factor is always 10 times greater for electrons than holes. Note that the induced charge variation over the device volume becomes much greater as the carrier extraction values decrease from high values. In both parts (a) and (b) of Fig. 13.24, the variation over the map of induced charge again becomes minimal for very small values for $v\tau^*/W$. Indeed, the trapping uniformity of the carriers appears to allow for spectrometers to be fabricated from material with very strong trapping. In practice, such devices are not practical for many reasons, which include the significantly lower induced charge and variations in the uniformity of trapping.

Irradiation from the cathode side is generally preferred for devices in which the hole extraction factor is poor. The total gamma-ray interaction distribution will be skewed toward the exposed or irradiated cathode surface, resulting in the spatial distribution of electron–hole pair excitation also being weighted toward the cathode. The induced charge signals then arise mainly from the motion of the more mobile electrons, and the poorer mobility of the holes is partially offset by their reduced average drift distance. The effect is most significant for X-rays or very low-energy gamma rays. As a result, a device with poor hole collection characteristics can be used to achieve reasonably good energy resolution for low-energy radiations.[53] Unfortunately, for moderate to high energy gamma rays, the resolution will suffer as predicted from Figure 13.24. Anode irradiation of planar devices with poor hole collection will inevitably suffer poor gamma-ray energy resolution.

B. Cadmium Telluride (CdTe) Detectors

Cadmium telluride (CdTe) combines relatively high atomic numbers (48 for Cd and 52 for Te) with a large enough bandgap energy (1.52 eV) to permit room temperature operation.

The probability of photoelectric absorption per unit pathlength is roughly a factor of 4 to 5 times higher in CdTe than in Ge, and 100 times larger than in Si for typical gamma-ray energies. Applications of this material therefore most often involve situations in which a high gamma-ray detection efficiency per unit volume is at a premium.

High purity CdTe crystals can be grown using a number of different fabrication techniques.[55] These include solution growth, variations of Bridgman growth, and the traveling heater method (THM). The latter is commonly used to produce detector grade CdTe, in that the technique allows for lower growth temperatures, reduced impurity concentrations, and lower defect concentrations than typical Bridgman methods.[55] THM CdTe is often purposely doped with Cl to compensate background impurities and native defects, resulting in high resistivity p-type material (up to 10^9 Ω-cm). The devices are fabricated by first slicing and lapping samples to the desired dimensions. The surfaces are generally polished with a chemomechanical technique followed by pure chemical etching. Contact formation varies, although a popular technique employs electroless metal deposition with solutions of $AuCl_3$ or $PtCl_4$. A chemical reaction takes place when the solution comes into contact with the CdTe surface, which deposits a film of the noble metal. Another configuration that has been demonstrated[56] consists of high-resistivity n-type CdTe with heavily doped p^+ and n^+ contacts grown by liquid phase epitaxy techniques. Such a configuration resembles the p^+-v-n^+ configuration common in HPGe detectors (see Fig. 12.2).

Because of rather poor collection efficiency of holes, energy resolution achievable in CdTe detectors is generally not comparable with that obtainable in Si or Ge. Some representative figures are given in Table 13.3, and an example of a measured spectrum is shown in Fig. 13.26. As a result of the charge carrier transport difficulties, most CdTe detectors are relatively small and are usually planar in design. Energy resolution is best for low-energy gamma-ray and X-ray irradiation on the cathode side. Although CdTe detectors can be relatively efficient for low-energy gamma rays, acceptable detection efficiency for high-energy gamma rays requires larger volumes, which unfortunately translates into poor spectroscopic performance. Electronic pulse rejection and correction[59] techniques have been successfully applied to CdTe detectors to improve energy resolution. Selection of pulses based on shape characteristics can dramatically improve gamma-ray energy resolution at the expense of gamma-ray sensitivity. Electronic methods that compensate for charge trapping based on measurement of the pulse

Table 13.3 Properties of Semiconductor Materials

Material	Z	Density (g/cm³)	Bandgap (eV)	Ionization Energy (eV per e-h pair)	Best Gamma-Ray Energy Resolution (FWHM)	Reference
Si (300 K)	14	2.33	1.12	3.61		65
(77 K)			1.16	3.76	400 eV at 60 keV	65
(77 K)					550 eV at 122 keV	65
Ge (77 K)	32	5.33	0.72	2.98	400 eV at 122 keV	65
					900 eV at 662 keV	65
					1300 eV at 1332 keV	57
CdTe (300 K)	48/52	6.06	1.52	4.43	1.7 keV at 60 keV	47, 58
					3.5 keV at 122 keV	58
HgI$_2$ (300 K)	80/53	6.4	2.13	4.3	3.2 keV at 122 keV	47, 72
					5.96 keV at 662 keV	72
Cd$_{0.8}$Zn$_{0.2}$Te (300 K)	48/30/52	6	1.64	5.0	11.6 keV at 662 keV	47, 80

Figure 13.26 Example of pulse height spectra from a CdTe detector. Cases are shown with and without a correction technique based on electronically compensating for hole trapping based on measured pulse rise time. (From Richter and Siffert.[59])

shape are capable of improving energy resolution without discarding many of the events, and a FWHM value of 1.13% at 662 keV has been obtained [59] with this technique.

If spectroscopic information is not required, CdTe detectors can be applied effectively to simple pulse counting in a variety of applications in which the high gamma-ray efficiency allows for a very compact detector size. CdTe can also be operated in current mode in high gamma-ray fluxes. This includes the *photovoltaic* mode (see p. 399) in which no external voltage is applied to the detector. In this case, the charge carriers created by gamma-ray interactions are collected through the internal field formed by the contact potential of the junction, much the same as in a solar cell.[60] CdTe detectors have also been used in the pulse mode spectroscopy of energetic charged particles,[61] but with an energy resolution inferior to that of silicon detectors.

A persistent problem in the use of some CdTe detectors is the phenomenon of polarization, which under certain conditions of operation leads to a time-dependent decrease in the counting rate and charge collection efficiency.[62] This polarization is related to the capture of electrons by deep acceptors within the material. The resulting buildup of space charge not only interferes with carrier collection, but also leads to a gradual decrease in the effective thickness of the detector active region (or depletion region).

Commercially available CdTe detectors range in size from 1 mm to over 1 cm in diameter, but are normally limited to thicknesses of a few millimeters or less. They are relatively rugged and stable in field use and can routinely be operated at temperatures up to 30°C without excessive noise. Current mode operation is possible to temperatures up to 70°C. Recent results with small-diameter CdTe detectors have demonstrated their usefulness in portable instruments for X-ray spectroscopy, in which the detector crystal is often chilled with a Peltier cooler to reduce the detector leakage current and to reduce noise from the preamplifier.[63] Some reviews of CdTe detector performance in a variety of applications are given in Refs. 47, 55, 58, 64 and 65.

C. Mercuric Iodide (HgI₂) Detectors

Mercuric iodide is another material that has attracted considerable attention as a potential detector for X-rays and gamma rays. HgI_2 is a semiconducting material with a large bandgap energy of 2.13 eV, which permits operation at room temperature while producing only very small thermally generated leakage current. The high photoelectric cross section (see Fig. 13.22) leads to a much greater low-energy gamma-ray absorption efficiency than in Ge, at some energies as much as 50 times greater. In this regard, HgI_2 is an even more efficient photon detector than CdTe. For example, 85% of 100 keV photons are absorbed within a 1-mm thickness of HgI_2, yet the same result would require 2.6 mm of CdTe and 10 mm of Ge.

The development of detectors from this material remains largely at the research stage. Problems remain with HgI_2 that restrict its widespread application, including low hole mobilities, short charge carrier mean free drift lengths, space charge polarization, and surface degradation. The buildup of trapped charges can contribute to polarization,[66] in which the electric field is distorted and efficient charge collection is further disturbed. Devices have often been observed to perform best after the operating voltage has been applied over a period of several weeks,[67] only to return to the prior state of performance after shutting off the device. The improvement in energy resolution and detection efficiency has been ascribed to the electromigration of mobile defects and impurities that may be slowly swept through the detector active volume. Also, the material must be encapsulated[68] with parylene or a similar material to reduce deterioration of the crystal surfaces over extended periods of time.

Despite these operational difficulties, there are some applications in which HgI_2 has been successfully applied. Crystals with thicknesses less than 1 mm can demonstrate reasonable spectral performance and have been applied successfully in the measurement of X-rays and low-energy gamma rays. Figure 13.27 shows the performance that can be obtained with a thin detector operated at 0°C for low-energy X-rays.

For good detection efficiency at higher gamma-ray energies, thicker detectors are needed. It has been possible to grow useable HgI_2 crystals with thicknesses up to 1.2 cm and volumes of about 10 cm^3. However, the best results tend to be obtained from detectors of smaller volume because of remaining difficulties with efficient charge collection. Energy resolution of 1.5% FWHM at 662 keV has been reported with a 1.7 mm thick, 2.2 cm^2 area detector,[67] but typically the spectroscopic performance for HgI_2 detectors of this size is much poorer. Large volume detectors with dimensions of 4 cm × 4 cm × 1.2 cm show a best energy resolution of only about 7% FWHM at 662 keV.[69]

Efforts have been carried out[70,71] to restore the energy resolution in thick HgI_2 through pulse processing techniques. Such electronic correction relies on identifying the separate pulse contributions from holes and electrons, and subsequently adding through artificial means a component to the output signal that corrects for hole charge carrier trapping losses. Since the electrons are collected much more rapidly than holes, an estimate of the separate hole and electron contributions can be made from the change in the leading

Figure 13.27 Example of a measured pulse height spectrum from a HgI_2 detector. (From Ponpon and Sieskind.[69])

edge slope of the signal pulse before and after the electrons have been collected. Electronic compensation techniques are capable of producing energy resolution figures of a few percent for detectors up to 5 mm thick.[70] Another approach is to employ unique detector structures in which the induced charge is due primarily to the flow of electrons rather than holes.[72] Achievable energy resolution is still far inferior to that of Ge-based detectors, but the very high volumetric detection efficiency, the compactness of size, and the room temperature operability of HgI_2 could be advantageous in some applications. There have also been demonstrations[73] of the successful implementation of HgI_2 detector arrays for position sensing or imaging applications.

The large bandgap energy (2.13 eV) and resulting low leakage current allow such detectors to be operated as simple conductivity counters, in which the creation of a depletion region is no longer necessary as is required with Si or Ge detectors. Therefore, rectifying contacts are not necessary to reduce leakage current, and the device current–voltage curves take on dissimilar characteristics to that of conventional junction or surface barrier detectors. (Rectifying junctions are necessary for materials that have low resistivity and high thermal carrier generation, neither of which is the case for wide bandgap materials such as HgI_2.) Consequently, these detectors can be operated with either positive or negative polarity voltage applied to either surface contact. The choice of polarity, however, is extremely important if soft radiation is involved that does not penetrate deeply into the detector. Much better performance can then be obtained by arranging the polarity so that holes are collected over the smaller distance (movement towards the entrance contact), leaving the much more mobile electrons to traverse the full thickness of the detector.

Although HgI_2 is primarily of interest as an X-ray or gamma-ray detector, measurements have also been made[74] of its response to charged particles. Because of its high density, HgI_2 has a stopping power for energetic protons of about twice that of Si, the usual semiconductor material used for charged particle detectors. A nearly linear energy response was observed, with an energy resolution of 5 to 15% FWHM for H and He ions fully stopped in the detectors. While this energy resolution is more characteristic of scintillators than semiconductor detectors, some improvements might be expected by employing the techniques mentioned above to compensate for charge loss resulting from hole trapping.

D. $Cd_{1-x}Zn_xTe$ ("CZT") Detectors

A more recent candidate for room temperature radiation spectrometers has been introduced within the last decade, namely, the various compositions of the ternary compound $Cd_{1-x}Zn_xTe$, where x is the blending fraction of ZnTe in CdTe. The acronym CZT is commonly used for this material. The extremes of Zn concentration commonly used range from $Cd_{0.96}Zn_{0.04}Te$ to $Cd_{0.8}Zn_{0.2}Te$, which have room temperature bandgap energies of 1.53 eV and 1.64 eV, respectively.[75] The gamma-ray absorption efficiency is similar to that of CdTe, yet the increased bandgap reduces the intrinsic free carrier concentration and the operating leakage current below that generally experienced with CdTe devices. The measured mobilities are 1350 $cm^2/V-s$ for electrons and 120 $cm^2/V-s$ for holes,[76] hence the drift velocity characteristics are significantly different. Similar to CdTe, the measured hole lifetime is much shorter than the electron lifetime, being documented as between 100 ns to several microseconds for electrons and from 50–300 ns for holes.[47,77]

High resistivity CdZnTe is most often grown through the unseeded high-pressure vertical Bridgman (HPVB) technique.[78] Unfortunately, HPVB CdZnTe ingots presently have numerous defects including metal inclusions, twins, grain boundaries, and in some cases severe polycrystallinity. Consequently, spectrometer quality samples are often painstakingly identified and separated from the bulk ingots, a process that ultimately increases the overall detector cost. Alternative crystal growth methods are under investigation.

Figure 13.28 Pulse height spectrum from a 5 mm × 5 mm × 2 mm thick planar CdZnTe detector. (From McGregor and Hermon.[47])

Despite these difficulties, CdZnTe-based gamma-ray spectrometers have shown significant improvement in both size and energy resolution over other room temperature operated gamma-ray spectrometers. A room temperature spectrum from a typical planar CdZnTe detector is presented in Fig. 13.28, showing the predictable low-energy skewing (or "tail") from severe hole losses during transport. Reducing the detector thickness enhances the spectroscopic performance. Additionally, the spectra are best for low-energy gamma rays (shown at 60 keV), whereas measurements of high-energy gamma rays demonstrate poorly resolved gamma-ray peaks and severe "tailing" effects. Unlike CdTe devices, polarization does not appear to be an issue with CdZnTe. Commercially available planar CdZnTe detectors have average dimensions ranging from 5 mm × 5 mm × 1 mm to 10 mm × 10 mm × 2 mm.

The energy resolution of CdZnTe spectrometers can be improved when techniques are applied that eliminate the sensitivity of the induced charge to the slow hole motion. These *single polarity charge sensing* configurations[79–84] employ specialized shapes or electrode configurations to accomplish a circumstance in which the induced charge derives primarily from only the motion of the conduction electrons, but is still independent of their position of creation within the detector active volume. Shown in Fig. 13.29 is a room temperature

Figure 13.29 Pulse height spectrum for 662-keV gamma rays taken with a 1-cm³ CdZnTe spectrometer operated with coplanar grid electrodes. (From He et al.[80])

gamma-ray spectrum of 662 keV gamma rays from a 1 cm^3 "coplanar grid" CdZnTe detector. The device uses the concept introduced by Frisch for gas ion chambers, except that the grid is fabricated on the device surface.[79,80] Other detector designs more closely resembling the Frisch grid are under investigation.[83] Another single polarity sensing approach utilizes a pixelated anode, in which charge induction on the small pixels occurs primarily when charge carriers are in their vicinity.[84] The "small pixel effect" (see Appendix D) weights the induced charge output signal strongly toward electron motion and the adverse effects of poor hole transport properties are significantly reduced. Single polarity sensing detectors are in the experimental stage of development, but they offer some promise for future improved performance from materials with poor hole transport.

E. Other Semiconductor Detectors

Gallium arsenide (GaAs) has been studied as a radiation detector since the early 1960s[85,86] and was the first compound semiconductor operated at room temperature that demonstrated good gamma-ray energy resolution.[87] The room temperature bandgap energy of GaAs is 1.43 eV and the average ionization energy is 4.3 eV/e-h pair, which indicates that acceptable energy resolution should be possible for room temperature operation.[86] The atomic numbers of Ga (31) and As (33) bracket that of Ge, so the expected gamma-ray interactions and detection efficiency per unit mass should be similar to germanium detectors. Early experimental GaAs spectrometers were fabricated from very high purity liquid phase epitaxial material, but excessive cost and repeatability issues have prevented their introduction as commercially viable detectors. Attempts at using bulk-grown GaAs material for charged particle and gamma-ray spectrometers have been only modestly successful.[47,88] Unfortunately, electric field distortions and charge carrier trapping effects, generally believed to be a consequence of native deep level and impurity defect centers,[89,90] have prevented the commercial realization of bulk GaAs radiation spectrometers. Unique applications for GaAs devices have seen success as high-speed photoconductors[91,92] (see p. 491). If spectroscopic information is not important (as in particle tracking), the good resistance of GaAs devices to radiation damage effects has sparked interest in their application in particle physics measurements at high-energy accelerators.

Lead iodide (PbI$_2$) remains interesting as an alternative semiconductor for gamma-ray spectroscopy. The large atomic numbers of lead (82) and iodine (53) indicate that the material will have very good gamma-ray interaction and detection efficiency. Its bandgap energy is 2.55 eV, implying that the material can be operated in environments well above room temperature. Although PbI$_2$ was first studied in the early 1970s as a radiation detector, material issues regarding its use remain.[93,94] Recent improvements have been realized,[95] yet poor charge carrier mobilities coupled with serious charge carrier trapping have continued to prevent PbI$_2$ from becoming a commercially viable semiconductor spectrometer.

There has also been some interest[96] in the use of amorphous silicon layers as radiation detectors. Silicon in amorphous form offers the potential of much lower cost and larger surface area compared with the single crystals conventionally used in silicon detector fabrication. Both the mobility and lifetime values for electrons and holes in amorphous material are many orders of magnitude below those observed in silicon crystals. As a result, only very thin layers (up to 15 μm) have been used successfully as detectors, and then only with fractional charge collection. Layers of this type, although not generally useful in energy spectroscopy, can function as simple charged particle counters for applications in which energy resolution is not important. Pulses from alpha particles have been measured[96] for layers as thin as 5 μm.

Some limited utilization of diamond as a detector material has been made in specialized applications. Diamond is an insulating material with a very large bandgap (5.6 eV) and

Table 13.4 Some Alternative Compound Semiconductor Materials

Material	Z	Density (g/cm^3)	Bandgap (eV)	Ionization Energy (eV per e-h pair)	Best Gamma-Ray Energy Resolution (FWHM)	Reference
GaAs	31/33	5.32	1.43	4.2	2.5 keV at 60 keV	86, 87
					2.6 keV at 122 keV	87
InP	49/15	4.79	1.35	4.2	—	47
CdSe	48/34	5.8	1.73	—	8.5 keV at 60 keV	108
PbI$_2$	82/53	6.2	2.55	4.9	1.83 keV at 60 keV	47, 95
GaSe	31/34	4.55	2.03	4.5	—	107

can be operated as a simple conduction counter by applying electrical contacts to opposite faces of the crystal. Detectors of this type, although obviously limited to small sizes, are characterized by a fast response time of several nanoseconds and superior long-term stability. They can be operated in either conventional pulse mode or in current mode in high radiation fields.[97] Measurements have demonstrated an energy resolution of 82 keV for room temperature spectroscopy of alpha particles.[98] Because of their wide bandgap, diamond detectors are well suited for operation at elevated temperatures, and their use at 250–300°C has been reported.[99,100] Thin films of diamond produced by chemical vapor deposition process (CVD diamond) have also been shown to be useable as radiation detectors.[101–104] Because of their high resistance to radiation damage, CVD diamond devices have found application in particle tracking for high-energy physics experiments and have been successfully fabricated as microstrip position-sensing detectors.[105]

Several other semiconductors have been studied with limited success, including InP,[106] GaSe,[107] and CdSe.[108] Table 13.4 lists some properties of various semiconductor materials for comparison. None of these materials has reached the point of commercial viability. Most are limited to very small sizes and there are often persistent problems with efficient charge collection. Development work continues, however, in the hope of producing an alternative to Si and Ge capable of room temperature operation. Armantrout et al.[109] have carried out a preliminary screening of many additional compound semiconductors, none of which has received much attention as a potential detector material because of the lack of availability in practical crystal sizes.

III. AVALANCHE DETECTORS

The normal semiconductor detector is the solid-state analogue of a conventional gas-filled ionization chamber. The objective in either case is simply to collect all charge carriers that are created by the incident radiation within the active volume. Under certain conditions, it is also possible to achieve charge multiplication in solid-state detectors. The resulting device is then the analogue of the proportional counter and is called an *avalanche detector*. Although not nearly as widely applied as gas proportional counters, they have achieved some degree of popularity for the detection of low-energy radiations.

Gain is achieved within the semiconductor material (normally silicon) by raising the electric field sufficiently high to enable the migrating electrons to create secondary ionization during the collection process. Avalanche detector designs must incorporate a device geometry that allows the generation of high fields within the volume of the detector without allowing the field at the surface to become so large as to create surface breakdown. This effect is sometimes accomplished by beveling or *contouring* the detector shape[110–112] as

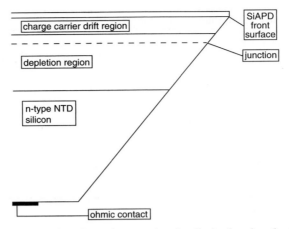

Figure 13.30 Cross section through an avalanche diode showing the beveled edge used to avoid surface breakdown. (From Farrel et al.[112])

illustrated in Fig. 13.30. An alternative structure shown in Fig. 13.31 is called the *reach through* configuration.[113–115] Gains of up to several hundred in the total collected charge are possible, but there has been some difficulty in achieving uniform multiplication across the entire entrance window of such detectors. The gain is also a strong function of applied voltage and temperature,[116] and stable operation requires careful attention to these factors.

Charge carriers created within the high-field region will be multiplied by a variable gain, depending on their position relative to the boundaries of the multiplication region. The internal configuration of avalanche detectors usually provides a normal *drift* region adjacent to the multiplication region where the electric field is too low to create multipli-

Figure 13.31 The "reach-through" configuration for an avalanche diode. (From Hauger et al.[115])

cation. If all incident radiations interact within the drift region, a more uniform multiplication of the charge can be expected because all electrons will then be subject to the same average multiplication. Nonetheless, the statistical variations in the avalanche process introduce considerable fluctuation into the multiplication process, and the energy resolution is typically inferior to that obtained in low-noise silicon detectors without gain.

Avalanche detectors offer some attractive properties both from the standpoint of small signal detection and in speed of response. The very high fields ensure that the transit time through the multiplication region is very short, and typical pulse rise times are of the order of 3 ns.[117] Typical time resolution is less than a nanosecond[118] and can be less than 0.1 ns under favorable circumstances.[115] This combination of properties permits use of avalanche detectors with good signal-to-background characteristics, even for the detection of very low-energy radiation or under conditions in which the detector temperature is elevated. For example, 1.5-keV X-rays can be detected with a 30–40% intrinsic efficiency at temperatures between 85 and 100°C while maintaining very low background levels.[117] Much of this sensitivity comes about because the electronic shaping times can be made very short due to the extremely fast rise time of pulses from these detectors. In many applications, the output signal of the avalanche detector is fed through a tunnel diode that is set with a discrimination point sufficiently high so as not to trigger on noise pulses. Because of the internal gain of the detector, X-rays with an energy as low as 0.6 keV can generate pulses that are above this discrimination point.

Avalanche detectors have found widespread applications in the detection of low-energy X-rays.[119–123] Other applications have involved use in portable gamma-ray monitors[124] and in the detection of low-energy beta particles from tritium.[125,126] Their use as photodiodes to detect the light from scintillators was described previously in Chapter 9.

IV. PHOTOCONDUCTIVE DETECTORS

A specialized type of solid state detector has proven useful in the measurement of short bursts of intense radiation. Under these conditions, many particles or photons interact in the active volume within the time required to drift the resulting charges to the electrodes. A *photoconductive detector* consists of a sample of semiconducting material fitted with two injecting contacts (or "ohmic" contacts) at opposite surfaces. When a voltage is applied, a measurable current flows (the *equilibrium current*) that is determined by the free carrier concentrations in the semiconducting material. In the discussion that follows, we will make the assumption that the usual conditions apply in which only one carrier (normally the electrons) dominate the conductivity of the material. The carrier concentration will be increased by irradiating the device with a pulse (or a continuous flux) of ionizing radiation, resulting in an increase in the material conductivity. The integrated *photocurrent* (the increase over the equilibrium current), or the photoconductive charge, is proportional to the energy deposited by the pulse of radiation. With true injecting contacts, the resistivity change is maintained in the sense that charge carriers that exit one contact are replaced by injected carriers that enter through the opposite contact.[127] For a short pulse of intense radiation, the current then decays with a characteristic time determined by the charge carrier lifetime until the observed current is restored back to the equilibrium current. Photocurrent gain (or "photoconductive gain") is defined as[127, 128]

$$G = \frac{\mu_C \tau_C^* \mathscr{E}}{W} = \frac{\tau_C^*}{t_C} \tag{13.6}$$

where μ_C is the charge carrier mobility, $\mathscr{E}$ is the applied electric field, W is the device thickness, τ_C^* is the charge carrier lifetime, and t_C is the time required for a carrier to drift across the detector. For large values of the carrier lifetime, the integrated charge can be many times greater than the initial charge created by the pulse of radiation, and the phrase *photoconductive gain* is therefore used to describe this increase. In the opposite extreme of short lifetimes, the observed pulse has a minimum time duration that is governed by the time width of the radiation pulse. While this extreme is sometimes chosen to promote optimum timing performance, it results in a much smaller integrated charge because the gain is now reduced below unity. The leakage current is very low in large bandgap or high resistivity materials, but they also require that special attention be paid in fabricating true injecting contacts. Without efficient charge injection, the conditions necessary for photoconductive gain are no longer satisfied.

Photoconductive gain for detectors intended for single radiation quanta is difficult to realize because a single particle or photon creates charge carriers in discrete locations within the device. Thus the bulk conductivity of the detector may not be perturbed enough to create a measurable change in the material conductivity. As a result, charge injection becomes less probable at the entrance contact, even though the opposite contact may allow for efficient carrier removal.

V. POSITION-SENSITIVE SEMICONDUCTOR DETECTORS

Detectors in which the position of interaction of the incident radiation is sensed together with its energy have application in a number of different areas. The two major types of detectors used for position sensing for charged particles are gas-filled proportional tubes (as described in Chapter 6) and silicon or germanium semiconductor diode detectors. The latter types are sometimes preferred because of their compactness and low bias voltage compared with gas-filled detectors. Semiconductor detectors, because of their greater stopping power, are also better suited for applications involving long-range radiations.

A. Resistive Charge Division

In its simplest form, the semiconductor position-sensitive detector consists of a one-dimensional strip of silicon or germanium for which one contact is made to have a significant series resistance. As shown in Fig. 13.32, one end of this resistive contact is grounded whereas the other leads to an amplifier for the derivation of the position signal. The resistive contact acts as a charge divider, and the amount of charge delivered to the position signal amplifier is proportional to x/L, where x is the distance of the interaction from the grounded end, and L is the strip length. A second signal (the E signal) proportional to the total charge deposited in the detector is derived from the normal conductive front electrode. If the position signal is divided by the E signal, a pulse is produced whose amplitude reflects the position of the interaction, independent of the amount of charge actually deposited. A review and analysis of pulse-processing methods for position sensing are given in Ref. 130.

Figure 13.32 Basic configuration of a position-sensitive semiconductor detector. The position signal (labeled P) is obtained by resistive charge division from the back layer. The E signal is proportional to the energy deposited by the particle in the detector active volume. (From Kaufman et al.[129])

Figure 13.33 A two-parameter display of events recorded from a position-sensitive detector of the type shown in Fig. 13.32. The E signal amplitude is displayed along the horizontal axis and the P signal along the vertical axis. A collimator with nine equidistant slits was placed in front of the detector. Incident particles were fission fragments from ^{252}Cf. (From Kaufman et al.[129])

The performance of one such detector is illustrated in Fig. 13.33. The contributions of nine different locations are cleanly separated for incident fission fragments with a wide spread in energy. Position resolution can be of the order of 0.5–1.0% FWHM at room temperature over active lengths of 50–100 mm.[131] The position resolution normally deteriorates as the energy of the incident particle decreases, because one of the determining factors is the statistical variation in the number of charge carriers collected at either end of the resistive strip.

Under some conditions, the undepleted portion of a normal silicon surface barrier can be used as the resistive back layer. More typically, the resistive contact is formed by metal evaporation or by ion implantation techniques. Ion-implanted layers are usually superior because they can be made uniform in resistance.[131,132] In other arrangements, the charge division is accomplished in an external resistive network.[133,134]

B. Microstrip Detectors

An alternative method of position sensing is to subdivide one of the electrodes into a number of independent segments or strips. Since electron–hole pairs created within the volume of the detector will travel along field lines to the corresponding electrode segment, a strong signal will be derived only from those segments that have collected appreciable charge carriers. If short-range particles are involved, then only one such signal will be generated. For longer-range particles, the "center of gravity" of the position of the track can be determined by interpolation of the signals from nearby segments. A silicon *microstrip* detector is one in which a series of narrow parallel strip electrodes have been fabricated on one surface using ion implantation and/or photolithography techniques.[135,136] For fine spatial resolution, such detectors have been produced with strip widths as small as 10 μm. To avoid the multiplicity of independent electronic channels that would be needed to measure the

Figure 13.34 Double-sided silicon strip detector. (From Hubbeling et al.[137])

signal from each strip independently, schemes have been developed to find the strip near-est the interaction using one-dimensional charge division along a line that interconnects all the strips. Using this approach, one can obtain a spatial resolution approaching 10 μm in one dimension for incident alpha particles.

Double-sided processing of the silicon wafer can produce configurations as illustrated in Fig. 13. 34 in which electrode strips are provided in orthogonal orientation on the oppo-site sides of the wafer. Since charges will be induced on both sets of strips by the motion of the electrons and holes, independent x and y coordinates can be obtained using the same techniques described above for single-sided microstrip detectors. In double-sided configu-rations, successfully fabricating strips on the n-side presents some special problems.[137] The silicon oxide layer on the surface, which is present to passivate the surface and reduce leak-age currents, inherently accumulates positively charged ions at the oxide–silicon interface. These charges produce a thin accumulation layer of mobile electrons in the silicon that rep-resents a conductive path between the electrode strips and prevents their independent operation. Several techniques are available[138] to remedy this potential problem, including the implantation of p-type strips (so called p-stops) between the electrodes to interrupt the electron layer[139] and restore good electrical isolation.

Silicon strip detectors have found widespread application in the tracking of high-energy particles in accelerator-based experiments and have also been applied in the imag-ing of X-rays. As one example,[140] a 300-μm-thick double-sided silicon strip detector with 50 μm pitch on both sides covered an imaging area of 32 mm × 32 mm. Each of the 640 strips on both sides were read out using multichannel integrated-circuit readout chips.

C. Pad or Pixel Detectors

The "brute force" approach to obtaining two-dimensional position information from a single-sided silicon detector is to fabricate the top electrode as a checkerboard pattern of indi-vidual small elements that are electrically isolated from each other. When the dimensions of the individual electrodes are a millimeter or larger, this type of device is usually called a pad detector. For electrode dimensions smaller than one millimeter, the common termi-nology is pixel detector. Electrical connection must be made to each individual electrode and separate electronic readout channels provided for each.

One advantage of this approach is that the small size of each individual electrode results in a relatively small capacitance and leakage current, and thus the electronic noise is reduced considerably from that observed from microstrip detectors of equivalent dimensions. In one example,[141] an equivalent noise charge (ENC, as defined on p. 630) of 100 electrons per pixel was observed from a pixel detector versus 1500 electrons for a similar strip detector. In addi-tion to resulting in a lower noise level, the lower leakage current may permit room tempera-ture operation when devices with larger electrode dimensions may require cooling.

Providing separate electrical contacts to each pixel presents a significant challenge. For small area detectors in which the pixels or pads are sufficiently large, wire bonding tech-

Figure 13.35 A "hybrid" pixel detector consisting of separate detector and electronics chips connected through indium bump bonds. (From Hallewell.[144])

niques can be used[142] to provide electrical connection from each pixel to the edge of the wafer. Alternatively, some double metal layer constructions allow contact to the interior pixels.[143] However, in the more common approach, illustrated in Fig. 13.35, a pixel detector chip is connected to a separate readout chip using flip chip solder bonding[145] or indium bump bonds. The readout chip is fabricated with exactly the same pitch as the detector pixels, so each bump provides an electrical connection between a single pixel and its corresponding readout electronics.

It is tempting to think about integrating the detector and readout electronics in a monolithic structure on a single silicon chip. Progress in this direction has been inhibited by several problems, including the basic incompatibility between the high resistivity silicon needed for substantial depletion regions for the detector and the lower resistivity silicon common in the fabrication of integrated circuits (ICs). There are also conflicts between the high temperature processes typical of ICs and the lower temperatures needed to preserve carrier lifetimes in the detector, and between the high voltages required for detector depletion and the lower voltages characteristic of IC operation.[146] Although there is continuing interest and effort to carry out this monolithic integration of the detector and electronics, present pixel or pad detectors are more often based on separate structures for each.

Pixel or pad detectors typically have active areas that are limited to a few square centimeters. Larger detector areas can be achieved by assembling individual modules into arrays, although at the expense of increasing complexity in what is already a complex device.

D. Semiconductor Drift Detectors

It has been demonstrated[147] that the drift time of charge carriers can be used to deduce the position of their formation in a silicon detector. In a *semiconductor drift detector,* a unique electrode configuration is used as illustrated in Fig. 13.36. Semiconductor junctions are formed at *both* faces of a large-area wafer, and each is reverse biased until the detector is fully depleted. Electrons created by ionizing radiation within the semiconductor are confined within an electric potential well and caused to drift in a direction parallel rather than perpendicular to the wafer surface. A collecting anode is fabricated near the edge of the wafer. The time required for these electrons to drift to the anode is then a linear measure of the distance between the anode and the position of the interaction. Tests of prototype drift detectors[149] indicate that a spatial resolution of 4 μm may be achieved over a drift distance of several millimeters. Detectors with active area as large as 2.5 cm × 2.5 cm have been operated[150]

Figure 13.36 Structure of a linear silicon drift detector. Electrons formed by the ionizing radiation are initially drawn to a potential minimum near the center of the wafer. They are then transported parallel to the surfaces by a potential gradient between the surface strips and collected at an anode near the edge of the wafer. (From Gramegna et al.[148])

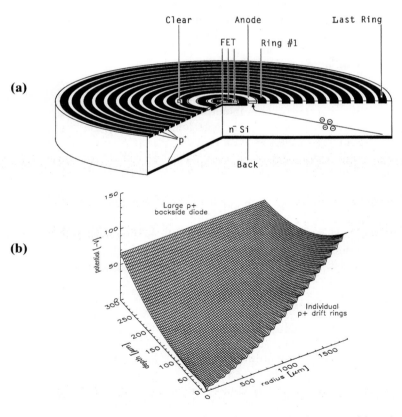

Figure 13.37 (*a*) Cross section through a cylindrical drift detector with an integrated JFET connected to the collecting anode. The nonstructured back surface of the detectors can provide a thin entrance window for weakly penetrating radiation. (*b*) Potential surface for electrons for the device of part (*a*). The central collecting anode is shown at the near corner of the surface. (From Lechner et al.[151])

with spatial resolution of less than 0.5 mm. Figure 13.36 shows a so-called linear drift device in which parallel strips are used to shape the field and create the potential gradient needed to drive the electrons to the collecting anode. In this case, the anode is segmented to allow determination of the second position coordinate in the dimension parallel to the strips.

Figure 13.37 shows a cylindrical drift geometry in which the collecting anode is located at the center of a series of circular rings that serve the purpose of shaping the electric field. In this case, all the ionization electrons are collected on a single small-area anode which is kept very small to minimize its capacitance. In the example shown, an on-chip junction field-effect transistor (JFET) has been integrated to minimize the added stray capacitance that connection to an external preamplifier would involve. Also shown in the figure is the resulting potential surface for electrons illustrating that electrons created anywhere within the bulk of the silicon are guided to the small area anode at the center. This configuration provides a flat back surface without surface structures that can serve as a thin entrance window for weakly penetrating radiations.

The drift detector configuration has an additional advantage that can be exploited[152,153] to improve energy resolution. Because the electrons can be drifted over large distances and collected on an electrode of very small size, the capacitance of the detector can be much smaller than that of an equivalent semiconductor diode of conventional design. The detector capacitance is always an important factor in determining the noise level of a spectroscopic system, and low capacitance is a significant advantage. Drift detectors have therefore gained considerable attention as X-ray spectrometers, and the device shown in Fig. 13.37 has a diameter of 3.5 mm and a fully depleted thickness of 300 μm. Because of the bulk leakage current that contributes to the measured charge, drift detectors tend to have the best energy resolution when cooled below room temperature. For example, the energy resolution reported[151] for a similar drift detector at 5.9 keV is 225 eV at room temperature, 170 eV at −10°C (263 K) (attainable by Peltier cooling), and 140 eV when cooled to 200 K. This latter figure is competitive with the lithium-drifted silicon detectors that have been the standard in this field for many years, and that are normally cooled to liquid nitrogen temperature (77 K) for their operation.

The small capacitance of drift or pixel detectors also leads to an advantage in high counting rate applications. Because the optimal shaping time for minimization of electronic noise (discussed in Chapter 17) shifts to lower values as the detector capacitance is decreased, shorter shaping times are possible that lead to a higher count rate capability for a given degree of pile-up. For example, shaping times for drift detectors can be as short as hundreds of nanoseconds, while optimum shaping times for lithium-drifted silicon detectors, because of their higher capacitance, are often in the tens of microseconds range.

While not needed for spectroscopic measurements, radial position sensing can also be performed in the ring geometry shown in Fig. 13. 37. The time of particle interaction can be sensed from the induced charge on the bottom surface electrode caused by the immediate flow of charges and the drift time measured to the time that the anode signal begins to rise.

E. Charge Coupled Devices

Charge coupled devices (CCDs) were first introduced in the early 1970s as silicon microcircuit devices capable of recording visible light images. Their application has grown rapidly, and CCDs are found in a wide range of optical components and cameras. These commercial applications remain the dominant economic force behind CCD development. However, a smaller subset of devices with similar properties, often called *scientific CCDs*, have emerged in the 1990s as extremely useful sensors for radiation detection and imaging. They have found widespread use in the tracking or imaging of high-energy minimum ionizing particles.[154] CCDs have also become a somewhat more complex but viable alternative to lithium-drifted silicon detectors for routine X-ray spectroscopy, especially at low energies.[155]

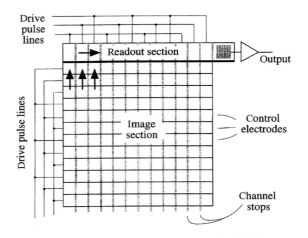

Figure 13.38 Layout of the surface of a CCD.

A schematic diagram of a simple CCD architecture is shown in Fig. 13.38. They are normally silicon structures produced by many of the standard microelectronics fabrication techniques used for integrated circuits. Typical scientific CCDs have square areas with dimensions of 1–2 cm on a side, and they are generally fabricated on a silicon wafer a few hundred microns thick. In the standard CCD, a depletion region is created immediately below the front surface of the device, and a potential minimum for electrons is created a few microns below the surface structure consisting of a metal-oxide-silicon (MOS) electrode structure. Pixel sizes tend to be on the order of 25 μm on a side. An alternative approach[156] is to replace the metal oxide electrode junction with *p-n* diode structures at the surface (see Fig. 13.39), with larger typical pixel dimensions of 150 μm on a side.[158] In some configurations, the silicon substrate is only partially depleted, leaving an undepleted silicon layer at the back as shown in Fig. 13.40. In other cases, the silicon wafer is fully depleted so that soft radiations can be incident from the back side of the CCD without encountering attenuation by the top surface structures.

The area of the CCD is subdivided into a large number of small pixels through the control electrodes and channel stops, illustrated in Fig. 13.38. When proper voltages are applied

Figure 13.39 Cross section through a transfer channel of *pn*-CCD. (From Soltau et al.[157])

Figure 13.40 Schematic representation of the diffusion of electrons following their interaction in a partially-depleted CCD. Electrons created in the depleted region are less likely to be spread among multiple pixels than those that are formed in the undepleted region because of their rapid collection caused by the strong electric field in the depleted region.

to the control electrodes, individual potential wells are created for each pixel, and any free electrons liberated by ionizing radiation within the dimensions of that pixel are accumulated at the corresponding potential well. Thus the CCD is inherently an integrating device, collecting charges in the pixels that simply accumulate over the exposure period.

The key to the popularity of the CCD lies in the simplicity of its serial readout technique. After the exposure period, the voltages on the control electrodes are modified by "drive pulses" that have the effect of tipping the potential minima and driving the accumulated electrons in a preferred direction. This is normally done using three drive lines for each pixel and adjusting the phase of the voltage between them. In the case illustrated in Fig. 13.38, the charges accumulated in the topmost row of the CCD are transferred into an adjacent readout section containing a number of pixels equal to that in one row. Again, the phases of the voltages on drive lines in the readout section are used to sequentially move the charges one at a time to an amplifier integrated on the CCD chip. Thus the first pixel to be read is the corner pixel in the upper right-hand corner. The second then is the pixel to its left in the original image. This process continues until the entire row has been read out sequentially. Then the drive line pulses in the imaging section are again adjusted to shift each row one step upward and the process is repeated. In this manner, all pixels from the imaging section are read out in serial fashion using a single readout node. The efficiency of the transfer of the charges from one potential well to the next is amazingly good in CCDs, and in the best devices, no appreciable charge is lost even when undergoing thousands of transfers.

The readout amplifier can have extremely low capacitance (on the order of 0.2 pF), and so even relatively small charge packets can be distinguished above the low noise level of the readout. For best noise suppression, shaping times of about 10 μs per pixel measurement are required,[159] limiting the maximum readout rate under these conditions to about 10^5 pixels per second. Since large area CCDs may contain as many as 10^6 or more pixels, readout times of more than a second are not uncommon in the "full frame" readout mode just described. To speed up the readout, a number of options have been developed. One approach involves rapid storage of the full pixel array in a second area of the CCD to allow resetting of the first image section, and a new accumulation period is started while the storage section is read out separately (called "frame storage" mode). Another approach is to divide the image section into quadrants and to provide four independent readout nodes at each corner.

With optimized systems,[160] the noise level of the readout can be as low as 2–3 electrons RMS equivalent noise, a level far below that of conventional stand-alone preamplifiers. Because of these very low noise levels, CCDs are capable of measuring the charge generated by a single particle or X-ray photon in a given pixel. Minimum ionizing particles such as fast electrons deposit about 80 electron–hole pairs per micron penetrated in silicon, so that thicknesses of tens of microns generate a thousand or more charges. While these signals would be difficult to process using conventional preamplifiers, the extremely small structures and low capacitance of the CCD readout allow their measurement with a reasonable signal-to-noise ratio.

In order for a CCD to be used to measure the energy of incident particles, the intensity must be low enough to have only a low probability of detecting more than one particle per pixel during the exposure period. Most of the pixels should then be empty during a full frame readout. Inevitably, there will be some charge sharing between pixels even for single events as a result of the diffusion of the electrons before they reach the potential minima (see Fig. 13.40). Also, for particles with an extended range, tracks may traverse the depletion region at an angle and thus create charges that would naturally be collected in more than one pixel. Thus, the best single particle spectroscopy with CCDs often is achieved by restricting recorded events to those in which only single pixels in a given neighborhood record a charge.

Throughout the exposure period, electrons will also be thermally generated within the silicon and the corresponding "leakage current" electrons will also be accumulated in all the pixels. The fluctuation in the number of these leakage electrons represents a noise source that can interfere with good energy resolution. The leakage becomes most important for thick depletion regions, large area pixels, and long exposure times. To reduce the influence of leakage current in spectroscopic measurements, it is almost universal to cool scientific CCDs by at least 50–100°C below room temperature. While this is an obvious operational complication, it is less demanding than the liquid nitrogen cooling required of thick Si(Li) detectors.

There is also a potential problem of "smearing" when a CCD is used to image incident radiation. During the exposure period, charges are recorded in the proper pixel corresponding to the position of incidence of the radiation. During the readout phase, however, these charges are in the process of moving across the face of the frame and any additional hits will be registered in the wrong position. For weakly penetrating radiations, a shutter can be used to block the incident radiation during the readout period. For more penetrating radiations where this solution is not viable, one would then want the readout time to be short compared with the exposure time to minimize the smearing effect. In those cases, it may be preferable to tolerate a higher noise level in the readout by raising the pixel sampling rate, to as high as 5×10^7 per second in some cases.[154] The recorded counting rate in spectroscopic measurements is thus limited by the simultaneous requirements of low pixel hit density per frame and a relatively long framing time required to suppress the smearing effect.

For the integrating mode of operation in which multiple hits per frame are expected in a typical pixel, some thought must be given to the charge storage capacity of a given pixel. In representative structures, numbers such as $1–5 \times 10^5$ electrons are quoted for the maximum electron storage per pixel.[159] Integration of the charge over an exposure period is a common method of operation in the imaging of various forms of radiation. For example, CCDs can be used in place of photographic emulsions in the recording of X-ray images or in carrying out autoradiography of tissue samples from the emitted beta particles.[161]

In spectroscopy mode where accepted events are limited to single hits per pixel, CCDs have demonstrated outstanding resolution for X-ray spectroscopy.[160] When noise levels in the readout are kept at a level of 2–3 electrons RMS, the noise becomes insignificant in measurements above about 1 keV. Then the energy resolution is limited only by the statistics of the charge carrier formation, a fundamental limit to all silicon devices. For example, an energy resolution value of 120 eV for 5.9 keV X-rays has been achieved,[155] which is very close to the estimated statistical limit.

The CCD and the drift chamber described earlier share many common properties. Both provide for transport of the electrons created by an ionizing particle over long distances parallel to the silicon wafer surface. The CCD accomplishes this in a synchronized sequential marching of rows of pixels toward the readout section, while the drift detector drives the charges continuously. The latter process is inherently faster and there is continuing interest[162] in combining the features of the two concepts to overcome some of the limitations in CCDs, primarily their slow readout. This is an area of active research[163] and new hybrid structures are likely to become available as this technology matures.

PROBLEMS

13.1 Give two reasons why the X-ray escape peak is less intense in silicon detectors compared with germanium detectors.

13.2 Estimate the maximum charge collection time (using data given in Chapter 11) for a 4-mm thick planar Si(Li) detector operated at 2000 V.

13.3 What must be the energy resolution (in percent) for an Si(Li) detector if it is to resolve separately the K-characteristic X-rays from copper and zinc?

13.4 What physical effects cause the detection efficiency of Si(Li) detectors to drop off at low (less than 5 keV) incident X-ray energies?

13.5 By using an NaI(Tl) well counter with absolute peak efficiency of 83%, a net of 146,835 counts was recorded under the 122 keV photopeak from a ^{57}Co source over a 15-min live time. The same source was then placed 10 cm from the face of an Si(Li) detector with 300-mm^2 surface area and a spectrum recorded over a 60-min counting period. If 730 counts were recorded under the 6.4-keV X-ray peak, what is the efficiency of the Si(Li) detector at this energy (see Table 13.2)?

13.6 Find the statistical limit for the energy resolution of a Si(Li) detector for the 59.5 keV gamma rays from ^{241}Am. Compare the result with the corresponding value for a HPGe detector at the same energy. What would the value be for a silicon drift detector?

13.7 Estimate the thickness of the following semiconductor materials that result in 50% of all incident 662 keV gamma ray photons undergoing at least one interaction: Si, Ge, CdTe, and HgI$_2$. In each case, also estimate the fractions of the initial interactions that are photoelectric absorption and Compton scattering.

13.8 A fully depleted silicon CCD is 300 μm thick and used to form a recorded image of incident X-rays whose energy is approximately 10 keV. It has an array of 256×256 pixels per frame, and is operated in a simple mode of alternating exposure and readout periods. The readout frequency is 100 kHz. The exposure time per frame is to be kept at least 20 times the total readout time per frame to minimize the effects of "smearing" of the recorded image of the incident X-ray pattern during the charge transfer period. The same measurement is also designed to measure the energy deposited by each individual X-ray photon, so one would like the probability to be less than 5% that more than one proton interacts in a typical pixel during the exposure time per frame.

(a) Determine the maximum X-ray interaction rate within the full image that can be accommodated under these conditions.

(b) Find the minimum required storage capacity in electrons of a typical pixel.

(c) If the charge due to leakage is to be kept less than 10% of the signal charge due to a single X-ray photon interaction in a pixel, estimate the maximum leakage current across the entire area of the CCD that can be tolerated.

REFERENCES

1. E. M. Pell, *J. Appl. Phys.* **31**, 291 (1960).
2. A. Coche and P. Siffert, "Lithium Drifted Silicon and Germanium Detectors," in *Semiconductor Detectors* (G. Bertolini and A. Coche, eds.). Elsevier-North Holland, Amsterdam, 1968.
3. P. E. Gibbons and P. Iredale, *Nucl. Instrum. Meth.* **53**, 1 (1967).
4. A. Lauber, *Nucl. Instrum. Meth.* **75**, 297 (1969).
5. M. Moszynski, W. Kurcewicz, and W. Przyborski, *Nucl. Instrum. Meth.* **61**, 173 (1968).
6. A. S. Antonov, *Solid State Phys.* **8**, 1325 (1966).
7. Z. Moroz and M. Moszynski, *Nucl. Instrum. Meth.* **68**, 261 (1969).

8. M. Moszynski and W. Przyborski, *Nucl. Instrum. Meth.* **64**, 244 (1968).
9. G. Lindstrom and R. Lisdat, *Nucl. Instrum. Meth.* **116**, 181 (1974).
10. P. Van Espen, H. Nullens, and F. Adams, *Nucl. Instrum. Meth.* **145**, 579 (1977).
11. R. E. Wood, P. Venugopala Rao, O. H. Puckett, and J. M. Palms, *Nucl. Instrum. Meth.* **94**, 245 (1971).
12. C. S. Rossington, R. D. Giauque, and J. M. Jaklevic, *IEEE Trans. Nucl. Sci.* **39**(4), 570 (1992).
13. J. L. Campbell, *Nucl. Instrum. Meth.* **B49**, 115 (1990).
14. J. L. Campbell et al., *Nucl. Instrum. Meth.* **A418**, 394 (1998).

15. T. He, R. P. Gardner, and K. Verghese, *Nucl. Instrum. Meth.* **A299,** 354 (1990).

16. P. Lechner and L. Strüder, *Nucl. Instrum. Meth.* **A354,** 464 (1995).

17. J. L. Campbell, *Nucl. Instrum. Meth.* **B109E110,** 71 (1996).

18. L. C. Alves, A. P. Jesus, and M. A. Reis, *Nucl. Instrum. Meth.* **B109E110,** 129 (1996).

19. M. L. Hildner, A. J. Antolak, and G. S. Bench, *Nucl. Instrum. Meth.* **A373,** 124 (1996).

20. G. Kalinka, *Nucl. Instrum. Meth.* **B88,** 470 (1994).

21. J. Felsteiner, S. Kahane, and B. Rosner, *Nucl. Instrum. Meth.* **118,** 253 (1974).

22. R. G. Musket, *Nucl. Instrum. Meth.* **117,** 385 (1974).

23. M. Hollstein, *Nucl. Instrum. Meth.* **82,** 249 (1970).

24. G. Bertolini, F. Cappellani, and G. Restelli, *Nucl. Instrum. Meth.* **112,** 219 (1973).

25. M. R. Khan, A. S. Lodhi, and D. Crumpton, *Nucl. Instrum. Meth.* **160,** 127 (1979).

26. G. Kalinka and K. Taniguchi, *Nucl. Instrum. Meth.* **B75,** 91 (1993).

27. A. H. F. Muggleton, *Nucl. Instrum. Meth.* **101,** 113 (1972).

28. J. L. Campbell and P. L. McGhee, *Nucl. Instrum. Meth.* **A248,** 393 (1986).

29. R. J. Gehrke and R. A. Lokken, *Nucl. Instrum. Meth.* **97,** 219 (1971).

30. J. L. Campbell and L. A. McNelles, *Nucl. Instrum. Meth.* **125,** 205 (1975).

31. W. J. Gallagher and S. J. Cipolla, *Nucl. Instrum. Meth.* **122,** 405 (1974).

32. B. Rosner, D. Gur, and L. Shabason, *Nucl. Instrum. Meth.* **131,** 81 (1975).

33. R. G. Musket and W. Bauer, *Nucl. Instrum. Meth.* **109,** 593 (1973).

34. J. L. Campbell, H. H. Jorch, and J. A. Thompson, *Nucl. Instrum. Meth.* **140,** 167 (1977).

35. H. I. Israel, D. W. Lier, and E. Storm, *Nucl. Instrum. Meth.* **91,** 141 (1971).

36. J. M. Jaklevic and F. S. Goulding, *IEEE Trans. Nucl. Sci.* **NS-18**(1), 187 (1971).

37. F. S. Goulding, *Nucl. Instrum. Meth.* **142,** 213 (1977).

38. I. Ahmad and F. Wagner, *Nucl. Instrum. Meth.* **116,** 465 (1974).

39. M. J. Berger et al., *Nucl. Instrum. Meth.* **69,** 181 (1969).

40. A. Damkjaer, *Nucl. Instrum. Meth.* **200,** 377 (1982).

41. M. J. Berger et al., *NBS Tech. Note* **489** (1969).

42. R. D. Von Dincklage and J. Gerl, *Nucl. Instrum. Meth.* **A235,** 198 (1985).

43. T. Frommhold et al., *Nucl. Instrum. Meth.* **A310,** 657 (1991).

44. R. Hofstadter, *Nucleonics* **4,** 2 (1949).

45. C. A. Klein, *J. Appl. Phys.* **39,** 2029 (1968).

46. H. L. Malm, *IEEE Trans. Nucl. Sci.* **NS-19,** 263 (1972).

47. D. S. McGregor and H. Hermon, *Nucl. Instrum. Meth.* **A395,** 101 (1997).

48. J. S. Iwanczyk, B. E. Patt, Y. J. Yang, and A. Kh. Khusainov, *Nucl. Instrum. Meth.* **A380,** 186 (1996).

49. W. Shockley, *J. Appl. Phys.* **9,** 635 (1938).

50. S. Ramo, *Proc. IRE* **27,** 584 (1939).

51. K. Hecht, *Z. Physik* **77,** 235 (1932).

52. W. Akutagawa and K. Zanio, *J. Appl. Phys.,* **40,** 3838 (1969).

53. G. F. Knoll and D. S. McGregor, MRS Proc. **302,** 3 (1993).

54. C. Scharager, P. Siffert, A. Holtzer, and M. Schieber, *IEEE Trans. Nucl. Sci.* **NS-27**(1), 276 (1980).

55. M. Hage-Ali and P. Siffert, *Semiconductors for Room Temperature Nuclear Detector Applications: Semiconductors and Semimetals,* **Vol. 43,** Chaps. 6–8, Academic Press, San Diego, 1995.

56. S. H. Shin et al., *IEEE Trans. Nucl. Sci.* **NS-32**(1), 487 (1985).

57. P. Siffert et al., *IEEE Trans. Nucl. Sci.* **NS-22**(1), 211 (1975).

58. P. Siffert, *MRS Proc.* **16,** 87 (1983).

59. M. Richter and P. Siffert, *Nucl. Instrum. Meth.* **A322,** 529 (1992).

60. R. J. Fox and D. C. Agouridis, *Nucl. Instrum. Meth.* **157,** 65 (1978).

61. R. A. Ristenen et al., *Nucl. Instrum. Meth.* **188,** 445 (1981).

62. P. Siffert et al., *IEEE Trans. Nucl. Sci.* **NS-23**(1), 159 (1976).

63. J. Pantazis, A. Huber, and P. Okun, *IEEE Trans. Nucl. Sci.* **NS-41,** 1004 (1994).

64. R. C. Whited and M. M. Schieber, *Nucl. Instrum. Meth.* **162,** 113 (1979).

65. E. Sakai, *Nucl. Instrum. Meth.* **196,** 121 (1982).

66. T. Mohammed-Brahim, A. Friant, and J. Mellet, *IEEE Trans. Nucl. Sci.* **NS-32**(1), 581 (1985).

67. V. Gerrish, *Nucl. Instrum. Meth.* **A322,** 402 (1992).

68. J. S. Iwanczyk et al., *IEEE Trans. Nucl. Sci.* **NS-37**(6), 2214 (1990).

69. J. P. Ponpon and M. Sieskind, *Nucl. Instrum. Meth.* **A380,** 173 (1996).

70. A. Beyerle, V. Gerrish, and K. Hull, *Nucl. Instrum. Meth.* **A242,** 443 (1986).

71. V. M. Gerrish, D. J. Williams, and A. G. Beyerle, *IEEE Trans. Nucl. Sci.* **NS-34**(1), 85 (1987).

72. B. E. Patt, J. S. Iwancyzk, G. Vilkelis, and Y. J. Wang, *Nucl. Instrum. Meth.* **A380,** 276 (1996).

73. B. E. Patt et al., *Nucl. Instrum. Meth.* **A366,** 173 (1995).

74. F. D. Bechetti et al. *Nucl. Instrum. Meth.* **213,** 127 (1983).

75. D. J. Olega, J. P. Faurie, S. Sivananthan, and P. M. Raccah, *Appl. Phys. Lett.* **47,** 1172 (1985).

76. Z. Burshtein et al., *Appl. Phys. Lett.* **63,** 102 (1993).

77. Z. He, G. F. Knoll, and D. K. Wehe, *J. Appl. Phys.* **84,** 5566 (1998).

78. J. F. Butler et al., *Mat. Sci. Engin.* **B16,** 291 (1993).

79. P. N. Luke, *IEEE Trans. Nucl. Sci.* **NS-42,** 207 (1995).

80. Z. He, G. F. Knoll, D. K. Wehe, and J. Miyamoto, *Nucl. Instrum. Meth.* **A388,** 180 (1997).

81. K.-L. Giboni and E. Aprile, *Nucl. Instrum. Meth.* **A416,** 319 (1998).

82. C. L. Lingren et al., *IEEE Trans. Nucl. Sci.* **NS-45,** 433 (1998).

83. D. S. McGregor and R. A. Rojeski, *IEEE Trans. Nucl. Sci.* **NS-46,** 250 (1999).

84. H. H. Barrett, J. D. Erskin, and H. B. Barber, *Phys. Rev. Lett.* **75,** 156 (1995).

85. A. Barraud, *Comptes Rendus* **257,** 1263 (1963).

86. D. S. McGregor and J. E. Kammeraad, *Semiconductors and Semimetals,* Vol. 43, Chap. 10, Academic Press, San Diego, 1995.

87. J. E. Eberhardt, R. D. Ryan, and A. J. Tavendale, *Nucl. Instrum. Meth.* **94,** 463 (1971).

88. A. D. Holland, A. D. T. Short, and T. Cross, *Nucl. Instrum. Meth.* **A346,** 366 (1994).

89. K. Berwick, M. R. Brozel, C. M. Buttar, M. Cowperthwaite, and Y. Hou, *Inst. Phys. Conf. Series* **135,** 305 (1993).

90. D. S. McGregor et al., *Nucl. Instrum. Meth.* **A343,** 527 (1994).

91. C. L. Wang, J. E. Flatley and M. D. Pocha, *Energy Tech. Rev.* Sept.-Oct., 15 (1991).

92. F. Foulon, B. Brullot, C. Rubbelynk, P. Bergonzo, and T. Pochet, *IEEE Trans. Nucl. Sci.* **NS-43**, 1372 (1996).
93. S. Roth and W. R. Willig, *Appl. Phys. Lett.* **18**, 328 (1971).
94. C. Manfredotti, R. Murri, A Quirini, and L. Vasanelli, *IEEE Trans. Nucl. Sci.* **NS-24**, 126 (1977).
95. V. Deich and M. Roth, *Nucl. Instrum. Meth.* **A380**, 169 (1996).
96. V. Perez-Mendez, J. Morel, S. N. Kaplan, and R. A. Street, *Nucl. Instrum. Meth.* **A252**, 478 (1986).
97. S. F. Kozlov et al., *IEEE Trans. Nucl. Sci.* **NS-24**(1), 235 (1977).
98. C. Canali et al., *Nucl. Instrum. Meth.* **160**, 73 (1979).
99. E. K. Konorova and S. F. Kozlov, *Sov. Phys. Semicond.* **4**, 1600 (1971).
100. F. Nava et al., *IEEE Trans. Nucl. Sci.* **NS-26**(1), 308 (1979).
101. E-K. Souw and R. J. Meilunas, *Nucl. Instrum. Meth.* **A400**, 69 (1997).
102. C. Manfredotti et al., *Nucl. Instrum. Meth.* **A410**, 96 (1998).
103. P. Weilhammer et al., *Nucl. Instrum. Meth.* **A409**, 264 (1998).
104. T. Behnke et al., *Nucl. Instrum. Meth.* **A414**, 340 (1998).
105. M. M. Zoeller et al., *IEEE Trans. Nucl. Sci.* **NS-44**(3), 815 (1997).
106. Y. Suzuki, Y. Fukuda, Y. Nagashima, and H. Kan, *Nucl. Instrum. Meth.* **A275**, 142 (1989).
107. H. Nakatani, E. Sakai, C. Tatsuyama, and F. Takeda, *Nucl. Instrum. Meth.* **A283**, 303 (1989).
108. M. Roth, *Nucl. Instrum. Meth.* **A283**, 291 (1989).
109. G. A. Armantrout, S. P. Swierkowski, J. W. Sherohman, and J. H. Yee, *IEEE Trans. Nucl. Sci.* **NS-24**(1), 121 (1977).
110. G. C. Huth, *IEEE Trans. Nucl. Sci.* **NS-13**(1), 36 (1966).
111. M. R. Squillante et al., *IEEE. Trans. Nucl. Sci.* **33**(1), 336 (1986).
112. R. Farrell, K. Vanderpuye, G. Entine and M. R. Squillante, *IEEE. Trans. Nucl. Sci.* **38**(2), 144 (1991).
113. P. P. Webb and A. R. Jones, *IEEE Trans. Nucl. Sci.* **NS-21**(1), 151 (1974).
114. P. P. Webb and R. J. McIntyre, *IEEE Trans. Nucl. Sci.* **NS-23**(1), 138 (1976).
115. J. A. Hauger et al., *Nucl. Instrum. Meth.* **A337**, 362 (1994).
116. I. Tapan et al., *Nucl. Instrum. Meth.* **A388**, 79 (1997).
117. G. C. Huth, "Avalanche Multiplying Diode," in *Semiconductor Nuclear Particle Detectors and Circuits,* Publication No. 1593, National Academy of Sciences, Washington, DC, 1969, p. 323.
118. A. Q. R. Baron and S. L. Ruby, *Nucl. Instrum. Meth.* **A343**, 517 (1994).
119. P. V. Hewka, G. C. Huth, and K. L. Swinth, *IEEE Trans. Nucl. Sci.* **NS-17**(3), 265 (1970).
120. P. J. Moldofsky and K. L. Swinth, *IEEE Trans. Nucl. Sci.* **NS-19**(1), 55 (1972).
121. A. Q. R. Baron, *Nucl. Instrum. Meth.* **A352**, 665 (1995).
122. A. Ochi, Y. Nishi, and T. Tanimori, *Nucl. Instrum. Meth.* **A378**, 267 (1996).
123. A. Q. R. Baron, R. Rüffer, and J. Metge, *Nucl. Instrum. Meth.* **A400**, 124 (1997).
124. A. R. Jones, *IEEE Trans. Nucl. Sci.* **NS-20**(1), 528 (1973).
125. R. A. Surette, *Nucl. Instrum. Meth.* **A337**, 588 (1994).
126. J. S. Gordon et al., *IEEE. Trans. Nucl. Sci.* **41**(4), 1494 (1994).
127. R. H. Bube, *Photoconductivity in Solids*, Wiley, New York, 1960.
128. A. Rose, *Concepts in Photoconductivity and Allied Problems*, Interscience Publishers, New York, 1963.
129. S. B. Kaufman, B. D. Wilkins, M. J. Fluss, and E. P. Steinberg, *Nucl. Instrum. Meth.* **82**, 117 (1970).
130. J. L. Alberi and V. Radeka, *IEEE Trans. Nucl. Sci.* **NS-23**(1), 251 (1976).
131. E. Elad and R. Sareen, *IEEE Trans. Nucl. Sci.* **NS-21**(1), 75 (1974).
132. E. Laegsgaard, F. W. Martin, and W. M. Gibson, *IEEE Trans. Nucl. Sci.* **NS-15**(3), 239 (1968).
133. J. E. Lamport, G. M. Mason, M. A. Perkins, and A. J. Tuzzolino, *Nucl. Instrum. Meth.* **134**, 71 (1976).
134. M. S. Gerber et al., *IEEE Trans. Nucl. Sci.* **NS-24**(1), 182 (1977).
135. E. H. M. Heijne et al., *Nucl. Instrum. Meth.* **178**, 331 (1980).
136. J. Yorkston, A. C. Shotter, D. B. Syme, and G. Huxtable, *Nucl. Instrum. Meth.* **A262**, 353 (1987).
137. L. Hubbeling et al., *Nucl. Instrum. Meth.* **A310**, 197 (1991).
138. M. Krammer, *Nucl. Instrum. Meth.* **A379**, 384 (1996).
139. J. Matheson et al., *Nucl. Instrum. Meth.* **A362**, 297 (1995).
140. M. Overdick et al., *Nucl. Instrum. Meth.* **A392**, 173 (1997).
141. T. Mouthuy, *Nucl. Instrum. Meth.* **A368**, 213 (1995).
142. P. Weilhammer et al., *Nucl. Instrum. Meth.* **A383**, 89 (1996).
143. W. T. Lin et al., *Nucl. Instrum. Meth.* **A389**, 415 (1997).
144. G. D. Haldwell, *Nucl. Instrum. Meth.* **A348**, 388 (1994).
145. G. Humpston and A. P. Needham, *Nucl. Instrum. Meth.* **A395**, 375 (1997).
146. G. Vanstraelen, I. Debusschere, C. Claeys, and G. Declerck, *Nucl. Instrum. Meth.* **A273**, 625 (1988).
147. E. Gatti, P. Rehak, and J. T. Walton, *Nucl. Instrum. Meth.* **226**, 129 (1984).
148. G. Gramegna et al., *IEEE Trans. Nucl. Sci.* **42**(5), 1497 (1995).
149. P. Rehak et al., *Nucl. Instrum. Meth.* **A248**, 367 (1986).
150. P. N. Luke, N. W. Madden, and F. S. Goulding, *IEEE Trans. Nucl. Sci.* **NS-32**(1), 457 (1985).
151. P. Lechner et al., *Nucl. Instrum. Meth.* **A377**, 346 (1996).
152. K. J. Rawlings, *Nucl. Instrum. Meth.* **A253**, 85 (1986).
153. J. Kemmer et al., *Nucl. Instrum. Meth.* **A253**, 378 (1987).
154. C. J. S. Damerell, *Rev. Sci. Instrum.* **69**(4), 1549 (1998).
155. R. P. Kraft et al., *Nucl. Instrum. Meth.* **A361**, 372 (1995).
156. L. Strüder et al., *SPIE* **982**, 129 (1988).
157. H. Soltau et al., *Nucl. Instrum. Meth.* **A377**, 340 (1996).
158. N. Meidinge, L. Strüder, H. Soltau, and C. V. Zanthier, *IEEE Trans. Nucl. Sci.* **42**(6), 2066 (1995).
159. D. H. Lumb and J. A. Nousek, *IEEE Trans. Nucl. Sci.* **39**(5), 1379 (1992).
160. A. D. Holland, *Nucl. Instrum. Meth.* **A377**, 334 (1996).
161. J. H. MacDonald, K. Wells, A. J. Reader, and R. J. Ott, *Nucl. Instrum. Meth.* **A392**, 220 (1997).
162. A. Castoldi and C. Fiorini, *Nucl. Instrum. Meth.* **A397**, 332 (1997).
163. A. Castoldi et al., *Nucl. Instrum. Meth.* **A409**, 379 (1998).

Chapter **14**

Slow Neutron Detection Methods

Neutrons are generally detected through nuclear reactions that result in prompt energetic charged particles such as protons, alpha particles, and so on. Virtually every type of neutron detector involves the combination of a target material designed to carry out this conversion together with one of the conventional radiation detectors discussed in earlier chapters. Because the cross section for neutron interactions in most materials is a strong function of neutron energy, rather different techniques have been developed for neutron detection in different energy regions. In this chapter, we discuss those methods that are of primary importance for the detection of neutrons whose energy is below the *cadmium cutoff* of about 0.5 eV. This is conventionally called the *slow neutron region* and is distinguished from intermediate and fast neutrons with energies above this value. Slow neutrons are of particular significance in present-day nuclear reactors and much of the instrumentation that has been developed for this energy region is aimed at the measurement of reactor neutron flux. Specific detector types that have evolved for this purpose are discussed at the end of this chapter.

We limit our discussion in this chapter to those methods that are intended to indicate only the detection of a neutron, with no attempt made to measure its kinetic energy. Devices that can measure slow neutron energies such as crystal spectrometers or mechanical monochromators are generally complex research-oriented instrumentation systems and are not covered here. In contrast, rather simple detectors can be used to measure the energy of neutrons of higher energy and these are discussed in the next chapter.

We also postpone discussion of *passive* neutron detectors, including activation foils, until Chapter 19. In this chapter we discuss only the common *active* detectors in which a pulse or current signal is produced by each neutron as it interacts in the device. General reviews of slow neutron detection devices and techniques may be found in Refs. 1–5. More detailed descriptions of those detectors developed specifically for reactor applications are given in Refs. 6 and 7.

I. NUCLEAR REACTIONS OF INTEREST IN NEUTRON DETECTION

In searching for nuclear reactions that might be useful in neutron detection, several factors must be considered. First, the cross section for the reaction must be as large as possible so that efficient detectors can be built with small dimensions. This is particularly important for detectors in which the target material is incorporated as a gas, of which we shall see several examples. For the same reason, the target nuclide should either be of high isotopic abundance in the natural element, or alternatively, an economic source of artificially enriched samples should be available for detector fabrication. In many applications, intense fields of gamma rays are also found with neutrons and the choice of reaction bears on the ability to discriminate against these gamma rays in the detection process. Of principal importance here is the Q-value of the reaction that determines the energy liberated in the reaction following neutron capture. The higher the Q-value, the greater is the energy given to the reaction products, and the easier is the task of discriminating against gamma-ray events using simple amplitude discrimination.

It is important to point out that all the common reactions used to detect slow neutrons result in heavy charged particles.[†] Possible reaction products are listed below:

$$\text{target nucleus} + \text{neutron} \rightarrow \begin{cases} \text{recoil nucleus} \\ \text{proton} \\ \text{alpha particle} \\ \text{fission fragments} \end{cases}$$

All the conversion reactions are sufficiently exothermic so that the kinetic energy of the reaction products is determined solely by the Q-value of the reaction and does not reflect the very small incoming energy of the slow neutron.

The distance traveled by the reaction products following their formation also has important consequences in detector design. If we are to capture the full kinetic energy of these products, the detector must be designed with an active volume that is large enough to fully stop the particles. If the detection medium is a solid, this requirement is easily achieved because the range of any of the reaction products shown does not exceed a few tenths of a millimeter in any solid material. If the detection medium is a gas, however, ranges of the reaction products (typically several centimeters) can be significant compared with detector dimensions and some may not deposit all their energy. If the detector is large enough so that these losses can be neglected, the response function will be very simple, consisting only of a single full-energy peak as shown in the sketch.

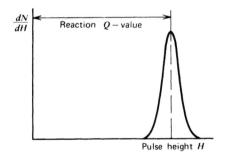

Under these circumstances the detector would exhibit a very flat counting plateau, and the ability to discriminate against low-amplitude events (such as gamma-ray-induced processes) would be maximized. If, on the other hand, a significant number of neutron-induced events do not deposit the full energy, a low-energy continuum is added to the pulse height distribution and the detector performance with respect to these criteria will suffer.

[†]Gamma rays produced by neutron capture are used in some specialized detectors, but these applications are relatively rare.

A. The ^{10}B(n, α) Reaction

Probably the most popular reaction for the conversion of slow neutrons into directly detectable particles is the ^{10}B(n, α) reaction. The reaction may be written

$$
{}^{10}_{5}\text{B} + {}^{1}_{0}\text{n} \rightarrow
\begin{cases}
{}^{7}_{3}\text{Li} + {}^{4}_{2}\alpha & 2.792 \text{ MeV (ground state)} \\
{}^{7}_{3}\text{Li*} + {}^{4}_{2}\alpha & 2.310 \text{ MeV (excited state)}
\end{cases}
\qquad (14.1)
$$

where the branching indicates that the reaction product ^{7}Li may be left either in its ground state or in its first excited state.[†] When thermal neutrons (0.025 eV) are used to induce the reaction, about 94% of all reactions lead to the excited state and only 6% directly to the ground state. In either case, the Q-value of the reaction is very large (2.310 or 2.792 MeV) compared with the incoming energy of the slow neutron, so that the energy imparted to the reaction products (^{7}Li and α) is essentially just the Q-value itself. Thus, the incoming kinetic energy of the neutron is submerged in the much larger reaction energy, and it is impossible to extract any information about its original value. Also, because the incoming linear momentum is very small, the reaction products must also show a net momentum of essentially zero. Consequently, the two reaction products must be emitted in exactly opposite directions, and the energy of the reaction will always be shared in the same manner between them. Individual energies of the alpha particle and lithium nucleus can be calculated simply by conservation of energy and momentum as follows:

$$E_{\text{Li}} + E_{\alpha} = Q = 2.31 \text{ MeV} \qquad (14.2)$$

$$
\begin{aligned}
m_{\text{Li}}v_{\text{Li}} &= m_{\alpha}v_{\alpha} \\
\sqrt{2m_{\text{Li}}E_{\text{Li}}} &= \sqrt{2m_{\alpha}E_{\alpha}}
\end{aligned}
\qquad (14.3)
$$

Solving Eqs. (14.2) and (14.3) simultaneously:

$$E_{\text{Li}} = 0.84 \text{ MeV} \quad \text{and} \quad E_{\alpha} = 1.47 \text{ MeV}$$

where the calculation has been carried out for the case of populating the excited state of ^{7}Li. A similar calculation would yield larger values by 21% for reactions leading to the ground state.

Figure 14.1 is a plot of cross sections versus neutron energy for a number of nuclear reactions of interest in neutron detection. The thermal neutron cross section for the ^{10}B(n, α) reaction is 3840 barns. The cross-section value drops rapidly with increasing neutron energy and is proportional to $1/v$ (the reciprocal of the neutron velocity) over much of the range. The utility of this reaction stems from its rather large and structureless cross section and from the fact that boron, highly enriched in its ^{10}B concentration, is readily available. The natural isotopic abundance of ^{10}B is 19.8%.

B. The ^{6}Li(n, α) Reaction

Another popular reaction for the detection of slow neutrons is the (n, α) reaction in ^{6}Li. Here the reaction proceeds only to the ground state of the product and is written simply as

$$
{}^{6}_{3}\text{Li} + {}^{1}_{0}\text{n} \rightarrow {}^{3}_{1}\text{H} + {}^{4}_{2}\alpha \qquad \begin{array}{c} Q\text{-value} \\ \overline{} \\ 4.78 \text{ MeV} \end{array}
$$

[†]The excited lithium nucleus quickly returns (half-life of $\sim 10^{-13}$ s) to its ground state with the emission of a 0.48 MeV gamma ray. We assume that this photon always escapes and does not contribute to the response of the detector.

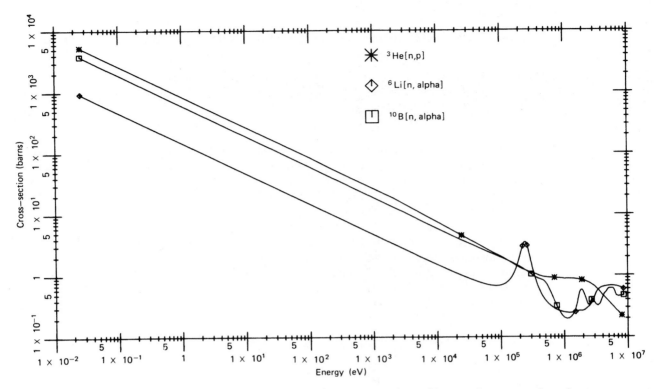

Figure 14.1 Cross section versus neutron energy for some reactions of interest in neutron detection.

Calculation of the reaction product energies for negligible incoming neutron energy yields the following:

$$E_{^3\text{H}} = 2.73 \text{ MeV} \quad \text{and} \quad E_{\alpha} = 2.05 \text{ MeV}$$

The alpha particle and triton produced in the reaction must be oppositely directed when the incoming neutron energy is low.

The thermal neutron cross section for this reaction is 940 barns. Figure 14.1 shows that the cross section remains below that for the ^{10}B reaction until the resonance region (> 100 keV). The lower cross section is generally a disadvantage but is partially offset by the higher Q-value and resulting greater energy given to the reaction products. ^{6}Li occurs with a natural isotopic abundance of 7.40% and is also widely available in separated form.

C. The ^{3}He(n, p) Reaction

The gas ^{3}He is also widely used as a detection medium for neutrons through the reaction

$$\underset{0.764 \text{ MeV}}{\overset{Q\text{-value}}{{}^3_2\text{He} + {}^1_0\text{n} \rightarrow {}^3_1\text{H} + {}^1_1\text{p}}}$$

For reactions induced by slow neutrons, the Q-value of 764 keV leads to oppositely directed reaction products with energies

$$E_{\text{p}} = 0.573 \text{ MeV} \quad \text{and} \quad E_{^3\text{H}} = 0.191 \text{ MeV}$$

The thermal neutron cross section for this reaction is 5330 barns, significantly higher than that for the boron reaction, and its value also falls off with a $1/v$ energy dependence (see Fig. 14.1). Although ^{3}He is commercially available, its relatively high cost is a factor in some applications.

D. The Gadolinium Neutron Capture Reaction

The cross section for thermal neutron capture of 255,000 barns in ^{157}Gd is among the highest nuclear cross sections found in any material. The isotope is 15.7% abundant in natural gadolinium, and neutron absorption results in an assortment of prompt reaction products that include gamma rays and conversion electrons. Because they are directly ionizing, it is the fast electrons that are the useful product in the application of this reaction for neutron detection and imaging. The most significant of these is a 72 keV conversion electron that is emitted in 39% of the capture reactions. Its range in typical Gd-containing layers is about 20 μm, so deposits of roughly this thickness can be employed as converters of incident neutrons into fast electrons that can be recorded in an adjacent detector. The conversion efficiency can be as high as 30% in such a layer, much larger than the 1% and 3–4% more typical of ^{6}Li- and ^{10}B-containing foils, respectively.[8] This reaction is frequently applied in neutron imaging, where the conversion electron serves to record the position of interaction of the neutron in an adjacent photographic film or other position-sensitive detector.

Liquid scintillators can also contain gadolinium in solution, typically at about 0.5% concentration, to make them highly sensitive to neutrons. Because the reaction products are fast electrons and gamma rays, gamma-ray backgrounds are more of a problem than for other conversion reactions in which heavily charged particles are produced and pulse shape discrimination techniques can be employed.

E. Neutron-Induced Fission Reactions

The fission cross sections of ^{233}U, ^{235}U, and ^{239}Pu are relatively large at low neutron energies and thus these materials can be used as the basis of slow neutron detectors. One characteristic of the fission reaction is its extremely large Q-value (approximately 200 MeV) compared with the reactions discussed previously. As a result, detectors based on the fission reaction can often give output pulses that are orders of magnitude larger than those induced from competing reactions or incident gamma rays, and very clean discrimination can be accomplished. Figure 14.2 shows a plot of fission cross sections of a variety of fissile nuclides, including some that are of primary use as fast neutron detectors. Almost all fissile nuclides are naturally alpha radioactive and consequently any detector that incorporates these materials will also show a spontaneous output signal due to decay alpha particles. The energy of the alpha particles, however, is always many times less than the energy given off in a fission reaction, and again these events can usually be discriminated easily on a pulse amplitude basis.

II. DETECTORS BASED ON THE BORON REACTION

A widely used detector for slow neutrons is the BF_3 proportional tube. In this device, boron trifluoride serves both as the target for slow neutron conversion into secondary particles as well as a proportional gas. A number of other boron-containing gases have been evaluated, but BF_3 is the near-universal choice because of its superior properties as a proportional gas, as well as its high concentration of boron. In nearly all commercial detectors, the gas is highly enriched in ^{10}B, resulting in an efficiency some five times greater than if the gas contained naturally occurring boron. Because the performance of BF_3 as a proportional gas is poor when operated at higher pressures, its absolute pressure in typical tubes is limited to about 0.5–1.0 atm.

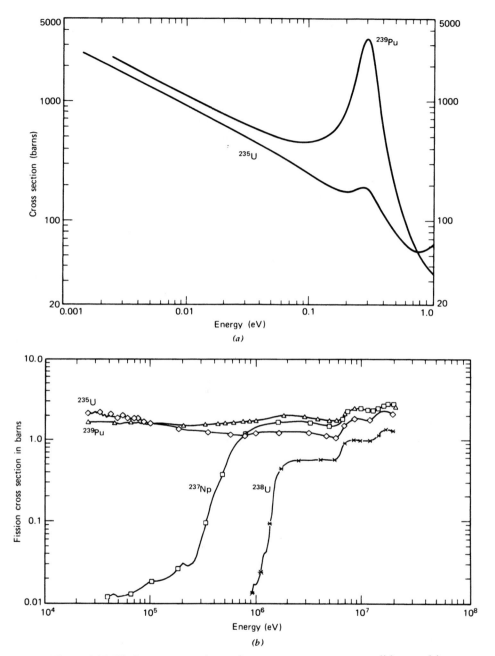

Figure 14.2 Fission cross sections of some common target nuclides used in fission chambers. (*a*) Slow neutron region where the cross sections shown are relatively large. (*b*) Fast neutron region. Chambers with ^{237}Np or ^{238}U are used as *threshold detectors* sensitive only to fast neutrons.

A. BF₃ Tube Pulse Height Spectra—The Wall Effect

Figure 14.3*a* shows the ideal pulse height spectrum expected from a BF$_3$ tube of very large dimensions. For a large tube, nearly all the reactions occur sufficiently far from the walls of the detector to deposit the full energy of the products within the proportional gas. In that event, all the energy of the reaction is deposited in the detector and the only variation is a result of the branching of the reaction between the excited state and ground state of the ^{7}Li product nucleus. The branching ratio for thermal neutrons is such that about 6% of the reactions lead to the ground state and 94% to the first excited state. Therefore, the areas under the peaks shown in Fig. 14.3*a* should be in the ratio 94:6 as illustrated.

Once the size of the tube is no longer large compared with the range of the alpha particle and recoil lithium nucleus produced in the reaction, some events no longer deposit the full reaction energy in the gas. If either particle strikes the chamber wall, a smaller pulse is produced. The cumulative effect of this type of process is known as the *wall effect* in gas counters. Because the range of the alpha particle produced in the reaction is on the order of 1 cm for typical BF$_3$ gas pressures, almost all practical tubes are small enough in diameter so that the wall effect is significant.

Figure 14.3*b* shows the differential pulse height spectrum expected from a tube in which the wall effect is important. The primary change from the spectrum shown in Fig. 14.3*a* is the addition of a continuum to the left of the peaks corresponding to partial energy deposition in the gas of the tube. The two steps or discontinuities in the continuum are an interesting feature of the spectrum and can be explained through the following argument.

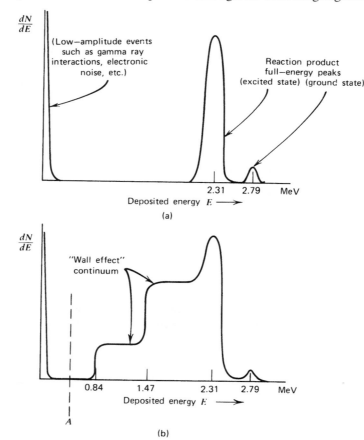

Figure 14.3 Expected pulse height spectra from BF$_3$ tubes. (*a*) Spectrum from a large tube in which all reaction products are fully absorbed. (*b*) Additional continuum due to the wall effect.

Because the incoming neutron carries no appreciable momentum, the two reaction products must be oppositely directed. If the alpha particle strikes the wall, the ^{7}Li recoil is therefore directed away from the wall and is very likely to deposit its full energy within the gas. Conversely, if the ^{7}Li recoil strikes a wall, the entire energy of the alpha particle from that same reaction is usually fully absorbed.

Thus, we expect to see wall losses for only one reaction product at a time. There are two possibilities: (1) the alpha particle hits a wall after depositing some fraction of its energy in the fill gas, whereas the ^{7}Li recoil is fully absorbed in the gas, or (2) the ^{7}Li recoil hits a wall after depositing part of its energy and the alpha particle is fully absorbed. Under case 1 above, the reaction could occur at a distance from the wall that might be anywhere between zero and the full alpha particle range. The amount of energy deposited in the gas can correspondingly vary from $(E_{Li} + 0)$ to $(E_{Li} + E_\alpha)$, as illustrated below.

Because all locations of the reaction are more or less equally probable, the distribution of deposited energy will be approximately uniform between these two extremes.

Parallel arguments can be made for case 2 to show that the energy deposited in the gas will vary from $(E_\alpha + 0)$ to $(E_\alpha + E_{Li})$.

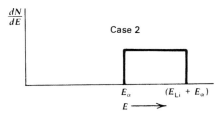

The combined energy deposition distribution of all events in which either reaction product strikes a wall will simply be the sum of the two cases.

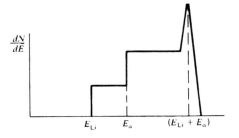

In addition to the wall effect events, the sketch above also shows the location of the full-energy peak that results from all those reactions from which both products are fully absorbed in the gas. The wall effect continuum extends from E_{Li}(0.84 MeV) up to the full-energy peak at $(E_{Li} + E_\alpha)$ (2.31 MeV). We have considered only those reactions leading to the ^{7}Li excited state because the wall effect continuum associated with the much less probable ground state is normally so small as to be submerged by the remainder of the spectrum.

The BF$_3$ tube is an example of a detector from which the differential pulse height spectrum tells us nothing about the energy spectrum of the incident radiation but is a function only of the size and geometry of the detector itself. In routine applications, there is consequently no motivation to record the pulse height spectrum from a BF$_3$ tube other than in an indirect manner. Instead, we are likely to seek a stable operating point or a counting plateau for which small drifts in operating parameters do not significantly affect the neutron sensitivity of the counter. That objective would be met by setting a fixed discriminator level at the point labeled A in Fig. 14.3b, which is below the minimum possible pulse amplitude from a neutron-induced reaction. From the arguments given in Chapter 4, we would expect a counting plateau to appear as the high voltage to the tube is varied, changing the internal gain of the proportional gas multiplication process. The flattest portion of that plateau should occur when the effective discrimination point is at the minimum in the differential pulse height spectrum, or point A. Under these conditions, all the neutrons will be counted, whereas low-amplitude events will be rejected. If the wall effect is eliminated or greatly suppressed by making the tube very large, a pulse height spectrum similar to that shown in Fig. 14.3a results. The counting plateau will then extend over a much greater range of applied voltage and thereby extend the useful operating range over which all neutron interactions are counted. A rather complete sample of spectra from a variety of BF$_3$ tubes is presented by Anderson and Malmskog.[9] A theoretical model has been developed by Cervellati and Kazimierski for the expected pulse height distribution from a BF$_3$ tube and is compared with experimentally measured distributions.[10]

B. BF$_3$ Tube Construction

The neutron detection efficiency can be increased and the wall effect suppressed by making the tube larger in dimension. Similar improvements can be achieved by raising the pressure of the BF$_3$ fill gas. Fowler[11] has reported the successful construction and operation of BF$_3$ tubes with diameter up to 15 cm and 180 cm long. Filling pressure ranged from 100 to 600 torr (approximately 13–80 kPa). Pressures in the range 200–300 torr (approximately 27–40 kPa) gave the best resolution in this work, whereas the full-energy peaks in the spectrum broadened considerably at higher pressure due to recombination and negative ion formation. In many counting situations, the poorer resolution is of no real consequence, and tubes with the higher gas pressure would be quite acceptable as long as a distinct counting plateau is maintained. Small-diameter tubes filled to several atmospheres pressure are commercially available, although pressures in the range 500–600 torr (approximately 67–80 kPa) are much more common.

In common with most proportional counters, BF$_3$ tubes are universally constructed using cylindrical outer cathodes and small-diameter central wire anodes. Aluminum is often the material of choice for the cathodes because of its low neutron interaction cross section. For low background applications, other materials such as stainless steel are preferred because aluminum normally shows a small amount of low-level alpha activity. With typical anode diameters of 0.1 mm or less, operating voltages tend to be about 2000–3000 V. Larger-diameter anode wires and/or higher fill gas pressures require higher applied voltages. Typical gas multiplication at operating voltage is on the order of 100–500.

BF$_3$ tubes of typical construction are normally limited to operating temperatures up to about 100°C, but tubes of special design can extend the operating range to as high as 150°C. However, the pulse amplitude decreases and the pulse height resolution decreases sharply[12] when operated well above room temperature. These changes may be related to the possible desorption of impurities from the counter wall or other components at elevated temperatures.

Because of the relatively high operating voltages, BF_3 tubes share some temperamental qualities with other proportional counters. Spurious pulses of about the same size as signal pulses can sometimes arise from fluctuations in leakage currents through insulators, especially under conditions of high humidity. Spurious counts can also arise in applications in which the counter is subject to vibration or shock.[13] These effects are attributed to detector microphonics and the influence of small particles of lint or dirt within the counter.

In common with other proportional counters, BF_3 tubes show significant effects of aging. In some cases[14] significant degradation in the performance is observed after operation of 10^{10}–10^{11} registered counts. This degradation is likely related to the contamination of the anode wire and cathode wall by molecular disassociation products produced in the avalanches. The same study indicates that ^3He tubes described later in this chapter are more resistant to these effects.

C. Gamma-Ray Discrimination

A very important consideration in many applications of BF_3 tubes is their ability to discriminate against gamma rays, which often are found together with the neutron flux to be measured. Gamma rays interact primarily in the wall of the counter and create secondary electrons that may produce ionization in the gas. Because the stopping power for electrons in gases is quite low, a typical electron will deposit only a small fraction of its initial energy within the gas before reaching the opposite wall of the counter. Thus, we should expect that most gamma-ray interactions will result in low-amplitude pulses that will lie in the tail to the left of point A in Fig. 14.3b. Simple amplitude discrimination can then easily eliminate these gamma rays without sacrificing neutron detection efficiency.

If the gamma-ray flux is sufficiently high, however, several complications can reduce the effectiveness of this amplitude discrimination. At high rates, pulse pile-up can result in apparent peak amplitudes for gamma rays which are considerably larger than any individual pulse. Brown[15] discusses the compromise that must then be struck in choosing the pulse-shaping time constant in the detector electronics. Short time constants are desirable to reduce the gamma-ray pile-up but may lead to reduction in the neutron-induced pulse amplitude due to incomplete charge integration. At very high gamma rates, there is evidence that chemical changes occur in the BF_3 gas caused by molecular disassociation, leading to degraded pulse height spectra from neutron-induced events.[16] If this degradation is sufficiently severe, it may no longer be possible to separate gamma- and neutron-induced events.[17] In extreme cases, the radiation-induced chemical changes can result in permanent damage to the tube. Verghese et al.[18] report successful discrimination against gamma rays at exposure rates as high as 12 R/h using a conventional BF_3 tube. Developmental tubes that employ activated charcoal within the tube to act as an absorbing agent for contaminants have been reported.[19] These tubes exhibit good operating characteristics in gamma-ray fluxes up to 1000 R/h.

D. Detection Efficiency of a BF_3 Tube

The detection efficiency for neutrons incident along the axis of a BF_3 tube is given approximately by

$$\epsilon(E) = 1 - \exp\left[-\Sigma_a(E)L\right] \tag{14.4}$$

where $\qquad$ $\Sigma_a(E)$ = macroscopic absorption cross

section of ^{10}B at neutron energy E

L = active length of the tube

Using Eq. (14.4), we find that the calculated efficiency for a 30-cm long BF$_3$ tube (96% enriched in ^{10}B) filled to 600 torr (80 kPa) is 91.5% at thermal neutron energies (0.025 eV) but drops to 3.8% at 100 eV. Thus, a BF$_3$ tube exposed to neutrons with mixed energies will respond principally to the slow neutron component. Equation (14.4) slightly overestimates the neutron counting efficiency because there usually are regions near the end of the tube in which charge collection is inefficient, resulting in reduced neutron response. The influence of these *dead spaces* is most severe for detectors whose length is small and has been the subject of experimental investigations that lead to a more precise prediction of detector efficiency.[20] *End window* designs are common in which the dead space and structural materials at one end of the tube are minimized.

Most practical BF$_3$ counters are filled with pure boron trifluoride enriched to about 96% in ^{10}B. However, because BF$_3$ is not ideal as a proportional counter gas, counters are sometimes manufactured using BF$_3$ with an admixture of a more suitable gas such as argon. This dilution causes a decrease in detection efficiency, but the pulse height spectrum from the tube generally shows sharper peaks and consequently a more stable counting plateau than tubes filled with pure BF$_3$.

E. Boron-Lined Proportional Counters

An alternate approach is to introduce the boron in the form of a solid coating on the interior walls of an otherwise conventional proportional tube. This configuration has the advantage that a more suitable proportional gas than BF$_3$ can now be used. Some applications, particularly those in which fast timing is important, are better served by introducing one of the common proportional gases discussed in Chapter 6. Also, the chemical degradation problems in BF$_3$ when exposed to high gamma ray fluxes can be greatly reduced by using alternative fill gases.

Because the maximum range of the alpha particles from the boron reaction is on the order of 1 mg/cm^2, the efficiency of boron-lined counters will improve only as the coating thickness is increased to about this value. Making the deposit any thicker will simply create layers in the coating that are too far from the filling gas to permit any reaction products to reach the gas, and the efficiency will actually begin to decrease slightly because of the added attenuation of the incident neutrons. Efforts have been made to increase the surface area available for coating by introducing boron-coated plates or baffles within cylindrical tubes, but these configurations have not achieved widespread popularity.

The pulse height spectrum to be expected from a boron-lined proportional chamber with a thick boron layer is sketched in Fig. 14.4. Because the interactions are now taking place in the wall of the chamber and the reaction products are oppositely directed, only one reaction product can be expected per interaction. If the alpha particle is directed toward the interior of the tube, the maximum energy it can deposit is its initial kinetic energy of 1.47 MeV. The actual alpha particle energy deposited in the gas will vary from this value down to zero as the possible location of the neutron interaction varies from the surface of the boron coating through those locations that are more than an alpha range away from the counter gas. Because all these locations are almost equally probable, the expected energy deposition distribution for alpha particles will be approximately rectangular in shape with a maximum at 1.47 MeV. This distribution is sketched in Fig. 14.4a as a dashed rectangle. A parallel argument can be made for the lithium recoil nucleus, with its maximum possible

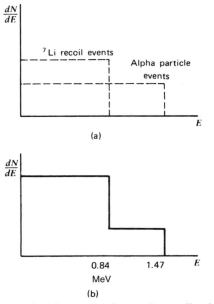

Figure 14.4 Idealized pulse height spectra from a boron-lined proportional tube. (*a*) Separate contributions of alpha particles and lithium recoil nuclei, which add to give the spectrum shown in plot (*b*).

deposited energy equal to 0.84 MeV. The sum of these two individual rectangular distributions is shown as the solid line in Fig. 14.4*b* and is a somewhat idealized indication of the expected pulse height spectrum from a boron-lined chamber with a thick (greater than 1 mg/cm^2) boron lining. From the discussion given in Chapter 4, a differential pulse height spectrum without a "valley" structure does not lead to a counting plateau. Thus, it would be expected that boron-lined chambers would be less satisfactory than BF_3 tubes in terms of long-term counting stability. Because the average energy deposited for neutron interaction is also considerably less than for BF_3 tubes, the gamma-ray discrimination ability of boron-lined chambers will be inferior to that of BF_3 tubes.

F. Boron-Loaded Scintillators

Because the output pulse from a BF_3 tube originates with reaction products created with a more-or-less random location and direction, typical pulses will have rise times that vary by as much as 3–5 μs for tubes of average size. A further disadvantage for neutron time-of-flight applications is that the point of interaction of the neutron cannot be defined more precisely than somewhere within the volume occupied by the BF_3 fill gas. Because typical tubes are as much as 10–20 cm long to provide reasonable interaction efficiency, pathlength uncertainties can be large.

In order to circumvent both of the above limitations, other types of boron-loaded detectors have been investigated. Scintillators made by fusing B_2O_3 and ZnS have found wide application in neutron time-of-flight measurements. These scintillators are usually kept quite thin (1–2 mm) because of the relative opaqueness of this material to its own scintillation light and also to minimize pathlength uncertainty. Boron-loaded plastic scintillators of large sizes are also available commercially containing up to 5% boron content, with a light output that is about 75% of that of the standard plastic. Higher boron concentrations are possible, but result in a further drop in the light yield.[21] These scintillators are much less effective at discrimination against gamma-ray backgrounds compared with BF_3 proportional tubes for several reasons. The secondary electrons from gamma-ray interactions tend to deposit all their energy in the solid scintillator, but they can only deposit a small fraction in the BF_3 gas. Furthermore, organic scintillators produce much more light per unit energy from electrons than from the heavy charged particles produced by neutrons,[†] while the ionization yield in gases is very nearly the same for each. Thus the inherent separation in amplitude between gamma-ray and neutron pulses from the BF_3 tube is no longer observed from the boron-loaded plastic scintillator. An alternative is to use a boron-loaded liquid scintillator that is commercially available with 4.4% boron enriched to 90% in ^{10}B. Then the pulse shape discrimination methods discussed in Chapter 8 that are generally applicable to liquid organic scintillators can be implemented[23] to discard most of the gamma-ray pulses while retaining those that are due to neutrons.

III. DETECTORS BASED ON OTHER CONVERSION REACTIONS

A. Lithium-Containing Slow Neutron Detectors

Because a stable lithium-containing proportional gas does not exist, a lithium equivalent of the BF_3 tube is not available. Nonetheless, the larger Q-value of the lithium reaction offers some real advantage whenever discrimination against gamma-ray pile-up and other low-amplitude events is at a premium. Also, because the lithium reaction goes exclusively to the ground state of the product nucleus, the same energy is always imparted to the reaction products for all slow neutron interactions. The resulting pulse height distribution in detectors that absorb all this energy is therefore a simple single peak. Although examples can be found[24] of the application of gas-filled detectors with solid lithium-based converters, the more common applications of this reaction employ the scintillation process to detect the products of the neutron-induced reaction.

Lithium-containing scintillators are quite common as slow neutron detectors. A logical choice, because of its chemical similarity to sodium iodide, is crystalline lithium iodide. If a small amount (less than 0.1 at. %) of europium is incorporated as an activator, light outputs of about 35% of the equivalent NaI(Tl) yield can be achieved. The scintillation mechanism is similar to that discussed earlier for sodium iodide. The scintillation decay time is approximately 0.3 μs.

Crystals of lithium iodide are generally large compared with the ranges of either of the reaction products from a neutron interaction. Therefore, the pulse height response will be free of wall effects and should be a single peak for all slow neutron interactions. The range of secondary electrons produced by gamma rays will not be large compared with typical crystal dimensions. The scintillation efficiency for lithium iodide is nearly the same for both electrons and heavy charged particles. (A 4.1 MeV electron will yield about the same light as the 4.78 MeV reaction products.) Therefore, a single gamma-ray interaction in lithium iodide is capable of producing a maximum pulse height approximately proportional to the energy of the gamma ray, whereas each neutron interaction will produce a pulse height equivalent to 4.1 MeV on the same scale. The gamma-ray rejection characteristics will therefore be inferior to that of typical gas-filled neutron detectors, in which a gamma ray can deposit only a small fraction of its energy.

[†]For example, the light output of a typical boron-loaded plastic scintillator (BC-454) for the reaction products from neutrons is reported[22] to be only 4% of that for electrons of the same energy.

Similar to sodium iodide, lithium iodide is highly hygroscopic and cannot be exposed to water vapor. Commercially available crystals are hermetically sealed in a thin canning material with an optical window provided on one face. Because of the high density of the material, crystal sizes need not be large for very efficient slow neutron detection. For example, a 10-mm thick crystal prepared from highly enriched ^{6}LiI remains nearly 100% efficient for neutrons with energy from thermal through the cadmium cutoff of 0.5 eV.

Other recipes for lithium-containing scintillators have achieved some popularity. One of these[25] consists of a lithium compound dispersed in a matrix of ZnS(Ag) with thickness of about 0.6 mm. These scintillators are commercially available and their efficiency is quoted as 25–30% for 0.1 eV neutrons. Thin layer combinations of LiF and ZnS are described in Refs. 26 and 27. Because of their small thickness, gamma discrimination is very effective since a large fraction of all secondary electrons created by gamma-ray interactions will escape without depositing their full energy.

Liquid scintillator loaded with ^{6}Li to 0.15% concentration is also available commercially.[28] This scintillator has a good light yield and transmission that make it suitable for use in large volumes. In common with other liquid scintillators, it permits pulse shape discrimination (see Chapter 8) against gamma rays while preserving nearly all of the neutron-induced pulses.

Lithium-containing glass scintillators have seen widespread application for the detection of fast neutrons, and a discussion of their principal properties is therefore postponed until Chapter 15. However, a unique form of these glass scintillators has emerged as a very useful detector of slow neutrons. These are the lithium-containing glass fibers that have been shown to provide detectable scintillation light from the interaction of a single neutron at the end of small-diameter fibers that are as long as several meters. This capability provides a flexibility in the design of slow neutron detectors that cannot be matched in other devices. Further discussion will be found beginning on p. 548.

B. The ^{3}He Proportional Counter

With a cross section even higher than that of the boron reaction, the ^{3}He(n, p) reaction is an attractive alternative for slow neutron detection. Unfortunately, because ^{3}He is a noble gas, no solid compounds can be fabricated and the material must be used in gaseous form.

^{3}He of sufficient purity will act as an acceptable proportional gas, and detectors based on this approach have come into common use. General properties of ^{3}He proportional tubes are surveyed in Ref. 29. In a large detector, one would expect each thermal neutron reaction to deposit 764 keV in the form of kinetic energy of the triton and proton reaction products. Because the range of these reaction products is not always small compared with the dimensions of the proportional tube, however, the wall effect discussed earlier for a BF$_3$ tube can also be important for ^{3}He proportional counters. The expected pulse height spectrum for a tube of typical size is illustrated in Fig. 14.5. Only a single full-energy peak should be expected for neutron energies that are small compared with 764 keV. The step structure to the left of the peak is similar to that shown in Fig. 14.3b for a BF$_3$ tube, except that the discontinuities will occur at energies corresponding to that of the proton (573 keV) and triton (191 keV).

The continuum in the pulse height spectrum due to the wall effect is detrimental from several standpoints. The voltage range over which an acceptable counting plateau will be observed is reduced, and the smaller pulse height for some neutron events will reduce the separation expected from low-amplitude, gamma-induced pulses. Consequently, consideration is often given in the design of ^{3}He tubes to minimize the wall effect. One obvious step is to build the counter with a diameter as large as possible so that most neutron interactions occur far away from the wall. Another is to increase the pressure of the ^{3}He gas to reduce the range of the charged particle reaction products. Because of the low atomic mass

Figure 14.5 Expected pulse height spectrum from a ^{3}He tube in which the wall effect is significant.

of ^{3}He, the ranges of the reaction products are unusually long and the wall effect is considerably more significant than for a BF$_3$ tube of the same size and fill gas pressure. One method of reducing the charged particle ranges is to add a small amount of a heavier gas to the ^{3}He to provide an enhanced stopping power. A detailed analysis of the wall effect in ^{3}He counters can be found in Ref. 30.

The rise time of the output pulse observed from either a BF$_3$ tube or a ^{3}He tube will depend both on the position of the neutron interaction and the orientation of the charged particle tracks with respect to the tube axis. Since both of these quantities are randomly variable, substantial differences are observed in the rise time of the pulses. In the example[31] shown in Fig. 14.6, the time to reach about 80% of final amplitude ranged from 0.3–1.6 μs. These variations are the result of differences in the drift times of the electrons from their formation along the particle tracks to the anode. The significant charges that give rise to the output pulse are produced only in the avalanches around the anode wire. If the electrons all have the same drift time, then the avalanches are triggered at nearly the same time. If the electron arrivals are staggered because of different drift times, however, the

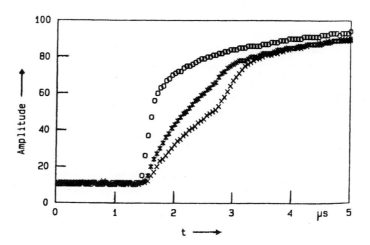

Figure 14.6 Variation in the digitized pulse leading edge from a ^{3}He tube due to differences in the orientation of the proton-triton tracks with respect to the anode wire. (From Dietz and Sosaat.[31])

avalanches are spread out in time and the pulse rise time is slower. In the same study, it was found that the full amplitude of the pulse was not realized until after about 50 μs. This slow component of the rise comes about because of the slow motion of the positive ions after they leave the immediate vicinity of the anode wire. In practical circumstances, a much shorter shaping time must be used to prevent pulse pile-up, and therefore conditions prevail in which there is a relatively large ballistic deficit. If the shaping times are chosen to be too small, the variability of the pulse rise time will also contribute to the ballistic deficit and will begin to broaden the peaks observed in the pulse height spectrum. If sufficiently extreme, these changes will reduce the length of the counting plateau and diminish the ability to discriminate against gamma rays.

Compared with BF_3 tubes, ^{3}He counters can be operated at much higher pressures with acceptable gas multiplication behavior and are therefore preferred for those applications in which maximum detection efficiency is important. The lower Q-value of the ^{3}He reaction, however, makes gamma-ray discrimination more difficult than for an equivalent BF_3 tube. When the gamma irradiation rate is high, the pile-up of the resulting pulses can raise their amplitudes to the point that a clean separation from the neutron-induced pulses is no longer possible. To minimize the gamma-ray pile-up, the choice of a gas additive such as CO_2 or Ar that speeds up the electron drift will allow the use of shorter shaping times in the pulse processing electronics. Other factors that influence the behavior of ^{3}He tubes in high gamma-ray environments include the choice of wall material and/or the use of an activated carbon coating on the inner tube wall to adsorb gas impurities.[32]

The acceptable operating temperature for ^{3}He tubes has been shown[33,34] to extend as high as 200–250°C. In general, the pulse amplitude increases and the pulse height resolution decreases with increasing temperature, while the pulse rise time shows little temperature dependence.

As with all proportional counters, purity of the gas is critical, and the most typical cause of failure is leakage of air into the tube over long periods of time. Another factor is the buildup of electronegative poisons in the gas with use. As in BF_3 tubes, a layer of activated charcoal within the tube has been shown to be effective in removing these poisons and can extend the useful lifetime of a ^{3}He detector.[35]

C. Fission Counters

The neutron-induced fission reaction can also serve as a means of converting a slow neutron into ionizing reaction products that can then be detected by conventional means. One outstanding characteristic of the fission reaction is the large amount of energy (200 MeV) liberated in the reaction, about 160 MeV of which appears as kinetic energy of the fission fragments. Therefore, the neutron-induced fission reactions can be expected to be of much larger magnitude in most detectors than any other competing reaction or other event due to background or counter contamination. Under these circumstances, extremely low background rates can be achieved and neutron counting can be practically carried out at very low counting rates. In some detector types (scintillators are often an extreme example) the output pulse does not always linearly reflect the energy of the exciting particle. Nonlinearities are particularly important for densely ionizing particles such as fission fragments, and therefore the size of the neutron-induced detector pulses may not be as large as would be calculated based on simple linearity with energy. In that event, the fission fragments still may be considerably larger than competing reactions, but not by the several orders of magnitude one might ordinarily expect.

The most popular form of fission detector is an ionization chamber that has its inner surfaces coated with a fissile deposit. Because of adverse chemical and physical properties, little success has been realized in trying to incorporate the fissile material in gaseous form as part of the counter gas. In this section we stress properties of fission chambers operated in pulse mode, which is typical of nonreactor applications. Fission chambers can also function in current mode, most often as reactor neutron flux monitors discussed later in this chapter.

The pulse height spectrum to be expected from a fission chamber depends primarily on the fissile deposit thickness and the geometric conditions under which the fission fragments are collected. For deposits that are very thin compared with fragment ranges, the familiar "double humped" fission fragment energy spectrum is observed, with the light and heavy fragment distributions peaking at about 100 and 70 MeV, respectively. If the deposit is made thicker to enhance detection efficiency, the energy loss of fragments within the deposit will reduce the average fragment energy and distort the shape of the measured distribution. Figure 14.7 illustrates the changes in spectrum shape to be expected as the thickness of a deposit of UO_2 is changed. These energy loss effects limit the practical deposit thickness to about 2–3 mg/cm². For a layer of highly enriched ^{235}U of this thickness, the corresponding detection efficiency in 2π counting geometry is about 0.5% at thermal energy, dropping to about 0.1% at 0.5 eV. Typical fission chambers employ a single layer, and are thus limited to an equivalent neutron detection efficiency. More complex fission chambers with higher detection efficiency can be designed,[37-40] by providing multiple layers of fissile deposits, and detecting the fragments in each segment of the chamber between the layers.

The dimensions of the counter tend to be similar to that of alpha particle detectors, because the average range of a fission fragment is approximately half the range of 5 MeV alpha particles. Furthermore, because the fission fragment starts out with a very large positive charge (on the order of 15 or 20 electronic charges) the energy loss at the beginning of the track is at its maximum, and the rate of energy loss continues to decrease as the fragment slows and additional charges are picked up. Just the reverse is true for most light charged particles, for which the rate of energy loss peaks near the end of the track. Therefore, in those detectors that do not fully stop the particle, fission fragments will yield a larger fraction of their total energy than would be expected for alpha particles or protons

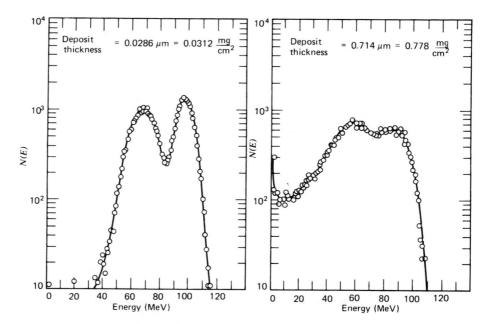

Figure 14.7 Energy spectra of fission fragments emerging from flat UO_2 deposits of two different thicknesses. A 2π detector is assumed which responds to fragments emitted in all directions from one surface of the deposit. (From Kahn et al.[36])

of the same range. For typical counter fill gases, the mean range is a few centimeters so that chambers of this dimension or larger can function reasonably as fission chambers.

The two fission fragments are always oppositely directed for slow-neutron-induced fission, and therefore detectors with a solid coating of fissionable material will respond only to the single fragment that is directed toward the active volume of the chamber. Some fission counters have been built with extremely thin backing material underneath a thin fissile deposit, so that both fragments can escape into opposite halves of a dual chamber, thereby permitting simultaneous detection of both fragments. Coincidence techniques can then discriminate against alpha particles and other background events. The very thin supports required for the fissile deposit are quite fragile, and consequently this type of fission chamber is not widely used in routine neutron detection applications.

The fissionable materials used in the construction of fission chambers are almost always alpha radioactive to some degree. As a result, typical fission chambers have an irreducible alpha-induced pulse rate that is an undesirable background. Because the average alpha particle energy is typically 5 MeV, whereas an average fission fragment energy is 10 times larger, these alpha pulses can often be simply discriminated on an amplitude basis. In those fissionable materials in which the alpha activity is relatively high (such as ^{239}Pu), problems often arise as a result of pile-up of alpha pulses to an amplitude that can overlap the lowest energy fission fragments. Because the maximum amplitude expected from pulse pile-up is an inverse function of detector resolving time, it can be suppressed through the use of fast fission detectors. Small methane-filled fission chambers have been reported[41] that show pulse widths as small as 10 ns. Extremely fast fission detectors can also be built by detecting fission fragments in a gaseous scintillator or by incorporating the fissionable material as part of fast solid scintillators.[42] An extensive review of various fission chamber designs may be found in Ref. 43.

IV. REACTOR INSTRUMENTATION

A. General Considerations

In thermal nuclear reactors, most of the power is generated through fission induced by slow neutrons. Therefore, nuclear sensors that are to be part of reactor control or safety systems are generally based on detectors that respond primarily to slow neutrons. In principle, many of the detector types discussed earlier in this chapter can be adapted for application to reactor measurements. However, the extreme conditions associated with reactor operation often lead to substantial design changes, and a category of slow neutron detectors designed specifically for this application has gradually evolved. Detailed reviews of reactor instrumentation are given in Refs. 6 and 7.

It is conventional to subdivide reactor instruments into two categories: *in-core* and *out-of-core*. In-core sensors are those that are located within narrow coolant channels in the reactor core and are used to provide detailed knowledge of the flux shape within the core. These sensors can be either fixed in one location or provided with a movable drive and must obviously be of rather small size (typically on the order of 10-mm diameter). Some examples are given in Ref. 44. Out-of-core detectors are located some distance from the core and thus respond to properties of the neutron flux integrated over the entire core. The detectors may be placed either inside or outside the pressure vessel and normally will be located in a much less severe environment compared with in-core detectors. Size restrictions are also less of a factor in their design.

The majority of neutron sensors for reactor use are of the gas-filled type. Their advantages in this application include the inherent gamma-ray discrimination properties found in any gas detector, their wide dynamic range and long-term stability, and their resistance to radiation damage. Detectors based on scintillation processes are less suitable because of the enhanced gamma-ray sensitivity of solid or liquid scintillators and the radiation-induced spurious events that occur in photomultiplier tubes. Semiconductor detectors are very sensitive to radiation damage and are never used in reactor environments.

Nearly all the gas-filled chambers, whether they are ionization or proportional counters, can be operated in a variety of modes. In pulse mode, each neutron interaction must be separated by sufficient time so that it may be resolved as an individual pulse. This mode is therefore limited to the lower ranges of neutron flux measurement but offers the benefit of gamma discrimination through simple amplitude selection of the output pulses. Pulse mode operation of most detectors is conventionally limited to rates below about 10^5 per second, although state-of-the-art techniques in chamber design and pulse-processing electronics can raise this limit to as high as 10^7 per second (see references cited in Ref. 45).

When flux levels become high enough so that pulse mode operation is no longer possible, neutron detectors are often operated in current mode. With proper ion chamber design, the range of operation can be extended to the maximum flux level of interest in reactors before serious nonlinear effects due to ion–electron recombination set in. Lower limits of current mode operation are usually determined by leakage currents that inevitably arise in the detector insulation and cable dielectric material. By operating the chamber in current mode, one sacrifices any chance of inherent gamma-ray discrimination because all pulses, whether large or small, add some contribution to the measured current.

One method of reducing gamma-ray sensitivity is to use the MSV mode of operation described in Chapter 4. This operational mode, commonly called the *Campbelling technique*, consists of deriving a signal that is proportional to the mean square of the current fluctuations from an ion chamber.[46–49] The mean square signal is seen from Eq. (4.8) to be proportional to the average pulse rate and the square of the ionization charge generated in each pulse. Because neutron-induced pulses result in much greater charge than pulses from gamma rays, the mean square signal will weight the neutron component by the square of the ratio of neutron- to gamma-ray-induced charge. This increase in neutron sensitivity is of greatest advantage in fission chambers. Although the MSV mode enhances the neutron sensitivity, it will not completely eliminate the gamma contribution. MSV mode operation has proved to be most useful in the intermediate reactor power ranges and in wide-range reactor control channels,[50–52] where one detector provides input to instrumentation that can operate in pulse mode, MSV mode, or current mode depending on the neutron flux level being measured.

A second approach for reducing the importance of the gamma-ray signal is to employ direct gamma-ray compensation in a specialized detector known as a compensated ion chamber (CIC). The CIC typically uses boron-lined ion chambers operating in current mode. Because of the much smaller Q-value of the neutron-induced reaction, neutron interactions in a boron-lined chamber result in an order of magnitude less charge than do neutron-induced events in a fission (uranium-lined) chamber. The effectiveness of the MSV mode of operation in discriminating against gamma rays is therefore reduced. The alternative approach of using a CIC is found to be more effective in reducing the gamma-ray contribution in boron-lined chambers than is the MSV mode of operation. The CIC, illustrated functionally in Fig. 14.8, employs a dual ion chamber from which two independent ion currents can be extracted separately. One chamber is boron-lined, whereas the construction of the second is nearly identical in terms of active volume and structural material, but without the boron lining. The current I_1 from the boron-lined chamber will consist of the sum of the current due to neutron interactions in the boron and the gamma-ray interactions in the chamber walls and gas-filled region. The current I_2 from the unlined chamber will reflect only the gamma-ray contribution. By taking the difference between these two currents, a signal current is derived which in principle will be proportional to only the neutron contribution. The two chambers are often constructed as concentric cylinders and consequently are not identical in their response to gamma rays. Therefore, the compensation is not exact and may vary slightly with gamma-ray flux. This variation may

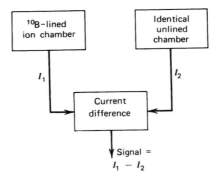

Figure 14.8 Operating principle of a compensated ion chamber (CIC).

require adjustments[53] in the balance between the two chambers to restore exact compensation for different gamma-ray flux levels. (See Refs. 6 and 7 for further discussion of CIC design and operation.)

B. Reactor Nuclear Instrument (NI) System Overview

Before proceeding with further discussion of the specialized detectors used in reactor safety and control systems, a general discussion of the distinctly different system approaches used in pressurized water reactors (PWR) and boiling water reactors (BWR) is instructive. The reader is referred to Ref. 52 for a more detailed description of reactor NI systems.

1. PRESSURIZED WATER REACTOR NUCLEAR INSTRUMENTATION

Detectors for the routine monitoring of reactor power in a PWR are located outside the reactor pressure vessel and are characterized by the following typical environmental conditions: neutron flux up to 10^{11} $n/cm^2 \cdot s$, gamma irradiation rates up to 10^6 R/h, and temperatures of approximately 100°C. Out-of-core sensors are the usual basis of reactor control and safety channels in a PWR. In choosing specific detector types, consideration must be given to the expected neutron signal level compared with noise sources, the speed of response of the detector, and the ability to discriminate against gamma-induced signals. Each of these criteria assumes different importance over various ranges of reactor power, and as a result multiple detector systems are usually provided, each designed to cover a specific subset of the power range.

Figure 14.9 illustrates a typical scheme for a PWR in which three sets of sensors with overlapping operating ranges are used to cover the entire power range of the reactor. The lowest range, usually called the *source start-up range*, is encountered first when bringing up reactor power from shut-down conditions. This range is characterized by conditions in

Figure 14.9 Typical ranges covered by out-of-core neutron detectors in a PWR.

which the gamma flux from the fission product inventory in the core may be large compared with the small neutron flux at these low power levels. Under these conditions, good discrimination against gamma rays is at a premium. Also, the expected neutron interaction rates will be relatively low in this range. Pulse mode operation of either fission chambers or BF_3 proportional counters is therefore possible, and the required gamma-ray discrimination can be accomplished by accepting only the much larger amplitude neutron pulses.

As the power level is increased, an intermediate range is encountered in which pulse mode operation is no longer possible because of the excessive neutron interaction rate. In this region the gamma-ray-induced events are still significant compared with the neutron flux, and therefore simple current mode operation is not suitable. The MSV mode of operation can reduce the importance of the gamma-ray signal in this range, but a more common method used in PWRs is to employ direct gamma-ray compensation using a CIC.

A third range of operation corresponds to the region near the full operating power of the reactor. The neutron flux here is usually so large that gamma-ray-induced currents in ion chambers are no longer significant, and simple uncompensated ion chambers are commonly used as the principal neutron sensor. Because these instruments are often part of the reactor safety system, there is a premium on simplicity that also favors uncompensated ion chamber construction.

In most PWR NI systems the geometric arrangement shown in Fig. 14.10 is employed. Two BF_3 proportional counters used in the source start-up range are placed on opposite sides of the core and two CIC detectors used in the intermediate range are placed in the same location or on the two opposing sides. Four power range monitors are then located at 90° intervals at positions between the BF_3 and CIC detectors. Each of the four power range monitors consists of two uncompensated ion chambers arranged end-to-end, resulting in a total detector length of 3–4 m. This arrangement provides both radial and axial neutron flux data for control and safety at full power as well as axial flux offset information needed for control of xenon oscillations.

Although PWR nuclear instrumentation employs out-of-core gas-filled detectors (primarily ion chambers) for control and safety channels, there remains a need for information on in-core spatial variations of the neutron flux. This information is necessary for fuel management and is provided by various types of detectors placed within the reactor core (see following descriptions).

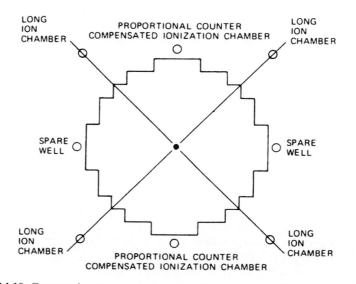

Figure 14.10 Geometric arrangement of out-of-core neutron detectors relative to the core of a PWR. (From Harrer and Beckerley.[52])

Figure 14.11 Typical ranges covered by in-core neutron detectors in a BWR.

2. BOILING WATER REACTOR NUCLEAR INSTRUMENTATION

The BWR NI system, like the PWR system, has three overlapping ranges as illustrated in Fig. 14.11. The three systems are called source, intermediate, and power range monitors. Unlike the PWR, which uses out-of-core neutron detectors, the neutron detectors are all located in-core. There are also many more detectors used in the BWR NI system than in the PWR system.

The source range monitoring system typically consists of four in-core fission chambers operating in pulse mode. Pulse mode operation provides good discrimination against gamma rays, which is necessary when measuring a relatively low neutron flux in the presence of a high gamma flux. A typical intermediate range monitoring system has eight in-core fission chambers operating in the MSV mode. The MSV mode promotes the enhanced neutron to gamma response required to provide a proper measure of neutron flux in the presence of gamma rays for both control and safety requirements. The power range monitoring system typically consists of 144–164 fission ion chambers distributed throughout the core. The fission chambers operate in current mode and are called *local power range monitors (LPRM)*. Current mode operation provides satisfactory neutron response at the high flux levels encountered between 2 and 150% full power. In a typical system, approximately 20 LPRMs are summed to provide input to one of the seven or eight average power range monitoring (APRM) systems. The APRM system provides input for both control and reactor protection systems. Details of the neutron monitoring systems used in the BWR are described in Ref. 52.

A diagram of a typical fission chamber used in BWR NI is shown in Fig. 14.12. The physical sizes of the detectors used in each range are similar. The fill-gas pressures and uranium loadings are different, depending on whether the detector is designed to operate in pulse, MSV, or current mode. The LPRM system typically employs a fission chamber with a mixture of both ^{235}U and ^{234}U. The advantage of this design for extended operation in a high neutron flux is described in the next section.

C. In-Core Detectors

There is often a need to place neutron sensors within the core of a nuclear reactor to provide information on the spatial variation of the neutron flux. Because of the small size (1–7 cm) of the channel in which these instruments must be located, emphasis is placed on compactness and miniaturization in their design. They may either be left in a fixed position or provided with a motorized drive to allow traverses through the reactor core. Some may provide a continuous readout, whereas others are interrogated only at periodic intervals. Typical operating conditions are: neutron flux at full power of 5×10^{13} n/cm$^2 \cdot$ s, gamma flux up to 10^8 R/h, operating temperature up to 300°C, and external operating pressure up to 2500 psi (17 MPa). In-core reactor measurements are the subject of a number of papers found in Ref. 44.

Figure 14.12 A typical in-core fission chamber used in BWR neutron monitoring systems.

1. FISSION CHAMBERS

Miniaturized fission chambers[49] can be tailored for in-core use over any of the power ranges likely to be encountered in reactor operation. Walls of the chamber are usually lined with highly enriched uranium to enhance the ionization current. These small ion chambers are typically made using stainless steel walls and electrodes, and operating voltage varies from about 50 to 300 V. Argon is a common choice for the chamber fill gas and is used at a pressure of several atmospheres. The elevated pressure ensures that the range of fission fragments within the gas does not exceed the small dimensions of the detector.

The gradual burn up of neutron-sensitive material is a serious problem for the long-term operation of in-core detectors. For example, a fission chamber using ^{235}U will show a sensitivity decrease of about 50% after exposure to an integrated neutron fluence of about 1.7×10^{21} n/cm^2 (Ref. 54). One method of reducing the effects of burnup in fission chambers is to combine fertile and fissile material in the neutron-sensitive lining of the chamber. Use of these *regenerative* chambers will gradually convert the fertile isotopes to fissile nuclei to help compensate for the burnup of the original fissile material present in the lining. The long-term response of fission chambers can greatly be improved with this method. For example, it is reported that fission chambers based on a mixture of ^{238}U and ^{239}Pu can maintain a sensitivity that does not change more than $\pm 5\%$ over an integrated neutron fluence of 4.8×10^{21} n/cm^2 (Ref. 54). Similar results have also been obtained with fission chambers based on a mixture of ^{234}U and ^{235}U.

Fission ion chambers that have been operated for long periods in high neutron fluxes will show a residual current or *memory* effect due to the buildup of fission products

Figure 14.13 Memory effect in fission chambers corresponding to various steady-state irradiation times. (From Roux.[55])

within the chamber. These fission products emit beta and gamma rays, which ionize the fill gas of the chamber and result in a significant ion current. Figure 14.13 shows the results to be expected if the current from a fission chamber is monitored following its removal from long-term exposure in a steady-state neutron flux. The residual current I is plotted as a fraction of the steady-state current I_0 observed during the neutron irradiation. One minute after removal about 0.1% of the signal current persists, whereas after 10 days the fission product activity has decayed sufficiently so that the residual current has dropped to about 10^{-5} of the steady-state signal.

One effect that can be important in ion chambers that must cover a wide range of irradiation rates is illustrated in Fig. 14.14. At lower rates the region of ion saturation is reached at a lower voltage than at higher rates. When the current is high, the density of ionization is correspondingly high and recombination will occur more readily than at lower currents. The electric field required to prevent recombination is therefore higher at high rates, evidenced by the increased voltage required to achieve ion saturation. It is important to select an operating voltage for these chambers at the highest irradiation rate or largest current that will be encountered. Although the change in current-voltage characteristics with increased neutron flux may be greater for in-core detectors than out of core detectors, a similar effect is observed in both the compensated and uncompensated ion chambers used in pressurized water reactors.

2. SELF-POWERED DETECTORS

A unique type of neutron detector that is widely applied for in-core use is the *self-powered detector*. These devices incorporate a material chosen for its relatively high cross section for neutron capture leading to subsequent beta or gamma decay. In its simplest form, the detector operates on the basis of directly measuring the beta decay current following capture of the neutrons. This current should then be proportional to the rate at which neutrons are captured in the detector. Because the beta decay current is measured directly, no external bias voltage need be applied to the detector, hence the name self-powered. Another

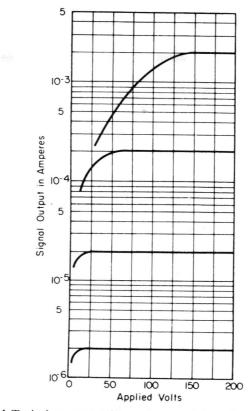

Figure 14.14 Typical current-voltage characteristics of an in-core fission chamber at different neutron irradiation levels. (Courtesy of Westinghouse Electric Corporation, Electronic Tube Division, Horseheads, NY).

form of the self-powered detector makes use of the gamma rays emitted following neutron capture. Some fraction of these gamma rays will interact to form secondary electrons through the Compton, photoelectric, and pair production mechanisms. The current of the secondary electrons can then be used as the basic detector signal.

The self-powered detector is also known by a variety of other names. In recognition of some of the early pioneering work done by J. W. Hilborn,[56,57] they are sometimes called *Hilborn detectors*. Other names that can be found in the literature include beta emission detectors, collectrons, electron emission detectors, and PENA (primary emission, neutron activation) detectors. Nonetheless, the *self-powered neutron (SPN)* detector remains the most common term applied to this family of devices.

Compared with other neutron sensors, self-powered detectors have the advantages of small size, low cost, and the relatively simple electronics required in conjunction with their use. Disadvantages stem from the low level of output current produced by the devices, a relatively severe sensitivity of the output current to changes in the neutron energy spectrum, and, for many types, a rather slow response time. Because the signal from a single neutron interaction is at best a single electron, pulse mode operation is impractical and self-powered detectors are always operated in current mode. Various types of SPN detectors are surveyed in Ref. 58, and considerations in their application to in-core measurements in a PWR are reviewed in Ref. 59.

Figure 14.15 Cross-sectional view of a specific self-powered detector design. (From Stevens.[60])

a. Detectors Based on Beta Decay

Figure 14.15 shows a sketch of a typical SPN detector based on beta decay. The heart of the device is the emitter, which is made from a material chosen for its relatively high cross section for neutron capture leading to a beta-active radioisotope. Ideally, the remainder of the detector does not interact strongly with the neutrons, and construction materials are chosen from those with relatively low neutron cross sections. Figure 14.16 illustrates some possible sequences that can contribute to the measured current. The principles of operation are very simple: The current corresponding primarily to the beta rays given off by the emitter is measured between the emitter and an outer shell, called the *collector*. The intervening space is filled with an insulator, which must be chosen to withstand the extreme temperature and radiation environment typically found in a reactor core. Various metallic oxides are often used as the insulator, with magnesium or aluminum oxide[61] most commonly used. The collector is typically high-purity stainless steel or Inconel. Great care must be taken in the fabrication of these detectors to keep the materials as clean as possible to avoid contamination by substances that might also become radioactive and add an interfering current to the measured signal. The small dimensions shown in Fig. 14.15 are necessitated by the small clearances available for instrument channels in typical reactor cores.

The key to the detector performance lies in the choice of emitter material. Factors to be considered in the selection of the emitter include its neutron capture cross section, together with the energy and half-life of the resulting beta activity. The capture cross section must be neither too high nor too low, because very low cross sections will lead to detec-

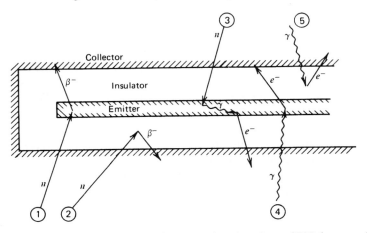

Figure 14.16 Representative events that can take place in an SPN detector. Events ① and ② are neutron capture followed by β−decay. Event ③ shows the interaction of a prompt gamma ray emitted upon neutron capture, giving rise to a fast secondary electron. Events ④ and ⑤ show interfering fast electrons arising from interactions of external gamma rays. In standard SPN detectors, event ① is the basis of its neutron response. In those with fast response, event ③ is the preferred mode of interaction.

Table 14.1 Properties of Emitter Materials for SPN Detectors Based on Beta Decay

Emitter Material	Nuclide of Interest and Percent Abundance[a]	Activation Cross Section at Thermal Energy[b]	Half-Life of Induced Beta Activity[a]	Beta Endpoint Energy[a]	Typical Neutron Sensitivity[c]
Vanadium	$^{51}_{23}$V (99.750%)	4.9 barns	225 s	2.47 MeV	$5 \times 10^{-23} \dfrac{A}{n/cm^2 \cdot s}$
Rhodium	$^{103}_{45}$Rh (100%)	139 barns	44 s		
		11 barns	265 s	2.44 MeV	$1 \times 10^{-21} \dfrac{A}{n/cm^2 \cdot s}$

[a]Data from Lederer and Shirley, *Table of Isotopes,* 7th ed., Wiley & Sons, New York, 1978.
[b]Data from BNL-325, 3rd ed., Vol. 1 (1973).
[c]Sensitivity quoted for emitter of 1-cm length and typical diameter.

tors with low sensitivity, whereas an excessively high cross section will result in rapid burn-up of the emitter material[62] in the high neutron fluxes associated with reactor cores. The beta rays produced should be of sufficiently high energy so that excessive self-absorption in the emitter or insulator is avoided, and the half-life of the induced activity should be as short as possible to permit the detector to respond quickly to rapid changes in neutron flux.

Based on these criteria, the two most popular choices for emitter material have been rhodium and vanadium. Table 14.1 summarizes some of the important properties of these materials when used as emitters in self-powered detectors. Vanadium produces a fairly simple beta decay with a half-life of 225 s, whereas rhodium gives rise to a more complex beta decay with an admixture of half-lives of 44 and 265 s. Figure 14.17 shows the resulting response of these materials to a step change in neutron flux level. Despite the fact that vanadium has a lower sensitivity and somewhat slower response than rhodium, vanadium emitters have become more common in reactor applications because the rate of burnup is significantly less, permitting use over periods of years in typical reactor fluxes.

In its simplest form, a self-powered detector with a single mode of induced activity and negligible burnup would behave as follows when exposed to a neutron flux for a period of time t:

$$I(t) = Cq\sigma N\varphi(1 - e^{-\lambda t}) \tag{14.5}$$

where C = dimensionless constant reflecting the specific geometry and collection efficiency of the detector
q = charge liberated (number of beta particles $\times$ e) per neutron absorbed
σ = activation cross section of the emitter material
N = number of emitter atoms
φ = neutron flux
λ = decay constant of activity produced in the emitter

At saturation, or when the detector has been exposed to the flux for a period of time that is long compared with the half-life of the induced activity, the steady-state current is given simply by

$$I_{sat} = Cq\sigma N\varphi \tag{14.6}$$

The saturated current is proportional to the neutron flux and consequently can serve as a corresponding monitor of the neutron flux level.

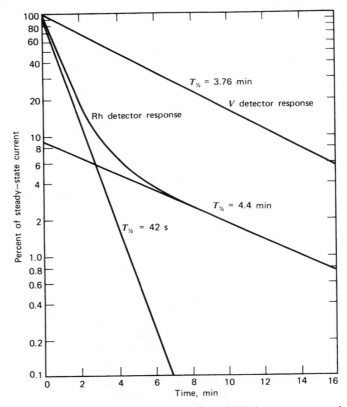

Figure 14.17 Response of rhodium and vanadium SPN detectors to an abrupt drop to zero of a steady-state neutron flux. (From Stevens.[60])

A more complex analysis of the output of these detectors must take into account a number of other factors, including neutron flux depression caused by emitter self-shielding, the Compton and photoelectrons produced from gamma rays that may accompany the beta decay process, and the self-absorption probability of the beta particles within the emitter. Some electrons are stopped in the insulator before reaching the collector, whereas others can be produced within the insulator material and travel to either the emitter or the collector. After some period of operation, an equilibrium is established in which the net amount of charge flowing into the insulator is exactly balanced by the charge flowing out. Monte Carlo calculations have shown that the current caused by processes taking place within the insulator is typically less than 15%[63] and consequently is not a dominant effect in determining detector response. Fairly detailed physical models have been described[63–66] that appear to give adequate predictions[67] of the response of SPN detectors, taking into account most of these effects.

b. Self-Powered Detectors Based on Secondary Electrons from Gamma Decay
One of the primary disadvantages of self-powered detectors based on beta decay is their relatively slow response time. Some efforts have been made to remedy this situation by electronic or digital processing of the signals,[68–70] but it would be preferable to speed up the inherent response of the detector itself. One method of accomplishing this objective is to rely on the secondary electrons produced by prompt capture gamma rays that can follow neutron capture in the emitter (see Fig. 14.16). These capture gamma rays are typically emitted within a very small fraction of a second, as opposed to the much slower decay of

typical neutron-induced beta activities. Even in vanadium and rhodium detectors, there is a prompt component of the signal that, although much smaller than the signal from the beta current, corresponds to the prompt capture gamma rays emitted upon neutron capture in these materials. The ratio of prompt to delayed signal in commercial vanadium detectors is reported to be about 6.5%.[68]

For fast self-powered detectors, it is more common to choose a specific emitter material that will optimize the signal arising from prompt capture gamma rays. Most experience has been gained using cobalt as a prompt emitter,[71–73] with cadmium also used in commercially available detectors. The behavior and sensitivity of other prompt emitter materials have been reported.[74,75] With some exceptions, the neutron sensitivity is substantially less for prompt detectors compared with those based on beta decay, but the much faster response can make them the detector of choice for certain applications.

The response to external gamma rays also can be quite significant for some emitter materials. Gamma rays incident on the detector give rise to secondary electrons, which can yield a discernible signal; the process is illustrated in Fig. 14.16. The gamma-ray-induced signal can be either positive or negative, in that the net flow of current may be either in the same or the opposite direction to the neutron-induced current. Which polarity prevails depends on whether more gamma-ray-induced electrons flow from the emitter to collector or vice versa. Either situation may exist, depending on the specific construction of the detector. The commonly used emitter materials for neutron-sensitive detectors (rhodium, vanadium, or cobalt) all have gamma-ray responses that typically are less than a few percent of the neutron response.[76] Detectors made with zirconium emitters have an almost pure gamma-ray response, whereas other materials such as platinum, osmium, or cerium will give a mixed response. Shields[76] first described the properties and application of a platinum detector for in-core flux measurements in power reactors, in which the combined gamma/neutron sensitivity provides a mix of prompt and delayed response. Platinum detectors have gained considerable popularity, and analyses of their response have been published in Refs. 77–80.

In all types of self-powered detectors, the effects of neutron and gamma-ray interactions in the connecting cable can be quite significant. As a result, considerable care must be taken in choosing the construction materials for the signal cable[81] that lead to the detector, especially in those regions in which these cables are subjected to high radiation fields. In order to aid in the suppression of false signals arising in the connection cables, twin signal leads are often used. One lead is connected through the cable to the emitter, whereas the other is included within the same cable but is terminated without electrical contact physically near the emitter. By electronically subtracting the unconnected lead signal from the current detected from the lead connected to the emitter, effects caused by cable interactions are approximately canceled out.

PROBLEMS

14.1 Discuss the feasibility of operating a BF_3 or ^3He tube in the ionization or Geiger regions rather than as a proportional tube. What practical considerations dictate the latter choice?

14.2 When operated at a gas multiplication factor of 1000, estimate the pulse amplitude produced by the interaction of a thermal neutron in a ^3He tube of 100 pF capacitance.

14.3 A BF_3 tube using natural boron shows a counting efficiency of 1% for 10 eV neutrons in a given application. By what factor can the efficiency be increased by using boron enriched to 96% ^{10}B?

14.4 Calculate the detection efficiency of a BF_3 tube (96% enriched in ^{10}B) filled to 600 torr (80 kPa) for incident thermal neutrons if their pathlength through the gas is 10 cm.

14.5 In which applications might one prefer to use a ^3He tube rather than a BF_3 tube for slow neutron counting?

14.6 In BF_3 tubes of small diameter, the "step" in the pulse height spectrum of Fig. 14.3b at 1.47 MeV has a finite positive slope that is much more noticeable than that for the step at 0.84 MeV. Give a physical explanation for this observation.

14.7 Sketch the pulse height spectrum expected from a boron-lined proportional tube for thermal neutrons if the boron thickness is small enough so that energy loss of the reaction products in the layer can be neglected.

14.8 Using the data given in Table 8.3, estimate the number of scintillation photons liberated by the interaction of a thermal neutron in ^{6}LiI(Eu) and in a typical Li glass scintillator.

14.9 Why is it not possible to increase the counting efficiency of a fission chamber indefinitely simply by increasing the thickness of the fissionable deposit?

14.10 The signal current from a typical application of a self-powered neutron detector is seldom more than a nanoampere. Find the equivalent number of beta particles transferred per second between the emitter and collector.

14.11 Find the fractional decrease in sensitivity (the "burn-up") of a rhodium self-powered detector if used in a neutron flux of $3 \times 10^{13}/\text{cm}^2 \cdot \text{s}$ over a period of 6 months.

REFERENCES

1. W. D. Allen, *Neutron Detection,* George Newnes, Ltd., London, 1960.
2. W. J. Price, *Nuclear Radiation Detection,* 2nd ed., Chap. 10, McGraw-Hill, New York, 1964.
3. H. Neuert, *Kernphysikalische Messverfahren,* Chap. 9, Verlag G. Braun, Karlsruhe, 1966.
4. A. Lorenz, *A Review of Neutron Detection Methods and Instruments,* UCID-16325 (1973).
5. G. Grosshoeg, *Neutron Ionization Chambers,* North-Holland, Amsterdam, 1979.
6. J. F. Boland, *Nuclear Reactor Instrumentation (In-Core),* Gordon & Breach, New York, 1970.
7. J. M. Harrer and J. G. Beckerley, *Nuclear Power Reactor Instrumentation Systems Handbook,* Vol. 1, Chaps. 2 and 3, TID-25952-Pl (1973).
8. D. A. Abdushukurov et al., *Nucl. Instrum. Meth.* **B84,** 400 (1994).
9. I. O. Anderson and S. Malmskog, "Investigation of the Pulse Height Distribution of Boron Trifluoride Proportional Counters," AE-84 (1962).
10. R. Cervellati and A. Kazimierski, *Nucl. Instrum. Meth.* **60,** 173 (1968).
11. I. L. Fowler, *Rev. Sci. Instrum.* **34,** 731 (1973).
12. E. Sakai et al., *IEEE Trans. Nucl. Sci.* **NS-30**(1), 802 (1983).
13. C. W. Peters, A. L. Snyder, and A. S. Gallia, Jr., *IEEE Trans. Nucl. Sci.* **NS-13**(1), 636 (1966).
14. T. Qian, P. Tonner, N. Keller, and W. J. L. Buyers, *IEEE Trans. Nucl. Sci.* **45**(3), 636 (1998).
15. D. P. Brown, *IEEE Trans. Nucl. Sci.* **NS-21**(1), 763 (1974).
16. T. Tomoda and S. Fukakusa, *Nucl. Instrum. Meth.* **224**(3), 557 (1984).
17. J. Csikai and M. Buczko, *Nucl. Instrum. Meth.* **8,** 73 (1960).
18. K. Verghese, J. R. Bohannon, and A. D. Kowalczuk, *Nucl. Instrum. Meth.* **74,** 355 (1969).
19. A. J. Stokes, T. J. Meal, and J. E. Myers, Jr., *IEEE Trans. Nucl. Sci.* **NS-13**(1), 630 (1966).
20. T. E. Sampson and D. H. Vincent, *Nucl. Instrum. Meth.* **95,** 563 (1971).
21. D. M. Drake, W. C. Feldman, and C. Hurlbut, *Nucl. Instrum. Meth.* **A247,** 576 (1986).
22. J. B. Czirr, G. M. MacGillivray, R. R. MacGillivray, and P. J. Seddon, *Nucl. Instrum. Method.* **A424,** 15 (1999).
23. H. P. Chou and C. Y. Horng, *Nucl. Instrum. Meth.* **A328,** 522 (1993).
24. O. P. Joneja, R. Hecker, and A. Mohsin, *Nucl. Instrum. Meth.* **193,** 563 (1982).
25. R. Stedman, *Rev. Sci. Instrum.* **31,** 1156 (1960).
26. R. H. Bossi and A. H. Robinson, *Trans. Am. Nuc. Soc.* **22,** 153 (1975).
27. F. Mantler-Niederstatter, F. Bensch, and F. Grass, *Nucl. Instrum. Meth.* **142,** 463 (1977).
28. S. Ait-Boubker et al., *Nucl. Instrum. Meth.* **A277,** 461 (1989).
29. W. R. Mills, Jr., R. I. Caldwell, and I. L. Morgan, *Rev. Sci. Instrum.* **33,** 866 (1962).
30. S. Shalev, Z. Fishelson, and J. M. Cuttler, *Nucl. Instrum. Meth.* **71,** 292 (1969).
31. E. Dietz and W. Sosaat, in *Physikalisch-Technische Bundesanstalt Annual Report* (R. Jahr and S. Jetzke, eds.), Braunschweig, 1989.
32. D. H. Beddingfield, H. O. Menlove, and N. H. Johnson, *Nucl. Instrum. Meth.* **A422,** 35 (1999).
33. E. Sakai, K. Kubo, and H. Yoshida, *IEEE Trans. Nucl. Sci.* **NS-27**(1), 776 (1980).
34. F. L. Glesius and T. A. Kniss, *IEEE Trans. Nucl. Sci.* **NS-35**(1), 867 (1988).
35. A. E. Evans, H. O. Menlove, R. B. Walton, and D. B. Smith, *Nucl. Instrum. Meth.* **133,** 577 (1976).
36. S. Kahn, R. Harman, and V. Forgue, *Nucl. Sci. Eng.* **23,** 8 (1965).
37. H. Ries et al., *Nucl. Instrum. Meth.* **185,** 373 (1981).
38. A. A. Bogdzel et al., *Nucl. Instrum. Meth.* **200,** 407 (1982).
39. K. H. Valentine et al., *IEEE Trans. Nucl. Sci.* **NS-32**(1), 384 (1985).
40. S. A. Wender et al., *Nucl. Instrum. Meth.* **A336,** 226 (1993).
41. N. W. Hill, J. T. Mihalczo, J. W. Allen, and M. M. Chiles, *IEEE Trans. Nucl. Sci.* **NS-22**(1), 686 (1975).
42. E. Catalano and J. B. Czirr, *Nucl. Instrum. Meth.* **143,** 61 (1977).
43. R. W. Lamphere, "Fission Detectors," in *Fast Neutron Physics,* Part I, p. 449 (J. B. Marion and J. L. Fowler, eds.), Interscience Publishers, New York, 1960.
44. *Proceedings of the Specialists' Meeting on In-Core Instrumentation and Reactor Assessment,* Nuclear Energy Agency (NEA), Organization for Economic Co-Operation and Development (OECD), 1984.
45. W. H. Ellis, J. L. Cooper, Jr., and G. H. Sanders, *IEEE Trans. Nucl. Sci.* **NS-20**(1), 639 (1973).
46. R. A. DuBridge, *IEEE Trans. Nucl. Sci.* **NS-14**(1), 241 (1967).
47. H. A. Thomas and A. C. McBride, *IEEE Trans. Nucl. Sci.* **NS-15**(1), 15 (1968).
48. N. R. Campbell and V. J. Francis, *JIEE* **93,** Part III (1946).
49. K. R. Prasad et al., *Nucl. Instrum. Meth.* **A418,** 420 (1998).
50. M. Oda, M. Wada, and S. Badono, *IEEE Trans. Nucl. Sci.* **NS-23**(1), 304 (1976).
51. S. Shirayama, T. Itok, C. C. Wimpee, and J. Sturz, Proceedings of the Specialists' Meeting on In-Core Instrumentation and Reactor Assessment, NEA/OECD, pp. 66–77 (1984).
52. J. M. Harrer and J. G. Beckerley, *Nuclear Power Reactor Instrumentation Handbook,* Vol. 2, Chaps. 15 and 16, TID-25952-P2 (1974).

53. H. S. McCreary, Jr., and R. T. Bayard, *Rev. Sci. Instrum.* **25**(2), 161 (1954).
54. H. Böck and E. Balcar, *Nucl. Instrum. Meth.* **124**, 563 (1975).
55. D. P. Roux, ORNL-3929 (1966).
56. J. W. Hilborn, *Nucleonics* **22**, 2, 69 (1964).
57. J. W. Hilborn, "Self-Powered Neutron Detector," U.S. Patent 3, 375, 370 (March 26, 1968).
58. N. P. Goldstein and W. H. Todt, *IEEE Trans. Nucl. Sci.* **NS-26**(1), 916 (1979).
59. D. P. Bozarth and H. D. Warren, *IEEE Trans. Nucl. Sci.* **NS-26**(1), 924 (1979).
60. H. H. Stevens, "Neutron Sensors—In-Core," Chap. 3 in *Nuclear Power Reactor Instrumentation Systems Handbook,* TID-25952-P1 (1973).
61. D. P. Bozarth and H. D. Warren, *Trans. Am. Nucl. Soc.* **23**, 517 (1976).
62. H. D. Warren, *Trans. Am. Nucl. Soc.* **23**, 460 (1976).
63. N. P. Goldstein, *IEEE Trans. Nucl. Sci.* **NS-20**(1), 549 (1973).
64. W. Jaschik and W. Seifritz, *Nucl. Sci. Eng.* **53**, 61 (1974).
65. H. D. Warren and N. H. Shah, *Nucl. Sci. Eng.* **54**, 395 (1974).
66. H. D. Warren, *Nucl. Sci. Eng.* **48**, 331 (1972).
67. P. S. Rao and S. C. Misra, *Nucl. Instrum. Meth.* **A253**, 57 (1986).
68. W. Seifritz, *Nucl. Sci. Eng.* **49**, 358 (1972).
69. J. M. Carpenter, R. F. Fleming, and H. Bozorgmanesh, *Trans. Am. Nucl. Soc.* **22**, 606 (1975).
70. L. A. Banda and B. I. Nappi, *IEEE Trans. Nucl. Sci.* **NS-23**(1), 311 (1976).
71. H. Böck and M. Stimler, *Nucl. Instrum. Meth.* **87**, 299 (1970).
72. H. Böck, *Nucl. Instrum. Meth.* **125**, 327 (1975).
73. J. C. Kroon, F. M. Smith, and R. I. Taylor, *Trans. Am. Nucl. Soc.* **23**, 459 (1976).
74. P. Gebureck et al., IAEA-SM-168/G-8, p. 783 (1973).
75. A. K. Mahant, P. S. Rao, and S. C. Misra, *Nucl. Instrum. Meth.* **A406,** 117 (1998).
76. R. B. Shields, *IEEE Trans. Nucl. Sci.* **NS-20**(1), 603 (1973).
77. G. F. Lynch, R. B. Shields, and P. G. Coulter, *IEEE Trans. Nucl. Sci.* **NS-24**(1), 692 (1977).
78. N. P. Goldstein, *IEEE Trans. Nucl. Sci.* **NS-25**(1), 292 (1978).
79. N. P. Goldstein, C. L. Chen, and W. H. Todt, *IEEE Trans. Nucl. Sci.* **NS-28**(1), 752 (1981).
80. D. S. Hall, *IEEE Trans. Nucl. Sci.* **NS-29**(1), 646 (1982).
81. R. Van Nieuwenhove, *Nucl. Instrum. Meth.* **A373,** 202 (1996).

Chapter 15

Fast Neutron Detection and Spectroscopy

In Chapter 14 on slow neutron detection, a number of neutron-induced reactions were discussed which can serve as the basis for the conversion of neutrons to directly detectable charged particles. In principle, all these reactions could be applied to detect fast neutrons as well. As shown in the cross-section plot of Fig. 14.1, however, the probability that a neutron will interact by one of these reactions decreases rapidly with increasing neutron energy. As a result, conventional BF_3 tubes have an extremely low detection efficiency for fast neutrons and consequently are almost never used for this purpose. For reasons to be discussed later in this chapter, the ^{3}He proportional counter is useful both for thermal neutron detection and for fast neutron spectroscopy. As a rule, however, fast neutron devices must employ a modified or completely different detection scheme to yield an instrument with acceptable detection efficiency.

The most important additional conversion process useful for fast neutrons is elastic neutron scattering. In this interaction an incident neutron transfers a portion of its kinetic energy to the scattering nucleus, giving rise to a *recoil nucleus*. The energy that can be transferred from a slow neutron is therefore very small, and the resulting recoil nuclei are too low in energy to generate a usable detector signal. Once the neutron energy reaches the keV range, however, recoil nuclei can be detected directly and assume a large importance for fast neutron detection. By far the most popular target nucleus is hydrogen. The cross section for neutron elastic scattering from hydrogen is quite large and its energy dependence is accurately known. More important, however, is the fact that an incident neutron can transfer up to its entire energy in a single collision with a hydrogen nucleus, whereas only a small fraction can be transferred in collisions with heavy nuclei. Therefore, the resulting *recoil protons* are relatively easy to detect and serve as the basis for a wide variety of fast neutron detectors.

An important distinction in the application of fast neutron detectors is whether an attempt is made to measure the energy of the incoming neutron. For all the slow neutron detection methods discussed in Chapter 14, the information on initial neutron kinetic energy is hopelessly lost in the conversion process, because the neutron energy is extremely small compared with the energy liberated in the conversion reaction itself (the Q-value). Once the incoming neutron energy is no longer negligible compared with the reaction Q-value (that means at least 10–100 keV for most of the reactions discussed in Chapter 14), the energy of the reaction products begins to change appreciably with changes in the neutron energy. An accurate measurement of the reaction product energies can then, in principle, be used to deduce the incoming neutron energy by simply subtracting the reaction Q-value. In elastic scattering the reaction Q-value is zero, so that neutron energies can begin to be measured by this technique at the point at which the resulting recoils have measurable kinetic energy. The collection of instruments and techniques applied to the measurement of fast neutron energy is conventionally included in the category of *fast neutron spectroscopy*.

In some instances, however, the purpose of the measurement is simply to record the presence of fast neutrons without a corresponding measurement of their energy. Such fast neutron counters can employ any of the methods discussed to convert neutrons to charged particles, and then simply record all pulses from the detector. Fast neutron counters of this type will have a severe variation in efficiency with neutron energy, but if the incident neutron energy is not likely to change greatly between measurements, they can provide useful information on the relative intensity of a fast neutron flux. Other applications in which the fast neutron spectrum may change considerably between measurements benefit from counters tailored to the application. We begin our discussion of fast neutron devices with counters of this type.

I. COUNTERS BASED ON NEUTRON MODERATION

A. General Considerations

The inherently low detection efficiency for fast neutrons of any slow neutron detector can be somewhat improved by surrounding the detector with a few centimeters of hydrogen-containing moderating material. The incident fast neutron can then lose a fraction of its initial kinetic energy in the moderator before reaching the detector as a lower-energy neutron, for which the detector efficiency is generally higher. By making the moderator thickness greater, the number of collisions in the moderator will tend to increase, leading to a lower value of the most probable energy when the neutron reaches the detector. One would therefore expect the detection efficiency to increase with moderator thickness if that were the only factor under consideration. A second factor, however, tends to decrease the efficiency with increasing moderator thickness: The probability that an incident fast neutron ever reaches the detector will inevitably decrease as the moderator is made thicker. Several effects are at work here, as illustrated in Fig. 15.1. As the detector becomes a smaller and smaller fraction of the total volume of the system, there will be a lower probability that a typical neutron path will intersect the detector before escaping from the surface of the moderator. Furthermore, a neutron may be absorbed within the moderator before it has a chance of reaching the detector. The absorption probability will increase rapidly with increasing moderator thickness because absorption cross sections generally are larger at lower neutron energies.

As a result of all these factors, the efficiency of a moderated slow neutron detector when used with a monoenergetic fast neutron source will show a maximum at a specific moderator thickness. Assuming that the moderator is the usual choice of a hydrogenous material such as polyethylene or paraffin, we find that the optimum thickness will range from a few centimeters for keV neutrons up to several tens of centimeters for neutrons in the MeV energy range.

If the thickness of the moderator is fixed at a fairly large value, the overall counting efficiency of the system versus incident neutron energy will also tend to show a maximum. Low-energy neutrons will not penetrate far enough into the moderator before they are likely to be captured in the moderator itself, whereas high-energy neutrons will not be adequately moderated for efficient detection. By careful choice of the diameter and composition of the moderator–detector system, its overall efficiency versus energy curve can often be shaped and tailored to suit a specific application.

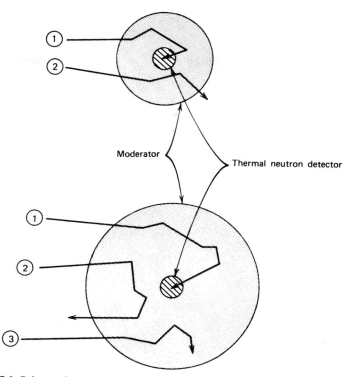

Figure 15.1 Schematic representation of neutron histories in moderated detectors. The small thermal neutron detector at the center is shown surrounded by two different thicknesses of moderator material. Histories labeled ① represent incident fast neutrons that are successfully moderated and detected. Those labeled ② are partially or fully moderated but escape without reaching the detector. History ③ represents those neutrons that are parasitically captured by the moderator. Larger moderators will tend to enhance process ③ while reducing process ②. See text for a discussion of the variation of process ①.

B. The Spherical Dosimeter

In an effort to develop a useful neutron spectrometer, Bramblett, Ewing, and Bonner[1] first investigated the properties of a small lithium iodide scintillator placed at the center of polyethylene moderating spheres of different diameters. The general behavior seen in this type of study is shown in Fig. 15.2. The difference in the shapes and position of the maxima in these response curves serves as the basis for using the set of spheres as a simple neutron spectrometer. By measuring the count rate with each sphere individually, an unfolding process can, in principle, provide some information about the energy distribution of the incident neutrons.[2-5]

A very useful type of counter has emerged from these studies. The response curve shown in Fig. 15.2 for the 12-in. sphere turns out to have a similar shape compared with the dose equivalent delivered per neutron as a function of energy (see Fig. 2.22). This resemblance is entirely coincidental, in that there is no fundamental relation between the probability of neutron detection at the center of the sphere and the magnitude of the dose delivered by that neutron in a biological medium. Nonetheless, this happy circumstance can be used to good advantage when it is desired to measure the dose equivalent due to

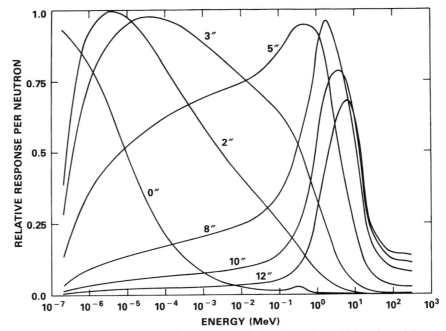

Figure 15.2 The energy dependence of the relative detection efficiencies of Bonner sphere neutron detectors of various diameters up to 12 inches. (From Johnson et al.[2])

neutrons with an unknown or variable energy spectrum. Because of the similarity of the two curves, the efficiency of the detector is high for those neutrons whose biological importance is high and is low for neutrons that deliver less dose. Therefore, the overall count from the detector in a polyenergetic spectrum will automatically include proper weighting factors for all energies and give a meaningful measure of the combined dose due to all the neutrons. The small size (typically 4 mm × 4 mm) of the LiI scintillator and the relatively high Q-value of the lithium capture reaction for neutrons allow for very effective discrimination against gamma rays, even in relatively large gamma-ray fields. The spherical geometry provides for a reasonably nondirectional detector response. Typical intrinsic efficiency for a 12-in. sphere, defined as the fraction of neutrons that strike the surface of the sphere which ultimately result in a count, is 2.5×10^{-4} at the peak of the 12-in. detector response. Translated into dose, the average response corresponds to about 3×10^3 counts/mrem. It is virtually the only monitoring instrument that can provide realistic neutron dose estimates over the many decades of neutron energies ranging from thermal energies to the MeV range. However, because the detection mechanism is not fundamentally related to radiation dose, substantial errors can arise when applied to widely different source conditions.[6]

These spherical moderator systems are generally known as *Bonner spheres* after one of the authors (T. W. Bonner) of the original paper describing its experimental investigation.[1] More recent work[7–11] has reexamined the response of spherical moderators used in combination with several different thermal neutron detectors, both experimentally and by neutron transport calculational techniques.

Hankins[12,13] first chose to study the response of a 10-in. diameter (25.40 cm) sphere used with a 4 mm × 4 mm lithium iodide scintillator. The moderator size was selected as that likely to give the closest fit between the efficiency and the dose equivalent per incident neutron over a wide range of neutron energy. Figure 15.3 shows the estimated response of the 10-in. sphere detector together with the dose per neutron curve. Obviously, the match is

Figure 15.3 Sensitivity of a 10 in. (25.4 cm) diameter moderating sphere surrounding a 4 mm × 4 mm LiI scintillator. Also shown is the relative dose per neutron labeled "inverse of RPG curve." (From Hankins.[12])

quite good over several decades of neutron energy. Below 100 keV, direct measurements of the detector efficiency are difficult. A neutron transport code can be applied to calculate the detector response over this energy range, and these results are included in Fig. 15.4. The calculations confirm the good match between the two curves above 100 keV and at thermal energy but show a sizable deviation in the intermediate energy range between 0.1 eV and 100 keV. The deviation is such as to lead to an overestimate of the neutron dose if the spectrum contains a significant component over this energy range. Although the deviation at 10 keV is as much as a factor of 5, measurements made when the neutron spectrum covers a broad range in energies will show a considerably lower average deviation.

An alternate version of the spherical neutron dosimeter developed by Leake[14] is shown in Fig. 15.5. This design has come to be widely implemented as the Harwell type 95/0075 neutron survey meter. A spherical ^{3}He proportional counter is substituted for the lithium iodide scintillator as the slow neutron detector. This substitution is made to

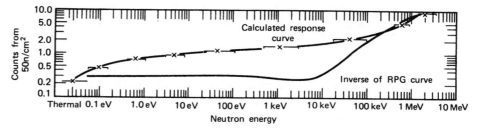

Figure 15.4 Calculated sensitivity of a 10-in. (25.4 cm) diameter moderating sphere, together with the relative dose per neutron (inverse of RPG curve). (From Hankins.[12])

Figure 15.5 A spherical neutron dosimeter based on a ^{3}He neutron detector. (From Leake.[14])

minimize the response of the detector to gamma rays, and Leake reports application of the dosimeter in gamma-ray fields as high as 20 R/h. Used with a simple 20.8-cm diameter polyethylene moderator, the energy response of the system to thermal and epithermal neutrons is higher than ideal. Therefore, a spherical cadmium absorber, perforated with holes, is placed around the ^{3}He detector to shape the response curve. Although the instrument still overresponds to neutrons in the keV energy range (by a factor of 4.9 at 10 keV), the response to broad neutron spectra typical of shielded fission sources does not deviate by more than ±40% for a very wide range of experimental and calculated spectra.[15,16] The response at high neutron energies drops off somewhat; the instrument underestimates the dose from 14 MeV neutrons shielded by concrete by about a factor of 2 (Ref. 15).

There have been continuing efforts to develop new designs of spherical or cylindrical neutron dosimeters[17–20] whose sensitivity more closely matches the dose per neutron curve over a wide energy range. Some of the challenge comes about because of evolving standards for quality factors (see Chapter 2) that are necessary inputs in calculating the desired curve shape. The incorporation of lead or other high-Z elements as heterogeneous components[21,22] increases the relative response to high-energy neutrons, where the response of a pure moderator tends to fall off. Secondary neutrons produced in (n,2n) and other multiplying reactions in these heavy elements at high energies can extend the useful sensitivity to several hundred MeV.

C. The Long Counter

A detector whose counting efficiency does not depend on the neutron energy can be a very useful device in many areas of neutron physics. For an ideal detector of this type, a graph of the detection efficiency versus neutron energy is a horizontal line, which has led to the name *flat response* detectors. Although no real detector exists with a perfectly flat response over the entire range of possible neutron energies, several designs have evolved that come close to this ideal.

Over the years, the most popular flat response neutron detector has been the *long counter*. Like the spherical neutron dosimeter, it is based on the principle of placing a slow neutron detector at the center of a moderating medium. For the long counter, however, the slow neutron detector is a BF$_3$ tube, and the system is designed to respond properly to neutrons only when they are incident from a specific direction.

The combination of a BF$_3$ tube and cylindrical moderator was first suggested as a flat response neutron detector by Hanson and McKibben.[23] A later design by McTaggart[24] is shown in Fig. 15.6a and has achieved fairly widespread acceptance as the standard long counter. The counter is designed to be sensitive only to neutrons incident on the right-hand face of the counter within the boron oxide shell. Those incident from other directions tend to be moderated by the outer annulus of paraffin and are subsequently captured in the boron layer without giving rise to a count. Neutrons incident on the front face parallel to the cylindrical axis will penetrate some distance before undergoing moderation. The

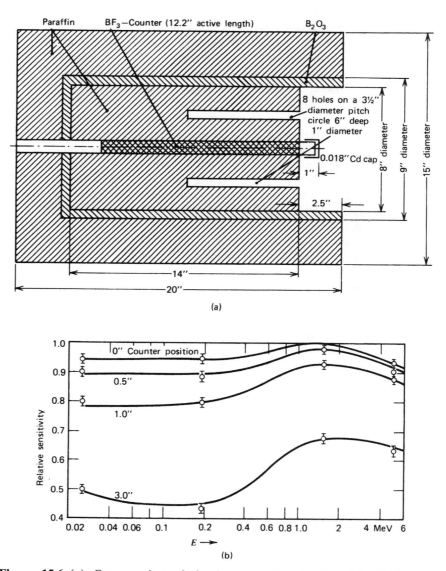

Figure 15.6 (*a*) Cross-section of the long counter developed by McTaggart. (*b*) Relative sensitivity of McTaggart long counter versus neutron energy. The parameter varied for the different curves is the distance the end of the BF$_3$ tube is shifted in from the front of the moderator face. The flattest response occurs when the tube is flush with the front face. (From K. H. Beckurts and K. Wirtz, *Neutron Physics.* Copyright 1964 by Springer-Verlag, Inc. Used with permission.)

average distance of penetration will increase as the neutron energy increases. If the BF_3 tube and cylindrical moderator are sufficiently long, then a typical cross section through the cylinder at the point of moderation will not be different for various energy neutrons. Therefore, the probability that the moderated neutron will find its way to the BF_3 tube and produce a count should not depend strongly on neutron energy. It is this property that leads to the flat energy response of the detector. The holes provided in the front surface prevent a fall-off in the efficiency at neutron energies below 1 MeV by allowing lower-energy neutrons to penetrate farther into the moderator, away from the front surface from which they might otherwise escape. Figure 15.6*b* shows a plot of the efficiency of a McTaggart long counter versus neutron energy for various axial positions of the BF_3 tube. With the BF_3 tube flush with the front surface, the efficiency does not change by more than 10% over the neutron energy range shown. A long counter of similar design by DePangher and Nichols[25] has also achieved some recognition as a standard in health physics measurements, and documentation of its flat response between about 2 keV and 6 MeV is given in Refs. 26–28.

Long counters derive many of their operational characteristics from the BF_3 tube on which their design is based. Sensitivity to relatively high levels of gamma rays can be eliminated by simple amplitude discrimination, while continuing to count all the neutron interactions in the tube. The long counter normally displays good long-term stability and traditionally has achieved widespread application as a neutron flux monitor in a wide variety of neutron physics experiments.

A modified long counter with some improved characteristics has been developed by East and Walton[29] and is shown in Fig. 15.7. It substitutes ^{3}He detectors for the BF_3 tube used in McTaggart design and provides five separate detectors near the center of the cylindrical moderator. By using high-pressure ^{3}He tubes, the multiple detector arrangement leads to a rather high overall neutron detection efficiency of 11.5%, compared with a standard long counter efficiency of about 0.25%. The 12 holes that penetrate the inner polyethylene cylinder are covered on the front face by a 19-mm thick ring of polyethylene to provide a geometry of moderation that best favors an overall flat response. The measured efficiency of this detector is plotted in Fig. 15.8 for a number of neutron sources with broad energy distributions. While the measured efficiency is essentially constant for these sources, greater fluctuation has been observed[30] for more nearly monoenergetic neutrons. These variations can arise because of narrow resonances in the cross sections of some of the constituent materials.

Figure 15.7 A high-efficiency long counter utilizing multiple ^{3}He tubes. (From East and Walton.[29])

Figure 15.8 Efficiency of the long counter shown in Fig. 15.7 versus the average energy of some neutron sources. The efficiency figures are for a point source located 1 m from the detector face. (From East and Walton. [29])

D. Other Detectors Based on Moderation

A number of detection systems have evolved that consist of a spherical assembly into which the neutron source is inserted. This approach can obviously be used for small portable neutron sources and can also be adapted for a highly collimated parallel beam of neutrons by providing a small-diameter entrance channel through which the beam can pass to reach the center of the assembly. A typical design consists of placing multiple ^{3}He or BF$_3$ counters in a hydrogenous moderator,[31–36] and in some sense is a spherical version of the long counter. Overall counting efficiencies on the order of 1% can be achieved with a response that is flat to within ±1% over a neutron energy range from 30 keV to 5 MeV.[31]

Other flat response detectors that also rely on neutron moderation have been developed. The *grey neutron detector* of Poenitz[37,38] uses a NaI(Tl) scintillator to detect the 2.2 MeV capture gamma rays produced when neutrons are thermalized in a hydrogenous sphere. To provide a faster response, the *black neutron detector* was also introduced,[39,40] which is based on the light produced in a hydrogenous spherical scintillator as the neutron is moderated. Both types of detectors can provide a very flat efficiency curve over several decades of neutron energy.

II. DETECTORS BASED ON FAST NEUTRON-INDUCED REACTIONS

The detectors described in the previous section rely on the slowing down of a fast neutron in a moderating material before its detection as a thermal neutron. The moderating process eliminates all information on the original energy of the fast neutron and normally cannot be used if an attempt is made to extract energy information. Furthermore, the detection process is relatively slow. In most designs, the neutron must undergo multiple collisions with moderator nuclei followed by diffusion as a thermal neutron (that may take tens or hundreds of microseconds) before the detection signal is generated. As a result, such detectors cannot provide a fast detection signal required in many neutron detection applications.

Both these limitations may be overcome if the fast neutron can be made to induce directly a suitable nuclear reaction without the moderation step. The reaction products will then have a total kinetic energy given by the sum of the incoming neutron kinetic energy and the Q-value of the reaction. Provided the neutron energy is not a hopelessly small fraction of the Q-value, a measurement of the reaction product energies will give the neutron energy by simple subtraction of the Q-value. Additionally, the detection process can potentially be fast because the incoming fast neutron will typically spend no more than a few

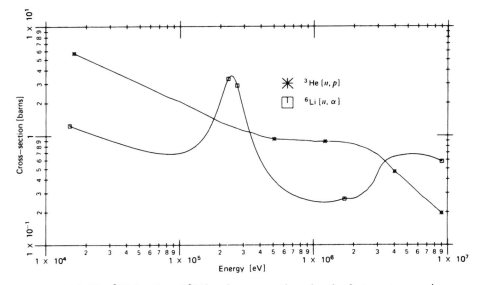

Figure 15.9 The ^{3}He(n, p) and ^{6}Li(n, α) cross sections for the fast neutron region.

nanoseconds in the active volume of the detector, and only a single reaction need occur to provide a detector signal. However, the cross sections for typical fast-neutron-induced reactions are orders of magnitude lower than the corresponding thermal neutron cross sections, and such detectors will inevitably show a much lower detection efficiency than their thermal neutron counterparts.

Excluding elastic scattering, which will be the topic of the next section, there are only two reactions of major importance in fast neutron spectroscopy: ^{3}He(n, p), and ^{6}Li(n, α). Both of these reactions were discussed at the beginning of Chapter 14. A plot of the cross-section variation with neutron energy for the fast region is shown in Fig. 15.9. We now emphasize the application of these reactions in neutron spectroscopy, where the neutron energy is inferred by measuring the energy of the reaction products. It should be clear that the same detectors can be used simply to detect the presence of fast neutrons by arranging to count all (or some fixed fraction) of the neutron-induced reactions in the detector. A third reaction, neutron-induced fission, is not of interest in spectroscopy because of the very high Q-value associated with the reaction. The fission process, however, can serve as the basis of a fast neutron counter if energy information is not required.

Finally, a class of fast neutron detectors (called *activation counters*) is based on detecting the radioactivity induced in certain materials. These detectors have proved to be useful when applied to sources producing short pulses of fast neutrons. Because these devices do not produce prompt signals, a discussion of their properties is postponed until Chapter 19.

A. Methods Based on the ^{6}Li(n, α) Reaction

As seen from Fig. 15.9, the cross section for the ^{6}Li(n, α) reaction drops off rather smoothly with increasing neutron energy, except for the pronounced resonance at a neutron energy of about 250 keV. The relatively large Q-value of 4.78 MeV is an advantage in thermal neutron detection but limits the application in fast neutron spectroscopy to neutrons with energy of at least several hundred keV. A competing reaction, ^{6}Li(n, n'd)^{4}He, has a Q-value of -1.47 MeV and becomes the dominant neutron-induced reaction at energies above about 2.5 MeV. Because this reaction leads to three products, one of which is a neutron that normally escapes, a continuum of deposited energy should be expected even for monoenergetic incident neutrons. Therefore, although it will contribute some neutron pulses, this additional reaction is generally an undesirable part of the response of any detector that attempts to measure the incident neutron energy.

If we neglect the continuum that may be introduced by this latter reaction, the response function of fast neutron detectors based on the lithium reaction should be a single peak located at an energy equal to the neutron energy plus the Q-value of the (n, α) reaction (4.78 MeV). In practical situations, an additional peak is often observed at 4.78 MeV from reactions induced by neutrons whose energy has been reduced to the thermal range by moderation in the laboratory walls, shielding, and any other material in the vicinity of the detector. Unless special care is taken to eliminate those low-energy neutrons, their large interaction cross section will result in many events, all of which deposit the same energy (the reaction Q-value) in the detector. The resulting peak is usually called the *epithermal peak* and can provide a convenient energy calibration point for the detector output.

1. THE LITHIUM IODIDE SCINTILLATOR

The lithium iodide (europium-activated) scintillation crystals discussed in Chapter 14 in connection with thermal neutron detection can also be applied to fast neutron spectroscopy.[41,42] However, the use of this material at room temperature is severely hampered by its nonlinear response to the tritons and alpha particles produced from the lithium reaction. This nonlinearity results in a resolution of about 40% for the full-energy peak caused by incident thermal neutrons. This very broad response function seriously limits the application of LiI(Eu) in fast neutron spectroscopy. It has been shown[43] that the nonlinearities in the crystal response can be substantially reduced by cooling the crystal to liquid nitrogen temperature, which improves the resolution of the full-energy peak to about 20%. The practical problems involved in cooling the crystal are significant, but fast neutron spectra measurements have been reported using this technique.[42]

2. LITHIUM GLASS SCINTILLATORS

Lithium can also be incorporated in other scintillation matrices and used as a fast neutron detector. Because of their relatively poor light output and nonlinearity of response, these scintillators have not been used directly for neutron spectroscopy in the manner described above for lithium iodide scintillators. In neutron time-of-flight spectroscopy, however, the detector need register only the arrival time of a neutron, and various lithium scintillators have evolved for this application. Granular scintillators consisting of mixtures of lithium fluoride and zinc sulfide have been developed[44,45] for time-of-flight use. Lithium-containing glass scintillators have become much more popular in these applications, however, because of the relatively fast response time and large areas that can easily be fabricated. Silicate glasses of various compositions, generally with cerium activation, are used as the scintillation medium.

General characteristics of glass scintillators are described in Refs. 46–51. Table 15.1 lists some properties of commercially available lithium glass scintillators and shows that lithium concentrations of up to 7.7% can be obtained. The low-background properties of NE912 are important for low-level neutron counting and are achieved through the use of materials that are low in natural thorium. In addition to the highly enriched ^{6}Li formulations shown in Table 15.1, equivalent scintillators are also available in which the lithium is present as natural lithium (7.4% ^{6}Li, 92.6% ^{7}Li) or depleted lithium ($> 99.9\%$ ^{7}Li). The latter are neutron insensitive and can be used to measure separately the gamma contribution in a mixed neutron–gamma irradiation.

The time resolution that can be achieved from these detectors depends somewhat on the pulse amplitude distributions produced by the incident neutrons but can be as low as a few nanoseconds. Unfortunately, glass scintillators show a much reduced light output per unit energy for heavy charged particles compared with electrons, and reaction products

Table 15.1 Properties of Some Commercially Available Lithium Glass Scintillators[a]

Manufacturer's Identification[b]	Type NE902	NE905	NE908	NE912
Density (g/cm^3)	2.6	2.48	2.674	2.55
Refractive index	1.58	1.55	1.57	1.55
Melting point (°C)	1200	1200	1200	1200
λ of emission maximum (nm)	395	395	395	397
Light output relative to anthracene	22–34%	20–30%	20%	25%
Decay constant (ns)	75	100	5 + 75	75
Content of Li	2.2 wt %	6.6 wt %	7.5 wt %	7.7 wt %
^6Li enrichment	95%	95%	95%	95%
Background α activity per 100 g of glass (/min)	100–200	100–200	100–200	20
Resolution expected for thermal neutrons (depends on glass thickness)	13–22%	15–28%	20–30%	20–30%

[a]Data from McMurray et al.[47]

[b]Commercial identification numbers are those used by Nuclear Enterprises, Ltd. Lithium glass scintillators of similar properties are also available through Koch-Light Laboratories, Ltd.

with 4.78 MeV energy will yield about the same light output as a 1.2 MeV gamma ray. The gamma-ray discrimination ability of these scintillators[52] is therefore not as good as other detectors in which the response is more uniform for all particles. The detection efficiency for thick lithium glasses is of considerable interest in many applications and is difficult to calculate accurately because of the influence of multiple scattering within the glass. Resonances in the scattering cross sections of various materials in the glass lead to sharp peaks in the detection efficiency at neutron energies above about 100 keV.

3. LITHIUM GLASS FIBER SCINTILLATORS

As introduced in Chapter 8, glass scintillators lend themselves to fabrication in the form of small-diameter optical fibers. When the glass contains lithium (normally enriched in ^6Li), these fibers provide a unique detection medium for neutrons. The alpha particle and triton produced by slow neutron reactions with ^6Li have ranges in glass of about 7 and 40 μm, respectively, so a fiber diameter of several times the sum of these ranges is needed to absorb all their energy in a large fraction of the neutron interactions. In contrast, typical secondary electrons produced by gamma rays have ranges of a millimeter or more, so choosing a small fiber diameter helps minimize their energy deposition. An optimum diameter of around 100 μm maximizes the amplitude difference between neutron-induced and gamma-ray-induced pulses to facilitate discrimination against gamma-ray backgrounds.

The cerium-doped glass scintillators of the types listed in Table 15.1 can be drawn into fibers with about this diameter. When provided with a thin glass cladding of lower refractive index, about 3.3% of the light emitted in neutron-induced reactions is captured in each direction and propagated along the length of the fiber.[53] Even though attenuation lengths as large as 2 meters have been reported,[53] other observations[54] show a much greater attenuation of the scintillation light. Some of the discrepancy comes about because of the overlap of the emission band of the Ce^{3+} fluorescence sites with the absorption of Ce^{4+} atoms that may be present in variable concentration because of cerium oxidation that takes place during the process of drawing the fiber.

To achieve better light transmission, there has been some interest in substituting terbium as the activator in the glass fibers.[54,55] Terbium activation results in several significant changes, including a shift in the peak emission wavelength to 550 nm, and a much slower fluorescent decay time of about 3 ms. Problems of optical self-absorption are greatly reduced so that longer attenuation lengths are easier to achieve than with cerium activated glass. However, the long decay time necessitates unconventional pulse processing techniques,[54] or restriction to applications in which the light yield is measured using current mode or charge integration techniques.

4. LITHIUM SANDWICH SPECTROMETER

Another way in which the lithium reaction has been widely used to measure fast neutron energies is outlined in Fig. 15.10. A thin layer of lithium fluoride or other lithium-containing material is prepared on a very thin backing and placed between two semiconductor diode detectors. When the neutron energy is low, the two reaction products are oppositely directed and coincident pulses should be observed from the two semiconductor detectors. Neglecting energy loss of the charged particles before they reach the active volume, the sum of the energy deposited in the two detectors should be equal to the incoming neutron energy plus the Q-value of the lithium reaction.

In practice, complications arise because of the necessity to use finite thicknesses for both the lithium deposit and the backing on which it is supported. Figure 15.10 illustrates that the energy loss in the target materials, which does not contribute to the detected signal, will vary as the angle of the emitted reaction products changes through all possible values. Particularly troublesome are those reaction products that are emitted near the plane of the target foil, which will have long pathlengths through the deposit and backing before escaping into the detectors. These energy loss effects can be minimized by making the deposit very thin, but only at the expense of a reduced counting efficiency. Alternatively, the paths near the plane of the foil can be eliminated by geometric collimation between the deposit and detectors, but again, the counting efficiency will be reduced.

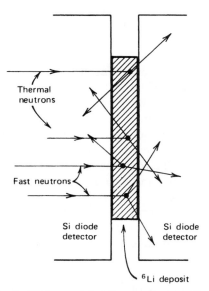

Figure 15.10 Elements of a lithium sandwich spectrometer. Reaction products from thermal neutrons are always oppositely directed, whereas fast neutron interactions will lead to some net forward momentum for the triton and alpha particle. If both are emitted in the forward direction, no coincidence will exist.

The employment of coincidence detection of the two reaction products greatly reduces the background in semiconductor sandwich spectrometers. Any background events that occur only in one detector will automatically be eliminated. If the neutron energy is significant compared with the Q-value, the reaction products must have some momentum in the direction of the incoming neutron and will not be exactly oppositely directed. Then, some fraction of all reactions will lead to two products, both of which strike the same detector and do not give rise to coincidences (see Fig. 15.10). The fraction of neutron events lost to this effect becomes more significant as the neutron energy increases.

Several methods for processing the data from semiconductor sandwich spectrometers have been described. The simplest is to record the sum signal from coincident pulses and deduce the neutron energy by subtracting the Q-value of the reaction. At neutron energies below several hundred keV, this method becomes quite sensitive to small errors and uncertainties. An alternative scheme first proposed by Maroni[56] is based on measuring the difference in energy between the triton and alpha particle. For reaction products that are collinear with the incoming neutron direction, this difference is very sensitive to small changes in neutron energy below about 100 keV. Rickard[57] discusses several other approaches to the analysis of the detector signals that can be advantageous in some applications. General discussions of the application of lithium sandwich detectors in fast neutron spectroscopy can be found in Refs. 58–64.

B. Detectors Based on the ^{3}He(n, p) Reaction

The ^{3}He(n, p) reaction discussed in Chapter 14 has also been widely applied to fast neutron detection and spectroscopy. The fast neutron cross section plotted in Fig. 15.9 falls off continuously with increasing neutron energy. Several competing reactions must be considered in any detector based on this reaction. The most significant of these is simple elastic scattering of the neutrons from helium nuclei. The cross section for elastic scattering is always larger than that for the (n, p) reaction, and this predominance becomes more pronounced as the neutron energy becomes larger. For example, the cross sections are about equal at a neutron energy of 150 keV, but elastic scattering is about three times more probable at 2 MeV. In addition, a competing (n, d) reaction on ^{3}He is possible at neutron energies exceeding 4.3 MeV, but the cross section is low for energies below about 10 MeV. The (n, p) reaction and elastic scattering therefore account for the important features of ^{3}He detector response for all but the highest neutron energies.

The pulse height spectrum from a detector based on the ^{3}He reaction should show three distinct features. Neglecting the wall effect, the full energy of the reaction products is always totally absorbed within the detector and a spectrum similar to that shown in Fig. 15.11 should be expected. The first feature is a full-energy peak corresponding to all the (n, p) reactions induced directly by the incident neutrons. This peak occurs at an energy equal to the neutron energy plus the Q-value of the reaction. Second, a pulse height continuum results from elastic scattering of the neutron and a partial transfer of its energy to a recoiling helium nucleus. The maximum energy of the continuum can be calculated from Eq. (15.4) (given later in this chapter) and is 75% of the incoming neutron energy for ^{3}He. Third, an epithermal peak appears in virtually every spectrum taken with a ^{3}He detector (as in detectors based on the lithium reaction) and corresponds to the detection of incident neutrons that have been reduced to the thermal range by moderation in external materials. All such neutron interactions deposit an energy equal to the Q-value, or 764 keV.

Wall effects arise whenever the dimensions of the detector are not large compared with the ranges of the secondary particles produced in these reactions. As described in Chapter 14 in connection with BF$_3$ tubes, the effect on the pulse height spectrum is to fill in the region to the left of the peaks shown in Fig. 15.11.

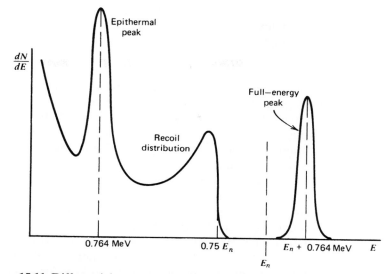

Figure 15.11 Differential energy spectrum of charged particles expected from fast neutrons incident on a ^{3}He detector.

1. THE ^{3}He PROPORTIONAL COUNTER

If a large ^{3}He-filled proportional counter is irradiated by fast neutrons, a spectrum similar to that shown in Fig. 15.11 will be observed. To minimize the wall effect in smaller counters, several atmospheres of pressure are usually used, and a second component consisting of a heavier gas such as krypton is often added to the ^{3}He to reduce the reaction product ranges. Specific designs and applications of ^{3}He proportional tubes are described in Refs. 65–70.

The undesirable influences of the wall effect and elastic scattering in ^{3}He proportional counters can be ameliorated considerably at the price of added complexity. The additional information carried by the rise time of the output pulses can be used to eliminate a large number of these unwanted events, while retaining virtually all the direct ^{3}He neutron capture events. In any proportional counter, the charge collection time depends on the radial distance at which the ions are formed by the original charged particle. Those tracks that are either very short or parallel to the axis of the proportional tube will generate ionization electrons, all of which are collected in about the same time. The resulting avalanches will thus be synchronized, and the corresponding pulse rise time will therefore be small. Tracks that cover a wide range of radii will generate electrons with widely different collection times and pulses with longer rise times. Rejection of the elastically scattered ^{3}He recoils is possible because their specific ionization is greater than protons of the same energy. As a result, recoils have a shorter range and are likely to be limited to a smaller variation in radii per track. They will therefore tend to produce pulses with a shorter rise time than the preferred proton events and can be discriminated on that basis. Because there is some overlap in these rise times, the rejection will also eliminate some true signals and will reduce the overall counting efficiency. With proper selection criteria, it has been possible to reject virtually all recoil events with no more than about a factor of 2 reduction in the (n, p) detection efficiency.[65]

There is also evidence[71] that wall effect pulses have an average rise time considerably slower than that of the full-energy pulses and can also be suppressed by eliminating long rise time events. As a bonus, long rise time rejection will also effectively eliminate gamma-ray background because fast electrons will generally travel completely across the active volume of the tube. More elaborate analysis including two-dimensional storage of both amplitude and rise time information for each pulse[72] or digital signal processing techniques[73] allow for a more selective choice of acceptance parameters but also introduce added complexity.

2. THE ³He IONIZATION CHAMBER

Although ³He filled chambers were first developed as proportional counters, there is some advantage in designing such chambers to operate instead as gridded ionization chambers of the type described beginning on p. 152. By avoiding the added fluctuations introduced by avalanche formation, gridded ion chambers can display significantly better pulse height resolution than the equivalent proportional counter. When applied to fast neutron spectroscopy, this advantage translates into superior energy resolution. Some examples of the design and application of ³He ionization chambers are given in Refs. 74–82.

Based on an original design by Shalev and Cuttler,[83] a widely used version of this type of chamber uses a mixture of ³He, argon, and methane at partial pressures of 3, 6, and 0.5 atm, respectively. The predominance of the heavy gas argon reduces the ranges of the reaction products and minimizes the complications of wall and end effects or partial energy loss in the gas. The FWHM of the full-energy peak identified in Fig. 15.11 ranges from about 12 keV for thermal neutrons to 20 keV for incident 1 MeV neutrons.[74] The slow charge collection time and small pulse amplitude from this type of detector create more severe problems compared with proportional tubes. Relatively long shaping times (5–10 μs) are necessary to fully develop the pulse, making such chambers susceptible to microphonic noise and pulse pile-up. Sensitive low-noise preamplifiers are also needed to preserve the good energy resolution of the chamber.

3. THE ³He SCINTILLATOR

As described previously in Chapter 8, the noble gases including helium can be used as scintillators. Pure helium has a rather poor light yield, but the addition of xenon with as little as a few percent concentration can enhance the light yield by as much as a factor of 5 (Ref. 84). The emitted light is relatively low in intensity compared with conventional scintillation materials, and appears mostly in the ultraviolet region of the spectrum. It is common to use additives in the gas (such as nitrogen) or wavelength shifting materials (such as p-terphenyl) as reflecting layers on the inner surfaces of the scintillation chamber to convert much of the ultraviolet to the visible band. Purity of the gas is very important, since trace amounts of oxygen or organic vapors are known to reduce the light yield significantly.

The decay time of the scintillation is only several nanoseconds, leading to a very fast risetime of the output pulse. This advantage in timing compared with ³He proportional or ion chambers is offset by a poorer pulse height resolution. One design[85] has shown a FWHM of 121 keV for the full energy peak from 2.5 MeV neutrons, limited largely by the light collection efficiency variations throughout the volume of the gas. In order to increase the neutron detection efficiency, ³He scintillation chambers have been designed[84] to withstand up to 150 atm pressure.

4. THE ³He SEMICONDUCTOR SANDWICH SPECTROMETER

The configuration of a neutron-sensitive target surrounded on both sides by semiconductor detectors is most commonly used with ⁶Li as the target. This type of neutron detector was discussed earlier as the lithium semiconductor sandwich spectrometer. Less attention has been given to the use of ³He as the target material, but some potential advantages have spurred efforts in this direction. These advantages can include a considerably higher detection efficiency for equivalent neutron energy resolution and a cross section that is smooth and well known. The gain in efficiency, which can be as much as a factor of 20–50 (Ref. 86), is due mostly to the lower specific energy loss of the proton and triton reaction products in helium compared with that for the alpha particle and recoil triton in lithium targets. Consequently, thicker targets can be used which can consist of pure elemental ^{3}He. Also from Fig. 15.9, the fast neutron cross section for the ^{3}He reaction is larger than that for the ^{6}Li reaction. Disadvantages include the lower Q-value of the ^{3}He reaction which makes discrimination against gamma rays much more difficult. Furthermore, the larger volume of pressurized helium gas that must be substituted for the solid lithium target makes efficiency calculations more complicated and subject to uncertainties. Descriptions of the design and application of ^{3}He semiconductor sandwich spectrometers can be found in Refs. 86–88.

III. DETECTORS THAT UTILIZE FAST NEUTRON SCATTERING

A. General Properties

The most common method of fast neutron detection is based on elastic scattering of neutrons by light nuclei. The scattering interaction transfers some portion of the neutron kinetic energy to the target nucleus, resulting in a *recoil nucleus.* Because the targets are always light nuclei, this recoil nucleus behaves much like a proton or alpha particle as it loses its energy in the detector medium. Hydrogen, deuterium, and helium are all of interest as target nuclei, but hydrogen is by far the most popular. The recoil nuclei that result from neutron elastic scattering from ordinary hydrogen are called *recoil protons,* and devices based on this neutron interaction are often referred to as *proton recoil detectors.*

The Q-value of elastic scattering is zero because the total kinetic energy after the reaction by definition is the same as the kinetic energy before. For all practical purposes the target nuclei are at rest, and therefore the sum of the kinetic energies of the reaction products (recoil nucleus and scattered neutron) must equal that carried in by the incident neutron. For single scattering in hydrogen, the fraction of the incoming neutron energy that is transferred to the recoil proton can range anywhere between zero and the full neutron energy, so that the average recoil proton has an energy about half that of the original neutron. Therefore, it is usually possible to detect preferentially fast neutrons in the presence of gamma rays or other low-energy background, but the discrimination becomes more difficult as the incoming neutron energy drops below a few hundred keV. By employing techniques such as pulse shape or rise time discrimination to eliminate gamma-ray-induced events, specialized proton recoil detectors can be used to a neutron energy as low as 1 keV. Recoil methods are insensitive to thermal neutrons except through any competing reactions that might be induced in the target material or other parts of the detector.

1. KINEMATICS OF NEUTRON ELASTIC SCATTERING

We first define some symbols to be used in the equations that follow:

A = mass of target nucleus/neutron mass

E_n = incoming neutron kinetic energy (laboratory system)

E_R = recoil nucleus kinetic energy (laboratory system)

Θ = scattering angle of the neutron in the center-of-mass coordinate system

θ = scattering angle of the recoil nucleus in the lab coordinate system

These definitions are illustrated graphically in Fig. 15.12.

For incoming neutrons with nonrelativistic kinetic energy ($E_n \ll 939$ MeV), conservation of momentum and energy in the center-of-mass coordinate system gives the following relation for the energy of the recoil nucleus:

$$E_R = \frac{2A}{(1 + A)^2}(1 - \cos \Theta)E_n \tag{15.1}$$

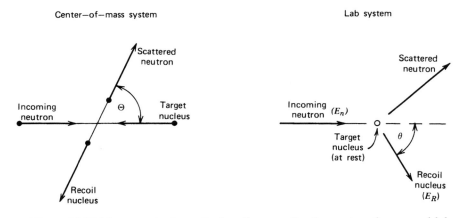

Figure 15.12 Neutron elastic scattering diagrams for the center-of-mass and laboratory coordinate systems.

To convert to the more familiar laboratory coordinate system in which the original target nucleus is at rest, we use the following transformation:

$$\cos \theta = \sqrt{\frac{1 - \cos \Theta}{2}} \tag{15.2}$$

which, when combined with Eq. (15.1), gives the following relation for the recoil nucleus energy in terms of its own angle of recoil:

$$E_R = \frac{4A}{(1 + A)^2} (\cos^2 \theta) E_n \tag{15.3}$$

From Eq. (15.3) we can see that the energy given to the recoil nucleus is uniquely determined by the scattering angle. For a grazing angle encounter in which the neutron is deflected only slightly, the recoil is emitted almost perpendicular to the incoming neutron direction ($\theta \cong 90°$), and Eq. (15.3) predicts that the recoil energy is near zero. At the other extreme, a head-on collision of the incoming neutron with the target nucleus will lead to a recoil in the same direction ($\theta \cong 0$), resulting in the maximum possible recoil energy,

$$E_R\Big|_{max} = \frac{4A}{(1 + A)^2} E_n \tag{15.4}$$

Table 15.2 Maximum Fractional Energy Transfer in Neutron Elastic Scattering

Target Nucleus	A	$\dfrac{E_R}{E_n}\Big|_{max} = \dfrac{4A}{(1 + A)^2}$
$^1_1$H	1	1
$^2_1$H	2	$8/9 = 0.889$
$^3_2$He	3	$3/4 = 0.750$
$^4_2$He	4	$16/25 = 0.640$
$^{12}_6$C	12	$48/169 = 0.284$
$^{16}_8$O	16	$64/289 = 0.221$

Table 15.2 lists the maximum fraction of the incoming neutron energy that can be transferred to a recoil nucleus in a single collision for a variety of target nuclei. As the target nucleus mass increases, the maximum fractional energy transfer decreases. Only in collisions with ordinary hydrogen can the neutron transfer all its energy in a single encounter. The trend shown in the table explains why only light nuclei are of primary interest in recoil detectors, with hydrogen assuming the predominant role.

2. ENERGY DISTRIBUTION OF RECOIL NUCLEI

We must also be concerned with the way in which the recoil energies are distributed between a minimum of zero and the maximum given in Table 15.2. Because all scattering angles are allowed, in principle, a continuum of possible recoil energies between these extremes should be expected. If we define $\sigma(\Theta)$ as the differential scattering cross section in the center-of-mass system, then, by definition, the probability that the neutron will be scattered into $d\Theta$ about Θ is

$$P(\Theta)\,d\Theta = 2\pi \sin \Theta\, d\Theta\, \frac{\sigma(\Theta)}{\sigma_s} \qquad (15.5)$$

where σ_s is the total scattering cross section integrated over all angles. We are more interested in the distribution in recoil nucleus energy and will let $P(E_R)dE_R$ represent the probability of creating a recoil with energy in dE_R about E_R. Now, because $P(E_R)dE_R = P(\Theta)\,d\Theta$, it follows that

$$P(E_R) = 2\pi \sin \Theta\, \frac{\sigma(\Theta)}{\sigma_s} \cdot \frac{d\Theta}{dE_R} \qquad (15.6)$$

Now, evaluating $d\Theta/dE_R$ from Eq. (15.1) and substituting, we obtain

$$P(E_R) = \frac{(1 + A)^2}{A}\, \frac{\sigma(\Theta)}{\sigma_s} \cdot \frac{\pi}{E_n} \qquad (15.7)$$

Equation (15.7) shows that the shape expected for the recoil energy continuum is just the same as the shape of the differential scattering cross section $\sigma(\Theta)$ as a function of the center-of-mass scattering angle of the neutron. For most target nuclei, the shape of $\sigma(\Theta)$ will tend to be somewhat peaked to favor forward and backward scattering as shown in Fig. 15.13.

A very important simplification holds if the scattering process is *isotropic in the center-of-mass-coordinate system*. Then, $\sigma(\Theta)$ does not change with Θ and is equal to a constant $\sigma_s/4\pi$. This fortunate circumstance is not generally the case but does hold for scattering from hydrogen over most of the energy range of interest ($E_n < 10$ MeV). Because hydrogen is by far the most important target nucleus, the simplifications that result are of real significance. *The expected proton recoil energy distribution is therefore a simple rectangle,* extending from zero to the full incident neutron energy, as sketched in Fig. 15.14. The response function of a detector based on simple hydrogen scattering should therefore have a correspondingly simple rectangular shape. The average proton energy is then one-half of the neutron energy. As discussed in the following sections, however, there are a number of complicating factors that can distort this simple rectangular response.

3. DETECTION EFFICIENCY

The detection efficiency of a device based on recoil protons or other recoil nuclei can be calculated from the scattering cross section σ_s. If nuclei of only one species are present in the detector, the intrinsic efficiency is given simply by

$$\epsilon = 1 - \exp(-N\sigma_s d) \qquad (15.8a)$$

where N is the number density of target nuclei, σ_s is the scattering cross section for these nuclei, and d is the pathlength through the detector for incident neutrons. Carbon often

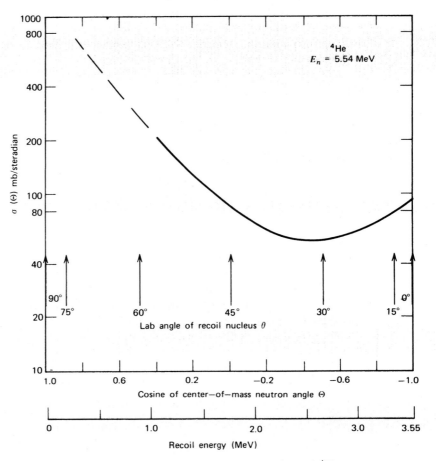

Figure 15.13 The differential scattering cross section of ^{4}He at a neutron energy of 5.54 MeV. Also indicated are the corresponding angle and energy of the helium recoil nucleus in the laboratory system.

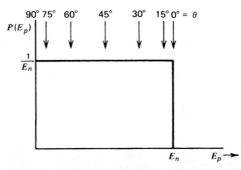

Figure 15.14 Energy distribution of recoil protons E_p produced by monoenergetic neutrons. Recoil energies are indicated for various values of the recoil emission angle θ as given by Eq. (15.3). For scattering from hydrogen, E_p is equivalent to E_R in Eqs. (15.1) through (15.7).

appears in combination with hydrogen in proton recoil detectors, and competing effects due to carbon scattering must then be taken into account. The counting efficiency, neglecting multiple scattering, is then given by

$$\epsilon = \frac{N_H \sigma_H}{N_H \sigma_H + N_C \sigma_C}\{1 - \exp[-(N_H \sigma_H + N_C \sigma_C)d]\} \tag{15.8b}$$

where the subscripts H and C refer to the separate hydrogen and carbon values for the quantities defined above.

Figure 15.15 (*a*) Elastic scattering cross sections for ^{1}H, ^{2}H, and ^{3}He. (*b*) Elastic scattering cross sections for ^{4}He and ^{12}C.

Plots of the scattering cross section for several materials of interest in fast neutron detectors are given in Fig. 15.15. An empirical fit to the hydrogen scattering cross section suggested by Marion and Young[89] is

$$\sigma_s(E_n) = \frac{4.83}{\sqrt{E_n}} - 0.578 \text{ barns} \tag{15.9}$$

where E_n is in MeV. This approximation is accurate to within about 3% over the neutron energy range of $0.3 < E_n < 30$ MeV.

B. Proton Recoil Scintillators

One of the easiest ways to use proton recoil in the detection of fast neutrons is through the application of hydrogen-containing scintillators. Fast neutrons incident on the scintillator give rise to recoil protons whose energy distribution should be approximately rectangular, ranging from zero to the full neutron energy. Because the range of the recoil protons is usually small compared with the dimensions of the scintillator, their full energy is deposited in the scintillator and the expected pulse height distribution is also approximately rectangular. Hydrogen-containing scintillation detectors are widely applied in many areas of fast neutron physics research, as indicated in the extensive review by Harvey.[90]

1. SCINTILLATION MATERIALS

Because scintillation materials that contain hydrogen are quite common, there is no shortage of candidates for use as fast neutron scintillators. Successful applications have been reported using organic crystals such as anthracene or stilbene, liquid scintillators that combine an organic scintillant dissolved in a hydrogen-containing organic solvent, and plastic scintillators in which an organic scintillant is incorporated in a bulk matrix of polymerized hydrocarbon. A general discussion of the basic properties of these scintillators can be found in Chapter 8.

Much of the early work in developing proton recoil scintillators was done using crystals of anthracene or stilbene. Anthracene has the largest light output of any organic scintillator, but attention gradually shifted to stilbene because of its superior gamma-ray rejection characteristics. However, both crystals are difficult and expensive to obtain in large sizes (greater than a few centimeters in dimension) and are also subject to damage from thermal and mechanical shock. A further disadvantage stems from the directional variation of the light output from such crystals, which depends on the orientation of the path of the charged particles with respect to the crystal axis. It is not unusual to observe variations as large as 25% as the charged particle orientation is varied. This effect, although not serious if the detector is used only to count neutrons, greatly complicates the job of unfolding an observed pulse height spectrum to derive the incident fast neutron energy spectrum.

It is far more common to use plastic and liquid scintillators for fast neutron applications, and their properties were summarized previously in Table 8.1. These materials are relatively inexpensive, can be tailored to a wide variety of sizes and shapes, and are totally nondirectional in their response. Plastics can be obtained in large sizes including rods, bars, and sheets, and are relatively simple to machine for custom applications. The attenuation length for the scintillation light in typical plastics can be several meters or more, an important consideration when scintillators of large size are required.

Liquid scintillators are sometimes preferred because of their pulse shape discrimination capability, discussed in Chapter 8, to suppress gamma-ray response. Some representative measurements of the nonlinear light yield of liquid scintillators as a function of recoil proton energy are given in Refs. 91–97. They are available commercially either in bulk or in sealed glass containers for convenient use. For specialized purposes, liquid scintillators are also available with deuterium substituted for the normal hydrogen content. Since fast neutron scattering from deuterium is not isotropic in the center-of-mass system, the energy distribution of the recoil deuterons no longer approximates the rectangular distribution expected for recoil protons. Instead, the measured pulse height spectra for mono-energetic neutrons[98] show a rise toward the maximum recoil energy that has some of the characteristics of a peak. This shape may, under some circumstances, be better suited to facilitate deconvolution calculations of the incident neutron spectrum. Other formulations based on hexafluorobenzene are totally free of hydrogen and can be used to measure separately the gamma-ray contribution in a mixed fast neutron/gamma-ray field.

2. SCINTILLATOR SIZE

In choosing a size for a recoil proton scintillator, several compromises must be struck. The first involves a trade-off of counting efficiency versus energy resolution. By making the scintillator thick, the efficiency calculated from Eq. (15.8) is obviously enhanced. For example, if the scintillator is made about 5 cm thick in the direction of the neutron path, interaction probabilities of at least 40% will hold for neutrons whose energy is less than 2 or 3 MeV. As the neutron energy increases, the detection efficiency will decrease, and consequently there will be strong motivation to make the scintillator larger. However, it is more difficult to achieve uniform light collection from a large-volume scintillator and the energy resolution will worsen. Another factor that often limits scintillator size is the pulse rate due to gamma rays interacting within the detector. In many applications, this rate exceeds that from fast neutrons, and the scintillator must be kept sufficiently small so that the pile-up of gamma-ray events is not a problem.

If the scintillator is used as a fast neutron spectrometer, other factors enter into the choice of size. In small crystals, a typical neutron is likely to scatter only once, and the energy spectrum of proton recoils will closely approximate the rectangular distribution discussed earlier. As long as the scintillator dimensions are larger than a few millimeters, escape of protons from the surface is unlikely, and the response function of the detector is simple and easily calculated. As the detector dimensions are increased, multiple scattering of the neutrons becomes more likely and the response function is more complicated and harder to predict. Because an accurate knowledge of the response function is critical for the unfolding process, one would like to keep the scintillator small so that these complicating effects do not introduce large uncertainties.

3. RESPONSE FUNCTION

For that subset of detectors for which the rectangular response function is a reasonable approximation, the task of deriving the incident neutron energy spectrum is particularly simple. Because the derivative with respect to energy of a rectangular distribution is zero everywhere except at the maximum, the derivative of the recoil proton spectrum will give a narrow peak located at the incident neutron energy. The derivative of the spectrum recorded from a complex source therefore gives an easily calculated representation of the incident neutron spectrum.

For response functions that are more complex, the general techniques of deconvolution or unfolding, discussed in Chapter 18, must be applied. Representative examples of the application of unfolding methods to the neutron response of organic scintillators are given in Refs. 99–103. Because of their complexity, unfolding calculations of this type are inevitably cast in the form of large computer codes.[104–106] An accurate knowledge of the detector response function is a necessary input to all these unfolding methods. We therefore list some of the factors that distort the simple rectangular distribution in organic scintillators and show the qualitative effect on the response function.

a. Nonlinear Light Output with Energy

As discussed in Chapter 8, the light output from most organic scintillators does not increase linearly as the deposited energy increases. This nonlinear behavior distorts the expected rectangular proton energy distribution of Fig. 15.16c into the pulse height

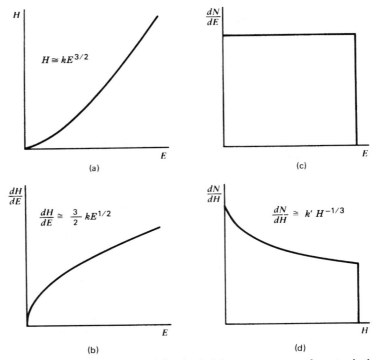

Figure 15.16 Part (*a*) shows a plot of pulse height versus energy for a typical organic scintillator. This nonlinear response leads to a distortion of the rectangular proton recoil spectrum of part (*c*) into the spectrum shown in part (*d*).

distribution shape sketched in Fig. 15.16*d*. For many organic scintillators, the light output H is approximately proportional to $E^{3/2}$, in which case

$$H = kE^{3/2} \tag{15.10}$$

and the pulse height distribution shape is given by

$$\frac{dN}{dH} = \frac{dN/dE}{dH/dE} = \frac{\text{constant}}{\frac{3}{2}kE^{1/2}} = k'H^{-1/3} \tag{15.11}$$

where k and k' are proportionality constants. This relation is only an approximation to the actual distortion caused by the nonlinearity, and a more detailed description of the light output versus energy is required as a basis for accurate spectrum unfolding.[107]

b. Edge Effect

If the scintillator is sufficiently small or the neutron energy very high so that the range of the recoil protons is not small compared with detector dimensions, some effects can be expected as a result of the escape of protons from the surface of the scintillator. The event is not lost but simply shifted to an energy lower than would normally have been observed. The effect on the response function will be to shift events from high pulse height to low pulse height, further increasing the slope of Fig. 15.16*d*.

c. Multiple Scattering from Hydrogen

For detectors that are not small, it is possible for an incident neutron to scatter more than once from hydrogen nuclei before escaping from the scintillator. Because all such events normally occur within a very short period of time compared with the pulse shaping time, the light from all recoil protons is summed and a pulse produced whose amplitude is proportional to the total light output. Multiple scattering will therefore increase the average pulse height and change the expected response function by adding events at large pulse heights at the expense of those at lower amplitudes.

d. Scattering from Carbon

All organic scintillators contain carbon as well as hydrogen. Because of the decreased scintillation efficiency for high dE/dx particles, carbon recoils produced by neutron elastic scattering do not contribute much to the detector output. However, scattering from carbon does affect the detector response function indirectly in that the scattered neutrons may still undergo a hydrogen scattering before escaping from the scintillator. Because the neutron energy has been decreased in the initial carbon scatter, the proton recoil spectrum produced from carbon-scattered neutrons will not extend to as high an energy as that from unscattered neutrons. Because the incident neutron can lose between 0 and 28% of its initial energy in a carbon scattering, the maximum energy of a subsequent recoil proton will vary between 100 and 72% of the original energy. The corresponding effect on the detector response function is shown in Fig. 15.17*b*.

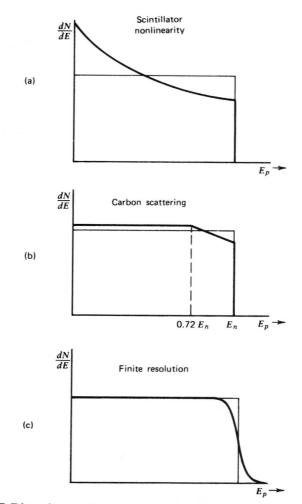

Figure 15.17 Distortions to the rectangular recoil proton energy spectrum due to three separate factors.

e. Detector Resolution

The discussion to this point has dealt with ideal detector response without considering the spread introduced by nonuniform light collection, photoelectron statistics, and other sources of noise. These sources of dispersion will tend to wash out some of the distinct structure expected in the response function, as illustrated in Fig. 15.17c.

f. Overall Response Function

The manner in which all these distorting effects combine to give the overall detector response function is illustrated in Fig. 15.18, for a 2.54 cm × 2.54 cm cylindrical stilbene crystal at a neutron energy of 2.6 MeV. This example is fairly typical of any organic or plastic scintillator response to neutrons whose energy lies below the threshold of competing reactions discussed in the following section.

(a)

(b)

Figure 15.18 (a) Calculated response function at a neutron energy of 2.6 MeV (for perfect detector resolution) for a 2.54 cm × 2.54 cm cylindrical stilbene crystal. Separate components are: a—single scattering with detector nonlinearity and edge effects; b—double scattering from hydrogen; c—the combination of single and double scattering from hydrogen, with the spectrum normalized to the same number of counts as in spectrum a; d—proton recoils from carbon-scattered neutrons; e—composite spectrum consisting of the sum of spectra c and d. (b) Measured pulse height spectrum (with significant zero offset) for the same crystal included in the calculation of part (a). The added influence of imperfect detector resolution is evident. (From Bormann et al.[108])

g. Competing Reactions at High Neutron Energies

Once the neutron energy exceeds 8 or 9 MeV, two competing reactions must be considered in the overall response of organic scintillators:

	Neutron Energy at Threshold
$^{12}C(n, \alpha)^9Be$	6.17 MeV
$^{12}C(n, n')3\alpha$	7.98 MeV

The combined effects of multiple scattering and the competing reactions can be seen in Fig. 15.19, which plots the counting efficiency of a liquid organic scintillator versus neutron energy. In the figure, the competing reactions $^{12}C(n, \alpha)$ and $^{12}C(n, n')3\alpha$ are seen to contribute to the detection efficiency at high neutron energies. The threshold energies for these reactions are 6.17 and 7.98 MeV, but they become significant only above about 9 MeV. An inhibiting factor is the lower light output per unit energy that most organic scintillators exhibit for alpha particles compared to recoil protons.

4. DETECTOR COUNTING EFFICIENCY VERSUS DISCRIMINATOR BIAS LEVEL

A common mode of application of proton recoil detectors is to set a fixed discrimination level to eliminate all pulse amplitudes below a given size. Some finite discrimination level is always required to eliminate inevitable noise pulses that arise spontaneously in the counting system, but gamma rays and other background events may dictate discrimination at a substantially higher level. Because the proton recoil energy distribution extends all the way to zero, some recoil events will inevitably be eliminated in the discrimination process. Therefore, the efficiency assuming that all pulses are counted is sometimes called the *zero bias* efficiency, and a real detector with a finite discrimination level will always have a somewhat lower efficiency, particularly at lower neutron energies.

Figure 15.19 The calculated efficiency of a NE213 liquid scintillator (1.9-cm radius, 10-cm length) for a discrimination level of 0.81 MeV. Identification of the symbols is as follows: ● combined efficiency from all processes; ○ single hydrogen scattering; × n-H, n-H double scattering; ⊙ n-C, n-H double scattering; △ n-C, n-H, n-H triple scattering; □ (n, α) and (n, n')3α reactions. (From Hermsdorf et al.[109])

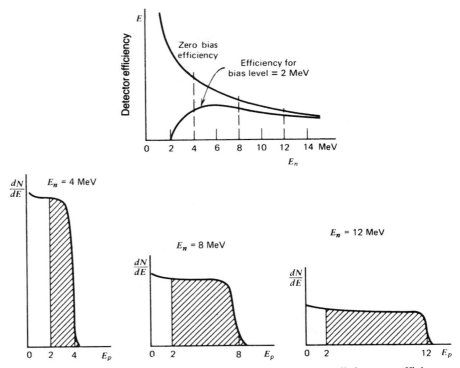

Figure 15.20 The effect of discrimination level on proton recoil detector efficiency. The upper graph gives both the zero bias efficiency and the reduced efficiency when a discrimination level corresponding to a proton energy of 2 MeV is used. See text for further elaboration.

The effect of discrimination level on counting efficiency is shown graphically in Fig. 15.20. Sketched along the bottom of the figure are three differential pulse height spectra representing the response of a scintillator to three different neutron energies. Neglecting more complex interactions, the total area under each of these curves is proportional to the number of neutron–hydrogen scattering events in the detector at that energy. This number is proportional to the zero bias efficiency, shown plotted versus neutron energy at the top of the figure.

When a discriminator is used, that area under the differential spectrum that lies to the right of the discrimination level is proportional to the number of recorded counts. For the example shown, this reduced efficiency is also plotted versus neutron energy on the upper graph. This efficiency obviously drops to zero for neutron energies below the discrimination level and approaches the zero bias efficiency at neutron energies that are large by comparison. More detailed analyses of detection efficiency of organic scintillators can be found in Refs. 110–116.

5. ENERGY CALIBRATION WITH GAMMA RAYS

The light output of organic scintillators for electrons is always higher per unit energy than for heavy charged particles. Therefore, the problem of gamma discrimination is made more severe because it may require a neutron with energy of 2 or 3 MeV to give the same light output as a 1 MeV gamma ray. Because the scintillation response to electrons is fairly linear, a gamma-ray source can be used to calibrate the energy scale of the detector output, with the result often quoted in units of MeVee (MeV electron equivalent—see p. 227).

The low Z-value of the constituents of organic scintillators (hydrogen, carbon, and perhaps oxygen) results in a very low photoelectric cross section, so that virtually all gamma-ray interactions are Compton scatterings. A gamma-ray spectrum taken with an organic scintillator will therefore show no photopeaks, and Compton edges are the only distinguishable features. Because there are no photopeaks, some point on the Compton edge must be selected and associated with the maximum energy of a Compton recoil electron. One study[117] of both a stilbene crystal and a NE102A plastic scintillator provides data to relate the measured features of the Compton continuum to the maximum energy of the Compton electrons. More detailed procedures that take into account variability of the detector resolution are described in Refs. 118 and 119.

6. PULSE SHAPE DISCRIMINATION AGAINST GAMMA RAYS

As discussed in Chapter 8, the relative intensity of the fast and slow components of the light yield of some organic scintillators depends on the specific ionization of the ionizing particle (see Fig. 8.5). Therefore, particles of different mass or charge will produce signal pulses with different time characteristics. Specifically, gamma-ray-induced fast electrons generate a larger fraction of their scintillation light in the prompt component as compared with recoil protons. The methods of pulse shape discrimination discussed in Chapter 17 can be very effective in rejecting gamma-ray pulses from organic scintillators, while retaining reasonable efficiency for fast neutrons.[120–123]

If the gamma-ray pulses are not simply rejected but instead are analyzed in a separate spectrum, simultaneous spectroscopy on both neutrons and gamma rays can be carried out. Because of the lack of full-energy gamma-ray peaks, the gamma-ray spectra must be unfolded using the same type of computation as is applied to the neutron spectrum. Both calculated[124,125] and measured[126–128] response functions to gamma rays for NE213 liquid scintillator have been published and can be used as input to the unfolding codes. Although the light yield versus energy characteristics for NE213 are much more nearly linear for electrons than for recoil protons, there is some indication[128] that a measurable nonlinearity is present for electron energies above 1.6 MeV.

In theory, there is another possible method to distinguish neutrons and gamma rays in detectors with sufficiently fast response. Both neutrons and gamma rays are likely to undergo multiple scattering if the detector volume is reasonably large compared with their mean free paths. The gamma rays travel at the speed of light between collisions, while fast neutrons travel more slowly (the velocity of a 1 MeV neutron is about 5% of the speed of light). Therefore signals from neutrons that scatter more than once will tend to have a greater time duration than those from gamma rays. The differences are subtle—the average time between neutron collisions in a typical plastic scintillator is only a few nanoseconds. Efforts to exploit this effect to separate neutrons from gamma rays in a fast plastic scintillator[129] have not yet proved to be practical because time spreads from other causes tend to obscure the small differences that are expected.

C. Gas Recoil Proportional Counters

1. GENERAL

As an alternative to the organic scintillator, gas proportional counters can also be used to measure fast neutrons through the recoil process. In these applications, the fill gas is usually hydrogen, a hydrogen-containing gas such as methane, or some other low-Z gas such as helium. In the case of hydrogen, the expected proton recoil spectrum should again be the simple rectangular shape described earlier. As in the case of organic scintillators, however, complicating effects often arise which distort this simple response function and make the task of unfolding observed pulse height spectra a great deal more complicated.

Because the detection medium is a gas with relatively low density, recoil proportional counters inevitably have a lower counting efficiency than typical organic scintillators. Equation (15.8) can again be used to estimate the neutron detection efficiency for simple hydrogen-filled counters, and typical values for neutrons in the MeV energy range will be less than 1%. This reduced interaction probability is not entirely lacking some compensating advantages. For example, the interaction probability for scattered neutrons will also be quite low, and thus the response function of gas detectors will be largely free of the multiple scattering complications discussed earlier for scintillators.

Recoil proportional counters are undoubtedly more sensitive to practical problems than organic scintillators. As discussed in Chapter 6, purity of the fill gas is of utmost importance and microscopic air leaks will ultimately lead to detector failure. Although recoil proportional counters are commercially available, there is a common tendency for experimenters to design their own counters specifically tailored to the application required. The reader should note that considerable attention to construction details such as surface preparation, high vacuum pumping before gas filling, and high voltage insulator design are necessary to have a successful detector. By comparison, scintillators are relatively easy to assemble and operate. The fact that recoil proportional counters are not large compared with the range of recoil nuclei in a gas means that the correction for wall and end effects, usually quite small in scintillators, becomes an important consideration in determining the response of these detectors. An extensive review of response functions for recoil proportional counters of different designs is given in Ref. 130.

2. GAMMA-RAY SENSITIVITY AND PULSE SHAPE DISCRIMINATION

Another consequence of the low-density gas detection medium is the behavior of the detector in the presence of gamma rays. Typical recoil scintillators will have dimensions that are large with respect to both the recoil nuclei produced by neutron interactions as well as to fast electrons created in gamma-ray interactions. Therefore, the full energy of each is almost always deposited within the detector active volume. Gas proportional counters, on the other hand, are often of a size that is comparable with recoil nuclei ranges, and usually they are much smaller than the range of gamma-ray-produced recoil electrons. Therefore, it is likely that neutron-induced events deposit all their energy in the gas, whereas gamma rays will deposit only a small part of their energy. Furthermore, the inherent response of organic scintillators is such that the light output may be as much as a factor of 2 or 3 times greater for electrons than for charged particles of the same energy. This extreme variation in response does not exist in proportional counters, in which the energy expended to create one ion pair changes only slightly with the nature of the ionizing particle.

A difference that works to the disfavor of the proportional counter is the nature of gamma-ray interactions in the detector.[131] For the organic scintillator, both neutrons and gamma rays must interact within the scintillator volume in order to give rise to detected pulses. In the gas counter, neutrons must interact within the fill gas, but gamma rays may interact either in the gas or, more likely, in the much denser walls and other construction materials of the counter, leading to secondary electrons that can escape into the gas volume. Therefore, if the fill gas of a recoil proportional counter with the same elemental composition as that for an organic scintillator is chosen, one would expect a considerably larger ratio in the number of gamma ray to neutron pulses in the proportional counter compared with the organic scintillator. In the proportional counter, however, the average gamma-ray pulse amplitude would be considerably smaller.

As in all proportional counters, the rise time of observed signal pulses depends primarily on the radial distribution of the ionization track that gives rise to that signal pulse. As discussed for ^{3}He tubes, this property can be used to count preferentially neutron-induced events in the presence of potentially interfering gamma radiation. Provided the neutron energy is fairly low, proton recoil tracks will tend to be rather short and consequently will be confined to a limited range in radius. Gamma-ray-induced fast electrons will almost always pass completely through the gas and, on the average, involve a much greater range of radii in the tube. As a result, neutron-induced pulses will tend to have shorter rise times than those induced by gamma rays, and pulse shape discrimination methods will serve to differentiate between the two. Extension of the proton recoil technique in proportional counters down to a neutron energy as low as 1 keV, as pioneered by Bennett,[132] is critically dependent on elimination of the majority of the gamma-ray pulses by this method.

3. COUNTER DESIGN AND CHOICE OF FILL GAS

Although some results have been reported for spherically shaped chambers,[133–136] the majority of recoil proportional counter designs incorporate cylindrical geometry similar to that discussed in Chapter 6 and illustrated in Fig. 6.6. As in all proportional counters, the axial anode wire must be very uniform and of small diameter to ensure uniform gas multiplication along its length. The field tubes at either end of the anode wire are common devices used to abruptly terminate the region in which gas multiplication can occur, so that the active volume of the counter is well defined. Nonetheless, there remains some nonuniformity (sometimes called the *tip effect*) due to the curvature of the electric field lines at the point of discontinuity between the field tube and anode wire. Bennett and Yule[137,138] report that the electric field distortions can be minimized by reducing the outer diameter of the cathode in these regions. Nonetheless, the distortion of the electric field in the end regions of the tube remains one of the important complicating factors that must be considered in analyzing the efficiency and response function of the detector.

By far the majority of all recoil detectors utilize hydrogen or some compound of hydrogen to make use of recoil protons produced by neutron elastic scattering. The simple rectangular energy distribution for proton recoils illustrated in Fig. 15.14 is distorted by several important effects in proportional counters. The most important of these is the effect of the finite size of the active volume of the chamber which leads to proton tracks that are truncated either in the walls or end of the tube. There are several approaches to this problem. One is simply to live with these losses and either attempt to calculate or measure their effects on the overall detector response function. Reasonably good success has been achieved with this approach under some circumstances,[139–141] but it would clearly be preferable to eliminate these events as much as possible. If those proton tracks that leave the active volume can be identified and discarded, a much simpler response function will result. A track that leaves the active volume at the cylindrical cathode can be recognized if a somewhat more complex chamber is constructed, in which the cathode consists of a cage of fine wire so that the proton can pass through and continue to produce ionization outside the cathode. A ring of counters around the central region can be used to detect this leaked ionization and thus to reject the original event by anticoincidence. Such a chamber has been described by Heiberg,[142] in which losses to the ends of the tubes were also eliminated through the use of signals derived from the field tubes. Tracks that overlap the active volume and end region of the tube will give coincident pulses on both the anode and field tubes and therefore can also be rejected through anticoincidence.

Although hydrogen is in many ways the ideal target for neutron scattering, its use as a fill gas in proportional counters is limited by its low density and relatively low stopping power for recoil protons. Methane is a more common choice, but complicating effects introduced by the carbon nuclei must then be taken into account. It has been estimated[137] that carbon ions will create approximately 75% as much ionization in the gas as the equivalent energy proton. Because the maximum energy of a carbon ion is 28% of the neutron energy, we would then expect that all carbon recoil pulses should lie below about 21% of the maximum proton recoil pulse amplitude. Helium is also widely used[143] as a fill gas, and typical recoil spectra are shown in Fig. 15.21. Because elastic scattering from helium is not

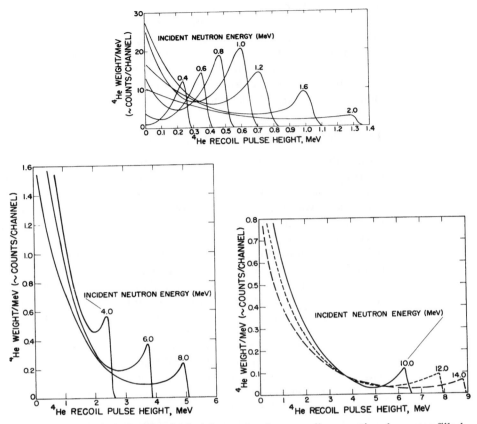

Figure 15.21 Calculated pulse height spectra for a recoil proportional counter filled with ^{4}He at 8 atm pressure. (From Atwater.[144])

isotropic in the center-of-mass system (see Fig. 15.13), these distributions will not be rectangular even in the absence of complicating effects.

4. ENERGY CALIBRATION AND LINEARITY

Because sources of monoenergetic neutrons are not commonly available, the energy calibration of a proton recoil proportional counter is not a straightforward process. Gamma rays cannot be used as previously described in the calibration of proton recoil scintillators because the secondary electrons created in gamma ray interactions generally cannot be stopped in the fill gas. One possible method is to incorporate a small amount of ^{3}He into the fill gas so that, when irradiated with thermal neutrons, proton and triton pairs sharing 764 keV kinetic energy (see p. 508) are generated internally. The corresponding peak in the recorded spectrum then provides an energy marker. Although this technique has proven very useful, errors in the calibration can result[145] because one of these particles is not a proton. Differences then can arise that are related to differences in the energy required to form an ion pair or due to disparities in the particle ranges. At low energies, the addition of a small amount of radioactive ^{37}Ar will result in a peak from the 2.82 keV X-rays that are emitted in its decay.[140] This approach has the disadvantage that the calibration peak is present in all measurements and may interfere with an interesting region of the spectrum. The calibration is also sensitive to potentially large differences in the energy loss behavior of the low energy photoelectrons generated by the X-rays compared with the recoil protons of interest.

Above about 10 keV, there seems to be little doubt that the ionization produced by recoil protons in standard proportional counters is essentially linear with the proton recoil energy. Stated another way, the W-value (defined as the energy expended per ion pair produced) in hydrogen is a constant for protons above 10 keV. Below this energy, however, there is evidence[146] that the value of W falls off and may be as low as about 70% of its high-energy value at a proton energy of 1 keV. These nonlinearities in the relation between energy and ionization must obviously be taken into account when unfolding the detector response to low-energy neutrons.

D. Proton Recoil Telescopes

In conventional organic scintillators or recoil proportional counters, one has little choice but to accept all angles of neutron scattering as they occur. Therefore, the response function of these detectors to monoenergetic neutrons incorporates all recoil proton energies up to the neutron energy and has the approximately rectangular shape discussed earlier. For spectroscopy purposes, however, it would be preferable if the response function were a simple narrow peak to avoid the problems of spectrum unfolding otherwise required. If only those proton recoils that occur at a fixed angle with respect to the neutron direction can be singled out, the recoil proton energy will be fixed for monoenergetic neutrons and the response function will approach the ideal narrow peak. Devices based on a narrow selection of recoil directions are generally known as *proton recoil telescopes* and have been applied to a wide variety of fast neutron measurements.

From Eq. (15.3), the energy of recoil protons observed at an angle θ with respect to the incoming neutron direction is given simply by

$$E_p = E_n \cos^2 \theta \tag{15.12}$$

Recoil proton telescopes can be applied only to situations in which the incoming neutron direction has been defined by collimation or other means.

A schematic diagram of a common form of recoil telescope is shown in Fig. 15.22. Neutrons are incident on a thin film, usually made from an organic polymer, whose thickness is kept small compared with the range of the lowest energy recoil proton to be measured. The angle θ at which recoil protons are observed is defined by positioning a detector some distance from the radiator, with the intervening space evacuated to prevent proton energy loss. Because of the $\cos^2 \theta$ fall-off of recoil proton energy, the detector is usually positioned at a small angle with respect to the neutron direction. Many designs put the proton detector at $\theta = 0$, but others choose a finite observation angle to avoid neutron-induced background events in the detector from the primary beam.

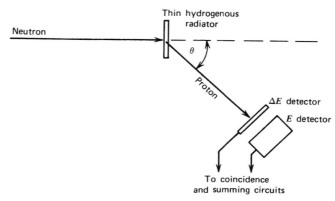

Figure 15.22 A proton recoil telescope.

Although a single detector can, in principle, be used to measure the proton energy and hence the neutron energy through Eq. (15.12), multiple detectors are often used in coincidence to reduce backgrounds from competing reactions and other unwanted events. The arrangement shown in Fig. 15.22 is a common one in which a very thin ΔE detector is placed in front of a thicker E detector, which fully stops the recoil protons. By operating the two detectors in coincidence, only particles incident from the direction of the radiator are recorded. If both detector responses are linear, the sum of the two signals will be proportional to the total proton energy. More elaborate particle identification schemes (see p. 396) can be applied to this arrangement by demanding that the ΔE signal be of proper size relative to the E signal to correspond to the energy loss expected from protons. This requirement will then effectively eliminate other charged particle events such as alpha particles produced in (n, α) reactions in the radiator or other structural components. Proton recoil telescopes of this design, which use semiconductor diode detectors to measure recoil proton energy, are described in Refs. 147 and 148. Other detector types, including gas proportional counters and scintillation detectors, also have been incorporated as the proton detector.

The dominant disadvantage of the proton recoil telescope is its extremely low detection efficiency (typically one count per 10^5 incident neutrons.) This low efficiency stems from two factors, neither of which can be improved without sacrificing energy resolution of the device. First, the radiator thickness must be kept small to avoid appreciable energy loss of the recoil protons before they leave the radiator. Usable radiator thicknesses lead to a probability of about 10^{-3} or 10^{-4} that the incident neutron undergoes a scattering event within the radiator. Second, the solid angle subtended by the recoil proton detectors must be kept relatively small to avoid including too large a spread in recoil angles, and consequently smearing the peak response function.

One of the attractive features of proton recoil telescopes is the fact that their detection efficiency can be calculated quite accurately. Because complications such as multiple scattering or wall effects are largely avoided, the probabilities of neutron scattering and subsequent proton recoil detection are quite easily calculated from the accurately known hydrogen scattering cross section and geometric evaluation of the detector solid angle. Useful analyses of these factors can be found in Refs. 149–152.

One way to improve the detection efficiency is to make the radiator thick and to record separately the energy loss of the protons before they escape. Then the radiator takes the form of a hydrogenous detector such as an organic scintillator[153,154] or methane-filled proportional counter.[155,156] Protons that originate at any depth up to their range can deposit part of their energy in the radiator-detector, and then leave its surface to deposit the remainder in a second detector located some distance away to define the recoil angle. By analyzing the coincident pulses observed from the two detectors, the neutron energy that gave rise to the recoil proton can be inferred.

E. Capture-Gated Neutron Spectrometer

As was illustrated in Fig. 15.18, the pulse height spectrum produced by monoenergetic neutrons in a typical organic or plastic scintillator shows a broad continuum that stretches from zero amplitude up to the equivalent of the full neutron energy. This continuum reflects the energy of single or multiple recoil protons produced by the fast neutron in its interactions within the detector. Because most fast neutrons enter the scintillator and then escape after having deposited a variable fraction of their energy, this broad continuum is the result.

The ideal pulse height spectrum is a single peak for monoenergetic neutrons. An alternative approach has been introduced using organic scintillators that comes closer to this ideal. The principle is illustrated in Fig. 15.23. Imagine that we have a reasonably large (say a diameter of 10 cm or more) plastic scintillator that has been loaded with a small percentage of ^{10}B. The larger the scintillator volume, the greater will be the probability that a

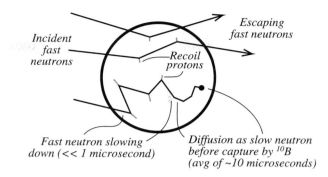

Figure 15.23 Principle of the capture-gated neutron spectrometer using a plastic scintillator loaded with ^{10}B.

neutron enters the scintillator as a fast neutron, but suffers enough collisions so that it does not escape but is captured as a slow neutron in the scintillator. In this case, a series of recoil protons will be produced whose energy must sum to be the incident neutron energy. Because the fast neutron is traveling at a significant fraction of the speed of light, all the recoil protons are produced within a short period of time, typically less than 50 ns. Because they appear in a short time compared with typical pulse shaping times, all the light from the recoil protons adds together to form a single output pulse. If a method is implemented to select only this type of event and discard all others in which the neutron escapes without having lost all its energy, we would then expect a single peak in the corresponding pulse height spectrum.

Once the neutron has lost its energy through these multiple collisions, it continues to diffuse as a thermal neutron within the scintillator. If no ^{10}B were present, it would eventually be captured by one of the hydrogen atoms in the plastic, giving rise to a 2.2 MeV capture gamma ray with a high probability of escape. However, if a sufficient boron concentration is present, the high cross section for capture in ^{10}B will predominate, and virtually all thermalized neutrons will be captured by boron nuclei. In most of these capture events, 2.3 MeV is released as kinetic energy of heavy charged particles (see Chapter 14) that will deposit their energy locally and give rise to a second pulse of light. The average time separation between the light produced by the recoil protons and the light produced by the boron capture products is the mean diffusion time of thermal neutrons in the boron-loaded scintillator, and might typically be around 10 μs.

Thus we have a unique signature that can be used to single out just those neutrons that lose all their energy in the detector. That signature is a pulse (produced by the multiple recoil protons) that is followed within 10–20 μs by a second pulse that corresponds to the capture in ^{10}B. To make the selection even more specific, pulse height selection criteria can be applied to the second pulse to require that its energy correspond to a window around 2.3 MeV. Neutrons that enter the detector, create one or more recoil protons, and escape are rejected because a second pulse does not occur as in the case of full energy deposition.

The expected sequence of pulses is illustrated on Fig. 15.24. When the incident rate of neutrons is low, there is only a small probability that an unrelated neutron will create a pulse within the gating period and create a "chance coincidence" that would be a false identification. Since this probability is uncorrelated with the occurrence of a previous pulse, the contribution of these chance events can be determined and later subtracted from the recorded spectrum by inserting a long delay before the onset of the gating period. At higher rates, the contribution of chance events becomes more significant and ultimately

Figure 15.24 Sequence of two pulses expected from a capture-gated neutron spectrometer when the incident fast neutron is thermalized and captured.

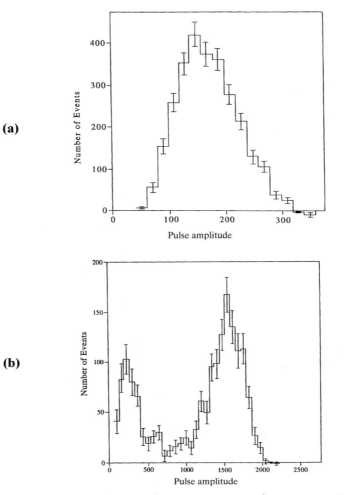

Figure 15.25 Measured spectra from a capture-gated neutron spectrometer for incident neutrons with energies of 1.2 MeV (part *a*) and 6.7 MeV (part *b*). (From Czirr and Jensen.[160])

limits the maximum practical counting rate when the chance rate overwhelms the true rate.

Spectrometers of this type can be constructed using, as in the example above, boron-loaded plastic,[157,158] boron-loaded liquid scintillator (which allows for pulse shape discrimination against incident gamma rays),[159] or heterogeneous combinations of plastic scintillator and ^{6}Li loaded glass scintillator plates.[160] In the latter case, the neutron capture leads to deposition of 4.78 MeV in the glass scintillator. The different decay times of the plastic scintillator and the lithium glass allow pulse shape discrimination techniques to be used to differentiate between the proton recoil and capture pulses, permitting a more specific identification of the full-energy absorption event.

One advantage of capture-gated spectrometers is that the neutron detection efficiency can be as high as 10% in the few MeV range,[160] dropping to about 1% at 14 MeV. These efficiencies obviously depend on the size and detailed design of the detector, but are much higher than those of the proton recoil telescope described in the previous section. While the measured pulse height spectra illustrated in Fig. 15.25 consist primarily of a single peak, the width of the peak is rather broad indicating a poorer energy resolution than might be expected based on the response of the scintillator to monoenergetic charged particles of the same energy. Some of this degraded energy resolution is caused by the nonlinear response typical of organic scintillators. When a neutron loses all its energy, the multiple recoil proton energies must sum to exactly the energy of the neutron. Because different combinations of proton energies will be involved for every individual case, however, nonproportionality of the energy-to-light conversion will mean that the light will not always sum to the same value. The greater the nonlinearity of the scintillator, the greater will be the spread in the light produced even by incident monoenergetic neutrons. In addition to this broadening, there is also a rise in the spectrum at low pulse amplitude in the response to neutrons of energies above 5 or 6 MeV. This rise is caused by the growing importance of inelastic scattering of the neutron from the carbon nuclei that are also present in the scintillator. In an inelastic event, some of the neutron energy may escape as a gamma ray and the recoil proton energies then fall short of representing the full energy of the neutron.

PROBLEMS

15.1 Calculate the efficiency of a 4-mm thick ^{6}LiI scintillator for incident 1 MeV neutrons. Repeat for thermal neutrons.

15.2 A thermal neutron detector is placed at the center of a spherical moderator that is exposed to a source of 5-MeV neutrons. If the moderator diameter is varied while holding all other conditions constant, sketch the corresponding expected variation of the counting rate. Offer physical explanations for the behavior of this curve at both large and small diameters.

15.3 A lithium iodide scintillator is often used as the central detector in the neutron spherical dosimeter. Sketch the expected pulse height spectrum from the scintillator in this application.

15.4 An incident fast neutron is moderated and then diffuses a total pathlength of 10 cm before being captured in the BF$_3$ tube of a long counter. Estimate the time delay between the time of neutron incidence and the leading edge of the output pulse.

15.5 Incident 3 MeV neutrons interact in a lithium sandwich spectrometer. Calculate the reaction product energies for the case in which the alpha particle is emitted in the forward direction at 0° and the triton at 180°.

15.6 Calculate the maximum proton energy from the ^{3}He(n, p) reaction when induced by 1.5 MeV neutrons.

15.7 Explain the physical origin of the epithermal peak observed in most pulse height spectra from ^{3}He proportional tubes when used with fast neutron sources.

15.8 Calculate the detection efficiency of a methane-filled proportional counter for incident 100 keV neutrons if the gas pressure is 1 atm and the neutron pathlength through the gas is 5 cm.

15.9 A 1 MeV neutron enters a plastic scintillator and undergoes two sequential scatterings from hydrogen nuclei before escaping. If the first scattering deflects the neutron at an angle of 40° with respect to its original direction and the scattering sites are 3 cm apart, calculate the time that separates the two events. If the PM anode time constant is 20 ns, will the two events be resolved?

15.10 Sketch the differential pulse height spectrum you would expect from a proton recoil detector if the incident neutron energy spectrum is known to have three very prominent and narrow peaks at 75, 150, and 300 keV.

15.11 Show that the angle (in the laboratory frame) between a recoil proton and the corresponding scattered neutron is always 90°.

15.12 Using the parameters listed in Table 6.1, estimate the maximum pulse amplitude expected if a methane-filled proportional counter with the following properties is irradiated by 1-MeV neutrons: gas pressure, 0.75 atm; applied voltage, 2000 V; anode radius, 0.005 cm; cathode radius, 2 cm; tube capacitance, 60 pF.

15.13 A silicon detector is irradiated by 1 MeV neutrons. Find the minimum and maximum energies expected for the silicon recoil nuclei produced in elastic scattering of the incident neutrons.

15.14 What basic physical difference leads to the observation that the recoil energy distribution from 5 MeV neutron scattering from hydrogen is uniform or rectangular shaped, while it is highly nonuniform for scattering from helium?

15.15 What factor limits increasing the detection efficiency of a proton recoil telescope by simply increasing the thickness of the hydrogenous radiator?

15.16 A capture-gated neutron spectrometer is based on the use of a plastic scintillator that is loaded with natural boron to 5% by weight (see properties of BC-454 in Table 8.1). Assume that an incident neutron is fully moderated at a position near the center of the scintillator, and begins to diffuse as a thermal (0.025 eV) neutron. Calculate the expected mean time from the start of the diffusion process to the time the thermal neutron triggers a capture reaction in ^{10}B. Offer a justification for why this value is likely to be smaller than the observed pulse pair separation times for capture-gated spectrometers with this composition.

REFERENCES

1. R. L. Bramblett, R. I. Ewing, and T. W. Bonner, *Nucl. Instrum. Meth.* **9**, 1 (1960).
2. T. L. Johnson, Y. Lee, K. A. Lowry, and S. C. Gorbics, *Proceedings of the American Nuclear Society Topical Meeting on Theory and Practices in Radiation Protection and Shielding,* April 1987.
3. G. J. H. Jacobs and R. L. P. van den Bosch, *Nucl. Instrum. Meth.* **175**, 483 (1980).
4. L. W. Brackenbush and R. I. Scherpelz, PNL-SA-11645, CONF-840202-13 (1983).
5. W. H. Miller and R. M. Brugger, *Nucl. Instrum. Meth.* **A236**, 333 (1985).
6. D. W. O. Rogers, *Health Phys.* **37**, 735 (1979).
7. N. E. Hertel and J. W. Davidson, *Nucl. Instrum. Meth.* **A238**, 509 (1985).
8. V. Mares, G. Schraube, and H. Schraube, *Nucl. Instrum. Meth.* **A307**, 398 (1991).
9. A. Aroua et al., *Nucl. Instrum. Meth.* **A321**, 298, 305, and 312 (1992).
10. V. Mares and H. Schraube, *Nucl. Instrum. Meth.* **A337**, 461 (1994).
11. V. Mares and H. Schraube, *Nucl. Instrum. Meth.* **A366**, 203 (1995).
12. D. E. Hankins, LA-2717 (1962).
13. D. E. Hankins and R. A. Pederson, LAMS-2977 (1964).
14. J. W. Leake, *Nucl. Instrum. Meth.* **63**, 329 (1968).
15. K. G. Harrison, *Nucl. Instrum. Meth.* **166**, 197 (1979).
16. J. W. Leake, *Nucl. Instrum. Meth.* **178**, 287 (1980).
17. C. Birattari et al., *Nucl. Instrum. Meth.* **A297**, 250 (1990).
18. C. Birattari et al., *Nucl. Instrum. Meth.* **A324**, 232 (1993).
19. M. Tan, C. M. Chen, Y. Q. Wen, and Z. Wang, *Nucl. Instrum. Meth.* **A339**, 573 (1994).
20. A. Klett and B. Burgkhardt, *IEEE Trans. Nucl. Sci.* **44**(3), 757 (1997).
21. C. Birattari et al., *Rad. Prot. Dosim.* **44**(1/4), 193 (1992).
22. H. H. Hsu, K. R. Alvar, and D. G. Vasilik, *IEEE Trans. Nucl. Sci.* **41**(4), 938 (1994).
23. A. O. Hanson and M. L. McKibben, *Phys. Rev.* **72**, 673 (1947).
24. M. H. McTaggart, AWRE NR/Al/59 (1958).
25. J. De Pangher and L. L. Nichols, BNWL-260 (1966).
26. J. B. Hunt and J. C. Robertson, *Proceedings of the First Symposium on Neutron Dosimetry in Biology and Medicine,* EUR 4896 d-f-e, 935 (1972).
27. D. R. Slaughter and D. W. Rueppel, *Nucl. Instrum. Meth.* **145**, 315 (1977).
28. J. B. Hunt and R. A. Mercer, *Nucl. Instrum. Meth.* **156**, 451 (1978).
29. L. V. East and R. B. Walton, *Nucl. Instrum. Meth.* **72**, 161 (1969).
30. A. E. Evans, *Nucl. Instrum. Meth.* **199**, 643 (1982).
31. R. F. Barrett, J. R. Birkelund, and H. H. Thies, *Nucl. Instrum. Meth.* **68**, 277 (1969).
32. B. K. Kamboj, M. G. Shahani, U. V. Phadnis, and D. Sharma, *Nucl. Instrum. Meth.* **148**, 57 (1978).
33. E. Hochhäuser and E. Schönfeld, *Nucl. Instrum. Meth.* **80**, 347 (1970).
34. K. K. Sekharan, H. Laumer, B. D. Kern, and F. Gabbard, *Nucl. Instrum. Meth.* **133**, 253 (1976).
35. E. A. Sokol et al., *Nucl. Instrum. Meth.* **219**, 336 (1984).
36. R. E. Mayer et al., *Nucl. Instrum. Meth.* **A324**, 501 (1993).
37. W. P. Poenitz, *Nucl. Instrum. Meth.* **58**, 39 (1968).
38. W. P. Poenitz, *Nucl. Instrum. Meth.* **72**, 120 (1969).
39. W. P. Poenitz, ANL-7915 (1972).
40. D. J. DeSimone et al., *Nucl. Instrum. Meth.* **A388**, 443 (1997).
41. J. R. P. Eaton and J. Walker, *Proc. Phys. Soc.* (*London*) **83**, 301 (1964).
42. D. R. Johnson, J. H. Thorngate, and P. T. Perdue, *Nucl. Instrum. Meth.* **75**, 61 (1969).
43. R. B. Murray, *Nucl. Instrum. Meth.* **2**, 237 (1958).
44. A. R. Spowart, *Nucl. Instrum. Meth.* **75**, 35 (1969).
45. A. R. Spowart, *Nucl. Instrum. Meth.* **82**, 1 (1970).
46. J. M. Neill, D. Huffman, C. A. Preskitt, and J. C. Young, *Nucl. Instrum. Meth.* **82**, 162 (1970).
47. W. R. McMurray, N. J. Pattenden, and G. S. Valail, *Nucl. Instrum. Meth.* **114**, 429 (1974).

48. A. R. Spowart, *Nucl. Instrum. Meth.* **135**, 441 (1976).

49. A. R. Spowart, *Nucl. Instrum. Meth.* **140**, 19 (1977).

50. E. J. Fairley and A. R. Spowart, *Nucl. Instrum. Meth.* **150**, 159 (1978).

51. S. Yamaguchi, *Nucl. Instrum. Meth.* **A274**, 573 (1989).

52. G. L. Jensen and J. B. Czirr, *Nucl. Instrum. Meth.* **205**, 461 (1983).

53. K. H. Abel et al., *Nucl. Instrum. Meth.* **A353**, 114 (1994).

54. G. B. Spector, T. McCollum, and A. R. Spowart, *Nucl. Instrum. Meth.* **A329**, 223 (1993).

55. G. Zanella et al., *Nucl. Instrum. Meth.* **A359**, 547 (1995).

56. C. Maroni, F. Russo, and E. Verondini, *Nucl. Instrum. Meth.* **74**, 256 (1969).

57. I. C. Rickard, *Nucl. Instrum. Meth.* **113**, 169 (1973).

58. M. G. Silk, *Nucl. Instrum. Meth.* **66**, 93 (1968).

59. G. B. Bishop, *Nucl. Instrum. Meth.* **62**, 247 (1968).

60. R. A. Wolfe and W. F. Stubbins, *Nucl. Instrum. Meth.* **60**, 246 (1968).

61. H. Bluhm and D. Stegemann, *Nucl. Instrum. Meth.* **70**, 141 (1969).

62. G. Koutzoukos and C. B. Besant, *J. Br. Nucl. Energy Soc.* **14**, 83 (1975).

63. P. J. Clements, *Nucl. Instrum. Meth.* **127**, 61 (1975).

64. T. Pinelli et al., *Nucl. Instrum. Meth.* **150**, 497 (1978).

65. A. Sayres and M. Coppola, *Rev. Sci. Instrum.* **35**, 431 (1964).

66. J. L. Friedes and R. E. Chrien, *Rev. Sci. Instrum.* **35**, 469 (1964).

67. T. Fuse, T. Miura, A. Yamaji, and T. Yoshimura, *Nucl. Instrum. Meth.* **74**, 322 (1969).

68. S. Nishino, T. Nakamura, and T. Hyodo, *Mem. Fac. Eng.* (*Kyoto Univ.*) **35**(3), 309 (1973).

69. E. Dietze et al., *Nucl. Instrum. Meth.* **A332**, 521 (1993).

70. N. Takeda and K. Kudo, *IEEE Trans. Nucl. Sci.* **41**(4), 880 (1994).

71. S. Izumi and Y. Murata, *Nucl. Instrum. Meth.* **94**, 141 (1971).

72. J. M. Cuttler, S. Greenberger, and S. Shalev, *Nucl. Instrum. Meth.* **75**, 309 (1969).

73. H. Takahashi et al., *Nucl. Instrum. Meth.* **A353**, 164 (1994).

74. H. Franz et al., *Nucl. Instrum. Meth.* **144**, 253 (1977).

75. J. G. Owen, D. R. Weaver, and J. Walker, *Nucl. Instrum. Meth.* **188**, 579 (1981).

76. W. A. Fisher, S. H. Chen, D. Gwinn, and R. R. Parker, *Nucl. Instrum. Meth.* **219**, 179 (1984).

77. A. E. Evans, *IEEE Trans. Nucl. Sci.* **NS-32**(1), 54 (1985).

78. K.-H. Beimer, G. Nyman, and O. Tengblad, *Nucl. Instrum. Meth.* **A245**, 402 (1986).

79. H. Ohm, K.-L. Kratz, and S. G. Prussin, *Nucl. Instrum. Meth.* **A256**, 76 (1987).

80. F. Hoenen and W. Bieger, *Nucl. Instrum. Meth.* **A259**, 529 (1987).

81. M. J. Loughlin, J. M. Adams, and G. Sadler, *Nucl. Instrum. Meth.* **A294**, 606 (1990).

82. T. Iguchi, N. Nakayamada, H. Takahashi, and M. Nakazawa, *Nucl. Instrum. Meth.* **A353**, 152 (1994).

83. S. Shalev and J. Cuttler, *Nucl. Sci. Eng.* **51**, 52 (1973).

84. A. E. Evans, Jr., "Development of a High-Pressure ^{3}He Neutron Scintillator Spectrometer," Los Alamos National Laboratory Program Technical Note, LA-Q2TN-82-109, Apr. 29, 1982.

85. M. S. Derzon, D. R. Slaughter, S. G. Prussin, *IEEE Trans. Nucl. Sci.* **NS-33**(1), 247 (1986).

86. H. Bluhm, *Nucl. Instrum. Meth.* **115**, 325 (1974).

87. M. G. Silk, "The Determination of the Fast Neutron Spectrum in Thermal Reactors Using ^{6}Li and ^{3}He Semiconductor Spectrometers," AERE-R-5183 (1966).

88. T. R. Jeter and M. C. Kennison, *IEEE Trans. Nucl. Sci.* **NS-14**(1), 422 (1967).

89. J. B. Marion and F. C. Young, *Nuclear Reaction Analysis,* North-Holland, Amsterdam, 1968.

90. J. A. Harvey and N. W. Hill, *Nucl. Instrum. Meth.* **162**, 507 (1979).

91. A. A. Naqvi et al., *Nucl. Instrum. Meth.* **A325**, 574 (1993).

92. A. Aksoy et al., *Nucl. Instrum. Meth.* **A337**, 486 (1994).

93. S. Mouatassim et al., *Nucl. Instrum. Meth.* **A359**, 530 (1995).

94. A. A. Naqvi, M. M. Nagadi, S. Shaheen, and A. Bari, *Nucl. Instrum. Meth.* **A356**, 330 (1995).

95. N. Colonna and G. Tagliente, *Nucl. Instrum. Meth.* **A416**, 109 (1998).

96. F. Arneodo et al., *Nucl. Instrum. Meth.* **A418**, 285 (1998).

97. J. H. Lee and C. S. Lee, *Nucl. Instrum. Meth.* **A402**, 147 (1998).

98. F. D. Brooks et al., *Nucl. Instrum. Meth.* **A410**, 319 (1998).

99. J. Devos et al., *Nucl. Instrum. Meth.* **135**, 395 (1976).

100. R. H. Johnson et al., *Nucl. Instrum. Meth.* **145**, 337 (1977).

101. D. Slaughter and R. Strout II, *Nucl. Instrum. Meth.* **198**, 349 (1982).

102. R. Ofek, A. Tsechanski, A. E. Profio, and G. Shani, *Nucl. Instrum. Meth.* **A278**, 513 (1989).

103. P. R. P. Coelho, A. A. Da Silva, and J. R. Maiorino, *Nucl. Instrum. Meth.* **A280**, 270 (1989).

104. M. J. Coolbaugh, R. E. Faw, and W. Meyer, "Fast Neutron Spectroscopy in Aqueous Media Using an NE213 Proton Recoil Spectrometer System," COO-2049-7 (1971).

105. M. E. Toms, *Nucl. Instrum. Meth.* **92**, 61 (1971).

106. P. Marinkovic, S. Avdic, M. Pesic, and N. Zavaljevski, *Nucl. Instrum. Meth.* **A321**, 333 (1992).

107. N. Sasamoto and S. Tanaka, *Nucl. Instrum. Meth.* **148**, 395 (1978).

108. M. Bormann, R. Kühl, K. Schäfer, and U. Seebeck, *Nucl. Instrum. Meth.* **88**, 245 (1970).

109. D. Hermsdorf, K. Pasieka, and D. Seeliger, *Nucl. Instrum. Meth.* **107**, 259 (1973).

110. J. L. Fowler et al., *Nucl. Instrum. Meth.* **175**, 449 (1980).

111. M. Drosg, D. M. Drake, and P. Lisowski, *Nucl. Instrum. Meth.* **176**, 477 (1980).

112. J. Cub et al., *Nucl. Instrum. Meth.* **A274**, 217 (1989).

113. K. Gul, A. A. Naqvi, and H. A. Al-Juwair, *Nucl. Instrum. Meth.* **A278**, 470 (1989).

114. T. Akimoto et al., *IEEE Trans. Nucl. Sci.* **38**(5), 1040 (1991).

115. S. Meigo, *Nucl. Instrum. Meth.* **A401**, 365 (1997).

116. M. A. Al-Ohali et al., *Nucl. Instrum. Meth.* **A396**, 388 (1997).

117. R. Cherubini et al., *Nucl. Instrum. Meth.* **A281**, 349 (1989).

118. G. Dietze and H. Klein, *Nucl. Instrum. Meth.* **193**, 549 (1982).

119. A. V. Hristova, E. I. Vapirev, L. T. Tsankov, and V. Jordanov, *Appl. Radiat. Isot.* **41**(9), 887 (1990).

120. R. E. Howe, *Nucl. Instrum. Meth.* **190**, 309 (1981).

121. A. G. Da Silva et al., *Nucl. Instrum. Meth.* **A264**, 381 (1988).

122. M. Moszynski et al., *Nucl. Instrum. Meth.* **A317**, 262 (1992).

123. M. Moszynski et al., *Nucl. Instrum. Meth.* **A350**, 226 (1994).

124. N. A. Lurie, L. Harris, Jr., and J. C. Young, *Nucl. Instrum. Meth.* **129**, 543 (1975).

125. C. Chen, J. A. Lockwood, and L. Hsieh, *Nucl. Instrum. Meth.* **138**, 363 (1976).

126. D. T. Ingersoll and B. W. Wehring, *Nucl. Instrum. Meth.* **147**, 551 (1977).

127. L. Büermann et al., *Nucl. Instrum. Meth.* **A332**, 483 (1993).

128. T. Novotny, L. Büermann, S. Guldbakke, and H. Klein, *Nucl. Instrum. Meth.* **A400**, 356 (1997).

129. P. L. Reeder et al., *Nucl. Instrum. Meth.* **A422**, 84 (1999).

130. V. V. Verbinski and R. Giovannini, *Nucl. Instrum. Meth.* **114**, 205 (1974).

131. P. K. Ray and E. S. Kenney, *Nucl. Instrum. Meth.* **144**, 579 (1977).

132. E. F. Bennett, *Nucl. Sci. Eng.* **27**, 16 (1967).

133. P. W. Benjamin, C. D. Kemshall, and J. Redfearn, *Nucl. Instrum. Meth.* **59**, 77 (1968).

134. E. Korthaus, EURFNR-1197; KFK-1994 (1974).

135. W. H. Miller, *Nucl. Instrum. Meth.* **A279**, 546 (1989).

136. K. Weise, M. Weyrauch, and K. Knauf, *Nucl. Instrum. Meth.* **A309**, 287 (1991).

137. E. F. Bennett and T. J. Yule, ANL-7763 (1971).

138. E. F. Bennett and T. J. Yule, *Nucl. Instrum. Meth.* **98**, 393 (1972).

139. N. L. Snidow and H. D. Warren, *Nucl. Instrum. Meth.* **51**, 109 (1967).

140. R. Gold and E. F. Bennett, *Nucl. Instrum. Meth.* **63**, 285 (1968).

141. D. W. Vehar and F. M. Clikeman, *Nucl. Instrum. Meth.* **190**, 351 (1981).

142. S. A. Heiberg, *Nucl. Instrum. Meth.* **63**, 71 (1968).

143. M. Weyrauch, A. Casnati, P. Schillebeeckx, and M. Clapham, *Nucl. Instrum. Meth.* **A403**, 442 (1998).

144. H. F. Atwater, *Nucl. Instrum. Meth.* **100**, 453 (1972).

145. I. R. Brearley, A. Bore, N. Evans, and M. C. Scott, *Nucl. Instrum. Meth.* **192**, 439 (1982).

146. H. Werle, G. Fieg, H. Seufert, and D. Stegemann, *Nucl. Instrum. Meth.* **72**, 111 (1969).

147. T. B. Ryves, *Nucl. Instrum. Meth.* **135**, 455 (1976).

148. M. Cambiaghi, F. Fossati, and T. Pinelli, *Nucl. Instrum. Meth.* **82**, 106 (1970).

149. H. Gotoh and H. Yagi, *Nucl. Instrum. Meth.* **97**, 419 (1971).

150. H. Gotoh and H. Yagi, *Nucl. Instrum. Meth.* **101**, 395 (1972).

151. D. Sloan and J. C. Robertson, *Nucl. Instrum. Meth.* **198**, 365 (1982).

152. B. R. L. Siebert, H. J. Brede, and H. Lesiecki, *Nucl. Instrum. Meth.* **A235**, 542 (1985).

153. K. N. Geller, D. Eccleshall, and T. T. Bardin, *Nucl. Instrum. Meth.* **69**, 141 (1969).

154. J. H. Osborne et al., *Nucl. Instrum. Meth.* **A345**, 308 (1994).

155. H. Borst, *Nucl. Instrum. Meth.* **169**, 69 (1980).

156. C. Mori et al., *Nucl. Instrum. Meth.* **A422**, 75 (1999).

157. W. C. Feldman, G. F. Auchampaugh, and R. C. Byrd, *Nucl. Instrum. Meth.* **A306**, 350 (1991).

158. E. A. Kamykowski, *Nucl. Instrum. Meth.* **A317**, 559 (1992).

159. T. Aoyama et al., *Nucl. Instrum. Meth.* **A333**, 492 (1993).

160. J. B. Czirr and G. L. Jensen, *Nucl. Instrum. Meth.* **A349**, 532 (1994).

Chapter 16

Pulse Processing and Shaping

In this chapter we begin discussion of the methods used to extract information from the pulses produced by radiation detectors. Many of the topics discussed in this and the following two chapters are more thoroughly covered in texts on nuclear electronics, of which Refs. 1 and 2 are examples. Leaving more detailed circuit descriptions to these specialized texts, we stress here the implementation of pulse-handling and processing methods from a *user's* rather than a designer's point of view. A functional or "blackbox" approach will be taken to most electronic components, with the exceptions limited only to those cases in which a more detailed description is necessary to permit an intelligent choice between alternatives or to allow optimization of adjustments normally carried out by the user.

I. DEVICE IMPEDANCES

A basic concept in the processing of pulses from radiation detectors is the impedance of the devices that comprise the signal-processing chain. A simplified representation of the input and output configurations of a typical component is shown in Fig. 16.1. Both the input and output impedances can in general involve capacitive or inductive components, but for the sake of simplicity, a purely resistive impedance will be assumed.

The input impedance Z_i represents the extent to which a device loads a given signal source. A high input impedance will draw very little current from the source and therefore present only a very light load. For example, the input impedance of an oscilloscope is always very high to avoid perturbing the signals that are being inspected. For most applications, input impedances of devices are kept high to avoid excessive loading, but other factors may sometimes dictate situations in which the input impedance must be low enough to load the source significantly.

The output impedance can be thought of as an internal resistance in series with a voltage generator representing the output stage of a given component (see Fig. 16.1). For most applications, one would ideally want this output impedance to be as low as possible to minimize the signal loss when the output is loaded by a subsequent component. In Fig. 16.1, the voltage V_L appearing across a loading Z_L is given by the voltage-divider relation

$$V_L = V_S \frac{Z_L}{Z_0 + Z_L} \tag{16.1}$$

The open-circuit or unloaded ($Z_L = \infty$) voltage appearing at the output of the device is simply V_S. To preserve maximum signal level, one normally wants V_L to be as large a fraction of V_S as possible. If the output impedance is low compared with the load ($Z_0 \ll Z_L$), then $V_L \cong V_S$ and essentially all the signal voltage is transmitted to the load. If the output

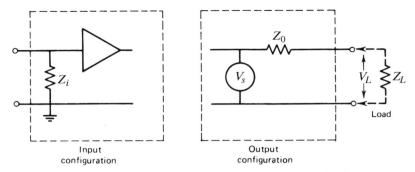

Figure 16.1 Idealized input and output configurations. For the input configuration, the triangle represents an ideal operational amplifier with infinite input impedance. Therefore, the effective device input impedance is Z_i. For the output configuration, Z_0 represents the output impedance in series with an ideal voltage generator V_S.

impedance is equal to the load $(Z_0 = Z_L)$, then $V_L = V_S/2$ and only half the unloaded output voltage of the device appears across the load. Output stages with low output impedance thus are often an important design goal. As an example, an emitter–follower configuration is a common type of output circuit that can have an output impedance of less than 1 ohm. It can therefore drive loads of 50–100 ohms or larger without significant signal attenuation.

When devices are interconnected in a signal chain, the load Z_L presented to a given component is the impedance of the following component. If all output impedances are low compared with input impedances, the maximum signal level is preserved throughout the chain. This condition is often realized in normal detector pulse-processing systems. If very fast pulses are being handled, however, considerations involving reflections in coaxial cables can dictate impedance-matching conditions in which the output/input impedance ratio is not always small, and some signal attenuation will occur.

II. COAXIAL CABLES

A. Cable Construction

Virtually all interconnection of components in a signal chain for radiation detector pulses is carried out using shielded coaxial cable. A diagram of a typical cable construction is shown in Fig. 16.2. The shielded construction is designed to minimize pickup of noise from stray electric and electromagnetic fields. To preserve the flexibility of the cable, the outer shield is usually made of braided strands of fine copper wire. The effectiveness against low-frequency electric fields is determined primarily by the tightness of the braided shield. High-frequency electromagnetic fields are shielded by virtue of the skin effect. At

Figure 16.2 Construction of a standard coaxial cable.

Table 16.1　Properties of Coaxial Cables[a]

	Insulating Material	Cable Diameter (cm)	Characteristic Impedance (ohms)	Signal[b] Propagation	HV Rating	Cable Capacitance (pF/m)	Signal Attenuation per Meter	
							MHz	dB
RG-8/U	Polyethylene	1.03	52	0.659	5000	96.8	100	0.066
							400	0.154
RG-11/U	Polyethylene	1.03	75	0.659	5000	67.3	100	0.066
							400	0.138
RG-58/U	Polyethylene	0.50	53.5	0.659	1900	93.5	100	0.135
							400	0.312
RG-58C/U	Polyethylene	0.50	50	0.659	1900	100.1	100	0.174
							400	0.413
RG-59/U	Polyethylene	0.61	73	0.659	2300	68.9	100	0.112
							400	0.233
RG-62/U	Semisolid polyethylene	0.61	93	0.840	750	44.3	100	0.102
							400	0.207
RG-174/U	Polyethylene	0.25	50	0.659	1500	101.0	100	0.289
							400	0.656
RG-178/U	TFE teflon	0.18	50	0.694	1500	95.1	400	0.951
			Double Shielded Coaxial Cables					
RG-9/U	Polyethylene	1.07	51	0.659	5000	98.4	100	0.062
							400	0.135
RG-223/U	Polyethylene	0.52	50	0.659	1900	101.0	100	0.157
							400	0.328

[a]Data derived in part from Coaxial Cable Catalog, Belden Corporation, Richmond, IN.

[b]Fraction of speed of light in a vacuum (3.00×10^8 m/s).

frequencies at which the skin depth is comparable to or smaller than the braid strand thickness (say, greater than 100 kHz), the shielding is quite effective but will become less so at lower frequencies. Under extreme conditions, it is sometimes necessary to surround the braid with a second shield to fully exclude the effects of very strong fields through which the cable must pass. Doubly shielded coaxial cables are commercially available (see Table 16.1) in which a second braided shield is provided, but a better solution for difficult cases is to run a conventional cable inside tubing made of a solid conductor. For most routine applications, single-shielded cable provides entirely satisfactory isolation from unwanted signal pickup.

B. Cable Properties

The velocity of propagation for pulses through a coaxial cable is a function only of the dielectric materials separating the central conductor and the outer shield and is inversely proportional to the square root of the dielectric constant. Cables using air or some other gas as a dielectric have a propagation velocity very close to the velocity of light in a vacuum (3.00×10^8 m/s). Virtually all general-duty signal cables utilize a solid such as polyethylene for the dielectric, in which case the velocity of propagation is about 66% of that of light in a vacuum. Some special-duty cables with polyethylene foam dielectric have a somewhat higher velocity of propagation. At the other extreme, special delay cables with helically wound central conductors can reduce the velocity of propagation by factors of 100 or more.

Table 16.1 lists important properties of coaxial cable types commonly used for nuclear instrumentation. The "RG/U" designation arose originally as a military specification and the numbers were assigned to various cable types in order of approval. Consequently, they bear no relation to specific cable properties. The RG/U designation implies adherence to certain quality standards not usually found in lower-cost coaxial cables intended for consumer electronics applications. The historical trend has been to replace the older large-diameter signal cables (such as RG-8/U) with smaller-sized cables such as RG-59/U or RG-62/U, or with miniature types such as RG-178/U. In signal cables, the important specifications are usually the characteristic impedance (see discussion later in this chapter) and the capacitance per unit length. In cables intended to carry bias voltage to detectors, the maximum voltage rating is also important.

No real cable is a perfect transmission line. There will always be dissipative losses caused by imperfect dielectrics and resistance of the center conductor that will result in some attenuation and distortion of the transmitted pulse, especially its high-frequency components. For most applications, these effects are small and can often be neglected for cables shorter than a few tens of meters. However, for demanding situations involving the transmission of fast rise time pulses, some attention should be paid to the high-frequency attenuation specifications of the cable to prevent deterioration of the fast characteristics of the transmitted pulse. As an example, the distortion of a 1 ns leading edge pulse after transmission through 3 m of RG-174/U cable is easily seen on a fast oscilloscope.[3] Specially made rigid transmission lines using solid copper tubing for both the shield and center conductor, which display superior high-frequency characteristics compared with conventional coaxial cables, are described in Ref. 4.

C. Noise Pickup and Component Grounding

The outer shield also serves to interconnect the chassis of each component with that of the next. When all components are mounted in the same instrument bin or rack, this electrical connection is often redundant. When components are physically separated, however, the shield will tend to establish a common ground potential for all components. If all chassis are not grounded internally to the same point, some dc current may need to flow in the shield to maintain the common ground potential. In many routine applications, this ground current is small enough to be of no practical consequence. However, if components are widely separated and internally grounded under widely different conditions, the shield current can be large and its fluctuations may induce significant noise in the cable. Under these conditions, such *ground loops* must be eliminated by ensuring that all components are internally grounded to a single common point for the entire system. It is therefore common practice to plug all equipment power lines (including oscilloscopes or other monitoring equipment) into a common power strip that connects to the main electrical lines only at a single outlet. Undesirable transient signals can also be induced in cable shields (especially loop configurations) if nearby equipment involves the fast switching of large currents. Computers are a potential source of high frequency noise pickup and may need to be kept at a distance from detectors and preamplifiers where signal levels are low. Brookshier[5] presents a general analysis of many sources of noise pickup in instrumentation systems that are interconnected using coaxial cables.

A technique known as *common mode rejection* is sometimes helpful in reducing the effects of noise pickup on interconnecting cables. The receiving device (often the linear amplifier) is designed with a differential input so that the signal is measured on one input relative to a reference voltage at the second input. Connection to the two inputs is made by identical cables that are run side-by-side to the signal source (often the preamplifier). Only the first cable is connected to the signal source, whereas the cable from the reference input is left open at the sending end. In principle, much of the pickup will appear on both cables and will therefore be eliminated by the differential input.

D. Characteristic Impedance and Cable Reflections

A general discussion of pulse transmission through coaxial cables is best divided into two extremes: cases in which low-frequency or slow pulses are transmitted, or those in which high-frequency or fast pulses are involved. The distinction of whether a given application involves fast or slow pulses depends on a comparison of the fastest pulse component (usually the rise time) with the transit time of the pulse through the cable. For cables with solid polyethylene dielectric, the transit time is about 5.1 ns/m. Pulses having rise times that are large compared with the transit time are slow pulses, whereas those having a rise time comparable to or shorter than the transit time are fast pulses. For cables of a few meters in length, only the pulses from very fast detectors will qualify as fast pulses, whereas even relatively slow detector outputs must be considered as fast pulses if the cable is several hundred meters long. However, most routine situations involve pulses which by this definition are slow.

For slow pulses, the cable acts much like a simple conductor interconnecting components. Its important properties then are simply its series resistance and capacitance to ground. The resistance of the central conductor is very small for cables less than a few hundred meters in length, so the most significant parameter usually is the cable capacitance. This capacitive loading will increase linearly with cable length, but is seldom a practical problem except between the detector and preamplifier. Because the noise characteristics of a typical preamplifier will deteriorate with increasing input capacitance, there is a premium on keeping the additional capacitive loading to a minimum. In most other aspects, the choice of cable for slow pulse applications is not critical, and virtually any of the cables listed in Table 16.1 will suffice for routine interconnection of components.

For fast pulses, a number of other considerations become important. Prominent among these is the *characteristic impedance* of the cable. It is a property that depends on the dielectric material and diameters of the inner conductor and outer shield of the cable but is independent of the cable length. To illustrate the physical meaning of characteristic impedance, imagine that the voltage generator shown in Fig. 16.3 is capable of generating a step voltage change from zero to some finite value V_0 at a time $t = 0$. If this voltage step is applied to the input of a coaxial cable, it will travel down the length of the cable with the velocity of propagation discussed in the previous section. During the time this voltage step is traveling down the cable, current is being drawn from the signal source because a finite charge per unit length of the cable is required to raise the central conductor voltage from zero to V_0. If the cable were infinitely long, this current drain would be continuous because the step voltage would propagate indefinitely and continue to draw current from the

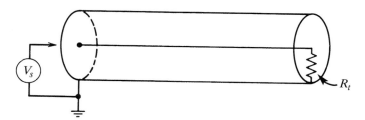

Figure 16.3 Application of a step voltage V_S to a coaxial cable terminated in R_t.

source. In this example, the characteristic impedance of the cable is simply given by the ratio of the step voltage V_0 divided by the current drawn by the infinitely long cable.

For cables of finite length, we must pay some attention to the conditions that exist at the far end of the cable. The cable is *terminated* by the effective resistance R_t that appears between the central conductor and outer shield of the cable at the far end. If the cable is connected to an electronic component, then the termination resistance is effectively just the input impedance of that component. If the cable is simply left unconnected, its termination resistance is infinite, or it may be terminated by connecting a resistor between the central conductor and the outer shield. When it is desired to terminate a cable that is connected to a component with high input impedance with a lower termination resistance, a resistor-to-ground (called a *shunt terminator*) can be inserted parallel to the input of the device so that the effective termination is the parallel combination of the input impedance and the shunt terminator resistance.

Let us next consider what happens when a cable is terminated in its own characteristic impedance. For the sake of example, assume that we have a long cable with 50-ohm characteristic impedance which is terminated by connecting a 50-ohm resistor across the far end. If we apply a 5 V step input to the sending end, then during the time the step is propagating along the length of the cable a current of 100 mA is drawn from the source. The current simply goes into charging successive segments of the cable as the voltage step propagates. During this period the 50-ohm resistor does not draw current because the center conductor voltage at the far end of the cable remains at zero. When the voltage step finally reaches the far end, the situation changes. Current no longer goes into charging up the cable because it is already fully charged over its entire length. However, the 50-ohm resistor now feels a 5 V potential and will therefore begin to draw a current of 100 mA. If the 5 V step is held by the signal generator at the sending end of the cable, this current through the 50-ohm resistor will continue indefinitely. Therefore, the signal generator must continuously supply a 100 mA current to the cable from the instant the step is applied, and it must maintain this current even after the step has propagated along the entire length of the cable. Therefore, as far as the signal generator is concerned, a *coaxial cable terminated in its own characteristic impedance behaves like an infinitely long cable of the same impedance.*

Other conditions prevail if the cable is not terminated in its characteristic impedance. There will then be an abrupt change in the properties of the medium through which the step is propagating, and reflections from the end of the cable will therefore be generated. If the cable is shorted (R_t is made equal to zero), then the step is inverted and reflected back down the cable toward the sending end with an amplitude equal to the original step. If the end of the cable is simply left unterminated (R_t is infinite), then a reflection of the same polarity and amplitude will propagate back toward the sending end. Only when the cable is terminated in its own characteristic impedance are reflections completely avoided.

Table 16.2 Reflection Conditions Created by Various Terminations at the End of a Coaxial Cable with Characteristic Impedance Z_0. Step Input Waveform with Amplitude A Is Assumed

Termination Resistance R	Reflected Step Amplitude
0	$-A$
Between 0 and Z_0	Between $-A$ and 0
Z_0	0
Between Z_0 and ∞	Between 0 and $+A$
∞	$+A$

Table 16.2 shows the reflection conditions for a step voltage of amplitude A for termination conditions between the extremes of zero and infinity.

When transmitting fast pulses through coaxial cables, these reflections can be very undesirable and may lead to distortions of the transmitted pulse form. In these applications, therefore, the experimenter must pay careful attention to the impedance of the cable and the termination conditions at each end. Because the majority of cables used for nuclear pulse applications have characteristic impedances of either 50 or 93 ohms (see Table 16.1), most of the commercial circuits intended for fast pulse applications are designed with input and/or output impedances of 50 or 93 ohms as well. It is usually sufficient to properly terminate only the receiving end of the cable because, in principle, all reflections are thereby avoided. If slight mismatches occur, however, a small reflected component will be propagated back to the sending end where the termination conditions now become important. The output impedance of the generating device determines this termination value, and if it is also equal to the characteristic impedance of the cable, further reflections are suppressed. Instrument systems in which all the input, output, and cable impedances are the same value can therefore be rather generally interconnected without fear of significant reflections.

If the device impedance does not match the cable impedance and reflections must be avoided, external termination resistors may be used at the point where the cable connects to the device in order to create proper termination conditions. A termination resistor added in series will add its resistance value to the device impedance at either end of the cable, while a shunt resistor added in parallel to the output or input will lower the device impedance to the equivalent resistance of the parallel combination (see Fig. 16.4). For example, a 50Ω cable connected to the high-impedance (say $>1000\Omega$) input of a device will be properly terminated by using a 50Ω shunt resistor, since the parallel combination of 50Ω and 1000Ω has an equivalent resistance value very close to 50Ω.

E. Useful Coaxial Cable Accessories

1. TERMINATORS

As described in the previous section, it is often necessary to convert a high-impedance input at the end of a coaxial cable to an impedance that matches the characteristic impedance of the cable to prevent reflections. Because of the widespread use of 50-ohm systems for fast pulse transmission, 50-ohm termination is a common requirement. Consequently, 50-ohm terminators are widely used in fast pulse work and consist of a compact cylindrical 50-ohm resistance to ground fitted with a standard coaxial connector. These can be of the

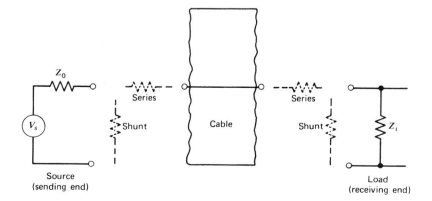

Figure 16.4 Methods of inserting terminators at either end of a coaxial cable.

simple one-ended type with a plug connection, or the "feedthrough" type with a plug and jack on opposite ends. The latter type can be connected directly to a high-impedance input to provide a shunt termination and a substitute jack for cable attachment.

2. PULSE ATTENUATOR

In some pulse-handling situations, the need arises to reduce the amplitude of a pulse to match the input requirements of a given component. For linear signals, this practice should be avoided if possible because attenuation will inevitably deteriorate the signal-to-noise characteristics of the pulses. At times, however, an attenuation step cannot be avoided.

A simple resistive voltage divider, as shown in Fig. 16.5a, will work as an attenuator, but some attention must be paid to impedance levels and high-frequency performance. To provide an invariant multiplication factor of

$$\frac{R_2}{R_1 + R_2}$$

the divider must be used with a source impedance $Z_0 \ll R_1$, and the attenuated pulse must be supplied to a following component whose input impedance $Z_i \gg R_2$. Because of resistor nonidealities and stray capacitance, a simple divider configuration is not usable for pulses with rise or fall times much less than about 100 ns, due to nonlinear attenuation of higher-frequency components and a resulting distortion of the pulse shape. Although the high-frequency performance can be improved with the addition of parallel compensating capacitors (e.g., see Ref. 6), attenuators of this type are not widely used for fast pulse work.

The configuration of Fig. 16.5b is known as a *T-section* attenuator and is more popular as a fast pulse attenuator. It has the advantage that symmetric and equal input and output impedances are realized for convenient matching to coaxial cables. Attenuators of this

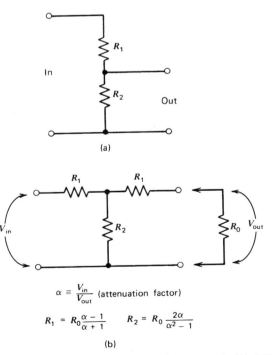

$$\alpha \equiv \frac{V_{in}}{V_{out}} \text{ (attenuation factor)}$$

$$R_1 = R_0 \frac{\alpha - 1}{\alpha + 1} \qquad R_2 = R_0 \frac{2\alpha}{\alpha^2 - 1}$$

(b)

Figure 16.5 (*a*) A simple voltage-divider attenuation network. (*b*) A T-attenuator network. The attenuation factor holds provided the network is loaded with the value of R_0 shown. The input and output impedances are then also equal to R_0.

Figure 16.6 A symmetric pulse splitter that can be used to drive two loads while maintaining matched impedance levels.

type, using thin-film resistors mounted in closely fitting metallic cylinders and provided with standard coaxial connections, have excellent high-frequency characteristics and can be used with nanosecond pulses.

3. PULSE SPLITTER

It is occasionally necessary to split the signal chain into two branches at a given point. For slow pulses for which impedance matching is not important, the familiar coaxial "tee" connector can be used without fear of pulse distortion. When impedance matching is required to prevent cable reflections, the configuration shown in Fig. 16.6 serves to distribute a pulse applied to any terminal to the other two while maintaining a constant impedance level. For example, in a 50-ohm splitter, the resistance values should each be 16.6 ohms so that the impedance looking into any terminal will be 50 ohms if the other two terminals are connected to 50-ohm loads. However, the signal level delivered to each load will be only half that if the load were directly coupled to the source.

4. INVERTING TRANSFORMER

When dealing with signal pulses of various types, it is sometimes desirable to invert the polarity to be compatible with the input requirements of a particular component. In general, an active circuit involving an amplifier stage is required to carry out a general reversal of the polarity of a pulse. However, in the special case that fast pulses with rise and fall times of a few tens of nanoseconds are involved, the *inverting transformer* shown in Fig. 16.7 is a useful, simple, and compact device. The inversion is carried out simply by grounding opposite extremes of the primary and secondary windings on a pulse transformer and will do a creditable job of inverting a pulse as long as its duration is rather short (say, less than 100 ns). For longer pulses, some nonlinearities and distortion of the output pulse are to be expected.

III. PULSE SHAPING

In dealing with signal pulses from radiation detectors, it is often desirable to change the shape of the pulse in some predetermined fashion. By far the most common application is

Figure 16.7 The basic configuration of an inverting pulse transformer.

Figure 16.8 The pulses with long tails shown in part (*a*) illustrate the apparent variation in amplitude due to pulse pile-up. These effects can greatly be reduced by shaping the pulses as in part (*b*).

in processing a train of pulses produced by a preamplifier. In order to ensure that complete charge collection occurs, preamplifiers are normally adjusted to provide a decay time for the pulse which is quite long (typically 50 μs). If the rate of interaction in the detector is not small, these pulses will tend to overlap one another and give rise to a pulse train that has the appearance shown in Fig. 16.8*a*. Because it is the amplitude that carries the basic information (the charge Q deposited in the detector), the "pile-up" of pulses on the tails of preceding pulses, which have not fully decayed to zero, can be a serious problem. Because the time spacing between nuclear pulses is random, each pulse can be superimposed on a different residual tail and the resulting amplitude no longer is a good measure of Q from that event.

The ideal solution is to *shape* the pulses in such a way as to produce a pulse train similar to that shown in Fig. 16.8*b*. Here all the long tails have been eliminated, but the information carried by the maximum amplitude of the pulse has been preserved. The pulses have been shaped in the sense that their total length has been reduced drastically but in a way that does not affect the maximum amplitude.

This type of pulse shaping is conventionally carried out in the linear amplifier element of a nuclear pulse signal chain. In the discussions that follow, we illustrate the basic shaping operations using simple analog passive networks. In practice, the equivalent shaping steps generally are implemented using more complex active circuits or by carrying out arithmetic operations on a digitized version of the input waveform. This latter approach, known as *digital pulse processing*, is described further in Chapter 17. Regardless of the actual method of implementation, the general properties of various shaping methods remain similar to those of the equivalent simple networks used here for illustration.

A. *CR* and *RC* Shaping

In general electrical circuits, the term *RC shaping* refers to the use of passive resistor–capacitor networks to carry out a desired alteration in pulse shape. When discussing nuclear pulse shaping, it is conventional to make a semantic distinction between *differentiator* or *CR* networks on one hand, and *integrator* or *RC* networks on the other. Both operations can also be thought of as filtering in the frequency domain, and one purpose of pulse shaping is to improve signal-to-noise ratio by limiting the response of the instrumentation to those frequency ranges in which the signal has useful components, while reducing as much as possible the transmission of frequency components from the various sources of noise.

In both types of network, the *time constant* given by the product of resistance and capacitance plays a critical role. In the analysis that follows, we represent this time constant as τ, or

$$\tau \equiv RC \tag{16.2}$$

(The units of τ are seconds if R is in ohms and C is in farads.)

1. CR DIFFERENTIATOR OR HIGH-PASS FILTER

A basic CR differentiator network is diagrammed in Fig. 16.9. From the circuit equations, the input voltage E_{in} and output voltage E_{out} are related by

$$E_{in} = \frac{Q}{C} + E_{out} \tag{16.3}$$

where Q represents the charge stored across the capacitor. Now, differentiating with respect to time,

$$\frac{dE_{in}}{dt} = \frac{1}{C}\frac{dQ}{dt} + \frac{dE_{out}}{dt} \tag{16.4}$$

$$\frac{dE_{in}}{dt} = \frac{1}{C}i + \frac{dE_{out}}{dt} \tag{16.5}$$

Noting that $E_{out} = iR$ and setting $RC = \tau$, we obtain

$$E_{out} + \tau\frac{dE_{out}}{dt} = \tau\frac{dE_{in}}{dt} \tag{16.6}$$

Now, if we make RC sufficiently small, we can neglect the second term on the left and

$$E_{out} \cong \tau\frac{dE_{in}}{dt} \tag{16.7}$$

Thus, in the limit of small time constant τ, the network acts to produce an output E_{out} that is proportional to the time derivative of the input waveform E_{in}—hence the name *differentiator*. In order to meet these conditions, the time constant should be small compared with the duration of the pulse to be differentiated.

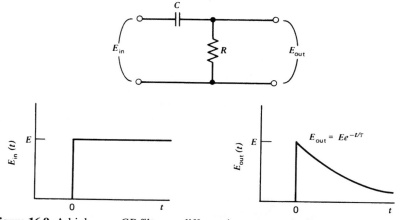

Figure 16.9 A high-pass CR filter or differentiator network. The response to a step function input is illustrated.

In the opposite extreme of large time constant, the first term on the left of Eq. (16.6) can be neglected and we have

$$\tau \frac{dE_{out}}{dt} \cong \tau \frac{dE_{in}}{dt} \tag{16.8}$$

and setting the constant of integration equal to zero

$$E_{out} \cong E_{in} \tag{16.9}$$

Therefore, if the conditions for differentiation are not met, the network will tend to pass the waveform without alteration.

We can solve Eq. (16.6) for arbitrary E_{in} waveforms. Let us state two specific results.

(a) *Sinusoidal E_{in}*
For

$$E_{in} = E_i \sin 2\pi f t \tag{16.10}$$

it can be shown that

$$\frac{E_{out}}{E_i} = |A| \sin(2\pi f t + \theta) \tag{16.11}$$

where

$$|A| = \frac{1}{[1 + (f_1/f)^2]^{1/2}} \qquad \theta = \tan^{-1}\left(\frac{f_1}{f}\right)$$

$$f_1 \equiv \frac{1}{2\pi\tau}$$

For high-frequency inputs, $f \gg f_1$, and $|A| \cong 1$. Hence, high frequencies are passed to the output with little attenuation, and we have a *high-pass filter*. Low frequencies are attenuated, because for $f \ll f_1$, $|A| \cong 0$. In the limit of $f = 0$ (constant voltage), no signal is transmitted and the network serves to block dc voltages because of the capacitor that is in series with the signal.

(b) *Step Voltage Input*
For

$$E_{in} = \begin{cases} E & (t \geq 0) \\ 0 & (t < 0) \end{cases}$$

the output is

$$E_{out} = Ee^{-t/\tau} \tag{16.12}$$

This case approximately represents the shaping of a fast-rising signal pulse with a long tail by a single *CR* differentiator. Note that the fast leading edge of the step is *not* differentiated because τ is not small compared with its rise time. The leading edge is therefore simply passed through, and the shaping consists of "differentiating away" the long tail. The amplitude E is not affected provided τ does not become small compared with the finite rise time of the actual signal pulse.

2. *RC* INTEGRATOR OR LOW-PASS FILTER

When configured as shown in Fig. 16.10, a passive *RC* network can also serve as an integrator. The circuit equation is now

$$E_{in} = iR + E_{out} \tag{16.13}$$

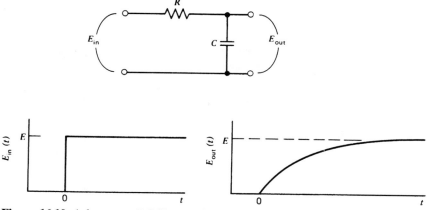

Figure 16.10 A low-pass RC filter or integrator network. The response to a step function input is illustrated.

The current i also represents the rate of charging or discharging of the capacitor.

$$i = \frac{dQ}{dt} = C\frac{dV_c}{dt} \tag{16.14}$$

or

$$i = C\frac{dE_{out}}{dt} \tag{16.15}$$

Now combining Eqs. (16.13) and (16.15) and setting $RC = \tau$, we obtain

$$E_{in} = \tau\frac{dE_{out}}{dt} + E_{out} \tag{16.16}$$

Rearranging, we have

$$\frac{dE_{out}}{dt} + \frac{1}{\tau}E_{out} = \frac{1}{\tau}E_{in} \tag{16.17}$$

Now, if RC is sufficiently large, only the first term on the left is significant, and

$$\frac{dE_{out}}{dt} \cong \frac{1}{\tau}E_{in}$$

or

$$E_{out} \cong \frac{1}{\tau}\int E_{in}\,dt \tag{16.18}$$

Hence, the name *integrator*. The network will integrate provided the time constant τ is large compared with the time duration of the input pulse.

In the opposite extreme of small time constant only the second term on the left of Eq. (16.17) is significant, and therefore

$$\frac{1}{\tau}E_{out} \cong \frac{1}{\tau}E_{in}$$

or

$$E_{out} \cong E_{in} \tag{16.19}$$

Thus, if the conditions for integration are not met, the network tends to pass the waveform without change.

Again we apply Eq. (16.17) to some specific input waveforms.

(a) *Sinusoidal E_{in}*
For

$$E_{in} = E_i \sin 2\pi f t$$

the solution is

$$\frac{E_{out}}{E_i} = |A| \; \sin(2\pi f t + \theta) \tag{16.20}$$

where

$$|A| = \frac{1}{[1 + (f/f_2)^2]^{1/2}} \qquad \theta = -\tan^{-1}\!\left(\frac{f}{f_2}\right)$$

$$f_2 \equiv \frac{1}{2\pi\tau}$$

Again, we examine the extremes of frequency response. If

$$f \gg f_2, \quad \text{then} \quad |A| \cong 0$$

and if

$$f \ll f_2, \quad \text{then} \quad |A| \cong 1$$

Therefore, the network blocks high frequencies, passes low frequencies without attenuation, and is thus a low-pass filter.

(b) *Step Voltage Input*
For

$$E_{in} = \begin{cases} E & (t \geq 0) \\ 0 & (t < 0) \end{cases}$$

the output is

$$E_{out} = E(1 - e^{-t/\tau}) \tag{16.21}$$

This response is also plotted in Fig. 16.10. Recall that the circuit performs as an integrator in the limit of large τ. The mathematical integral of the step input should be a linearly increasing ramp. The actual response starts out as a linear ramp, but over a sufficiently long time scale the circuit time constant is no longer large by comparison, and the integration begins to fail. The output voltage then approaches the input step as a limit.

3. CR-RC SHAPING

The output of a single differentiating network shown in Fig. 16.9 is not a very attractive waveform for pulse analysis systems. The sharply pointed top makes subsequent pulse height analysis difficult because the maximum pulse amplitude is maintained only for a very short time period. Furthermore, because the differentiation allows all high-frequency components of any noise mixed with the signal to be passed by the network, the signal-to-noise characteristics of the network in practical applications are usually very poor. If a stage of *RC* integration is added following the differentiation, however, both of these drawbacks are considerably improved. The combination of a single stage of differentiation followed by a single stage of integration is in fact a common method of shaping preamplifier pulses.

Figure 16.11 shows the elements of the basic *CR-RC* shaping network. An ideal unity-gain operational amplifier (with infinite input impedance and zero output impedance)

Figure 16.11 A shaping network consisting of sequential differentiating and integrating stages, sometimes denoted a *CR-RC* network.

separates the two individual networks for impedance isolation so that neither network influences the operation of the other. The general solution of the response of the combined network to a step voltage of amplitude E at $t = 0$ is

$$E_{out} = \frac{E\tau_1}{\tau_1 - \tau_2} (e^{-t/\tau_1} - e^{-t/\tau_2}) \tag{16.22}$$

where τ_1 and τ_2 are time constants of the differentiating and integrating networks, respectively. Plots of this response for several different combinations of τ_1 and τ_2 are shown in Fig. 16.12.

In nuclear pulse amplifiers, *CR-RC* shaping is most often carried out using equal differentiation and integration time constants. In that event, Eq. (16.22) becomes indeterminant, and a particular solution for this case is

$$E_{out} = E \frac{t}{\tau} e^{-t/\tau} \tag{16.23}$$

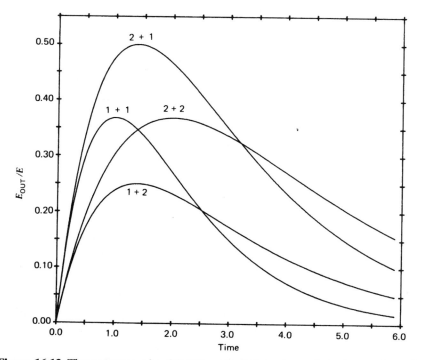

Figure 16.12 The response of a *CR-RC* network to a step voltage input of amplitude E at time zero. Curves are shown for four pairs of differentiator + integrator time constants. Units of the time constants and time scale are identical.

Plots of this step input response for two different values of time constant τ are also shown in Fig. 16.12.

One factor in the choice for the time constant of the shaping circuits is the charge collection time in the detector being used. In the interests of reducing pile-up, one would like to keep these time constants short so that the shaped waveform can return to the baseline as quickly as possible. On the other hand, once the shaping time constants become comparable with the rise time of the pulse from the preamplifier, the input to the network no longer appears as a step voltage and some of its amplitude is lost. This loss is called the *ballistic deficit* (see p. 628) and can be avoided only by keeping the time constants long compared with the charge collection time in the detector. At the same time, the signal-to-noise characteristics of the pulses are also influenced by the choice of shaping time. As described in Chapter 17, the interplay of the sources of series and parallel noise leads to an optimum choice of shaping time when minimization of the electronic noise contribution to energy resolution is the sole objective. This noise level may be unimportant in detectors such as scintillators where the inherent detector resolution is poor, but it can become a predominant consideration in the choice of shaping times in other cases (such as silicon X-ray spectrometers) in which the suppression of electronic noise is critical. The proper choice for shaping time for a given circumstance thus becomes a complex balance between factors involving ballistic deficit, electronic noise, and pulse pile-up, which is explored further in Chapter 17.

4. GAUSSIAN OR CR-(RC)ⁿ SHAPING

If a single CR differentiation is followed by several stages of RC integration, a pulse shape that approaches a mathematical Gaussian is realized. If the differentiation and n integration time constants are all the same value τ, the particular solution of the corresponding circuit equation is

$$E_{\text{out}} = E\left(\frac{t}{\tau}\right)^n e^{-t/\tau} \tag{16.24}$$

In practice, four stages of integration ($n = 4$) are sufficient so that the difference between the resulting pulse shape and a true Gaussian is negligible. The time required for the shaped pulse to reach its maximum amplitude (often called the *peaking time*) is equal to $n\tau$.

For equal time constants throughout, a $CR\text{-}(RC)^4$ network results in a peaking time that is a factor of 4 longer than that for a simple $CR\text{-}RC$ network. However, if the time constants are adjusted to give equal peaking times for the two methods, the more symmetric shape of the Gaussian pulse results in a faster return to the baseline. Pulse pile-up at high counting rates is thereby reduced. Gaussian shaping also has the advantage of better signal-to-noise characteristics for individual pulses compared with $CR\text{-}RC$ shaping (see Fig. 17.15). For these two reasons, Gaussian shaping has become a popular choice in gamma-ray spectroscopy systems employing high-resolution germanium detectors.

5. ACTIVE FILTER PULSE SHAPING

Earlier in this chapter, it was shown that the processes of differentiation and integration correspond to filtering steps in the frequency domain. The CR differentiator functions as a high-pass filter, while the RC integrator is a low-pass filter. Alternative methods are available to the circuit designer to carry out similar operations using active circuits that incorporate elements such as transistors and diodes. In the design of the shaping networks of modern linear amplifiers, most designers have now substituted these active filters for the passive RC networks described earlier in this chapter. Their function, however, remains similar to that of the corresponding passive network. For example, one of the most common choices is the combination of a high-pass filter (differentiator) followed by several stages of active low-pass filters (integrators) to produce the Gaussian pulse shaping discussed above.

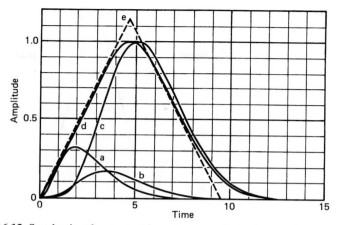

Figure 16.13 Synthesis of a near triangular waveform (curve *d*, normalized to unity) through the summation of active integrator output waveforms *a, b,* and *c.* Curve *e* is a symmetrical triangle for comparison. Time is in units of the initial differentiation time constant. (From Goulding et al.[7])

6. TRIANGULAR SHAPING

The discussion of signal-to-noise behavior of different pulse-shaping schemes in linear amplifiers given in Chapter 17 points out the theoretical advantage that a symmetrical triangular shape has over the Gaussian shape (see Fig. 17.15). It is practically impossible to achieve a triangular shape using passive circuit elements alone. If active filter shaping elements are provided, however, it has been shown[7] that a waveform approximating a symmetrical triangle can be synthesized. After an initial differentiation, the outputs from active integrators with different characteristics are weighted and mixed as shown in Fig. 16.13. Similar schemes for producing a triangular pulse shape are offered as a feature of some commercially available linear amplifiers.

7. TRAPEZOIDAL SHAPING

For detectors in which the charge collection time is variable depending on the position of the radiation interaction, the potential problem of ballistic deficit (see p. 628) favors pulse shaping methods that lead to a shaped pulse with a flat top. One such shape is a trapezoid that can be synthesized using either analog[8] or digital[9] techniques. The benefits are illustrated in Figure 16.14, where it is assumed that the input to the shaper is a sequence of three tail pulses, each arising from the same total charge but having different charge collection times. The effect of a trapezoidal shaper for these pulses is seen to produce output pulses all with the same maximum amplitude, whereas the shaping methods described previously without a flat top would lead to pulses with slightly differing amplitudes. Using trapezoidal or other flat-topped shaping methods can avoid this problem of variable ballistic deficit that may degrade energy resolution, for example from large volume coaxial germanium detectors.

B. Pole-Zero Cancellation

Thus far our comments on pulse shaping have assumed that the input pulse from the preamplifier consists of a step voltage. Although the decay of the preamplifier pulse is usually long, it is not infinite and the finite decay will have a measurable effect on the response of the networks discussed above. For example, a basic *CR-RC* differentiator–integrator will no longer produce a strictly unipolar response if the input pulse has the finite decay shown

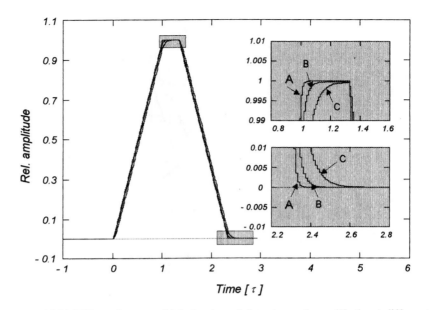

Figure 16.14 Effect of trapezoidal shaping of detector pulses with three different rise times that increase in the sequence A, B, and C. In the case shown, if the trapezoidal flat top is longer than the variation in rise time, the shaped pulses reach the same amplitude and ballistic deficit is avoided. (From Jordanov and Knoll.[9])

in Fig. 16.15. Instead, there will be a slight zero crossover or *undershoot* of the pulse, which then recovers back to zero with a time characteristic of the preamplifier decay time. Because preamplifiers have long decays (of the order of 50 μs), the undershoot persists for a relatively long time. If another pulse arrives during this period of time, it will be superimposed on the undershoot and an error will be induced in its amplitude. The problem is particularly severe for very large signal pulses that overload the amplifier and consequently lead to rather large following undershoots.

Our previous analysis of the response of a *CR* differentiator to a step input [Eq. (16.12)] showed that the output was a simple exponential decay with no undershoot. The transfer function for this network is $\tau_1 s/(1 + s\tau_1)$, where τ_1 is the time constant and s is the Laplace variable. Assume that this output is now supplied to a second *CR* differentiator with time constant τ_2 and transfer function $\tau_2 s/(1 + s\tau_2)$. The overall transfer function is then $\tau_1 \tau_2 s^2/(1 + s\tau_1)(1 + s\tau_2)$, and it converts a step input to an output pulse with undershoot. The poles in the denominator of the transfer function ensure that a simple exponential decay will not be possible.

The above picture represents the origin of the undershoot if a preamplifier pulse with finite decay is presented to a *CR* differentiator in a shaping network. The term *pole-zero cancellation*[10] describes a technique in which the network is modified as in Fig. 16.15 to again restore the simple exponential output without undershoot. A resistance R_{pz} is added in parallel with the capacitor of the *CR* network, resulting in a modified transfer function of the form

$$\frac{\tau_1(1 + sR_{pz}C_1)}{(1 + s\tau_2)(R_{pz}C_1 s\tau_1 + R_{pz}C_1 + \tau_1)}$$

If the value of R_{pz} is now chosen such that

$$R_{pz} = \frac{\tau_2}{C_1}$$

the transfer function reduces to

$$\frac{1}{\tau_2(s+k)} \quad \text{where} \quad k \equiv \frac{\tau_1 + \tau_2}{\tau_1 \tau_2}$$

This result, with its single pole in the denominator, ensures that the network once again produces a simple exponential decay for a step input.

A somewhat more heuristic way of describing the function of the modified network of Fig. 16.15 is to note that R_{pz} allows an attenuated replica of the input pulse to pass through to the output. If the input is a positive step with finite decay, then the slight undershoot from a normal *CR* differentiator can be cancelled by this slight admixture of the input pulse. A proper choice for the resistance value depends on the value of the input decay time from the particular preamplifier in use, and therefore in practice it is often chosen empirically. The pole-zero network is simply adjusted while the output pulse is inspected on an oscilloscope and adjustments are made to eliminate the long undershoot by visual observation. Alternatively, some commercially-available systems carry out an automatic pole-zero adjustment by sensing the undershoot electronically.

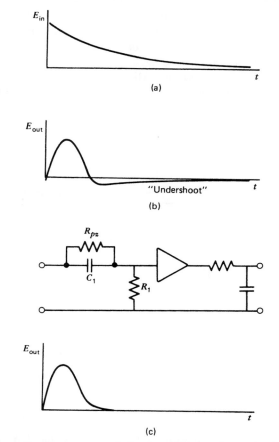

Figure 16.15 Application of pole-zero cancellation to eliminate the undershoot (*b*) normally generated by a *CR-RC* shaping network for an input step with finite decay time. By adding an appropriate resistance R_{pz} to the differentiator stage, a waveform without undershoot (*c*) can be obtained.

C. Baseline Shift

1. THE ORIGIN OF THE PROBLEM

The discussion to this point has dealt with the shaping of single isolated pulses. In any radiation detector system, pulses will occur in time sequence with many other pulses. This pulse train leads to a potential problem not encountered with single pulses. In Fig. 16.16a, we show the response of the *CR-RC* network to a train of regularly spaced rectangular pulses. Because capacitors cannot conduct direct current, the average dc voltage of any point to the right of the capacitor of the differentiating network in Fig. 16.11 must be zero. If the dc voltage were nonzero, a finite dc current must flow through the resistor to ground. Any such current would have to come through the capacitor, which is a physical impossibility. Therefore, the baseline on which these rectangular pulses are superimposed must be suppressed below the true zero level such that the areas enclosed by the output waveform above and below the zero axis are equal. The amount by which this apparent baseline is depressed below true zero is called the *baseline shift* and will obviously become more severe as the average spacing between pulses is made smaller.

In nuclear pulse analysis systems, the amplitude of the pulse carries much of the basic information. Because the amplitude is measured relative to a true zero baseline, the presence of any baseline shift will reduce the apparent pulse amplitude. For regular periodic pulses such as those shown in Fig. 16.16a, the problem would not be too severe because the baseline shift would be constant and appropriate compensation could be made. However, for pulses from radiation detectors, both the amplitude and spacing are variable and the situation will appear more like that shown in Fig. 16.16b. Now the degree of baseline shift is not constant and no adequate compensation can be carried out.

In principle, baseline shifts can be eliminated if the pulse shape is made to be *bipolar* rather than *monopolar*. Bipolar pulses contain both positive and negative lobes, as shown in Fig. 16.17, and differ from monopolar pulses, which are restricted to voltages only on one

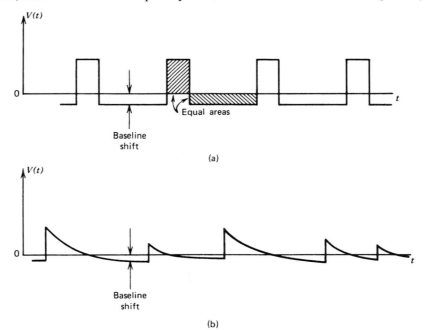

Figure 16.16 Illustrating baseline shift that must take place in ac-coupled circuits. In part (*a*) uniform and regular pulses lead to fixed baseline shift, but the random pulses of part (*b*) give rise to variable shift.

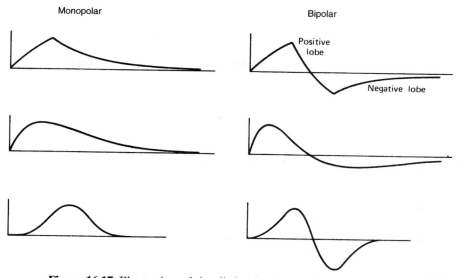

Figure 16.17 Illustration of the distinction between monopolar and bipolar signal pulses of various shapes.

side of the zero axis. If the positive and negative lobes of a bipolar pulse are of equal area, its average dc value is zero and it can be passed by a capacitor without alteration of the baseline. As discussed in Chapter 17, all pulse shaping operations, whether resulting in monopolar or bipolar pulses, affect both the signal pulse and the noise that inevitably is also present. The signal-to-noise ratio is generally poorer after bipolar shaping compared with monopolar shaping that results in the same pulse width. Thus a logical choice is to use monopolar shaping for better signal-to-noise characteristics at low counting rates and to convert to bipolar shaping only when required for baseline restoration performance at high count rates. This choice is often made available to the user through a front panel switch on standard amplifiers.

Effects of baseline shifts could also be avoided if all elements in the signal chain were dc coupled rather than interconnected by coupling capacitors. Some pulse-processing components are in fact dc coupled for this reason, but it is impractical to dc couple some elements such as a linear amplifier. One reason is that very small shifts in the input zero level to a high-gain amplifier will, if dc coupled, give rise to very large offsets at the output. Furthermore, if *CR* differentiation is used at any stage in the amplifier, it automatically becomes ac coupled. Therefore, ac coupling is the usual rule, and baseline shift must be countered at high pulse rates through the use of bipolar pulse shaping or through active baseline restoration.

2. BASELINE RESTORATION

One method to avoid baseline shift is to use shaping networks (described later) that produce bipolar pulses. Inevitably, the added shaping step required to produce the bipolar shape deteriorates the signal-to-noise characteristics of the system compared with simple monopolar shaping. Therefore, it is sometimes attractive to retain monopolar pulses and instead make use of an active electronic circuit to eliminate the resulting baseline shift. Such a *baseline restorer* circuit has as its primary purpose the return to true zero of the baseline between pulses in as short a time as possible. As a bonus, the use of baseline restoration also greatly reduces the deleterious effects of low-frequency disturbances, such as power-line hum and vibrational microphonics, which may be present along with the signal.

Figure 16.18 Equivalent circuit of a baseline restorer. R_0 is the output impedance of the operational amplifier, normally with gain $A = 1$. The time constant is given by the product $(R + R_0)C$.

The equivalent circuit of a baseline restorer is diagrammed in Fig. 16.18. In principle, the switch is open only during the duration of each pulse, and its closing restores the output voltage to zero, with a time constant given by product of $(R + R_0)$ and the coupling capacitance C, where R is the equivalent series resistance of the switch and R_0 is the output impedance of the operational amplifier. In actual circuits, the role of the switch is carried out by diodes[11] or by more complex nonlinear circuitry.[12–14] Other restorer designs are described in Refs. 15–18. Although the use of these circuits introduces some degree of additional noise into the system, the benefits derived normally predominate, and most high-resolution spectrometry systems now employ some type of active baseline restoration.

To be effective, baseline restoration must take place near the end of the signal chain so that no further ac coupling takes place between the restorer and the point at which the pulses are analyzed or measured. If another capacitively coupled component does exist after the active restoration, the carefully restored baseline level will again shift in the manner illustrated in Fig. 16.16. Baseline restorers are therefore most commonly found at the

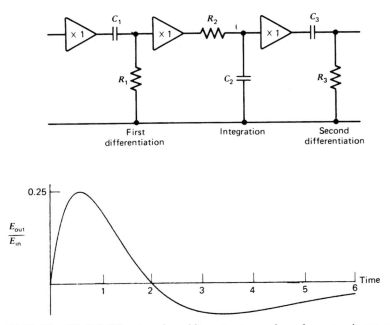

Figure 16.19 The *CR-RC-CR* network and its output waveform for a step input when all three time constants are equal. Units of the time scale are in values of this time constant.

output stage of linear amplifiers, which are intended to supply pulses directly to the analog-to-digital converter (ADC) of a multichannel analyzer.

D. Other Pulse Shaping Methods

1. DOUBLE DIFFERENTIATION, OR CR-RC-CR SHAPING

In order to achieve a bipolar pulse shape, a second state of differentiation is sometimes added to the *CR-RC* configuration discussed previously. Again, the most common choice is to make all three time constants approximately the same, which results in the pulse form sketched in Fig. 16.19. The bipolar shape makes baseline shifts much less severe, but because the two lobes of the pulse are not of exactly equal area, some baseline shift will remain. This type of shaping is most useful at high counting rates, but at lower rates the signal-to-noise characteristics are usually inferior to single-stage differentiation and integration.

2. SINGLE DELAY LINE (SDL) SHAPING

The properties of coaxial cables discussed earlier can also be applied to carry out pulse shaping. In particular, recall that a coaxial cable that is shorted at the receiving end will give rise to a reflection when a step voltage reaches that end of the cable. The reflection is a step moving back toward the sending end of the cable, with an amplitude equal to the initial step but opposite in polarity.

A configuration in which a shorted transmission line can be used to shape a step input is shown in Fig. 16.20. The transmission line or coaxial cable is assumed to be long enough so that the propagation time through its entire length is long compared with the rise time of the step voltage applied to the input of the network. The unity-gain operational amplifier

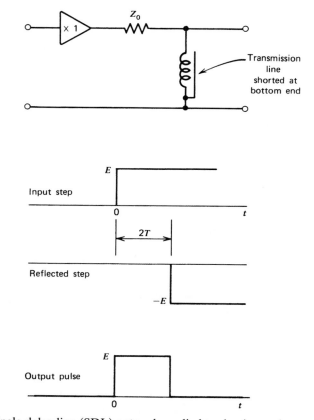

Figure 16.20 Single delay line (SDL) network applied to shaping an input step waveform.

simply provides impedance isolation at the input. Because the amplifier is assumed to have zero output impedance, the resistor Z_0 terminates the cable in its own characteristic impedance at the sending end. The far end of the cable is assumed to be shorted, or terminated in zero resistance.

The waveforms shown in Fig. 16.20 illustrate the sequence of events if a step voltage is applied to the input of the network. The initial positive step is assumed to be applied to the network at time $t = 0$ and to persist for a long period of time. The step propagates down the cable, is reflected from the shorted end, and travels back to the sending end as an inverted or negative step, where further reflections are prevented by the impedance-matching resistor Z_0. The voltage observed at the output of the network is simply the algebraic sum of the original and reflected waveforms, or the rectangular pulse shown in the figure. The time width of this shaped pulse is the down-and-back propagation time through the length of cable used for the shaping. For many applications, this propagation time should be of the order of a microsecond. In order to avoid excessively long cables, special delay lines with much reduced propagation velocity are often used for this purpose. Hence, the process just described is often known as *single delay line* (SDL) shaping.

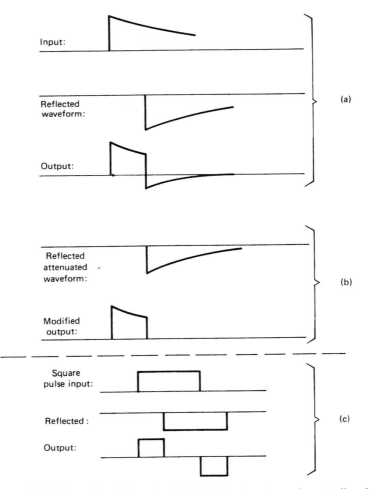

Figure 16.21 (*a*) Effect of applying simple SDL shaping to an input tail pulse with decay time comparable with the propagation time of a delay line. (*b*) Remedy to the undershoot of part (*a*) accomplished by partially attenuating the reflected waveform. (*c*) Effect of SDL shaping on a rectangular shaped input pulse whose length exceeds the delay line down-and-back time.

The propagation time of the delay line must always be larger than the rise time of the pulse from the preamplifier to avoid the ballistic deficit discussed under *RC* shaping. Also, if the decay time of the preamplifier pulse is not many times larger than the propagation time, a situation will result as illustrated in Fig. 16.21*a*, in which there is an undesirable undershoot following the shaped pulse. If the preamplifier decay time is always a fixed value, this undershoot can virtually be eliminated by slightly attenuating the reflected pulse as shown in the figure. This attenuation can be accomplished by raising the termination resistance at the reflecting end of the cable from zero to a small value that is less than the cable characteristic impedance.

Similar shaping is also commonly used to reduce the length of pulses with very fast (of the order of nanoseconds) leading edges. In this case the role of the delay line is carried out by a normal coaxial cable of a few meters length, commonly referred to as a *shorted stub*. As is shown in Fig. 16.21*c*, this type of shaping applied to a rectangular input pulse gives rise to two shaped pulses of opposite polarity separated in time by the length of the original rectangular pulse.

3. DOUBLE DELAY LINE (DDL) SHAPING

Because the output of a single delay line shaping network is a unipolar pulse for typical preamplifier pulses, the same comments apply as were made earlier for single differentiation networks. At high counting rates, it may be desirable to substitute a shaping network that results in bipolar pulses to eliminate baseline shift. Bipolar pulses can be produced if two delay lines are used in the configuration shown in Fig. 16.22, known as *double delay line* (DDL) shaping. Both delay lines should have equal propagation time, and the resulting pulse will have positive and negative lobes of equal amplitude and duration. Therefore,

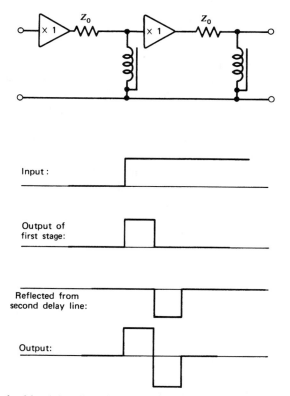

Figure 16.22 A double delay line (DDL) shaping network with equal delay times applied to a step input waveform.

an average dc level of zero can accurately be maintained, and baseline shift in subsequent ac-coupled circuits will virtually be eliminated. Although DDL shaping therefore has excellent high counting rate characteristics, it applies no high-frequency filtering to the signal and consequently is usually inferior from a signal-to-noise standpoint compared with methods based on *RC* networks. Therefore, one seldom sees delay line shaping used in conjunction with detectors with inherently high energy resolution (such as germanium detectors), but with poorer resolution detectors (such as scintillators) the difference in overall system resolution can be negligible. In these circumstances the better high counting rate characteristics of DDL shaping may dictate its choice.

PROBLEMS

16.1 Find the transit time of a pulse through 15 m of RG-59/U coaxial cable.

16.2 Describe a convenient method of measuring the characteristic impedance of a coaxial cable by varying its termination conditions.

16.3 For each instance cited below, determine whether termination to prevent cable reflections is potentially needed.

(a) Transmission of 0.5 μs rise time pulses through 20 m of RG-59/U cable.

(b) Transmission of a 10 ns rise time pulse through 10 m of RG-62/U cable.

16.4 (a) Fast pulses are to be transmitted from a source with output impedance of 50 Ω to another component with input impedance of 1 kΩ. Choose one of the coaxial cable types listed in Table 16.1, and show the termination conditions required to avoid cable reflections at either end.

(b) If the output pulse amplitude from the source without any load attached is 5 V, find the amplitude that appears across the input of the second component after the termination required in part (a).

16.5 Find the resistance values R_1 and R_2 in the T-attenuator network of Fig. 16.5 if an attenuation factor of 10 is needed while preserving a 50-Ω impedance level.

16.6 Prove that the resistance values in the pulse splitter of Fig. 16.6 should be 16.6 Ω in order to distribute a pulse to two 50-Ω loads while maintaining a 50-Ω impedance level.

16.7 Pulses from a preamplifier are produced with exponential tails with 50-μs time constant. What is the minimum

spacing between adjacent pulses so that amplitude change due to pile-up is less than 1%?

16.8 An input voltage of the form $V(t) = E[1 - \exp(-t/k)]$, where E and k are constants, is supplied to a simple *RC* integrator circuit. Derive the form of the output voltage.

16.9 A differentiator circuit has component values of $C = 500$ pF and $R = 500$ Ω. Find the frequency of a sinusoidal input voltage that will be attenuated by a factor of 2 by this circuit.

16.10 A step voltage of 1 V amplitude is applied to the input of a *CR-RC* differentiator–integrator network with equal time constants. What is the amplitude of the shaped pulse?

16.11 Derive Eq. (16.22) for the response of a *CR-RC* network to a step voltage input by using Eq. (16.12) as the input waveform to the *RC* stage.

16.12 What is the principal advantage of bipolar shaped pulses over monopolar shaped pulses?

16.13 What property of the detector is reflected in the rise time of the tail pulse obtained by a large time-constant collection circuit?

16.14 Pulses from a detector are shaped to produce monopolar pulses of approximately triangular shape with 10 V amplitude and width of 5 μs. Find the *average* value of the baseline shift after these pulses are passed through a capacitor if the pulse rate is 100/s. Repeat for a rate of 50,000/s.

16.15 Sketch the result of shaping a rectangular monopolar pulse of 200 ns width by using a shorted stub of RG-59/U of 10-m length.

REFERENCES

1. E. Kowalski, *Nuclear Electronics,* Springer-Verlag, New York, 1970.
2. P. W. Nicholson, *Nuclear Electronics,* Wiley, New York, 1974.
3. F. A. Kirsten, *IEEE Trans. Nucl. Sci.* **NS-20**(5), 22 (1973).
4. K. Porges, W. Corwin, L. Burkel, and E. Lewandowski, *Rev. Sci. Instrum.* **41,** 138 (1970).
5. W. K. Brookshier, *Nucl. Instrum. Meth.* **70,** 1 (1969).
6. J. Millman and H. Taub, *Pulse, Digital, and Switching Waveforms,* McGraw-Hill, New York, 1965.
7. F. S. Goulding, D. A. Landis, and N. W. Madden, *IEEE Trans. Nucl. Sci.* **NS-30**(1), 301 (1983).
8. F. S. Goulding et al., *IEEE Trans. Nucl. Sci.* **41**(4), 1140 (1994).
9. V. T. Jordanov and G. F. Knoll, *Nucl. Instrum. Meth.* **A345,** 337 (1994).
10. C. H. Nowlin and J. L. Blankenship, *Rev. Sci. Instrum.* **36,** 1830 (1965).
11. L. B. Robinson, *Rev. Sci. Instrum.* **32,** 1057 (1961).

12. R. L. Chase and L. R. Poulo, *IEEE Trans. Nucl. Sci.* **NS-14**(1), 83 (1967).
13. E. A. Gere and G. L. Miller, *IEEE Trans. Nucl. Sci.* **NS-14**(1), 89 (1967).
14. F. S. Goulding, D. A. Landis, and R. H. Pehl, UCRL-17560 (1967).
15. M. Bertolaccini and C. Bussolati, *Nucl. Instrum. Meth.* **100,** 349 (1972).
16. E. R. Semple, *IEEE Trans. Nucl. Sci.* **NS-19**(1), 445 (1972).
17. N. Karlovac and T. V. Blalock, *IEEE Trans. Nucl. Sci.* **NS-22**(1), 457 (1975).
18. E. Fairstein, *IEEE Trans. Nucl. Sci.* **NS-22**(1), 463 (1975).

Chapter **17**

Linear and Logic Pulse Functions

I. LINEAR AND LOGIC PULSES

In any pulse-processing system, it is important to distinguish between two types of signal pulse. A *linear* pulse is defined as a signal pulse that carries information through its amplitude, and sometimes by its shape as well. A sequence of linear pulses may therefore differ widely in size and shape characteristics. On the other hand, a *logic* pulse is a signal pulse of standard size and shape that carries information only by its presence or absence, or by the precise time of its appearance. Virtually all radiation detector signal chains start out with linear pulses and, at some point, a conversion is made to logic pulses based on some predetermined criteria.

A. Linear Pulses

Although there are a number of exceptions, linear pulses in radiation detector signal chains are usually one of three general types. *Fast linear pulses* are those that correspond to collecting the output current of a radiation detector using a collection circuit whose time constant is small (see discussion in Chapter 4). The rise and fall characteristics of these pulses are therefore determined primarily by the time history of the current generated directly by the interaction of the radiation with the specific detector.[†] Because these times are typically a few microseconds or less in most radiation detectors, fast linear pulses are seldom longer than several microseconds and may be much narrower for very fast detectors. For reasons outlined in Chapter 4, the signal-to-noise properties of fast linear pulses are always far inferior to those of the corresponding tail pulse derived by integrating the charge output of the detector across a large time constant collection circuit (see the following discussion). However, the rapid rise and fall can often offset this disadvantage when timing information and high counting rates are more important than amplitude resolution. The polarity of fast linear pulses often depends on the polarity of the bias voltage applied to the detector and can, in general, be either positive or negative.

A linear *tail pulse* is the type of linear pulse generated when the output of a radiation detector is collected across an equivalent circuit of large time constant. As discussed in Chapter 4, the leading edge of such a pulse corresponds to the time over which the charge produced by the detector is integrated across the capacitance of the collection circuit. Therefore, the time characteristics of the leading edge are determined exclusively by the charge collection time within the detector. The decay or return to zero of the pulse is determined by the time constant of the collection circuit. In order to ensure complete charge collection or to avoid ballistic deficit, this time constant must be large compared with the

---

[†]For this reason, such pulses are sometimes also called *current pulses*.

detector charge collection time. Therefore, such pulses have a long tail compared with their leading edge, giving rise to the name tail pulse. From the arguments given in Chapter 4, the amplitude reached by a tail pulse in a given situation will always be many times larger than that of the equivalent fast linear pulse. Polarity can again be either positive or negative, but negative tail pulses are generally encountered more frequently in common applications.

For linear tail pulses, the most significant parameter besides amplitude is the *rise time* of the leading edge of the pulse. It should be pointed out that the term *rise time* always refers to the leading edge of pulses whether they are positive or negative polarity, even though the leading edge may have a negative slope for negative polarity pulses. The most widely used definition of rise time is the interval between the times at which the pulse reaches 10% and 90% of its final amplitude. For fast linear pulses, the full width at half maximum amplitude (FWHM) is often quoted as a measure of the overall pulse width.

A *shaped* linear pulse is a tail pulse whose time width has drastically been reduced by one of the shaping methods discussed in Chapter 16. Rather than the 50 or 100 μs decay times of tail pulses, the time width of a shaped linear pulse is typically only several microseconds. Shaped linear pulses are usually produced by a linear amplifier and consequently, within limits, can have arbitrary amplitude and polarity. Their dynamic range or "span" can often be chosen to be consistent with the dynamic range of circuits to which the linear pulse will be supplied. The Nuclear Instrument Module (NIM) standard (outlined in Appendix A) recommends that positive polarity linear pulses be used with a span of either 0–1 or 0–10 V, with the latter choice by far the more popular.

B. Logic Pulses

There are two classes of logic pulses specified in the NIM standard. *Standard logic pulses* are used in general applications where performance at extremely high counting rates or fast switching times are not required. They are to be positive in polarity and within the amplitude limits specified in Appendix A. Their width is not standardized but is usually of the order of 1 μs. Pulse shape tends to be a unipolar square pulse to allow the most abrupt switching between logic states. *Fast logic pulses* arise primarily in circumstances in which fast timing information is of paramount importance. The rise time of these logic pulses should be as short as possible and is usually of the order of a few nanoseconds. Because most timing is carried out from the leading edge of the pulse, their width is not of primary importance and is a factor only in determining the maximum practical counting rate. Because of the importance of cable reflection when dealing with short rise time pulses, fast logic pulses are usually handled in systems in which the input, output, and cable impedances are all set to be 50 ohms. In this way, simple interconnection of the components ensures undistorted transmission of fast logic pulses from one unit to the next. The NIM standard calls for negative polarity for fast pulses, and more detailed specifications are given in Appendix A.

Although they are technically not logic pulses, *gate pulses* share some common properties. A gate pulse is usually rectangular in shape and, when applied to the gate pulse input of a linear gate, will cause the gate to open for a period of time equal to the width of the gate pulse. Its amplitude and polarity must correspond to the transition between closed and open conditions of the linear gate and are usually positive and of the order of several volts. Its width, however, may be variable and is usually adjusted to the requirements of the particular application.

II. INSTRUMENT STANDARDS

Before the widespread use of solid-state circuitry, most electronic instrumentation used for nuclear measurements consisted of vacuum tube circuits housed in a self-contained chassis and required only connection to a power line. To facilitate mounting in a vertical stack, much of this equipment was built to fit a frame commonly known as the standard *19-in. relay rack.* Individual chassis are screw mounted one above the other in a relay rack, and widespread use persists to this day for the convenient mounting of nuclear electronic components. Some bulky units such as high-voltage power supplies continue to be manufactured in the full 19-in. width, but the advent of solid-state circuitry has led to more compact packaging in which only a fraction of the full width need be occupied by a single electronic unit.

It has become common practice to manufacture most nuclear electronics in standard modules, a number of which can fit into a housing called a *bin* or *crate,* which occupies the full 19-in. width. Two important international standards, the NIM and the CAMAC (Computer Automated Measurement and Control) standards, have come to be widely accepted, and now virtually all modular nuclear electronics are manufactured according to one of them. These systems have resulted in a great simplification for the user, and one can quite generally intermix components and bins from different manufacturers with assurance that those adhering to a common standard will be mutually compatible. The NIM system is usually better suited for small-scale linear pulse processing normally encountered in the routine application of radiation detectors. The more costly CAMAC system, which is strongly oriented toward large digital systems and computer interfacing, is usually necessary only when large-scale signal processing involving many detectors or logical operations is required.

As a self-contained system, each of these standards shares the following attributes:

1. The basic dimensions for the bin and modules are specified so that all modules and bins within a given system are mechanically interchangeable.
2. The philosophy generally adopted is that only the bin (or crate) will be connected to the laboratory ac power, and it will contain the necessary power supplies to generate all the dc supply voltages required by all modules contained within that bin. Individual modules therefore do not contain their own power supplies, but draw power from the bin in which they are located.
3. The connector interface between the module and bin must therefore be standardized both electrically and mechanically, so that any standard module can draw its required power when plugged into any one of the available bin locations.
4. Specifications are included for the polarity and span of both logic and linear pulses of various types.

Details of both the NIM and CAMAC standards are included in Appendix A.

III. APPLICATION SPECIFIC INTEGRATED CIRCUITS (ASICs)

Beginning in the 1980s, many of the functions traditionally carried out in self-standing electronic units such as NIM or CAMAC modules began to appear in the form of silicon microcircuits, generally known as *application specific integrated circuits* (ASICs). Microcircuits that carry out basic functions such as amplification, voltage comparison, signal encoding, and digital accumulation can be combined on special purpose ASICs that function as preamps, discriminators, timing triggers, and counters. Because of the small scale of the integrated circuits, many parallel channels of independent pulse processing units can be laid out on a single chip. The designer of custom circuits or those assembling large scale electronics for experiments involving many detectors can use these building blocks to assemble compact custom pulse processing units. For a number of years, the major emphasis was in applications in large-scale, high-energy physics experiments in which particle tracking is the major objective. In this case, spectroscopic information is not important, and as a result many of the early ASICs that were developed were intended for counting only and did not preserve the pulse height information.

In the 1990s, ASICs began to become available that were designed for applications in which amplitude information or energy resolution was also important. Commercial suppliers now provide a variety of ASICs with a rapidly growing set of features that can be applied in spectroscopic measurements from detectors that produce multiple output signals. Prime examples are silicon microstrip detectors, in which each electrode strip must be connected to a separate readout channel, or pixelated semiconductor detectors where many independent channels of spectroscopic readout are required. Great progress has been made in keeping noise levels sufficiently low so that multichannel readout of 128 or more separate signal channels can be provided without a great compromise in the energy resolution that could be obtained using traditional self-standing pulse processing units. The availability of these ASICs has facilitated the development of multielement detectors of all types and will certainly have a major impact on their future development. Because evolution of this technology remains a rapidly changing area, the reader is advised to consult current literature for an update on capabilities that can be provided in the integrated circuit format.

IV. SUMMARY OF PULSE-PROCESSING UNITS

In almost all applications of radiation detectors operated in pulse mode, the output of the detector is converted to a linear pulse whose amplitude and shape may carry information of interest to the user. This linear pulse may be processed in the signal chain before its properties are recorded in some way, or logic pulses may be derived to convey other information. Of the various electronic units in the signal chain, some are linear devices intended to perform operations on a linear pulse input to provide a linear pulse output. Other modules are strictly logic units which examine logic pulse inputs using set selection criteria to determine whether a logic pulse will be produced at the output. A third general category are linear-to-logic converters, in which logic output pulses are produced only when the input linear pulse meets predetermined conditions. In some circumstances, supplying a linear pulse to a device intended for logic pulse inputs may cause the device to function in some manner, but the variable size and shape of the linear pulse will usually result in unreliable operation. It is therefore important to maintain a clear distinction between linear and logic pulse functions, and to carry out linear-to-logic conversion only at a definite point in the signal chain.

To provide a general overview, Table 17.1 lists the summarized properties of some of the more popular types of nuclear instrumentation modules. Each of these is discussed in greater detail in the applications sections that follow. These modules represent the basic building blocks from which common pulse-processing chains are assembled. Although we treat each as a separate module, some of the functions shown are often combined into a single NIM module for convenience and versatility (e.g., *timing single-channel analyzers* are widely available which combine the time pick-off and SCA functions). All the functions shown are commonly available in NIM format, and most of the logic functions are available in CAMAC as well.

The largely digital operations that are the speciality of CAMAC require modules designated as scalers, coincidence units, and analog-to-digital converters. The common digital manipulations involving registers, buffers, encoders, and so on are explored in texts on digital methods (e.g., Ref. 1) and are not discussed here.

The remainder of this chapter is divided into sections centering on application of various modules in different signal chains. Each is intended to typify a commonly encountered situation in radiation measurements, but no attempt is made to provide an exhaustive review of the wide assortment of applications to which the modules may be put. Most basic

Table 17.1 Summary of Common Pulse-Processing Functions

(A) Linear–Linear	In	Out
PREAMPLIFIER	Linear charge pulse from the detector	Linear tail pulse
LINEAR AMPLIFIER	Linear tail pulse	Amplified and shaped linear pulse
BIASED AMPLIFIER	Shaped linear pulse	Linear pulse proportional to amplitude of input pulse that lies above input bias level
PULSE STRETCHER	Fast linear pulse	Conventional shaped linear pulse of amplitude equal to input pulse
SUM AMPLIFIER	Two or more shaped linear pulses	Shaped linear pulse with amplitude equal to the sum of coincident input pulses
DELAY	Fast linear or shaped linear pulse	Identical pulse after a fixed time delay
Linear input — LINEAR GATE / Gate input	(1) Shaped linear pulse (2) Gate pulse	Linear pulse identical to linear input if gate pulse is supplied in time overlap

(B) Linear–Logic	In	Out
INTEGRAL DISCRIMINATOR	Shaped linear pulse	Logic pulse if input amplitude exceeds discrimination level
DIFFERENTIAL DISCRIMINATOR (SINGLE-CHANNEL ANALYZER)	Shaped linear pulse	Logic pulse if input amplitude lies within acceptance window
TIME PICK-OFF (TRIGGER)	Fast linear or shaped linear pulse	Logic pulse synchronized with some feature of input pulse

(C) Logic–Linear	In	Out
START STOP — TIME-AMPLITUDE CONVERTER	Logic start and stop pulses separated by time Δt	Shaped linear pulse with amplitude proportional to Δt

(D) Logic–Logic	In	Out
COINCIDENCE	Logic pulses at two or more inputs	Logic pulse if pulses appear at all inputs within a time interval τ (resolving time)
ANTI-COINCIDENCE	Logic pulses at two inputs	Logic pulse only if pulse appears at one input *without* pulse at second input within time τ
SCALER	Logic pulses	One logic pulse for every N input pulses

module types are detailed under the application in which their use is most critical. However, the discussion begins with one element common to nearly all applications (the preamplifier) and two ancillary elements (the detector bias supply and the test pulse generator) that are also important in virtually any instrumentation system.

V. COMPONENTS COMMON TO MANY APPLICATIONS

A. Preamplifiers

1. GENERAL

As outlined in Chapter 4, the fundamental output of all pulse-type radiation detectors is a burst of charge Q that is liberated by the incident radiation. For G-M tubes and many scintillation counter applications, Q is sufficiently large so that a fairly large voltage pulse is produced by integrating this charge pulse across the summed capacitance represented by the detector, connecting cable, and input of the recording circuitry. For most other detectors, however, the charge is so small that it is impractical to deal with the signal pulses without an intermediate amplification step. The first element in a signal-processing chain is therefore often a *preamplifier* (or "preamp") provided as an interface between the detector and the pulse-processing and analysis electronics that follow.

The preamplifier is usually located as close as possible to the detector. From a signal-to-noise standpoint, it is always preferable to minimize the capacitive loading on the preamp, and therefore long interconnecting cables between the detector and preamp should be avoided if possible. One function of the preamp is to terminate the capacitance quickly and therefore to maximize the signal-to-noise ratio. Because of convenience or safety considerations, the components that follow in the pulse processing chain often are located at some distance from the detector and preamp. Thus another requirement is for the preamp output stage to be capable of driving its signal into the large capacitance represented by the long interconnecting cable, or that it have a low output impedance.

The preamplifier conventionally provides no pulse shaping, and its output is a linear tail pulse. The rise time of the output pulse is kept as short as possible, consistent with the charge collection time in the detector itself. The decay time of the pulse is made quite large (typically 50 or 100 µs) so that full collection of the charge from detectors with widely differing collection times can occur before significant decay of the pulse sets in.

2. VOLTAGE- AND CHARGE-SENSITIVE CONFIGURATIONS

Preamplifiers can be of either the *voltage-sensitive* or *charge-sensitive* type. Historically, the voltage-sensitive type is the more conventional in many electronic applications and consists simply of a configuration that provides an output pulse whose amplitude is proportional to the amplitude of the voltage pulse supplied to its input terminals. A schematic diagram of a voltage-sensitive configuration is shown in Fig. 17.1. If the time constant of the input circuit (the parallel combination of the input capacitance and resistance) is large

Figure 17.1 Schematic diagram of a simplified voltage-sensitive preamplifier configuration. R_2 is the feedback resistance.

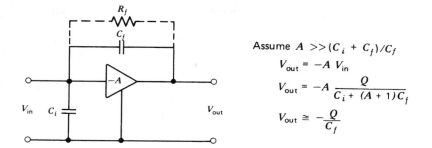

Assume $A \gg (C_i + C_f)/C_f$

$V_{out} = -A \, V_{in}$

$V_{out} = -A \dfrac{Q}{C_i + (A+1)C_f}$

$V_{out} \cong -\dfrac{Q}{C_f}$

Figure 17.2 Simplified diagram of a charge-sensitive preamplifier configuration. If the conditions indicated are met, the output pulse amplitude becomes independent of the input capacitance C_i. The time constant given by the product $R_f C_f$ determines the decay rate of the tail of the output pulse.

compared with the charge collection time, then the input pulse will have an amplitude equal to

$$V_{max} = \frac{Q}{C} \tag{17.1}$$

where C is the input capacitance. For most detectors, the input capacitance is fixed so that the output pulse produced by a voltage-sensitive preamp is proportional to the charge Q liberated by the incident radiation. If the input capacitance were to change, however, this desirable constant proportionality would no longer hold. In semiconductor diode detectors, for example, the detector capacitance may change with operating parameters. In these situations, a voltage-sensitive preamp is undesirable because the fixed relationship between V_{max} and Q is lost.

The elements of a charge-sensitive configuration that can remedy this situation are shown in Fig. 17.2. For this circuit, the output voltage is proportional to the total integrated charge in the pulse provided to the input terminals, as long as the duration of the input pulse is short compared with the time constant $R_f C_f$. Changes in the input capacitance no longer have an appreciable effect on the output voltage. Although originally developed for use with semiconductor diode detectors, this charge-sensitive configuration has proved its superiority in a number of other applications, so that preamplifiers used with other detectors in which the capacitance does not necessarily change are also often of the charge-sensitive design.

Ideally, the rise time of the pulse produced from the preamplifier is determined only by the charge collection time in the detector and is independent of the capacitance of the detector or preamp input. That is generally the case, but for fast detectors the rise time may also be influenced by time constants arising from several factors.[2] Prominent among these is any series resistance from undepleted regions in semiconductor detectors or imperfect electrical contact to the detector active volume. This resistance couples with capacitance at the input of the preamplifier to define a time constant that can slow the rise of the pulse if it is comparable with or exceeds the charge collection time.

3. NOISE CHARACTERISTICS

Probably the most important specification for a preamplifier is its noise figure. This specification is normally quoted as the FWHM of the response function of the system that is due only to the preamplifier noise. The figure is normally given as the equivalent energy spread in the type of detector for which the preamplifier is designed. The noise figure is a strong function of the capacitance with which the preamp input is loaded. For example, a good quality preamplifier used with silicon diode detectors may have a noise figure of 1.6 keV with zero input capacitance, but this figure may double if the input is loaded with 100 pF. The input capacitance arises from both the inherent detector capacitance and from the connecting cable between the detector and preamplifier. It is therefore important to keep the interconnecting cable as short as possible and to choose a detector whose inherent capacitance is no larger than necessary.

For a wide assortment of applications, the noise level of commercially available preamplifiers is sufficiently low so that their contribution to the FWHM of the system response is small compared with the inherent contributions of the detector itself. This is not the case for very-low-energy measurements, however, and, as an example, the preamplifier noise contribution is very significant in the measurement of energy deposition in a semiconductor diode detector below a few keV. Therefore, a strong motivation remains to reduce the inherent preamplifier noise for these low-energy measurements.

One of the significant contributors to preamplifier noise is the Johnson noise associated with the feedback resistor (R_f in Fig. 17.2). The noise is made smaller by increasing the resistance value, but the longer time constant leads to very long tails on the output pulses. Problems can then arise with respect to overload recovery and pile-up in the preamplifier (see below). An alternative approach is to eliminate the feedback resistor altogether. Two preamplifier designs that do not require resistive feedback are called *pulsed optical feedback* and *transistor reset* preamplifiers. Their operation is described beginning on p. 614. In both cases, the elimination of the feedback resistor and its Johnson noise permits a lower preamplifier noise contribution compared with conventional resistive feedback preamplifiers.

In demanding situations, the noise generated in the preamplifier input stage can also be reduced by cooling. The practical problems involved generally make this approach unattractive except in the case of applications in which the detector itself is operated at reduced temperature. Most cooled preamplifier applications are therefore in conjunction with germanium or Si(Li) detectors, for which the first stage of the preamp is held near liquid nitrogen temperature inside the detector cryostat, or with semiconductor X-ray detectors that are used with Peltier electrical cooling.

4. OVERLOAD RECOVERY AND PILE-UP

The charge-sensitive preamplifier is relatively susceptible to saturation when very large pulses are supplied to its input. The overload recovery properties of the preamplifier are therefore an important specification for those applications in which frequent large pulses might obscure the signal to be measured. Methods to reduce the sensitivity of charge-sensitive preamplifiers to overloading pulses are described in Refs. 3 and 4.

Because the output of a preamplifier is a tail pulse with a rather long decay time, the pile-up of pulses is inevitable, except at very low signal rates. Although the effects of this pile-up are largely removed by subsequent shaping (see Chapter 16), one potential effect cannot be dealt with so easily. In Fig. 17.3, the piled-up output of a preamplifier at high rates is sketched. The average level of this waveform will increase with rate and may approach the limit of linear operation of the preamplifier. In that event, some piled-up pulses may drive the preamplifier into saturation and thus will be seriously distorted. Choosing small values for the feedback resistor R_f will minimize this effect by ensuring rapid decay of the pulse, but at the expense of an increased noise contribution.

The pulse rate at which saturation occurs depends on the type of connection between the preamp and the detector (see Fig. 17.5). In a dc-coupled preamplifier (the type commonly built into germanium detectors), the saturation behavior is determined by its feedback resistance value R_f (see Fig. 17.2) and its saturation output level V_m. When saturation conditions are reached, the current flowing through R_f is simply given by V_m/R_f. This current will be supplied by the charges produced by radiation interactions in the detector. For

Figure 17.3 The pile-up of pulses within the preamplifier at high rates. If the saturation level of the preamplifier is exceeded, some pulses can be seriously distorted.

a single pulse, the charge will be Ee/ϵ, where E is the average energy deposited in the detector, e is the charge on the electron, and ϵ is the energy required to form one charge carrier pair. When multiplied by the rate at which events are occurring, a current is obtained which must equal the current flowing through R_f. Thus, if r_m is defined as the pulse rate at saturation

$$\frac{Eer_m}{\epsilon} = \frac{V_m}{R_f}, \quad \text{or} \quad r_m = \frac{V_m \epsilon}{EeR_f}$$

Since V_m and R_f are constants for a preamplifier of given design, the product Er_m, called the *energy rate limit,* can be specified to describe the conditions under which it will go into saturation

$$Er_m = \frac{V_m \epsilon}{eR_f}$$

The value of ϵ is appropriate to the type of detector with which the preamp is designed to operate. For example, a preamp with $R_f = 10^9 \ \Omega$ and V_m of 10 V used with a germanium detector ($\epsilon = 2.96$ eV/electron–hole pair) would have an energy rate limit given by the above expression as 1.85×10^5 MeV/s. Thus pulses with an average energy of 1 MeV would saturate the preamplifier at a rate of 185,000/s, while smaller pulses with an average energy of 0.5 MeV would not saturate until a rate of 370,000/s.

Stand-alone preamplifiers intended for use with a variety of different detectors normally are ac coupled, and as shown in Fig. 17.5, a coupling capacitor isolates the input of the preamplifier from the detector. This coupling ensures that the average dc voltage to the right of the coupling capacitor must be zero. However, when randomly occurring pulses are produced by the detector, a fluctuating voltage will now appear at this point. If the maximum voltage fluctuation is large enough, it can also exceed the saturation level of the preamplifier output.

From the discussion of the Campbell theorem in Chapter 4, the RMS value of these fluctuations depends on the product of the rate and the square of the average amplitude [see Eq. (4.8)]. Therefore the saturation specifications for ac coupled preamplifiers must be made in terms of the product of the rate multiplied by the square of the average energy per pulse. This quantity is now defined as the *energy-squared rate limit* and will be given as one of the preamplifier characteristics. For example, a specification of 3×10^6 MeV^2/s would permit a maximum counting rate for pulses with average amplitude of 1 MeV of 3×10^6 per second. This saturation rate would be lowered by a factor of 4 if the average amplitude were increased from 1 to 2 MeV.

5. ACTIVE RESET TECHNIQUES

There are two potential advantages to be gained if the feedback resistance R_f is eliminated from the charge-sensitive preamplifier configuration shown in Fig. 17.2. One is the improvement in the noise level as a result of removing the Johnson noise associated with the resistor. The second relates to the performance at very high pulse rates, where the pile-up and overload problems discussed in the preceding section limit the maximum rate at which a preamplifier with resistive feedback can be used.

Without the feedback resistor, the pulses of charge from the detector are simply accumulated on the feedback capacitance (C_f in Fig. 17.2). The output voltage then grows in the random staircase fashion shown in Fig. 17.4, with each upward step corresponding to a separate pulse. This staircase voltage corresponds to the sequence of pulses illustrated in Fig. 17.3, where the tails now have infinite length. (As before, the tails are later removed by the shaping process in a following linear amplifier, producing a series of shaped pulses with amplitudes proportional to each individual step.)

Some method must now be provided to reset the preamplifier voltage to zero when the staircase approaches some maximum voltage short of the saturation level. One method[5–7] is to discharge the feedback capacitor by momentarily illuminating the drain–gate junction of the input stage FET with light. A light pulse of a few microseconds duration is produced by a light-emitting diode connected to the output stage. This *pulsed optical feedback* avoids any physical connection to the input stage and does not add to the input capacitance, keeping noise levels very low. It was developed for use in demanding situations where minimum noise from the preamplifier is critical. One common application is in X-ray spectroscopy with Si(Li) detectors, where the preamplifier contribution to the system noise can be as low as 100 eV equivalent.[6]

An alternative reset method generally called *transistor reset* is often applied[8,9] in preamplifiers intended for high-rate applications. Here the voltage is reset using an active circuit with a transistor connected to the input stage. While this approach cannot match the low capacitance and noise figures of the pulsed light feedback technique, it can provide adequately low noise performance when used with detectors whose inherent capacitance is not extremely small.[8]

In both the pulsed optical and transistor reset techniques, the sudden large drop in the output voltage during the reset can cause overloading problems if processed by a following linear amplifier. Therefore, such preamplifiers provide an "inhibit" output signal that is turned on just before the reset occurs and is maintained until the reset is completed. The inhibit signal is then supplied to a gate input on the amplifier to block signals during its dura-

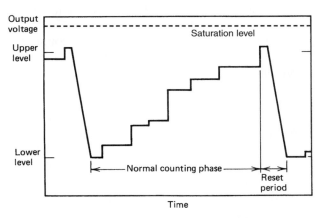

Figure 17.4 The output voltage of an active reset preamplifier. The reset phase is triggered when the voltage crosses the upper-level limit during the normal counting phase. Each upward step is a separate signal pulse.

tion. This period represents a dead time that can become significant at high rates and for large amplitude pulses when frequent resets are required. One advantage of transistor reset compared with pulsed optical techniques of present design is that this dead time can be made shorter, allowing higher-rate operation. Reset dead times of a few microseconds are typical.

Another advantage inherent to both pulsed optical and transistor reset is that the pole-zero cancellation adjustment (see p. 593) in the following amplifier is no longer required. The need for the adjustment is normally brought about by the finite decay time of the tails of pulses produced by resistive feedback preamplifiers. This adjustment is critical to achieving good resolution performance, especially at high rates. The active reset methods produce pulses that have infinite decay time, removing the pole-zero cancellation requirement.

6. DETECTOR BIAS VOLTAGE

Another function normally carried out by the preamplifier is to provide a means for supplying bias voltage to the detector. Arrangements can always be made to supply voltage to the detector independent of the preamplifier, but it is usually convenient to do so through the preamplifier. In Fig. 17.5, two configurations are shown in which the bias voltage is supplied through a load resistance R_L. A single cable between the preamplifier and detector provides both the voltage to the detector and the signal pulse to the input of the preamplifier. The load resistance, together with the input capacitance, determines the time constant across which the detector current is collected. From the standpoint of minimum noise,

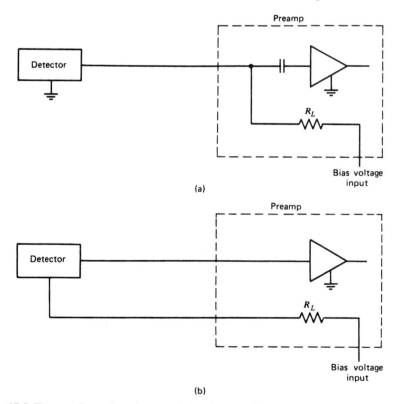

Figure 17.5 Two configurations for supplying detector bias through a preamplifier. (*a*) An ac-coupled arrangement, in which a coupling capacitor is provided between the detector and preamplifier circuits. This allows interchanging different values of R_L without affecting the preamplifier input. (*b*) A dc-coupled configuration that eliminates the coupling capacitor and generally leads to better noise performance for critical applications. Now the detector must be isolated from ground, and changing the bias resistor R_L may affect the input stage characteristics.

R_L should be as large as possible. Practical limitations always dictate that its value be no more than a few thousand megohms because any dc signal or leakage current drawn by the detector must also flow through this resistor. Leakage currents are especially troublesome for semiconductor diode detectors and, in those cases in which R_L is large, can lead to a substantial dc voltage drop across R_L. In that event, the voltage actually applied to the detector is less than the supply voltage, and experimenters must be aware of the magnitude of the leakage current so that they can compensate for the voltage drop across R_L by raising the supply voltage. In some preamplifiers, it is recommended that R_L be changed by the user to suit the specific application to which the preamplifier is put.

Note that when a preamplifier is dc-coupled to the detector, all the leakage current produced from the detector must be accommodated by the preamp. In a resistor-feedback configuration, the current must flow through the feedback resistance (R_f in Fig. 17.2) and the resulting voltage that appears across R_f will cause problems if it becomes large enough. In dc-coupled preamps using alternative active reset methods, leakage current will be integrated on C_f between pulses, causing a positive slope in the "horizontal" portion of the stairstep output. This extra accumulated charge will cause more frequent resets, increasing the system dead time. For these reasons, ac-coupling in which the dc current from the detector is blocked by the coupling capacitor becomes the necessary choice for detectors with high leakage current.

The best noise performance in present designs results from the use of a field effect transistor (FET) as the input stage. FETs are notoriously sensitive to overvoltage and are easily damaged by the transients that can be generated by switching a detector bias supply in coarse steps or abruptly disconnecting or turning off the voltage. As a result, many commercial preamplifiers are provided with overvoltage protection circuits. For the ultimate in low noise performance, however, it is often necessary to switch out the protection circuits, and in such circumstances the bias supply must be changed only gradually and in a continuous fashion.

7. TEST PULSE INPUT

Most preamplifiers are also provided with an input labeled *test pulse,* which is intended to receive the output of a pulse generator for system test purposes. Figure 17.6 shows one means of applying this test pulse to the input stage of a charge-sensitive configuration. If the pulser is sufficiently stable, its output should be resolved into a single channel by a pulse height analyzer connected to the system output. Any broadening of this response into more than one channel can be attributed to the inherent electronic noise of the measurement system. It is often interesting to measure this inherent electronic noise width because any measured detector response can never be better resolved than this electronic limit. It is important to make this measurement with the detector attached to the input of the

Figure 17.6 A charge-sensitive preamplifier that has been provided with a test pulse input. If a step voltage pulse of amplitude V_t is applied to this input, a charge equal to $V_t C_t$ is supplied to the preamplifier input stage. R_t is a small-value termination resistance.

preamplifier because its capacitance is often a significant factor in the overall noise characteristics of the preamplifier. The test pulse input provides a convenient means of carrying out this noise determination as well as simply checking the overall function of the signal-processing system prior to a measurement. In more sophisticated systems, a test pulse input may also be required during the course of a measurement for purposes of dead time determination or gain stabilization (see Chapter 18).

8. CONSIDERATIONS IN THE USE OF PREAMPLIFIERS

The signal-processing section of preamplifiers intended for either solid-state diode detectors or gas-filled ionization or proportional counters is quite similar. The primary differences are in the value chosen for the load resistor R_L in Fig. 17.5 and in the degree of isolation provided along the high-voltage path through the preamplifier. Preamplifiers intended for proportional counter and HPGe detector applications must be designed to withstand several thousand volts, whereas silicon junction detectors seldom require bias supplies of more than a few hundred volts. The value of the load resistance R_L can be higher (to minimize its noise contribution) for gas-filled detectors because their leakage current is normally lower than that of solid-state detectors.

Preamplifiers intended for scintillation counters are usually quite different. Because the signal level from a photomultiplier (PM) tube is rather high, the gain and noise specifications required of a scintillation preamplifier are relatively undemanding. Although it is quite possible to use a scintillation counter without a preamplifier in many applications, it is convenient to include a preamplifier simply to avoid the changes that can occur in the time constant of the equivalent anode circuit of the PM tube if it is connected directly to a linear amplifier or other measuring circuits. Without the preamplifier, the length of the cable used and the input impedance of the measuring circuits both affect the time constant and can lead to situations in which nonoptimum time constants result from equipment rearrangement. The high voltage to scintillation counters is also supplied somewhat differently as compared with semiconductor diode or gas-filled detectors. The signal from a PM tube is derived from the anode and is independent of the high voltage, which is supplied directly to the resistor divider string in the tube base. Therefore, scintillation counter preamplifiers do not have a bias voltage supply input; instead, the high voltage is supplied directly to the scintillation counter base.

Reflecting the usual interests of the user, most commercial preamplifiers are optimized with respect to the pulse height resolution for pulses derived from a specific detector. In some applications, however, information on the timing of the pulse is more important than an accurate measure of its amplitude, and modifications must be made in the way the basic detector pulse is handled. Sherman and Roddick[10] discuss the methods that can be used to derive a fast timing signal from preamplifiers, while at the same time making minimum compromises in energy resolution.

It is conventional to supply the low voltage power required for the operation of the preamplifier through a multiconductor cable connected to the chassis of a subsequent electronic component, most often the linear amplifier. The type of connector and specific pin assignments for this power cable are covered in the NIM standard; see Appendix A. When there is a large separation between the preamplifier and subsequent electronics, this long interconnecting power cable is occasionally the source of noise problems as a result of pickup and/or ground loops. In that event, it is better policy to provide a separate local preamplifier power supply.

B. Detector Bias and High-Voltage Supplies

With few exceptions (such as the self-powered neutron detector), virtually all radiation detectors require the application of an external high voltage for their proper operation.

This voltage is conventionally called *detector bias,* and high-voltage supplies used for this purpose are often called detector bias supplies.

Some characteristics of detector bias supplies that can be important in specific applications are the following:

1. The maximum (and minimum) voltage level and its polarity.
2. The maximum current available from the supply.
3. The degree of regulation against long-term drifts due to changes in temperature or power line voltage.
4. The degree of filtering provided to eliminate ripple at power line frequency or other low-frequency noise.

The sophistication required of the bias supply varies greatly with the detector type. For detectors that draw very little current (such as an ion chamber) the bias supply can be as simple as a dry cell battery. On the other hand, supplies that must simultaneously provide high voltage and relatively high current involve a substantial amount of design engineering and can be among the heaviest and bulkiest of the equipment normally found in a nuclear instrumentation system.

The most demanding common application is to provide high voltage for the operation of PM tubes in connection with scintillation counters (see Chapter 9 for design details). Typical scintillation high-voltage supplies must be capable of providing up to 3000 V with a current of a few milliamperes. The output must also be well regulated to prevent gain shifts in the PM tube that are due to drifts in the high-voltage level. Bias supplies for proportional counters must also supply relatively high voltages, but the current demands are considerably less. However, the degree of regulation and filtering is again important because any high-voltage fluctuations appear superimposed on the signal. Similarly, high voltage supplies for germanium detectors may need to provide up to 5000 V for large-volume coaxial configurations, but only modest currents are required. Semiconductor diode detectors draw relatively little current and the voltage demands seldom exceed 1000 V. The high-voltage supply for G-M tubes is the least demanding of all in several respects. Although the voltage may be as high as several thousand volts, the current requirements are very small. Furthermore, stability can be very poor because the plateau characteristics of the tube ensure that the counting rate will be relatively independent of bias voltage perturbations.

The voltage level on most bias supplies is adjustable either through switching in steps or by means of a continuously adjustable helipot. When used with preamplifiers with FET input stages, continuous adjustment avoids switching transients that may be potentially damaging to the FET. Some designs provide alternate protection by limiting the rate of change of the voltage between steps.

C. Pulse Generators

An electronic pulse generator is indispensable in the initial setup and calibration of virtually any nuclear instrumentation system. Furthermore, some methods of gain stabilization and dead time determination require the output of a pulse generator to be mixed with signal pulses during the course of a measurement. Pulsers are therefore a very common element in most radiation instrumentation systems.

A tail pulse generator with adjustable rise and decay times is probably the most useful of all pulser types. Its output is conveniently fed to the test pulse input on preamplifiers or used directly in place of the preamplifier output. In its simplest form, the pulser output provides for convenient adjustment of system parameters such as shaping time constants, pole-zero parameters, and various timing and delay adjustments. If the output amplitude is constant with a high accuracy, a measurement of the amplitude distribution recorded by the pulse analysis system determines the electronic noise level present in the system.

Most pulse generators also provide a front panel adjustment for the amplitude of the pulse. In some designs, often called *precision pulsers,* this adjustment is accurately controlled and can be set to a given level with high precision. Such pulsers can then be used to check the integral linearity of a pulse-handling system simply by recording the output amplitude for several different settings of the input pulse amplitude. The output of most precision pulsers is sufficiently stable so that, in the absence of system electronic noise, all pulses of constant amplitude would be resolved into a single channel in a multichannel analysis system.

Some designs can also provide a pulse source of other than constant amplitude. One of the more useful is called a *sliding pulse generator,* which produces pulses whose amplitude is uniformly distributed between zero and some set maximum. Recording these pulses in a multichannel pulse analysis system gives a direct measure of the differential linearity of the multichannel analyzer (see Chapter 18).

For purposes of determining the pulse resolving time of a given measurement system, a *double pulse generator* can be very useful. Here, two pulses of fixed amplitude are produced periodically with a variable time spacing between the two pulse leading edges. The resolving time can be determined simply by increasing the time spacing between the double pulses and observing the transition point at which the system begins to recognize two distinct pulses rather than only one.

For normal pulse generators, the interval between pulses is uniform and periodic. However, for determining some system parameters such as pile-up behavior and other time-dependent phenomena, a periodic source does not adequately represent the random time spacing encountered from actual radiation detector pulses. Therefore, designs have evolved that provide a source of randomly spaced pulses of constant amplitude. In such *random pulsers,* the noise signal from an internal component is often used to trigger randomly the time at which a pulse is produced at the output. Another method of producing randomly spaced pulses is to use the triggering feature provided on many pulse generators. In this mode, the pulse generator will produce an output pulse only when provided with an external trigger pulse. If this trigger pulse is derived from a random source such as a separate radiation detector, the output of the pulser can again be an accurate simulation of the random pulse spacing normally encountered in radiation detector signal chains.

Pulsers other than tail pulse generators can also play a useful role in detector pulse processing. A *gate generator* produces a square pulse of variable amplitude and width when triggered by an external source, usually a standard logic pulse. As its name implies, it is most useful in providing the gate pulse necessary to open or close a linear gate based on external logic criteria. Other square pulse generators, especially those that can be triggered and that provide a variable delay between the trigger pulse and output pulse, can be very useful in providing logic pulses to various points in the counting system under variable conditions.

VI. PULSE COUNTING SYSTEMS

The signal chain shown in Fig. 17.7 represents a basic measurement scheme in which only the number or rate of pulses from a radiation detector are to be recorded. The tail pulse output of the preamplifier typically has an amplitude of a few tens or hundreds of millivolts and is too small to be counted directly. Furthermore, the pile-up of these long pulses at high rates could cause stability problems. Therefore, the next step is normally to process the pulses through a linear amplifier. Here a voltage gain of 1000 or more can be provided so that the shaped linear pulse at its output can easily cover a span of 0–10 V. The shaping requirements in a simple counting system are usually not severe, and only at relatively high counting rates must one pay close attention to the specific method of shaping chosen. On

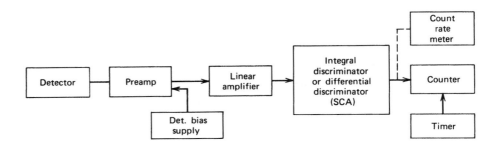

Figure 17.7 Elements of a typical signal chain for pulse counting.

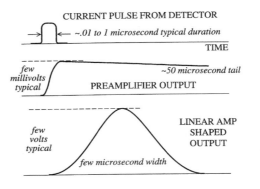

Figure 17.8 Intercomparison of typical pulses at sequential points along the signal processing chain.

the other hand, spectroscopy systems described later place much more emphasis on shaping, and a detailed description of this and other functions of linear amplifiers is therefore postponed until the following section on pulse spectroscopy.

The interrelationship between the various signal pulses at different points in a typical pulse processing system is illustrated in Fig. 17.8. The current pulse from the detector is integrated on a long time constant circuit in the preamplifier, producing a tail pulse output. The linear amplifier then shapes this pulse to a much shorter width, and increases its amplitude by a factor given by its gain.

A. Integral Discriminator

In order to count the pulses reliably, the shaped linear pulses must be converted into logic pulses. The *integral discriminator* is the simplest unit that can be used for this conversion and consists of a device that produces a logic output pulse only if the linear input pulse amplitude exceeds a set discrimination level.[†] If the input pulse amplitude is below the discrimination level, no output appears. This selection process is illustrated in Fig. 17.9a. Unless specifically designed otherwise, the logic pulse is normally produced shortly after the leading edge of the linear pulse crosses the discrimination level. This *leading edge*

[†]The discrimination level is sometimes referred to as the *pulse height bias* level of the counting system. This terminology is somewhat unfortunate because of the potential confusion with the phrase *detector bias,* commonly used to describe the external dc voltage supplied to the detector, and is not used here.

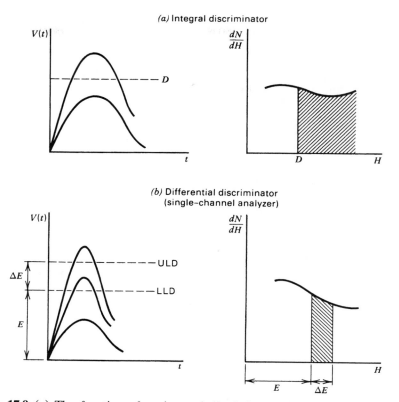

Figure 17.9 (*a*) The function of an integral discriminator. Of the two input pulses shown, only the larger clears the discrimination level *D* and produces a logic pulse output. The effect is to select only the area to the right of the amplitude *D* in the differential pulse height spectrum. (*b*) The function of a differential discriminator or single-channel analyzer. For the three input pulses shown, only the intermediate amplitude lies within the acceptance window and leads to an output logic pulse. In the pulse height spectrum, only the pulses falling within the cross-hatched area are accepted.

timing is compared with other schemes of generating the logic pulse in the later section on time pick-off methods.

The discrimination level is normally adjustable by a front-panel control. In many counting situations, the level is set just above the system noise so that the maximum sensitivity for counting detector pulses of all sizes is realized. Other situations may call for a higher discrimination level to count selectively only events above a set minimum size. For example, much of the background may be limited to relatively low pulse amplitudes so that some finite discrimination level may greatly enhance the signal-to-background counting ratio.

Integral discriminators must be designed to accept shaped linear input pulses of a specific amplitude span (usually 0–10 V positive). The stability and linearity of the discriminator adjustment are usually adequate for routine applications but may become important specifications for demanding situations.

B. Differential Discriminator (Single-Channel Analyzer)

Another linear-to-logic converter in widespread use involves two independent discrimination levels. As illustrated in Fig. 17.9*b*, a *differential discriminator* or *single-channel analyzer* (SCA) produces a logic output pulse only if the input linear pulse amplitude lies between the two levels. The action of the unit is therefore to select a band of amplitudes or window in which the input amplitude must fall in order to produce an output pulse.

Several systems of nomenclature and adjustment persist for SCAs. In some units, the *lower-level discriminator* (LLD) and *upper-level discriminator* (ULD) are independently adjustable from front-panel controls. In others, the lower level is labeled the *E* level, and the window width or difference between levels is labeled ΔE and can be varied separately without affecting the *E* level.

In counting systems, the SCA can serve to select only a limited range of amplitudes from all those generated by the detector. A common example is one in which the window is set to correspond only to those events in the detector that deposit the full energy of an incident radiation. In this way, one type or energy of radiation often can be measured selectively in the presence of other radiations.

In normal SCAs, the time of appearance of the logic pulse is not closely coupled to the actual event timing, and use of these logic pulses in timing measurements will often give imprecise results. If one of the time pick-off methods discussed later in this chapter is incorporated into the SCA design, the logic pulse can be much more closely correlated with the actual event time. Modules with this feature are often called *timing SCAs* and are widely used in coincidence applications or other timing measurements.

Most SCAs provide the option of switching out the upper-level discriminator and using the unit as a simple integral discriminator controlled by the lower level. The input linear pulses are intended to be shaped with typical 0.5–10 μs widths and a pulse height range that most commonly is 0–10 V positive. Bipolar pulses with positive leading edges are also normally acceptable. Other specifications that can be important in some applications include the linearity of the discriminator level adjustments, the stability of the levels with respect to temperature changes, and the SCA dead time (usually 1 or 2 μs larger than the input pulse width).

C. Scalers or Counters

As the final step in a counting system, the logic pulses must be accumulated and their number recorded over a fixed period of time. The device used for this purpose may be a simple digital register that is incremented by one count each time a logic pulse is presented to its input. In nuclear pulse counting applications, such devices are sometimes called *scalers* as a historic anomaly that dates from the time when digital registers of reasonable size were not widely available. Then it was common to use a scaling circuit to divide the input pulse repetition rate by a fixed factor such as 100 or 1000 so that the rate would be low enough to be directly recorded by an electromechanical register. These systems have been replaced by all-electronic digital registers. The scaling function persists only in the sense that an overflow output pulse is often provided when the maximum content of the register is exceeded (usually no less than 10^5 or 10^6 counts). We henceforth refer to such units as *counters* because that term more adequately describes the actual function.

Counters are commonly operated in one of two modes: preset time and preset count. In the preset time mode, the counting period is controlled by an internal or external timer. This timer may be built as part of a common chassis with the counter, or separate timers can be obtained as individual modules. In the preset count mode, the counter will accumulate pulses until a specified total has been achieved, at which point the counting period is terminated. If the period of time over which these counts have been accumulated can be recorded independently, the counting rate can be determined. The preset count mode has the advantage that a given statistical precision can be specified before the start of the measurement, since the duration of counting will be prolonged until enough counts have been accumulated to guarantee the desired statistical precision.

Counters can also be of the *blind* or *display* type. In a display counter, the contents of the register are continuously displayed on a front-panel numeric indicator. Display counters provide the advantage that the progress of a measurement can be monitored visually and malfunctions can often be detected quickly by aberrant behavior of the digital display. Blind counters do not provide a visual display but, instead, can generate a coded logic readout of the register content when triggered by an external command. Because blind counters are less expensive than the display type, they have found favor in large-scale systems in which many independent counts must simultaneously be accumulated. In that event, the blind counters are often part of a CAMAC system in which the interrogation and readout take place over the dataway.

A *printing counter* is one in which an interface has been provided to generate the proper readout signals to drive a conventional line printer or other device. Other features found in some counters include an internal input gate that can be controlled by a gate pulse supplied to the unit, or a built-in integral discriminator to eliminate any noise that may appear along with the input pulses. Other specifications of importance are the minimum time separation between the leading edge of two logic pulses in order that they be counted as separate events (the *pulse pair resolving time*) and the maximum counting rate at which the counter may be driven.

D. Timers

The function of a timer is simply to start and stop the accumulation period for an electronic counter or other recording device. Obviously its most important property is the precision to which the time interval is controlled. Two general methods of control are commonly encountered. The simplest and least expensive method is to base the timing interval on the frequency of the alternating current of the power line to which the unit is connected. The precision of the timing is therefore determined solely by the accuracy and stability of the power line frequency. Utility companies usually do a good job of controlling the accuracy of the power line frequency when integrated over a day or more in order to maintain the accuracy of clocks also synchronized to the line frequency. On the other hand, the frequency may wander substantially over short periods of time, and timers based on power line synchronization thus may give rise to substantial timing interval errors if the interval is less than a few hours. In order to guarantee better accuracy, timers based on internal crystal-controlled circuits are preferred for more exacting measurements. The most important specification in this case then becomes the stability of the timing frequency to changes in temperature.

E. Counting Rate Meter

In some situations it is advantageous to have a visual indication of the rate at which pulses are being counted in the system. This function can sometimes be provided by a display counter in which the experimenter visually observes the rate at which counts are accumulated. Because of the random spacing between nuclear events, small changes in counting rate are difficult to observe in this way, particularly at low counting rates.

A *counting rate meter* provides a more direct means of indicating the rate at which pulses are being accumulated. In its most common form, a rate meter can be represented by the diode pump circuit outlined in Fig. 17.10. The output stage of the logic device driving the rate meter is represented by the voltage generator and series output impedance R_f. Each logic pulse, as it enters the circuit, deposits a small fixed amount of charge on the storage capacitor C_t. This capacitance is also continuously discharged by a current that flows through the resistance R. If the rate of arrival of logic pulses is constant, an equilibrium will eventually be established in which the rate of charge deposition on the capacitor is just equal to the rate of its discharge through the resistance. Equilibrium is reached after several values of the time constant of the circuit have elapsed following an increase or decrease in the rate. This time constant is given simply by the product of the capacitance C_t and the parallel resistance R.

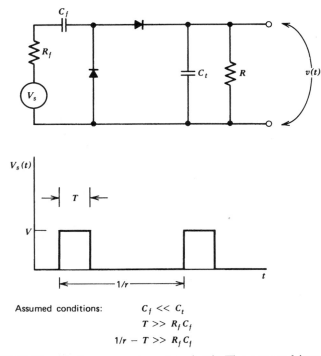

Assumed conditions:

$$C_f \ll C_t$$
$$T \gg R_f C_f$$
$$1/r - T \gg R_f C_f$$

Figure 17.10 The diode pump rate meter circuit. The assumed input voltage waveform V_s is also shown.

If the conditions shown in the figure are met, the average voltage appearing at the output of the circuit is

$$\bar{v} = iR = QrR = C_f VrR \tag{17.2}$$

where r is the average rate at which pulses are supplied to the circuit, and Q is the charge deposited per pulse given by the product of the coupling capacitor C_f and the pulse amplitude V. This output voltage is therefore proportional to the rate of arrival of the input logic pulses. If the input pulses were regularly spaced in time, the output voltage would have the appearance sketched in Fig. 17.11a. Longer time constants result in a more nearly constant signal, but the response to abrupt changes in rate will be slower. The full-scale range of the meter is normally varied by selecting the value of R with a front-panel control. Other rate meter circuits have been developed[11] that provide an output proportional to the logarithm of the count rate. These meters allow compression of the counting rate scale so that several decades may be monitored without the inconvenience of switching between scales.

When dealing with events from a radiation detector, the spacing between pulses is irregular and fluctuations in the output voltage arise as a result of the random variation in spacing. The rate meter signal then has an appearance typical of that sketched in Fig. 17.11b. The standard deviation σ of this signal can be defined as the square root of the variance of the values derived by sampling the signal many times at random and may be derived as follows.

The differential contribution to the output voltage produced by a rate r during the time between t and $t + dt$ is $(Qr/C_t) \, dt$. Because C_t is continuously discharged through R, the importance of this contribution decreases exponentially with time and at a later time t_0 is multiplied by a factor $\exp[-(t_0 - t)/RC_t]$. Similarly, the differential contribution to the standard deviation is $(Q/C_t)\sqrt{r} \, dt$ with an importance that also decreases exponentially with time. If the mean rate r does not change, the total variance of the output voltage can

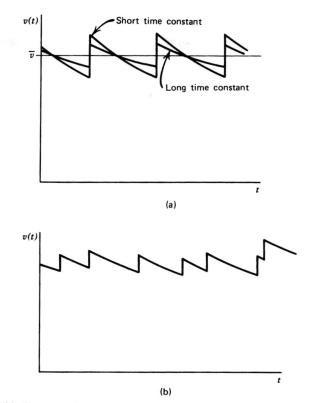

Figure 17.11 Output voltage from the diode-pump rate meter for (*a*) periodic pulses and (*b*) random pulses.

be obtained by integrating the independent weighted contributions of all prior time intervals:

$$\sigma_v^2 = \left(\frac{Q}{C_t}\right)^2 r \int_{-\infty}^{t_0} \exp\left(-\frac{2(t_0 - t)}{RC_t}\right) dt \qquad (17.3)$$

$$= \frac{Q^2 R}{2C_t} r \qquad (17.4)$$

The standard deviation is the square root of the variance:

$$\sigma_v = Q \sqrt{\frac{Rr}{2C_t}} \qquad (17.5)$$

The fractional standard deviation is usually of more interest and can be derived by combining Eqs. (17.2) and (17.5),

$$\frac{\sigma_v}{\bar{v}} = \frac{1}{\sqrt{2rRC_t}} \qquad (17.6)$$

which is often expressed as a percentage. The fractional standard deviation is usually selectable through a front-panel switch that varies the value of C_t. To achieve small fluctuations, C_t must be large, but the resulting long time constant limits the rate meter speed of response to rate changes.

An interesting digital equivalent of the rate meter circuit has been described by White.[12] It is based on the observation that in the diode pump circuit (Fig. 17.10) the amount of charge removed from the capacitor per unit time is always a fixed fraction of the total charge on the capacitor. By analogy, the digital rate meter consists of a register into which input pulses are gated for a fixed period of time. At the end of this time, a fixed fraction of the register content is subtracted from the accumulated content. The cycle of accumulation and fixed fraction subtraction is then repeated continuously. As in the analog rate meter, an equilibrium is exponentially approached in which the rate at which pulses are added to the register becomes equal to the rate at which they are subtracted. The equivalent time constant is given by

$$\tau = TF$$

where T is the accumulation gating time and F is the fraction of pulses subtracted at each step. With this substitution for the time constant $\tau = RC_t$, the standard deviation predicted by Eq. (17.6) for the analog rate meter also applies to the digital equivalent.

It should be emphasized that the output of a rate meter circuit, although it represents an analog quantity that can be displayed on a meter, is not the equivalent of operating the detector in current mode. For the rate meter indication, all pulses from the detector that clear a discrimination level and generate a logic pulse contribute equally to the output. If the same detector is operated in current mode, the importance of each pulse is weighted by its amplitude or energy. Keeping all other factors constant, both the rate meter indication and the current output would double if the source intensity were increased by a factor of two. However, if the energy spectrum were to change, the effect on the current output would be not be closely coupled with the change in the rate meter indication.

F. Dead Time in Counting Systems

An important consideration in many counting applications is the loss of events due to the dead time of the system. For some detectors (notably the G-M tube) the detector mechanism itself limits the minimum interval between events for which two distinct pulses can be counted. More often, however, the detector will be capable of producing pulses that are separated by a time that is less than the dead time inherent in the operation of an electronic component in the signal chain, and therefore it will be this component that determines the system dead time. In the simple counting systems shown in Fig. 17.7, this limiting component is usually the integral discriminator or the SCA. Although there are many exceptions, the dead time of a discriminator or SCA is typically related to the width of the linear pulse presented to its input and is characteristically a microsecond or two larger than this width.

The validity of the corrections for dead time losses discussed in Chapter 4 depends on the assumption that the dead time is constant for all events. The inherent dead time of electronic units can sometimes vary with the amplitude or shape of the input pulse, so that steps to set the dead time artificially are warranted in some critical applications. In this approach, the dead time is standardized by an element such as a linear gate, which is held closed for a fixed period of time following each pulse. This time is chosen to be larger than the dead time of any component in the system, so that accurate corrections can be made, even under conditions in which wide variations in pulse amplitude are encountered. The dead time of this element may also be measured conveniently by direct observation on an oscilloscope.

VII. PULSE HEIGHT ANALYSIS SYSTEMS

A. General Considerations

Next to the simple counting of pulses, the most common procedure in nuclear measurements involves recording the amplitude distribution of pulses produced by a radiation detector. Most often the object is to deduce properties of the incident radiation from the position of peaks in the recorded spectrum, although other aspects of the spectrum may be of interest in different situations. The performance required of the instrument system used to record the pulse height spectrum is largely dependent on the inherent energy resolution of the detector. If the detector energy resolution is relatively poor, the requirements of the recording system are undemanding and easy to meet. On the other hand, detectors with good energy resolution require careful attention to the pulse-processing system to ensure that additional degradation of the resolution is minimized.

A simple pulse height analysis system is shown in Fig. 17.12a. The key element in this signal chain is the linear amplifier, which shapes the pulses from the preamplifier and provides enough amplification to match the input range of pulse amplitudes for which the multichannel analyzer has been designed. It is the shaping function of the linear amplifier that often dominates the performance of the pulse-processing system. For low-resolution detectors, virtually any of the shaping techniques discussed in Chapter 16 can be used to reduce the long tails of the preamplifier output pulses. For high-resolution detectors, however, consideration must be given to the effect of various shaping methods on the pulse properties with regard to signal-to-noise and pile-up. Also shown in Fig. 17.12b is a pulse spectrometry system to which several components have been added to mitigate the effects of pulse pile-up and baseline shift.

The strategies employed in choosing parameters for the pulse-processing system change drastically with the expected counting rate. If the counting rate is low (say 100 counts/s or less), the system can be optimized with regard to processing each individual pulse without much concern for interfering effects between pulses.[†] The problem becomes much more

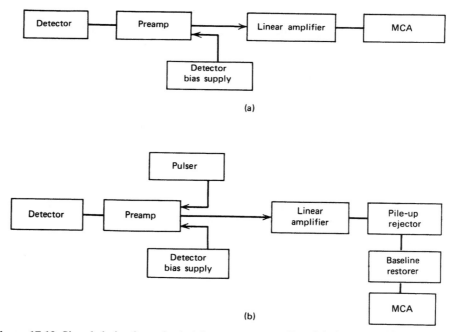

(a)

(b)

Figure 17.12 Signal chains for pulse height spectrometry. Part (a) shows a rudimentary system for noncritical applications, whereas system (b) includes some added functions that can improve performance for high-resolution detectors when operated at high pulse rates.

[†]A quick estimate of the importance of rate-related effects can be made by calculating the *duty cycle* obtained by multiplying the effective width of the shaped pulse by the rate. If this product is less than about 10^{-3}, these effects should be minimal and can often be neglected. A duty cycle of 10^{-2} is a moderate rate, whereas in high-rate situations it may approach 10^{-1}. For a typical pulse width of 5 μs, the corresponding rates are 200, 2000, and 20,000 per second.

complicated at high rates, where instrument parameters chosen to minimize pulse pile-up and to ensure rapid return of the pulse waveform to the baseline often conflict with the choices that would be made on a signal-to-noise basis alone. Thus, it is in the high-resolution–high-rate situation that the demands are greatest, and the proper choices are often difficult to predict in advance. Decisions are then frequently made on an empirical basis by varying the system parameters to find the best combination for each individual situation.

The most critical choices involve the method chosen for pulse shaping within the linear amplifier. The following section on linear amplifier operation therefore stresses the various compromises that must be struck in choosing the type and time characteristics of the pulse shaping. These choices necessarily involve the general concepts of ballistic deficit, signal-to-noise, and pile-up of signal pulses.

B. Ballistic Deficit

The rise time of the pulse from the preamplifier normally corresponds to the charge collection time in the detector itself. If the full amplitude of the preamplifier pulse is to be preserved through the shaping process, the shaping time constants must be large compared with the preamplifier pulse rise time. Because the shaping time constants cannot always be chosen as arbitrarily large, the amplitude of the shaped pulse can sometimes be slightly less than that attainable with very long time constants. The degree to which the infinite time constant amplitude has been decreased by the shaping process is called the *ballistic deficit* (see Fig. 17.13).

In detectors with a constant charge collection time, relatively large ballistic deficits can often be tolerated because a constant fraction of the amplitude for each pulse is lost. Should the charge collection time vary, however, a variable amount of each pulse will be lost, which can lead to resolution degradation. In these cases, one must then choose longer time constants than might be optimum on the basis of signal-to-noise or pile-up considerations.

This problem is obviously most severe for those detectors with the largest variation in charge collection time. Rise time variations are often important in proportional counters, where the radial variation of the primary ionization determines the spread in the charge collection time. Perhaps the biggest variations occur in coaxial germanium detectors, where large differences in collection time correspond to different radial locations of the radiation interaction in the detector. Optimum shaping times then tend to be many times greater than the average collection time.

In the common pulse-shaping methods, events leading to the largest ballistic deficit will also produce the longest peaking time of the shaped pulse. Because of this correlation, it is theoretically possible to correct for ballistic deficit on a pulse-to-pulse basis by estimating its magnitude by sensing the peaking time. A corresponding voltage can then be added to the shaped pulse to compensate for the amplitude loss. Ballistic deficit variations can be a

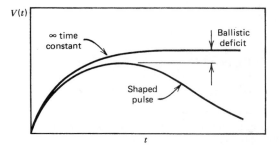

Figure 17.13 Definition of the ballistic deficit. If the shaping times are fixed, the ratio of the deficit to the amplitude of the shaped pulse will be constant for pulses of the same leading edge shape but will vary if this shape changes.

dominant contributor to the energy resolution of gamma-ray spectra taken under high rate conditions where short shaping times are needed to minimize pulse pile-up. Goulding and Landis[13] have shown that use of a correction circuit based on this principle can significantly improve energy resolution under these conditions, and this approach has been made available in at least one commercially available shaping amplifier. A comparison of the Goulding–Landis method and alternative approaches using gated integrators for minimizing the effects of ballistic deficit in large germanium detectors can be found in Ref. 14.

C. Signal/Noise Considerations

By definition, noise is any undesired fluctuation that appears superimposed on a signal source. Fig. 17.14 gives a graphical representation of the way in which a random noise component can degrade the amplitude information carried by pulses from a radiation detector. A general analysis of noise sources in radiation systems is beyond the scope of this discussion but may be found in texts such as Ref. 15. At this point we outline only some of the more important considerations regarding noise sources and the practical steps that can be taken to minimize their influence on measured pulse height spectra.

The important sources of noise occur near the beginning of the signal chain where the signal level is at a minimum. Noise generated at this point undergoes the same amplification as the signal, whereas noise generated further along the signal chain is usually much smaller than the signal. Therefore, discussions of electronic noise sources generally center on the preamplifier and, most importantly, its input stage. As outlined by Radeka,[16] the noise sources can conveniently be categorized into those that are effectively in parallel with the input and those that are in series with the signal source. For example, sources of parallel noise include the fluctuations in the leakage current within the detector itself and in the gate-source current of the FET in the input stage of the preamplifier. Series noise sources involve the contribution of Johnson noise associated with series resistances and the thermal noise of the input FET. The frequency spectrum of both series and parallel

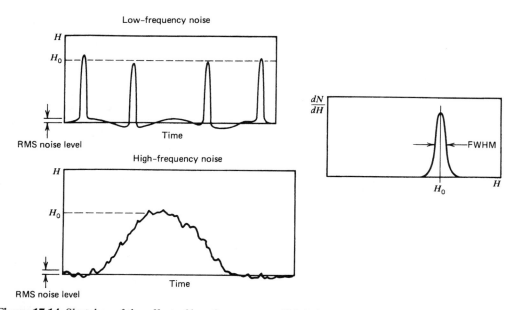

Figure 17.14 Sketches of the effect of low-frequency and high-frequency noise on signal pulses of constant amplitude. The effect is to broaden the peak recorded in the differential pulse height spectrum for these pulses. The FWHM of the peak of Gaussian-distributed noise will be equal to the RMS noise level multiplied by a shape factor of 2.35, provided noise is the only factor in broadening the peak.

noise is very broad, and for some analytical purposes it is often assumed that a "white" or uniform distribution in frequency holds.

In contrast, the frequency spectrum of the signal is confined to a much narrower band. For example, the signal from a detector whose charge collection time is about a microsecond can never contain useful information on a time scale of nanoseconds. Therefore, low-pass filtering of the output of such a detector will eliminate the contribution of high-frequency noise but will not affect the information-carrying components of the signal. Similarly, low-frequency power line pick-up is a noise source that can potentially degrade the signal, but high-pass filtering can remove this component without substantially affecting the signal pulse shape from most detectors. Therefore, the pulse-shaping role traditionally carried out by the linear amplifier actually amounts to selective filtering to remove as much broad spectrum noise as possible without severely attenuating the useful signal components. The pulse shaping is normally carried out through a combination of differentiating and integrating circuits, but from the discussion of Chapter 16, it is equally valid to regard these operations as high-pass and low-pass filtering, respectively.

The amount of noise added by the preamplifier–amplifier combination is often expressed in terms of the *equivalent noise charge* (ENC). This is defined as the amount of charge that, if applied suddenly to the input terminals of the system, would give rise to an output voltage equal to the RMS level of the output due only to noise. Formally, the ENC is expressed in absolute units of charge, or coulombs. However, it has become commonplace to divide this value by the unit charge on an electron (1.6×10^{-19} C) to express it instead in units of electron charges. For example, the electronic noise in a given system may be expressed as "100 electrons," with the understanding that the ENC is equal to 1.6×10^{-17} C.

For a specific type of detector, the contribution of electronic noise to the FWHM of peaks in the energy spectrum can be calculated from the ENC. From Fig. 17.14, the FWHM is equal to 2.35 multiplied by the RMS value of the noise level, or 2.35 times the ENC value. To convert to energy units one must also multiply by ϵ, the deposited energy required to create one charge carrier. In silicon at room temperature, $\epsilon = 3.62$ eV, so the relationship becomes:

$$\text{FWHM (eV in silicon)} = 2.35 \cdot 3.62 \text{ eV} \cdot \text{ENC} = 8.51 \cdot \text{ENC (electrons)}$$

Thus the previous example of an ENC of 100 electrons predicts a peak width contribution (FWHM) caused by electronic noise in silicon detectors of 851 eV.

The effect of pulse shaping on signal-to-noise ratios in nuclear pulse systems has been the subject of extensive theoretical studies. The reader is referred to detailed discussions of this topic in Refs. 17 and 18, and only a summarized statement of the results is given here. These analyses are generally based only on the assumption that the noise is broadly distributed in frequency and deal with the problem from the standpoint of signal pulses that are widely separated in time. As we shall see, the interfering effects of signal pulses at high rates can sometimes override the considerations of signal-to-noise ratios for single isolated pulses given below.

It can be shown[18] that the best possible signal-to-noise ratio is achieved if the signal pulses are shaped to the form of an infinite cusp (see Fig. 17.15). As a practical matter, the pulse shape must have finite width, and the finite cusp has the best signal-to-noise properties if pulse shapes of finite width are compared. Because it has been shown[19] that no other type of pulse shaping (including time-dependent or nonlinear processes) can yield a pulse shape superior to the cusp in its signal-to-noise properties, this pulse shape has become the standard by which the performance of other methods of pulse shaping are compared.

The cusp is not particularly practical for several reasons. The top is sharply pointed, which makes pulse amplitude measurement difficult unless pulse-stretching methods are employed. Furthermore, the return to the baseline is rather slow, which is undesirable from a pile-up standpoint. The cusp shape is also difficult to achieve with practical shaping

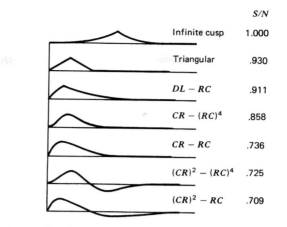

	S/N
Infinite cusp	1.000
Triangular	.930
$DL - RC$	.911
$CR - (RC)^4$	.858
$CR - RC$	.736
$(CR)^2 - (RC)^4$	.725
$(CR)^2 - RC$	.709

Figure 17.15 Various pulse shapes and their signal-to-noise ratio (S/N) relative to the infinite cusp. Various time constants are chosen to yield minimum noise, and pulse shapes are normalized to constant height. (Reproduced with permission from *Nuclear Electronics,* by P. W. Nicholson. Copyright 1974, by John Wiley & Sons Ltd.)

circuits. For all these reasons, other methods of pulse shaping are more commonly used that are inferior to the cusp in signal-to-noise properties but are preferable because some of the above drawbacks are avoided.

Fig. 17.15 lists some of the common methods of pulse shaping generally found in linear pulse amplifiers, together with the signal-to-noise performance compared with the infinite cusp. All those shown in the figure can be achieved using the methods of passive shaping discussed in Chapter 16. More complicated methods of pulse shaping involving active circuits are useful in systems in which the ultimate in signal-to-noise must be attained. Of these, active filters that approximate a Gaussian or triangular shape (see p. 593) have been incorporated into commercially available linear amplifiers.

A specific type of noise, called *microphonics,* can sometimes be a problem in practical detector systems. Mechanical vibrations transmitted to the detector–preamplifier input stage can produce small fluctuations in capacitance that cause a modulation of the output signal. The effects are most pronounced for systems such as ionization chambers or semiconductor spectrometers for which the small charge per pulse requires keeping the input stage capacitance at a minimum. For example, it has been demonstrated[20] that a capacitance change of only 5×10^{-7} pF between the FET gate and the high-voltage bias of a Si(Li) system corresponds to a microphonic signal equivalent to the deposition of 10 eV in the detector. For stray capacitance values on the order of 1 pF, such a change corresponds to a mechanical motion of only one part in 10^7. Fortunately, much of this noise will occur at low frequencies and will lie outside the frequency band passed by the shaping networks in the linear amplifier. In those cases in which microphonic noise is a problem, choosing short shaping times (corresponding to raising the low-frequency cutoff) will help minimize this contribution to the system noise level.

D. Dependence of Noise on Shaping Time and Capacitance

The importance of various sources of electronic noise in the measured signal-to-noise ratio depends on choices made in the filtering or shaping operation. All of the shaping methods illustrated in Fig. 17.15 involve a choice of parameters that determine the time characteristics of the shaping network and the resulting time-to-peak of the shaped pulse. One of the

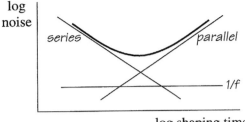

Figure 17.16 Variation of the contributions of series and parallel noise sources as a function of the shaping time in the pulse processing electronics. The minimum in the total occurs at the shaping time that results in equal contributions from series and parallel noise. The $1/f$ noise contribution is unaffected by the shaping time.

important choices presented to the user in spectroscopic amplifiers is a wide range of possible shaping times.

As illustrated in Fig. 17.16, the contribution of series noise mentioned in the previous section tends to become less important as the shaping time is increased. On the other hand, sources of parallel noise become more important. There is yet another category of noise sources that are generally grouped together under the term *1/f noise* that includes, for example, the effects of capture and release of charges in the input FET. This latter category contributes to the electronic noise in a manner that does not change with choice of the shaping time. Considering the effect of the combination of all three sources of noise (which will combine in quadrature), the overall electronic noise contribution will go through a minimum as the shaping time is increased from a small value. This optimum shaping time will generally occur at the point where the series and parallel noise contributions are equal, and might typically be in the range from 1–20 μs for preamplifiers associated with the common silicon or germanium semiconductor detectors.

The contribution of series noise as a fraction of the signal increases with detector capacitance, while the relative contribution of parallel noise to the signal remains independent of this capacitance. For this reason, the optimum shaping time among similar types of detectors tends to shift toward longer times for those with larger capacitance. An experimental optimization of the shaping time, such as that illustrated later in Fig. 17.24, is an important procedure the user should carry out as part of the initial set up of the detector system if electronic noise contribution is to be kept to a minimum.

E. Pile-Up

1. THE PROBLEM

The fact that pulses from a radiation detector are randomly spaced in time can lead to interfering effects between pulses when counting rates are not low. These effects are generally called *pile-up* and can be minimized by making the total width of the pulses as small as possible. Other considerations of ballistic deficit and signal-to-noise prevent reduction of the pulse width beyond a certain point, and therefore the effects of pulse pile-up at high rates are often very significant.

Pile-up phenomena are of two general types, which have somewhat different effects on pulse height measurements. The first type is known as *tail pile-up* and involves the superposition of pulses on the long-duration tail or undershoot from a preceding pulse (see Fig. 17.17). Under conditions described earlier, tails or undershoots can persist for

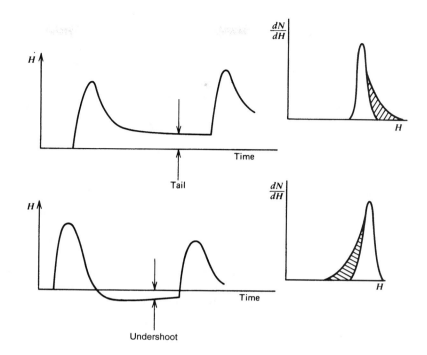

Figure 17.17 Pile-up from the tail or undershoot of a preceding pulse. Both are usually categorized as tail pile-up. The effect on the differential pulse height spectrum for constant-amplitude pulses is shown as the cross-hatched area at the right.

relatively long periods of time so that tail pile-up can be significant even at relatively low counting rates. The effect on the measurement is to worsen the resolution by adding wings to the recorded peaks in the pulse height spectra as shown in the figure. The remedy for tail pile-up is to eliminate residual tails or undershoots through the use of pole-zero cancellation or active baseline restoration techniques discussed in Chapter 16. Pulse shapes that return quickly to the baseline will also help eliminate the irreducible tail pile-up caused by the normal decay of the shaped pulse.

A second type of pile-up, generally called *peak pile-up,* occurs when two pulses are sufficiently close together so that they are treated as a single pulse by the analysis system. As shown in Fig. 17.18, the superposition of pulses with relatively flat tops will lead to a combined pulse with an apparent amplitude equal to the sum of the two individual amplitudes. Lesser degrees of overlap will give a combined pulse with an amplitude somewhat less than the sum. Not only does this type of pile-up lead to distortions in the recorded spectrum, including an occasional sum peak, but it also interferes with quantitative measurements based on measuring the area under full-energy peaks. The pile-up of two full-energy pulses effectively removes both from the proper position in the pulse height spectrum, and the area under the full-energy peak in the spectrum will no longer be a true measure of the total number of full-energy events. Because peak pile-up leads to the recording of one pulse in place of two, the total area under the recorded spectrum is also smaller than the total number of pulses presented to the system during its live time.

The seriousness of peak pile-up for a given situation can be estimated from the counting rate and the effective width of the signal pulses. The effective width is difficult to define, except for rectangular pulses, but can be approximated by the FWHM of the first lobe of

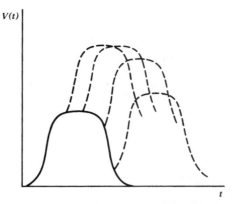

Figure 17.18 Peak pile-up, in which two closely spaced signal pulses combine to form one distorted pulse. Several different cases are sketched with increasing overlap between the first and second pulse.

the shaped pulse.[21] If we neglect second-order effects, the interval distribution [Eq. (3.71)] can be applied to estimate the degree of pile-up to be expected. The probability of observing an interval greater than τ is given by

$$P\left(>\tau\right) = \exp\left(-n\tau\right) \tag{17.7}$$

where n is the true rate of signal pulses. For pile-up to be avoided, the interval following each pulse must be greater than the effective pulse width. For $n = 20{,}000$ s^{-1} and an effective width of 5 μs, Eq. (17.7) predicts that 90.5% of all intervals will be greater than this width. However, two pulses are affected if a pile-up event takes place, so that only about 81% of all true events escape pile-up and are properly recorded. At high rates, higher-order pile-up events must be considered in which more than two pulses may overlap simultaneously.

If all signal pulses were rectangular in shape with constant amplitudes, the spectral effects of pile-up would be relatively simple. At very low rates, only a single peak should appear in the recorded pulse height spectrum because pile-up is then negligible. At higher rates, any pile-up that occurs can result only in a total amplitude that is an integral multiple of the single pulse amplitude. Therefore, the effect of pile-up would be to introduce sum peaks that would appear only at uniformly spaced positions in the recorded spectrum. Of these, the *double sum peak* corresponding to the simple pile-up of two pulses will be the most intense and will appear at a position corresponding to twice the basic amplitude.

A more realistic representation must account for the variable amplitude and nonrectangular shape of most signal pulses.[22] The pile-up spectra will then have the continua shown in Fig. 17.19, which fill in the space between expected sum peaks and the single amplitude peak. These events correspond to the partial pile-up of pulses with leading and trailing edges that are not vertical, and to pile-up of individual pulses with different amplitudes.

Pile-up can also influence measurements other than the pulse height distribution. In particular, timing measurements can be affected by the resulting changes in pulse shape that accompany pile-up. For example, Refs. 24 and 25 describe the quantitative losses that can occur in coincidence experiments based on crossover timing when pile-up distorts the pulse shape. Wilkinson[26] also points out the errors that can occur in time interval measurements when variable counting rates are encountered in the course of the measurement.

2. PILE-UP REJECTION

One method of avoiding the pulse height spectrum effects arising from a peak pile-up is to use some means of *pile-up rejection*. Most commercial linear amplifiers are provided with active circuitry that can discard pulses that are expected to be affected by pile-up. Some

Figure 17.19 Spectral effects of peak and tail pile-up. The dashed curve shows a ^{55}Fe spectrum taken at a low counting rate at which pile-up is negligible. The solid curve shows a high-rate spectrum and illustrates the sum peak and continuum caused by peak pile-up. The low-energy tail added to the primary peak by overshoot or tail pile-up is also observed. (From Wielopolski and Gardner.[23])

practical methods of pile-up rejection are reviewed in Refs. 27–29. Circuits of this type can dramatically reduce the spectral distortions due to pile-up,[23] but the quantitative losses must be dealt with in the same manner as dead time losses. (See analysis below.) An example of the degree of pile-up rejection that can be achieved is shown in Fig. 17.20.

Because the shape of a piled-up pulse differs from that of a normal pulse not affected by pile-up, the techniques of pulse shape discrimination discussed later in this chapter can also be used to reject pile-up.[31] This approach has the advantage of including other benefits of pulse shape discrimination, including the rejection of "defective" pulses resulting from incomplete charge collection from semiconductor detectors. The use of digital pulse processing has broadened the scope of techniques that can be used to reject pile-up events and introduces the possibility of recovering the information lost in the pile-up process (see p. 658). However, most common systems for pulse processing rely only on the selective rejection of piled-up pulses to clean up the recorded spectrum without attempting to separate them.

A common technique employed in many pile-up rejectors is to pass the signal through two parallel branches—a *fast* and a *slow* branch. In the fast branch, one of the time pick-off techniques described later in this chapter is used to generate a very short logic pulse as quickly as possible. In the slow branch, the event is processed using conventional (and somewhat slower) shaping techniques designed to best measure its amplitude. This shaped pulse is then passed through a linear gate only if a second pulse is *not* triggered in the fast branch during the processing time in the slow branch. In this way, the output pulses should not be contaminated by events that occur during the processing time, and the corresponding spectrum is, in principle, free of pile-up distortions.

Figure 17.20 Pulse height spectra recorded with and without a pile-up rejector. The true event rate n was $2.6 \times 10^5/\text{s}$ and the rejector resolution time τ was 300 ns. Note that the contributions to the spectrum from pile-up are greatly reduced by the rejector, while the numbers of counts in the primary peaks (free of pile-up) are unaffected. (From Goulding and Landis.[30])

In reality, however, the fast–slow pile-up rejection technique only reduces the problem but cannot totally eliminate it. There is still some possibility that two or more events will occur within the pulse resolution time of the fast branch, and only one fast pulse will be produced. The slow branch will then process an amplitude that is some combination of the multiple events, and a pulse resulting from pile-up is again recorded. The likelihood of this occurrence depends on the time behavior of the fast branch, which can be approximated using one of the two models for dead time behavior introduced in Chapter 4. It is therefore instructive to look in somewhat greater detail at the nature of the events lost in dead time processes, because these same combinations of closely spaced events will also contribute to pile-up.

3. STATISTICAL ANALYSIS OF PILE-UP EVENTS

We begin by defining the terms that will be used throughout this analysis. An *event* is the consequence of a radiation interaction in the detector that should lead to a recorded pulse, in the absence of dead time or pile-up. A determination of the amplitude spectrum of true events is often the object of our measurement. A *count* is a pulse as actually registered by the recording system. Because of dead time and/or pile-up, fewer counts are recorded than the number of true events. Also, some recorded counts correspond to the pile-up of more than one event, and the recorded amplitude distribution is distorted compared with the true event spectrum.

The analysis also assumes that any inherent dead time of the detector/preamplifier is small compared with the pile-up resolution time τ of the pulse-processing system. This time τ is now defined as the minimum time that must separate two events so that they do not pile up. Thus, events arrive at the amplifier as a Poisson distributed random process and are assumed to pile up if they occur with a time spacing less than τ following a previous event. The discussion of dead time losses given previously in Chapter 4 assumed that events were simply lost under these conditions, but we now extend the analysis to examine the categories of counts that may result. True events are again assumed to occur at a rate n. Due to pile-up, the recording system will perceive counts at a lower rate m. We now seek to classify these counts according to the number of true events that contribute to each count. The results will depend on whether the system behaves in a paralyzable or nonparalyzable manner, as defined in Chapter 4.

a. Nonparalyzable System

In Chapter 4 [see Eq. (4.24)], it was shown that the event rate n and counting rate m are related by

$$n = \frac{m}{1 - m\tau} \quad \text{or} \quad m = \frac{n}{1 + n\tau} \tag{17.8}$$

We now derive the probability that a typical count corresponds to the pile-up of two or more true events. Over the time τ, the average number of true events is simply $n\tau$. From the Poisson distribution [Eq. (3.24)] we can write the probability that exactly x events occur over this time as

$$P(x) = \frac{\bar{x}^{x} e^{-\bar{x}}}{x!} = \frac{(n\tau)^{x} e^{-n\tau}}{x!} \tag{17.9}$$

Each of these time intervals is started by a true event. If *no* additional events occur over the following time τ, a count is recorded that is free of pile-up. The probability that a given *count* falls in this category is

$$P(0) = e^{-n\tau} \tag{17.10}$$

Note that n is the true event rate, not the observed counting rate. Similarly, the probability that a given count represents the pile-up of exactly two events is

$$P(1) = n\tau e^{-n\tau} \tag{17.11}$$

In general, the probability that $(x + 1)$ true events pile up to produce a given count is

$$P(x) = \frac{(n\tau)^{x} e^{-n\tau}}{x!} \tag{17.12}$$

Since the Poisson distribution is normalized,

$$\sum_{x=0}^{\infty} P(x) = 1$$

and we have accounted for all counts.

As a cross check, we note that a count corresponds to $(x + 1)$ events with a probability of $P(x)$. Therefore, the average number of true events per count is

$$
\begin{aligned}
\langle x \rangle &= \sum_{x=0}^{\infty} (x + 1)\, P(x) = \sum_{x=0}^{\infty} (x + 1)\, \frac{(n\tau)^{x} e^{-n\tau}}{x!} \\
&= e^{-n\tau} \sum_{x=0}^{\infty} (x + 1)\, \frac{(n\tau)^{x}}{x!} = e^{-n\tau} e^{n\tau} (n\tau + 1) \\
&= n\tau + 1
\end{aligned}
\tag{17.13}
$$

But, by definition, $\langle x \rangle = n/m$, so

$$\frac{n}{m} = n\tau + 1 \quad \text{or} \quad m = \frac{n}{1 + n\tau}$$

Since this expression is identical to Eq. (17.8) above, we have correctly accounted for all true events as well.

b. Paralyzable System

From Eq. (4.27), we have previously shown that for a paralyzable system

$$m = ne^{-n\tau} \tag{17.14}$$

The arguments leading to Eq. (17.10) are still valid for a paralyzable system, so the probability that a given *count* is free of pile-up remains

$$P(0) = e^{-n\tau}$$

Two (and only two) events will pile up under paralyzable conditions if the following sequence occurs: a true event at $t = 0$, with a second event sometime in the interval $0 < t < \tau$ that is followed by an event-free interval of length τ. The probability that a given count originates in this way can be written as

$$P(1) = \int_0^\tau \left(\begin{array}{c} \text{probability of} \\ \text{no event } 0 \to t \end{array} \right)\left(\begin{array}{c} \text{probability of} \\ \text{event in } dt \end{array} \right)\left(\begin{array}{c} \text{probability of} \\ \text{no event } t \to (t + \tau) \end{array} \right)$$

$$= \int_0^\tau e^{-nt}\, n\, dt \cdot e^{-n\tau}$$

$$= e^{-n\tau}(1 - e^{-n\tau})$$

We can extend this analysis to predict the probability that a count results from the pile-up of *three* true events. In this case, following an initial event at $t = 0$, one additional event must occur with time spacing less than τ, followed by the sequence described above for two-event pile-up. Thus, we can write

$$P(2) = \int_0^\tau \left(\begin{array}{c} \text{probability of} \\ \text{no event } 0 \to t \end{array} \right)\left(\begin{array}{c} \text{probability of} \\ \text{event in } dt \end{array} \right) \cdot P(1)$$

$$= \int_0^\tau e^{-nt}\, n\, dt \cdot e^{-n\tau}(1 - e^{-n\tau})$$

$$= e^{-n\tau}(1 - e^{-n\tau})^2$$

To generalize, the probability that a given count is formed from the pile-up of $(x + 1)$ events is

$$P(x) = e^{-n\tau}(1 - e^{-n\tau})^x \tag{17.15}$$

Note that

$$\sum_{x=0}^\infty P(x) = e^{-n\tau} \sum_{x=0}^\infty (1 - e^{-n\tau})^x = e^{-n\tau}e^{+n\tau} = 1$$

so that we have accounted for all counts.

To check the accounting of true events, we first note that each recorded count results from $(x + 1)$ true events with a probability $P(x)$. The average number of events per count is therefore

$$\langle x \rangle = \sum_{x=0}^\infty (x + 1)\, P(x) = e^{-n\tau} \sum_{x=0}^\infty (x + 1)(1 - e^{-n\tau})^x \tag{17.16}$$

$$= e^{-n\tau}e^{+2n\tau} = e^{+n\tau}$$

This expression is consistent with the result obtained directly from Eq. (17.14):

$$\langle x \rangle = \frac{n}{m} = \frac{n}{ne^{-n\tau}} = e^{+n\tau}$$

so the check on true events is verified.

The pulse height spectrum measured in the presence of pile-up can be thought of as the sum of a number of individual spectra caused by different categories of counts. One of these is the original event spectrum that is free of pile-up, and its relative contribution to the measured spectrum is given by $P(0)$ in Eq. (17.10). In addition, the spectrum created by the pile-up of two events will be added with a relative contribution equal to $P(1)$. Spectra that are due to higher-order pile-up will also become significant as the event rate increases and will add individual spectra caused by the pile-up of $(x + 1)$ events with a relative contribution given by the expressions for $P(x)$ in Eqs. (17.12) and (17.15). Each of these contributions is plotted as a function of true event rate in Fig. 17.21, both for paralyzable and nonparalyzable systems. In the pulse height spectrum previously shown in Fig. 17.20, the individual contributions are evident for the pile-up of up to four true events [corresponding to $P(3)$ above].

If no pile-up rejection is used, then τ in the previous expressions represents the effective width of the shaped pulse at the amplifier output (see Fig. 17.18). Typical values for germanium detectors are in the 3–10 μs range. The system is most likely to behave as a paralyzable one, since the arrival of a true event during a previous pulse will extend the period for pile-up by another time τ. When pile-up rejection is applied, the effect is to greatly suppress the pile-up contributions in the recorded spectrum by imposing a much smaller value for τ. This time now becomes the pulse resolution time in the fast branch of the rejector, and typical values are fractions of a microsecond. The system behavior (whether paralyzable or nonparalyzable) will depend on the specific design of the rejection circuits.

One of the effects of pile-up rejection is to ensure that partially overlapped pulses are efficiently discarded, leaving only pile-up corresponding to the nearly complete overlap of two pulses. Thus, the lower spectrum in Fig. 17.20 more cleanly separates the contributions of each order of pile-up that remains. Tenney[32] has demonstrated how individual pile-up spectra can be calculated under certain assumed conditions and gives examples for some simple true event pulse height spectra. Other discussions of spectral distortions caused by pile-up are given in some detail in Refs. 23 and 33–39.

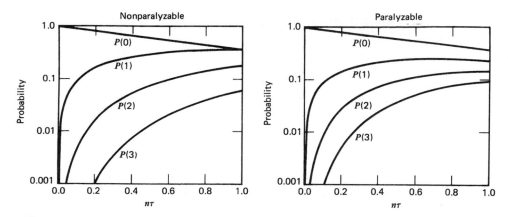

Figure 17.21 Plots of the probabilities that a recorded count is free of pile-up [$P(0)$] or due to ith-order pile-up [$P(i)$]. The abscissa is the product of true event rate n and the pile-up resolution time τ. The left plot is from Eq. (17.12) for nonparalyzable systems, and the right plot is from Eq. (17.15) for paralyzable systems.

It should be emphasized that the results derived above are based on recorded counts. The throughput of the system is better gauged by considering the effect of pile-up on the fraction of true events that are properly recorded free of pile-up. For the nonparalyzable case, recall that the average number of true events per count is, from Eq. (17.13),

$$\langle x \rangle = n\tau + 1$$

The probability (per count) of recording a pile-up-free event is $P(0) = e^{-n\tau}$. Therefore, the fraction of true events that escape pile-up is

$$f_e\big|_{\text{nonpara}} = \frac{P(0)}{\langle x \rangle} = \frac{e^{-n\tau}}{n\tau + 1} \tag{17.17}$$

For $n\tau \ll 1$, the first-order expression is

$$f_e\big|_{\text{nonpara}} \cong \frac{1 - n\tau}{1 + n\tau} \cong 1 - 2n\tau$$

This result can be compared with the first-order expression for the fraction of *counts* that are free of pile-up:

$$f_c = P(0) = e^{-n\tau} \cong 1 - n\tau$$

Thus, the first-order losses to pile-up are twice as great for true events compared with those for recorded counts. This comparison is in agreement with the heuristic observation made earlier that, at low count rates, two events are lost for every pile-up count.

In the case of a paralyzable system, Eq. (17.16) gives the average number of events per count as

$$\langle x \rangle = e^{n\tau}$$

so the fraction of true events that escape pile-up is now

$$f_e\big|_{\text{para}} = \frac{P(0)}{\langle x \rangle} = \frac{e^{-n\tau}}{e^{n\tau}} = e^{-2n\tau} \tag{17.18}$$

Again, the first-order expression for $n\tau \ll 1$ is

$$f_e\big|_{\text{para}} \cong 1 - 2n\tau$$

which is identical to the nonparalyzable case. Plots of $f_e\big|_{\text{nonpara}}, f_e\big|_{\text{para}}$, and f_c are shown in Fig. 17.22.

Finally, the rate at which true events are recorded free of pile-up (r_{pf}) is of primary interest to the user. In systems with an effective pile-up rejector, r_{pf} is approximately the observed output rate, where τ represents the processing time in the slow branch. This rate is given by the product of the true rate n and the fraction of true events that escape pile-up (f_e). For the nonparalyzable case, we use Eq. (17.17) for this fraction and find

$$r_{pf}\big|_{\text{nonpara}} = n f_e\big|_{\text{nonpara}} = \frac{n e^{-n\tau}}{1 + n\tau} \tag{17.19}$$

As a check, this rate should also be given by the product of the counting rate m and the probability (per count) of escaping pile-up $P(0)$:

$$r_{pf}\big|_{\text{nonpara}} = m P(0) = \frac{n}{1 + n\tau} e^{-n\tau}$$

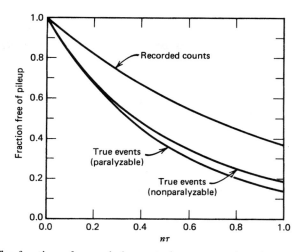

Figure 17.22 The fraction of recorded counts (upper curve) and true events (lower two curves) that escape pile-up as a function of true event rate n.

and we obtain the same result. As the true rate is increased, r_{pf} goes through a maximum. Increasing the true rate further will continue to increase the total counting rate, but the growing importance of pile-up actually decreases r_{pf}. This behavior is illustrated in Figure 17.23. The maximum value of r_{pf} is given by $0.206/\tau$, and it is reached at a true event rate of $0.618/\tau$.

For the paralyzable case, the same type of analysis leads to

$$r_{pf}\big|_{\text{para}} = nf_e\big|_{\text{para}} = ne^{-2n\tau} \tag{17.20}$$

Now r_{pf} reaches a maximum value of $0.184/\tau$ at a true event rate of $0.500/\tau$. Figure 17.23 also plots this case, together with the total counting rate (without pile-up rejection) for both models. Recall (see Fig. 4.8) that the counting rate for the paralyzable model also goes through a maximum given by $0.368/\tau$ at a true rate of $1/\tau$. At that point, Eq. (17.18) predicts that only 13.5% of the true events are recorded free of pile-up.

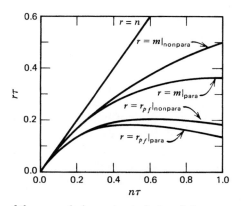

Figure 17.23 Variation of the recorded count rate (m) and the rate at which true events are recorded free of pile-up (r_{pf}) for both paralyzable and nonparalyzable systems. The true event rate is n and the pile-up resolution time is τ.

Finally, let us use these results to predict a straightforward quantity, the rate at which counts due to pile-up are observed, or r_{pu}. That rate is given by the product of the overall counting rate m multiplied by the probability that a given count involves pile-up. Thus

$$r_{pu} = m(1 - P(0)) \qquad (17.21)$$

where $P(0)$ is the previously defined probability that a given count is free of pile-up. For both the paralyzable and nonparalyzable models, $P(0) = e^{-n\tau}$, where n is the true event rate. Therefore

$$r_{pu} = m\,(1 - e^{-n\tau}) \qquad (17.22)$$

We now examine this expression in the limit of low rates where the following approximations can be used

$$e^{-n\tau} \cong 1 - n\tau \quad \text{and} \quad m \cong n(1 - n\tau)$$

Thus

$$r_{pu} \cong n(1 - n\tau)\,n\tau$$

and to first order

$$r_{pu} \cong n^2\tau \qquad (17.23)$$

This simple and widely used result [for example, see Eq. (10.13)] shows that, in the limit of small pile-up contributions, the rate of counts involving pile-up is proportional to the resolving time τ, but depends on the square of the pulse rate. Thus the ratio of recorded counts that are due to pile-up to those that are free of pile-up is simply given by $n\tau$, and increases linearly with pulse rate. The relative prominence of the distortions caused by pile-up in the pulse height spectrum shown in Fig. 17.19 will therefore increase approximately in proportion to the pulse rate at which the spectrum is recorded. Also note that, in this same low-pile-up limit, the fraction of true events that are involved in pile-up is given by $2n\tau$, twice the fraction of recorded counts that develop from pile-up.

4. QUANTITATIVE CORRECTION FOR PILE-UP EFFECTS

The most effective technique of experimentally correcting for the quantitative errors in peak areas introduced by pile-up is to employ the *pulser method*. The output of a fixed amplitude pulse generator is mixed with the pulses from the detector, normally at the pre-amplifier stage. The amplitude of the pulses from the generator is adjusted to create an artificial peak in the recorded pulse height spectrum at a location that does not interfere with peaks or other spectral features from the signal pulses. Since the number of pulses injected from the generator can be accurately determined, a measurement of the area under the artificial peak in the spectrum yields the fraction of those pulses that have escaped pile-up. The assumption is then made that this same fraction also applies to the signal pulses to allow correction of their measured peak areas for pile-up effects. This assumption is rigorously valid only for a pulse generator that produces pulses that are randomly distributed in time, and *random pulse generators* are commercially available for such applications. Periodic pulse generators may also be used provided that the pulse rate is kept small compared with the signal pulse rate. (Differences arise because random pulses can pile up with each other, but periodic pulses cannot. Monte Carlo simulations[40] show that a pulser rate of less than 10% of the signal rate generally results in a negligible difference.) An extensive analysis of the correction of pulse height spectra for pile-up effects for both periodic and random pulse generators can be found in Ref. 36, and a useful discussion of the issues involved in pile-up and dead time losses in gamma-ray spectroscopy are presented in

Ref. 41. This latter reference shows that, with careful attention to methods for detecting and correcting for rate-related losses, quantitative measurements with germanium detectors can be carried out with very small errors even for changing counting rates that can be as high as tens of thousands per second.

F. Linear Amplifier

Two primary functions are conventionally provided by the linear amplifier element in the pulse-processing chain: pulse shaping and amplitude gain. The unit accepts tail pulses as an input, often of either polarity, and produces a shaped linear pulse with standard polarity and output pulse amplitude range (the NIM standard is positive polarity, 0–10 V amplitude). The amplification factor or gain required varies greatly with application but is typically a factor between 100 and 5000. The gain is normally adjustable over a wide range through a combination of coarse and fine controls. If the product of input amplitude and gain exceeds the maximum design output amplitude, the amplifier will *saturate* or *limit* and produce a distorted output pulse with a flat top at the amplitude at which saturation occurs (typically 10 V). Linear amplification will be realized only for those pulses that are short of this saturation level. Amplifier specifications normally include a figure for maximum nonlinearity over the design range, which is usually small enough to have little practical consequence.

Of all the elements in a pulse height analysis system, the linear amplifier presents the user with the greatest variety of operational characteristics and, as a result, the most difficult choices. The conflicting requirements placed on pulse shaping from consideration of signal-to-noise, ballistic deficit, and pulse pile-up must be weighed for each potential application. Fundamental to these considerations is the inherent trade-off between counting rate and resolution. High rates favor bipolar pulses with small width, whereas optimum shaping for best resolution points toward unipolar pulses with relatively large widths. On many commercial linear amplifiers, the shaping times are adjustable. Then the choice may be optimized for a given application by measuring the pulse height resolution of the system as the shaping times are varied. Of course, it is important to carry out these measurements at a counting rate equivalent to that expected in the application itself.

Examples of this optimization are given in Fig. 17.24, where the effects of varying the shaping time in the amplifier are shown for two different pulse rates. In both cases, there is

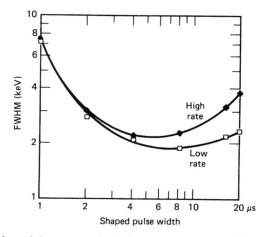

Figure 17.24 Variation of the energy resolution of a typical germanium detector system as the amplifier shaping time is changed. The horizontal axis is the width of the shaped pulse at half maximum. The upper curve is for a relatively high rate (62,000/s) and the lower curve is for a low rate (2000/s). (Data from Fairstein.[42])

an optimum shaping time that minimizes the FWHM of the measured gamma-ray peak from a germanium detector. For the low-rate case, pile-up and baseline shift effects are negligible. As the shaping time is increased from small values, the FWHM decreases (resolution improves) for two reasons: The ballistic deficit decreases (see p. 628) and the contribution of series noise (due primarily to fluctuations in the channel current of the input stage FET) is reduced. At the other extreme, increasing the shaping time causes a gradual increase in the FWHM due to a greater sensitivity to parallel noise (originating primarily from fluctuations in detector leakage current). At an optimum shaping time between these extremes, the observed FWHM goes through a minimum value. For the high-rate case, the minimum occurs at a smaller value of the shaping time, and the corresponding best FWHM is somewhat larger than for the low rate. Now the effects of pile-up and baseline shift are important for all but the smallest values of shaping time near the left of the plot. For larger values of the shaping time, these effects increasingly contribute to the measured FWHM and are reflected in the steeper rise to the right of the minimum. In general, one would expect the optimum shaping time to move to smaller values as the counting rate is increased and the effects of pile-up and baseline shift become more severe.

Most amplifiers in use today employ one of the pulse-shaping methods discussed in Chapter 16. For low-resolution systems (such as scintillation detectors), virtually any method of pulse shaping will be acceptable at low rates, since the detector itself will dominate the observed resolution. For high rates, double delay line shaping can be useful because of its relatively narrow pulse width and almost complete freedom from baseline shift due to the symmetrical bipolar pulse shape. For high-resolution systems (such as germanium gamma-ray spectrometers), shaping choices are more critical. Delay line shaping is not used because of its relatively inferior signal-to-noise characteristics. At low counting rates, the unipolar near-Gaussian pulse shape produced by one differentiation followed by several integration networks is a common choice. Alternatively, triangular shaping (see p. 593) can offer a slightly better noise suppression. At high rates, the baseline shift that accompanies unipolar pulses requires either the substitution of a network that results in bipolar shape (provided as an alternative output on many units), or the use of some type of baseline restoration. Increasingly, active baseline restoration is included as a standard feature in the design of modern linear amplifiers.

More elaborate shaping schemes are sometimes warranted in demanding circumstances, often involving large-volume germanium detectors. Here the large variations in pulse rise times can result in resolution loss through ballistic deficit effects. Shaping techniques that result in a flat-top pulse can offset these effects (see Fig 16.14), so shaping that results in a trapezoidal output or a cusp with flat top has some advantage. These shapes are not easily implemented in standard analog shaping networks, but can be synthesized using digital filters in digital pulse-processing systems (see p. 655).

Performance at high rates can also benefit from use of a *gated integrator* at the output stage of the amplifier shaping networks. Its function is outlined in Fig. 17.25. The unipolar pulse from a Gaussian shaping network is integrated on a capacitance that is part of an active circuit. The integration is continued for a time interval that is typically 8–10 times the shaping time used in the Gaussian network (2–3 times the peaking time). If an additional event occurs during this period, the process is aborted by the pile-up rejection circuits in the amplifier. At the end of this interval, the capacitance is abruptly discharged by closing a switch. The resulting pulse amplitude is now proportional to the area rather than the peak value of the shaped pulse and is therefore much less sensitive to changes in ballistic deficit caused by variable charge collection times in the detector. Consequently, the shaping times can be made much shorter than could be tolerated without the gated integrator to minimize the deleterious effects of pile-up on energy resolution. An example of the high-rate performance of a specific system is shown in Fig. 17.26. Here the pile-up rejection causes the output rate to pass through a maximum as a function of input rate (see Fig. 17.23). Very

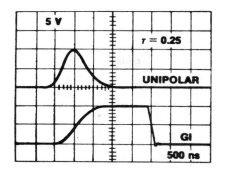

Figure 17.25 Waveforms in the operation of a gated integrator. The unipolar output of a standard Gaussian shaping network is integrated in an active circuit to give the waveform shown in the lower plot. (From Britton et al.[9])

high throughput systems for germanium detectors have been demonstrated[43] that employ active reset preamplifiers to avoid saturation, a gated integrator to minimize the shaping time needed in the amplifier, and fast ADCs in the multichannel analyzer to minimize storage time of the pulses. Reasonable spectroscopic performance has been shown for input rates in the 300,000–400,000/s range, with the recording of pile-up-inspected output events at rates approaching 100,000/s.

The response of a linear amplifier to very large overloading pulses can also be an important factor in determining its overall performance, especially at high rates. The overload recovery time is usually included in a list of amplifier specifications. The output pulse shape may return quickly to zero for normal pulses, but strongly saturating pulses can substantially extend the time required to re-establish true zero baseline. Even for bipolar shaped pulses, the two lobes may not saturate in a symmetric way. Then equal areas above and below the zero level are no longer maintained, and some baseline shift must occur. Thus, active baseline restoration can be beneficial for high-rate operation with bipolar as well as unipolar pulse shapes.

Figure 17.26 An example of the throughput of a pulse-processing system at high input rates. In this case, a gated integrator allows the use of a short (0.25 μs) Gaussian shaping time. The top curve is for the amplifier alone. The remaining plots show the additional losses due to the conversion time of the ADC of the multichannel analyzer used to record the pulses. (From Britton et al.[9])

In summary, a linear amplifier is subject to performance criteria that often require trade-offs depending on the particular detector in use and the details of its application. Important amplifier functions include:

1. Signal amplification.
2. Pulse shaping that promotes accurate measurement of collected charge with maximum immunity from ballistic deficit and variations in shape of the current–time profile. (This consideration favors long shaping times.)
3. Pulse shaping to minimize pile-up and overload and therefore maximize performance at high counting rates. (This factor argues for short shaping times.)
4. Pulse shaping to provide for best signal-to-noise ratio in the analysis of individual pulse amplitude. (This criterion favors the shaping methods listed in Fig. 17.15 that show the best noise immunity.)
5. For demanding applications, the provision of active circuits for pile-up rejection and baseline restoration to help preserve performance at high rates.

G. Biased Amplifier

The function of a biased amplifier is outlined in Fig. 17.27. The input is usually a shaped linear pulse as produced at the output of a normal linear amplifier. A biased amplifier, however, produces no output pulse for any input pulse that lies below a given *bias level,* which normally can be adjusted by a front panel control. The entire range of the biased amplifier output is assigned to the range of input pulses between the bias level and the specified maximum input level. The effect is to magnify this limited range of input amplitudes by spreading it out over the entire output range. This function is especially useful if a small region of a pulse height spectrum must be examined in great detail. The small region can then be spread over the total number of amplitude channels available in the multichannel analyzer.

Figure 17.27 The function of a biased amplifier is to provide linear amplification for only that portion of each input pulse that lies above the bias level.

Because the biased amplifier responds only to that portion of the input pulse that lies above the bias level, the output pulse is usually significantly narrower than the input pulse width. Consequently, some cases may require the use of a pulse stretcher (see below) to restore the pulse width necessary for subsequent amplitude measurement.

H. Sum or Difference Amplifier

In some measurements it may be advantageous to sum the outputs of several detectors to provide a single pulse for later analysis. Sum amplifiers, which can accept multiple inputs (usually the shaped outputs of linear amplifiers) and provide an output pulse that is the algebraic sum of all coincident inputs, are commercially available. A sum amplifier can be converted into a difference amplifier by inverting one of the pulses before the summation.

I. Linear Gate

In principle, a linear gate functions as a switch for linear pulses. When the gate is open, linear pulses pass through the gate without change or distortion. While the gate is closed, all pulses are blocked. The opening of the gate is controlled by the presence or absence of a gate pulse at a second input at the time the linear pulse is presented to the linear input. The gate pulse is a standard logic pulse whose width is usually adjustable and chosen to be somewhat longer than the linear pulse. Proper time overlap may require delay of the linear pulse to allow the gate pulse to arrive first. The *pedestal* of a gate is an undesirable dc level that can be added to the linear pulse in the process of passing through the gate. Ideal linear gates should have zero pedestal, or a means should be provided for the adjustment of the pedestal to zero under experimental conditions. Other desirable properties are low feedthrough under closed gate conditions, low nonlinearity, and fast switching between open and closed states.

J. Pulse Stretcher

A stretcher unit produces a linear pulse output whose shape is set to a standard compatible with following components and whose amplitude is equal to the maximum amplitude of the input pulse. Its most common application is to accept fast linear pulses or those that have been shaped with a sharply peaked waveform and convert them into a nearly flat-topped pulse of the same amplitude, which suits the ADC requirements of most multichannel analyzers.

VIII. DIGITAL PULSE PROCESSING

The types of analog pulse-shaping techniques outlined in Chapter 16 and earlier in this chapter represented the only choices available in the design of pulse processing systems for many decades. Beginning in the early 1990s, however, the development of high-speed analog-to-digital converters (ADCs) with good resolution opened a new possibility for the digital processing of pulses from detectors. A number of commercially available spectroscopic systems now incorporate digital processing in place of the traditional analog approach, and they demonstrate significant advantages over analog systems in some circumstances. They have proven especially useful in the processing of pulses from large-volume germanium detectors at high counting rates, where the conflicting demands for good energy resolution and minimum pulse pile-up are particularly sensitive to the pulse shaping process that is chosen. These commercial systems also digitally implement other spectroscopy amplifier functions such as automatic pole-zero adjustment, baseline restoration, and gain control in a user-friendly overall package.

With sufficiently fast and accurate ADCs, all the information in a single analog pulse from a detector and preamplifier can be preserved by taking multiple samples of the pulse amplitude during the duration of the pulse and converting the samples into a sequence of digital values. Many samples are required during the charge collection time of the detector if the full shape information is to be preserved, so very fast ADCs are needed. The pulse characteristics are conveyed by a string of numbers that can then be manipulated using standard digital operations such as addition, multiplication, and so on. Combinations of these operations can emulate the standard shaping (or filtering) steps of differentiation and integration, and produce pulse shapes that are similar to those that have proven useful in analog shaping. These digital shapers (or filters) have the same effect as the equivalent analog processes in determining the overall width of the shaped pulse as well as in enhancing its signal-to-noise characteristics.

There are a number of potential advantages to digital signal processing. It offers an unlimited flexibility in the choice of shaping parameters since these are only inputs to the software or firmware routines that carry out the digital operations. For example, changing the shaping time in a conventional analog shaper requires switching physical circuit components and is limited to a predetermined range of discrete choices. In a digital processor, shaping times are just software parameters that can take on any arbitrary values. Furthermore, shaping to produce triangular, trapezoidal, or cusp-like shapes that have potential advantage in some circumstances is difficult to implement in analog circuits (see Fig. 16.13) and results in only an approximation to the ideal. These same shapes are easily produced digitally, and their shaping parameters can be changed arbitrarily to suit the application. Other pulse shapes that are impossible to achieve in analog circuits, such as a flat-top pulse with cusp-like rise and fall, have been shown to have some favorable characteristics when implemented in digital systems.

Another advantage of the digital approach is stability. After the conversion to digital data, there can be no further change in that information, and subsequent processing will be totally immune from the drifts that can take place in analog processors because of temperature or line voltage changes. In contrast with analog circuits, digital manipulations introduce no further noise into the system and are perfectly linear. It is also possible to introduce time delays in the pulse processing system in the form of "first in, first out" (FIFO) registers that, in contrast to analog delays, introduce absolutely no distortions in the digitized pulse shape. One potential disadvantage arises when fast timing information is important. Unless complex operations are carried out, digital systems are limited in timing accuracy to the nearest period of the sampling frequency. Although this period is typically only a few nanoseconds, analog timing methods will generally yield more precise timing information when very fast detectors are involved.

A. Analog-to-Digital Converters

The first and critical step in digital pulse processing is the conversion of the analog pulse waveform into digital data. ADCs have undergone a rapid evolution in connection with the widespread growth of digital technology, and much of the progress that has taken place is driven by strong economic forces from applications in communications and imaging. Their use in nuclear pulse processing is a minor market by comparison, but it has benefited greatly from the intense effort that has gone into ADC development.

We will introduce the general principles of ADCs in this section, and then focus on the types known as *flash* and *subranging* ADCs that are of most importance in the digital processing of pulses. Later in Chapter 18, we will return to the topic to discuss the successive approximation and linear ramp ADCs that are of greater interest for use in multichannel analyzers.

1. GENERAL PROPERTIES

The basic function of an ADC is to produce a digital code (or number) at its output that is proportional to an analog voltage supplied to its input. In the general type of ADC considered in this chapter, conversions are carried out continuously at a fixed clock frequency. A 100-MHz clock will produce 100 megasamples per second (MSPS), or one sample every 10 ns. In an ideal ADC, each conversion of input voltage to output code is independent, perfectly linear, and occurs instantly. As we will see, nonidealities limit both the maximum sampling frequency and the linearity and accuracy of the conversion. Regardless of ADC type, some general properties are defined in the following sections.

2. LINEARITY

The ideal ADC would perform a perfectly linear conversion of analog voltage to code value. Under these conditions, a plot of input voltage versus code number would be a simple straight line. Because of the discrete nature of the output code, there will actually be a small range of input voltages for each output code value, so the midpoint of this range is technically the quantity that should be plotted versus the code number. The *integral linearity* specification for the ADC is generally quoted as the maximum deviation of the measured curve from the best fit straight line, expressed as a percentage of the full range value.

A more sensitive measure of linearity can be carried out using a voltage generator that produces a slowly varying output that follows a perfect ramp or triangle shape. Then the time that the input voltage is within the range corresponding to any given output code should be constant, and the same number of conversions should take place for every code value. A plot of the number of recorded conversions versus the code number should then be constant across the entire output code scale. Deviations from the average constant value are then quoted as the *differential nonlinearity* (DNL), and the maximum deviation sometimes quoted in manufacturer's specifications in units of *least significant bits* (LSBs).

The differential nonlinearity can also be defined[44] on a channel-to-channel basis as the code width $W(k)$ for channel k, defined in the following section, with respect to the ideal width Q

$$DNL(k) = (W(k) - Q)/Q$$

so that a value of 1 indicates that the specific code width is twice the ideal, a value of 0 indicates the ideal width, and −1 indicates that the width is zero and the code is totally missing. A plot of the $DNL(k)$ defined in this way for a typical 8-bit flash ADC is shown in Fig. 17.28. Note the wide channel-to-channel variation in the effective code width, not unusual for flash ADCs. A single value for differential nonlinearity of the ADC is sometimes quoted as the $DNL(k)$ with the largest absolute value over the entire range.

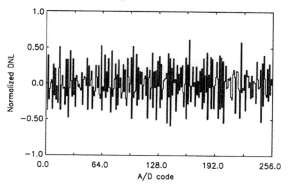

Figure 17.28 The differential nonlinearity (as defined in the text) plotted versus ADC code for an 8-bit flash ADC. (From Crawley et al.[44])

Figure 17.29 Histogram of the frequency of occurrence of measured code widths for a representative flash ADC. (From Crawley et al.[45])

Another way to quantify the differential nonlinearity is to quote the root-mean-square (RMS) value of all the code width variations from the average width. This value then gives a more global measure of the width nonuniformity. A histogram of measured code widths for a specific flash ADC is plotted in Fig. 17.29, and the RMS deviation is related to the width of the peak in this histogram.

When ADCs are used in multichannel analyzers as described in Chapter 18, their performance is much more sensitive to differential nonlinearities than when applied in digital pulse processing. For conventional pulse height analysis, only a single sample of the pulse amplitude is taken, so nonlinearities have a proportional effect on channel nonuniformity of the analyzer. In digital pulse processing, multiple samples are produced over the duration of a single pulse, and there will be some tendency to average out the effects of sampling errors caused by differential nonlinearity.

3. CHANNEL PROFILE AND CODE WIDTH

Assume that an analog voltage of perfectly controllable value is supplied to an ADC. As this voltage is smoothly varied, the output code should switch abruptly between adjacent values. A plot of the output versus input should look like the sketch below:

The vertical scale is shown in increments of one LSB of the output code. As a result of noise and other nonideal behavior, the transitions between output codes are not exactly vertical as idealized above. If we focus on a single output code, then a precise calibration will reveal the following typical result for a "channel profile":

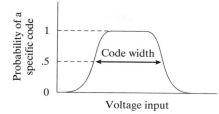

The "code width" is defined as the range of analog input voltages beginning and ending with the 0.5 probability level on the above plot. Ideally, all code widths are the same, and any variations are directly evidenced as departures from constant differential linearity. Codes (or channels) that are "too wide" will collect more conversions than those that are "too narrow" in the test involving ramp or triangular inputs, and the plot of differential linearity versus channel number should mimic a plot of code width versus channel number.

4. ADC RESOLUTION OR BIT DEPTH

An 8-bit ADC will generate 2^8, or 256, discrete codes over its input range. A 12-bit ADC will extend the number of codes to 4096. This number defines the maximum number of increments (or channels) into which the input voltage can be sorted. Thus an 8-bit ADC will offer a "digital resolution" of $1/256$, or just under 0.4% for voltages that lie near the maximum of the input range, while the corresponding value for a 12-bit ADC is less than 0.025%. The number of bits needed in a specific case is largely dependent on the signal-to-noise ratio of the input. If the noise level is several percent, then choosing an ADC with more than 8 bits will only result in digitizing meaningless noise fluctuations. Conditions with higher signal-to-noise ratios will demand greater bit depth.

One can define a "quantization error" ϵ associated with an ideal conversion process that becomes smaller with larger bit depth. As shown in Fig. 17.30, a zero value for ϵ

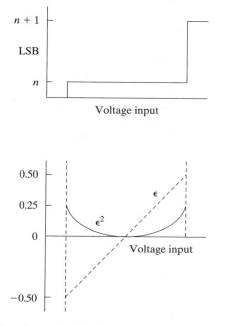

Figure 17.30 The top plot shows the ideal conversion between analog voltage input and the output code of an ADC. The quantization error ϵ (in units of code width) is plotted below, together with the square of its value ϵ^2.

corresponds to an analog voltage at the midpoint of a channel, and ϵ becomes $\pm$ half of the code width at the extreme edges of the channel. The RMS value of this ideal quantization error can be calculated from the integration

$$\langle \epsilon^2 \rangle = 2 \int_0^{0.5} x^2 \, dx = \frac{1}{12}$$

$$\epsilon \equiv \sqrt{\langle \epsilon^2 \rangle} = \frac{1}{\sqrt{12}} \tag{17.24}$$

in units of the code width.

5. EFFECTIVE NUMBER OF BITS

There are sources of noise that are associated with any ADC that introduce some fluctuations in the conversion process. If the noise is at a high enough level, it can reduce the significance of the LSBs in the converted code. Thus the concept of "effective number of bits" E has been introduced[44] with the following definition:

$$E = N - \log_2(\text{RMS}_{\text{noise}}/\epsilon) \tag{17.25}$$

where N is the number of digitized bits and ϵ is the ideal quantization error given above. $\text{RMS}_{\text{noise}}$ is the measured RMS deviation of the converted code from a precisely known but variable input voltage that is swept over the ADC range. At a minimum, this measurement will include the ideal quantization error, so the logarithm term is zero if no other sources of noise are significant. Then $E = N$, and a 10-bit ADC would deliver 10 fully significant bits. However, if additional noise is present to the extent that the $\text{RMS}_{\text{noise}}$ is twice ϵ, then the same ADC really produces only nine effective bits. The noise level, and thus the value of E, can be expected to vary with operating parameters such as sampling frequency and the rate of change of the input voltage.

6. THE FLASH ADC

In many ways, the flash ADC is the simplest type of ADC design, representing something of a brute force approach to digitizing the signal. As diagrammed in Fig. 17.31, the basic layout consists of a stack of threshold comparators, each set to a progressively higher voltage level. These reference voltages are conventionally provided via a resistive voltage divider with equal resistance values between each comparator. An n-bit flash ADC requires 2^n comparators, and the increasing complexity for large values of n limits contemporary flash ADCs to no more than 8 to 10 bits (256 to 1024 codes).

The input voltage is presented in parallel to all comparators, and all those with reference voltage below that of the input will produce a logic 1 output. Those comparators with reference voltage above the input produce a logic 0. The resulting pattern of logic levels is read by the latch register and converted to a corresponding binary number by the encoder. The readout frequency is controlled by clock pulses and is as high as 1000 MHz in the sample of present-day designs shown in Table 17.2.

Because of the possible drift and instability in the large number of analog voltage comparisons that are required, the differential nonlinearity of flash ADCs is often fairly poor. In the extreme, some output codes may actually be missing. To improve this weakness, some applications will employ ADCs with higher resolution than actually required so that the LSBs can be discarded.

7. SUBRANGING ADCs

The limitations of flash ADCs have led to the widespread use of an alternative architecture for fast ADCs applied in digital pulse processing. The *subranging ADC* defines a series of expanding scales to allow a greater resolution and lower power consumption while

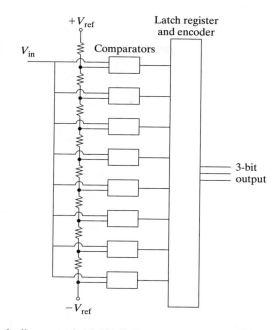

Figure 17.31 Block diagram of a 3-bit flash ADC. Each additional one bit of resolution doubles the required number of comparators.

Table 17.2 Some Examples of Fast Analog-to-Digital Converters

Manufacturer	Model	Resolution [bits]	DNL [typ] [LSB]	Speed [MHz]	Power Consumption [mW]
Flash ADCs					
Harris Semiconductor	HI3246	8	0.5	120	340
	HI1166	8	0.3	250	1400
	HI1276	8	0.3	500	2800
Signal Processing Technology	SPT7750	8	0.9	500	5500
	SPT7760	8	0.9	1000	5500
Analog Devices	AD9048	8	0.3	35	550
	AD9020	10	1	60	2800
Sub-ranging ("Multi-Pass or Pipelined") ADCs					
Analog Devices	AD9283	8	0.5	100	90
	AD9051	10	0.75	60	250
	AD9224	12	0.33	40	415
Burr-Brown	ADS931	8	0.7	30	69
	ADS823	10	0.25	60	265
	ADS807	12	0.5	53	324
Harris Semiconductor	HI5714/8	8	0.4	80	325
	HI5766	10	0.5	60	320
Signal Processing Technology	SPT7861	10	0.5	40	160
	SPT7935	12	0.6	20	80

Figure 17.32 Analogy to the operation of a subranging ADC in which magnifiers are used in sequence to expand the analog voltage scale. For the example shown, a sequence of three 3-bit analog-to-digital conversions produces the 9-bit binary code of 011101010, or 234 on a decimal scale of 0–511.

retaining the high-frequency sampling characteristic of flash ADCs. Figure 17.32 illustrates by analogy the process that is implemented. At the first stage, a coarse determination is made with a flash ADC (perhaps to a resolution of 3 bits) of the most significant bits of the eventual converted digital value. This initial measurement determines the position of the "magnifying glass" that is placed over the region including the voltage signal to be measured. In the case of the example, the lens has a magnifying power of $\times 8$ that allows a closer examination of the "subrange" of the first scale. Now the voltage is digitized on a finer scale, again to 3 additional bits in the example. Once more, the result of this digitization repositions the "magnifying glass" to allow another digitization to an additional 3 bits, now involving only a very small range of the analog voltage. The result from this last digitization provides the least significant bits of the conversion. Combining the results from the three stages produces a 9-bit code that is the output of the unit. The subranging approach allows a considerable reduction in the number of comparators that are needed to carry out the conversion. In this example, three 3-bit flash ADCs were used to complete a 9-bit conversion. Since each of these ADCs involves 2^3 (or 8) comparators, a total of 24 are needed. If the same process were done in a single flash ADC, it would require 2^9 (or 512) comparators. This reduction leads to a decreased power consumption in subranging units (see Table 17.2) that is beneficial to promote stability and to facilitate use in battery-powered portable instrumentation.

Figure 17.33a shows the architecture of a module that carries out one stage of the process described above. At each clock cycle, the input voltage is momentarily held by a track-hold (T/H) circuit. A flash ADC carries out a digitization to n bits ($n = 3$ in the example above), and that value serves as the digital output of the stage. That same value is supplied to a digital-to-analog converter (DAC) and subtracted from the input voltage. The analog difference is sent through an amplifier with gain of G corresponding to the magnification factor ($G = 8$ in the example above), and the amplified analog voltage is passed on to the next stage. A number of such modules can now be combined in a "multipass" or "pipelined" arrangement illustrated in Fig. 17.33b, with all stages synchronized by a common clock. The digital outputs of all the modules are combined in a logic block that typically includes functions of bit alignment, error correction, and self-calibration to form the overall digital result. Because of the serial operation of the individual stages, the final result is produced a number of clock cycles following the appearance of the corresponding analog input. This "latency time" is the same for all samples and introduces some delay

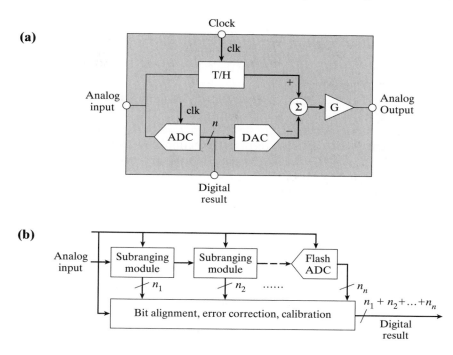

Figure 17.33 (*a*) The elements of a single subranging stage or module. T/H is a track-hold circuit, ADC is a flash ADC, DAC is a digital-to-analog converter, Σ is an analog summing unit, and G represents the gain of an analog amplifier. (*b*) The combination of multiple subranging modules in a fast pipelined ADC.

between input and output, but it does not affect the frequency of the conversions. Examples of some representative commercially available multipass or pipelined ADCs are given in Table 17.2. Units have been developed with resolutions of up to 16 bits and with conversion frequencies of up to a few hundred MHz.

B. Digital Shaping or Filtering

1. LINEAR TIME-INVARIANT FILTERS

The processes of shaping the signal pulse from a radiation detector can be thought of as a filtering operation. For example, we have seen that the CR differentiator is equivalent to a high-pass filter in the frequency domain. Most of the pulse shaping operations involve what are known as "linear time-invariant filters." Their definition incorporates the restriction that the properties of the filter are fixed and do not change with time. A "transversal filter" or "finite impulse response" filter is, by definition, limited to using input data over the restricted time range $L < t < 0$, where L is the length of the filter. This time range reaches backward from the present time simply because only data that have already occurred can contribute to the output at a specific instant. The mathematical expression of the filtering process then takes the form of the convolution integral

$$S(t) = \int_{t-L}^{t} V(t')\, H(t - t')\, dt' \tag{17.26}$$

where $H(t)$ is called the *impulse response function* of the system. This function reflects the operations carried out by the analog shaping networks in conventional linear amplifiers.

The definitions given above apply to cases in which the input data appear as an analog or smoothly varying function. That would be the case if we were dealing with the standard analog voltage output produced by a radiation detector. However, the fast analog-to-digital converters described in the previous section can be applied to convert this analog voltage into a series of digital samples. Once this conversion to digital data has taken place, it is then possible to apply digital, rather than analog, filtering techniques to carry out the shaping of these pulses.

For the digital case, we start by assuming that the input signal has been sampled into a series of values $V(i)$ that starts at $i = 0$. The time period between each sequential value of i simply represents the sampling period of the ADC. For the filtering process, the equivalent of the impulse response function now becomes a discrete set of weighting factors $H(0) \ldots H(L)$. We will assume that this filter is of finite length L so that the values of $H(i)$ are 0 for values of $i < 0$ or $i > L$. The convolution process equivalent to Eq. (17.26) above then can be written

$$S(j) = \sum_{i=j-L}^{i=j} V(i)H(j - i) \tag{17.27}$$

where $S(j)$ now represents the discrete series of output values produced by the digital filtering process.

To give a more graphic picture of this filtering process, let us choose an example for which the filter consists of four weighting factors, $H(0) \ldots H(3)$, so the length of the filter L is equal to 3. Values of $H(i)$ are by definition 0 outside this range. The input consists of the digital samples $V(i)$, where we assume that the data begin at $i = 0$. Evaluating the sum given in Eq. (17.27) for this example then gives, starting with the first nonzero output value ($j = 0$)

$$S(0) = \sum_{i=-3}^{i=0} V(i) H(-i) = V(0) H(0)$$

The next output value will be for $j = 1$

$$S(1) = \sum_{i=-2}^{i=1} V(i)H(1 - i) = V(0) H(1) = V(1) H(0)$$

The process continues producing output values for each succeeding value of j, the number of terms in the summation reaching four at $j = 3$, and remaining at four for all higher values of j until the end of the input data sequence is approached.

This convolution process is illustrated graphically in Fig. 17.34. For the example shown, the weighting factors are written along the upper line in sequence from right to left.

S(0) = V(0)H(0)

S(1) = V(0)H(1) + V(1)H(0)

S(2) = V(0)H(2) + V(1)H(1) + V(2)H(0)

Figure 17.34 An illustration of a digital transversal filter. The filter elements are shown (right to left) along the top line of each step, with the digitized data along the lower line. The output is formed by multiplying factors that are aligned and summing.

Directly below, the sampled input values are written starting at $V(0)$. The output is then given by multiplying each of these input voltages by the weighting factor shown directly above it and summing over the length of the filter. The first nonzero value of the output occurs for the example of $j = 0$. Each successive value of the output is obtained by displacing the string of input values one step sequentially to the left and again carrying out the multiplication and summation. This process can be visualized as "pulling" the set of weighting functions through the sequence of input data (or "pushing" the data through the filter), advancing one step at a time to produce the sequential output values. When the filter is well past the start of the input data, each output value consists of the sum of four terms, and only the input data that are within three steps of the most recent data point being processed contribute to the output at that corresponding point.

Simple transverse filters can easily carry out the basic pulse shaping steps of differentiation and integration. They can also be combined in more complex ways to produce triangular, trapezoidal, or other useful shapes.[46,47] These shaping filters share the common property that they simply implement a series of digital weighting factors that remain perfectly stable unless intentionally changed. The time required to carry out the digital operations is short enough so that these schemes can be implemented on a pulse-by-pulse basis in real time, in much the same manner as analog shaping.

2. ADAPTIVE FILTERING

Digital pulse-processing systems, since they are controlled through software, easily allow the filter weighting functions to be customized to suit the circumstances of the measurement. In our previous discussion of pulse shaping, it was illustrated that each particular shaping choice results in a somewhat different theoretical signal-to-noise ratio (see Fig. 17.15). The choice of the optimum shaping method depends on the magnitude and nature of the noise that is present with the signal. Systems have been designed[48–51] that sample the detailed character of the noise present with the signal and then choose a weighting function that is optimal for the measured results. This is one example of *adaptive* filtering that exploits the flexibility of the digital system to easily adjust the shaping parameters.

Another example of adaptive filtering is applied to systems in which the counting rate changes substantially during a measurement. In many cases, one would like to use shaping methods that lead to wider pulses at low counting rates than can be tolerated at high counting rates as a result of excessive pile-up. An adaptive filtering approach that changes the width of the shaping function as a function of counting rate can adjust automatically to these changing conditions.

C. Pulse Shape Analysis

Once the input waveform from the detector has been digitized, measurements of its detailed shape also become straightforward. We have seen many examples of detectors in which the detailed time profile of the current pulse reveals additional information about the nature of the event in the detector. Examples are in distinguishing one type of radiation from another, measuring the spatial position of an event, or distinguishing pulses that arise from multiple gamma-ray interactions from those that are due to a single interaction only. Sophisticated algorithms can be used to inspect the string of digital data representing each individual pulse so that rather subtle differences among them can be distinguished. As these operations become more complex, however, it is increasingly difficult to carry them out entirely in real time at high counting rates because of the limited time available between signal pulses. In such cases, the digital data can be sent to memory and the processing completed off-line after the measurement.

D. Digital Baseline Restoration

In digital pulse-processing systems, the baseline restoration process described on p. 597 can be accomplished by digitally sampling the baseline between pulses and then subtracting an appropriate value from the measured pulse amplitude.[52] Independent baseline estimates can be obtained for each pulse that is processed or, alternatively, a common estimate applied to a series of pulses between more widely spaced samples of the baseline. Ideally, one would like multiple digital measurements of the baseline to be taken for a given sample to enhance its accuracy, requiring a relatively long time interval between pulses. Long intervals occur less frequently than short ones, so there must be a trade-off between the accuracy of the baseline measurement and the frequency with which they can be made. Digital filters that can be applied to these baseline samples to optimize the measurement accuracy are described in Ref. 53.

E. Deconvolution of Piled-Up Pulses

The availability of digital pulse-processing systems opens the possibility of doing more than simply tolerating the losses associated with pile-up. If a detailed digital sampling is made of each output pulse shape from a pulse processing system, it should in principle be possible to distinguish those that originate from a single event from those involving pile-up in which more than one event have contributed. The possibility then arises of deconvolving the piled-up pulses into their separate components through a detailed analysis of their shape.

Figure 17.35 shows an example of a piled-up pulse, sampled at 10 ns intervals, that actually arises from two events in the detector. Since the shaping process produces pulses of a predictable shape when operating on isolated single events, piled-up pulses depart from this simple profile and can therefore be selected for further analysis. Combinations of the expected shape from two events with varying amplitude and time spacing can be fitted to the observed pulse shape, following an iterative process. Once the optimal fit is obtained, then the individual amplitudes of each event are obtained from the parameters of the fit, and the mutual interference of the two events is eliminated. Not only is the corresponding distortion removed from the recorded pulse height spectrum, but now the full information

Figure 17.35 Demonstration of the deconvolution of piled-up pulses from the output of a scintillation counter. The sampled data from the ADC operating at 100 MHz are shown as the series of points in both plots. In the bottom plot, an analytic fit to these data using the assumption that a single pulse shape should suffice results in a poor fit. In the top plot, an iterative procedure was used to fit two pulse components that allow the separation of both from the piled-up recorded data. (From Komar and Mak.[54])

carried by the two separated pulses can be obtained. Although the utility of this type of pile-up deconvolution has been demonstrated in off-line systems,[54,55] the required processing times are still sufficiently long that it is not yet routine in standard spectroscopy systems in which the pulse analysis is carried out in real time.

IX. SYSTEMS INVOLVING PULSE TIMING

In a large number of applications, information on the precise arrival time of a quantum of radiation in the detector is of particular interest. When timing information is the major purpose, detector pulses are often handled quite differently than when accurate pulse height measurement is the object. The accuracy with which timing can be performed depends both on the properties of the specific detector and the type of electronics used to process the signal. The best timing performance is obtained for the fastest detectors, that is, those in which the signal charge is collected most rapidly. For detectors with equal charge collection time, those that generate the greatest number of information carriers (ion pairs, electron–hole pairs, etc.) per pulse will be influenced least by the "graininess" of the signal and therefore will demonstrate superior timing properties. The timing characteristics of a given system depend greatly on the dynamic range (ratio of maximum to minimum pulse height) of the signal pulses. If all the signal pulses are confined to a narrow range of amplitudes, many different timing schemes can give good results. If the pulse amplitude must cover a wide range, however, some sacrifice in timing accuracy nearly always results.

A. Time Pick-Off Methods

The most fundamental operation in timing measurements is the generation of a logic pulse whose leading edge indicates the time of occurrence of an input linear pulse. Electronic devices that carry out this function are called *time pick-off* units or *triggers*.

Factors that lead to some degree of uncertainty in deriving the timing signal are always present. Sources of timing inaccuracy are conveniently divided into two categories. Those that apply when the input pulse amplitude is constant are usually called sources of *time jitter,* whereas those effects that derive primarily from the variable amplitudes of input pulses are grouped together in a category called *amplitude walk* or *time slewing.* Relative importance of these two categories depends on the dynamic range expected in the input pulse amplitude. The best timing performance will be achieved if the input pulses are confined to a very narrow range in amplitude, because then only sources of time jitter contribute to uncertainty. More often, however, practical applications require that pulses of different amplitudes be processed, and the additional contribution of walk will worsen the overall time resolution of the system.

An important source of time jitter is the random fluctuations in the signal pulse size and shape. These fluctuations can arise from several sources. One is the electronic noise added by components that process the linear pulse prior to the time pick-off. Another is the discrete nature of the electronic signal as generated in the detector. When the number of information carriers that make up the signal is low, statistical fluctuations in their number and time of occurrence will also be reflected in the size and shape fluctuations of the pulse. This effect will obviously be greatest for small-amplitude pulses and for detectors that generate relatively few information carriers such as scintillation counters.

1. LEADING EDGE TRIGGERING

The easiest and most direct time pick-off method is to sense the time that the pulse crosses a fixed discrimination level. Such *leading edge* timing methods are in common use and can be quite effective, especially in situations in which the dynamic range of the input pulses is not large.

Figure 17.36 The time jitter in leading edge triggering arising from random noise. An envelope is shown of many repeated signal pulses of the same amplitude and shape, but with a random contribution of noise.

The effect of time jitter on leading edge timing is shown in Fig. 17.36. The random fluctuations superimposed on signal pulses of identical size and shape may cause the generation of an output logic pulse at somewhat different times with respect to the centroid of the pulse. The timing errors will be approximately symmetrical and will increase if the slope of the leading edge of the pulse is decreased.

The amplitude walk associated with a leading edge trigger is graphically demonstrated in Fig. 17.37. The two pulses shown have identical true time of origin but give rise to output logic pulses that differ substantially in their timing. Under extreme conditions, the amplitude walk can amount to the full rise time of the input pulse and is often unacceptable in situations where accurate timing is needed over a wide amplitude range.

Even if the input amplitude is constant, walk can still take place if changes occur in the shape of the pulse. Detectors with variable charge collection time, such as germanium detectors, produce output pulses with variable rise time as illustrated in Fig. 17.38. Changes that occur in the pulse shape before the discrimination point will affect the timing and can constitute another source of timing uncertainty. The sensitivity of leading edge triggering to timing walk because of amplitude and shape variations is minimized by setting the discrimination level as low as possible. However, the discrimination point should be in a region of steep slope on the pulse leading edge to minimize uncertainties due to jitter. Practical compromises in these somewhat conflicting requirements often lead to optimum time resolution for levels that are set at about 10–20% of the average pulse amplitude.

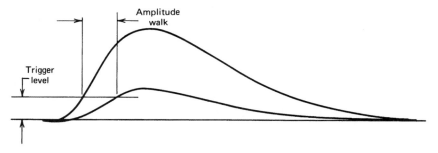

Figure 17.37 Amplitude walk in leading edge triggering. Two pulses with identical shape and time of occurrence but different amplitude are seen to cross the trigger level at different times.

Figure 17.38 Time walk in leading edge triggering due to variations in pulse shape or rise time. Changes in either the overall pulse rise time or the shape of the leading edge for pulses with the same rise time can be responsible for time walk.

2. CROSSOVER TIMING

When leading edge triggering is applied to pulses with a wide amplitude range, amplitude walk often results in large timing uncertainties. Another time pick-off method, known as *crossover timing,* can greatly reduce the magnitude of amplitude walk but requires the input pulse to have bipolar shape. Figure 17.39 shows two pulses with different amplitudes as shaped by a *CR-RC-CR* double differentiating network. Although the pulse amplitudes differ widely, the time at which the waveform crosses from the positive to the negative side of the axis is theoretically independent of the amplitude and depends only on the shaping time constant chosen for the network. Double delay line shaping produces the same effect and is usually preferred in crossover timing. Compared with leading edge timing, crossover methods greatly reduce amplitude walk, but only at the expense of increased jitter. The noise introduced by the required shaping stage and the increased susceptibility of the crossover point to statistical fluctuations in the signal contribute to the increased jitter.

A related method known as *fast crossover* can be implemented for the specific case of fast scintillator–photomultiplier tube signals. A fast bipolar pulse is produced directly at the PM tube anode using a "tee" connection and a shorted length of coaxial cable. This *shorted stub* technique is described in Chapter 16, and the resulting waveform is shown in Fig. 16.21a. Because the time to the zero crossover point is determined by the length of the shorted cable, it is also independent of the signal pulse amplitude. However, the pulse shape must not vary if this freedom from walk is to be preserved.

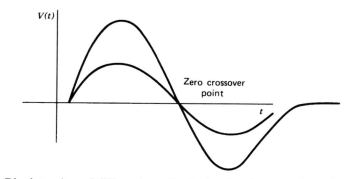

Figure 17.39 Bipolar pulses of different amplitude showing the same time of zero crossover.

3. CONSTANT FRACTION TIMING

If the input dynamic range is small, the lower jitter of leading edge timing results in superior timing performance compared with crossover timing. Furthermore, it is empirically found that the best leading edge timing characteristics are obtained when the timing discriminator is set at about 10–20% of the pulse amplitude. These observations have led to the development[56] of a time pick-off method that produces an output signal a fixed time after the leading edge of the pulse has reached a *constant fraction* of the peak pulse amplitude. This point is then independent of pulse amplitude for all pulses of constant shape. Therefore, pulses over a wide dynamic range can be accepted with much the same freedom from amplitude walk as exhibited by crossover timing, but with lower jitter. The electronic shaping steps required to carry out constant fraction timing are diagrammed in Fig. 17.40. The process involves taking the preamplifier output [part (*a*)] and multiplying it by the fraction *f* that is to correspond to the desired timing fraction of full amplitude. The input waveform is also inverted and delayed for a time greater than the pulse rise time to give the waveform shown in part (*c*). The sum of waveforms (*b*) and (*c*) is then taken to give the pulse that is sketched in part (*d*). The time that this pulse crosses the zero axis is independent of the pulse amplitude and corresponds to the time at which the pulse reaches the fraction *f* of its final amplitude. Detailed descriptions of constant fraction discriminator circuits and their use in several timing applications can be found in Refs. 57–60, and examples implemented in the form of integrated circuits are described in Refs. 61–63.

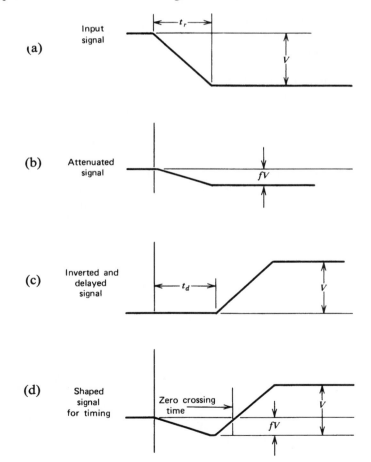

Figure 17.40 Waveforms in the constant fraction time pick-off method. For clarity, only the leading edge of the pulse is shown.

4. AMPLITUDE AND RISE TIME COMPENSATED (ARC) TIMING

In situations where the shape or rise time of the pulse can change, even constant fraction timing methods cannot eliminate walk. To meet the need for accurate timing for germanium detectors, in which the rise time variations are large, a system known as ARC timing, first described by Chase,[64] has gained considerable popularity.[65,66] The scheme amounts to basing the timing signal on a fixed fraction of only the early portion of the pulses and is therefore unaffected by shape changes that may occur at a late point in the waveform.

The pulse-processing methods are similar to those used in constant fraction timing and are shown in Fig. 17.41. The only difference is that the delay time illustrated in the figure is chosen to be as small as practical (of the order 5–10 ns). The delay must be sufficiently long so that the initial portion of the output pulse is allowed to significantly exceed the noise level, so that a simple discriminator signal can be generated that will actuate the circuits that sense the crossing of the waveform back across zero. Using this method, the time of zero crossing is given by $t_d/(1 - f)$, provided that the input pulse rises linearly to that point. Therefore, this crossover time again occurs at a constant fraction of the amplitude of all input pulses and will furthermore be insensitive to any slope change in the rise of the input pulse that occurs after the zero crossing point.

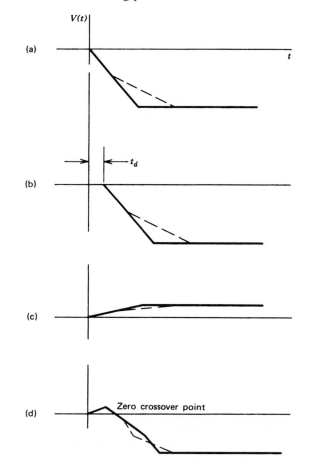

Figure 17.41 Waveforms in the ARC time pick-off method. The original waveform (*a*) is delayed (*b*), inverted, and attenuated (*c*). The sum of (*b*) and (*c*) gives waveform (*d*) with a zero crossover point used to generate the timing signal. The dashed waveform has a slower rise time than the solid, but both give the same timing signal.

5. EXTRAPOLATED LEADING EDGE (ELET) TIMING

As an alternative to ARC timing, an extrapolation principle has also been applied success-fully to the almost walk-free derivation of timing signals from germanium detectors with variable rise time.[67-69] The method is similar to ARC timing in that it is based only on the initial portion of the pulse rise and is not affected by shape variations that may follow that initial portion. ELET timing uses two independent discriminators set at different discrim-ination levels to carry out an extrapolation of the pulse waveform back to its time of ori-gin. For purposes of illustration, assume that the second discrimination level is set at twice the level of the first. For pulses with a linear rise, the time difference between the discrim-ination points should be equal to the time delay between the true start of the pulse and the time at which the first discriminator is crossed. The technique employs time-to-amplitude converters to carry out an effective extrapolation back to zero, so that a timing pulse cor-responding to the true time of origin of the input pulse is generated. The two discrimina-tion levels are ideally set as low as possible, so that the assumption of linearity need be met for only a short portion of the leading edge of the pulse.

6. FIRST PHOTOELECTRON (FPET) TIMING

A unique method for providing timing information from scintillation detectors can be applied when the conditions of the experiment limit the possible triggering time to a rela-tively small time interval. If this interval is small compared with the average time spacing between photomultiplier tube noise pulses, excellent timing can be carried out simply by sensing the arrival of the photomultiplier signal corresponding to a single photoelectron. This is, in effect, leading edge timing with a trigger level that is as low as physically possible. Because photomultiplier tube noise would also trigger at this level, the probability of a noise pulse within the time range of interest must remain small. This condition has been made practical by low-noise bialkali photomultiplier tubes in which noise rates may be as low as tens or hundreds per second. The use of FPET was pioneered by Lynch[70] and is par-ticularly advantageous for timing from scintillators when low-energy gamma rays are involved. Further studies of FPET with sodium iodide scintillators are given in Ref. 71.

7. COMPARISON OF TIME PICK-OFF METHODS

As a general rule, leading edge triggering gives the best time resolution for signal pulses whose amplitudes are restricted to a narrow range and whose shape characteristics do not vary. When pulses with a wide range of amplitude are processed, leading edge methods show large amplitude walk. Constant fraction timing methods are very effective at reduc-ing amplitude walk when the pulse shape does not change, and this technique has largely replaced crossover timing originally introduced for the same purpose. For example, scintil-lation detectors produce pulses of fixed shape when a given type of radiation is involved. A useful comparison of leading edge and constant fraction pick-off techniques applied in timing measurements using scintillators can be found in Ref. 72. For detectors that produce pulses with variable rise time (such as germanium detectors), both ARC and ELET timing methods can be used to good advantage.[73-75] Alternatively, more elaborate methods that sense the rise time (often using two independent discriminators set at different levels) can be employed to make corrections in the timing signal[76,77] to reduce the influence of vari-able rise time.

The time resolution achievable with any of these pick-off methods varies greatly with detector type and is determined primarily by the characteristics of the charge collection process that are reflected in the output pulse rise time. A discussion of the time behavior of each major detector type is included with its description given in other chapters. The best timing is attainable from detectors with fast and nonvariable rise time. Among detectors with equal rise time, those producing the largest signal will generally result in better tim-ing performance. For example, plastic scintillators can produce a best time resolution of around 100 ps (defined as the FWHM of the prompt coincidence peak, as described in the following section) for large amplitude pulses, but the resolution will worsen for smaller pulses. A useful figure of merit for scintillator timing performance is the total light yield divided by the decay time, which gives the initial photon emission rate. Slower scintillators such as NaI(Tl) and BGO can show time resolution values of 1–2 ns if pulses of nearly constant amplitude are involved but will have poor timing performance if pulses covering a broad range of amplitude are processed. One study[78] of 14 different germanium detector systems measured timing resolutions ranging from about 2 to 10 ns using a version of the constant fraction triggering technique. Proportional counters generally lead to poorer timing resolution because of the slower charge collection and variability of the pulse rise time.

B. Measurement of Timing Properties

1. MULTICHANNEL TIME SPECTROSCOPY

We begin our discussion of pulse timing systems by considering some basic concepts of time interval measurement. The time-to-amplitude converter (TAC) is a device (discussed in greater detail later in this chapter) that produces an output pulse with an amplitude proportional to the time interval between input "start" and "stop" pulses. The differential amplitude distribution of the output pulses as recorded by a multichannel analyzer is thus a measure of the distribution of time intervals between start and stop pulses and is often called a *time spectrum*.

The time spectrum bears a close relation to the pulse height spectrum (or differential pulse height distribution) discussed in Chapter 4. The abscissa, rather than pulse height, is the time interval length T. The ordinate is dN/dT, the differential number of intervals whose length lies within dT about the value T. The area under the time spectrum between any two limits is the number of intervals of length between the same limits. The discrete approximation to the ordinate, $\Delta N/\Delta T$, is the number of intervals of length within a finite increment ΔT, normalized to the length of that increment. During the discussion that follows, it is also convenient to divide the number of recorded intervals by the measurement time to obtain the corresponding rate. The ordinate then becomes dr/dT for the time spectrum, to be interpreted as the differential rate of occurrence of intervals whose length lies within dT about T.

A simple system for recording a time spectrum is shown in Fig. 17.42. For the sake of illustration we assume that a single source of pulses such as a radiation detector provides a split output that is simultaneously sent down two signal branches. A time pick-off unit in each branch provides timing logic pulses that are supplied to the start and stop inputs of a TAC. A fixed time delay is also present in the stop branch. The multichannel analyzer (MCA) records the number of time intervals that lie within many contiguous increments or channels of width ΔT, ranging from zero to some maximum interval length set by the range of the TAC. The maximum time range of the TAC is assumed to be small compared with the average spacing between signal pulses, so that the probability of more than one signal pulse per TAC interval is small. Alternatively, the output of a pulse generator may be substituted for the random pulse source and adjusted to have a period between pulses which is larger than the TAC maximum range.

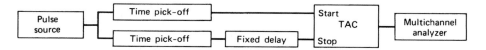

Figure 17.42 A simplified system for recording multichannel time spectra from a split common source.

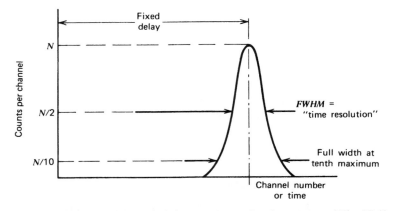

Figure 17.43 The multichannel time spectrum for the system of Fig. 17.42.

Under these conditions, the time spectrum recorded by the MCA is exceedingly simple. In the absence of time jitter or walk, each timing pulse is produced at precisely the same time in each branch, and therefore the start and stop pulses are always separated by the fixed delay value. The TAC always produces a constant-amplitude output that is therefore stored in a single channel of the MCA. If the delay is made smaller, the interval between start and stop pulses also decreases, producing a TAC pulse of smaller amplitude which will be stored in a lower-numbered channel.

A more realistic time spectrum shown in Fig. 17.43 takes into account the fact that any time pick-off method will always involve some degree of timing jitter and walk. If these timing uncertainties in each branch are independent of each other, the time spectrum will display a distribution as shown in the sketch. Because the time pick-off in each branch is likely to behave similarly with regard to time walk caused by amplitude or shape variations, most of the distribution spread measured in this conceptual experiment would be caused by those random sources of uncertainty or jitter that are independently generated in each branch.

The full width at half maximum of the time distribution is often used as a measure of the overall timing uncertainty in the measurement system and is called the *time resolution*. Another widely quoted specification is the full width at $\frac{1}{10}$ maximum, which more fairly accounts for tails sometimes observed at either extreme of the distribution. The recording of the time interval distribution in the manner just described is often called *multichannel time spectroscopy*.

2. PROMPT AND CHANCE COINCIDENCE SPECTRA

We now imagine the system to be reconfigured as shown in Fig. 17.44. Two independent detectors are irradiated by a common radioisotope source that is assumed to emit at least two detectable quanta in *true coincidence*; that is, both radiations arise from the same nuclear event within the source. We further assume that for all true coincidences, the nuclear decay scheme is such that there is no appreciable time delay between the emission of both radiations.

The time spectrum taken under these conditions will have the general appearance shown in Fig. 17.45. Some fraction of all the true coincidence events will give rise to radiations that are detected simultaneously in both detectors. These true coincidence events will appear in the same region of the time spectrum as did the split pulse output in the previous example, producing a *prompt coincidence* peak. If there were no delay difference between the two branches, this peak would be centered about time zero, and therefore only about half of its shape would be measured. Introducing the fixed delay into the stop chan-

Figure 17.44 A simplified system to record multichannel time spectra from a radioisotope source emitting coincident radiation.

nel moves the entire time spectrum to the right by an amount equal to the delay and allows both sides of the prompt coincidence peak to be recorded. The area under the peak, after subtraction of the continuum discussed later in this section, gives the total number of detected coincidences.

The width of the prompt distribution is likely to be somewhat greater than in the previous example because separate detectors are now involved. Amplitude walk, for example, will no longer be identical in both branches and therefore will also contribute to the width of the distribution together with the time jitter. If detectors, timing electronics, and triggering conditions are nearly identical in both branches, then all sources of time jitter and walk should be symmetric. Under these conditions the prompt coincidence peak should also be symmetric with a width that indicates the total contribution of all sources of time uncertainty. If one branch differs substantially from the other, then asymmetric prompt coincidence peaks will often result. For example, the effect of amplitude walk in leading edge triggering is to produce a small number of timing pulses that occur substantially later than the majority. If a system subject to amplitude walk is used in the stop branch in conjunction with a system with little time uncertainty in the start branch, the measured prompt coincidence distribution will have an asymmetric tail in the direction of longer intervals.

In addition to the coincident events, each detector will typically produce a much larger number of pulses that correspond to the detection of one quanta for which there may not be a corresponding coincident emission, or for which the coincident radiation escapes detection in the opposite detector. For these events there can be no true coincidence. Because of their random distribution in time, however, some sequences will occur by chance in which a stop pulse will be generated within the TAC time range following an

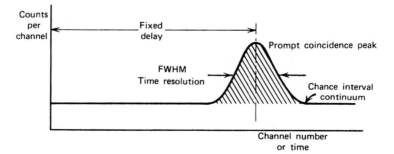

Figure 17.45 The multichannel time spectrum for a radioisotope source emitting some radiation in prompt coincidence. The cross-hatched area gives the total number of recorded coincident events. The time resolution of the system is conventionally defined as the FWHM of the prompt coincidence peak.

unrelated start pulse. These events are called *chance intervals*, and their intensity depends on the rates at which pulses are generated in either branch leading to the TAC. If these rates, often called *singles rates*, are not high compared with the reciprocal of the time range of the TAC, the chance interval distribution will be uniform over the entire time range, as shown in Fig. 17.45.

The amplitude of the chance distribution can be derived as follows: Let r_1 and r_2 represent the rates of arrival of uncorrelated start and stop pulses, respectively, at the TAC inputs. For typical applications, r_1 and r_2 will be much larger than the true coincidence rate and thus are essentially equal to the singles rate in either branch. (If r_1 is not small compared with the inverse of the TAC time range, account must be taken of those start pulses that are lost because the TAC is busy when the start pulse arrives. We ignore such losses here.) After each start, the probability that an interval of length T will elapse without a stop pulse is simply e^{-Tr_2}. The differential probability of a stop pulse arriving within the next differential time dT is just $r_2\, dT$. Because both independent events must occur, the overall differential probability of generating an interval within dT about T is simply $r_2 e^{-Tr_2}\, dT$. The differential rate of these intervals is then simply the product of the rate of arrival of start pulses multiplied by this probability, or $r_1 r_2 e^{-Tr_2}\, dT$. Now as long as r_2 is not large compared with the reciprocal of the TAC time range, $r_2 T$ will be small and the exponential can be approximated by unity. The differential distribution dr/dT is then constant and equal to $r_1 r_2$. If the output of the TAC is recorded by an MCA with a time width of ΔT per channel, then the chance interval rate per channel is simply $r_1 r_2 \Delta T$.

In a multichannel time spectroscopy measurement, one would normally like to maximize the true coincidence peak compared with the chance interval background. For equal areas, peaks with the narrowest width will be most prominent, so that improvements in the timing accuracy of either branch, which diminish the overall timing uncertainty, are always beneficial. Other experimental factors can also help improve the ratio of the prompt peak to the chance continuum. For example, use of energy selection criteria in each branch can limit r_1 and r_2 by discarding any events that cannot correspond to true coincidences. The chance continuum will therefore be reduced without affecting the area under the true coincidence distribution. The true rate also scales linearly with source activity, whereas the chance rate is proportional to the product of r_1 and r_2 or the square of the source activity. Consequently, using as low a source activity as permitted by reasonable counting statistics will also enhance the prominence of the true coincidence peak. Varying the rates by changing the counting geometry, however, affects both the true and chance rates by an equal factor.

3. MEASUREMENTS USING A COINCIDENCE UNIT

An equivalent single-channel method is available to carry out the type of time spectroscopy just described, but at the price of a somewhat increased total measurement time. The single-channel method, in effect, consists of setting a time window to accept only those sequences in which the interval between start and stop pulses lies within a narrow band. The situation is somewhat analogous to pulse height analysis, where both single-channel and multichannel methods are used. As discussed earlier in this chapter, the unit used to perform narrow band pulse height selection is called a single-channel analyzer.

For time spectroscopy, the *coincidence unit* performs the equivalent function and selects from all intervals only those for which the time difference between inputs is less than a circuit parameter, known as the *resolving time*. A coincidence unit, in its simplest form, has two identical logic inputs. Whereas the TAC must receive a start pulse and stop pulse within the time range in that specific order, a coincidence unit will produce a logic output if pulses arrive at either input within the resolving time of a second pulse at the opposite input. The order of the arrival is not significant.

The system shown in Fig. 17.46 illustrates the application of a coincidence unit to time spectroscopy. The fixed delay is assumed to be the same as in the previous example, and a

Figure 17.46 A simplified system to record coincidence-delay curves from a radioisotope source emitting coincident radiation.

variable delay has been inserted into the opposite branch. We initially assume that the coincidence unit is set with a resolving time equal to one-half the time width of one channel in the multichannel time spectroscopy example. We now record the rate from the coincidence unit as a function of the variable delay value. If the delay is set to a value that corresponds to a time interval at the midpoint of one of the channels in the multichannel spectrum, then the rate measured by the coincidence unit will be exactly the same as the rate recorded by that specific channel. Under these conditions, the coincidence unit produces an output for predelay intervals that range from $(t_v - t_f - \tau)$ to $(t_v - t_f + \tau)$ where t_f and t_v are the fixed and variable delay times, and τ is the coincidence resolving time. If a series of measurements is now made in which the coincident rate is measured as t_v is varied in increments of 2τ, then a curve that is exactly equivalent to the multichannel time spectrum of Fig. 17.45 will be generated by plotting this rate versus the delay setting. This type of plot is often called a *coincidence-delay curve* and can be used to reproduce the multichannel time spectrum as just described, provided the resolving time τ is small compared with the overall time resolution of the system.

In common coincidence measurements, however, the object is not to map out fully the time interval spectrum but rather simply to record the number of true coincidence events. The coincidence resolving time is therefore chosen to be larger than the system time resolution, so that the acceptance time window can fully encompass all true coincidences.

Figure 17.47 graphically illustrates the interrelation between the measured coincidence rate and the differential distribution dr/dT versus T for the case of a relatively large resolving time τ. This differential distribution is the same as the multichannel spectrum of Fig. 17.45, normalized by the measurement time so that the distribution now is in units of rate. At a given delay setting t_v the measured coincidence rate corresponds to the area under the differential spectrum between the time window limits of $t_v - \tau$ and $t_v + \tau$.

Ideally, coincidence measurements are carried out with delay t_v adjusted to the point indicated as ① on the figure. Here, the acceptance time window is centered around the prompt coincidence peak in the spectrum. The measured coincidence rate corresponds to the cross-hatched area and consists of two additive terms. The *true coincidence rate* corresponds to the net area under the prompt coincidence peak, whereas the *chance coincidence rate* corresponds to the area of the flat chance continuum on which the peak is superimposed.

If the delay is changed to a point well away from the prompt peak (such as point ② on Fig. 17.47), only chance coincidences are measured. The rate can be deduced from the results of the previous section, where it was shown that the amplitude of the differential distribution dr/dT for chance intervals is the product of the two singles rates $r_1 r_2$. Therefore, the area under the differential spectrum in the chance continuum region is this

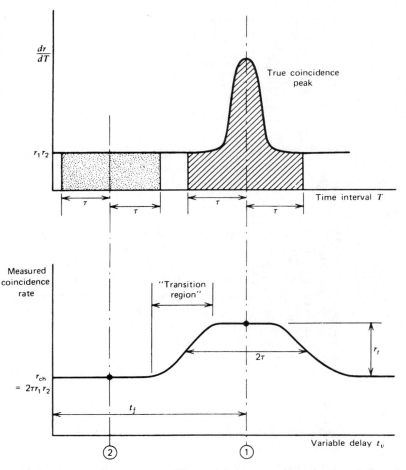

Figure 17.47 Relation between the differential time spectrum (upper plot) and the coincidence-delay curve recorded with the system of Fig. 17.46 (lower plot). The recorded coincidence rate at any specific delay value t_v is equal to the area under the differential spectrum between the limits $t_v - \tau$ and $t_v + \tau$. In the lower plot, r_{ch} represents the chance coincidence rate, and r_t is the true coincidence rate.

amplitude multiplied by the width of the time window 2τ. The general result for any twofold coincidence unit is therefore that the chance coincidence rate from uncorrelated inputs at rates r_1 and r_2 is given by

$$r_{ch} = 2\tau r_1 r_2 \qquad (17.28)$$

This chance contribution must be subtracted from the rate measured at point ① to derive the net true coincidence rate.

The remainder of the coincidence-delay curve shown at the bottom of Fig. 17.47 can be traced out by measuring the total coincidence counting rate as t_v is varied. This is a very useful calibration procedure during the initial setup of a standard coincidence measurement. Starting at ①, assume that the delay is changed very gradually. For a while nothing changes because the amount of area under the chance distribution that is lost by moving one limit of the acceptance window is made up by that area gained at the opposite limit. However, when the delay is changed sufficiently so that some portion of the true coincidence peak begins to be lost, the measured coincidence rate starts to drop off. When the delay has been changed sufficiently, the entire peak is excluded and only the chance rate remains. The transition region at either side extends over a range of delay equal to the full width (at its base) of the true coincidence peak. The midpoint of a transition region corresponds to the delay setting at which one edge of the time window exactly bisects the prompt coincidence peak. The time difference between these two midpoints (as illustrated on the figure) is therefore a measure of the time window width and is equal to 2τ, twice the coincidence unit resolving time.

Coincidence-delay curves generated by assuming different values for τ are shown in Fig. 17.48. The minimum value of τ for which all true coincidences can be recorded is half the total width (at its base) of the true coincidence peak. The curve labeled C in the figure corresponds to this case. With the resolving time set to this minimum, only one specific value of the delay will lead to counting all true coincidences. Should any time drift occur in either branch, true coincidence events would begin to be lost. Therefore, in most coincidence measurements one would like to have the resolving time τ somewhat larger than this minimum to allow some leeway for such drifts or other timing changes. On the other hand, the chance coincidence rate increases linearly with τ, and the resolving time should be kept as small as possible to maximize the true-to-chance coincidence ratio. The usual compromise between these conflicting considerations is one in which τ is chosen to be several times the system time resolution as in curve B. The coincidence-delay curve then has a flat top or plateau, which represents the range in delay settings that can be tolerated without losing true coincidences. In the initial calibration of the coincidence system, delays are then adjusted to choose an operating point near the middle of this plateau. Because of slight differences in the inherent delay of pulse-processing components, this point may correspond to some apparent delay difference between the two branches.

In practice, the fixed delay t_f in Fig. 17.46 is usually omitted and the coincidence-delay curve is obtained by using only the variable delay t_v. The curve is then approximately centered about zero, and only the right half is recorded with t_v in the lower signal branch. The remainder of the curve is then obtained by shifting t_v to the upper branch, where its value represents "negative delay."

Coincidence measurements are not necessarily confined to two signal branches but can, in general, involve inspecting any number of signals for true coincidence. For such multiple inputs, all signals must arrive within a total time interval corresponding to the resolving time of the unit for an output logic pulse to be produced. The measured rate will again

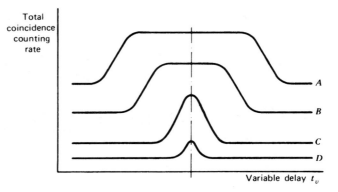

Figure 17.48 Coincidence-delay curves for different values of the coincidence resolving time τ. Curve A corresponds to the largest value of τ, curve D to the smallest. Values of τ larger than half the width (at its base) of the prompt coincidence peak in the differential time spectrum of Fig. 17.47 lead to the flat-topped plateau shape of curves A and B. In curve C, τ is just equal to this value and the full true coincidence rate is obtained only at one specific delay value. In curve D, τ is too small to obtain all the true coincidences at any delay value.

be a mixture of true and chance coincidences, but correction for the latter is more difficult, as discussed in the following section.

4. CHANCE COINCIDENCE CORRECTIONS

In standard coincidence measurements it is essential to correct the recorded coincidence rate for the contribution due to chance events to derive the rate due to true coincidences alone. For twofold coincidence systems, Eq. (17.28) allows calculation of the chance coincidence contribution if each singles rate is measured and the resolving time of the coincidence unit is known. Alternatively, the chance rates can be measured directly by temporarily inserting a large delay in either branch of the system so that the true coincidence peak occurs well away from the acceptance time window of the coincidence unit. The latter approach is usually preferable because it can be applied more easily to situations in which the singles counting rate may not be constant over the period of the measurement. Several electronic schemes have been suggested[79-81] for simultaneously measuring the total coincidence and chance coincidence rates throughout the course of the measurement by periodically excluding the true coincidences in a similar manner.

It is possible that there may be some confusion in attempting to reconcile the result just obtained for the two-fold chance coincidence rate with the results for pile-up rates obtained earlier in this chapter. Figure 17.49 illustrates the comparison. The input rates r_1 and r_2 are assumed to be uncorrelated Poisson random processes. The top diagram shows the result for the (first order) chance coincidence rate from Eq. (17.28) as $2\tau r_1 r_2$, where τ is the coincidence resolving time. In the lower diagram, we imagine combining the same two input lines with the same singles pulse rates into a common line. We then ask what the pile-up rate will be, assuming that the pile-up and coincidence resolving times are the same. The first-order result for the rate of piled-up counts from Eq. (17.23) is τr^2, where τ is the pile-up resolving time. For this example, the predicted pile-up rate is $\tau(r_1 + r_2)^2$, which is clearly different from the chance coincidence rate in the top diagram.

There is a fundamental difference between the cases of chance coincidence and pile-up that resolves this apparent contradiction. In the case of the coincidence unit, pulses from *both* branches are required to occur within the resolving time. For pile-up, it is possible for pulses from either branch to generate a pile-up event with a pulse from the *same* branch, as well as to generate a pile-up event between the two branches. By expanding the expression for the pile-up rate, we see that it is consistent with the sum of three separate rates: the pile-up of pulses from line 1, the pile-up of pulses from line 2, and the pile-up of pulses from opposite lines. This cross term $(2\tau r_1 r_2)$ does indeed follow the prediction of the chance coincidence rate, and therefore both representations are consistent.

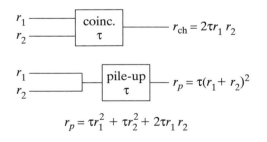

$$r_p = \tau r_1^2 + \tau r_2^2 + 2\tau r_1 r_2$$

Figure 17.49 At the top, a coincidence unit with resolving time τ results in a chance coincidence rate of $2\tau\, r_1 r_2$. In the bottom diagram, two input lines with equivalent singles rates are combined in a system with pile-up resolving time of τ. Expansion of the resulting pile-up rate shows three components: pile-up of pulses in line 1, pile-up of pulses in line 2, and the pile-up involving pulses from opposite lines.

In multiple coincidence systems, it is usually not possible to make an analytic correction for chance contributions. Instead, supplemental measurements must be carried out in which the separate twofold coincidences between various inputs are individually determined and applied in a more complex analysis, such as that given in Refs. 82 and 83. For the example of a triple coincidence unit, the chance rate from totally uncorrelated inputs can be calculated from the resolving time and individual singles rates as $3\tau^2 r_1 r_2 r_3$. If this were the only source of chance events, the correction could be made analytically as in the case of twofold coincidences. However, the added complexity arises because of partially correlated events in which a true coincidence in two branches is accompanied by a chance event in the third branch within the system resolving time. Because the probability of these events depends on the specifics of each experiment, they must, in general, be experimentally evaluated for each case.

5. DETERMINATION OF COINCIDENCE RESOLVING TIME

Several methods are available to the experimenter for the determination of the resolving time of a coincidence circuit. One is to provide totally uncorrelated inputs and to measure simultaneously both the singles rates and the resulting chance coincidence rate. The resolving time can then be calculated from Eq. (17.28). In setting up such a measurement, care must be taken to exclude any possibility of true coincidences between the two branches. The source must be incapable of generating coincident radiations that can interact in both detectors, or alternatively, separated sources with adequate shielding should be used to prevent true coincidences. Consideration should also be given to the possibility that true coincidences may arise from the scattering of radiation from one detector to the other.

The resolving time can also be measured by recording the coincidence-delay curve. As shown in Fig. 17.47, the FWHM of the portion of the curve that corresponds only to true coincidences is equal to twice the resolving time. For this method to be practical, a source with a sufficiently high probability of true coincidence emission must be used to ensure that the true coincidence rate stands out well above the chance background.

6. DELAYED COINCIDENCE AND OTHER INTERVAL MEASUREMENTS

In the previous example, the time spectrum was illustrated for a source that emitted prompt coincidence radiation. One way of defining prompt coincidence is to include any events that are separated by a delay time that is small compared with the instrumental time resolution. There are many occasions in which radiations are emitted in the same nuclear decay but are separated in time because of an intermediate nuclear state of finite lifetime. The time distribution should then show an exponential tail to the right of the prompt peak shown previously in Fig. 17.45. By measuring the time constant of this tail, the decay constant of the intermediate state can be deduced. These measurements can be carried out either with the multichannel technique using a TAC or as a series of single-channel measurements using a coincidence unit and variable delay. The latter method has historically been known as the *delayed coincidence* technique, and before the days of multichannel time spectroscopy was widely used to measure time interval distributions.

Multichannel time spectroscopy can be applied quite generally to any situation in which a time interval is to be measured. For example, in neutron time-of-flight spectroscopy, the start pulse is supplied from a detector that senses the time at which the neutron is generated, whereas the stop pulse is taken from a detector in which the neutron interacts after traveling some distance. The time interval between these two events is then a measure of the flight time of the neutron from which its energy can be calculated. There are many other examples of physical measurements in which the time interval distribution is important, and methods originally developed for nuclear measurements have been applied to a large assortment of determinations in other scientific fields.

7. MEASUREMENT OF ABSOLUTE SOURCE ACTIVITY USING COINCIDENCE TECHNIQUES

As an illustration of one of the common applications of coincidence circuits, we consider the problem of measuring the absolute activity of a given radioisotope source. If the source emits two coincident radiations that are distinguishable, methods can be applied that eliminate the need to know absolute detector efficiencies in order to calculate the source activity. Because detector efficiencies are often uncertain and hard to determine, use of these methods can provide a more accurate means of source activity determination than otherwise available.

Assume that the source activity (disintegrations per second) is given by S. Further assume that each such disintegration gives rise to two coincident radiations with no correlation of the angle of emission of one with respect to the other. We now arrange two detectors, such that detector 1 records pulses only from radiation 1, whereas detector 2 records pulses from radiation 2 only. The output of these detectors after appropriate pulse processing is fed to a coincidence unit with a resolving time τ. By separately measuring the two singles rates and the coincidence counting rate, the source activity can be determined as follows.

The singles counting rate in branch 1 (r_1), corrected for background and dead time losses, can be written as the product of the source activity S multiplied by some overall efficiency ϵ_1. This efficiency includes the solid angle subtended by the detector, the interaction probability within the detector, and the fraction of detector pulses accepted by the subsequent circuitry. A similar relation can be written for branch 2. Therefore, we have

$$r_1 = \epsilon_1 S \tag{17.29}$$
$$r_2 = \epsilon_2 S \tag{17.30}$$

The true coincidence rate r_1 can be predicted by noting that two independent events must occur: Radiation 1 must be detected in branch 1, whereas at the same time radiation 2 must be detected in branch 2. The independent probabilities (they are independent if there is no angular correlation) of these two events are ϵ_1 and ϵ_2. Therefore, the probability that both occur is simply their product $\epsilon_1 \epsilon_2$. The true coincidence rate is thus the product of this combined probability and the source activity S:

$$r_t = \epsilon_1 \epsilon_2 S \tag{17.31}$$

The measured coincidence rate r_{12} is the sum of the true coincidence rate and the chance coincidence rate. Therefore,

$$r_{12} = r_t + r_{ch}$$
$$r_{12} = \epsilon_1 \epsilon_2 S + r_{ch} \tag{17.32}$$

Now we can solve the three equations (17.29), (17.30), and (17.32) simultaneously and thereby eliminate any two variables. By eliminating the efficiencies ϵ_1 and ϵ_2, we can write

$$S = \frac{r_1 r_2}{r_{12} - r_{ch}} \tag{17.33}$$

This expression gives the source activity in terms of directly measured rates and the chance coincidence rate which can be measured using the methods described earlier. In many applications, the two coincident radiations selected are beta (β) and gamma (γ) rays emitted in the decay of a given isotope, and the method is often called β-γ coincidence. The requirement of no angular correlation can be relaxed if one of the radiations is detected over a 4π solid angle. Therefore, a common implementation of the method is to use a 4π proportional counter as the beta detector in connection with a gamma-ray detector subtending a smaller solid angle. Accuracies approaching 1% can be obtained with this method, whereas activity measurements that rely on a prior knowledge of detector efficiency are nearly always less precise.

The analysis above is somewhat oversimplified, because it is often difficult to meet the condition that each detector responds only to one of the two radiations. In β-γ coincidence measurements, it is normally quite easy to eliminate the β sensitivity of the γ detector by interposing an absorber that is thicker than the maximum distance of penetration of the beta particles involved. However, most beta particle detectors show some response to gamma rays. Two gamma rays of distinguishable energies are also sometimes used as the two radiations (γ-γ coincidence) by selecting each photopeak separately in two gamma-ray detectors. It is then difficult to avoid including some of the higher-energy gamma rays in the lower-energy branch because of the contribution of the Compton continuum from the higher-energy gamma rays. Also, the complications of true and chance summing (see p. 322) that may occur in either detector must be considered. For a more detailed discussion of these and other complications in absolute activity measurements using coincidence methods, see Ref. 84.

C. Modular Instruments for Timing Measurements

Pulse-processing systems composed of readily available nuclear instrumentation modules are shown in Fig. 17.50 for simple coincident measurements and time spectroscopy. If there

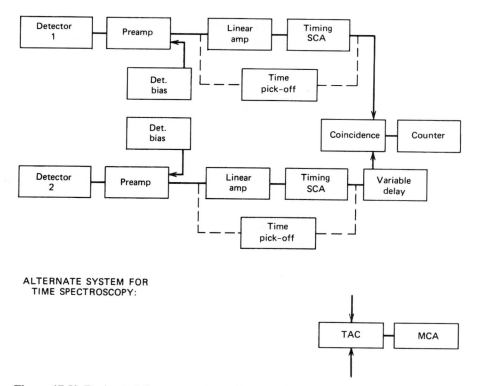

Figure 17.50 Basic modular system for twofold coincidence measurements. The use of a time pick-off in place of the linear amplifier and timing SCA, as indicated by the dashed lines, is warranted if no pulse amplitude selection is desired. An alternative arrangement for multichannel time spectroscopy is indicated at the bottom, in which the coincidence unit and counter are replaced by a TAC and MCA.

Figure 17.51 A simple fast–slow pulse-processing system in which a slow amplitude branch is used to select only those fast timing pulses that correspond to events of a predetermined amplitude set by the SCA window.

is no interest in pulse height information, these systems can be designed exclusively for their timing properties, using procedures that often compromise the pulse height information. As an example, the time constant for the output pulses from a scintillation counter can be chosen to be very small so that fast linear pulses are produced. These pulses follow the time profile of the collected electrons from the photomultiplier tube and provide the best performance in leading edge time pick-offs. However, as discussed in Chapter 4, fast linear pulses have inferior pulse height information when compared with the integrated tail pulse, which is usually provided when pulse height information is desired.

In applications in which both timing and pulse height information must be extracted, it has become common to arrange a *fast–slow* instrumentation scheme as outlined in Fig. 17.51. In the example shown, it is desired to accept timing pulses only for those detector pulses of a specific amplitude. In this case, separate signals for timing and amplitude are processed through the fast and slow branches so that optimum choices on pulse shaping, and so on, may be made independently to optimize the performance with regard to timing or amplitude information. The output from the amplitude branch is then used to operate a gate that can accept or reject the corresponding timing pulse. The time delays through the slower amplitude branch are such that it is normally required that a fixed delay be added to the timing branch before the two are combined.

1. TIME PICK-OFF MODULES

For optimum performance, the time pick-off is often best located directly at the detector. This is seldom done in practice, however, because of the practical problems in preserving good amplitude resolution at the same time. One exception is the scintillation counter, in which both the anode and a preceding dynode of the photomultiplier tube can be tapped to provide output signals. By choosing the anode load resistance to be small (say, 50 ohms), the anode circuit time constant will normally be smaller than the time over which electrons are collected and a fast linear pulse will therefore result. A preceding dynode can also be used to generate a signal by inserting a load resistance between that dynode and the voltage divider or other source of dynode dc voltage. If the value of this load resistor is high, the resulting time constant can be large and tail pulses will be generated. These tail pulses then serve as the primary signal for amplitude processing or the slow branch, whereas the fast linear pulses are used for timing purposes. (Note that dynode pulses are of opposite polarity from the anode pulse because more electrons leave each dynode than are collected from the preceding stage.)

It is more likely that the time pick-off follows the preamplifier and operates on the leading edge of the preamplifier output pulse. Commercial preamplifiers are available in which a premium is placed on fast rise time performance, so that the output leading edge reflects the charge collection time from the detector as closely as possible.

Conventional crossover methods of time pick-off require a shaped linear pulse to provide the bipolar shape on which the crossover method is based. Such pick-offs must then obviously follow the shaping networks in the linear amplifier. Other time pick-off methods are also occasionally built into single-channel analyzers or integral discriminators intended to accept a shaped linear pulse. These units are a convenience in timing situations in which the ultimate in performance is not warranted and can be used with virtually no sacrifice in the separate amplitude resolution of the pulses.

2. COINCIDENCE UNITS

Many common coincidence units are based on an overlap principle in which the width of the input pulses directly determines the resolving time of the coincidence circuit. Other types of circuits are sensitive only to the leading edge of the input pulse, and the resolving times can be chosen independently. The latter type is more flexible for situations in which detectors with different time properties may be interchanged.

Most coincidence units provide multiple inputs (often up to four) that can independently be switched in or out of the circuit. When only one input is switched in, every logic pulse presented to that input is simply routed through to the output of the coincidence unit. This mode of operation provides a convenient way of recording the singles rate in any input branch. When two inputs are switched in, the unit functions as a simple twofold coincidence, three switched inputs lead to threefold coincidence, and so on.

Coincidence units often provide for at least one *anticoincidence* input. Pulses provided to this input within a predetermined time act to cancel the normal output of the unit. For example, if one normal input is selected, an output pulse will appear for each input pulse that is not accompanied by an anticoincidence pulse within the resolving time of the circuit.

3. TIME-TO-AMPLITUDE CONVERTER (TAC)

The great popularity of TACs for the measurement of time intervals stems from the wide availability of multichannel analyzers in most measurement laboratories. By converting the time interval to a proportional pulse amplitude, the TAC allows the use of well-developed methods (discussed in Chapter 18) for the analysis and storage of pulse amplitudes as a substitute for the direct measurement of the time interval.

One of the more important properties of the TAC is the linearity of its time interval to amplitude conversion. In order to test the linearity, means must be provided for introducing fixed delays of known magnitude between start and stop pulses. For timing periods up to a few hundred nanoseconds, this can be accomplished by using variable lengths of coaxial cables. Other methods applicable for testing TAC linearity over wider time ranges are given in Ref. 85.

TACs are generally of two distinct types. The *overlap* type is based on supplying start and stop pulses of standard rectangular shape to the converter and measuring the area of overlap between the two. If the two pulses are coincident, the overlap will be complete, whereas if they are separated by a pulse width, there will be no overlap. Therefore, if an output pulse is generated whose amplitude integrates the area of overlap, the time to amplitude conversion is carried out. The principal merit of the overlap scheme is that it is very fast compared with other methods. Unfortunately, it tends to have poor linearity and accuracy specifications, and therefore it is used mainly in those applications in which maximum counting rates are of primary interest.

In the *start–stop* type of TAC, the start pulse initiates some circuit action, such as the charging of a capacitor by a constant-current source. This action continues until terminated by the appearance of the stop pulse. The constant current generates a linear ramp voltage, which is stopped at an amplitude proportional to the interval between the start and stop pulses. Designs of this type tend to have better linearity characteristics than overlap types and are more commonly encountered in routine time spectroscopy measurements.

4. DIRECT TIME-TO-DIGITAL CONVERTER (TDC)

In the time spectroscopy system of Fig. 17.44 the time interval between start and stop pulses is first converted to a pulse amplitude. In the multichannel analyzer, this pulse amplitude is digitized and recorded. It is ironic that the digitization is often carried out by converting the amplitude back into a time interval over which clock oscillator pulses are accumulated (see Chapter 18). Therefore, it is only logical to consider eliminating the step of converting the time interval to an amplitude and instead to digitize the interval directly by using it to gate the output of a constant-frequency clock. Application of this method, however, is limited by the highest frequency at which clock pulses can be reliably accumulated and counted. The present upper limit[86] is about 1 GHz, corresponding to a period of 1 ns. Thus intervals of the order of 100 ns or more can be measured to within 1% in this manner, but the digitization of shorter intervals suffers from the coarseness of the time scale imposed by the period of the clock frequency. Instead, more complex interpolation techniques have been developed[87–89] to stretch the basic interval by typical factors of 256 or more. A "multistop" TDC is one that has a number of parallel channels, all triggered by a common start signal. Each of the channels is controlled by an independent stop pulse to allow the timing of many events in parallel. The general principles of time interval measurement outlined in Ref. 90 are often incorporated in modern designs on a single application-specific integrated circuit.

5. TIME DELAYS

In many timing applications it is necessary to introduce delays at one or more points in the signal chain for timing adjustment or calibration purposes. On a nanosecond scale, the length of coaxial cables used to interconnect the various modules can be varied to provide differences in transit time. Beyond about 100 ns, cable lengths become excessively long (> 30 m) for convenient use. Use of delay lines with helically wound center conductors can greatly reduce the physical length of the cable for a given transit time, but they have relatively poor high-frequency transmission characteristics and will badly distort fast pulses. When used with shaped linear pulses that have already been high-frequency filtered, the distortions are less serious, and passive delay lines of various designs can provide adjustable delays up to several microseconds. Linear amplifiers often incorporate a delay line to provide a supplemental delayed output for convenience in setting up pulse timing systems.

Delay elements for logic pulses are less demanding because the shape of the pulse carries no information. One type of logic pulse delay unit is based on generating a ramp voltage and sensing the point at which the ramp exceeds a discrimination level. The ramp is initiated by the input logic pulse, and its termination produces a similar logic pulse after a delay time that can be continuously adjusted by varying the discrimination level.

6. WIDEBAND AND TIMING FILTER AMPLIFIERS

When the need for timing information is paramount, the detector output is often chosen to be a fast linear pulse of short duration. This type of pulse is commonly supplied to a leading edge time pick-off unit to provide a prompt timing signal. If the amplitude of the fast linear pulse is insufficient to reach a convenient trigger level in the time pick-off, it is necessary to provide some prior amplification. The linear amplifiers discussed previously will not suffice for this application because their shaping circuits are designed to eliminate the high-frequency components in the signal for optimum amplitude resolution. In order to preserve the fast linear pulse shape, an amplifier with linear response to as high a frequency as possible is desired. From the standpoint of baseline stability, it can also be desirable to dc couple all stages of the amplifier and thereby extend the frequency response to zero. Such amplifiers are called *wideband amplifiers*, and because they filter neither high nor low frequencies from the signal, they provide no shaping to the input pulse. When applied to nuclear pulse processing, they are more conventionally called *fast amplifiers* and ideally produce an output pulse that is a faithful replica of the input pulse, but with amplitude gain. A common application is in the processing of pulses from the anode of a photomultiplier tube where the signal levels are high enough to make electronic noise considerations relatively unimportant.

There are also some applications in which some degree of filtering of the signal from a preamplifier is useful to suppress noise, but timing information is still paramount. One example is the use of pulses from a germanium detector–preamplifier for timing purposes. A unit generally known as a *timing filter amplifier* can be useful in these circumstances. It offers some amplitude gain and adjustable differentiation and integration time constants that are typically much shorter (a few hundred nanoseconds or less) than those common in linear amplifiers. The resulting pulses generally have faster rise times and smaller widths than those from linear amplifiers, and are well-suited for the input of the time pick-off units described earlier in this chapter. The amplitude information will be noisier than that produced by conventional shaping in a linear amplifier, but that is not a major concern in measurements in which timing is the important objective.

X. PULSE SHAPE DISCRIMINATION

The usual information carried by a linear pulse is its amplitude (and time of occurrence). There are occasions in nuclear measurements, however, when the shape of the pulse also assumes some importance. Most shape differences arise because of differences in the time profile of the current produced at the detector by a radiation interaction. If a fast linear pulse is generated with a short collection time constant at the detector, the shape of the pulse is a reproduction of the time history of this current and will therefore directly reproduce these differences. In the more common application, the linear tail pulse obtained by collecting this current across a large time constant will show changes only in its leading edge characteristics. Therefore, *pulse shape discrimination* (PSD) methods designed to sense the difference between such events are sometimes also called *rise time discrimination* methods.

Sensing differences in pulse shape can serve a useful purpose when applied to the output of a number of different detector types. Some applications have been outlined in previous chapters and include (1) discrimination against gamma-ray background in organic scintillators used as fast neutron detectors; (2) separation of various particle types in scintillators such as CsI(Tl); (3) discrimination between short-range and long-range particles in proportional counters; (4) discarding defective pulses from silicon or germanium detectors to improve resolution; (5) separation of radiations of different range in the phoswich detector; (6) rejection of piled-up pulses, which can occur in any signal chain at high pulse rates.

There are two general approaches that have been used to carry out pulse shape discrimination measurements. The first is based on electronic methods of sensing the differences in the rise time of an output pulse, while the second derives a signal based on integrating the total charge over two different time periods. Both of these techniques are in widespread use, and a number of intercomparisons[91–93] show that there are applications in which either may be preferable.

The rise time measurement can be implemented in a number of ways including using two constant fraction timing discriminators to sense when the pulse crosses levels that are different percentages of its final amplitude. A more common implementation is the so-called crossover method, which is based on passing the pulse through a shaping network to produce a bipolar shape. This can be either a *CR-RC-CR* network or double delay line shaper, as discussed in Chapter 16. In either case, the time at which the bipolar pulse crosses zero in theory does not depend on pulse amplitude but instead is a function of the pulse shape and rise time. The time interval between the beginning of the pulse and the zero crossover point will therefore be an indication of the differences in pulse shape prior to the shaping network. Conventionally, the time difference is measured between a leading edge trigger set as low as possible at the beginning of the pulse and a second trigger that senses the crossover point. This time difference is then converted by a TAC into a pulse amplitude. Systems of this type have been described in Refs. 94–97. A representation of a typical multichannel analysis of the TAC output is shown in Fig. 17.52. The TAC output is normally fed to a single-channel analyzer to select only those time intervals that correspond to the radiation of interest. This step is equivalent to selecting a band of channels that includes only one of the peaks in Fig. 17.52. The SCA output can then be used to gate the pulse-recording system only when the acceptance criteria are met. More elaborate implementations involve the simultaneous recording of the pulse shape and pulse height information in a two-parameter multichannel analyzer, as described in Chapter 18.

A second major approach to the design of pulse shape discriminators is based on independent measurements of the integrated charge over two different time regions of the pulse. The ratio of these signals will be approximately constant for pulses of common shape, independent of pulse amplitude. The distribution of ratio values will have the appearance shown in Fig. 17.52 if pulses of two specific shapes are involved. Such measurements can either be carried out with analog circuitry (for example, as in Refs. 98–100) or using digital techniques.[101–107] The performance of pulse shape discriminators deteriorates at high pulse rates due to pile-up effects, but can be improved through the rejection of closely spaced pulsed pairs at the PSD input.[108,109]

Several specifications are of importance in discussing PSD circuits. One is the figure of merit M defined in Fig. 17.52. This factor, as introduced in Ref. 110, is a measure of the separation that can be achieved between different types of events in a given application. The figure of merit is likely to depend on the dynamic range of the input pulses, defined as the ratio between the maximum and minimum amplitude pulses acceptable by the system. Effective pulse shape discrimination systems will operate over a dynamic range of 100 or more.

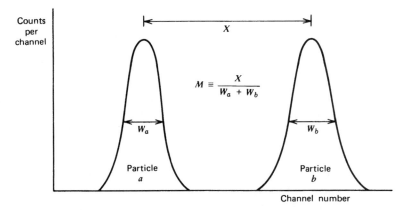

Figure 17.52 Definition of the figure of merit M for pulse shape discrimination applications. In the most common PSD method, the abscissa corresponds to the crossover time of a doubly differentiated input pulse.

PROBLEMS

17.1 Derive the expression $V_{out} \cong -(R_2/R_1)\,V_{in}$ for the voltage-sensitive preamplifier configuration of Fig. 17.1. (Note that the input current must flow through both R_1 and R_2.)

17.2 Derive the expression $V_{out} \cong -Q/C_f$ for the charge-sensitive preamplifier configuration of Fig. 17.2. (Note that the input charge is divided between C_i and C_f.)

17.3 Why is it preferable to locate a preamplifier as close as possible to the detector with which it is used?

17.4 Sketch a configuration consisting of two integral discriminators and an anticoincidence unit that will perform the function of a single-channel analyzer.

17.5 A rate meter circuit can be made to respond to rapid changes more quickly by reducing its time constant. What penalty is normally also associated with this same change?

17.6 A given application involves processing pulses at a low average rate from a high-resolution detector. If the objective is to preserve the pulse height resolution as much as possible, indicate which of the options given below is the better choice:

(a) Short or long shaping times.

(b) Monopolar or bipolar shaping.

(c) With or without active baseline restoration.

17.7 Pulses corresponding to a particular full-energy peak occur at a rate of 8000/s in a system in which the total pulse rate is 25,000/s. Estimate the fraction of the full-energy events that are lost due to pile-up, if the effective pulse width is 4 μs.

17.8 A given application involves signal pulses with constant shape and only a small variation in amplitude. Which of the time pick-off methods discussed in the text is likely to give the best timing performance?

17.9 List two independent methods of pulse shaping that can yield the bipolar shape needed for crossover timing.

17.10 A single-channel analyzer produces an output logic pulse for any input pulse that lies within a narrow *acceptance window* in the full pulse height spectrum. What electronic unit carries out the analogous function within the time spectrum?

17.11 Random and uncorrelated pulses are supplied at rates r_1 and r_2 to the inputs of an anticoincidence unit with resolving time τ. Taking into account the effects of chance coincidences, what should be the observed output rate?

17.12 Give two independent methods of determining the chance coincidence contribution to a standard twofold coincidence measurement.

17.13 What length of conventional coaxial cable is necessary to shift the position of the prompt coincidence peak in a time spectrum by 100 ns?

17.14 A sample of ^{60}Co will show a gamma-ray spectrum with peaks at 1.17 and 1.33 MeV, plus a β^- spectrum with endpoint energy of 0.31 MeV. Draw two independent decay schemes that are consistent with the above observations, and outline an experiment that could help decide between the alternatives.

17.15 Two sodium iodide scintillators are arranged to count one each of two gamma rays emitted in coincidence with no angular correlation by a radioactive source. In each of the cases listed below, describe the effect of the indicated change on the ratio of true coincidence to chance coincidence counting rates, and indicate the practical limits to which each of the changes may be carried in order to improve this ratio.

(a) Moving the detector positions to change the solid angle subtended by each.

(b) Varying the amount of source material.

(c) Changing the coincidence circuit resolving time.

(d) Varying the pulse height window width used to select a given gamma energy.

17.16 The following data were obtained for the coincidence-delay curve in a coincidence experiment:

(a) What is the resolving time of the coincidence unit?

(b) What is the width of the prompt coincidence peak in the time spectrum?

(c) What is the singles rate, assuming that it is about the same in both branches supplied to the coincidence unit?

17.17 A 12-bit subranging ADC can consist of 3 stages of 4 bits each, or 4 stages of 3 bits each. Which configuration requires the smaller total number of comparators? If both configurations were operated at the same clock frequency, which would have the lower latency time?

17.18 The digital equivalent of a step input voltage consists of sampled data $V(i) = 0$ for $i < 0$ and $V(i) = 1$ for $i \geq 0$. Find the first five nonzero terms for the output of a digital filter applied to these data that consists of the following weighting factors:

(a) A "moving average" or "boxcar filter" of $H(0) = H(1) = H(2) = H(3) = 0.25$

(b) A "differentiator" with $H(0) = 1, H(1) = -1$

(c) A "differentiator with memory" of $H(0) = H(1) = 0.5$; $H(2) = H(3) = -0.5$

(d) An "exponential integrator" with $H(i) = 0$ for $i < 0$ and $i > 5$; $H(i) = \exp(-i/2)$ for $0 \leq i \leq 5$

REFERENCES

1. J. Millman, *Microelectronics*, McGraw-Hill Book Co., New York (1979).
2. J. W. T. Dabbs and F. J. Walter, eds., *Semiconductor Nuclear Particle Detectors*, Publication No. 871, Washington DC: NAS-NRC, 237 (1961).
3. M. M. Satterfield, G. R. Dyer, and W. J. McClain, *Nucl. Instrum. Meth.* **75**, 312 (1969).
4. V. Radeka, *IEEE Trans. Nucl. Sci.* **NS-17**(1), 269 (1970).
5. D. A. Landis, F. S. Goulding, R. H. Pehl, and J. T. Walton, *IEEE Trans. Nucl. Sci.* **NS-18**(1), 115 (1971).
6. K. Kandiah, A. J. Smith, and G. White, *IEEE Trans. Nucl. Sci.* **NS-22**(5), 2058 (1975).
7. D. A. Landis, N. W. Madden, and F. S. Goulding, *IEEE Trans. Nucl. Sci.* **NS-26**(1), 428 (1979).
8. D. A. Landis, C. P. Cork, N. W. Madden, and F. S. Goulding, *IEEE Trans. Nucl. Sci.* **NS-29**(1), 619 (1982).
9. C. L. Britton, T. H. Becker, T. J. Paulus, and R. C. Trammell, *IEEE Trans. Nucl. Sci.* **NS-31**(1), 455 (1984).
10. I. S. Sherman and R. G. Roddick, *IEEE Trans. Nucl. Sci.* **NS-17**(1), 252 (1970).
11. J. J. Eichholz, ANL-6968, Argonne National Laboratory, (1966).
12. G. White, *Nucl. Instrum. Meth.* **125**, 313 (1975).
13. F. S. Goulding and D. A. Landis, *IEEE Trans. Nucl. Sci.* **NS-35**(1), 119 (1988).
14. M. Moszynski and G. Duchene, *Nucl. Instrum. Meth.* **A308**, 557 (1991).
15. A. B. Gillespie, *Signal, Noise and Resolution in Nuclear Counter Amplifiers*, Pergamon Press, London, 1953.
16. V. Radeka, *Nucleonics* **23**(7), 53 (1965).
17. E. Kowalski, *Nuclear Electronics*, Springer-Verlag, New York, 1970.
18. P. W. Nicholson, *Nuclear Electronics*, Wiley, London, 1974.
19. M. Bertolaccini, C. Bussolati, and E. Gatti, *Nucl. Instrum. Meth.* **41**, 173 (1966).
20. A. H. F. Muggleton, *Nucl. Instrum. Meth.* **101**, 113 (1972).
21. S. L. Blatt, *Nucl. Instrum. Meth.* **128**, 277 (1975).
22. D. W. Datlowe, *Nucl. Instrum. Meth.* **145**, 365 (1977).
23. L. Wielopolski and R. P. Gardner, *Nucl. Instrum. Meth.* **133**, 303 (1976).
24. S. L. Blatt, *Nucl. Instrum. Meth.* **49**, 235 (1967).
25. M. Furrer, J.-J. Gostely, and P. Lerch, *Nucl. Instrum. Meth.* **226**, 455 (1984).
26. D. H. Wilkinson, *Nucl. Instrum. Meth.* **134**, 149 (1976).
27. S. L. Blatt, J. Mahieux, and D. Kohler, *Nucl. Instrum. Meth.* **60**, 221 (1968).
28. C. W. Williams, *IEEE Trans. Nucl. Sci.* **NS-15**(1), 297 (1968).
29. J. Bartosek, J. Masek, F. Adams, and J. Hoste, *Nucl. Instrum. Meth.* **104**, 221 (1972).
30. F. S. Goulding and D. A. Landis, *IEEE Trans. Nucl. Sci.* **NS-25**(2), 896 (1978).
31. K. A. Sjöland and P. Kristiansson, *Nucl. Instrum. Meth.* **B94**, 333 (1994).
32. F. H. Tenney, *Nucl. Instrum. Meth.* **219**, 165 (1984).
33. R. Gold, ANL-6949, (1965).
34. R. P. Gardner and L. Wielopolski, *Nucl. Instrum. Meth.* **140**, 289 (1977).
35. D. W. Datlowe, *Nucl. Instrum. Meth.* **145**, 379 (1977).
36. P. C. Johns and M. J. Yaffe, *Nucl. Instrum. Meth.* **A255**, 559 (1987).
37. Q. Bristow, *IEEE Trans. Nucl. Sci.* **39**(4), 712 (1992).
38. Q. Bristow, *Radioact. Radiochem.* **5**(1), 22 (1994).
39. T. Fazzini, G. Poggi, P. Sona, and N. Taccetti, *Nucl. Instrum. Meth.* **A356**, 319 (1995).
40. M. Wiernick, *Nucl. Instrum. Meth.* **96**, 325 (1971).
41. R. M. Lindstrom and R. F. Fleming, *Radioact. Radiochem.* **6**(2), 20 (1995).
42. E. Fairstein, *IEEE Trans. Nucl. Sci.* **NS-32**(1), 31 (1985).
43. M. L. Simpson et al., *IEEE Trans. Nucl. Sci.* **38**(2), 89 (1991).
44. H. B. Crawley et al., *Nucl. Instrum. Meth.* **A342**, 578 (1994).
45. H. B. Crawley et al., *IEEE Trans. Nucl. Sci.* **41**(4), 1181 (1994).
46. V. T. Jordanov and G. F. Knoll, *Nucl. Instrum. Meth.* **A345**, 337 (1994).
47. V. T. Jordanov, G. F. Knoll, A. C. Huber, and J. A. Pantazis, *Nucl. Instrum. Meth.* **A353**, 261 (1994).
48. G. Bertuccio, A. Fazzi, A. Geraci, and M. Sampietro, *Nucl. Instrum. Meth.* **A353**, 257 (1994).
49. E. Gatti, A. Geraci, and G. Ripamonti, *Nucl. Instrum. Meth.* **A381**, 117 (1996).
50. A. Geraci, G. Ripamonti, and A. Pullia, *Nucl. Instrum. Meth.* **A403**, 455 (1998).
51. E. Gatti, A. Geraci, and G. Ripamonti, *Nucl. Instrum. Meth.* **A417**, 131 (1998).
52. A. Pullia and G. Ripamonti, *Nucl. Instrum. Meth.* **A376**, 82 (1996).
53. A. Pullia, G. Gritti, and G. Ripamonti, *IEEE Trans. Nucl. Sci.* **44**(3), 331 (1997).
54. R. J. Komar and H.-B. Mak, *Nucl. Instrum. Meth.* **A336**, 246 (1993).
55. V. Drndarevic, P. Ryge, and T. Gozani, *Nucl. Instrum. Meth.* **A277**, 532 (1989).
56. D. A. Gedcke and W. J. McDonald, *Nucl. Instrum. Meth.* **55**, 377 (1967).
57. B. Leskovar and C. C. Lo, *Nucl. Instrum. Meth.* **123**, 145 (1975).
58. J. D. McGervey, J. Vogel, P. Sen, and C. Knox, *Nucl. Instrum. Meth.* **143**, 435 (1977).
59. M. O. Bedwell and T. J. Paulus, *IEEE Trans. Nucl. Sci.* **NS-25**(1), 86 (1978).
60. K. Rytsölä, *Nucl. Instrum. Meth.* **199**, 491 (1982).
61. M. Tanaka, H. Ikeda, M. Ikeda, and S. Inaba, *IEEE Trans. Nucl. Sci.* **39**(5), 1321 (1992).
62. D. M. Binkley, *IEEE Trans. Nucl. Sci.* **41**(4), 1169 (1994).
63. M. L. Simpson, C. L. Britton, A. L. Wintenberg, and G. R. Young, *IEEE Trans. Nucl. Sci.* **42**(4), 762 (1995).
64. R. L. Chase, *Rev. Sci. Instrum.* **39**, 1318 (1968).
65. J. J. Kozyczkowski and J. Bialkowski, *Nucl. Instrum. Meth.* **137**, 75 (1976).
66. M. A. El-Wahab and A. El-Arabi, *IEEE Trans. Nucl. Sci.* **40**(2), 147 (1993).
67. J. P. Fouan and J. P. Passerieux, *Nucl. Instrum. Meth.* **62**, 327 (1968).
68. S. Bose, C. C. Dey, B. K. Sinha, and R. Bhattacharya, *Nucl. Instrum. Meth.* **A295**, 219 (1990).
69. A. R. Frolov, T. V. Oslopova, and Yu. N. Pestov, *Nucl. Instrum. Meth.* **A356**, 447 (1995).
70. F. J. Lynch, *IEEE Trans. Nucl. Sci.* **NS-13**(3), 140 (1966).
71. C. Hohenemser, R. Reno, and A. P. Mills, *IEEE Trans. Nucl. Sci.* **NS-17**(3), 390 (1970).
72. T. J. Paulus, *IEEE Trans. Nucl. Sci.* **NS-32**(3), 1242 (1985).
73. P. Ryge and R. R. Borchers, *Nucl. Instrum. Meth.* **95**, 137 (1971).

74. Z. H. Cho and R. L. Chase, *Nucl. Instrum. Meth.* **98**, 335 (1972).

75. H. Engel, H. Schneider, and R. Spitz, *Nucl. Instrum. Meth.* **142**, 525 (1977).

76. D. C. S. White and W. J. McDonald, *Nucl. Instrum. Meth.* **115**, 1 (1974).

77. J. Kasagi, H. Ohnuma, and N. Ohyama, *Nucl. Instrum. Meth.* **193**, 557 (1982).

78. T. J. Paulus, T. W. Raudorf, B. Coyne, and R. Trammell, *IEEE Trans. Nucl. Sci.* **NS-28**(1), 544 (1981).

79. A. E. Blaugrund and Z. Vager, *Nucl. Instrum. Meth.* **29**, 131 (1964).

80. R. Avida and S. Gorni, *Nucl. Instrum. Meth.* **52**, 125 (1967).

81. Z. H. Cho and T. R. Gerholm, *Nucl. Instrum. Meth.* **73**, 67 (1969).

82. C. H. Vincent, *Nucl. Instrum. Meth.* **127**, 421 (1975).

83. R. L. Chase, *IEEE Trans. Nucl. Sci.* **NS-23**(1), 244 (1976).

84. NCRP Report No. 58, *A Handbook of Radioactivity Measurements Procedures*, 2nd ed., National Council on Radiation Protection and Measurements. Bethesda, MD, 1985.

85. S. Cova and M. Bertolaccini, *Nucl. Instrum. Meth.* **77**, 269 (1970).

86. S. Veneziano, *Nucl. Instrum. Meth.* **A409**, 363 (1998).

87. J. A. Harder, BNL 39840, May 1987.

88. B. Turko, *IEEE Trans. Nucl. Sci.* **NS-26**(1), 737 (1979).

89. B. T. Turko, R. D. Macfarlane, and C. J. McNeal, *Int. J. Mass Spectrom. Ion Phys.* **53**, 353 (1983).

90. D. I. Porat, *IEEE Trans. Nucl. Sci.* **NS-20**(5), 36 (1973).

91. G. Ranucci, *Nucl. Instrum. Meth.* **A354**, 389 (1995).

92. D. Wolski et al., *Nucl. Instrum. Meth.* **A360**, 584 (1995)

93. Z. Cao and L. F. Miller, *Nucl. Instrum. Meth.* **A416**, 32 (1998).

94. L. J. Heistek and L. Van der Zwan, *Nucl. Instrum. Meth.* **80**, 213 (1970).

95. G. W. McBeth, J. E. Lutkin, and R. A. Winyard, *Nucl. Instrum. Meth.* **93**, 99 (1971).

96. D. W. Glasgow, D. E. Velkley, J. D. Brandenberger, and M. T. McEllistrem, *Nucl. Instrum. Meth.* **114**, 535 (1974).

97. P. Plischke et al., *Nucl. Instrum. Meth.* **136**, 579 (1976).

98. J. M. Adams and G. White, *Nucl. Instrum. Meth.* **156**, 459 (1978).

99. J. R. M. Annand, *Nucl. Instrum. Meth.* **A262**, 371 (1987).

100. V. T. Jordanov, J. A. Pantazis, and A. C. Huber, *Nucl. Instrum. Meth.* **A380**, 353 (1996).

101. Z. W. Bell, *Nucl. Instrum. Meth.* **188**, 105 (1981).

102. T. Kumahara and H. Tominaga, *IEEE Trans. Nucl. Sci.* **NS-31**(1), 451 (1984).

103. J. H. Heltsley et al., *Nucl. Instrum. Meth.* **A263**, 441 (1988).

104. M. Moszynski et al., *Nucl. Instrum. Meth.* **A317**, 262 (1992).

105. M. Moszynski et al., *Nucl. Instrum. Meth.* **A336**, 587 (1993).

106. M. N. Al-Haddad, C. Lin, W. H. Miller, and R. Berliner, *IEEE Trans. Nucl. Sci.* **41**(5), 1765 (1994).

107. V. T. Jordanov and G. F. Knoll, *IEEE Trans. Nucl. Sci.* **42**(4), 683 (1995).

108. R. B. Piercey, J. E. McKisson, M. A. Herath Banda, and M. R. Shavers, *IEEE Trans. Nucl. Sci.* **NS-34**(1), 82 (1987).

109. J. Bialkowski, M. Moszynski, and D. Wolski, *Nucl. Instrum. Meth.* **A275**, 322 (1989).

110. R. A. Winyard, J. E. Lutkin, and G. W. McBeth, *Nucl. Instrum. Meth.* **95**, 141 (1971).

Chapter **18**

Multichannel Pulse Analysis

A measurement of the differential pulse height spectrum from a radiation detector can yield important information on the nature of the incident radiation or the behavior of the detector itself and is therefore one of the most important functions to be performed in nuclear measurements. By definition (see Chapter 4), the differential pulse height spectrum is a continuous curve that plots the value of dN/dH (the differential number of pulses observed within a differential increment of pulse height H) versus the value of the pulse height H. The ratio of the differentials can never be measured exactly, but rather all measurement techniques involve a determination of $\Delta N/\Delta H$ (the discrete number of pulses observed in a small but finite increment of pulse height H). The increment in pulse height ΔH is commonly called the *window width* or *channel width*. Provided ΔH is small enough, a plot of $\Delta N/\Delta H$ versus H is a good discrete approximation to the continuous curve that represents the actual differential pulse height spectrum. As a practical matter, a distinction is seldom made between the continuous distribution and its discrete approximation, and all such plots are generally referred to as differential pulse height distributions or *pulse height spectra*.

I. SINGLE-CHANNEL METHODS

The differential discriminator (single-channel analyzer [SCA]) described in the previous chapter can be used to record a steady-state pulse height spectrum. The window is set to a small width ΔH and the number of output pulses produced over a measurement period is recorded as ΔN. In a somewhat tedious process, this window can then be moved stepwise over the pulse height range of interest. Sequential measurements of $\Delta N/\Delta H$ plotted at the midpoint H value of the window will then trace out the shape of the differential distribution. This serial process is inefficient, in that most pulses are ignored during a given measurement since they lie outside the specific window chosen. However, before the days of multichannel analyzers, this manual procedure was often the only method available to measure pulse height spectra.

A better approach is to employ multiple SCAs as in Fig. 18.1. Here the measurement is converted from one that is serial to one that is parallel, and every pulse can now contribute to the measured spectrum. All the inputs are connected together and each output fed to a separate counter. The lower level of the SCA at the bottom of the stack is set to zero, and that for the top SCA is set to correspond to the largest pulse height of interest. The lower level of the intermediate SCAs are arranged at equal intervals between these extremes. The window width of each SCA is identical and is set equal to the spacing between adjacent discrimination levels. This arrangement thus provides a series of contiguous pulse height windows of equal width, as illustrated in Fig. 18.2.

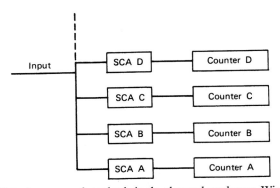

Figure 18.1 An array of stacked single-channel analyzers. Windows A, $B, C, \ldots$ are assumed to be contiguous and of equal width ΔH, with A at the bottom of the pulse height scale.

An input pulse presented to this array will fall into one and only one of the multiple windows set by the SCAs. Therefore, each input pulse results in an increment of one count added to the corresponding SCA counter. One can therefore view the overall process as the sorting of each input pulse into the proper window and incrementing the content of that counter by one. A small pulse will correspond to a window near the bottom of the stack, whereas a large pulse will fall into a window near the top. At the end of a measurement period, the sum of all the counters will simply be the total number of pulses presented to the input.

If we now plot the number of recorded pulses ΔN in each counter divided by the window width ΔH versus the average pulse height for each window, we derive a discrete representation of the differential pulse height distribution. In this context, each window is conventionally called a *channel* and is numbered in increasing order from left to right. The lowest channel corresponds to the pulse height window at the bottom of the range and records only those pulses whose amplitudes are very small. The largest channel numbers are plotted at the right of the horizontal axis and record only the pulses of largest amplitudes.

This process of sorting successive signal pulses into parallel amplitude channels is commonly called *multichannel pulse height analysis*. As a practical matter, schemes based on

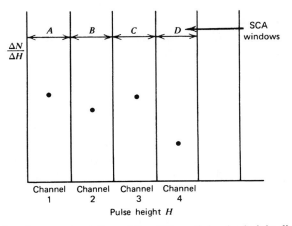

Figure 18.2 A pointwise representation of the differential pulse height distribution obtained from the stacked SCA array of Fig. 18.1 by plotting the content of each counter ΔN (normalized to the window width ΔH) versus the midpoint of the corresponding SCA window.

stacked independent SCAs are seldom attractive because of complications introduced by drifts in the various discrimination levels and window widths. These drifts can lead to over-lapping or noncontiguous channels whose width may also not be constant. As a result, other approaches have evolved for accomplishing the same purpose. The standard device designed to carry out this function is known as a *multichannel analyzer* (MCA), and the following sections discuss some general properties and functions of these instruments.

II. GENERAL MULTICHANNEL CHARACTERISTICS

A. Number of Channels Required

In any pulse height distribution measurement, two factors dominate the choice of the number of channels that should be used for the measurement: the degree of resolution required and the total number of counts that can be obtained. If an arbitrarily large number of counts can be accumulated, there is no disadvantage in making the number of channels as large as one wishes. By providing a large number of channels, the width of any one channel can be made very small and the resulting discrete spectrum will be a close approximation to the continuous distribution. For a faithful representation, the true distribution should not change drastically over the width of one channel. If peaks are present in the spectrum, this requirement translates into specifying that at least four or five channels should be provided over a range of pulse height corresponding to the FWHM (full width at half maximum) of the peak. Figure 18.3b shows a hypothetical differential distribution taken under conditions in which the number of channels is too small to meet this criterion. The resulting distortions and loss of resolution in the spectrum are obvious.

The channel requirements can also be expressed in terms of detector resolution R. For a peak with a mean pulse height H

$$R \equiv \frac{\text{FWHM}}{H} \quad \text{or} \quad H = \frac{\text{FWHM}}{R} \quad (18.1)$$

We can express both H and FWHM in terms of numbers of channels, and furthermore, we now require that at least five channels be provided over the FWHM of the peak. The position of H in units of channels is therefore

$$\text{peak position } H = \frac{5 \text{ channels}}{R} \quad (18.2)$$

and at least this number of channels must be provided. A detector whose energy resolution is 5% therefore requires a minimum of 100 channels, and a detector with 0.2% resolution would require 2500 channels. This argument is valid only in those cases in which the full range of pulse amplitude is recorded with constant channel width ranging from zero to the maximum pulse height. The channel requirement can be reduced by selectively recording only a portion of the spectrum with a large value of *zero offset* discussed in the next section.

The above arguments would suggest that one should always use the maximum number of channels possible. A second factor arises, however, when the available measurement time limits the total number of pulses that contribute to the recorded spectrum. Because the number of events that fall within any one channel will vary in proportion to its width, the content of a typical channel varies inversely with the total number of channels provided over the spectrum. Choosing a larger number of channels will consequently cause the relative statistical uncertainty of each content to increase, and the channel-to-channel fluctuation of the data over smooth portions of the spectrum will become more noticeable. If these fluctuations are large enough, they can begin to interfere with the ability to discern small features in the spectrum. Very small peaks can become lost in statistical noise. These effects are illustrated in Fig. 18.3d.

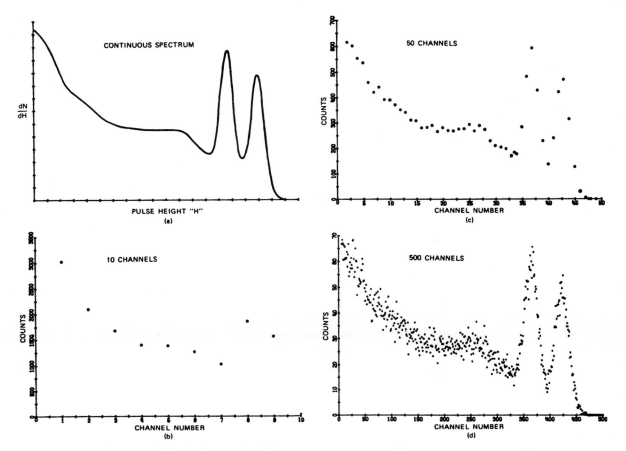

Figure 18.3 An illustration of the effect of varying the number of channels used to record the differential distribution at the top left. A total of 15,000 counts were accumulated for each of the three multichannel spectra. In (*b*) the number of channels is too small to show sufficient detail, in (*c*) the choice is about right, and too many channels were used in (*d*). The low average number of counts per channel in spectrum (*d*) leads to large statistical fluctuations that could obscure small additional peaks if they were present.

Another factor in choosing the number of channels across the width of typical peaks in the spectrum is the behavior of the software used to locate the peak and quantify its properties. Even if the underlying data were to follow a perfect Gaussian distribution, the "binning" effect of sorting the data into channels with finite width results in a series of points that do not fall on a perfect Gaussian curve because of the averaging effect that takes place within each channel. Neglecting statistical fluctuations, the contents of the channels would become closer to a true Gaussian as a larger number of channels are provided across the peak width. Thus if the analysis software assumes that the data in the region of a peak (after subtraction of any underlying continuum) follow a true Gaussian or some variant thereof (as most do), it may be better to allocate 9 to 12 channels across the FWHM to minimize the errors that are introduced due to the binning effect.[1]

B. Calibration and Linearity

The ideal MCA would perform a perfectly linear conversion of pulse height to channel number. Under these conditions, a plot of pulse height versus channel number would be a

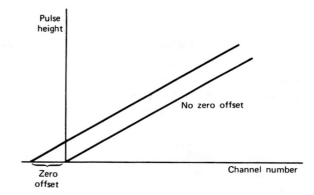

Figure 18.4 Typical calibration plot for a linear MCA with and without zero offset.

simple straight line, as illustrated in Fig. 18.4. In addition, it is usually convenient to introduce some *zero offset* shift of the origin, such that a nonzero amplitude is required for storage in the first channel. The zero offset is sometimes desirable to suppress high count rates from very small noise pulses, which may appear with the signal, or to assign the available channels only to the upper portion of a spectrum to be recorded. The zero offset is adjustable through front panel controls or as a software selection on many commercial MCAs.

In many situations, the signal pulses are first sent through a linear amplifier with variable gain. Then the slope of the calibration plot can also be varied by changing the gain factor. For example, Fig. 18.5 shows the calibration plots for an ideal MCA for three different values of amplifier gain. The same effect can also be achieved if the MCA allows selection of the conversion gain of its analog-to-digital converter (see next section).

The user is generally interested in an initial calibration of an MCA that will determine the energy scale of the pulse height distribution. Assuming the analyzer is sufficiently linear, two parameters need to be determined: the slope and intercept of the calibration line shown in Fig. 18.4. The easiest calibration method is to place sources of known energy at the detector and record the channel number into which the centroid of the resulting full-energy peak falls. Because two points completely determine a straight line, only two energies are, in principle, required. However, other sources are often used to provide additional points along the line for confirmation purposes and a test of linearity. If peaks of known energy cannot conveniently be provided, a pulse generator may also be used to

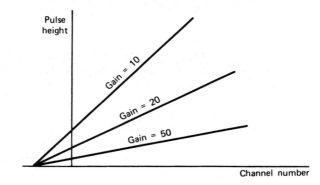

Figure 18.5 Different MCA calibration plots for three values of amplifier gain.

provide points for a calibration plot. Only pulses of known *relative* amplitude are required to test linearity and determine the zero offset, but some other means must then be used to relate independently the pulse height scale to energy.

The linearity of the MCA can be measured or quoted in several different ways. The most direct method is to make a measurement of the channel number in which pulses of known amplitude are stored, and then make the plot of pulse amplitude versus channel number shown in Fig. 18.6. The maximum deviation of the measured curve from a best-fit straight line is a measure of the *integral linearity* and is conventionally quoted as a percentage of the full MCA range. Nonlinearities are most often observed at either pulse height extreme and are typically less than 0.1% for well-designed analyzers.

A more sensitive method of detecting nonlinearities is to employ a source of pulses with a uniform distribution in amplitude. If the MCA is connected to accumulate these pulses over a period of time, a uniform or flat distribution of counts in all channels should result (see Fig. 18.6). Sufficient counting time must be allowed so that the statistical fluctuations in channel content are small compared with the desired measurement accuracy. Such a calibration measures the *differential linearity* of the MCA. Deviations from uniformity of a few percent are typical of good MCAs.

The integral and differential calibrations are interrelated, because the differential plot will be simply proportional to the slope of the integral plot at any point. In Fig. 18.6, the two calibrations are intended to be consistent and illustrate the same nonlinearity. The differential calibration, although more difficult to set up, is obviously a more sensitive test of MCA linearity.

The differential nonlinearity of an MCA is not a serious issue in most applications if it is small and varies smoothly from channel to channel. More significant problems can be created in pulse height analysis if the differential nonlinearity plot of the type shown in Fig. 18.6 has abrupt discontinuities or "bumps" that sometimes occur. These artificial features will appear superimposed on smooth regions of all pulse height spectra recorded by the analyzer and can easily be misinterpreted as true structure in the spectrum. Such discontinuities will show up in the measurement using a uniformly distributed pulse source,

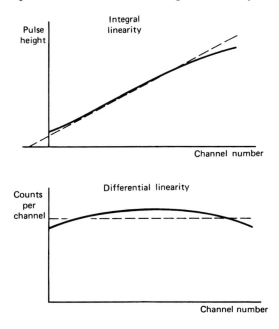

Figure 18.6 Examples of integral and differential linearity measurements for an MCA.

but they can also be detected in a much simpler test.[2] The pulse height distribution from many types of detectors is known to be smoothly varying over an extended amplitude interval (for example, a plastic scintillator responding to a gamma-ray source). The amplifier gain can be adjusted to spread this interval over the full range of the MCA, and a large number of counts are recorded to minimize the channel-to-channel statistical fluctuations. Even though the resulting recorded spectrum will not be uniform over the full range, it will still show any sudden discontinuities or periodic structures that are the most troublesome features of differential nonlinearity.

Although not as widely used as the nonlinearity tests, measurements of the channel profile discussed in Chapter 17 can also be applied in the performance testing of MCAs. Precision pulse generators whose amplitude can be continuously varied with a very fine vernier control are used to trace out the type of profile illustrated on p. 651. The probability that a typical pulse is stored in a preselected channel is plotted versus the pulse amplitude as it is varied from below to above the channel range. Ideally, these channel profiles have sharply defined limits, and there are no tails extending substantial distances beyond the nominal channel boundaries. MCAs with overly broad channel profiles that overlap with neighboring channels have a reduced resolution, and the effective number of channels is actually smaller than the nominal value. Any differential nonlinearity of the analog-to-digital converter will also be reflected in a corresponding departure from constant channel width over its range as measured by individual channel profiles.

III. THE MULTICHANNEL ANALYZER

A. Basic Components and Function

The multichannel analyzer (MCA) is comprised of the basic components illustrated in Fig. 18.7. Its operation is based on the principle of converting an analog signal (the pulse amplitude) to an equivalent digital number. Once this conversion has been accomplished, the extensive technology available for the storage and display of digital information can be brought to bear on the problem of recording pulse height spectra. As a result, the analog-to-digital converter (ADC) is a key element in determining the performance characteristics of the analyzer.

The flash and subranging ADCs previously discussed in Chapter 17 produce a continuous series of digital output values at a fixed clock frequency. In contrast, the ADCs intended for use in MCAs operate in a different mode. They are designed to produce only a single output value for each pulse presented to their input that is proportional to the peak amplitude of that pulse. They are therefore often called *peak sensing* ADCs. The input circuitry

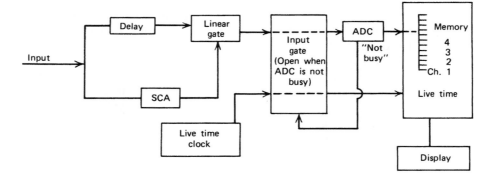

Figure 18.7 Functional block diagram of a typical MCA.

for these ADCs must include a capability to sense the arrival of an input pulse and to sample and hold the maximum amplitude of that pulse for a time needed to carry out the conversion to a digital value. The output of the ADC appears in a register that is used to address a standard digital memory that has as many addressable locations as the maximum number of channels into which the spectrum can be subdivided. The number of memory locations is normally made a power of 2, with memories of 1024 to 8192 being common choices. The maximum count capacity of any one channel is set by the bit depth of the memory, and in commercially available units can be as high as 2^{32}, or over 4×10^9 counts.

The basic function of the MCA involves only the ADC and the memory. For the purposes of illustration, we imagine the memory to be arranged as a vertical stack of addressable locations, ranging from the first address (or channel number 1) at the bottom through the maximum location number (say, 1024) at the top. Once a pulse has been processed by the ADC, the analyzer control circuits seek out the memory location corresponding to the digitized amplitude stored in the address register, and the content of that location is incremented by one count. The net effect of this operation can be thought of as one in which the pulse to be analyzed passes through the ADC and is sorted into a memory location that corresponds most closely to its amplitude. This function is identical to that described earlier for the stacked single-channel analyzers illustrated in Fig. 18.1. Neglecting dead time, each input pulse increments an appropriate memory location by one count, and therefore the total accumulated number of counts over all memory is simply the total number of pulses presented to the analyzer during the measurement period. A plot of the content of each channel versus the channel number will be the same representation of the differential pulse height distribution of the input pulses as discussed earlier for the stacked single-channel analyzers.

A number of other functions are normally found in an MCA. As illustrated in Fig. 18.7, an *input gate* is usually provided to block pulses from reaching the ADC during the time it is "busy" digitizing a previous pulse. The ADC provides a logic signal level that holds the input gate open during the time it is not occupied. Because the ADC can be relatively slow, high counting rates will result in situations in which the input gate is closed for much of the time. Therefore, some fraction of the input pulses will be lost during this dead time, and any attempt to measure quantitatively the number of pulses presented to the analyzer must take into account those lost during the dead time.

To help remedy this problem, most MCAs provide an internal clock whose output pulses are routed through the same input gate and are stored in a special memory location. The clock output is a train of regular pulses synchronized with an internal crystal oscillator. If the fraction of time the analyzer is dead is not excessively high, then it can be argued that the fraction of clock pulses that is lost by being blocked by the input gate is the same as the fraction of signal pulses blocked by the same input gate. Therefore, the number of clock pulses accumulated is a measure of the *live time* of the analyzer or the time over which the input gate was held open. Absolute measurements can therefore be based on a fixed value of live time, which eliminates the need for an explicit dead time correction to the data. Further discussion of the dead time correction problem for MCAs is given later in this chapter.

Many MCAs are also provided with another linear gate that is controlled by a single-channel analyzer. The input pulses are presented in parallel to the SCA and, after passing through a small fixed delay, to the linear pulse input of this gate. If the input pulse meets the amplitude criteria set by the SCA, the gate is opened and the pulse is passed on to the remainder of the MCA circuitry. The purpose of this step is to allow rejection of input pulses that are either smaller or larger than the region of interest set by the SCA limits. These limits, often referred to as the LLD (lower-level discriminator) and ULD (upper-level discriminator), are chosen to exclude very small noise pulses at the lower end and very large pulses beyond the range of interest at the upper end. Thus, these uninteresting pulses never reach the ADC and consequently do not use valuable conversion time, which would otherwise increase the fractional dead time. If an MCA is operated at relatively high fractional dead time (say, greater than 30 or 40%), distortions in the spectrum can arise because of the greater probability of input pulses that arrive at the input gate just at the time it is either opening or closing. It is therefore often advisable to reduce the counting rate presented to the input gate as much as possible by excluding noise and other insignificant small-amplitude events with the LLD, and if significant numbers of large-amplitude background events are present, excluding them with an appropriate ULD setting.

The contents of the memory after a measurement can be displayed or recorded in a number of ways. Virtually all MCAs provide a CRT display of the content of each channel as the Y displacement versus the channel number as the X displacement. This display is therefore a graphical representation of the pulse height spectrum discussed earlier. The display can be either on a linear vertical scale or, more commonly, as a logarithmic scale to show detail over a wider range of channel content. Standard recording devices for digital data, including printers and storage media, are commonly available to store permanently the memory content and to provide hard copy output.

Because of the similarity of many of the MCA components just described to those of the standard personal computer (PC), there is a widespread availability of plug-in cards that will convert a PC into an MCA. The card must provide the components that are unique to the MCA (such as the ADC), but the normal PC memory, display, and I/O hardware can be used directly. Control of the MCA functions is then provided in the form of software that is loaded into the PC memory. Some compromises in performance of the plug-in boards are often necessary because the noisy electronic environment inside the PC produced by the many digital switching operations is somewhat hostile to the sensitive analog operations required in ADCs. Thus there are also units in which the ADC operations are housed within an external NIM module that communicates with the PC through an interface cable. In some cases, the MCA is incorporated in a computer-based spectroscopy system that allows software control of the MCA functions as well as other settings such as detector voltage supply and parameters of the shaping amplifier such as gain, shaping time, pile-up rejection, and spectrum stabilizer operation (see later section in this chapter).

B. The Analog-to-Digital Converter

1. GENERAL SPECIFICATIONS

The job to be performed by the ADC is to derive a digital number that is proportional to the amplitude of the pulse presented at its input. Its performance can be characterized by several parameters:

1. The speed with which the conversion is carried out.
2. The linearity of the conversion, or the faithfulness to which the digital output is proportional to the input amplitude.
3. The resolution of the conversion, or the "fineness" of the digital scale corresponding to the maximum range of amplitudes that can be converted.

The nominal value of the resolution depends on the number of bits provided by the ADC, and is specified as the maximum number of addressable channels. Thus a 12-bit ADC will provide 2^{12}, or 4096 channels of resolution. From the previous discussion of ADC properties in Chaper 17, the effective resolution may be less than this value if electronic noise or instability result in typical channel profiles that are overly broad. For the types of ADCs generally chosen for use in MCAs, the effective resolution should not deviate greatly from the nominal value.

The voltage that corresponds to full scale is arbitrary, but most ADCs designed for nuclear pulse spectroscopy will be compatible with the output of typical linear amplifiers. Zero to 10 V is thus a common input span. Shaping requirements will also usually be specified for the input pulses, and most ADCs require a minimum pulse width of a few tenths of a microsecond to function properly.

The *conversion gain* of an ADC specifies the number of channels over which the full amplitude range will be spread. For example, at a conversion gain of 2048 channels, a 0-V to 10-V ADC will store a 10 V pulse in channel 2048, whereas at a conversion gain of 512, that same pulse would be stored in channel 512. At the lower conversion gain, a smaller fraction of the MCA memory can be accessed at any one time. On many ADCs, the conversion gain can be varied for the purposes of a specific application. The resolution of an ADC must be at least as good as the largest conversion gain at which it will be used.

The conversion speed or dead time of the ADC is the critical factor in determining the overall dead time of the MCA. Therefore, a premium is placed on fast conversion, but practical limitations restrict the designer in speeding up the conversion before linearity begins to suffer. The fastest ADCs, the flash or subranging type discussed in Chapter 17, are rarely used in MCAs because of their poor differential linearity. Two other types dominate in contemporary MCAs: linear ramp converters and successive approximation ADCs. The first of these, although the slowest, generally has the best linearity and channel profile specifications compared with the other types and has gained the most widespread application in MCAs. Successive approximation ADCs offer faster conversion times, but generally with poorer linearity and channel uniformity.

2. THE LINEAR RAMP CONVERTER (WILKINSON TYPE)

The linear ramp converter is based on an original design by Wilkinson[3] and is illustrated in Fig. 18.8. The input pulse is supplied to a comparator circuit that continuously compares the amplitude with that of a linearly increasing ramp voltage. The ramp is conventionally generated by charging a capacitor with a constant-current source that is started at the time the input pulse is presented to the circuit. The comparator circuit provides as its output a gate pulse that begins at the same time the linear ramp is initiated. The gate pulse is maintained "on" until the comparator senses that the linear ramp has reached the amplitude of the input pulse. The gate pulse produced is therefore of variable length, which is directly proportional to the amplitude of the input pulse. This gate pulse is then used to operate a

Figure 18.8 Block diagram of a linear ramp (Wilkinson type) ADC.

linear gate that receives periodic pulses from a constant-frequency clock as its input. A discrete number of these periodic pulses pass through the gate during the period it is open and are counted by the *address register*. Because the gate is opened for a period of time proportional to the input pulse amplitude, the number of pulses accumulated in the address register is also proportional to the input amplitude. The desired conversion between the analog amplitude and a digital equivalent has therefore been carried out.

Because the clock operates at a constant frequency, the time required by a Wilkinson-type ADC to perform the conversion is directly proportional to the number of pulses accumulated in the address register. Therefore, under equivalent conditions, the conversion time for large pulses is always greater than that for small pulses. Also, the time required for a typical conversion will vary inversely with the frequency of the clock. In order to minimize the conversion time, there is a premium on designing circuits that will reliably handle clock pulses of as high a frequency as possible. Clock frequencies of 100 MHz are representative of present-day commercial designs.

The Wilkinson-type ADC leads to contiguous pulse height channels, all ideally of the same width. Because the linear ramp generation can be very precise, this design is characterized by good linearity specifications, accounting for its widespread popularity in MCAs. Variants of the Wilkinson design that lead to some improvement in speed are described by Nicholson,[4] and a clock frequency of up to 400 MHz can be achieved in advanced designs.

3. THE SUCCESSIVE APPROXIMATION ADC

The second type of ADC in common use is based on the principle of successive approximation. Its function can be illustrated by the series of logic operations shown in Fig. 18.9. In the first stage, a comparator is used to determine whether the input pulse amplitude lies in the upper or lower half of the full ADC range. If it lies in the lower half, a zero is entered in the first (most significant) bit of the binary word that represents the output of the ADC. If the amplitude lies in the upper half of the range, the circuit effectively subtracts a value equal to one-half the ADC range from the pulse amplitude, passes the remainder on to the second stage, and enters a one in the most significant bit. The second stage then makes a

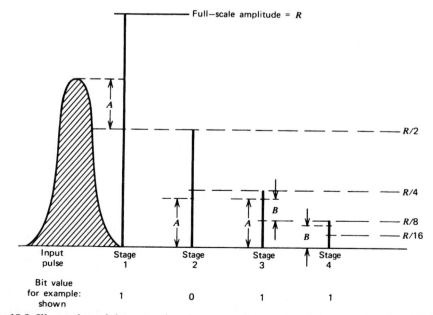

Figure 18.9 Illustration of the operational sequence for a successive approximation ADC. Four stages are shown which generate the four-bit word shown at the bottom as the digital output.

similar comparison, but only over half the range of the ADC. Again, a zero or one is entered in the next bit of the output word depending on the size of the remainder passed from the first stage. The remainder from the second stage is then passed to the third stage, and so on. If 10 such stages are provided, a 10-bit word will be produced that will cover a range of 2^{10} or 1024 channels.

In its most common circuit implementation, the successive approximation ADC makes multiple use of a single comparator that has two inputs: one is the sampled and held input voltage and the other is produced by a digital-to-analog converter (DAC). For the first-stage comparison, the input to the DAC is set to a digital value that is half the ADC output range. Depending on the result of the initial comparison, the second-stage comparison is then carried out with the digital input to the DAC set to either 25 or 75% of the range, and so on. In this way, analog subtractions are avoided but the functional operation is equivalent to that described above.

For a given successive approximation ADC, the conversion time is constant and independent of the size of the input pulse. For typical converters with a 10-bit output, the conversion time can be a few microseconds or less. Adding stages to increase the resolution only increases the conversion time in proportion to the number of total stages or in proportion to the logarithm of the maximum number of channels. Their speed advantage is therefore most pronounced when the number of addressed channels is large. The major disadvantage of typical successive approximation ADCs is a more pronounced differential nonlinearity compared with linear ramp ADCs. Some refinements to their design to help overcome this limitation are described in Refs. 5–7.

4. THE SLIDING SCALE PRINCIPLE

The linearity and channel width uniformity of any type of ADC can be improved by employing a technique generally called the *sliding scale* or *randomizing* method. Originally suggested in 1963 by Gatti and co-workers,[8] the method has gained popularity (e.g., Refs. 9 and 10) through its implementation using modern IC technology. It has been particularly helpful in improving the performance of both successive approximation and flash ADCs.

Without the technique, pulses of a given amplitude range are always converted to a fixed channel number. If that channel is unusually narrow or wide, then the differential linearity will suffer in proportion to the deviation from the average channel width. The sliding scale principle is illustrated in Fig. 18.10. It takes advantage of the averaging effect gained by spreading the same pulses over many channels. A randomly chosen analog voltage is added to the pulse amplitude before conversion and its digital equivalent subtracted after the conversion. The net digital output is therefore the same as if the voltage had not

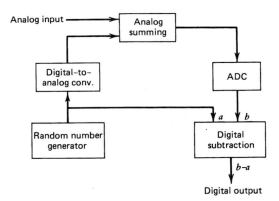

Figure 18.10 Functional diagram of the sliding scale principle for ADCs.

been added. However, the conversion has actually taken place at a random point along the conversion scale. If the added voltage covers a span of M channels, then the effective channel uniformity will improve as $\sqrt{M}$ if the channel width fluctuations are random. The implementation of Fig. 18.10 derives the added voltage by first generating a random digital number and converting this number to an analog voltage in a DAC. The same digital number is then subtracted after the conversion.

One of the disadvantages of the technique is that the original ADC scale of N channels is reduced to $N - M$. If a pulse occurs that would normally be stored in a channel number near the top of the range, the addition of the random voltage may send the sum off scale. Other designs[10,11] avoid this limitation by using either upward averaging (as described above) or downward averaging (by subtracting the random voltage) depending on whether the original pulse lies in the lower or upper half of the range. The choice of M can then be as large as $N/2$ to maximize the averaging effect without reducing the effective ADC scale.

Because the sliding scale method involves converting a fixed pulse amplitude through different channels whose width may vary, a potential disadvantage is a broadening of typical channel profiles. If this broadening is severe enough, it will compromise the resolution of the ADC. Another potential problem is that, if the addition and subtraction steps are not perfectly matched in scale factor, periodic structures can be generated in the differential nonlinearity that appear as artifacts in recorded spectra.

C. The Memory

The memory section of an MCA provides one addressable location for every channel. Any of the standard types of digital memory can be used, but there is sometimes a preference for "nonvolatile" memory, which does not require the continual application of electrical power to maintain its content. Then, data acquired over long measurement periods will not be lost if the power to the MCA is accidentally interrupted.

Most MCAs make provisions for subdividing the memory into smaller units for independent acquisition and storage of multiple spectra. In this way a 4096 channel analyzer can be configured as eight separate 512 channel memory areas for storing low-resolution spectra, or as a single 4096 channel memory for a high-resolution spectrum.

In most analyzers, provision is made for the negative incrementing of memory content as well as additive incrementing. In this "subtract" mode, background can conveniently be taken away from a previously recorded spectrum by analyzing for an equal live time with the source removed.

D. Ancillary Functions

1. MEASUREMENT PERIOD TIMING

Virtually all MCAs are provided with logic circuitry to terminate the analysis period after a predetermined number of clock pulses have been accumulated. One often has the choice between preset *live time* or *clock time*, which are distinguished by whether the clock pulses are routed through the input gate (see Fig. 18.7). Normally, quantitative comparisons or subtraction of background are done for equal live time periods, and this is the usual way of terminating the analysis period.

2. MULTISCALING

Multichannel analyzers can be operated in a mode quite different from pulse height analysis, in which each memory location is treated as an independent counter. In this *multiscaling* mode, all pulses that enter the analyzer are counted, regardless of amplitude. Those that arrive at the start of the analysis period are stored in the first channel. After a period of time known as the *dwell time*, the analyzer skips to the second channel and again records pulses of all amplitudes at that memory location. Each channel is sequentially allocated one such dwell time for accumulating counts, until the entire memory has been addressed. The dwell time can be set by the user, often from a range as broad as from 1 μs to several minutes. The net effect of this mode is to provide a number of independent counters equal to the number of channels in the analyzer, each of which records the total number of counts over a sequential interval of time. This mode of operation can be very useful in studying the behavior of rapidly decaying radioactive sources or in recording other time-dependent phenomena.

3. COMPUTER INTERFACING

Stand-alone MCAs share many features with general purpose computers. In its most basic form, the MCA can only increment and display the memory, but more elaborate operations can be carried out if it is provided with some of the features of a small computer. For example, one of the most useful functions is to allow summation of selected portions of the spectrum, generally called *regions of interest* (ROIs). Cursors are generated whose position on the displayed spectrum can be used to define the upper and lower bounds of the channel numbers between which the summation is carried out. This operation has obvious practical use for simple peak area determination in radiation spectroscopy. Other operations, such as addition or subtraction of two spectra or other manipulations of the data can also be provided.

More complex computer-based systems are also widely available that are based on PCs with an appropriate ADC under software control. In this approach, the functions mentioned above can be duplicated through software routines that may be modified or supplemented by the user. Useful operations of this type can range from simple smoothing of the spectra to damp out statistical fluctuations, to elaborate spectrum analysis programs in which the position and area of apparent peaks in the spectrum are identified and measured. Current manufacturer's specification sheets are often the best source of detailed information in this rapidly evolving area.

4. MULTIPARAMETER ANALYSIS

The simplest application of multichannel analysis is to determine the pulse height spectrum of a given source. This process can be thought of as recording the distribution of events over a single dimension–pulse amplitude. In many types of radiation measurements, additional experimental parameters for each event are of interest, and it is sometimes desirable to record simultaneously the distribution over two or more dimensions.

One example is in the case in which not only the amplitude of the pulse carries information, but also its rise time or shape. Categories of events can often be identified based on unique combinations of amplitude and shape, while a clean separation might not be possible using either parameter alone. In the example shown in Fig. 18.11, both the amplitude and shape (measured from the rise time) of each pulse from a liquid scintillation counter are derived in separate parallel branches of the pulse processing system. The object will now be to store this event according to the measured values of both these parameters.

In any unit designed for multiparameter analysis, at least two separate inputs with dedicated ADCs must be provided, together with an associated coincidence circuit. The memory now consists of a two-dimensional array in which one axis corresponds to pulse amplitude and the other to pulse shape. Because both parameters are derived from the same event, they appear at the two inputs in time coincidence. The multiparameter analyzer recognizes the coincidence between the inputs and increments the memory location corresponding to the intersection of the corresponding pair of digitized addresses. As data accumulate, the intensity distribution then takes the form of a two-dimensional surface with local peaks representing combinations of amplitude and shape that occur most frequently. The data are

Figure 18.11 Example of a two-parameter analysis of the pulses from a photomultiplier tube viewing a liquid scintillator vial. Events build up into a surface represented by the contour lines on the display of the two-dimensional memory.

sometimes displayed as a surface contour plot, or as an isometric view of the surface from a perspective that can be changed using software routines for best viewing.

Multiparameter analyzers generally require a much larger memory than single-parameter analyzers because, for equal resolution, memory requirements for two parameters are the square of the number of channels required for only one parameter. Often, however, one of the two parameters need not be recorded with the same degree of resolution so that nonsquare (rectangular) memory configurations will suffice. Three or more parameters require multidimensional memory allocations that can be impractical in size. In experiments in which many parameters are of interest for each event (for example, from multiple detectors recording coincident events), it may no longer be possible to dedicate a large enough memory to sort all the information in real time. An alternative mode of data recording called *list mode acquisition* can then be employed. Each of the multiple parameters is digitized with a separate ADC in real time, and the results are then quickly written to memory. A clock may also be read to provide a "time stamp" to record the time of occurrence of the event. As the measurement proceeds, the collection of data grows with each observed event but requires a memory whose size is only the product of the total number of events multiplied by the number of recorded parameters. After the conclusion of the measurement, the data can then be sorted off-line based on arbitrary criteria involving any or all of the recorded parameters.

E. MCA Dead Time

The dead time of an MCA is usually comprised of two components: the processing time of the ADC and the memory storage time. The first of these was discussed earlier and, for a Wilkinson-type ADC, is a variable time that is proportional to the channel number in which the pulse is stored. The processing time per channel is simply the period of the clock oscillator. For a typical clock frequency of 100 MHz, this time is 10 ns per channel. Once the pulse has been digitized, an additional fixed time of a few microseconds is generally required to store the pulse in the proper location in the memory. Thus, the dead time of an MCA using an ADC of this type can then be written

$$\tau = \frac{N}{v} + B \qquad (18.3)$$

where v is the frequency of the clock oscillator, N is the channel number in which the pulse is stored, and B is the pulse storage time. The analyzer control circuits will hold the input gate closed for a period of time that equals this dead time. A *dead time meter* is often driven by the input gate to indicate the fraction of time the gate is closed, as a guide to the experimenter. One normally tries to arrange experimental conditions so that the fractional dead time in any measurement does not exceed 30 or 40% to prevent possible spectrum distortions.

The automatic live time operation of an MCA described earlier is usually quite satisfactory for making routine dead time corrections. Circumstances can arise, however, in which the built-in live time correction is not accurate. When the fractional dead time is high, errors can enter because the clock pulses are not generally of the same shape and duration as signal pulses. One remedy[12,13] is to use the pulser technique described on p. 642 to produce an artificial peak in the recorded spectrum. If introduced at the preamplifier, the artificial pulses undergo the same amplification and shaping stages as the signal pulses. The fraction that are recorded then can account for both the losses due to pileup and the analyzer dead time. Several authors[14,15] have reviewed the live time correction problem and suggested conditions under which the pulser method is not accurate. To avoid potential problems, the pulse repetition rate must not be too high, and the use of a random rather than periodic pulser is preferred. Under these conditions, the pulser method can successfully handle virtually any conceivable case in which the shape of the spectrum does not change during the course of the measurement.

Additional complications arise if spectrum shape changes occur during the measurement, which lead to distortions and improper dead time corrections with the pulser method. A better method first suggested by Harms[16] can accommodate spectrum changes but requires a nonstandard mode of MCA operation. In this method, the analysis is run for a fixed *clock time*. If a pulse is lost because the analyzer is dead (this can be sensed externally), compensation is made immediately by assigning a double weight to the next pulse and incrementing the corresponding memory location by two. Spectra that change during the course of the measurement are properly accommodated because the correction automatically takes into account the amplitude distribution of signal pulses at the time of the loss. This correction scheme has been tested by Monte Carlo simulation for a wide variety of spectrum shapes and time variations and has proved to be quite accurate in all cases.[17] At higher rates, however, the assumption breaks down that only one pulse was lost during the dead time, and the Harms method begins to undercorrect for losses. One remedy is to modify the correction process by first calculating the expected number of counts lost during a dead period from a running measurement of the input pulse rate. The memory location corresponding to the next converted pulse is then incremented, not by two counts as above, but by one plus the calculated number of lost pulses. These artificial counts do not have the same statistical significance as the same number recorded normally, but they do maintain the total spectrum content as if there had been no losses. Normally called *loss free counting*,[18] this technique is applicable in situations in which the dead fraction of the MCA is as high as 80%.[19] It is particularly helpful in measurements from short-lived radioisotopes from which the counting rate may change by many orders of magnitude over the measurement period.

IV. SPECTRUM STABILIZATION AND RELOCATION

A. Active Spectrum Stabilization

Any drifts that occur along the signal chain during the course of the measurement will cause peaks in the recorded spectrum to broaden or become otherwise distorted. These drifts can come about through changes in temperature of the detector or associated electronics, gradual changes in voltage levels, or variations of the gain of the various active elements in the signal chain. Despite the best efforts to control temperature and other environmental conditions, spectra taken over long periods of time with high-resolution detectors often suffer an apparent loss of resolution due to these drifts. In some detectors (notably scintillation counters), large changes in counting rate can also lead to apparent gain changes over fairly short periods of time.

A *spectrum stabilizer* is a device that in some way senses the position of a peak in the measured spectrum during the course of the measurement, compares its position with a known reference, and generates an error signal that can be used in a feedback loop to change the gain of the system to restore the peak to its correct position. The element in the signal chain to which the feedback is applied can be any component whose gain can be conveniently changed, including the photomultiplier tube for a scintillation counter, or the linear amplifier in other spectroscopic systems.

The most common method of sensing the peak position during the measurement is illustrated in Fig. 18.12. Two SCA windows are set on opposite sides of a clearly resolved symmetrical peak and connected to circuitry that can measure the difference in count rate between the two windows. At the start of the measurement, the gains are adjusted so that the peak is precisely centered between the two windows, and the count rate difference is therefore zero. If the gain does drift, a difference in count rate between the two windows will be detected and will serve as the error signal. In this scheme, the window positions must be extremely stable because any changes in the SCA window positions will be interpreted as system gain changes. Ultimate stability is therefore limited by questions of

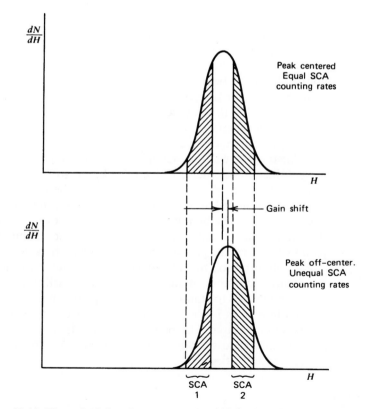

Figure 18.12 The principle of a spectrum stabilizer based on peak sensing. An error signal is derived from the counting rate difference in the two SCA windows.

SCA window stability. Methods have been developed[20,21] to generate extremely stable SCA windows for this application.

For systems based on MCAs, the function of the windows can be performed by preselected groups of channels. In that way, the peak is held constant in the recorded spectrum and questions of window stability are no longer applicable. Digital stabilizers are available as commercial units that communicate with the MCA and adjust the gain of a preceding amplifier according to the error signals sensed by the stabilizer.

The gain correction can be implemented in a number of different ways. In one extreme, a small gain adjustment is made following almost every pulse that is recorded near the vicinity of the peak centroid. The edges of the two windows are brought together so that, within the region of interest, only a pulse that is stored in the preset peak centroid channel does not trigger an adjustment. If a pulse falls in the lower window, the gain is increased; and if the pulse falls in the upper window, the gain is decreased. Because of the normal statistical fluctuation in pulse amplitudes, this process results in slight "wandering" of the gain even in the absence of any real gain shifts. If the gain steps used in the adjustments are small enough, the broadening effect of this gain wandering can generally be kept to negligible levels.[22]

An alternative approach is to collect a control pulse height spectrum (perhaps restricted to the channels near the peak of interest) in parallel with the primary spectrum in a separate section of memory. At periodic intervals, a software fit to the control peak is made and the current position of the centroid is derived. If this value is significantly different from the reference value set at the start of the measurement, then an appropriate gain adjustment is implemented. The control memory is then cleared and the process repeated at intervals throughout the measurement.

In common gain stabilizers, two peaks in the spectrum are usually monitored—one near the top and the other near the bottom of the spectrum range. In that way, corrections to drifts in both the slope and zero offset of the MCA calibration may be made if the amplifier allows adjustment of both its gain and zero offset.

If the spectrum to be recorded does not have readily identifiable peaks, other approaches must be adopted. A common remedy is to place a radioisotope "check source" near the detector that will produce its own peak or peaks in the spectrum. One possible drawback is that this source may also add undesirable background to other regions of the recorded spectrum. An alternative procedure is to place the lower window in a flat region of the spectrum and the upper window in a region where the spectrum drops off rapidly.[23] By adjusting the window width, a net zero counting rate difference can be set at the beginning of the measurement. Software techniques can also be applied[24] to sense changes in the shape of the recorded spectrum to serve as the basis for gain adjustments. In scintillation counters, methods based on the apparent position of a reference light pulser peak are also in common use. The light pulser must be incorporated into the scintillation assembly, where it produces pulses that do not interfere with the region of the spectrum in which the signal peaks of interest lie. This method of stabilization is described in more detail in Chapter 10.

Another option to stabilize spectra without peaks involves artificially creating a peak by mixing the output of an electronic pulser with the signal pulses. The standard methods can then be used to stabilize the pulser peak. The pulses should be introduced at the earliest possible point in the signal chain because any drifts that occur prior to that point will not be compensated. The pulser technique is similar to that described earlier (see p. 642) for determining losses caused by pile-up and dead time, and the same peak can also be used for gain stabilization.

B. Spectrum Alignment

An alternative to the active gain stabilization methods described above is to simply subdivide the measurement period into a number of small time intervals and record a separate pulse height spectrum for each. If the individual time intervals are small enough, the gain drifts can be negligible within any single interval. However, if some drift does occur over the course of the full measurement, then the individual recorded spectra will not be aligned exactly. The degree of misalignment can be deduced from the location of peaks in the spectra, preferably using several that are widely spaced along the pulse height axis.

For best statistical accuracy, it is generally desirable to combine all the data into a single spectrum, but then some method of alignment must be implemented. This process is not as simple as it might first appear, since each spectrum will, in general, have different gain and/or zero offset. For alignment of multiple spectra, the raw data must therefore be "rebinned" or sorted into a new set of channels of a "master" spectrum, with each new channel having different width and edge locations with respect to the original spectra. The situation is illustrated in Fig. 18.13. The question then arises of how the number of counts in an original channel should be redistributed among the new channels that overlap its position.

Any measured pulse height spectrum is a discrete sampling from the continuous differential distribution dN/dH versus H that characterizes the source of the pulses. The number of counts in any given channel is just proportional to the area under the continuous distribution between the channel edges. To rebin the original data, it would be best to resample the original continuous distribution into the new channels. Because of the finite width of each channel, some information is lost in the original sampling process, and a perfect knowledge of the continuous distribution is not available. However, if the channel widths are small enough so that the continuous distribution changes only slowly over the width of a channel, then an analytical curve that is fitted to the points representing the counts in each channel (plotted at the channel midpoint) will be an acceptable approximation to the original continuous distribution. The rebinning process can then proceed as illustrated in Fig. 18.13, in the following steps:

1. An analytical fit is made to the data in the vicinity of the relocated channels to approximate the original continuous distribution. The functional form of the fit can be a polynomial of degree M that is fitted to the $M + 1$ channels that are centered about the channel of interest. Some factors of importance in choosing the value of M are discussed in Ref. 25.

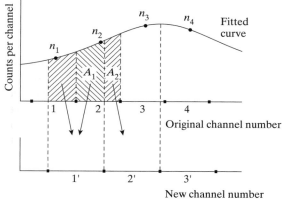

Figure 18.13 Schematic representation of the "rebinning" process to relocate multichannel spectra. The original spectrum is shown at the top as a series of points plotted at the midpoints of the channels 1, 2, 3, ... A continuous curve has been fit to these points. At the bottom is the new channel scale showing channels 1', 2', 3', ... In the deterministic process mentioned in the text, the original number of counts n_2 in channel 2 are redistributed to new channels 1' and 2' with fractional assignments of $A_1/(A_1 + A_2)$ and $A_2/(A_1 + A_2)$, respectively.

2. The area under this fitted curve is determined between the edges of the nearest relocated channel, and the fractions of this area that overlap the original channel edges are determined.

3. These fractions are then used as weighting factors in reassigning the counts in an original channel to one, two, or more new channels, depending on the degree of overlap. The fraction of counts assigned to any new channel is proportional to the area under the curve calculated for the range of overlap (see Fig. 18.13).

If this reassignment of counts is carried out using the deterministic process described above, all counts in the original spectrum are reassigned to the new spectrum and the process conserves areas or absolute numbers of counts. Since the counts in any one original channel are distributed over multiple new channels, however, the statistical behavior of the relocated data is changed. Even if the counts in an original channel follow Poisson statistics (as is the usual case), the relocated counts will generally display a smaller expected variance. Statistical tests applied to the rebinned or combined spectra will therefore show results that depart from Poisson behavior. An alternative stochastic relocation process[25] can be applied that preserves the Poisson statistics of the relocated data. The number of counts relocated to a specific new channel is obtained by a random sampling from a binomial distribution whose average is the deterministic value. The channel-to-channel statistical fluctuations of the relocated spectrum will then mimic those of the original,[26] and standard goodness-of-fit tests that are usually part of spectrum analysis software can be applied to the relocated data.

V. SPECTRUM ANALYSIS

A. Deconvolution or Unfolding

The differential pulse height spectrum dN/dH that is recorded from any radiation detector is the convolution of its inherent response function and the energy distribution of the incident radiation. We represent the incident energy distribution as $S(E)$, where $S(E)dE$ is the differential number of incident quanta with energy within dE about E. Information about this incident spectrum $S(E)$ is often the object of our measurement. The measured pulse height spectrum results from the convolution:

$$\frac{dN}{dH} = \int R(H, E)\, S(E)\, dE \qquad (18.4)$$

Here, $R(H, E)\, dH\, dE$ is the differential probability that a quantum of energy within dE about E leads to a pulse with amplitude within dH about H. In the special case that the incident radiation consists of only a single energy E_0, the spectrum $S(E)$ can be represented by the product of the number of quanta S_0 and the Dirac delta function:

$$S(E)\, dE = S_0\, \delta(E - E_0)\, dE$$

Putting this representation into the equation above, we observe

$$\left.\frac{dN}{dH}\right|_{E=E_0} = S_0 R(H, E_0)$$

Under these conditions, the measured pulse height distribution dN/dH is given by the *response function* $R(H, E_0)$ multiplied by the number of recorded pulses S_0. The response function, introduced earlier in Chapter 4, is characterized by noting that $R(H, E_0)\, dH$ is the differential probability of recording a pulse of amplitude within dH about H given an incident quantum of energy E_0. The shape of the response function will generally be different for incident quanta of different energy E_0. It may also depend on a number of other possible variables, including the operating conditions of the detector, the source–detector geometry, and the counting rate. In most cases, it is assumed that all these variables can be controlled sufficiently so that the only significant parameter is the radiation energy. When the spectrum is recorded by an MCA, Eq. (18.4) takes the discrete form

$$N_i = \sum_j R_{ij} S_j \tag{18.5}$$

where N_i is the recorded count in the ith channel, R_{ij} is the response matrix coupling the ith pulse height interval with the jth energy interval, and S_j is the radiation intensity in the jth energy interval.

The objective of many radiation measurements is to deduce the energy distribution of the incident radiation, or all values of the source elements S_j. We assume that the source distribution is discretized into L such intervals. The recorded spectrum is made up of M values of N_i, one for each channel. We can therefore write M simultaneous equations of the form of Eq. (18.5). If we assume that the entire response matrix R_{ij} is known, it is then theoretically possible to solve these equations for all the S_j elements, provided $M \geq L$. This process is generally called *spectrum deconvolution* or *spectrum unfolding*.

If the response function $R(H, E_0)$ is a narrow peak or mathematical delta function at each energy (the response matrix R_{ij} is then diagonal), a one-to-one correspondence exists between the measured pulse height and radiation energy. Such is very nearly the case, for example, for semiconductor diode detectors used to detect heavy charged particles (see Chapter 11). Then, the measured pulse height spectrum can be interpreted directly in terms of the incident radiation spectrum, and no unfolding is required. More often, however, the response function is more complex and may include both secondary peaks and a general continuum. The measured pulse height spectrum is then much more complicated, and unfolding may be necessary for a complete evaluation of the incident energy spectrum.

Two general problems arise in the unfolding process. The response functions may be subject to some uncertainty in that they may not be measured experimentally for all energies of interest. Furthermore, the detector operating conditions may change between the calibration and measurement steps, leading to corresponding changes in the response function. A second difficulty is due to the statistical nature of the recorded data. The content of each channel is subject to a statistical variance that can be estimated from the discussions given in Chapter 3. These statistical uncertainties will propagate through the unfolding calculation and give rise to corresponding variances in the calculated energy spectrum. Unless a large number of counts are collected over the measured spectrum, these statistical uncertainties can result in unacceptably large fluctuations in the derived energy spectrum.

Because of these imperfections, an exact set of solutions S_j cannot in general be obtained from the simultaneous equations. Instead, approximate solutions are sought, which are in some sense a best estimate of the incident energy spectrum. Many computer programs developed for this purpose are based on seeking a minimum in the weighted sum of residuals

$$\epsilon^2 = \sum_i W_i \left(N_i - \sum_j R_{ij} S_j \right)^2 \tag{18.6}$$

where the weighting factors W_i are often chosen to be inversely proportional to the statistical variance of each data point. The process of minimizing this sum of residuals amounts to a least-square fitting to the measured data. To reduce the effects of statistical uncertainties, many unfolding codes include some form of *data smoothing*, in which the content of each channel is replaced by a weighted average over a number of adjacent channels. To avoid loss of energy resolution, the interval over which the smoothing is carried out must always be less than the corresponding interval over which the response function changes rapidly.

The form of Eq. (18.5) also assumes simple linear superposition of all individual response functions. In most detectors, this assumption will be valid in the absence of complicating effects caused by pulse pile-up. Because these effects are accentuated at high count rates, unfolding methods that perform adequately at low counting rates may become unsatisfactory for spectra recorded at higher rates.

Although unfolding methods can generally be applied to the output of all radiation detectors, there are two main application areas where the techniques have been most thoroughly developed. One of these deals with the spectra recorded by proton recoil detectors used with fast neutrons. Here, the response functions consist only of continua that, to first approximation, are almost rectangular in shape (see Fig. 15.14). Techniques referenced in Chapter 15 for unfolding of these spectra are variants of the general deconvolution approach described above. The other major application concerns spectra recorded by gamma-ray spectrometers employing either scintillators or germanium detectors. Because of the presence of pronounced peaks in the response functions, the methods employed in this area tend to be somewhat different and are described later in this section.

The difficulty of the tasks involved in spectrum unfolding also depends on the nature of the incident radiation energy spectrum. If the source emits only monoenergetic radiations of a few separate energies, S_j in Eq. (18.5) has only a few nonzero terms. The recorded spectrum is therefore the simple superposition of several response functions that can, in principle, be measured in advance. On the other hand, if the incident radiation spectrum covers a continuum of energies, there will be a large number of S_j terms, and it is less practical to provide a set of experimentally measured response functions. In this event, calculations or analytic models are often used to provide the needed response matrix or to interpolate between measured functions.

B. Spectrum Stripping

If the incident spectrum consists only of a few discrete energies, a method known as *spectrum stripping* can sometimes be applied. The response functions $R(H, E_0)$ for each of the anticipated energies are stored in the memory of an MCA or associated computer. These response functions may be generated either by direct measurements using monoenergetic calibration sources or by calculation. The recorded spectrum is then "unpeeled" by starting with the largest anticipated energy. The corresponding response function is multiplied by a variable factor and subtracted from the recorded spectrum until the upper end of the spectrum is reduced to zero. The next lower energy is then chosen and the subtraction process repeated. Through this procedure, each succeedingly lower energy can be "stripped" until the entire recorded spectrum is reduced to zero. The factor by which each response function was multiplied then gives the relative intensity of the corresponding component. If the number of possible energies is sufficiently small, the spectrum stripping process can be carried out manually, with a simple visual inspection determining the point at which each subtraction process is adequate.

C. Analysis of Spectra with Peaks

Although in principle it is possible to use general unfolding procedures for gamma-ray spectra, a full deconvolution of the spectrum is seldom carried out. One reason is that the response functions for scintillators or germanium detectors are quite complicated, and uncertainties quickly propagate over the entire spectrum. A more important reason, however, is that a full deconvolution is usually unnecessary. Gamma-ray spectra almost always consist of a number of discrete gamma-ray energies, and the response functions show a predominant peak. One would therefore expect a corresponding peak in the pulse height spectrum for every discrete energy in the incident radiation that can be resolved, and the spectrum analysis can consist of simply locating and quantifying these peaks. (If continuous gamma-ray spectra are involved, individual peaks will not appear in the recorded spectrum and there is no recourse but to carry out a full deconvolution.)

Computer programs are continually being developed and improved for the automatic localization and quantification of peaks in multichannel spectra. Many are available as commercial packages supplied through the vendors of multichannel analysis equipment. Others are available as public domain software. Reviews of some of these programs as applied to germanium detector spectra[†] can be found in Refs. 27–29, and an intercomparison of their performance with test spectra is published in Ref. 30. Two representative examples (SAMPO and GAUSS) are described in Refs. 31 and 32. The programs serve two main purposes: (1) to locate the centroid position of all statistically significant peaks appearing in the spectrum and (2) to determine the net area under each peak that lies above the continuum or background on which the peak is superimposed.

1. PEAK LOCALIZATION

The task of picking out statistically significant peaks that rise above a smoothly varying continuum appears deceptively simple. When implemented in software, the search process must be relatively sophisticated to avoid the sensing of false peaks that results from statistical fluctuations in the background, and at the same time remain highly efficient for finding true peaks of low intensity. The problem is further complicated by the fact that two closely lying peaks may not be fully separated in the spectrum, and the program must be able to distinguish such partially resolved doublets from simple single peaks.

One feature of a smooth symmetric peak is that its second derivative shows a large negative excursion at the center of the peak position. Therefore, many peak search routines are based on sensing the second differences in channel-to-channel data that have been smoothed to minimize the effects of statistical fluctuations. Once a tentative peak has been identified by this procedure, additional criteria are normally tested to check its acceptability. The data in the local area may be fitted with an analytic function (usually a Gaussian or a variant thereof), and the width may be required to match that of an appropriate calibration peak. Peaks with unusual width are normally investigated as candidates for possible unresolved doublets by methods that often involve an iterative procedure, in which varying doublet spacing and relative intensities are tried. For each detected peak, the parameters of the best-fit analytic function are then used to define the position of its centroid. The effectiveness with which a typical peak search routine can locate many peaks in a single spectrum is shown by Fig. 18.14.

2. DETERMINATION OF PEAK AREA

Once each peak has been located, the next step normally involves a sequential examination of the data in the immediate vicinity of individual peaks to determine their area. Various automated techniques for determining peak areas in multichannel spectra have been compared in Refs. 34–37. The methods can be subdivided into two groups: those that obtain the area from a fitted analytic function and those that carry out a direct summation of the data points between prescribed limits. In either approach, the contribution of the continuum on which the peak is superimposed must be subtracted. The data in the channels on either side of each peak are normally used to define the continuum, and an assumed linear or quadratic curve is fit to produce an estimated continuum in the region under the peak. Channel-by-channel subtraction of the continuum then produces corrected data for the following steps.

In the direct summation method, some criteria must be preselected for the number of channels over which the data will be summed. In principle, the summation should extend over all channels that are significantly above the continuum. In practice, it is normally

---

[†]Although many of these same techniques can be applied to analyze spectra from scintillators, we limit this specific discussion to procedures that are most useful for spectra in which the peaks are relatively narrow as a result of the good energy resolution of the detector.

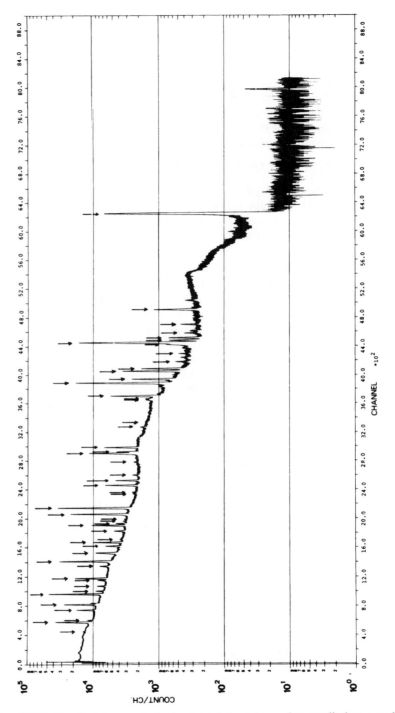

Figure 18.14 Results of a computerized peak search routine applied to a pulse height spectrum from a germanium gamma-ray spectrometer. Arrows indicate the locations of peaks found by the program. (From Sasamoto et al.[33])

limited to a given number of channels on either side of the peak centroid, depending on the detector resolution. A number of alternative schemes are discussed in Ref. 35.

A more common procedure is to use the area obtained from an analytic fit to the data points comprising the peak. Most assumed shapes involve a primary Gaussian peak with a small additive component (often exponential) to represent tailing on the low-energy side of the peak, caused by incomplete charge collection within the detector. Least-square methods are normally used to fit the function, and the area is then derived from the fitted shape parameters. Examples of a number of complex fitting functions applied to germanium detector photopeaks are given in Refs. 38–43.

The fitting process produces numerical values for the three most important parameters: the centroid of the peak, its width, and its area. Each of these parameters has an uncertainty that arises from the statistical fluctuations in the channel-to-channel data. For a pure Gaussian peak with no underlying background, it is not surprising that the expected standard deviation in the value A for the area under the peak, or the total number of counts, is just $\sqrt{A}$, and the fractional standard deviation is $1/\sqrt{A}$. This result is simply a reaffirmation of the most basic prediction of counting statistics discussed in Chapter 3. Statistical uncertainties also enter the fitted values for the centroid and width of the peak. It can be shown[1] that the expected variance (square of the standard deviation) of the fitted centroid value is simply given by σ^2/A, where σ is the standard deviation characterizing the width of the underlying Gaussian. A similar analysis shows that the variance in the fitted value (call it W) for the Gaussian FWHM is given by $W^2/2A$. Thus the expected fractional uncertainty due to counting statistics in all three parameters shows the same proportionality to $1/\sqrt{A}$, where A for a steady-state source is simply proportional to the total counting time.

PROBLEMS

18.1 How many pulse height channels should be provided to represent adequately the full spectrum from a system with 0.3% pulse height resolution?

18.2 Two peaks in a recorded pulse height spectrum are separated by 24 channels. Assuming a perfectly linear system, by how many channels will this separation change if the gain of the amplifier supplying the MCA is decreased from 1000 to 750, and the zero offset of the MCA is increased from 10 to 15 channels?

18.3 A Wilkinson-type ADC has a conversion gain of 2048 channels and a maximum conversion time of 25 μs. At what frequency must the oscillator operate?

18.4 How many stages are required in a successive approximation ADC to achieve a conversion gain of 4096 channels?

18.5 Why is it not possible to apply the formulas for dead time corrections developed in Chapter 4 to correct for losses in an MCA using a linear ramp ADC?

18.6 An MCA using a Wilkinson-type ADC operating at 80 MHz has a pulse storage time of 2.5 μs.

(a) What is the analyzer dead time for pulses stored in channel number 300?

(b) What will be the fractional dead time for a true pulse rate of 5000/s if the average pulse amplitude falls in channel number 220? Repeat for a true rate of 50,000/s.

(c) If the analyzer is set to record for a live time of 10 min, how much actual time will elapse under the conditions of part (b)?

18.7 Equally spaced pulses of fixed amplitude are generated at an adjustable frequency in an electronic pulser. They are supplied to an MCA with a dead time of 90 μs for the size of pulse involved. Sketch a plot of percent dead time losses versus pulser frequency over the range from 10 to 30 kHz.

18.8 The coincident outputs from two detectors are to be recorded in a two-dimensional multiparameter analyzer. If the detectors have pulse height resolutions of 0.5 and 2.5%, estimate the total number of memory locations needed to record faithfully the coincident output of both detectors over their full span.

18.9 A pulse height spectrum is recorded in an MCA by counting a source for a given live time, and then removing the source and subtracting background for an equal live time. In the upper region of the spectrum, the source contributes nothing so the average channel content after subtraction should be zero. Due to the influence of counting statistics, however, some channel-to-channel fluctuation about zero is observed. If the background is about 300 counts/channel over this region of the spectrum, what deviation from zero should be expected in a typical channel?

18.10 One commonly used method to sense the position of a peak superimposed on a continuous background is to take second differences in the multichannel data. To illustrate the behavior in the vicinity of a peak, make a plot of the second derivative of a Gaussian peak across its full width.

18.11 The following is a portion of a gamma-ray pulse height spectrum recorded using a germanium detector and MCA:

Channel Number	Counts	Channel Number	Counts
711	238	720	1625
712	241	721	1739
713	219	722	1412
714	227	723	901
715	242	724	497
716	280	725	308
717	409	726	256
718	736	727	219
719	1190	728	230

It can be assumed that the data consist of a constant background plus a Gaussian-shaped peak.

(a) Plot the data and estimate the constant background level. Find the net number of counts under the peak by direct summation. Estimate the centroid location of the peak (to the nearest tenth of a channel). Estimate the FWHM of the peak.

(b) Fit the net counts in the region of the peak with a Gaussian function using an available software fitting routine. Obtain the resulting peak area, centroid, and standard deviation, and compare with the results of part (a).

REFERENCES

1. J. D. Valentine and A. E. Rana, *IEEE Trans. Nucl. Sci.* **43**(5), 2501 (1996).
2. H. Spieler, "Pulse Processing and Analysis," lecture notes from IEEE/NPSS Short Course on Radiation Detection and Measurement, Albuquerque (1997).
3. D. H. Wilkinson, *Proc. Cambridge Philos. Soc.* **46**, Pt. 3, 508 (1950).
4. P. W. Nicholson, *Nuclear Electronics*, Wiley-Interscience, London, 1974.
5. M. Brendle, *Nucl. Instrum. Meth.* **144**, 357 (1977).
6. S. G. Gobbur, D. A. Landis, and F. S. Goulding, *Nucl. Instrum. Meth.* **140**, 405 (1977).
7. D. E. Carter and G. Randers-Pehrson, *Nucl. Instrum. Meth.* **199**(3), 497 (1982).
8. C. Cottini, E. Gatti, and V. Svelto, *Nucl. Instrum. Meth.* **24**, 241 (1963).
9. C. B. A. Correia and C. A. N. Conde, *Nucl. Instrum. Meth.* **A235**, 536 (1985).
10. X. Xianjie and P. Dajing, *Nucl. Instrum. Meth.* **A259**, 521 (1987).
11. A. Nour, *Nucl. Instrum. Meth.* **A363**, 577 (1995).
12. O. U. Anders, *Nucl. Instrum. Meth.* **68**, 205 (1969).
13. H. H. Bolotin, M. G. Strauss, and D. A. McClure, *Nucl. Instrum. Meth.* **83**, 1 (1970).
14. M. Wiernik, *Nucl. Instrum. Meth.* **96**, 325 (1971).
15. E. J. Cohen, *Nucl. Instrum. Meth.* **121**, 25 (1974).
16. J. Harms, *Nucl. Instrum. Meth.* **53**, 192 (1967).
17. C. F. Masters and L. V. East, *IEEE Trans. Nucl. Sci.* **NS-17**(3) 383 (1970).
18. G. P. Westphal, *Nucl. Instrum. Meth.* **146**, 605 (1977).
19. G. P. Westphal, *Nucl. Instrum. Meth.* **163**, 189 (1979).
20. M. Yamashita, *Nucl. Instrum. Meth.* **114**, 75 (1974).
21. P. J. Borg, P. Huppert, P. L. Phillips, and P. J. Waddington, *Nucl. Instrum. Meth.* **A238**, 104 (1985).
22. D. O'Connor, G. F. Knoll, and R. Ziff, *Nucl. Instrum. Meth.* **A353**, 291 (1994).
23. M. Matoba and M. Sonoda, *Nucl. Instrum. Meth.* **92**, 153 (1971).
24. J. Lu and C. C. Watson, *IEEE Trans. Nucl. Sci.* **39**(5), 1407 (1992).
25. S. P. Tsai, E. R. Mucciolo, and O. Helene, *Nucl. Instrum. Meth.* **A345**, 538 (1994).
26. O. Helene and M. Moralles, *Nucl. Instrum. Meth.* **A378**, 624 (1996).
27. C. M. Lederer, in *Radioactivity in Nuclear Spectroscopy* (J. H. Hamilton and J. C. Manthuruthil, eds.), Gordon & Breach, New York, 1972.
28. I. De Lotto and A. Ghirardi, *Nucl. Instrum. Meth.* **143**, 617 (1977).
29. W. Westmeier, *Nucl. Instrum. Meth.* **180**, 205 (1981).
30. S. P. Nielsen and S. E. Palsson, *Nucl. Instrum. Meth.* **A416**, 415 (1998).
31. M. J. Koskelo, P. A. Aarnio, and J. T. Routti, *Computer Phys. Commun.* **24**, 11 (1981).
32. A. E. Egger et al., *IEEE Trans. Nucl. Sci.* **42**(4), 267 (1995).
33. N. Sasamoto, K. Koyama, and S. Tanaka, *Nucl. Instrum. Meth.* **125**, 507 (1975).
34. P. A. Baedecker, *Anal. Chem.* **43**, 405 (1971).
35. L. Kokta, *Nucl. Instrum. Meth.* **112**, 245 (1973).
36. J. Hertogen, J. De Donder, and R. Gijbels, *Nucl. Instrum. Meth.* **115**, 197 (1974).
37. G. Kennedy, *Nucl. Instrum. Meth.* **A299**, 349 (1990).
38. B. L. Roberts, R. A. J. Riddle, and G. T. A. Squier, *Nucl. Instrum. Meth.* **130**, 559 (1975).
39. L. A. McNelles and J. L. Campbell, *Nucl. Instrum. Meth.* **127**, 73 (1975).
40. Y. Takeda, M. Kitamura, K. Kawase, and K. Sugiyama, *Nucl. Instrum. Meth.* **136**, 369 (1976).
41. H. Baba, S. Baba, and T. Suzuki, *Nucl. Instrum. Meth.* **145**, 517 (1977).
42. H. H. Jorch and J. L. Campbell, *Nucl. Instrum. Meth.* **143**, 551 (1977).
43. O. Ciftcioglu, *Nucl. Instrum. Meth.* **174**, 209 (1980).

Chapter 19

Miscellaneous Detector Types

I. CHERENKOV DETECTORS

A category of radiation detectors is based on the light that is emitted by a sufficiently fast charged particle passing through an optically transparent medium with index of refraction greater than 1. An extensive coverage of the origins and applications of this *Cherenkov light* can be found in Ref. 1. The light is emitted whenever the velocity of a charged particle exceeds that of light in the medium through which it is passing, or

$$\beta n > 1 \tag{19.1}$$

where n is the refractive index of the medium and β is the ratio of the velocity of the particle in the medium to that of light in a vacuum. Detectors based on sensing the Cherenkov light are widely used in high-energy physics experiments but have only limited use for the lower-energy radiations of interest in this text. For particles in the tens of MeV range or lower, electrons are the only category to achieve sufficiently high velocity to emit Cherenkov light in available materials. Applications therefore involve either fast primary electrons such as beta particles or energetic secondary electrons produced in gamma-ray interactions. Examples are given in Refs. 2 and 3.

Cherenkov detectors bear some similarity to common scintillation detectors, in that the emitted light is converted into an electrical signal by a photomultiplier tube in optical contact with the Cherenkov medium. However, several important properties are quite different.

1. As indicated by Eq. (19.1), a minimum particle velocity is required in a given medium in order to generate any Cherenkov light. Therefore, Cherenkov detectors have an inherent discrimination ability that is unique among radiation detectors. Their response is limited to electrons whose energy exceeds a minimum or threshold given by

$$E_{\text{th}} = m_0 c^2 \left(-1 + \sqrt{1 + \frac{1}{n^2 - 1}} \right) \tag{19.2}$$

where $m_0 c^2$ represents the electron rest-mass energy (0.511 MeV). The threshold electron energy is plotted in Fig. 19.1 as a function of the index of refraction n. Also plotted is the minimum energy gamma ray that can produce Compton electrons of this energy by 180° scattering. Because photoelectric absorption is negligible at these energies in low-Z materials, this latter curve also represents a practical energy threshold for the detection of gamma rays. This inherent energy discrimination can be a very useful feature in situations in which a high rate of low-energy events

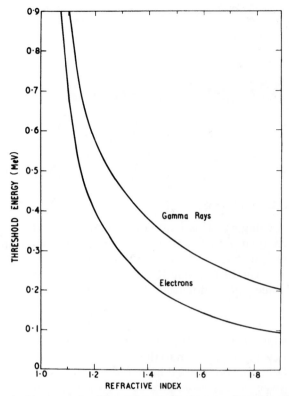

Figure 19.1 The threshold energy for Cherenkov radiation as a function of the index of refraction of the detection medium. Curves are shown both for electrons and for gamma rays that can yield, by 180° Compton scattering, an electron of the threshold energy. (From Sowerby.[2])

would otherwise be recorded. Potential problems arising from pulse pile-up or excessive rates can be avoided if these unwanted events are eliminated at the start. By using media with different indices of refraction, various detection thresholds can be chosen.

2. The light is emitted over the very short time required for the electron to slow from its initial velocity to below the threshold velocity. This time will typically be of the order of picoseconds in solids or liquids, so that Cherenkov detectors have the potential of being exceptionally fast. As a practical matter, their timing properties are normally limited by the photomultiplier tube.

3. The most important drawback of Cherenkov detectors is the low level of light produced. As shown in Fig. 19.2, the number of photons emitted per electron in common Cherenkov media is only several hundred per MeV. This corresponds to a conversion of only about 10^{-3} of the particle energy into visible light, about a factor of 100 smaller than the corresponding figure for an efficient scintillator (see Chapter 8).

4. In contrast to scintillation light, which is emitted isotropically, Cherenkov photons are emitted preferentially along the direction of the particle velocity. The light is confined to the surface of a cone with vertex angle θ, where

$$\cos \theta = \frac{1}{\beta n}$$

(19.3)

Figure 19.2 Calculated yield of Cherenkov photons in the 300–600 nm wavelength region for several detection media. (From Sowerby.[2])

The cone of light, when projected onto a planar surface, defines an ellipse or ring that is recorded in ring-imaging Cherenkov (RICH) detectors[4,5] developed for applications in high-energy particle physics experiments.

5. The yield of Cherenkov photons per unit wavelength λ is proportional to $1/\lambda^2$. The emission is therefore concentrated in the short-wavelength region of the spectrum. At extremely short wavelengths, the yield eventually drops off since the refractive index as a function of wavelength approaches unity.

Choice of the Cherenkov medium is made from materials with good optical transmission properties and, ideally, no scintillation component. By proper choice of the material, the index of refraction can vary from 1 to about 1.8. Silica aerogels have been demonstrated to be useful as solid Cherenkov media, covering the range from about 1.01 to 1.07 in refractive index depending on their density. Gases at various pressures can cover the range from 1 to about 1.2. Convenient materials with index between 1.2 and 1.33 are scarce, but liquified gases can be used for this purpose. The region between 1.33 and 1.47 is conveniently covered by mixtures of glycerin and water. Above an index of 1.47, many transparent solids such as lucite (perspex), glasses, and crystalline materials are common choices.

The Cherenkov effect can also generate light in the glass face plate of photomultiplier tubes exposed to gamma rays. This light can be an undesirable complication in the response of ordinary scintillation detectors under conditions of high gamma-ray background.

II. GAS-FILLED DETECTORS IN SELF-QUENCHED STREAMER MODE

A mode of operation has been developed for gas multiplication detectors that is distinct from the traditional proportional and Geiger regimes described in Chapters 6 and 7. It is called the *limited streamer* or *self-quenched streamer* (SQS) mode and has found application primarily in position-sensitive multiwire chambers. In common with proportional or Geiger–Mueller detectors, SQS devices rely on the formation of Townsend avalanches in a gas to multiply the amount of charge collected from ion pairs formed along the track of incident ionizing particles. However, the SQS mode of operation differs in the way in which these avalanches are allowed to propagate through the gas.

The discussions in Chapter 7 stressed the role that ultraviolet (UV) photons play in the spread of avalanches in the Geiger mode. These photons are emitted by excited atoms formed by electron collisions during the progression of an avalanche. In the proportional mode, the spread of avalanches is prevented either by keeping the avalanche small or by adding a "quench" gas that absorbs the UV photons without additional electron release. Only one avalanche is produced per original ion pair, and therefore the size of the output pulse remains proportional to their number. In the Geiger mode, the UV photons emitted in one avalanche propagate throughout the gas volume and create additional avalanches that eventually spread along the entire length of the anode wire. The process is terminated by the buildup of space charge around the anode, and the large output pulse size no longer depends on the number of original ion pairs.

In the SQS mode, the avalanches are also allowed to propagate, but only in a controlled fashion. By using gas mixtures that strongly absorb the UV photons, the formation of additional avalanches far from the original site is prevented. Instead, new avalanches are limited to the immediate vicinity of the original. It is observed experimentally[6] that, under proper conditions, the new avalanches will grow in the form of a narrow "streamer" that extends radially away from the anode wire surface (see Fig. 19.3). These streamers are 150–200 μm thick and extend a few millimeters from the anode. They are self-limiting and terminate with a final length that increases with applied voltage. If the original ion pairs are formed over a limited axial distance, only a single streamer is formed. For more extended tracks, multiple streamers are observed. At sufficiently high values of the voltage, a single electron can trigger a streamer.

One model of this behavior[6] is based on the local distortions in the applied electric field that are created within an avalanche. At high values of the gas multiplication, the positive ions form a cone-shaped cloud of space charge that creates a local dipole field. Its effect is to

Figure 19.3 Photographs of individual SQSs for two different applied voltages. The anode position is indicated by the line segments at the bottom. (From Atac et al.[6])

reduce the electric field near the sides of the cone but to reinforce it at its tip. Electrons formed near this region by UV photons will preferentially create additional avalanches near the cone tip where the field is the highest. The streamer thus grows radially outward after beginning near the wire surface. It terminates when the lower applied field at larger radii is no longer high enough to sustain further avalanches. This model is sometimes called the *photon-mediated* mechanism for SQS formation. It is clear that the creation of streamers involves a complex interplay of avalanche formation and changing electric fields, together with photon emission and absorption that change in importance with conditions. An alternative process has been proposed[7] for heavily ionizing radiations such as alpha particles that involves the staggered arrival of electrons at the anode from different locations along the particle track. The earliest electron to arrive at the anode triggers an avalanche, and subsequent electrons drifting inward from the track arrive at slightly later times. If the track is short, many of these electrons arrive at the tip of the positive ion cone that exists from previous avalanches and cause it to grow outward as a streamer. This process is called the *ionization track-induced SQS*,[8] and under some conditions may coexist with the photon-mediated mechanism. Other models for the formation of SQS behavior can be found in Refs. 9–11.

Figure 19.4 shows the transition from proportional to SQS mode in a gas-filled detector. Pulses from streamer formation appear with an amplitude that is typically an order of

Figure 19.4 Variation of the avalanche size with applied voltage in both the proportional and SQS modes of operation. Results are shown for methane and several argon–methane mixtures. (From Koori et al.[12])

magnitude larger than those from simple proportional operation. There is usually a range of applied voltage over which both proportional and SQS modes are observed. Streamer formation is promoted by using an anode wire of relatively large diameter (25–100 μm) and is observed only in certain gas mixtures.

The SQS mode combines some features of both Geiger and proportional mode operation. As in a Geiger tube, the additional avalanches that are created in the streamer increase the pulse size and reduce demands placed on pulse-processing electronics. This internal amplification is particularly helpful in position-sensitive detectors where the larger signal reduces position uncertainties compared with proportional operation.[13] However, as in a Geiger tube, the amplitude of the signal from an individual streamer is no longer a measure of the size of the initiating event. The advantages of proportional operation for radiation spectroscopy are therefore lost. By preventing the spread of avalanches along the anode wire, SQS operation avoids the long dead time of Geiger tubes. The rise time of typical output pulses has been measured[6] as 30–40 ns, so the SQS mode can result in good timing precision as well.

III. HIGH-PRESSURE XENON SPECTROMETERS

Xenon gas at pressures near its critical point (58 atm, 17°C, density of 1.1 g/cm³) represents an attractive medium for gamma-ray spectroscopy. The high atomic number (54) of xenon promotes the photoelectric absorption process that is important for full conversion of the incident gamma-ray energy to charge. Gaseous detectors are not limited to small volumes by constraints of crystal growth or other fabrication considerations. While there have been some examples of the use of high-pressure xenon in proportional counters[14] or proportional scintillation detectors,[15] principal interest has been in its use in pulse mode ion chambers. Because of the short mean free path of electrons in the high density gas, typical drift velocities are limited to around 10^5 cm/s for reasonable electric field values. It has been demonstrated[16] that for a density of 0.5 g/cm³, the free electron lifetime is about 5 ms for an electric field of 2×10^5 V/m, sufficiently long to allow over 99% charge collection efficiency in chambers of typical dimensions. Achieving this long lifetime, however, requires the strictest attention to eliminating impurities that might attach free electrons. Chambers must be constructed using ultrahigh vacuum techniques to avoid the inclusion of any organic materials, and electronegative impurities must be maintained below about 0.5 ppm.

Xenon at high pressures is far from an ideal gas and has a very high compressibility near its critical point, as illustrated in Fig. 19.5. For example, at a density of 0.8 g/cm³, the

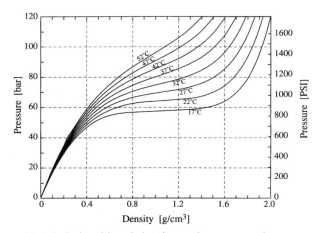

Figure 19.5 Relationship of density and pressure of xenon at the constant temperatures shown. (From Mahler et al.[16])

pressure will increase from 56 to 88 atmospheres with a temperature change from 17 to 52°C. Provided that the pressure boundary is designed to sustain these high pressures, xenon chambers have the added virtue of being usable over wide temperature ranges without significant deterioration in their performance.[17,18] The chamber wall must be several millimeters thick in order to safely contain the pressurized gas. Using typical materials such as titanium or stainless steel, there will be severe attenuation of low-energy incident gamma-ray photons below about 100 keV.

Most successful demonstrations of the use of high-pressure xenon as a spectrometer medium[16,19] have used ion chambers fitted with a Frisch grid (see Chapter 5). The W-value of xenon at low pressure is 21.9 eV per ion pair. This value is observed to decrease gradually with increasing xenon density, and it is lower by about 15% when a density of 1.7 g/cm^3 is reached.[17] Using a Fano factor of 0.13, the energy resolution limit set by charge carrier statistics is about 0.5% at 662 keV. In practice, typical high-pressure gridded ion chambers show much poorer energy resolutions of 2–4%. The difference represents the added contributions of electronic noise and microphonics because of the very low-level signals that are produced. Relatively long shaping times must be used in processing electronics because of the slow drift velocity of the electrons, allowing more of the microphonic signal to affect the chamber resolution. To help speed the electron drift, it has been found that the addition of about 1% of hydrogen gas to the xenon enhances the drift velocity and promotes a faster output signal.[20]

Xenon also is a scintillator and there will be prompt scintillation light emitted when the incident particle interacts in the gas. This light is in the ultraviolet region of the spectrum centered around 170 nm, and appears with a mixture of decay times of 2.2 and 27 ns. The appearance of this scintillation light can provide a "time zero" signal if the light can be conducted to an ultraviolet-sensitive photomultiplier tube outside the pressure boundary.[21] By timing the difference of the appearance of the scintillation light and the collection of the electrons at the anode, the drift distance can be deduced yielding the depth of the particle interaction within the gas.

IV. LIQUID IONIZATION AND PROPORTIONAL COUNTERS

The conventional detectors in which charges created by ionizing radiation are directly collected employ either a gas or a semiconductor material as the detection medium. The great advantage afforded by the high density of solids or liquids has stimulated the search for additional materials suitable for use in radiation detectors. The liquid and solid phases of noble gases are condensed dielectrics in which electrons remain free and for which sufficient purification is possible to eliminate electronegative impurities. Therefore, active development continues in the incorporation of these cryogenic materials into detectors that operate under principles similar to conventional ionization or proportional counters.

Some properties of liquified noble gases of interest in detector applications are given in Table 19.1. Of these, liquid xenon has received the most attention and has been applied successfully both as a conventional ion chamber, where the charges are simply collected, and as a proportional counter, in which charge multiplication occurs at high values of the electric field. The electron drift velocity saturates at 3×10^3 m/s for electric field values greater than about 10^5 V/m (Refs. 23 and 24). The onset of multiplication is observed to occur at about 10^8 V/m (Ref. 25).

Successful operation of these counters requires careful attention to purity of the liquified gas, and some system of continuous recycling through a purifier is usually incorporated into the design. Liquified gases require that the counter be maintained at cryogenic temperatures, which is an obvious operational handicap for most applications. However, the low values for W and F in Table 19.1 hold out the promise of excellent energy resolution (see Chapter 6) if charge collection can be made efficient. Furthermore, the high atomic

Table 19.1 Properties of Some Condensed Media for Ionization Chambers

	Z	$\left(\dfrac{\text{g}}{\text{cm}^3}\right)$	Boiling Point	W (eV/ion pair)		F (Fano factor)	
				Calculated[a]	Experimental[a]	Model 1[b]	Model 2[c]
Liquid Ar	18	1.41	87 K	23.3	23.6	0.107	0.116
Liquid Kr	36	2.15	120 K	19.5		0.057	0.070
Liquid Xe	54	3.52	166 K	15.4	15.6	0.041	0.059
			Melting point				
Solid Ar	18	1.62	84 K				

[a]From T. Doke et al.[22]

[b]Based on W. Shockley, *Czech. J. Phys.* **B11**, 81 (1961).

[c]Based on G. D. Alkhazov et al., *Nucl. Instrum. Meth.* **48**, 1 (1967).

number of xenon (54) is extremely attractive for applications involving gamma-ray spectroscopy. A liquid xenon ionization chamber has the potential of providing the good gamma-ray detection efficiency of NaI, with an improved energy resolution intermediate between that of a scintillator and a germanium detector.

If the impurity concentrations are kept very low (less than one part per billion of oxygen equivalent), the distance over which electrons can be drifted in liquid argon or xenon approaches a meter or more.[26] Chambers with large active volume are therefore feasible, and position-sensing can be carried out by measuring the electron drift time.[27] The energy resolution in small gridded ionization chambers is typically 5–6% for gamma rays,[28] but values of 2.7% in liquid argon[29] and 4.3% in liquid xenon[30] have been reported for 976 keV conversion electrons. These figures are still much poorer than predicted from theory, and it is thought that the degradation in resolution is caused by variations in recombination that is due to a high density of ions and electrons along the tracks of delta rays (low-energy electrons) produced in large numbers along the primary path.[31–33] An ultraviolet photon is often emitted in the recombination process, and efforts have been made to "reclaim" the lost ionization by mixing an organic dopant with low photoionization threshold into the liquid. There have been some reported results[30] that indicate an improved energy resolution with this approach.

Liquid argon is also a scintillator, as described in Chapter 8. Most of the scintillation light derives from excited molecules that are formed by the recombination of ion–electron pairs. Applying an electric field near 10^6 V/m to collect the free electrons also suppresses recombination, and the scintillation intensity drops to one-third of its zero field value. Both scintillation and ion chamber signals can be measured simultaneously from the same sample of liquid argon,[34] and it is observed that the sum of the two signals is approximately independent of electric field strength.

It is also possible[35,36] to operate liquid xenon chambers in the proportional-scintillation mode (see p. 194). This choice has the potential advantage of more stable operation than one in which charge multiplication is required. Again, both the scintillation light and the collected charge can serve as independent measures of the energy deposition. This approach also lends itself to applications in which sensing the position of interaction is important.[37]

Some attention has been given to solid rather than liquid argon.[38] The electron mobility is actually higher in the solid phase,[23] but problems related to poisoning and polarization of the solid have inhibited further development.

There is also interest in some nonpolar liquids as ionization detector media at room temperature. For example, tetramethylsilane has been shown to be an acceptable liquid filling for ionization chambers operated both in pulse[39] and current[40] modes. As in liquified noble gases, stable operation requires careful attention to purity.[41]

V. CRYOGENIC AND SUPERCONDUCTING DETECTORS

A unique class of radiation detectors has emerged that rely on the unique properties of materials near absolute zero temperature. A useful review of these low temperature cryogenic devices can be found in Ref. 42. Some of these rely on the phenomena of superconductivity, while others take advantage of the exceptionally small thermal capacity of materials at low temperature.

It has become routine to operate germanium or lithium-drifted silicon detectors at liquid nitrogen temperature of 77 K. Supplies of liquid nitrogen are commonly available at most laboratories, and achieving these temperatures is only a matter of filling a liquid nitrogen storage dewar. The devices described in this section are operated at much lower cryogenic temperatures, generally below 1 K. Cooling to the boiling point of helium of 4 K is straightforward using a supply of the condensed liquid. To reach lower temperatures, more complex cryostats are required that rely on so-called *dilution refrigerators* based on the thermodynamics of ^{3}He-^{4}He mixtures. These devices, while somewhat costly, have become commercially available and are capable of achieving routine operating temperatures of 0.1 K, adequate for operation for all the devices described in the following sections.

A. Cryogenic Microcalorimeters or Bolometers

Any substance when exposed to ionizing radiation will in principle show an increase in temperature caused by the energy absorbed from the incident particles or photons. Instruments known as *bolometers* have been used for many years for thermal or infrared radiation in which the incident flux on a target is measured through the temperature rise as sensed by a thermistor. For most common conditions with ionizing radiation, this temperature rise is too small to be measured unless the sample is exposed to a very high flux of radiation. For example, an absorbed dose of 1 gray (100 rad) represents by definition the absorption of 1 joule per kilogram (J/kg) of the absorber. The flux of incident radiation needed to deposit this energy is rather large, corresponding, for example, to the full absorption of over 10^9 5 MeV alpha particles per gram of the absorber. Converting energy units, one finds that the adiabatic absorption of that energy in water results in a temperature rise of only 2.39×10^{-4} °C. Consequently, calorimetric measurements in which the temperature rise in a thermally isolated sample is measured directly have traditionally been limited to relatively intense radiation fields.

The sensitivity of such calorimetric methods can be made many orders of magnitude better if the temperature rise is measured in miniature samples of material maintained at very low temperature. The heat capacity of dielectric materials or crystals at low temperatures is proportional to T^3, where T is the absolute temperature. Therefore, the lower the temperature, the greater will be the temperature increase per unit of absorbed energy. This temperature rise must be measured against the background of the statistical fluctuations in temperature caused by random fluctuations in the energy content of the absorber. Such fluctuations are due to the random variations in the flow of *phonons* across the thermal coupling between the absorber and the surrounding medium. It can be shown[43] that the mean square value of these fluctuations is proportional to T^5. Reducing the temperature therefore also greatly suppresses the background against which the signal must be measured.

Figure 19.6 Operational principle of the cryogenic bolometer.

By reducing temperatures to below 1 K, it is possible to sense the momentary temperature rise *caused by a single photon or incident particle*. A schematic sketch of the resulting temperature pulse is shown in Fig. 19.6. This temperature pulse can be measured using a thermistor in contact with the absorber providing an electrical signal that is analogous to the signal pulse from more conventional pulse mode detectors. What is unique is the fact that the pulse does *not* consist of discrete charges (such as ion pairs or electron–hole pairs) that are collected but rather is only a change in the thermal state of the absorber. Thus, one completely avoids the limits on energy resolution set by charge carrier statistics in virtually every other type of radiation detector. Ultimate limits are now set only by thermodynamic fluctuations at the phonon level in the absorber. Since phonon energies are typically several orders of magnitude below the energy required to create an electron–hole pair in a semiconductor, the theoretical energy resolution is many times better for the microcalorimeter. Analyses[44,45] show that fundamental thermodynamic limits could result in energy resolution figures as low as several eV or less.

The target chosen for this type of microcalorimeter is generally a crystalline or semiconductor material of very small volume to minimize the total heat capacity. Typical areas are less than 1 mm^2 with thicknesses from 25–500 μm. The absorber is held by supports of small cross section that provide the thermal link to the low temperature cryostat. Figure 19.7 shows an array of 12 silicon targets that produced the excellent energy resolution also illustrated in the figure. The FWHM of 7.3 eV is almost a factor of 20 better than that obtained from typical Si(Li) X-ray spectrometers. This outstanding energy resolution capability has stimulated interest in their application to X-ray measurements in which the small target size and high cost of the calorimeter are not detriments, and very active development of these devices is continuing.

Because this type of thermal detector has a response that is ultimately limited by the speed of phonon (sound) propagation, the rise time observed from microcalorimeters is very long (hundreds of microseconds or more) compared with standard pulses from charge collection devices. They are therefore limited to relatively low counting rates before excessive pile-up limits the measurement. If arrays of many microcalorimeters with separate readouts are employed, the overall counting rate can reach higher values while also providing position information on the incident radiation.

B. Transition-Edge Sensors

The microcalorimeters described in the previous section typically use dielectric insulators and semiconductors for the absorber material because their heat capacity becomes very small at low temperature. Thus the temperature rise caused by a single particle or photon can then be measured using conventional thermistors and preamplifiers. Unfortunately, the deposited energy in these materials goes into phonon excitations in the crystal that relax to a thermal distribution in a slow and complex manner. Therefore the temperature pulse as measured by the thermistor has very slow components of a millisecond or more (see Fig. 19.7), and the counting rate at which spectroscopic measurements can be made is severely limited.

Figure 19.7 Array of 12 microcalorimeter elements, each 0.25 mm × 1.0 mm × 15 μm thick. The corresponding energy spectrum from a ^{55}Fe source is shown in both a normal and expanded scale. A typical pulse profile is shown in the insert. (From McCammon et al.[46])

An alternative temperature measurement device has been developed known as a *superconducting transition edge sensor* that greatly improves upon these limitations.[47,48] The scheme is illustrated in Fig. 19.8. Again, a small absorber is coupled to a low-temperature source through a thermal link. Now, however, the temperature sensing is carried out through a superconducting film that is in contact with one of the absorber surfaces. The low-temperature sink is operated at a temperature well below the transition temperature required for the film to revert from normal conduction to the superconducting state. A voltage is placed across the film that causes a small current to flow while it is in its normal conducting state. This current causes a small amount of ohmic heating given by

$$P = IV = \frac{V^2}{R}$$

If we now imagine cooling the assembly starting from a temperature at which the film is normally conducting, the transition temperature region shown in Fig. 19.8b is reached where there is an abrupt drop in the resistivity as the film converts to the superconducting state. As the temperature drops through this transition, the current increases abruptly and thus the ohmic heating also increases. An equilibrium condition will be established in which the temperature stabilizes at the point at which the ohmic heating is just balanced by the heat flow through the absorber and to the low temperature sink. By choosing the voltage and film resistivity properly, this stable operating temperature can be set on the steep portion of the transition edge. This is sometimes referred to as a condition of *electrothermal feedback*, in that any tendency for the temperature of the film to rise is strongly countered by an increase in the film resistance and a corresponding drop in the ohmic

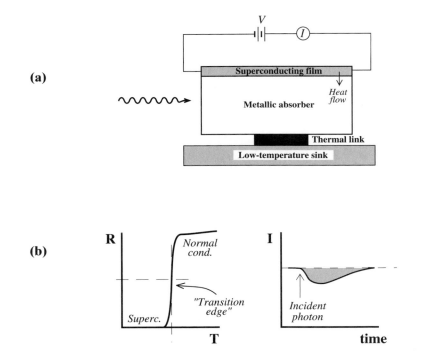

Figure 19.8 Principle of the superconducting transition edge sensor shown schematically in part (a) of the figure. Conditions are established in which the temperature of the superconducting film is in the transition range between superconducting and normal as shown on the left in part (b). When energy is deposited by a particle, the momentary drop in current is shown on the right.

heating. Likewise, any trend to reduce the film temperature would be offset by an increase in the ohmic heating. The temperature of the film becomes self-regulating at a value that is on the steep portion of the transition.

If a particle or photon deposits energy in the absorber, it will tend to raise the absorber and film temperature. However, the electrothermal feedback will strongly limit the temperature rise by causing a decrease in the current flowing through the film. Thus the total heat flow through the thermal link remains essentially constant, with the heat generated by the particle or photon being compensated by a momentary drop in the ohmic heating. A measurement of the area of the drop in current, shown as a shaded portion on Fig. 19.8*b*, is thus proportional to the decrement in ohmic heating caused by the particle or photon. When multiplied by the applied voltage, the integrated drop in current is an absolute measurement of the energy deposited in that event, independent of any calibration. In practice, very sensitive current measurements are made using superconducting quantum devices (SQUIDs) to enhance the accuracy and sensitivity of the measurement. The electrothermal feedback leads to a thermal response time that can be as much as a factor of 100 faster than the normal thermal relaxation time[49] experienced with thermistor readout.

Microcalorimeters based on transition edge sensors, because of their sensitivity, can be used with absorbers of substantially greater heat capacity than those using thermistor temperature measurement. As a result, metals can be chosen for the absorber material in which the thermalization time of the absorbed energy is much faster than in semiconductors. Normal metal absorbers diffuse the thermal energy more quickly and completely than semiconductors. It is predicted[49] that the lower limit on response time can be of the order of 1 microsecond, permitting much higher rate operation than for microcalorimeters based on thermistor readout. Transition-edge microcalorimeters have been shown to yield at least as good energy resolution as those observed for slower microcalorimeters, with a best value to date[50] of 7.2 eV measured at 5.9 keV X-ray energy.

C. Superconducting Grains and Strips

The very small heat capacity of materials near absolute zero temperature can be exploited in another way to detect single particles or photons. Specifically, the transition between the normal conducting and superconducting state that occurs in some materials such as tin or indium at low temperatures that was illustrated in the previous section can be exploited directly for the detection process. In a very small grain of such material with diameter of a few micrometers, the energy deposited by an energetic particle can be sufficient to transform the grain from the superconducting to the normal state. The energy required for this transition can be as small as hundreds of eV if the grain mass is small and it is maintained in a metastable state very close to the transition temperature.

As illustrated in Fig. 19.9, when an external magnetic field is present, the transition temperature decreases as the field strength is raised. By cooling a superconductor and raising the magnetic field strength, it is possible to create conditions in which a grain is superconducting

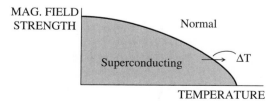

Figure 19.9 The superconducting state is limited to the shaded area on this plot of temperature and magnetic field strength. Superconducting grain detectors are based on observing the transition across this boundary.

but metastable, in the same sense that a supersaturated vapor is metastable to the formation of liquid droplets. It then requires deposition of only a small amount of energy to trigger the transition of the entire grain from the superconducting to the normal state. Since this transition takes place in a magnetic field, an abrupt change takes place in the field configuration when the grain "flips" from superconducting to normal. As sketched in Fig. 19.10, the Miesner effect excludes the magnetic field lines from the interior of the grain while it is in the superconducting state, but after the transition to the normal state the magnetic field lines penetrate the volume of the normally conducting metal. This sudden change in the magnetic field configuration generates an electromagnetic pulse that can be sensed in a readout loop at some distance through the voltage induced by the changing flux. The signal strength is proportional to the product of the volume of the granule and the strength of the magnetic field. Again, SQUID detectors have been shown[51–53] capable of detecting the electromagnetic signal created by the transition of a single grain of diameter ranging from 5–15 μm.

Detectors with dimensions of several centimeters have been constructed by dispersing a large number of superconducting grains in a dielectric filler material such as paraffin. Sensors placed at various points on the surfaces of this "colloid" detector can not only sense the transition of a single grain anywhere within its volume, but can also provide information on the position of the interaction through triangulation techniques. For example, a spatial resolution of 300 μm has been demonstrated using these techniques in a 1-cm^3 colloid detector.[54]

Colloid detectors have demonstrated sensitivity to 6 keV X-rays[55] and nuclear recoils with energy down to 1 keV.[56] A colloid detector with tin granules of 22-μm diameter filled to 16% volume fraction demonstrated 100% detection efficiency for minimum ionizing particles in a 2-mm overall thickness.[57] Colloid detectors using boron powder or borated paraffin as a matrix have demonstrated sensitivity to neutrons as well.[58,59]

The flipping of the grains generally takes place in a time period of tens or hundreds of nanoseconds following the triggering particle, and timing resolution of as good as 10 ns has been obtained experimentally.[60] If a particle has a range that is long enough to intersect multiple grains, then the larger superimposed signal can be used to separate long range particles from those of shorter range in which only a single grain is involved.[61]

Colloid detectors are most often applied as simple counters in which no energy information on the triggering event is provided. However, after the grain has "flipped" from the superconducting to the normal state, there is a natural tendency for it to "flop" back to its original superconducting condition as the original temperature is restored. This transition can also be sensed externally, and the time interval between the "flip and flop" correlated with the total energy deposited in the grain by the particle.[62]

The transition from the superconducting to normal state caused by an energetic particle can also be sensed in a strip of superconductor typically a few μm or less in width. A constant current is first caused to flow through the strip, and kept just below the critical current that would cause transition of the strip from its superconducting to normal state. If a particle deposits energy at some point along the strip, the local "hot spot" with normal conductivity will expand by thermal diffusion, forcing the current density in the remaining superconducting region to rise above the critical value. At that point, a "bridge" of normal

Superconducting   **Normal**

Magnetic field lines

Figure 19.10 On the left, external magnetic fields are excluded from the superconductor volume by the Miesner effect. Once the grain undergoes a transition to normal conductivity, the field lines now penetrate the volume.

conductivity is formed across the full width of the strip, and a voltage drop will now appear across the length of the strip. The strip can be reset back to its original superconducting state if a strong thermal link is in place to a low temperature sink. Arrays of these superconducting strips are of potential interest in the tracking of energetic particles in high energy physics experiments.[63]

D. Superconducting Tunnel Junctions

In a superconducting material, many of the electrons are bound together in a loosely associated pair known as a *Cooper pair*. It is this association that helps explain the absence of electrical resistance in superconductors. The binding energy of Cooper pairs is very small, approximately 0.5×10^{-3} eV. Other than at absolute zero, some Cooper pairs are spontaneously broken by thermal energy and appear as "quasiparticles" or electrons that are no longer bound together. Thus there is an energy structure for electrons in a superconductor that is sketched in Fig. 19.11. The energy band containing Cooper pairs is separated from that for quasiparticles by an energy gap equivalent to the Cooper pair binding energy. This band structure is analogous to that introduced in Chapter 11 for electrons in semiconductors. A major difference is that, while the bandgap in semiconductors is of the order of one eV, the energy gap for superconductors is over a factor of a thousand times smaller.

When a charged particle interacts in a superconductor, it loses some of its energy through the breaking of Cooper pairs and the formation of quasiparticles. Thus if a means can be devised to measure the number of quasiparticles formed, a measurement of the particle energy can be carried out. The number of information carriers per event, in this case quasiparticles, will be three orders of magnitude higher than in semiconductor detectors. One would therefore anticipate that the statistical limits on energy resolution would be smaller, roughly by a factor of the square root of 1000 or 30. Complicating factors such as the fraction of energy converted and the effective Fano factor will modify this estimate, but nonetheless one can anticipate outstanding energy resolution if the quasiparticles formed by the incident particle can be accurately quantified.

A device that can do just that is illustrated in Fig. 19.12. It is known as a superconducting tunnel diode, and consists of thin films of superconductors separated by an insulating layer that is thin enough to allow the quantum mechanical process of barrier tunneling of electrons or quasiparticles from one side of the junction to the other. The device finds applications in electronic circuits because of the unique current voltage characteristics also illustrated in Fig. 19.12. The effect of applying a voltage to the junction is to raise the potential diagram on one side of the junction relative to the other by an amount equal to the electronic charge multiplied by the applied voltage. The sudden rise in the current that is observed from the diode occurs at an applied voltage that corresponds to the width of the energy gap. At that point, tunneling is greatly enhanced because of the strong concentration of electrons on one side of the junction and the availability of energy states on the other to which they can tunnel.

Figure 19.11 Energy structure for electrons in a superconductor.

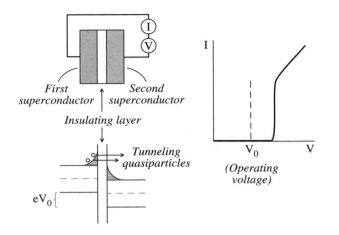

Figure 19.12 Principle of the superconducting tunnel diode. The operating voltage when used as a detector for quasiparticles is V_0.

When used as a radiation detector, the tunnel diode is operated at a somewhat lower bias of V_0 as indicated on Fig. 19.12. Here the operating current (or leakage current of the device) is very low since the only source is those tunneling quasiparticles that are created by thermal excitation. However, when an energetic particle creates a burst of quasiparticles, the sudden increase in tunneling is measurable as an increase in this current for the short time until the quasiparticles tunnel or recombine. The decay time of the quasiparticles tends to be on the order of a few microseconds, and the duration of the observed pulse is therefore of this order of magnitude.

Tunnel diodes are thin film devices that, by definition, have small volume. The restriction on thickness is necessary so that, when the quasiparticles are trapped in the layer, they have the opportunity to make many attempts to tunnel across the junction during their lifetime. On the other hand, one would often like larger volumes or areas to serve as efficient radiation detectors. A step in this direction is permitted by the process of "quasiparticle trapping"[64,65] in which a second superconducting material with larger band gap is placed in thermodynamic contact with the thin film superconductor on one side of the junction. An example of quasiparticle trapping commonly employed is to use an aluminum thin film coupled to a niobium larger volume superconductor. The wide-gap superconductor serves primarily as the absorber of the radiation, and it is in its volume that the majority of the quasiparticles are formed. Since this volume is not in contact with the insulating layer, these quasiparticles no longer have an opportunity to tunnel directly. However, in their normal diffusion, many will arrive at the interface between the two superconducting materials. At this point, there is a natural tendency for the quasiparticles to enter the thin film superconductor with narrower gap, and they are then trapped there because crossing the boundary in the opposite direction is energetically disallowed. Thus the larger-volume superconductor serves as a source of quasiparticles that subsequently tunnel after being trapped in the thin-layer superconductor.

Even with quasiparticle trapping employed, devices of this type remain very small (surface areas of the order of 1 mm^2), but they have found application in high-resolution X-ray spectroscopy. The best results reported to date are an energy resolution of 29 eV at 5.9 keV.[66] While this does not match the spectacular energy resolution observed from microcalorimeters, it is still a factor of about 5 better than the common commercial instrument used for X-ray spectroscopy, the lithium-drifted silicon detectors described in Chapter 13.

VI. PHOTOGRAPHIC EMULSIONS

The use of photographic film to record ionizing radiations dates back to the discovery of X-rays near the turn of the century but has remained an important technique through the present time. A thorough presentation of the uses of photographic emulsions for ionizing radiation is given in the text by Herz.[67]

Ordinary photographic film consists of an emulsion of silver halide grains (consisting mainly of silver bromide) suspended in a gelatin matrix and supported with a backing of glass or cellulose acetate film. The action of ionizing radiation in the emulsion is similar to that of visible light, in that some of the grains will be "sensitized" through interaction of the radiation with electrons of the silver halide molecules. The process forms semi-stable clusters consisting of a few neutral atoms of silver at the surface of some of the grains. The sensitized grains remain in this state indefinitely, thereby storing a latent image of the track of the ionizing particle through the emulsion. In the subsequent development process, the entire sensitized grain is converted to metallic silver, vastly increasing the number of affected molecules to the point that the developed grain is visible. Following development, the emulsion is fixed by dissolving away the undeveloped silver halide grains, and a final washing step removes the processing solutions from the developed emulsion.

Radiation applications of photographic emulsions are conveniently divided into two categories: those in which a general darkening of the emulsion is recorded due to the cumulative effects of many interactions, and those in which single particle tracks are individually recorded. The first category includes the broad field of *radiography*, in which an image is recorded of the transmitted intensity of a beam of radiation. The specialized films used for this purpose do not differ radically from ordinary photographic films. Single-particle tracks, however, are best recorded in *nuclear emulsions*, which are much thicker and differ in composition from photographic emulsions.

A. Radiographic Films

A sketch of a typical radiographic measurement is given in Fig. 19.13. The incident radiation is either in the form of a point source or a parallel beam, which will therefore cast a definite "shadow" of the object on the recording emulsion. Although the most common radiations used in radiography are X-rays generated by standard tubes, higher-energy gamma rays from radioisotope sources or high-energy bremsstrahlung from electron linear accelerators also are applied in industrial radiography of thick objects. Neutrons can also be used in radiography, as discussed below.

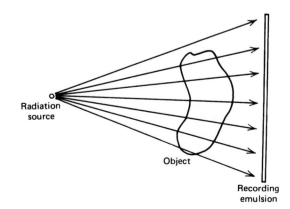

Figure 19.13 The elements in the formation of a radiographic image.

Radiographic emulsions are typically 10–20 μm thick with grains up to 1 μm in diameter. The silver halide concentration amounts to about 40% by weight. To increase their sensitivity, some radiographic films are "double sided," where the emulsion is applied to both surfaces of the base film.

In X- or gamma-ray radiography, the transmitted photon must interact to form a secondary electron if it is to be recorded. The direct interaction probability of a photon of typical energy within the emulsion itself is usually no more than a few percent at best. Therefore, films applied in the direct imaging of X-rays or gamma rays are relatively insensitive. A typical sensitivity curve is shown in the sketch below, where D is defined as the density of the developed film as measured by an optical densitometer. Three regions of this curve are readily identifiable. At low exposures, too few developed grains are produced to measurably increase the film density above the inherent "fog" level and the emulsion is underexposed. At the other extreme, high exposures result in a dense concentration of developed grains and a resulting overexposure. Intermediate exposures result in a density that varies approximately linearly with exposure, and this is the region of normal operation. The dynamic range (ratio between the largest and smallest exposures that result in useful density changes) is limited to no more than about a factor of 50–100 for typical films.

In medical radiography or other applications in which the incident radiation intensity is limited, steps are often taken to enhance the sensitivity of the emulsions. One method is to sandwich the film between foils made from materials with high atomic number. Photoelectric or Compton interactions within these converter foils may then contribute secondary electrons, which add to those created within the emulsion itself. Alternatively, *intensifier screens* may be placed next to the emulsion which consist of light-emitting phosphors of high atomic number, such as calcium tungstate. Interactions within the screen create visible light through the normal scintillation process, which then leads to additional sensitization of the emulsion. Because the light travels in all directions, a compromise must be struck between sensitivity and spatial resolution in choosing the thickness of the screen. In typical situations, the sensitivity of films to X-rays can be increased by a factor of 10 through the use of such intensifier screens.

Neutron radiography can also be carried out using emulsions, provided the appropriate type of converter foil is used. Slow neutrons can be imaged by sandwiching the emulsion between thin layers containing gadolinium (see p. 509). The conversion electrons emitted upon neutron absorption can enter the emulsion and lead to its sensitization. If the gadolinium layer and emulsion are kept thin, then the darkening of the emulsion will occur only in the immediate vicinity of the neutron absorption, and spatial resolution of the image is preserved.

In the technique of *autoradiography*, the radiation source to be imaged exists within the sample itself. For example, biological processes can be studied by tagging a given substance with tritium or carbon-14. If a thin layer of the sample is then placed in contact with a radiographic emulsion over a sufficient exposure time, the beta particles emitted in the radioisotope decay will be recorded as an image on the developed emulsion. Using these techniques, the detailed spatial distribution of radioisotope within the sample can easily be recorded.

B. Nuclear Emulsions

When the object of the measurement is to record individual particle tracks, specialized formulations known as *nuclear emulsions* are conventionally used. Here, the thickness of the emulsion is increased to as much as 500 μm to allow the recording of the entire track of many particles. To enhance the density of developed grains along the track, the concentration of silver halide within the emulsion is also increased to as much as 80%. The unusual thickness and composition of nuclear emulsions require the use of more elaborate procedures to ensure their uniform development. The track of ionizing particles is then visible under microscopic examination as a trail of developed silver grains that may become nearly continuous under certain conditions. An example is reproduced in Fig. 19.14. The length of the track can often be used as a measure of the particle range or energy, and the density of the track can serve to distinguish between radiations of different dE/dx. In some emulsions, a minimum value of dE/dx is required for the development of a visible track, so that an inherent discrimination ability against lightly ionizing radiations is achieved. Other formulations will develop tracks for low dE/dx particles, including fast electrons.

The theory of the response of nuclear emulsions to ionizing radiations is reviewed in Refs. 68 and 69. Under different conditions, the sensitization of grains can be considered a *single hit* process in which only a single encounter with an ionizing particle is required, or it may be a *multihit* process in which more than one such encounter is required. In the latter case, a definite threshold in ionization density is expected.

Although obviously limited to situations in which a prompt detector signal is not required, nuclear emulsions offer a number of advantages compared with conventional detectors. Reference 70 reviews some typical applications. No associated equipment is needed, so that very simple emulsion detectors may be used in remote experiments where conventional methods may not be applicable. Under proper ambient conditions of low temperature and humidity, the track may be stored indefinitely as a latent image within the emulsion and, once developed, is a permanent record. Emulsions loaded with specialized target nuclei such as boron or uranium can be made sensitive to thermal neutrons. Fast-neutron-induced tracks can be registered through recoil protons generated within the emulsion itself.

Figure 19.14 Tracks of fast heavy ions in Ilford K.5 nuclear emulsion. The ions are all incident from the left with initial energy of 10 MeV per nucleon. The track density is seen to decrease as the ion slows down and loses charge by electron pickup. (Photomicrographs courtesy of Prof. E. V. Benton, University of San Francisco.)

C. Film Badge Dosimeters

For many decades, the standard method of monitoring radiation exposure for personnel involved the use of photographic emulsions. Although increasingly replaced by other methods, *film badge dosimeters* consisting of a small packet of film in a light-tight wrapping are still found in a number of radiation monitoring applications. Emulsions offer a passive method of recording the results of radiation exposure over very long periods of time that is virtually free of the gradual loss of information or "fading" that sometimes is a problem in competing passive techniques such as the TLD detectors described in the next section of this chapter. An evaluation of the dose accumulated over the course of exposure is carried out by comparing the density of the developed film with that of an identical film exposed to a calibrated dose. In this way, variations in emulsion sensitivity and developing procedures are canceled out.

The film holder plays an important part in the film badge response because it conventionally contains filters that are held in contact with the film and alter its response. As an example of one function of such a filter, Fig. 19.15 shows the response of a typical photographic emulsion to equal doses of gamma rays with different energies. The sensitivity of the film is greatest for photons of low energy, so that an unfiltered film will tend to overestimate the dose of soft X-rays and gamma rays to soft tissue compared with those of higher energy. However, by covering the film with a combination filter composed of tin and lead, the response can be made quite flat over a wide range of photon energy. The density

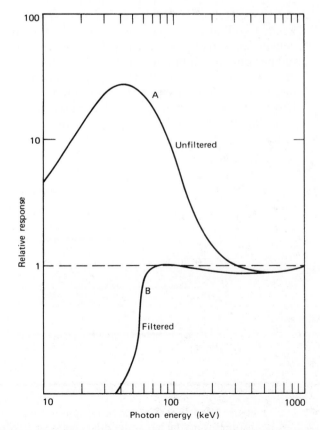

Figure 19.15 Relative response of a typical film badge dosimeter with and without a compensation filter. (From R. H. Herz, *The Photographic Action of Ionizing Radiations*, Copyright 1969 by John Wiley & Sons, Inc.)

of the film can then be used as a measure of gamma-ray exposure, independent of the details of the incident spectrum.

The film badge holder also normally contains a set of small filters confined to local regions of the film. These filters are made of different materials of variable thickness, so that differences in the corresponding film density can be used to sort out various components of the radiation exposure. For example, thin filters of low-Z material will effectively stop soft beta particles without seriously affecting transmitted gamma rays. Other metallic filters can help to separate out the contribution of different components of the gamma-ray spectrum.

A cadmium or gadolinium-containing converter foil can be used to evaluate the exposure of the film to thermal neutrons. The secondary radiation (fast electrons and capture gamma rays) emitted in these materials upon capture of a thermal neutron will lead to some incremental exposure of the film. By comparing the density behind such a filter to that behind a lead–tin filter of similar photon absorption properties, the additional darkening due to thermal neutrons can be estimated. Exposure of films to typical doses of fast neutrons does not lead to sufficient density to be measured directly. However, individual tracks due to recoil protons generated within the emulsion can be observed under a microscope and can be counted to give an estimate of the dose due to fast neutrons.

VII. THERMOLUMINESCENT DOSIMETERS AND IMAGE PLATES

A. The Thermoluminescence Mechanism

The inorganic scintillation materials discussed in Chapter 8, when exposed to ionizing radiation, emit light in the form of prompt fluorescence. The scintillation photons are given off when the electron–hole pairs that were formed by the incident radiation recombine at an activator site. These materials are purposely kept free of other impurities and defects in order to maximize the yield of prompt scintillation light.

A different class of inorganic crystals, known as *thermoluminescent dosimeters* (TLDs), are based on a somewhat opposite approach. Instead of promoting the quick recombination of electron–hole pairs, materials are used which exhibit high concentrations of trapping centers within the bandgap. As illustrated in Fig. 19.16, the desired process is now one in which electrons are elevated from the valence to the conduction band by the incident radiation but are then captured at various trapping centers. If the distance of the trap energy level below the conduction band is sufficiently large, there is only a small probability per unit time at ordinary room temperatures that the electron will escape the trap by being thermally excited back to the conduction band. Therefore, exposure of the material to a continuous source of radiation, although not resulting in a significant yield of prompt scintillation light, leads to the progressive buildup of trapped electrons.

Holes can also be trapped in an analogous process. An original hole created by the incident radiation may migrate through the crystal until reaching a hole trap with energy somewhat above the top of the valence band. If this energy difference is large enough, the hole will not migrate further and is then locked in place unless additional thermal energy is given to the crystal. A sample of TLD material will therefore function as an integrating detector in which the number of trapped electrons and holes is a measure of the number of electron–hole pairs formed by the radiation exposure.

After the exposure period, the trapped carriers can be measured through a process also illustrated in Fig. 19.16. The TLD sample is placed in a stream of heated gas or on a heated support, and its temperature is progressively raised. At a temperature that is determined by the energy level of the trap, the trapped electrons can pick up enough thermal energy so that they are re-excited back to the conduction band. Assuming that this temperature is lower than that required to free the trapped holes, the liberated electrons then migrate to near a trapped hole, where they can recombine with the emission of a photon.

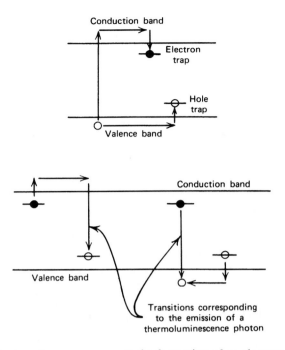

Figure 19.16 The top diagram represents the formation of an electron–hole pair in a TLD material, leading to population of electron and hole traps. The bottom diagram illustrates the two possible modes of recombination when the temperature is raised, which lead to the emission of a thermoluminescence photon.

Alternatively, if the holes are released at a lower temperature, they may migrate to a trapped electron and their recombination also results in a radiated photon. In either case, if the magnitude of the energy difference is about 3 or 4 eV, the emitted photons are in the visible region and are the basis of the TLD signal. Ideally, one such photon is emitted per trapped carrier. Therefore, the total number of emitted photons can be used as an indication of the original number of electron–hole pairs created by the radiation.

TLD systems thus derive a signal by using a heater in which the sample can be viewed by a photomultiplier tube. The light yield is recorded as a function of sample temperature in a *glow curve* of the type illustrated in Fig. 19.17. The basic signal related to the radiation exposure is the total number of emitted photons, or the area under the glow curve. If the sample is raised to a relatively high temperature, all the traps are depleted and the exposure record of the sample is "erased." TLD materials therefore have the very practical advantage of recyclability, and a single sample may be reused many times.

B. Thermoluminescent Materials

A thorough description of the properties and applications of different TLD materials is presented in Ref. 71. Some popular materials consist of crystals to which a small concentration of impurity has been added as an activator (e.g., $CaSO_4 : Mn$, where manganese is the activator). Others do not require the addition of an activator, but the traps are created by the inherent impurities and defects in the crystal. The choice among TLD materials must take into account considerations of trap depth and atomic number of the material. If the energy levels of the traps are very near the edge of the bandgap (as in $CaSO_4 : Mn$), the number of trapped carriers per unit exposure can be very large. Therefore, this material can be made sensitive to exposure as low as about 2×10^{-5} rads (0.2 μGy). The shallow traps

Figure 19.17 Typical thermoluminescent glow curves normalized to the same maximum intensity. Materials are A —$CaSO_4 : Mn$; B —LiF: Mg, Ti; C —CaF_2; and D — $CaF_2 : Mn$. The details of these curves will depend on preirradiation annealing procedures, radiation exposure level, and heating rate during readout. (From Fowler and Attix, in *Radiation Dosimetry,* 2nd ed., Vol. II (F. H. Attix and W. C. Roesh, eds.). Copyright 1966 by Academic Press. Used with permission.)

are somewhat unstable even at ordinary room temperature, however, and therefore the material will show a considerable "fading," which can lose as much as 85% of the trapped carriers over a few day's time.[72] Therefore, other materials such as $CaF_2 : Mn$ and LiF: Mg, Ti, with somewhat deeper traps, are better suited for longer-term exposures even though their sensitivity is several orders of magnitude less.

Of the many TLD materials, those based on LiF have been the most widely exploited. One of the most popular (known commercially as TLD100) has a concentration of about 400 ppm of Mg added to serve as the primary trapping centers. In addition, about 8 ppm of Ti are also added to provide *luminescent recombination centers* at which the detrapped holes and electrons recombine after being released from the trapping centers during the readout process (as opposed to their direct recombination shown in Fig. 19.16). The emitted light then has the characteristic spectrum of the Ti luminescence site, and reabsorption within the bulk of the material is minimized (as in an activated scintillation crystal). The emitted spectrum is a good match to the standard blue-sensitive photomultiplier tubes used for recording the glow curves. The trapped charges are quite stable at room temperature, and its fading rate is very low. A more recent TLD material (known as TLD200) is also based on LiF, but with Mg, Cu, and P activators. It shows a light output that is 10 times larger than TLD100 and is therefore applicable to lower exposures.

One of the major reasons for the popularity of LiF-based materials is the close match between the atomic numbers of its constituents and those of soft tissue. The energy deposited in LiF is therefore closely correlated with the gamma-ray exposure or dose equivalent over a wide range of gamma-ray energy. For TLD materials with higher atomic number, the enhanced photoelectric interaction probabilities exaggerate the response to low-energy X-rays or gamma rays. In Fig. 19.18 the relative response to 1 roentgen of gamma-ray exposure is plotted for a number of different TLD materials. Only the LiF curve remains reasonably constant over the wide range of photon energies plotted. Small wafers of TLD100 have thus come into widespread use as personnel dosimeters. These systems have replaced photographic film badges in most situations, because exposures may be measured directly by the user without the need for photographic developing of film, and TLDs may be reused many times by annealing at elevated temperatures. Because there is a great deal of variability between samples and methods of heating, absolute values of the radiation dose are normally determined by exposing a parallel sample to a known gamma-ray exposure. The minimum sensitivity of TLD100 is about 10^{-2} rads (100 μGy), and the signal remains linearly related to dose up to about 400 rads (4 Gy). At higher doses, this and other TLD materials display a nonlinear increase in response per unit exposure, a behavior known as *supralinearity*.[73-79]

Because natural lithium contains 7.4% ^{6}Li, TLD detectors made from LiF will be somewhat sensitive to slow neutrons through the (n, α) reaction described in Chapter 14. This response can be enhanced by using lithium enriched in ^{6}Li, or suppressed by using lithium consisting entirely of the ^{7}Li isotope. Some representative measurements of the response of such TLD materials to slow neutrons are reported in Refs. 80 and 81. TLD detectors sensitive to fast neutrons have also been fabricated by covering samples of a conventional material such as Al_2O_3 with polyethylene.[82, 83]

Specialized texts on the extensive subject of TLD detectors are readily available, and Refs. 84–87 are examples.

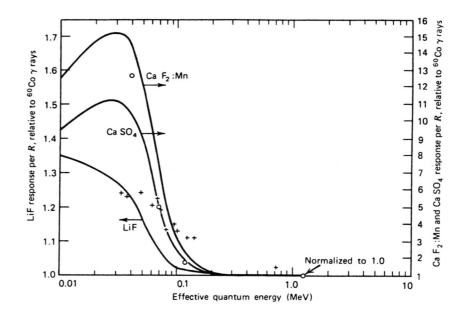

Figure 19.18 Variation with gamma-ray energy of the thermoluminescence response per roentgen. (Note the greatly different scale for LiF compared with the other materials.) Curves are calculated from gamma-ray interaction probabilities, assuming constant response per unit absorbed dose in the TLD material. Circles are experimental measurements for CaF_2 : Mn, crosses are measurements for LiF. (From Fowler and Attix, in Radiation Dosimetry, 2nd ed., Vol. II (F. H. Attix and W. C. Roesh, eds.). Copyright 1966 by Academic Press. Used with permission.)

C. Optically Stimulated Luminescence (OSL)

An alternative readout method can be applied to stimulate the luminescence from TLD materials. Instead of heating the material as in conventional TLD readout, a strong light source such as a laser or light-emitting diode can add energy to the trapped charges and cause their de-excitation through luminescence states. This technique makes use of the resulting *optically stimulated luminescence* (OSL), and has been exploited since the 1980s for the geological or archeological dating of materials through a measurement of their accumulated dose from natural sources of radiation. OSL has been demonstrated for a number of natural materials such as quartz and feldspar, or synthetic materials such as porcelain.[88] Beginning in the mid-1990s, similar optical techniques were introduced as readout techniques for other materials applied in personnel dosimetry.

In common with standard TLD readout, the technique consists of using a photomultiplier tube to measure the light that is emitted while energy is added to the sample. In the simplest form of optical readout, a continuous excitation source, such as a CW laser, illuminates the sample and the total light yield is measured over a period of many seconds. Because the excitation light is at a different wavelength from the emitted luminescence, optical filters can be used to shield the photomultiplier tube from the direct or reflected laser light used in this type of readout.

Alternatively, the readout laser or other light source can be pulsed. In this case the sample is illuminated for a brief time, and the subsequent luminescence is measured as it decays following the excitation pulse. The characteristics of the luminescence decay are generally complex, consisting of a mix of components with differing decay times. The relative intensities of these components can depend on the specific sample of material, temperature, absorbed dose level, and also the intensity of the readout light.[89] Multiple pulses of light are frequently employed, with the luminescence measurement made between successive pulses. If the pulse duration and observation time are relatively long, the process is sometimes called *delayed OSL* and will involve primarily the luminescence centers with longer lifetime. The readout time in this case might typically be 5 or 10 seconds. Alternatively, narrower light pulses can be used which are then followed by short duration observation times. This method, commonly known as *pulsed OSL,* is repeated through many excitation observation cycles and is primarily sensitive to the short-lived luminescent sites. This latter technique tends to be the most sensitive method of readout, and it can be accomplished in a shorter total time period than those described previously.

Some of the advantage of optically stimulated luminescence readout stems from the high sensitivity that is achievable. Using aluminum oxide,[90,91] often doped with carbon, sensitivity levels down to a few μGy (a few tenths of a mRad) have been demonstrated, well below that of standard TLD materials. The population of trapped charges also is not seriously depleted during pulsed optical readout, so an independent measurement of the absorbed dose can be made if desired at a later time. The samples can be annealed by exposure to intense blue light (of sufficient photon energy to detrap the charges) to eliminate nearly all the trapped charges and allow re-use of the sample. An additional advantage of OSL readout is that the sample temperature is not elevated significantly, so materials such as plastic substrates can be used that would not be compatible with the heating readout cycle required in standard TLDs.

D. Image Plates

The concept of optically stimulated readout of TLD materials has been extended in a unique form of position-sensitive radiation detector known as an *image plate.* (Other names sometimes used in the literature are *memory phosphors, storage phosphors, or photostimulable phosphors.*) Here the detector consists of a large-area plate (up to 40 cm on a side) that contains a layer of a phosphor that exhibits the same type of charge storage properties as found in TLD materials. The phosphor is generally mixed in an organic binder and deposited with a total layer thickness of up to 200 μm on a plastic substrate. The most common selection for the phosphor has been BaFBr: Eu^{2+} with a relatively high effective atomic number that helps promote its detection efficiency for incident radiations. For example, a 200-μm layer is about 50% efficient at detecting 50 keV X-rays, well above that of many alternative methods of X-ray imaging. Typical image plates are essentially 100% efficient for X-ray energies up to about 20 keV.[92]

Ionizing radiation interacting within the phosphor layer produces trapped electrons and holes very near the interaction site, forming a latent image of stored charges that represents the pattern of the incident radiation. This latent image shares the property of fading seen in many TLD materials, and as much as half of the stored charge can be lost as a result of spontaneous thermal detrapping over periods of hours. After the exposure, readout is accomplished by scanning a laser beam across the full area of the plate. The resulting luminescence light is conducted via optical fibers to a photomultiplier tube to measure its intensity as a function of the position of the laser spot. The output of the photomultiplier tube is processed by an analog-to-digital converter, and a digital image is recorded that corresponds to the original pattern of the incident radiation. The plate can be erased by exposure to intense light to allow its repeated use. The detailed physical mechanisms involved in the stimulation and luminescence processes are reviewed in Refs. 93–95.

A useful property of image plates is the large dynamic range that can be covered in linear fashion. From the weakest radiation that can be recorded up to the most intense where saturation occurs, at least five decades of exposure are normally possible, and eight decades have been demonstrated under some experimental conditions.[96] These figures can be contrasted with a value of about two decades for the dynamic range of radiographic film. The spatial resolution of the resulting image is limited primarily by the scattering and diffusion of the laser light[97] within the thickness of the phosphor layer. As a result, the best spatial resolution is obtained with thin layers, but at the sacrifice of detection efficiency. Typical image plates that are available commercially show a spatial resolution in the range of 100–200 μm.

Although used principally for X-ray imaging, image plates will also respond to other types of ionizing radiation. Applications have been demonstrated in which alpha particles[98] and fast electrons[99] are imaged. They have also been applied to thermal neutron imaging by using an external gadolinium converter[100] in contact with the plate, or by incorporating gadolinium or lithium as a component of the phosphor layer itself.[101,102] Image plates have also been shown to be useful in the autoradiography of very weak radioactivity using exposure times that can be extended over many days or weeks.[103,104]

VIII. TRACK-ETCH DETECTORS

A. The Track Registration Process

When an ionizing charged particle passes through a dielectric material, the transfer of energy to electrons results in a trail of damaged molecules along the particle track. In some materials, the track can be made visible upon etching in a strong acid or base solution. The entire surface of the material is attacked, but those points at which particle tracks have entered are etched about 10 times faster. The tracks can thus be made to form pits on the surface that are large enough to be easily visible through a conventional microscope. Materials used to detect particles in this manner are called *track-etch detectors*,[105] and their properties and applications have been thoroughly reviewed in a comprehensive book by Fleischer et al.[106]

In common with other passive detectors such as photographic emulsions or neutron activation foils, track-etch detectors have the advantages of simplicity and low cost. They also possess a very useful inherent threshold, in that there is a minimum value of the specific energy loss $(-dE/dx)$ required of the particle before the damage is severe enough to lead to an etchable track. The corresponding energy range for various charged particles over which tracks will be registered is illustrated in Figure 19.19. The threshold is always well above the specific energy loss of an electron track, so that track-etch materials are inherently insensitive to fast electron or gamma-ray interactions. Most materials also do not respond to lightly ionizing charged particles, such as protons or deuterons, and will therefore also be insensitive to the recoil protons produced by fast neutron interactions. However, some dielectrics will register proton tracks[107] over a certain range in energy.

Track-etch detectors share with nuclear emulsions the disadvantage of requiring individual track counting. When done by hand, the counting step is tedious at best and is very time consuming if a large number of tracks are recorded for statistical precision. Automatic counting systems have been developed but involve a sizable investment in cost and complication. Although some differentiation can be made of the particle type and energy based on appearance of the etched pit, many of the details of the original track are lost in the etching process. These detectors are therefore not nearly so well suited to the measurement of individual track properties as are nuclear emulsions.

The damage created by the incident particle can be through collisions of the particle itself or from the energetic delta rays (see Chapter 2) created along its track. The range of the delta rays may extend approximately 5 nm in any direction away from the particle position, so that the radius of the primary damage track is thought to be about this dimension.[106] After etching, the tracks are greatly enlarged to a diameter up to 10–20 μm. A photograph of etched tracks in polycarbonate film is shown in Fig. 19.20.

Etchable tracks are formed in a variety of materials. All are electrical insulators, although some wide-bandgap semiconductors are also known to record tracks. The materials fall into two main categories: inorganic solids, such as crystals and glasses, and organic solids, such as polymers. In the first category, the most popular materials are mica and flint glass, whereas polycarbonate and polyester films are the most common organic track-etch detectors. A list of the most useful materials is given in Table 19.2.

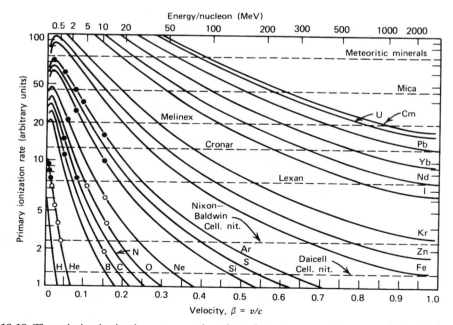

Figure 19.19 The relative ionization rate as a function of energy per nucleon or velocity is shown for various heavy ions. Horizontal dashed lines indicate the minimum damage rate required for 100% registration of tracks on some popular track-recording materials. The solid dots are experiments using accelerated ions on Lexan which gave 100% registration, and the open circles gave zero registration. (From R. L. Fleischer, P. B. Price, and R. M. Walker, *Nuclear Tracks in Solids.* Copyright 1975 by The Regents of the University of California; reprinted by permission of the University of California Press.)

Figure 19.20 Photograph of etched fission fragment tracks in polycarbonate film. The smallest scale division represents 10 μm. (Photo courtesy of D. M. Gilliam, National Bureau of Standards.)

Table 19.2 Commonly Used Track Etch Materials

	Atomic Composition	Least Ionizing Ion Seen
Inorganic Materials		
Quartz	SiO_2	100 MeV ^{40}Ar
Phlogopite mica	$KMg_2Al_2Si_3O_{10}(OH)_2$	
Muscovite mica	$KAl_3Si_3O_{10}(OH)_2$	2 MeV ^{20}Ne
Silica glass	SiO_2	16 MeV ^{40}Ar
Flint glass	$18SiO_2 : 4PbO : 1.5Na_2O : K_2O$	2–4 MeV ^{20}Ne
Organic Materials		
Polyethylene terephthalate (Cronar, Melinex)	$C_5H_4O_2$	36 MeV ^{16}O
Bisphenol A-polycarbonate (Lexan, Makrofol)	$C_{16}H_{14}O_3$	0.3 MeV ^4He
Polymethylmethacrylate (Plexiglas, Lucite, Perspex)	$C_5H_8O_2$	3 MeV ^4He
Cellulose triacetate (Cellit, Triafol-T, Kodacel TA-401 unplasticized)	$C_3H_4O_2$	
Cellulose nitrate (Daicell)	$C_6H_8O_9N_2$	0.55 MeV ^1H

Source: R. L. Fleischer, P. B. Price, and R. M. Walker, *Nuclear Tracks in Solids.* Copyright 1975 by The Regents of the University of California; reprinted by permission of the University of California Press..

A completely satisfactory model for the mechanisms that lead to the registration of tracks does not yet exist. Past attempts have been made to link the etchability of tracks to the density of atomic displacements or the creation of a "thermal spike" of molten material along the particle track. Another hypothesis, known as the *ion explosion spike* model,[106] is based on the momentary electric field created along the particle track by charge imbalance, caused by the ejection of many electrons from the immediate region of the track. None of these approaches can fully account for some of the observed differences in etching behavior. The threshold of track etching is best correlated with the *restricted energy loss*, which is tied to the energy deposition rate only along the primary particle path, excluding that of long-range delta rays.[108] Because the energy imparted to the delta rays is one component of the more traditional specific energy loss dE/dx, the restricted value is somewhat smaller, particularly at high particle energies.

B. Track Etching

A qualitative picture of how tracks are revealed by etching is sketched in Fig. 19.21a. A track is assumed to exist perpendicular to the surface of the medium, which is exposed to the etching solution. As a simplified model, we assume that the undamaged surface is

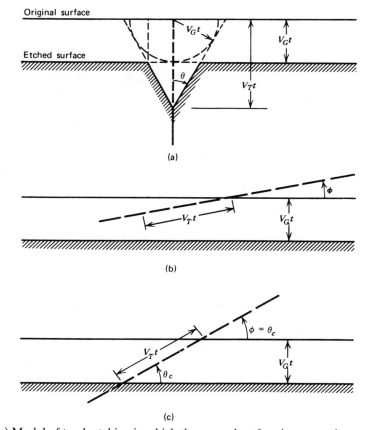

Figure 19.21 (*a*) Model of track etching in which the normal surface is removed at a velocity V_G and the damaged track at a velocity V_T, leading to a cone-shaped pit. (*b*) Tracks formed at an angle ϕ less than the critical angle θ_c are not revealed because the normal surface advances faster than the etch rate along the track. (*c*) The case in which the particle enters at the critical angle θ_c. Tracks entering the surface at an angle greater than the critical angle will be visible after etching. (From R. L. Fleischer, P. B. Price, and R. M. Walker, *Nuclear Tracks in Solids*. Copyright 1975 by The Regents of the University of California; reprinted by permission of the University of California Press.)

eroded away at a velocity V_G perpendicular to the surface. We further assume that the etching velocity along the damaged track is a greater value V_T. Under these conditions, the sketches illustrate that a cone-shaped pit is formed with an axis along the damaged track. Although this model is crude because it does not account for changes in etching rate with depletion of the etching solution or differences in the degree of damage along the track, it is adequate to explain many features of the observed behavior. For example, tracks that make a small angle with the surface can be erased by the etching process. Using the above model, it can be shown that the angle of incidence must exceed a critical angle θ_c in order to avoid its disappearance due to the progressive etching of the normal surface. This critical angle is given by

$$\theta_c = \arcsin\left(\frac{V_G}{V_T}\right) \tag{19.4}$$

For example, in polyester materials, the critical angle for track registration is about 5–15°. Because of variability of the etching conditions and the energy loss threshold, the etching behavior should be demonstrated in advance of any measurement, using a calibrated source of the same type and energy of particles involved in the measurement. A number of environmental factors can have a significant effect on the etching behavior. For example, the sensitivity of certain plastic track-etch detectors depends on the presence or absence of oxygen during the exposure.[106] The formation of tracks may be enhanced through the application of an electric field[109] or through exposure to ultraviolet radiation.[110]

The temperature history of the track-etch material is also important. By holding the material at elevated temperatures for some period of time, latent tracks may be annealed to the point that etched tracks will no longer be formed. In this way, track-etch detectors can be "erased" prior to the start of a given measurement. However, the potential also exists for the inadvertent loss of desired tracks if the time and/or temperature between an exposure and etching are too great.[111] Because the annealing properties of track-etch materials vary greatly, even between different samples of the same material, the fading properties of tracks should be investigated experimentally. It is empirically observed that the time required for a given fraction of the tracks to disappear can be described by a relation of the type

$$t = A \exp\left(\frac{U}{kT}\right) \tag{19.5}$$

where A and U are constants characteristic of the material, and kT is the Boltzmann temperature.

C. Track Counting and Applications

Manual counting procedures are normally required if a high degree of counting precision is needed, or if detailed information on the shape and length of the track is desired. Track densities under about $10^4/\text{cm}^2$ can manually be counted using a simple microscope or, in a somewhat less tedious manner, through the use of a projection microscope by which images of the etched tracks are displayed on a large screen. The etched material, if optically transparent, can also be placed in a photographic enlarger, and counting can be done from a photographic print.

A number of automatic schemes have also been developed for the counting of etched tracks. Many are based on optical methods in which a computer-controlled microscope stage allows examination of each individual track using a photomultiplier tube. One version[112] incorporates automatic scanning and self-focusing and is about 98% accurate in counting fission fragments in polycarbonate. Somewhat less accurate machines can be adapted from commercially available optical counting systems.

For some thin film track-etch detectors, automatic counting can be carried out through spark counting.[113] The tracks are etched to the point that they completely penetrate the thin film used for registration. When the film is slowly moved between high-voltage electrodes, a spark is formed whenever an etched hole passes between the electrodes. The number of these discharges can be recorded electronically as an indication of the total number of etched tracks. Other methods, based on the transmission of charged particles through the etched holes, have also been demonstrated.[114]

If individual tracks need not be counted, the overall track density can be indicated by several other means. Simple attenuation of a collimated light beam through the etched film provides one overall indicator of track density. Another method is based on measuring the total resistivity of an etched film over a local area of its surface. These latter methods can be useful as an indicator of overall exposure of the film to charged particles for dosimetry purposes.

Track-etch detectors can also be applied to neutron detection through the use of a converter such as ^{6}Li or ^{10}B, in conjunction with a material that registers alpha tracks. Alternatively, a foil of fissionable material can be applied to generate fission fragments that are easily registered in all track-etch materials. If the converter foil and registration material are kept in close contact, an image of the detected neutron distribution can be formed after etching, a technique that has proved to be useful in neutron radiography.

IX. SUPERHEATED DROP OR "BUBBLE DETECTORS"

A. General Principles

A unique type of radiation detector sketched in Fig. 19.22 is based on a suspension of many small droplets of a fluid held within an inert matrix that is immiscible with the fluid. For example, droplets of freon can be held in a polymer or gel matrix. Typical detectors consist of a few cubic centimeters of this emulsion that contain several tens of thousands of droplets, each with a diameter of 100 μm or less. Optical properties of the materials are chosen so that the emulsion appears virtually transparent while the droplets are suspended in their liquid state. Pressure and temperature conditions are adjusted so that the liquid droplets become metastable in a "superheated" or "overexpanded" state. The liquid would normally want to flash into a vapor, but some sort of nucleation trigger is required for the small droplets to undergo this phase transition.

That trigger can be provided by the charged particles produced by interactions of incident neutrons. According to models of this process,[115–118] the deposition of energy along the track of these charged particles produces local regions of high energy density (or a

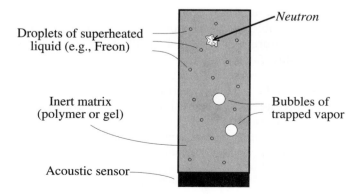

Figure 19.22 Principle of the bubble detector for neutrons. The acoustic sensor converts the device into an active detector.

"thermal spike") that causes local vaporization. If the bubble of vapor around the track reaches a certain critical size, it will trigger further vaporization and the entire droplet will flash into a bubble of vapor. This bubble is typically up to a millimeter in diameter, and is now easily seen by simple visual inspection. In general, only heavy charged particles will produce the high energy density needed to reach this critical vaporization stage, and the droplets are insensitive to the lower dE/dx electrons produced by incident gamma rays. The bubbles may remain fixed within the matrix, or in some versions can be collected by their gravity-driven motion to a specific location within the vial that contains the emulsion.

Because of the critical energy deposition requirement, there is a threshold in the energy of the charged particle needed to trigger bubble nucleation. It can be shown[119] that this minimum energy is proportional to the inverse square of the difference between the fluid vapor pressure and the local pressure within the vial. Fast neutrons can produce recoil nuclei with energies above this minimum by elastic scattering within the fluid droplets. At high degrees of superheat, the sensitivity increases because of the additional contribution of recoil protons from the surrounding gel that enter liquid droplets (the droplets themselves are normally free of hydrogen).

Thermal or intermediate energy range neutrons do not create recoils with sufficient energy. However, bubble detectors can be made sensitive to these neutrons as well by incorporating an element in the fluid that produces energetic charged particles through a neutron-induced reaction. The common example is to choose a chlorine-containing fluid that will then respond to low-energy neutrons through the reaction $^{35}Cl(n,p)^{35}S$, which will produce 17 keV sulfur recoils with a high energy density along their tracks. This capture reaction has its highest cross section for slow neutrons. It is therefore often triggered by neutrons that have entered the surrounding matrix and undergone a number of moderating collisions before diffusing to the droplet as a low energy neutron.[120]

B. Passive Bubble Detectors

In their simplest form, bubble detectors consist of a vial of a few cubic centimeters of emulsion in which the pressure has been lowered below the vaporization point of the fluid by means of an adjustable plunger or piston. It then operates as a passive device, and bubbles accumulate as neutron interactions cause the vaporization of individual droplets. At the end of the measurement, the number of bubbles can simply be counted manually or they can be automatically quantified using an optical scanner. The maximum number of counts for a passive detector is limited by the optical overlap of bubbles, making their quantitative count increasingly difficult beyond a few hundred. After the measurement, the detector can be recycled by applying pressure to recondense the bubbles back into liquid droplets, and the process repeated.

Because of the strong dependence of the triggering mechanism on the degree of superheat, bubble detectors show a strong inherent dependence of their sensitivity on ambient temperature. Under typical conditions, the number of bubbles produced by a given neutron exposure can change by 5% per °C. One method to reduce this temperature dependence is to incorporate a volatile liquid in the free space within the vial above the emulsion surface. As the temperature rises, the vapor pressure of this additive also increases, increasing the pressure on the emulsion. The higher pressure tends to reduce the change in superheat and can keep the temperature dependence at a much lower level. Another solution[120,121] is to use a temperature sensor to control heating strips in contact with the insulated vial to maintain a constant temperature somewhat above ambient room temperature.

C. Active Bubble Detectors

Bubble detectors can also be operated as active devices by using an acoustic sensor that is placed in contact with the vial containing the emulsion (see Fig. 19.22). As bubbles are formed, the sound that is produced can be picked up by the sensor and converted into an electrical pulse. This pulse tends to be a damped sinusoid with a typical duration of several milliseconds,[122] so that these applications are limited to lower rates than are typical of other electronic neutron detectors. Impact or vibration can also produce acoustical signals and potentially can cause false counts. For that reason, a second transducer is often incorporated into active instruments that is not in direct contact with the vial, and digital pulse shape analysis is carried out on the output of both sensors in order to distinguish only the bubble nucleation events and discard signals caused by vibration.

D. Applications in Neutron Dosimetry

The principal application of bubble detectors is in the measurement of fast neutron dose equivalent. For this purpose, ideally one would like a detector that produces one bubble per unit neutron dose, independent of the incident neutron energy. Then the number of bubbles accumulated over an exposure time can be a measure of the total dose, without knowledge of the incident neutron spectrum. This type of behavior requires that the detection efficiency of the device as a function of neutron energy match the curve of dose delivered by neutrons as a function of energy. It is found experimentally[120,121] that bubble detectors of the right composition can approximate this behavior. They respond efficiently at high neutron energies (where the dose per incident neutron is high) through direct recoils, and are less efficient for the detection of intermediate and slow neutrons (through the sulfur capture reaction) where the dose implication of these neutrons is lower. Typical sensitivities are in the range of a few bubbles/μSv (tens of bubbles per mRem), a level that is useful for routine personnel monitoring.

The threshold in energy for neutrons to produce bubbles by direct recoils varies with the degree of superheat, with this threshold decreasing as the degree of superheat is increased. Thus by properly adjusting either the temperature or pressure, a set of vials can be produced with response curves similar to those sketched in Fig. 19.23. These could be identical vials in which the pressures are different or vials operated at different temperatures. The set can then be used to do a form of fast neutron spectroscopy similar to that described in Chapter 15 for Bonner spheres. Knowing the shape of the response curves, unfolding codes can take as input the number of bubbles recorded in each vial and produce a calculated version of the fast neutron spectrum.[118,123,124]

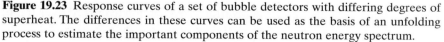

Figure 19.23 Response curves of a set of bubble detectors with differing degrees of superheat. The differences in these curves can be used as the basis of an unfolding process to estimate the important components of the neutron energy spectrum.

E. Other Considerations

As any bubble detector is exposed to neutrons, the available number of droplets decreases as they flash into bubbles. Under constant irradiation conditions, this depletion will lead to an exponential decrease in the sensitivity as a function of time. In active detectors, this efficiency change can be compensated by an electronic correction that is based on knowledge of the initial number of droplets and the number of recorded counts. Electronic compensation for sensitivity changes that are the result of temperature variation can also be incorporated.

While the principal application has been for the detection of neutrons, some versions of bubble detectors[125] have been investigated that can respond to low dE/dx particles produced by gamma rays. These gamma-sensitive bubble detectors have not reached the point of widespread application.

It has also been demonstrated[122,126] that bubble detectors can be made to be position-sensitive by placing acoustic sensors at both ends of a long vial. The time difference in the arrival of the acoustic signal at each of the sensors then provides a measure of the one-dimensional position of the neutron interaction within the vial. The relatively slow propagation velocity of sound allows good localization of the event, and a position resolution of 1 mm has been achieved using this approach.

X. NEUTRON DETECTION BY ACTIVATION

The neutron detectors discussed in Chapters 14 and 15 produce a prompt output pulse for each detected neutron. Neutron measurements can also be carried out indirectly through the radioactivity that is induced in some materials by neutron interactions. A sample of such a material can be exposed to a flux of neutrons for a period of time and then removed so that the induced radioactivity may be counted, using any of the conventional methods discussed earlier in this text. The measured radiations can then be used to deduce information about the number and/or energy distribution of the neutrons in the original field. The materials used in this way are often called *activation detectors*, and their applications are widely discussed in texts on neutron physics, such as Ref. 127.

Because neutron reaction cross sections are highest at low neutron energies, activation detectors are most commonly applied to the measurement of slow neutrons. To achieve a high degree of sensitivity, materials are chosen that have a large cross section for a neutron-induced reaction, which leads to a measurable form of radioactivity. Because the mean free path of neutrons in materials of high cross section is quite small, the thickness of the material must be kept small to avoid perturbing the neutron flux under measurement. Hence, the common geometric form of the material is that of a thin foil or small-diameter wire.

A. Activation and Decay

In the simplest case, the foil or wire is so thin that the probability of an interaction is small for any specific neutron. Then, the neutron flux remains unperturbed, and the rate R at which activation interactions occur within the foil is given by

$$R = \varphi \Sigma_{\text{act}} V \qquad (19.6)$$

where
φ = neutron flux averaged over the foil surface

Σ_{act} = activation cross section averaged over the neutron spectrum

V = foil volume.

Therefore, the rate of activation per unit mass is a direct indicator of the neutron flux magnitude.

As the foil is irradiated, the radioactive nuclear species that is formed also undergoes radioactive decay. The rate of decay is given simply by λN, where λ is the decay constant and N is the total number of radioactive nuclei present. The rate of change in N is given by the difference between the rate of formation and rate of decay

$$\frac{dN}{dt} = R - \lambda N \tag{19.7}$$

We assume that R is a constant, implying that the neutron flux does not vary during the exposure, and neglecting any "burnup" or decrease in the number of target nuclei over the measurement. The solution to Eq. (19.7) for the condition $N = 0$ at $t = 0$ is

$$N(t) = \frac{R}{\lambda}(1 - e^{-\lambda t}) \tag{19.8}$$

The activity A of the foil is given by λN, or

$$A(t) = R(1 - e^{-\lambda t}) \tag{19.9}$$

This induced activity therefore builds up with time, as shown in Fig. 19.24, and approaches an asymptote or *saturated activity* for infinitely long irradiation times given by

$$A_\infty = R = \varphi \Sigma_{\text{act}} V \tag{19.10}$$

Exposure times of three or four values of the half-life of the induced activity are sufficient to bring the foil activity to within 6–12% of the saturated value.

We assume that the irradiation has proceeded for a time t_0 at which time the foil is removed with an activity A_0:

$$A_0 = A_\infty(1 - e^{-\lambda t_0}) \tag{19.11}$$

After exposure to the neutron flux, the foil is transferred to an appropriate radiation counter for measurement of its activity. Because the activity is continuously decaying during this stage, careful account must be made of each of the times involved. If the counting is carried out over an interval between t_1 and t_2, the number of counts will be (see Fig. 19.24)

$$C = \epsilon \int_{t_1}^{t_2} A_0 e^{-\lambda(t - t_0)}\, dt + B$$

$$= \epsilon \frac{A_0}{\lambda} e^{\lambda t_0}(e^{-\lambda t_1} - e^{-\lambda t_2}) + B \tag{19.12}$$

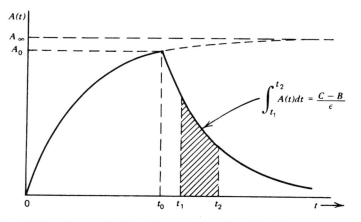

Figure 19.24 The activity of an activation detector after insertion in a constant neutron flux at time = 0 and removal at time = t_0. The measured number of counts is proportional to the area under the decay curve between t_1 and t_2. All times are measured from the start of irradiation at $t = 0$.

where ϵ is the overall counting efficiency (including any self-absorption effects) and B is the number of background counts expected in $t_2 - t_1$.

By combining Eqs. (19.11) and (19.12), we obtain for the saturated activity

$$A_\infty = \frac{\lambda(C - B)}{\epsilon(1 - e^{-\lambda t_0})e^{\lambda t_0}(e^{-\lambda t_1} - e^{-\lambda t_2})} \tag{19.13}$$

from which the neutron flux magnitude may be calculated from Eq. (19.10).

Activation foils are integrating detectors and therefore can provide no information about any time variation of the neutron flux over the course of the exposure. They have the advantages of small size, insensitivity to gamma radiation, and low cost. They can also tolerate exposure to extreme environments where other detectors might fail, and they require no electrical connections to the outside world. Activation foils are thus widely used for mapping the spatial variation of steady-state neutron fluxes in reactor cores, where the extreme temperature, pressure, and limited space severely constrain the type of conventional detector that may be used (see Chapter 14).

B. Activation Detector Materials

In choosing the material for an activation detector, a number of nuclear and physical properties need to be considered.

1. SHAPE OF THE CROSS SECTION

The material will obviously respond preferentially to those neutron energies for which its cross section is high. Radiative capture or (n, γ) reactions typically have largest cross sections at or near thermal energies, and therefore materials in which these reactions predominate are preferentially chosen for slow neutron detectors. Other reactions such as (n, p), (n, α), and (n, 2n) often require a minimum or threshold energy. Materials in which these reactions predominate can therefore be candidates as fast neutron detectors, which will be inherently insensitive to slow neutrons.

2. MAGNITUDE OF THE CROSS SECTION

As shown by Eq. (19.10), the saturated activity is linearly proportional to the average cross section for the activation reaction. Greatest sensitivity is therefore achieved by selecting materials with highest activation cross section, although some of this advantage may be offset by the need to keep the neutron absorption probability small enough to avoid perturbing the flux under measurement.

3. DECAY CONSTANT OF THE INDUCED ACTIVITY

The half-life of the induced activity should be neither too short nor too long, and for many applications, a value of a few hours is near the optimum. Long half-lives require the use of long irradiation times in order to approach saturation, and the specific activity becomes smaller with increasing half-life. Very short half-lives can lead to difficulties in transferring the foil to the counter without excessive delay time. The corresponding high activities may also create problems due to dead time effects within the counter.

4. PURITY AND INTERFERING ACTIVITIES

Very high purity of the material is often required to avoid interference from other neutron-induced reactions. If the half-life of the contaminant activity is sufficiently different from that of the primary, however, it may not pose a practical problem. Interfering activities with short half-lives can be eliminated simply by waiting some time before beginning the counting period, whereas those with long half-lives may not build up to a significant level if the irradiation period is short by comparison.

5. NATURE OF THE INDUCED ACTIVITY

The decay of the product nuclei produced in the activation reactions normally involves the emission of either beta particles or gamma rays. Gamma counting is often preferred because the penetrating nature of the radiation minimizes the effects of self-absorption within the sample. Furthermore, it is far easier to perform energy spectroscopy with gamma rays than beta particles, so that interfering activities and background can be discriminated more easily. In some cases, however, only beta activity is produced so that there is then no recourse but to carry out some type of beta counting. Continuous flow proportional counters are then often used in either 2π or 4π geometry.

6. PHYSICAL PROPERTIES

The properties of the activation materials play an obvious part in determining the physical environment in which they may be exposed. Materials in gaseous or liquid form are very difficult to apply except through elaborate encapsulation techniques that may interfere with the subsequent counting of the activity. Therefore, almost all activation detectors consist of metallic foils or wires that can be cut to the desired dimensions.

Table 19.3 lists a number of materials that are useful as detectors of slow neutrons. In each case, the activity is induced by an (n, γ) reaction that leads to the indicated radioactive product. In the thermal region, these materials have activation cross sections that vary approximately as $1/v$, where v is the neutron velocity. Many also show significant resonances in the cross section at certain neutron energies between about 1 and 1000 eV. The observed activity then corresponds to a mixture of activation caused by thermal neutrons, plus an additional component for neutrons with energy in the resonance region.

The contributions of the thermal and resonance neutrons can be separated through a technique known as the *cadmium difference* method. Cadmium is a metal whose radiative

Table 19.3 Materials Useful as Slow Neutron Activation Detectors

Element	Isotope (Abundance in Percent)	Thermal Activation Microscopic Cross Section (in 10^{-28} m^2)	Induced Activity	Half-Life
Manganese	^{55}Mn (100)	13.2 ± 0.1	^{56}Mn	2.58 h
Cobalt	^{59}Co(100)	16.9 ± 1.5	^{60m}Co	10.4 min
		20.2 ± 1.9	^{60}Co	5.28 y
Copper	^{63}Cu(69.1)	4.41 ± 0.20	^{64}Cu	12.87 h
	^{65}Cu(30.9)	1.8 ± 0.4	^{66}Cu	5.14 min
Silver	^{107}Ag(51.35)	45 ± 4	^{108}Ag	2.3 min
	^{109}Ag(48.65)	3.2 ± 0.4	^{110m}Ag	253 d
Indium	^{113}In(4.23)	56 ± 12	^{114m_1}In	49 d
		2.0 ± 0.6	^{114}In	72 s
	^{115}In(95.77)	160 ± 2	^{116m_1}In	54.12 min
		42 ± 1	^{116}In	14.1 s
Dysprosium	^{164}Dy(28.18)	2000 ± 200	^{165m}Dy	1.3 min
		800 ± 100	^{165}Dy	140 min
Gold	^{197}Au (100)	98.5 ± 0.4	^{198}Au	2.695 d

Source: K. H. Beckurts and K. Wirtz, *Neutron Physics*. Copyright 1964 by Springer-Verlag. New York. Used with permission.

capture cross section behaves in a very convenient manner. The cross section is very large for neutron energies below about 0.4 eV, where it drops abruptly and remains low for higher neutron energies. Thicknesses of the order of 0.5 mm of cadmium therefore act as a selective filter, blocking neutrons with energy below 0.4 eV but passing higher-energy neutrons with little attenuation. If one of the materials in Table 19.3 is made into two identical foils, one of which is covered with cadmium, the uncovered foil will respond both to the thermal and resonance neutrons, whereas the covered foil will indicate only the resonance contribution. By taking the difference between the two activations, the thermal contribution can be determined. The method does require some corrections for the nonideality of the cadmium filter, in that cadmium thicknesses that are large enough to fully stop all neutrons below 0.4 eV can also have a measurable effect on neutrons with higher energies. However, for most activation materials, corrections for the attenuation of the resonance neutrons are less than a few percent. By dividing the activity of the uncovered foil by that of the cadmium-covered one, the *cadmium ratio* is derived, which is often taken as an indication of the degree to which a given neutron field has been thermalized.

The activation method can be extended to the measurement of higher energy neutrons through the use of materials such as those listed in Table 19.4. The useful reactions in this group are threshold reactions, which require neutrons above a minimum energy in order to take place at all. Each material with a different threshold will respond to a somewhat different range of neutron energy. By exposing a set of these threshold activation foils to a given neutron field, the known differences in the shape of the cross sections can serve as a basis of

Table 19.4 Materials Useful as Threshold Activation Detectors

Material	Reactions of Interest	Isotopic Abundance (at %)	Half-Life	γ Energy (MeV)	γ Abundance (%)	Threshold (MeV)
F	$^{19}F(n, 2n)^{18}F$	100.0	109.7 min	0.511^+	194°	11.6
Mg	$^{24}Mg(n, p)^{24}Na$	78.7	15.0 h	1.368	100	6.0
Al	$^{27}Al(n, \alpha)^{24}Na$	100.0	15.0 h	1.368	100	4.9
Al	$^{27}Al(n, p)^{27}Mg$	100.0	9.46 min	0.84–1.01	100	3.8
Fe	$^{56}Fe(n, p)^{56}Mn$	91.7	2.56 h	0.84	99	4.9
Co	$^{59}Co(n, \alpha)^{56}Mn$	100.0	2.56 h	0.84	99	5.2
Ni	$^{58}Ni(n, 2n)^{57}Ni$	67.9	36.0 h	1.37	86	13.0
Ni	$^{58}Ni(n, p)^{58}Co$	67.9	71.6 d	0.81	99	1.9
Cu	$^{63}Cu(n, 2n)^{62}Cu$	69.1	9.8 min	0.511^+	195°	11.9
Cu	$^{65}Cu(n, 2n)^{64}Cu$	30.9	12.7 h	0.511^+	37.8°	11.9
Zn	$^{64}Zn(n, p)^{64}Cu$	48.8	12.7 h	0.511^+	37.8°	2.0
In	$^{115}In(n, n')^{115m}In$	95.7	4.50 h	0.335	48	0.5
I	$^{127}I(n, 2n)^{126}I$	100.0	13.0 d	0.667	33	9.3
Au	$^{197}Au(n, 2n)^{196}Au$	100.0	6.18 d	0.33–0.35	25–94	8.6
Li	$^{7}Li(n, \alpha n')t$	92.58	12.3 y	$0–0.019^\times$	$100^\times$	3.8

$^+$Annihilation radiation.

°Yield of annihilation photons assuming all positrons are stopped.

$^\times \beta$ particle energy and percent abundance.

Source: Kuijpers et al.[128]

an unfolding of the neutron energy distribution. Computer codes have been developed to carry out this unfolding process and have achieved a considerable degree of success.[128,129]

Accurate application of the activation technique in the quantitative measurement of neutron fluxes requires the application of a number of other correction factors. These involve the effects of the foil itself on the neutrons that are being measured. In a diffusing medium, the neutron flux in the immediate vicinity of the foil will be depressed due to the fact that some neutrons have been removed in their passage through the foil. Furthermore, the effective neutron flux at the center of the foil will be somewhat less than that at its surface due to attenuation or *self-shielding*. These effects are minimized by keeping the foil very thin, but the induced activity may then be so low as to create measurement difficulties. A complete analysis and discussion of the techniques required to account for flux depression and self-shielding of activation foils is presented in the text by Beckurts and Wirtz.[127]

C. Activation Counters

The discussion of activation materials in the previous section has emphasized techniques in which the induced activity is measured in a separate counting facility. A related class of instruments, commonly called *activation counters*, consists of a sample of the activation material placed in close proximity to a detector capable of responding directly to the induced activity. Activation counters have proved to be very useful in the measurement of fast neutrons produced in short bursts, typical of conditions encountered at accelerator facilities. The burst duration may be only a few nanoseconds or less, so pile-up precludes the use of pulse mode detectors such as the organic scintillators discussed in Chapter 15. At the same time, the neutron bursts may be infrequent enough so that the activity induced in standard activation foils is too small to be useful. By using appropriate materials coupled to efficient detectors, adequate sensitivity may be possible using the counters described below.

There are two general approaches to the design of activation counters. One is to employ slow neutron activation materials such as silver or rhodium (see Table 19.5) inside

Table 19.5 Some Neutron-Induced Reactions of Interest in Activation Counters

Percent Abundance	Reaction	Induced Activity	Half-Life	Thermal Cross Section	Application Reference
		SLOW NEUTRON REACTIONS			
51.8%	^{107}Ag(n, γ)^{108}Ag	1.49 MeV β$^-$	2.3 min	30 b ⎫	
48.2%	^{109}Ag(n, γ)^{110}Ag	2.24 or 2.82 MeV β$^-$	24.2 s	110 b ⎬	130–132
100%	^{108}Rh(n, γ)^{109}Rh	⎧ 2.47 MeV β$^-$	44 s	139 b ⎫	
		⎩ 2.44 MeV β$^-$	265 s	11 b ⎭	133

Percent Abundance	Reaction	Induced Activity	Half-Life	Threshold Energy	Application Reference
		THRESHOLD REACTIONS			
100%	^{75}As(n, n′)^{75m}As	0.30 MeV γ	17 ms	0.3 MeV	134
21.7%	^{207}Pb(n, n′)^{207m}Pb	24–304 keV γ	810 ms	1.6 MeV	135, 136
99.8%	^{16}O(n, p)^{16}N	⎧ 6.13 MeV γ	7.2 s ⎫	10.2 MeV	137
100%	^{19}F(n, d)^{16}N	⎩ 4.27–10.4 MeV β	⎭	~3 MeV	138
100%	^{23}Na(n, α)^{20}F	5.4 MeV β$^-$, 1.63 MeV γ	11.0 s	~7 MeV	139
100%	^{9}Be(n, α)^{6}He	3.51 MeV β$^-$	800 ms	~2 MeV	140

a moderating structure used to reduce the average energy of the incident fast neutrons. An example is the *silver counter* consisting of silver-wrapped (or silver-walled) G-M counters placed within a paraffin or polyethylene moderator.[130,131] The beta particles emitted by the activation products appear with half-lives of 24 s and 2.3 min and are detected with good efficiency by the G-M tube. The total number of counts recorded after many half-lives have elapsed is proportional to the neutron burst intensity. Other designs (see Fig. 19.25) have incorporated layers of silver with alternating plastic scintillator slabs to detect the beta particles, or a foil of rhodium in contact with a plastic scintillator.[133]

An alternative approach is to use threshold activation materials of the type listed in Table 19.4 and to rely on direct activation by the fast neutrons without moderation. This choice has the advantage of immunity from the effects of thermalized or low-energy neutrons often present as a background. However, now the reaction cross sections are much smaller than those for typical slow neutron activation materials. To boost the sensitivity during a short activation period, there is an advantage in choosing materials for which the induced activity has a much smaller half-life than those shown in Table 19.4. Then the measured activity can more closely approach the saturated value (A_∞ in Fig. 19.24). The half-life must be kept large enough to spread out the counts sufficiently so that dead time and pile-up effects in the detector are not serious. Attention has therefore focused on reactions that lead to half-lives of a few seconds or less, still long compared with typical time spacing between neutron bursts. The integrated counts following each burst remain, under these conditions, a linear measure of the total neutron intensity of each burst.

Figure 19.25 Example of a specific design for a silver activation counter. When used with fast neutrons, the counter is normally placed within a polyethylene moderator. The plot shows the counting rate observed for a period from 450 to 800 s following exposure to a fast neutron flux. (From Slaughter and Pickles.[132])

Some examples of threshold activation reactions that have been applied in such counters are listed in Table 19.5. In several designs,[134-140] the activation material is placed in contact with a plastic scintillator to count the beta or gamma rays emitted by the product. The oxygen reaction has been exploited[137] by using a liquid scintillator as both the source of oxygen and to detect the resulting ^{16}N gamma rays. Similarly, a sodium iodide scintillator has been used[139] to provide sodium nuclei for the ^{23}Na(n, α) reaction as well as to detect the beta and gamma rays emitted in the decay of the ^{20}F product. The fluorine reaction has been conveniently applied[138] by fitting a teflon cap over a germanium detector used to record the resulting gamma activity. In this case, one must bear in mind the sensitivity of germanium detectors to radiation damage caused by fast neutrons (see p. 451).

XI. DETECTION METHODS BASED ON INTEGRATED CIRCUIT COMPONENTS

A. Direct Ion Storage Dosimeter

The *direct ion storage* (DIS) dosimeter is based on the coupling of a gas-filled ion chamber with a semiconductor nonvolatile memory cell. These silicon-based devices (known as EPROMs, EEPROMs, or Flash-Memories) came into prominence in the early 1990s as a means to store analog information and have been widely applied in voice-recording consumer electronics. In contrast with standard memory cells that can only store a bit of digital information, EEPROMs have the capability of storing a variable analog voltage for indefinite times. In its standard configuration, the EEPROM consists of the structure shown in Fig. 19.26a. The information is stored as a charge on a "floating gate" that is completely surrounded by highly insulating silicon oxide. The charge is initially placed on the floating gate by injecting electrons by the tunneling process through the oxide. This charge will remain in place for years or longer since the electrons have no conductive discharge path. The magnitude of the stored charge can be measured nondestructively by monitoring the current between the source and drain.

For use in the DIS dosimeter, the EEPROM is modified in that the control gate is absent, and an opening in the oxide layer is provided to expose the top surface of the

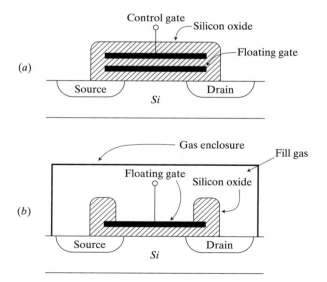

Figure 19.26 (*a*) Configuration of a non-volatile memory cell (EEPROM). (*b*) The direct ion storage dosimeter (DIS) consisting of the combination of a gas-filled ion chamber and an EEPROM.

floating gate. This surface is in contact with a gas that is contained within a conductive outer casing, creating a small ionization chamber (see Fig. 19.26b). If an initial voltage is placed on the floating gate, it will create an electric field throughout the gas volume that will serve to separate the ion pairs formed by ionizing radiation. The collected ionization charge serves to discharge the floating gate, and the change in the source–drain conduction can be measured in an electronic readout unit.

The voltage on the floating gate is generally limited to 30 V or less. In order to reach high enough electric field values to suppress recombination within the gas, the dimensions of the chamber must be kept no larger than a few millimeters. The uniformity of charge collection can be improved by connecting a conductive mesh in contact with the floating gate into the gas volume.[141] The low capacitance of the small structures results in an efficient conversion of charge to voltage, and excellent sensitivity of the DIS to well below 10 mSv has been demonstrated.[142]

B. MOSFET Dosimeters

The structure of a simple *metal oxide semiconductor field effect transistor* (MOSFET) is sketched in Fig. 19.27. It was suggested as early as 1974[143] that this type of structure could function as a means to measure cumulative exposure to ionizing radiation. While small in size and relatively insensitive compared with other methods of measurement, MOSFET devices have found useful applications as integrating dosimeters for applications in space and for the measurement of medical exposures to X- and gamma rays.[144] They can also be made sensitive to fast neutrons through the use of an adjacent hydrogenous converter layer to generate recoil protons.[145]

The principle of the measurement technique consists of monitoring the threshold voltage to the gate that just allows current to flow between the source and drain. During exposure to ionizing radiation, electron–hole pairs are formed in the insulating silicon oxide layer immediately below the gate. If a positive bias voltage is applied to the gate during the exposure period, there is a tendency for these charges to separate. The electrons will move toward the gate, and the holes toward the SiO_2–silicon interface where they tend to be trapped and form a fixed positive charge. The effect of this charge is to cause a shift to more negative values in the threshold gate voltage, and the amount of this voltage shift is a reasonably linear measure of the integrated dose. The higher the bias voltage, the greater will be the fraction of the charges collected, resulting in higher sensitivity. Typical MOSFET structures have an effective area of less than 0.1 mm^2 and show sensitivities in the range of 10–1000 mV/Gy. With a readout capable of measuring changes of the order of 0.5 mV, the corresponding minimum measurable dose would be 5×10^{-2} Gy (5 rad) to 5×10^{-4} Gy (50 mrad). Greater sensitivity can be achieved by stacking combinations of units or by custom designs with very thick oxide layers.

A simple MOSFET will show considerable temperature dependence of the threshold voltage, but this drawback can largely be overcome by using a structure with dual MOSFETs on the same silicon chip. The pair are operated with different bias voltages, and their temperature dependence offset by using the difference in their responses to indicate the measured dose.

Figure 19.27 Configuration of a metal oxide semiconductor field effect transistor (MOSFET).

C. Memory Cells as Detectors

It has been realized for several decades that the ionization charge created by an energetic particle can cause the state of a digital memory cell to change. These "single event upsets" can be caused either by alpha particles from trace radioactive impurities in the materials used to fabricate the cell, or by externally incident energetic charged particles that are components of the cosmic-ray background in space-based applications. A single alpha particle can create an ionization charge of over a million electron–hole pairs along its track in silicon (see Chapter 11). As the density of memory arrays has increased, a corresponding decrease has occurred in the size of the charge packets that are stored to constitute the digital information, increasing the susceptibility of the cells to these spontaneous upsets. As one consequence, the silicon and encapsulating materials used in the fabrication of memory chips are closely screened for low-level alpha-radioactive contamination to minimize the probability of upsets from this cause.

The dimensions of a typical *dynamic random-access memory* (dRAM) cell are on the order of a few μm. Its elements consist of a storage capacitor and an associated transistor, and its binary state is indicated by the presence or absence of a charge packet stored on the capacitor. Connected to each cell are address lines or "bit lines" that connect to a sense amplifier. The architecture is capable of addressing and reading the state of any cell or of changing that state. Because of the leakage of charge from the capacitors, each cell must have its contents refreshed on a cycle of several milliseconds when the cell charge is read, amplified, and rewritten back to the cell (hence the word *dynamic* in the above description).

As early as 1980, it was proposed[146] to use dRAMs as particle detectors. A charged particle can change the state of a memory cell if the ionization charge that it generates is located in a sensitive region of the cell.[147,148] Commercial memory chips normally have a protective layer over the surface, and this layer must be removed so that weakly penetrating particles can reach the sensitive regions. The efficiency for the detection of 5-MeV alpha particles can be as high as 50% over the area of the storage capacitors.[149] Other types of chips known as *static random-access memories* (SRAMs) have also been incorporated as particle detectors,[150,151] including a position-sensitive detector capable of providing direct images of a low-intensity ion beam.[152] The possibility of combining a neutron converter foil with a memory array for use in neutron imaging potentially is also of interest.[149,153]

PROBLEMS

19.1 (a) Find the minimum energy that an electron must have to produce Cherenkov radiation in quartz with an index of refraction of 1.47.

(b) Find the minimum energy for a gamma-ray photon to produce an electron of the energy calculated in part (a) through Compton scattering.

19.2 Compare the number of Cherenkov photons expected from a 2 MeV electron interacting in water to the number of scintillation photons expected from the same energy electron interacting in NaI(Tl).

19.3 Using the data given in Fig. 6.16, estimate the thickness of *liquid* xenon required to cause 50% of incident 30 keV photons to interact in the xenon. The densities of gaseous

and liquid xenon are 5.85 g/liter (at STP) and 3.52 kg/liter, respectively.

19.4 Explain the purpose of the intensifier screens that are used in conjunction with radiographic film for most clinical X-ray imaging.

19.5 Explain why the existence of electron trapping sites in a crystalline material is desirable if the material is used as a thermoluminescent dosimer, but undesirable if the material is used as a conventional scintillator.

19.6 From Fig. 19.19, select a track-etch material that will respond to neutron-induced fission fragments from a ^{235}U foil but will be insensitive to the alpha particles emitted in the spontaneous decay of the same material.

19.7 **(a)** A neutron flux is to be measured through the use of silver foil activation detectors and counting of the induced ^{108}Ag activity. In all cases, the foils are removed from the flux and counted for a 10-min period following a 2-min delay. Find the ratio of counts expected from two identical foils exposed in the same steady-state flux if one foil is exposed for 10 min and the other for 20 min.

(b) Find the factor by which the number of recorded counts could theoretically be increased by extending the counting period indefinitely.

19.8 What is the purpose of the Frisch grid in high-pressure xenon ion chambers?

REFERENCES

1. J. V. Jelly, *Cerenkov Radiation and its Applications,* Pergamon Press, London, 1958.
2. B. D. Sowerby, *Nucl. Instrum. Meth.* **97,** 145 (1971).
3. W. J. Gelsema, C. L. de Ligny, J. B. Luten, and F. G. A. Vossenberg, *Int. J. Appl. Radiat. Isotopes* **26,** 443 (1975).
4. J. Seguinot and T. Ypsilantis, *Nucl. Instrum. Meth.* **A343,** 1 (1994).
5. T. Ypsilantis and J. Seguinot, *Nucl. Instrum. Meth.* **A343,** 30 (1994).
6. M. Atac, A. V. Tollestrup, and D. Potter, *Nucl. Instrum. Meth.* **200,** 345 (1982).
7. A. Nohtomi et al., *IEEE Trans. Nucl. Sci.* **42**(4), 552 (1995).
8. A. Nohtomi, T. Sakae, M. Matoba, and N. Koori, *IEEE Trans. Nucl. Sci.* **44**(3), 691 (1997).
9. F. E. Taylor, *Nucl. Instrum. Meth.* **A289,** 283 (1990).
10. N. Koori et al., *Nucl. Instrum. Meth.* **A307,** 581 (1991).
11. H. Li, *IEEE Trans. Nucl. Sci.* **42**(5), 1505 (1995).
12. N. Koori et al., *IEEE Trans. Nucl. Sci.* **NS-33**(1), 395 (1986).
13. H. Ohgaki et al., *IEEE Trans. Nucl. Sci.* **NS-33**(1), 381 (1986).
14. R. K. Sood, Z. Ye, and R. K. Manchanda, *Nucl. Instrum. Meth.* **A344,** 384 (1994).
15. A. I. Bolozdynya et al., *IEEE Trans. Nucl. Sci.* **44**(6), 2408 (1997).
16. G. J. Mahler et al., *IEEE Trans. Nucl. Sci.* **45**(3), 1029 (1998).
17. A. Bolotnikov and B. Ramsey, *Nucl. Instrum. Meth.* **A396,** 360 (1997).
18. V. V. Dmitrenko et al., *Nucl. Instrum. Meth.* **A422,** 326 (1999).
19. G. Tepper, J. Losee, and R. Palmer, *Nucl. Instrum. Meth.* **A413,** 467 (1998).
20. G. Tepper and J. Losee, *Nucl. Instrum. Meth.* **A356,** 339 (1995).
21. G. Tepper and J. Losee, *Nucl. Instrum. Meth.* **A368,** 862 (1996).
22. T. Doke et al., *Nucl. Instrum. Meth.* **134,** 353 (1976).
23. L. Miller, S. Howe, and W. Spear, *Phys. Rev.* **166,** 871 (1968).
24. E. Shibamura et al., *Nucl. Instrum. Meth.* **131,** 249 (1975).
25. J. Prunier et al., *Nucl. Instrum. Meth.* **109,** 257 (1973).
26. E. Aprile, K. L. Giboni, and C. Rubbia, *Nucl. Instrum. Meth.* **A241,** 62 (1985).
27. K. Masuda, T. Doke, and T. Takahashi, *Nucl. Instrum. Meth.* **188,** 629 (1981).
28. A. S. Barabash et al., *Nucl. Instrum. Meth.* **A236,** 69 (1985).
29. E. Aprile, W. H.-M. Ku, J. Park, and H. Schwartz, *Nucl. Instrum. Meth.* **A261,** 519 (1987).
30. H. Ichinose et al., *Nucl. Instrum. Meth.* **A322,** 216 (1992).
31. K. L. Giboni, *Nucl. Instrum. Meth.* **A269,** 554 (1988).
32. E. Aprile, W. H.-M. Ku, and J. Park, *IEEE Trans. Nucl. Sci.* **NS-35**(1), 37 (1988).
33. J. Thomas, D. A. Imel, and S. Biller, *Phys. Rev.* **A38,** 5793 (1988).
34. H. J. Crawford et al., *Nucl. Instrum. Meth.* **A256,** 47 (1987).
35. A. Lansiart, A. Seigneur, J.-L. Moretti, and J. P. Morucci, *Nucl. Instrum. Meth.* **135,** 47 (1976).
36. K. Masuda et al., *Nucl. Instrum. Meth.* **160,** 247 (1979).
37. T. Doke, *Nucl. Instrum. Meth.* **A327,** 113 (1993).
38. J. H. Cobb and D. J. Miller, *Nucl. Instrum. Meth.* **141,** 433 (1977).
39. K. Masuda et al., *Nucl. Instrum. Meth.* **A241,** 607 (1985).
40. H. Jungblut and W. F. Schmidt, *Nucl. Instrum. Meth.* **A241,** 616 (1985).
41. K. Masuda, M. I. Lopes, and T. Doke, *Nucl. Instrum. Meth.* **A261,** 598 (1987).
42. A. Barone, *Nuclear Physics B (Proc. Suppl.)* **44,** 645 (1995).
43. D. McCammon et al., *IEEE Trans. Nucl. Sci.* **NS-33**(1), 236 (1986).
44. S. H. Moseley et al., *IEEE Trans. Nucl. Sci.* **NS-32**(1), 134 (1985).
45. G. W. Fraser, *Nucl. Instrum. Meth.* **A256,** 553 (1987).
46. D. McCammon et al., *Nucl. Instrum. Meth.* **A326,** 157 (1993).
47. K. D. Irwin, G. C. Hilton, D. A. Wollman, and J. M. Martinis, *Appl. Phys. Lett.* **69**(13), 1945 (1996).
48. K. D. Irwin, G. C. Hilton, J. M. Martinis, and B. Cabrera, *Nucl. Instrum. Meth.* **A370,** 177 (1996).
49. K. D. Irwin, G. C. Hilton, D. A. Wollman, and J. M. Martinis, *Jounal of Applied Physics* **83**(8), 3978 (1998).
50. D. A. Wollman et al., *Journal of Microscopy* **188**(3), 196 (1997).
51. D. Hueber, C. Valette, and G. Waysand, *Nucl. Instrum. Meth.* **167,** 201 (1979).
52. A. K. Drukier, *Nucl. Instrum. Meth.* **201,** 77 (1982).
53. M. Legros et al., *Nucl. Instrum. Meth.* **A263,** 229 (1988).
54. G. Meagher et al., *Nucl. Instrum. Meth.* **A348,** 696 (1994).
55. J. Seco et al., *Nucl. Instrum. Meth.* **A370,** 20 (1996).
56. M. Abplanalp et al., *Nucl. Instrum. Meth.* **A360,** 616 (1995).
57. G. Czapek et al., *Nucl. Instrum. Meth.* **A306,** 572 (1991).
58. A. K. Drukier, J. Igalson, and L. Sniadower, *Nucl. Instrum. Meth.* **154,** 91 (1978).
59. G. Meagher et al., *Nucl. Instrum. Meth.* **A370,** 8 (1996).
60. M. Furlan and K. Schiemann, *Nucl. Instrum. Meth.* **A374,** 413 (1996).
61. M. Furlan et al., *Nucl. Instrum. Meth.* **A370,** 17 (1996).
62. M. Abplanalp, *Nucl. Instrum. Meth.* **A370,** 11 (1996).
63. A. Gabutti, K. E. Gray, G. M. Pugh, and R. Tiberio, *Nucl. Instrum. Meth.* **A312,** 475 (1992).
64. N. E. Booth, *Appl. Phys. Lett.* **50**(5), 293 (1987).
65. N. E. Booth et al., *Nucl. Instrum. Meth.* **A315,** 201 (1992).
66. C. A. Mears et al., *Nucl. Instrum. Meth.* **A370,** 53 (1996).
67. R. H. Herz, *The Photographic Action of Ionizing Radiations,* Wiley-Interscience, London, 1969.
68. R. Katz and F. E. Pinkerton, *Nucl. Instrum. Meth.* **130,** 105 (1975).
69. R. Katz, L. Larsson, F. E. Pinkerton, and E. V. Benton, *Track Detec.* **1,** 49 (1977).
70. J. R. Erskine, *Nucl. Instrum. Meth.* **162,** 371 (1979).

71. J. R. Cameron, N. Suntharalingam, and G. N. Kenney, *Thermoluminescent Dosimetry,* University of Wisconsin Press, Madison, 1968.

72. J. F. Fowler and F. H. Attix, "Solid State Integrating Dosimeters," in *Radiation Dosimetry,* Vol. II, 2nd ed. (F. H. Attix and W. C. Roesh, eds.), Academic Press, New York, 1966.

73. E. Piesch, B. Burgkhardt, and S. Kabadjova, *Nucl. Instrum. Meth.* **126,** 563 (1975).

74. L. Larsson and R. Katz, *Nucl. Instrum. Meth.* **138,** 631 (1976).

75. B. Burgkhardt, D. Singh, and E. Piesch, *Nucl. Instrum. Meth.* **141,** 363 (1977).

76. B. Burgkhardt, E. Piesch, and D. Singh, *Nucl. Instrum. Meth.* **148,** 613 (1978).

77. M. P. R. Waligórski and R. Katz, *Nucl. Instrum. Meth.* **172,** 463 (1980).

78. R. A. Tawil et al., *Nucl. Instrum. Meth.* **A353,** 411 (1994).

79. M. Angelone, M. Chiti, and A. Esposito, *Nucl. Instrum. Meth.* **B117,** 428 (1996).

80. S. Tanaka and Y. Furuta, *Nucl. Instrum. Meth.* **133,** 495 (1976).

81. S. Tanaka and Y. Furuta, *Nucl. Instrum. Meth.* **140,** 395 (1977).

82. F. Spurny, M. Kralik, R. Medioni, and G. Portal, *Nucl. Instrum. Meth.* **137,** 593 (1976).

83. J. Henniger, B. Horlbeck, K. Hübner, and K. Prokert, *Nucl. Instrum. Meth.* **204,** 209 (1982).

84. M. Oberhofer and A. Scharmann (eds.), *Applied Thermoluminescence Dosimetry* (Ispra courses), Adam Hilger Ltd., Bristol, 1981.

85. Y. S. Horowitz, *Thermoluminescence and Thermoluminescent Dosimetry,* Vols. I, II, and III, CRC Press, Boca Raton, FL, 1984.

86. S. W. S. McKeever, *Thermoluminescence of Solids,* Cambridge University Press, Cambridge, U.K., 1985.

87. S. W. S. McKeever, M. Moscovitch, and P. D. Townsend, *Thermoluminescence Dosimetry Materials: Properties and Uses,* Nuclear Technology Publishing, Ashford, Kent, England (1995).

88. L. Botter-Jensen and S.W.S. McKeever, *Rad. Prot. Dos.* **65**(1-4), 273 (1996).

89. S. W. S. McKeever, L. Botter-Jensen, N. A. Larsen, and G. A. T. Duller, *Rad. Meas.* **27**(2), 161 (1997).

90. B. G. Markey, L. E. Colyott, and S. W. S. McKeever, *Rad. Meas.* **24**(4), 457 (1995).

91. S. W. S. McKeever, M. S. Akselrod, and B. G. Markey, *Rad. Protect. Dos.* **65** (1-4), 267 (1996).

92. M. Thoms, *Nucl. Instrum. Meth.* **A378,** 598 (1996).

93. A. Harrison et al., *Nucl. Instrum. Meth.* **A310,** 220 (1991).

94. H. von Seggern, *Nucl. Instrum. Meth.* **A322,** 467 (1992).

95. A. Winnacker, *Phys. Media* **9**(2-3), 95 (1993).

96. M. Thoms, *Nucl. Instrum. Meth.* **A389,** 437 (1997).

97. R. H. Templer et al., *Nucl. Instrum. Meth.* **A310,** 232 (1991).

98. C. Mori and A. Matsumura, *Nucl. Instrum. Meth.* **A312,** 39 (1992).

99. M. Takebe et al., *Nucl. Instrum. Meth.* **A363,** 614 (1995).

100. T. Bücherl, C. Rausch, and H. von Seggern, *Nucl. Instrum. Meth.* **A333,** 502 (1993).

101. J. Hofmann and C. Rausch, *Nucl. Instrum. Meth.* **A355,** 494 (1995).

102. N. Niimura et al., *Nucl. Instrum. Meth.* **A349,** 521 (1994).

103. C. Mori et al., *Nucl. Instrum. Meth.* **A339,** 278 (1994).

104. C. Mori et al., *Nucl. Instrum. Meth.* **A353,** 371 (1994).

105. R. L. Fleischer, P. B. Price, R. M. Walker, and E. L. Hubbard, *Phys. Rev.* **133,** A1443 (1964).

106. R. L. Fleischer, P. B. Price, and R. M. Walker, *Nuclear Tracks in Solids,* University of California Press, Berkeley, 1975.

107. H. B. Lück, *Nucl. Instrum. Meth.* **119,** 403 (1974).

108. E. V. Benton and W. D. Nix, *Nucl. Instrum. Meth.* **67,** 343 (1969).

109. H. Crannel, C. J. Crannell, and F. J. Kline, *Science* **166,** 606 (1969).

110. G. Siegmon, K. Bartholoma, and W. Enge, *Report of the Institute for Pure and Applied Nuclear Physics,* Christian Albrecht University, Kiel, Germany, 1FKK1 (1975).

111. E. Piesch and A. M. Sayed, *Nucl. Instrum. Meth.* **119,** 367 (1974).

112. R. Gold and C. E. Cohn, *Rev. Sci. Instrum.* **43,** 18 (1972).

113. J. Jasiak and E. Piesch, *Nucl. Instrum. Meth.* **128,** 447 (1975).

114. S. R. Dolce, *IEEE Trans. Nucl. Sci.* **NS-23**(1), 206 (1976).

115. S. C. Roy, R. E. Apfel, and Y.-C. Lo, *Nucl. Instrum. Meth.* **A255,** 199 (1987).

116. C.-K. C. Wang, *Health Phys.* **62**(3), 215 (1992).

117. M. J. Harper and J. C. Rich, *Nucl. Instrum. Meth.* **A336,** 220 (1993).

118. W. Lim and C. K. Wang, *Nucl. Instrum. Meth.* **A335,** 243 (1993).

119. B. White, D. Ebert, and F. Munno, *Health Phys.* **60**(5), 703 (1991).

120. F. d'Errico et al., *Rad. Protec. Dos.* **65**(1-4), 397 (1996).

121. F. d'Errico and W. G. Alberts, *Rad. Protec. Dos.* **54**(3-4), 357 (1994).

122. W. Lim and C. K. Wang, *Nucl. Instrum. Meth.* **A370,** 568 (1996).

123. T. Cousins, K. Tremblay, and H. Ing, *IEEE Trans. Nucl. Sci.* **37**(6), 1769 (1990).

124. F. d'Errico et al., *Rad. Protec. Dos.* **61**(1-3), 159 (1995).

125. M. T. Rashid, N. Ahmad, K. Kudo, and N. Takeda, *Nucl. Instrum. Meth.* **B72,** 491 (1992).

126. W. Lim and C. K. Wang, *Rev. Sci. Instrum.* **66**(12), 5442 (1995).

127. K. H. Beckurts and K. Wirtz, *Neutron Physics,* Springer-Verlag, New York, 1964.

128. L. Kuijpers et al., *Nucl. Instrum. Meth.* **144,** 215 (1977).

129. W. N. McElroy et al., AFWL-TR-67-41, Vol. 1, 2, 4 (1967).

130. F. J. Mayer and H. Brysk, *Nucl. Instrum. Meth.* **125,** 323 (1975).

131. A. Gentilini et al., *Nucl. Instrum. Meth.* **172,** 541 (1980).

132. D. R. Slaughter and W. L. Pickles, *Nucl. Instrum. Meth.* **160,** 87 (1979).

133. F. C. Young, *IEEE Trans. Nucl. Sci.* **NS-22**(1), 718 (1975).

134. E. L. Jacobs, S. D. Bonaparte, and P. D. Thacher, *Nucl. Instrum. Meth.* **213,** 387 (1983).

135. L. Ruby and J. B. Rechen, *Nucl. Instrum. Meth.* **15,** 74 (1962).

136. C. E. Spencer and E. L. Jacobs, *IEEE Trans. Nucl. Sci.* **NS-12**(1), 407 (1965).

137. R. H. Howell, *Nucl. Instrum. Meth.* **148,** 39 (1978).

138. A. Wolf and R. Moreh, *Nucl. Instrum. Meth.* **148,** 195 (1978).

139. V. E. Lewis and T. B. Ryves, *Nucl. Instrum. Meth.* **A257,** 462 (1987).

140. M. S. Rowland and J. C. Robertson, *Nucl. Instrum. Meth.* **224,** 322 (1984).

141. J. Kahilainen, *Rad. Protect. Dosim.* **66**(1-4), 459 (1996).

142. C. Wernli and J. Kahilainen, Proceedings of the 1996 International Congress on Radiation Protection, IRPA9, Vienna, p. 4-260.

143. A. G. Holmes-Siedle, *Nucl. Instrum. Meth.* **121,** 169 (1974).

144. M. W. Bower and D. E. Hintenlang, *Health Phys.* **75**(2), 197 (1998).

145. S. Kronenberg and G. J. Brucker, *IEEE Trans. Nucl. Sci.* **42**(1), 20 (1995).

146. G. Cerofolini, G. Ferla, and A. Foglio Para, *Nucl. Instrum. Meth.* **169,** 125 (1980).

147. D. G. Darambara and N. M. Spyrou, *Nucl. Instrum. Meth.* **A353,** 67 (1994).

148. D. G. Darambara and N. M. Spyrou, *Nucl. Instrum. Meth.* **A392,** 456 (1997).

149. J. C. Lund, F. Sinclair, and G. Entine, *IEEE Trans. Nucl. Sci.* **33**(1), 620 (1986).

150. G. A. Soli, B. R. Blaes, and M. G. Buehler, *IEEE Trans. Nucl. Sci.* **39**(5), 1374 (1992).

151. G. A. Soli, H. B. Garrett, and E. R. Fossum, *IEEE Trans. Nucl. Sci.* **43**(3), 1516 (1996).

152. T. Hamano et al., *Nucl. Instrum. Meth.* **B130,** 280 (1997).

153. D. G. Darambara and N. M. Spyrou, *Nucl. Instrum. Meth.* **A348,** 491 (1994).

Chapter 20

Background and Detector Shielding

Because of the cosmic radiation that continuously bombards the earth's atmosphere and the existence of natural radioactivity in the environment, all radiation detectors record some background signal. The nature of this background varies greatly with the size and type of detector and with the extent of shielding that may be placed around it. The background counting rate can be as high as many thousands of counts per second for large-volume scintillators, to a few counts per day in some specialized applications. Because the magnitude of the background ultimately determines the minimum detectable radiation level, it is most significant in those applications involving radiation sources of low activity. However, background is often important enough in routine usage so that the majority of radiation detectors are provided with some degree of external shielding to effect a reduction in the measured level. A second purpose of detector shielding is to provide a degree of isolation in laboratories where other radiation sources may be used or moved about during the course of a measurement.

The greatest value of keeping background small is encountered when extremely weak sources of radiation must be detected. The term *low-level counting* is often applied to such applications. Reference 1 provides a useful coverage of the techniques that can be applied to reduce background in low-level counting and details many of the background sources and shielding properties that are summarized in the remainder of this chapter.

I. SOURCES OF BACKGROUND

Background radiations are conveniently grouped into five categories:

1. The natural radioactivity of the constituent materials of the detector itself.
2. The natural radioactivity of the ancillary equipment, supports, and shielding placed in the immediate vicinity of the detector.
3. Radiations from the activity of the earth's surface (*terrestrial radiation*), construction materials of the laboratory, or other far-away structures.
4. Radioactivity in the air surrounding the detector.
5. The primary and secondary components of cosmic radiation.

A. Radioactivity of Common Materials

The radioactivity of ordinary construction materials is, in large part, due to low concentrations of naturally radioactive elements often contained as an impurity. The most important components are potassium, thorium, uranium, and the members of the long decay chains of uranium and thorium. Natural potassium contains 0.012% ^{40}K, which decays with a 1.26×10^9 year half-life through the decay scheme sketched below.

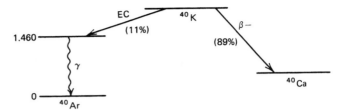

Radiations emitted are a beta particle with 1.314 MeV endpoint energy (89% yield), a gamma ray of 1.460 MeV energy (11%), and characteristic Ar X-rays following the electron capture. The high-energy gamma rays often lead to a recognizable peak in the background spectra from gamma-ray detectors, because potassium is a widespread component in concrete and other building materials.

Thorium, uranium, and radium are all members of long decay chains involving daughter products that emit a mixed spectrum of alpha, beta, and gamma rays. In the terrestrial gamma-ray spectrum shown in Fig. 20.1, the following daughter activities can be identified: in the thorium series, ^{228}Ac, ^{224}Ra, ^{212}Bi, ^{212}Pb, and ^{208}Tl; in the uranium series, ^{226}Ra, ^{214}Pb, and ^{214}Bi. The long-lived natural activities of ^{235}U and ^{40}K are also evident. A small peak caused by ^{7}Be which is continuously produced by cosmic-ray interactions in the upper atmosphere can also be detected.

In addition to the naturally occurring activities, background also consists of some fission-product activities that originated with atmospheric fallout from past weapons testing. In Fig. 20.1, the most prominent contributor is ^{137}Cs, but detectable amounts of ^{95}Zr, ^{95}Nb, ^{106}Ru, ^{125}Sb, and ^{144}Ce are also seen.

Table 20.1 lists the measured specific activity for some common materials used in the construction of detector systems. Certain materials such as pyrex glass contain potassium, thorium, and/or uranium as a normal constituent and therefore have a rather high background level. Most of the other examples include these components only as impurities, and their radioactivity is minimized by choosing highly purified samples. For this reason, copper or magnesium that has been electrolytically prepared is preferred when used in the construction of low-background counters. Stainless steel normally shows low background levels, but common sources of aluminum have a sufficient uranium and/or radium impurity level to be objectionable in demanding applications. Brass is generally of low activity, provided its lead content is low. Electrical solder[4,5] and some circuit board materials can be relatively high in radioactivity, so that attention must be given to the possible contribution of electronic components to the background rate of gamma-ray detectors. Results of a large set of measurements of the thorium, uranium, and potassium concentrations in common construction materials can be found in Ref. 6.

For scintillation counters, the glass envelope of the photomultiplier tube and the tube base or socket materials are potential sources of background. At some premium in cost, tubes that are made from quartz rather than glass can be obtained that will be of significantly lower activity. Bases for tubes intended for scintillation counting are also available with a special low-background formulation. In the past, potassium impurities in sodium iodide were a source of some detectable background, but modern fabrication techniques have now led to the availability of sodium iodide with negligible potassium contamination. Bismuth germanate (BGO) can show appreciable background[7] from ^{207}Bi, a radionuclide with 38-year half life most likely produced by cosmic ray proton transmutation of ^{206}Pb in lead contained in the same ore. It has been shown[7] that BGO obtained from lead-free ores does not carry this radioactive contamination. For germanium detector systems, the

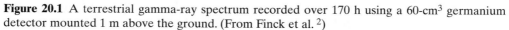

Figure 20.1 A terrestrial gamma-ray spectrum recorded over 170 h using a 60-cm^3 germanium detector mounted 1 m above the ground. (From Finck et al. [2])

Table 20.1 Levels of Activities from Natural Sources in Common Construction Materials

Material	Disintegrations/min per gram of Material		
	^{232}Th(583 keV)	^{238}U	^{40}K
Aluminum (6061 from Harshaw)	0.42	0.04	< 0.05
Aluminum (1100 from Harshaw)	0.24	< 0.017	< 0.06
Aluminum (1100 from ALCOA)	0.08	< 0.026	< 0.11
Aluminum (3003 from ALCOA)	0.10	< 0.026	0.56
Stainless steel (304)	< 0.006	< 0.007	< 0.06
Stainless steel (304-L)	< 0.005	< 0.005	< 0.02
Magnesium (rod)	0.06	< 0.04	0.1
Magnesium (ingot)	< 0.01	< 0.002	< 0.02
Magnesium (4 in. $\varnothing$ × 4 in. from Dow)	< 0.005	< 0.002	< 0.02
Magnesium (from PGT)	< 0.05	< 0.03	< 0.05
Beryllium copper alloy	< 0.02	< 0.06	< 0.2
Copper (sheet)	< 0.05	< 0.06	< 0.2
Pyrex window	0.45	0.27	3.8
Quartz window	< 0.018	< 0.018	< 0.07
Molecular sieve	4.4	3.0	9.0
Neoprene	< 0.008	< 0.01	0.36
Rubber	0.12	1.0	2.0
Apiezon Q	4.5	4.5	2.7
Electrical tape—3M	< 0.04	< 0.06	< 0.1
Cement (Portland)	0.25	1.3	4.5
Epoxy	0.006	0.01	0.19
Lacquer	0.002	0.005	0.04

Source: Camp et al.[3]

extreme degree of purity required of the germanium for its acceptability as detector material ensures that its inherent radioactivity is low.

Ironically, the shielding placed around a detector to reduce background from cosmic rays or terrestrial radiation may itself introduce a significant background from its inherent low-level radioactivity. The relative activity levels expected are discussed later in this chapter in connection with specific shielding materials. Shield surfaces can also be the source of significant activities as a result of airborne particle deposits or accidental surface contamination. The inner surfaces of low-background shields are therefore often sandblasted prior to installation to remove any surface radioactivity.

B. Airborne Radioactivity

A measurable amount of background can originate with radioactivity carried by the ambient air, either in the form of trace amounts of radioactive gases or dust particles. Radon (^{222}Rn) and thoron (^{220}Rn) are short-lived radioactive gases that originate as daughter products in the decay chains of the uranium and thorium present either in the soil or construction materials of the laboratory. Their concentration in the atmosphere can vary significantly[8] depending on the time of day and meteorological conditions. To eliminate the influence of radon on the background, the volume around the detector can be made air-tight and purged with a radon-free gas. Some laboratories use the boil-off nitrogen gas from the liquid nitrogen dewar used with germanium detectors for this purpose. Airborne dust can contain radioactivity from soil or radon decay products that can largely be eliminated through effective filtration of the air supply to the counting room.

C. Cosmic Radiation

A significant component of detector background arises from the secondary radiations produced by cosmic-ray interactions in the earth's atmosphere, as illustrated in the sketch below.

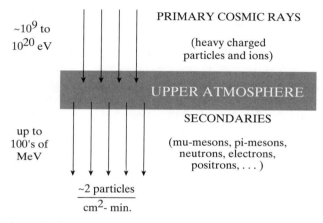

The primary cosmic radiation, which can be either of galactic or solar origin, is made up mainly of protons, plus some helium nuclei and heavy ions, with extremely high kinetic energies. In their interaction with the atmosphere, a large assortment of secondary particles is produced, including pi and mu mesons, electrons, protons, neutrons, and electromagnetic photons, with energies that extend into the hundreds of MeV range. Some of these radiations reach the earth's surface and can create background pulses in many types of detectors. At sea level, muons make up about 80% of the charged particle secondary flux, with an intensity of about one muon per cm^2 per minute.

Because of their very high kinetic energy, the cosmic primary and secondary particles have a relatively low specific energy loss $(-dE/dx)$ comparable with that of fast electrons. For example, the energy loss of muons ranges from about 1.1 to 1.8 MeV per g/cm^2 in absorber materials with high to low atomic number, respectively. Thus, the corresponding pulse amplitude is small in "thin" detectors designed to stop typical charged particles but not electrons. However, in solid detectors of large thickness, such as NaI(Tl) scintillators or germanium detectors, the deposited energy may be many MeV and the corresponding pulses can then be large compared with typical signal pulses.

The various secondary components differ in their hardness or ability to penetrate matter, and some gains are achieved even with modest amounts of shielding. Other components persist through many meters of common materials. At the earth's surface, the cosmic secondary radiations are directed primarily downward, so that shielding against cosmic background is most effective when located above the detector. For example, the muon flux follows an angular dependence of approximately $\cos^2 \theta$, where θ is the angle with respect to the vertical.

Secondary particles can also be created in the shielding surrounding the detector by high-energy cosmic components. For example, high-energy protons will produce energetic neutrons, especially in high-Z materials such as lead. Pions and muons can also trigger showers of energetic electrons and bremsstrahlung photons. These secondary radiations can offset some of the gains expected from making a shield thicker to provide more attenuation of other background sources. For example, increasing the thickness of lead shielding beyond 10–15 cm will have minimal effect in further reducing counts from external gamma rays, but may actually increase the background as a result of the influence of these secondary radiations. Fast neutrons from cosmic interactions (or any other source) can also create secondary gamma rays when a neutron is moderated and absorbed. If the shield contains hydrogen (e.g., as in concrete), the capture gamma ray at 2.22 MeV can sometimes be identified in the background spectrum.

II. BACKGROUND IN GAMMA-RAY SPECTRA

A. Relative Contributions

The importance of various components of the background changes greatly with the circumstance. In gamma-ray detectors without shielding, the cosmic-ray component is normally dominant. When significant shielding is provided, both the cosmic flux and the background caused by ambient sources of gamma rays are decreased, and radioactive contamination of structural and shielding materials around the detector becomes an important fraction of the remainder. Underground locations can help reduce the cosmic background, but tens or hundreds of meters of earth are needed to eliminate the hardest components of the meson flux.

Figure 20.2 shows the various components of the background spectrum recorded by a 7.62 cm × 7.62 cm NaI(Tl) detector within a massive shield consisting of 10 tons of lead and 160 kg of paraffin and boric acid mixture.[9] The various background components above an energy of 100 keV in the spectrum are tabulated in Table 20.2.

Figure 20.2 Various contributions to the background spectrum from a 7.62 cm × 7.62 cm NaI(Tl) scintillator inside a massive lead and borated paraffin shield. (From Stenberg and Olsson.[9])

Table 20.2 Components of a NaI(Tl) Scintillation Counter Background

Outside shield	29,200 counts/min
Inside shield	
Cosmic ray mesons	116.4
Cosmic ray neutrons	19.4
^{222}Rn daughters	25.9
^{40}K	8.6
Remaining background	33.1
Total	203.4 counts/min

Source: Stenberg and Olsson.[9]

The behavior of the background in a number of large-volume Ge(Li) detectors has been studied by Kaye et al.[10] Using shielding of the type shown in Fig. 20.3, a typical detector background is made up of a 30% contribution from cosmic radiation, 60% from the radioactive contamination of shielding materials, and 10% from radioactivity within the detector and unidentified sources. Typical background spectra from this configuration are shown in Fig. 20.4. Other representative studies of the background in low-level germanium spectroscopy systems can be found in Refs. 11–14.

The background in gamma-ray detectors can be expected to increase roughly as the detector volume. Therefore, in critical situations where low background is at a premium, it is important to select a detector size that is not larger than necessary to give a reasonable counting efficiency for the samples to be counted. As shown in Chapter 3, a statistically based selection criterion is to choose a detector size that maximizes the ratio of S^2/B, where S is the counting rate due to the source alone and B is the counting rate due to background. This same figure of merit can be used in the selection of other operating parameters, such as discrimination level, in setting up a low-level counting experiment.

Figure 20.3 A low-background shield configuration for a germanium detector. (From Kaye et al.[10])

Figure 20.4 The top curve shows the background spectrum recorded for an 85-cm^3 germanium detector in a normal laboratory using the shielding shown on Fig. 20.3. The bottom curve shows the gains achieved by relocating the counter deep inside the massive concrete structure of a dam and increasing the thickness of the lead shielding. (From Kaye et al.[10])

B. Variations in the Background Rate

The background rate from a typical radiation detector, although expected to be nearly steady state in time, may show a perceptible variation over periods of hours or days.[15] In experiments in which the signal rate is high, small background fluctuations are typically much lower than the inherent statistical fluctuation of the measurement itself, and consequently they are of no real importance. In such situations, a single background determination (repeated every day or so) is a sufficient measurement on which to base background subtraction of the measured counting rates.

When low-level activities are counted, the fluctuations in the background may be of the same order as the source strength and therefore must carefully be considered. The component of the background that results from radioactivity of the detector is virtually constant, but small, long-term variations can be observed[16] due to possible lack of equilibrium among the various members of the alpha decay chains of impurities such as thorium. Processing of materials can selectively remove some daughter products that gradually build back into secular equilibrium with a longer-lived parent. In that case, changes may take place in times on the order of many years. In general, more significant variations in the background arise from changes in either the cosmic ray intensity or the airborne radioactivity. Much of the observed variation in sensitive gamma-ray counters appears to be correlated with airborne activity in the form of the decay products of ^{222}Rn (Refs. 8 and 17). The importance of this background source depends greatly on the specific geometry and ventilation conditions around the detector, but variations of the order of a few percent in the background rate should not be unexpected.

In low-level counting experiments, it is therefore prudent to carry out a background determination near the time of the actual measurement itself. Background counts both before and after the measurement will help detect any changes in the background level. In critical situations, the background rate can be monitored during the course of the measurement by using a second detector whose properties are identical with those of the detector used in the measurement itself. In this way, any temporal changes in the background rate will affect both detectors and are thereby canceled.

C. Source-Related Background

In gamma-ray spectroscopy, some additional interfering radiation can be observed as a result of the interaction of primary gamma rays from the source with structural and shielding materials around the detector. The origins of these source-related background events were discussed previously in Chapter 10 and are illustrated schematically in Fig. 10.6. Important processes include Compton backscattering of the primary gamma rays, and the generation of secondary annihilation photons and characteristic X-rays through pair production or photoelectric absorption. These effects can be eliminated only by removing all materials from the immediate vicinity of the detector, but the practical demands of shielding usually do not permit this approach. Therefore, the potential source regions for these secondary radiations, particularly the inner surface of the radiation shield, are often designed with materials intended to reduce their importance.

To illustrate, Fig. 20.5 shows the low-energy portion of a gamma-ray spectrum from ^{22}Na as recorded by a NaI(Tl) scintillation detector. This portion of the spectrum shows the

Figure 20.5 A plot of the low-energy portion of the ^{22}Na gamma-ray spectrum (indicated on the insert) as recorded by a 7.62 cm × 7.62 cm NaI(Tl) scintillator. The effects of different shield liners on the fluorescent X-ray and backscatter peaks are illustrated. (From Stenberg and Olsson.[9])

perturbing effects of both a backscatter peak at about 170 keV and a peak at 77 keV arising from lead fluorescent X-rays generated by photoelectric absorption of the primary gamma rays within the surrounding lead shield. To reduce the latter component, lead shields are commonly lined with thin layers of low-Z material to absorb these lead X-rays, and which themselves emit much softer characteristic X-rays. Figure 20.5 illustrates the degree of suppression that can be achieved in the corresponding X-ray peak with linings of copper or cadmium. These gains are partially offset by an increase in the intensity of the backscatter peak caused by the greater probability of Compton scatter in these materials relative to photoelectric absorption.

III. BACKGROUND IN OTHER DETECTORS

A. Gas-Filled Counters

The background expected from gas-filled detectors such as proportional counters or Geiger tubes depends greatly on the minimum ionization required for registration of a pulse. In the Geiger tube, this minimum is, in principle, only one ion pair, so that any interaction, regardless of its amplitude, results in a registered pulse. All the previously mentioned sources of background can therefore contribute to the background rate observed from a Geiger tube. At sea level, an unshielded G-M tube of typical dimensions will show a background rate of several counts per second, which can greatly be reduced by a minimal shielding of several centimeters of lead.

Proportional counters can be operated with a finite discrimination level so that lightly ionizing events do not contribute to the measured counting rate. Because the fill gas is relatively transparent to fast electrons, the discrimination level can readily exceed the ionization that will be created along the track of an electron. When operated in this mode, the proportional counter is sensitive only to heavy charged particles such as alpha particles, and the operating point corresponds to the *alpha plateau* region discussed in Chapter 6. Under these circumstances, many of the possible sources of background are eliminated, and only those that deposit their energy in the form of densely ionizing charged particles can contribute to the measured background counting rate. Then, the most significant source is often the inherent alpha radioactivity of the construction materials of the detector itself. Table 20.3 shows the inherent alpha background activity of some materials common to gas-filled detectors. With some care as to the choice of construction materials, the background level for proportional counters operating on the alpha plateau is less than 1 count/min.

For gas-filled detectors operated in current mode, there can be no inherent pulse amplitude discrimination. Therefore, all the sources of background radiation mentioned previously will contribute to the background current measured from an ion chamber, for example. Ionization caused by cosmic rays at sea level amounts to about two ion pairs per second, per cubic centimeter of STP air.[18] Background currents caused by terrestrial radiation and radioactivity of surrounding materials can be about the same order of magnitude, whereas the current from natural alpha activity of common wall materials is normally an order of magnitude lower.[18]

B. Semiconductor Charged-Particle Detectors

As in germanium detectors, the inherent radioactivity of the high-purity semiconductor material itself is negligible for silicon diode detectors. Because typical surface areas are only a few square centimeters, and the contact electrodes are extremely thin, there is very little other material that can contribute background alpha activity. The depletion depth of silicon surface barriers is normally too small to develop significant pulses from low dE/dx radiations from cosmic rays or gamma-ray interactions, so the residual background is restricted to alpha particles from nearby surfaces. With properly chosen low-activity mate-

Table 20.3 Alpha Particle Emission Rates from Various Materials

Material	Alphas per cm² per hour		
	(J. A. B.)[a] E_T^d 250 keV	(S. & H.)[b] E_T 100 keV	(A. & J.)[c] E_T 1 MeV
Machined copper	0.09		
Commercial brass	0.05	0.2	0.13
Mild steel (stainless)	0.03	0.05	0.01–0.03
Commercial aluminum	0.31	0.2	0.27
Solder	28		
Aquadag	0.07		
Nickel		0.03	
Lead	60		
Perspex		0	
Zinc sulfide in Perspex sandwich (10 mg/cm²)		0.1	
Air from room	32/100 cm³ · h		
Cylinder nitrogen	0		
Cylinder argon		0	

[a]J. A. B.; J. A. Bearden, *Rev. Sci. Instrum.* **4**, 271 (1933).

[b]S. & H.; J. Sharpe and P. Holton, unpublished measurements.

[c]A. & J.; B. Al-Bataina and J. Janecke, *Nucl. Instrum. Meth.* **A255**, 512 (1987).

[d]E_T represents the detection threshold.

Source: Data in first two columns from J. Sharpe, *Nuclear Radiation Detectors*, 2nd ed., Methuen & Co., Ltd., London, 1964.

rials, the resulting background rate can be less than one per hour in commercially available counting systems. In order to retain this low background condition over time, however, precautions must be taken. The detector entrance window and nearby surfaces must be kept completely free of alpha-emitting contamination. There is a tendency for alpha particle sources to create such contamination through the escape of recoil nuclei during some alpha decays. To prevent these low-energy recoils from contaminating surfaces, it is common to use a low pressure of air in the counting chamber rather than a hard vacuum.

IV. SHIELDING MATERIALS

A. Conventional Materials for Low-Background Shields

1. LEAD

Because of its high density and large atomic number, lead is the most widely used material for the construction of detector shields. The photoelectric absorption cross section predominates up to gamma-ray energies as high as 0.5 MeV, and therefore even relatively hard gamma rays from external background sources (such as the 1.46 MeV gamma ray from ^{40}K) can be absorbed efficiently. Because of its high density, thicknesses of just a few centimeters of lead will provide a large reduction in the background of typical gamma-ray detectors. Lead is reasonably effective at removing many of the cosmic-ray components of the background, although thicknesses beyond about 10 cm do not result in an appreciable decrease in the resulting counting rate because of the buildup of secondary radiations, which are due to cosmic interactions within the lead itself.[19]

Lead is widely used in the form of rectangular "lead bricks" in the construction of simple gamma-ray shields. Potential problems that result from streaming through cracks between the bricks can be overcome by building the shield with multiple layers or by using specially shaped bricks with interlocking surfaces. Lead is also cast relatively easily into solid shapes, although some care must be taken in the casting process to avoid porosity or voids in the solidified shields. As an alternative to casting, a container of the proper size and shape can be filled with lead shot to form a shield with somewhat lower density. Other shielding materials are also commercially available that incorporate a high percentage of lead into plastic or epoxy compositions, which can be more readily molded and shaped.

Ordinary lead normally contains a significant amount of natural activity caused by low-level contaminants, and therefore lead that is either specially refined or reclaimed from very old sources is preferred in the construction of shields for low-background applications. When freshly refined, lead can contain significant amounts of ^{210}Pb, a daughter product of the decay of ^{226}Ra. This isotope decays with a half-life of 20.4 years, so that samples of lead that are many decades old will be relatively free of this activity. However, there is some indication[1,20,21] that properly refined lead from some sources may not contain troublesome amounts of ^{210}Pb. The presence of ^{210}Pb is a source of background primarily through the characteristic lead X-rays and bremsstrahlung created by high energy beta particles (endpoint energy of 1.2 MeV) emitted by its decay product ^{210}Bi. A very low yield of 45 keV gamma rays is normally negligible.[1]

Recently refined lead may also show detectable radioactivity caused by impurities from atmospheric fallout. Other background may arise from trace amounts of thorium daughters, or radioactive impurities common in antimony, which is a normal alloy of lead. Some lead samples have shown activities as high as 1.5 Bq/g (Ref. 22), but activities of one or two orders of magnitude lower are more typical of high-purity lead. Samples of lead with the lowest radioactivity are those from ancient sources. In one extreme example, 2000-year-old lead salvaged from a sunken Roman ship[23] showed no activity above a measurement threshold of 0.02 Bq/g.

2. STEEL

Iron or steel is also a common gamma-ray shielding material and is often used in situations where the size or configuration of the shield would make its construction from lead alone too expensive. In such circumstances, an outer layer of steel with an inner lead lining is often an effective compromise. Again, there is a premium on obtaining old material because steel fabricated after about 1950 is of noticeably higher activity than prewar steel. (Steel salvaged from World War II naval ships has been used in the construction of a significant number of low-background shields.) Some of the low-level activity found in more recently fabricated steel is traceable to radioactive materials inadvertently introduced in the recycling of scrap steel.

Because both the atomic number and density of steel are considerably lower than those of lead, thicknesses of several tens of centimeters are normally required for very low background applications.

3. TUNGSTEN

With its large atomic number of 74 and high density of over 19 g/cm^3, tungsten is an attractive material to consider for small shields or collimators. The pure metal is difficult to cast or machine, but powder metallurgy processes can produce a sintered form of tungsten. Its density is only slightly below that of the pure metal, and it is much more amenable to machining. It has found application in shielding conditions that require a high degree of radiation attenuation in a limited thickness.[24] Other possibilities include alloys of tungsten with nickel/iron or nickel/copper that retain densities of 86 to 97% of pure tungsten, but also have excellent machinability.

4. CONCRETE

Because of its low cost, concrete is often used in the construction of large-volume shields. However, its activity is relatively high due to ^{40}K, uranium, and fallout products included in its composition. It is therefore most commonly used as the outer constituent of a shield, with its own activity shielded by an inner layer of steel, lead, or other shielding material of lower activity.

The effectiveness of various thicknesses of concrete on the attenuation of the cosmic component of background is illustrated in Fig. 20.6. Although few experimenters have the

Figure 20.6 The background spectra from a 7.62 cm × 7.62 cm NaI(Tl) scintillator within a paraffin and 10.2 cm thick lead shield. The indicated locations are at different depths within the massive concrete structure of Grand Coulee Dam. The purpose of the paraffin was to indicate the importance of the neutron component by the relative prominence of the hydrogen capture gamma-ray peak at 2.22 MeV. (From Kaye et al.[19])

luxury of using Grand Coulee Dam as a radiation shield as in this study, significant benefits can be achieved by locating the counting laboratory in the basement of a multistory building, in which the concrete floors of upper stories can significantly attenuate the cosmic component.[3]

Although ordinary concrete consists mainly of water and low-Z elements, a special formulation known as *barytes concrete* contains a significant percentage of heavy components and is therefore much more effective in gamma-ray shielding.

B. Neutron Shielding

The shielding of neutrons is important from several aspects. Any detector designed for neutron counting must obviously be shielded against external sources of background neutrons to enhance its signal-to-background ratio. However, other types of detectors, including the low-background gamma-ray counting systems discussed in the previous section, can also be influenced by a fast neutron background. Much of this sensitivity arises because of the capture gamma rays created upon absorption of the neutrons within the detector or nearby materials. The shielding of the detectors against neutron backgrounds is obviously most important around neutron-producing facilities, but a measurable neutron component also exists in the natural background due to secondary products of cosmic-ray interactions.

Completely different principles apply to the selection of neutron shielding materials as compared with those for gamma rays. It is most important to quickly moderate the neutron to low energies, where it can readily be captured in materials with high absorption cross sections. The most effective moderators are elements with low atomic number, and therefore hydrogen-containing materials are the major component of most neutron shields. In this application, water, concrete, and paraffin are all inexpensive sources of bulk shielding. Because mean free paths of fast neutrons typically are tens of centimeters in such materials, thicknesses of 1 m or more are required for effective moderation of almost all incident fast neutrons.

Once the neutron has been moderated, it can be eliminated through an appropriate capture reaction. This absorption may be in the hydrogen already present for moderating purposes, although the capture cross section is relatively low. The thermal neutron may thus diffuse an appreciable distance before capture, reducing the effectiveness of the shield. Furthermore, capture in hydrogen leads to the liberation of a 2.22 MeV capture gamma ray, which, because of its high energy, is particularly undesirable in many types of detectors. Therefore, a second component is normally used in neutron shields, either homogeneously mixed with the moderator or present as an absorbing layer near its inner surface. This additive is chosen to have a high neutron capture cross section, so that the moderated neutrons will preferentially undergo absorption within this material.

Some reactions with large capture cross sections for slow neutrons have already been discussed in Chapter 14 as applied to slow neutron detectors. These same reactions are also useful in the shielding application, and therefore boron and lithium are common components of neutron shields. The ^{10}B(n, α) reaction has a high capture cross section at low energies, and boron can readily be incorporated into polyethylene and other moderating materials. The heavy charged particles from the neutron-induced reactions are absorbed in solids within less than 0.1 mm and therefore do not contribute additional background. The majority of the capture reactions, however, lead to an excited state in the product nucleus, which subsequently decays by the emission of a 0.48 MeV gamma ray. Applications that are potentially sensitive to gamma-ray backgrounds are therefore better served by substituting lithium for the boron, because the ^{6}Li(n, α) reaction proceeds directly to the ground state of the product and no gamma rays are emitted.

The effectiveness of some commercially available combinations of polyethylene and boron or lithium in the shielding of fast neutrons is compared with those of graphite and

Figure 20.7 The relative effectiveness of different shielding materials for 2.6 MeV neutrons as indicated by the thermal neutron flux for various thicknesses. (From Gujrathi and D'Auria.[25])

paraffin in Ref. 25. Figure 20.7, taken from this reference, shows that the mixtures of polyethylene with the thermal absorbers are considerably more effective per unit thickness than the remaining materials.

Cadmium is also widely applied as a thermal neutron absorber because thin sheets of the material are essentially opaque to thermal neutrons due to its very high cross section. Layers as thin as 0.5 mm are very effective absorbers of thermal neutrons, but the subsequent (n, γ) reaction will also add a secondary gamma-ray background.

V. ACTIVE METHODS OF BACKGROUND REDUCTION

A. Anticoincidence Shielding

The cosmic-ray component of detector background can be removed only through massive amounts of conventional shielding. An alternative arrangement is sketched in Fig. 20.8, in which the highly penetrating cosmic radiations are eliminated through the use of an *anticoincidence shield* or *guard counter*. The primary detector is surrounded by a second detector (or an array of detectors), and the output of the primary detector is accepted only when it is not accompanied by a coincident pulse in the outer detector. The source to be counted is oriented and shielded so that it produces interactions only in the primary detector. Therefore, in the simplest case, no pulses are affected that correspond to the complete absorption of the source radiation within the primary detector. However, the cosmic

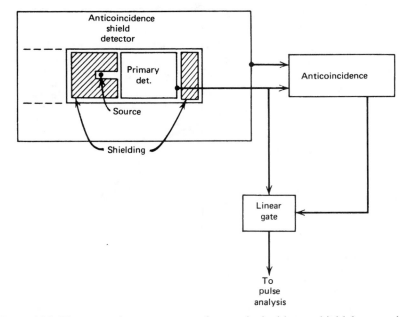

Figure 20.8 The general arrangement of an anticoincidence shield for cosmic background reduction.

radiations will likely penetrate both detectors and therefore can be eliminated from the output of the primary detector through the anticoincidence arrangement. As a bonus, the anticoincidence shield will also suppress the Compton continuum in the recorded spectrum, because a Compton-scattered gamma ray from the primary detector may also interact within the surrounding detector. This mode of operation for germanium detectors is discussed further in Chapter 12. The degree to which the anticoincidence technique can reduce the background spectrum in a germanium detector, when surrounded by an annular sodium iodide scintillator, is illustrated in Fig. 20.9. The anticoincidence detector can also be a large plastic or liquid scintillator,[3,26,27] a ring of Geiger tubes,[28] or virtually any other type of detector. It has become commonplace to provide guard counters as part of many types of low background detector systems. For example, most commercial proportional counter systems intended for general monitoring of samples for possible alpha or beta activity include a guard counter above the sample detectors. The guard counter often consists of a flat, large-area multiwire proportional counter whose output is used to eliminate many of the cosmic ray-induced events in the sample chambers by anticoincidence.

B. Coincidence Counting

If a radioisotope that emits more than one distinguishable radiation in coincidence is to be counted, the effective background can greatly be reduced through the use of coincidence techniques. A common example is that of beta decay to a daughter product, which promptly emits at least one gamma ray. A suitable arrangement can then consist of separate beta particle and gamma-ray detectors, with the source mounted to maximize the corresponding interaction rates in each detector. By demanding coincidence between the two detector outputs, many background events will be eliminated because they will occur in only one detector at a time.

As an example, Fig. 20.10 shows a double detector coincidence spectrometer in which the beta radiation is detected by a split 4π gas-flow proportional counter. This detector is fitted within the well of a 12.7 cm $\times$ 12.7 cm NaI(Tl) scintillator, which serves as the gamma-ray

Figure 20.9 The bottom spectrum shows the degree to which background in a 40-cm^3 germanium detector may be suppressed through the use of an annular NaI(Tl) scintillator in anticoincidence. A comparison of the two middle spectra illustrates the increase in background with detector active volume. The top curve shows the recorded spectrum using a sample of molecular sieve (a part of some germanium vacuum systems) as a source of natural background radiation. (From Kaye et al.[10])

spectrometer. By mounting the source between the two halves of the proportional counter, the counting efficiency for beta particles is about 70%. Because the scintillator will also be very efficient for gamma rays in this geometry, coincident pulses are obtained in 50–65% of the cases in which coincident radiations are emitted. This slight decrease in the overall counting efficiency is more than offset by the very large reduction in the background counting rate illustrated on the spectra in Fig. 20.10. For typical conditions, the improvement in the minimum detectable sample activity can amount to one or two orders of magnitude compared with a similarly shielded simple gamma detector.[29] The example shown also incorporates anticoincidence guard counters around the detector assembly to reduce the background level further.

Most applications in which extremely weak radioisotope sources must be counted take advantage of coincidence methods to suppress background as much as possible. Examples that contain references to earlier work can be found in Refs. 3, 10, and 30.

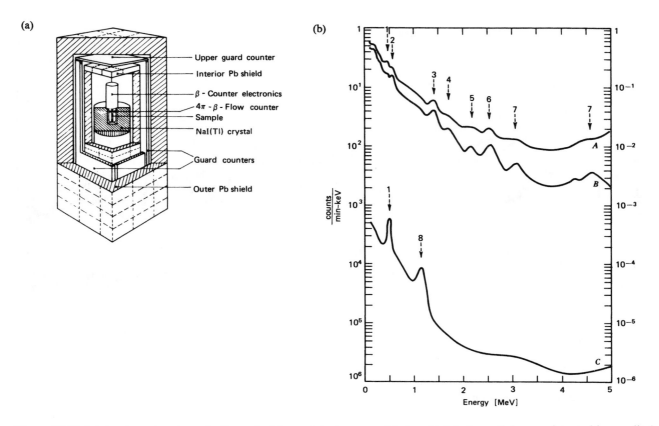

Figure 20.10 (*a*) A schematic view of a β-γ coincidence detector assembly, in which beta particles are detected by a split 4 π flow counter, and the gamma rays by a surrounding well-type NaI(Tl) scintillator. Anticoincidence guard detectors are also employed. (*b*) Background spectra from the same detector assembly. *A*: NaI(Tl) spectrum without β-γ coincidence or anti-coincidence shielding. *B*: Same, but with anticoincidence shielding. *C*: As in *B*, but with β-γ coincidence between the two detectors. Identified peaks probably correspond to 1—annihilation radiation; 2— [208]Tl; 3— [40]K; 4— [214]Bi, [212]Bi; 5— [214]Bi, [212]Bi; 6— [208]Tl; 7—alpha particles from detector components; 8—not specifically identified. (From Roedel.[29])

PROBLEMS

20.1 One potential source of background counts from sodium iodide scintillators is [40]K from trace-level potassium impurity in the crystal. Find the maximum potassium concentration (in ppm) if the corresponding background rate from a 7.62 cm × 7.62 cm cylindrical crystal is not to exceed 1 count/s.

20.2 Explain the presence of naturally occurring ^{14}C when its half-life (5600 years) is so short compared with the age of the universe.

20.3 Explain the following observation: For gas proportional tubes operated in pulse mode, the average size of cosmic-ray-induced pulses is usually much smaller than that of typical signal pulses, but the opposite is true for typical scintillation detectors used in gamma-ray spectroscopy.

20.4 Gas proportional counters operated on the alpha plateau can show background counting rates that are often less than 1 count/min. Point out the origin of the background

events that are observed and explain why this rate is so much less than usually encountered in many other detectors.

20.5 The background gamma-ray spectrum inside massive concrete shielding normally shows a prominent 2.22 MeV line. Explain the origin of these gamma rays.

20.6 Using a simple gamma-ray counting system, a sample containing ^{137}Cs results in a gross counting rate of about 10 counts/s. Removing the source results in a background rate of around 8 counts/s. The same sample is then taken to a β-γ coincidence counting system where the gross coincidence rate from source and background is about 3/s. Find the maximum value of the background coincidence rate with the source removed that will result in a lower expected fractional standard deviation in the measurement of the net rate due to the source alone, assuming that equal total measurement time is optimally divided between counting the source and background in both cases (see Ch. 3).

REFERENCES

1. P. Theodórsson, *Measurement of Weak Radioactivity*, World Scientific, Singapore, 1996.
2. R. R. Finck, K. Liden, and R. B. R. Persson, *Nucl. Instrum. Meth.* **135**, 559 (1976).
3. D. C. Camp, C. Gatrousis, and L. A. Maynard, *Nucl. Instrum. Meth.* **117**, 189 (1974).
4. K. Bunzl and W. Kracke, *Nucl. Instrum. Meth.* **A238**, 191 (1985).
5. R. L. Brodzinski, J. H. Reeves, F. T. Avignone, and H. S. Miley, *Nucl. Instrum. Meth.* **A254**, 472 (1987).
6. P. Jagam and J. J. Simpson, *Nucl. Instrum. Meth.* **A324**, 389 (1993). See also erratum published as *Nucl. Instrum. Meth.* **A334**, 657 (1993).
7. T. A. Lewis, *Nucl. Instrum. Meth.* **A264**, 534 (1988).
8. S. Okabe, T. Nishikawa, M. Aoki, and M. Yamada, *Nucl. Instrum. Meth.* **A255**, 371 (1987).
9. A. Stenberg and I. U. Olsson, *Nucl. Instrum. Meth.* **61**, 125 (1968).
10. J. H. Kaye, F. P. Brauer, J. E. Fager, and H. G. Rieck, Jr., *Nucl. Instrum. Meth.* **113**, 5 (1973).
11. G. Heusser, *Nucl. Instrum. Meth.* **B58**, 79 (1991).
12. J. Verplancke, *Nucl. Instrum. Meth.* **A312**, 174 (1992).
13. F. Sánchez et al., *Nucl. Instrum. Meth.* **A339**, 297 (1994).
14. R. Nunez-Lagos and A. Virto, *Appl. Radiat. Isot.* **47**(9/10), 1011 (1996).
15. K. S. Parthasarathy, *Nucl. Instrum. Meth.* **134**, 591 (1976).
16. U. Kilgus and R. Kotthaus, *Nucl. Instrum. Meth.* **A365**, 588 (1995).
17. K. S. Parthasarathy, *Nucl. Instrum. Meth.* **136**, 585 (1976).
18. M. H. Shamos and A. R. Liboff, *Rev. Sci. Instrum.* **39**, 223 (1968).
19. J. H. Kaye, F. P. Brauer, R. E. Connally, and H. G. Rieck, *Nucl. Instrum. Meth.* **100**, 333 (1972).
20. O. G. Bartels, *Health Phys.* **28**, 189 (1975).
21. R. J. Arthur, J. H. Reeves, and H. S. Miley, *IEEE Trans. Nucl. Sci.* **NS-35**(1), 582 (1988).
22. B. Grinberg and Y. Le Gallic, *Int. J. Appl. Radiat. Isotopes* **12**, 104 (1961).
23. A. Alessandrello et al., *Nucl. Instrum. Meth.* **B83**, 539 (1993).
24. S. Kobayashi, N. Hosoda, and R. Takashima, *Nucl. Instrum. Meth.* **A390**, 426 (1997).
25. S. C. Gujrathi and J. M. D'Auria, *Nucl. Instrum. Meth.* **100**, 445 (1972).
26. S. R. Lewis and N. H. Shafrir, *Nucl. Instrum. Meth.* **93**, 317 (1971).
27. J. H. Reeves, W. K. Hensley, and R. L. Brodzinski, *IEEE Trans. Nucl. Sci.* **NS-32**(1), 29 (1985).
28. A. Stenberg, *Nucl. Instrum. Meth.* **96**, 289 (1971).
29. W. Roedel, *Nucl. Instrum. Meth.* **61**, 41 (1968).
30. E. L. Grigorescu, *Nucl. Instrum. Meth.* **A369**, 574 (1996).

Appendix A

The NIM and CAMAC Instrumentation Standards

In Chapter 17, some of the general features of nuclear instrumentation standards were introduced. In this appendix, some details are provided for the two most common standards in general use: the NIM and CAMAC standards

A. The NIM (Nuclear Instrument Module) System

The NIM bin is designed to fit into the standard 19-inch relay rack and is subdivided into 12 individual module positions across its width. A NIM module occupies a unit width of 34.4 mm, although integral multiples of this width are permitted corresponding to modules of double width, triple width, and so on. Each of the 12 bin locations is provided with a 42-pin connector that mates with a corresponding connector at the back of each module. Pin assignments and functions are illustrated in Fig. A.1. Primary dc supply voltages provided by the bin are ± 12 V and ± 24 V. Some NIM bins also provide ± 6 V, mostly for modules using integrated circuits, but these voltages are not strictly required by the NIM rules. The NIM bin and modules can be of two standard heights—$5\frac{1}{4}$ inches (133 mm) or $8\frac{3}{4}$ inches (222 mm)—but the larger of these sizes is by far the more common.

Although some limited logic and switching operations can be performed through designated connector pins, the primary means of transmitting linear and logic pulses between NIM modules is by coaxial cables connected to appropriate jacks on either the back or front panel of the module. BNC connectors are specified for signal jacks, whereas SHV connectors are standard for high-voltage connections.

The NIM standard also recommends that shaped linear pulses correspond to one of three specific dynamic ranges:

1. 0 to +1 V (primarily for integrated circuits).
2. 0 to +10 V (primarily for transistor-based circuits).
3. 0 to +100 V (primarily for vacuum-tube-based circuits and largely obsolete).

NIM modules designed for the processing of linear pulses are signal compatible only if they share a common classification for the dynamic range of the signals.

Logic signal levels are also specified in the NIM standard. Standard logic levels for logic states and the transmission of digital data are given in Table A.1. Fast logic pulses in 50-ohm impedance systems are separately specified in Table A.2.

The table in the figure reads:

PIN	FUNCTION
1	RESERVED
2	RESERVED
3	SPARE
4	RESERVED
5	COAXIAL
6	COAXIAL
7	COAXIAL
8	+200 VOLTS D.C.
9	SPARE
✦ 10	+6 VOLTS
✦ 11	-6 VOLTS
12	RESERVED
13	SPARE
14	SPARE
15	RESERVED
✦ 16	+12 VOLTS
✦ 17	-12 VOLTS
18	SPARE
19	RESERVED
20	SPARE
21	SPARE
22	RESERVED
23	RESERVED
24	RESERVED
25	RESERVED
26	SPARE
27	SPARE
✦ 28	+24 VOLTS
✦ 29	-24 VOLTS
30	SPARE
31	SPARE
32	SPARE
✦ 33	117 VOLTS A.C. (HOT)
✦ 34	POWER RETURN GND
35	RESET
36	GATE
37	SPARE
38	COAXIAL
39	COAXIAL
40	COAXIAL
✦ 41	117 VOLTS A.C./NEUTRAL
✦ 42	HIGH QUALITY GND
G	GROUND GUIDE PIN

✦ MUST BE BUSSED TO ALL BIN CONNECTORS PG1B THROUGH PG12B

BIN CONNECTOR — MODULE CONNECTOR — REAR VIEW

GROUND GUIDE SOCKET GGS — GROUND GUIDE PIN GGP

NOTES:—
A. CONNECTORS ARE IN ACCORDANCE WITH DRAWING NO. ND-519.
B. RESERVED PINS ARE FOR FUTURE ASSIGNMENT BY THE COMMITTEE AND SHALL NOT BE USED UNTIL SUCH ASSIGNMENTS ARE MADE.
C. GP-1 = GUIDE PIN
GGP = GROUND GUIDE PIN
GS-1 = GUIDE SOCKET   } PER ND-519
GGS = GROUND GUIDE SOCKET
D. SEE ALSO SECTION F OF THIS REPORT

NOTE C
GS-1 — GP-1 — GS-1 — GP-1
GGS — GP-1 — GS-1 — GGP
BIN CONN.   MOD. CONN.
FRONT FACES
SHOWING GUIDE PINS & SOCKETS

Figure A.1 Pin assignments for the NIM standard connector between bin and module. (Adapted from National Bureau of Standards Photo 74-08-4044.)

Detailed specifications for NIM systems are contained in USAEC Report TID-20893, which was first issued in July 1964. This report has subsequently undergone revision every few years, and the interested user should seek out the latest revision.

The NIM standard does not adapt easily to situations in which large volumes of digital data must be processed. Furthermore, the basic module width is needlessly large for many digital units that do not require large numbers of front panel controls. These considerations, together with the desire for standard interfacing with digital computers, has led to the development of the CAMAC standard described next.

Table A.1 NIM Standard Logic Levels

	Output (Must Deliver)	Input (Must Respond to)
Logic 1	+4 to +12 V	+3 to +12 V
Logic 0	+1 to −2 V	+1.5 to −2 V

Table A.2 NIM Fast Logic Levels for 50 ohm Systems

	Output (Must Deliver)	Input (Must Respond to)
Logic 1	−14 to −18 mA	−12 to −36 mA
Logic 0	−1 to +1 mA	−4 to +20 mA

B. The CAMAC (Computer Automated Measurement and Control) Standard

The CAMAC standard is based on a *crate*, which also fits the standard 19-inch relay rack but which is subdivided into 25 individual module stations spaced 17.2 mm apart. Electrical connection between each module and the crate is made by a printed circuit board edge connector with 86 contacts. The width of the CAMAC station neatly accommodates many modules consisting of a single printed circuit board with mounted integrated circuits. When modules must involve bulkier equipment or more than one board, multiples of the basic width can be used, and the double-width or triple-width module will occupy two or three standard stations.

Within the crate, each connector provides access to the *dataway*, which is a data highway consisting of conductor busses for digital data, control signals, and power. One of the basic design features of CAMAC is that digital communication between plug-in modules within a crate occurs over this dataway. It replaces external interconnection of modules for many digital functions, but some coaxial cable connections must be retained for linear signals and other purposes. These signals can be coupled to either the front or back of a module using the type 50CM connector recommended in the CAMAC standard.

The extreme right-hand station within the crate is different from the remaining 24 and is called the *control station*. It is intended that this station will be occupied by a *crate controller* plug-in module. The crate controller is usually a double-width module that occupies one normal station (usually the 24th) as well as the control station. The crate controller provides all the control functions necessary for the transfer of data between modules in a crate and serves as the interface between the crate and any external equipment. No CAMAC system is complete without a crate controller, and none will function without one, except on an individual module basis. In order to communicate properly over the dataway, each individual module must have sufficient internal coding and decoding to read digital data from the dataway and to supply such data from its own internal circuits when requested by the crate controller.

A diagram of the dataway is shown in Fig. A.2. The power busses are connected to all 25 stations and provide ±6 V and ±24 V for module power. Although not required by CAMAC, ±12 V may also be available, depending on the specific crate design, and connections are reserved for these voltages. Except for the control station, all normal stations are also interconnected by 24 parallel *read* lines and 24 *write* lines. These lines can be used to transmit 24 parallel bits of data from a module along the read lines and to a module along the write lines. The control station is connected to each of the normal stations by separate private lines. Twenty-four of these are *station* lines (N lines), one of which corresponds to each normal station and must be activated to communicate with that specific

Figure A.2 Diagram of the CAMAC dataway.
(National Bureau of Standards Photo 72-534.)

station. An additional 24 are *look-at-me* lines (L lines), one for each station, by means of which each individual normal station may signal to the control station that it requires attention.

Modules cannot communicate directly with each other but must do so only through the crate controller. In providing a command to a given station, the controller, in addition to activating the corresponding N line, must also complete the command by providing a coded subaddress and function. These are communicated via five *function* lines (F lines) and four *subaddress* lines (A lines), which are fully bussed to all stations. Two *strobe* lines (S1 and S2) are also fully bussed and must be activated to initiate operations or to transfer data. The remaining busses shown in Fig. A.2 have fairly descriptive names that indicate their functions.

In addition to managing the internal module-to-module communication within a crate, the crate controller must also serve to interface the dataway with external equipment that is to be part of the overall system. Because CAMAC systems are of greatest interest when large volumes of digital information must be handled in a given application, the crate controller is often required to interface with the I/O structure of a computer. Standard crate controllers can be obtained for interfacing the more common types of laboratory computers, as well as to interface with the branch highway (type A controller) and serial highway (type L controller) described below.

If the number of CAMAC modules that must be used in a given application exceeds the room available in a single crate, or if several crates in different remote areas must be linked to a common data system, then a method must be provided to couple together more than one CAMAC crate. The arrangement can be either a parallel interconnection (a *branch highway*) consisting of 66 pairs of signal and ground wires, or a series chain (a *serial highway*) in which only two pairs of conductors are used. In the latter case, data are sent serially and it is therefore slower, but costs are often much lower than for a branch highway interconnection.

Detailed specifications for CAMAC are spelled out in IEEE Documents 583-1975, 596-1976, and 595-1976. A useful collection of introductory articles can be found in Ref. 1. A book that includes both a tutorial introduction to CAMAC and complete specifications has been published as Ref. 2.

REFERENCES

1. CAMAC Tutorial Issue, *IEEE Trans. Nucl. Sci.* **NS-20**, No. 2 (1973).

2. *CAMAC Instrumentation and Interface Standards*, IEEE Document No. SH06437, distributed by Wiley-Interscience, New York, 1976.

Appendix **B**

Derivation of the Expression for Sample Variance in Chapter 3

$\mathbf{F}$rom the discussion on p. 68, the definition of the residual for any data point is

$$d_i \equiv x_i - \bar{x}_e$$

Now by adding and subtracting the true mean value $\bar{x}$, we can relate the residual to the deviation

$$d_i = (x_i - \bar{x}) + \bar{x} - \bar{x}_e$$

$$= \epsilon_i + \frac{1}{N} \cdot N\bar{x} - \frac{1}{N} \cdot N\bar{x}_e$$

$$= \epsilon_i + \frac{1}{N} \sum_1^N \bar{x} - \frac{1}{N} \sum_{j=1}^N x_j$$

$$= \epsilon_i - \frac{1}{N} \sum_{j=1}^N (x_j - \bar{x})$$

$$d_i = \epsilon_i - \frac{1}{N} \sum_{j=1}^N \epsilon_j$$

Squaring both sides

$$d_i^2 = \epsilon_i^2 - \frac{2}{N} \epsilon_i \sum_{j=1}^N \epsilon_j + \frac{1}{N^2} \left(\sum_{i=1}^N \epsilon_i \right)^2$$

Summing over all data points

$$\sum_{i=1}^N d_i^2 = \sum_{i=1}^N \epsilon_i^2 - \frac{2}{N} \sum_{i=1}^N \epsilon_i \sum_{j=1}^N \epsilon_j + \frac{1}{N^2} \sum_1^N \left[\sum_{i=1}^N \epsilon_i \right]^2$$

$$= \sum_{i=1}^N \epsilon_i^2 - \frac{1}{N} \left(\sum_{i=1}^N \epsilon_i \right)^2$$

$$= \sum_{i=1}^N \epsilon_i^2 - \frac{1}{N} \left(\sum_{i=1}^N \epsilon_i^2 + \sum_{\substack{i,j=1 \\ i \neq j}}^N \epsilon_i \epsilon_j \right)$$

The last term in the equation above will sum to zero since ϵ_i and ϵ_j are uncorrelated, and there will be an equal number of positive and negative contributions. We are then left with

$$\sum_{i=1}^{N} d_i^2 = \sum_{i=1}^{N} \epsilon_i^2 - \frac{1}{N} \sum_{i=1}^{N} \epsilon_i^2$$

$$= \frac{N-1}{N} \sum_{i=1}^{N} \epsilon_i^2$$

Thus

$$\frac{1}{N} \sum_{i=1}^{N} \epsilon_i^2 = \frac{1}{N-1} \sum_{i=1}^{N} d_i^2$$

And substituting prior definitions from Chapter 3

$$s^2 = \frac{1}{N-1} \sum_{i=1}^{N} (x_i - \bar{x}_e)^2$$

This is now shown in Chapter 3 as Eq. (3.9).

Appendix C

Statistical Behavior of Counting Data for Variable Mean Value

A. Generalized Poisson Distribution

The Poisson distribution introduced in Chapter 3 strictly holds only for processes that are characterized by a constant mean value $\bar{x}$. When measuring radioactive sources, this condition translates into requiring that the observation time be short compared with the half-life of the source. Circumstances can arise, however, in which the true rate of events is expected to change over the course of the observation. The case of short-lived radioactive sources would be an example, as would counting events created by an accelerator beam whose intensity changes over the measurement time. An example is sketched in Fig. C.1a. We will represent this time-dependent true rate as $\mu(t)$. The average true rate over an observation time from 0 to T is then simply:

$$\bar{\mu} = \frac{1}{T} \int_0^T \mu(t)\, dt$$

It can be shown[1] that the expected number of events observed over an extended time interval, even when the true rate of events is changing, follows a Generalized Poisson distribution whose mean value is given by the average number over the interval, equal to $\bar{\mu}\,T$. Thus the probability of observing exactly x events is:

$$P(x) = \frac{(\bar{\mu}\,T)^x\, e^{-\bar{\mu}\,T}}{x!} \tag{C.1}$$

Therefore if the measurement consists of simply recording a single count over the full measurement period, Poisson statistics still describe the expected variance of the data if the identical measurement (including the same variation in the mean value) were repeated many times. In contrast, if the measurement instead consists of many short-duration counts that extend over the same total observation time, the distribution of the collection of recorded counts will no longer be expected to follow Poisson behavior. This case is treated in the following section of this appendix.

B. Distribution of Counts for Multiple Measurements with Varying Mean ("Modified Poisson Distribution")

A different circumstance arises if a series of measurements are made over a total period in which the mean is varying, as sketched in Fig. C.1b. For example, a 1028-channel

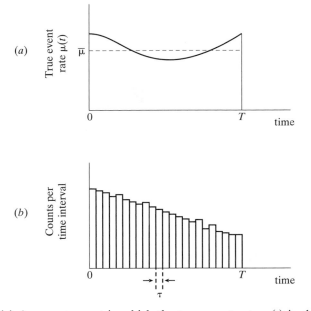

Figure C.1 (*a*) A measurement in which the true event rate $\mu(t)$ is changing as a function of time. The quantity of interest is the expected distribution of the total number of events (or counts) recorded over the time period from 0 to T. The average true rate over this time period is $\overline{\mu}$, and the mean number of events is $\overline{\mu}T$. (*b*) A series of counts, each taken using a short counting time τ, extended over a time range from 0 to T when the true rate is changing. The quantity of interest is now the expected distribution in the collection of data consisting of all the recorded counts.

multiscaling measurement might be carried out with dwell time of 1 second per channel for a source that is decaying with a 500-second half life. Over the full 1028-second duration of the measurement, the source will have decayed substantially, but the change in expected mean value over a single 1-second channel will be negligible. In the discussion of part A of this appendix, we were concerned only with the result obtained by adding together the counts from all the channels. We now will instead examine the separate question of the expected data distribution function for the many counts that have been recorded over the total observation time. More specifically, we would like to predict the probability of observing a given value x from the collection of data consisting of the counts appearing in all the channels.

The analysis begins by first recognizing that the Poisson distribution will predict the count distribution expected for the first interval provided the appropriate mean value is used. The next interval will have a slightly different mean, so a different Poisson will predict its count distribution. The distribution of counts observed from all the intervals will then be predictable from the distributions combined over the full measurement period. Assuming that there are n such intervals, each of equal length τ, then the total measurement time is given by $T = n\tau$. The distribution of interest will be

$$P(x) = \frac{1}{n} \sum_{j=1}^{n} P_j(x)$$

where $P_j(x)$ is the probability of observing x events in an interval of length τ, when the mean value for that interval is $\mu_j \tau$. We can then predict the probability of observing a given value x from the collection of all of the measurements as

$$P(x) = \frac{1}{n} \sum_{j=1}^{n} \frac{(\mu_j \tau)^x}{x!} e^{-\mu_j \tau} \tag{C.2}$$

This expression is sometimes called the *Modified Poisson Distribution*. Given a known variation of $\mu(t)$, the indicated summation can be carried out numerically.

An analytical solution can be obtained for one important case: that of an exponentially-decaying source without background, or $\mu(t) = \mu_0 e^{-\lambda t}$. By assuming a large number of intervals (n), each of a very short duration, and replacing the summation with an integration, it can be shown[2] that

$$P(x) = \frac{(\mu_0\tau)^x}{x!} e^{-\mu_0\tau} \cdot \frac{1}{T} \int_0^T \exp[-x\lambda t + \mu_0\tau(1 - e^{-\lambda t})] \, dt \qquad \text{(C.3)}$$

This equation is of the form

$$P(x) = P_0(x) \cdot C(x) \qquad \text{(C.4)}$$

where $P_0(x)$ is the Poisson distribution characterizing the first interval, and $C(x)$ is a "correction factor" that depends on the total observation time T and the source decay constant λ. Analytic expressions for $C(x)$ involving exponential integral functions and incomplete gamma functions are presented by Müller.[2]

Figure C.2 shows an example of this calculated distribution for a case in which T is equal to one value of the half-life of the source, or $T = \ln 2/\lambda$. The value of μ_0 was chosen to fit a set of measured data shown as points with error bars on the same figure. The experimental data were obtained from a 4096-channel multiscaler using 2 ms channel width (dwell time) and a source with a half-life of 16.06 s. The source activity resulted in about 50 counts per channel at the start of the measurement, dropping to about 25 counts per channel at the end. The mean over the entire measurement was about 36 counts per channel. Also shown is a standard Poisson distribution with the same mean value. It is clear that the Modified Poisson description is necessary for an adequate representation of the data in this case, since the mean is changing substantially over the full observation time. Additional examples are given in Ref. 3.

In any circumstance for which the true mean value changes significantly as a sequence of multiple counts is recorded, the Modified Poisson approach must replace the simple Poisson to predict the distribution of the recorded data. Some assumption must then be made about the true variation of the mean, and the simple exponential decay of the example given above is a rare case that can be treated analytically. More generally (as in the case of a time-dependent accelerator beam), an assumed model for the mean variation must be used to generate values of μ_j for Eq. (C.2) that can then be evaluated numerically to predict the expected distribution.

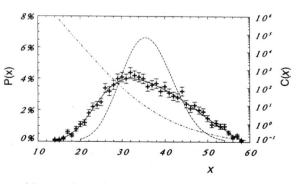

Figure C.2 The points with error bars show the frequency of observance of the value x within a short counting interval from a decaying source as detailed in the text. The solid line is a plot of Eq. (C.3) for this case in which the total measurement time is twice the half-life of the source. The dashed line is a plot of the standard Poisson distribution matched to the same mean value. The line shown with dashes and dots is the value of the correction factor $C(x)$ as defined in Eq. (C.4). (From Salma and Zemplén-Papp.[3])

REFERENCES

1. A. Papoulis, "Probability, Random Variables, and Stochastic Processes," McGraw-Hill Book Company, New York (1965).

2. J. W. Müller, *Jour. Radioanalytical Chem.* **61**, #1-2, 345 (1981).

3. L. Salma and É. Zemplén-Papp, *Nucl. Instrum. Meth.* **A312**, 591 (1992).

Appendix D

The Shockley-Ramo Theorem for Induced Charge

Throughout this text, it has been emphasized that signals from detectors arise because of the *motion* of charge carriers after they are formed by the incident radiation. This statement generally applies to gas-filled ion chambers or proportional counters as well as to semiconductor detectors such a silicon diodes or germanium spectrometers. Those unfamiliar with the process sometimes have a naive (and incorrect) notion that the detector signal is formed only when the charge carriers deposit their electrical charges on the electrodes. That picture would imply a delay before the onset of the pulse because of the time required to transport the carriers from their point of formation to the collecting electrode. As we will see below, no such delay exists, and the output pulse begins to form immediately when the carriers start their motion to the electrodes. (Because the pulse begins as soon as the incident particle deposits its energy in the detector, the timing of detection events can be much more accurate than if a delay were to exist.) Once the last of the carriers arrives at its collecting electrode, the process of charge induction ends and the pulse is fully developed. The time evolution of the signal is of fundamental importance in understanding the timing properties of detectors as well as in predicting the effects of changes in the location of the radiation interactions on the shape of the pulse.

The derivation of the time history of the induced charge in a pulse-type ion chamber that was given in Chapter 5 was for an idealized case of a one-dimensional planar detector without boundaries or edges. For more complex geometries, a more fundamental approach is needed to carry out the analysis. There are also classes of detectors in which electrodes are subdivided into separate strips or pixels. The task of predicting the types of signals expected on a specific electrode segment also requires that a more basic approach be taken to predict induced charges. The analysis that follows will be generally applicable to all types of detectors in which charge carriers formed by radiation are caused to move in an electric field within the detector volume, including the common cases in which the active volume is either a gas or a semiconductor.

A. Electric Potential, Electric Field, and Charge Carrier Motion

The first part of this appendix reviews the general approach to calculate the electric potential and electric field in a detector with arbitrary shape and orientation of its electrodes. The Poisson equation is the starting point for these calculations. It can be written

$$\nabla^2 \varphi = \rho / \epsilon \tag{D.1}$$

where φ is the electric potential, ρ is the charge density (which may vary with position), and ϵ is the dielectric constant for the detector medium. In the absence of trapped charges (which is usually the case for common detectors), $\rho = 0$ and the Poisson equation reduces to the Laplace equation:

$$\nabla^2\varphi = 0 \qquad (D.2)$$

One must now chose the Laplacian operator ∇^2 that is appropriate for the geometry under study. In orthogonal coordinates,

$$\nabla^2 = \frac{\partial^2}{\partial x^2} + \frac{\partial^2}{\partial y^2} + \frac{\partial^2}{\partial z^2}$$

Boundary conditions needed in the solution are chosen based on the operating conditions of the detector. If an external voltage V is applied between two electrodes, then one boundary condition is that the value of the potential φ must change by V between these electrodes. Analytical solutions are possible for some simple cases (see the development for the abrupt semiconductor junction beginning on p. 373). For complex geometries or detector shapes, solutions can be obtained only by using numerical techniques in the form of computer codes. Several commercial code packages are available that are capable of solving the Poisson or Laplace equation for complex geometries and electrode configurations. The result is a numerical solution for the electric potential φ everywhere within the detector volume.

The electric field $\mathscr{E}$ at any point can be obtained simply by taking the gradient of the electric potential

$$\mathscr{E} = -\text{grad } \varphi \qquad (D.3)$$

Neglecting the effects of diffusion, charge carriers generated within the detector will follow the electric field lines (or direction of maximum gradient in the potential) from their point of formation to the collecting electrode. If an assumption is made about their velocity as a function of electric field (for example, a proportional relationship indicating a constant value of the mobility), then the position of the charge as a function of time can be uniquely determined.

B. The Induced Charge

The general method to calculate induced charge on electrodes due to the motion of charge carriers in a detector makes use of the Shockley-Ramo Theorem[1,2] and the concepts of the *weighting field* and *weighting potential*. The theorem states that the instantaneous current induced on a given electrode is equal to

$$i = q\vec{v} \cdot \vec{E}_0 \qquad (D.4)$$

where q is the charge of the carrier, $\vec{v}$ is its velocity, and $\vec{E}_0$ is called the weighting field. Another way of stating the same principle is that the induced charge on the electrode is given by the product of the charge on the carrier multiplied by the difference in the weighting potential φ_0 from the beginning to the end of the carrier path:

$$Q = q\Delta\varphi_0 \qquad (D.5)$$

To find this weighting potential φ_0 as a function of position, one must solve the Laplace equation for the geometry of the detector, but with some artificial boundary conditions:

1. The voltage on the electrode for which the induced charge is to be calculated is set equal to unity.
2. The voltages on all other electrodes are set to zero.

3. Even if a trapped charge is present within the detector volume, it is ignored in the calculation (i.e., the Laplace equation, Eq. (D.2), is used rather than the Poisson equation).

The solution under these conditions gives the weighting potential, and its gradient is the weighting field. The weighting potential is *not* the actual electric potential in the detector, but instead serves as a convenience that allows simple determination of the induced charge on the electrode of interest by taking differences in the weighting potential at the start and end of the carrier motion. The path of the carrier must still be determined from the actual electric field lines. If the position of the carrier as a function of time is determined as previously described, then the time profile of the induced charge (or the induced current) can also be traced out to determine the shape of the output pulse.

C. Illustration: The Pixelated Detector

In Fig. D.1a, the configuration of a planar detector is sketched in which the x-y surface on the left is completely covered with a conventional continuous electrode. The opposite surface has an electrode that is subdivided into a checkerboard pattern of individual pixels. Each pixel is an independent square electrode, and small gaps are present between adjacent pixels. Separate electrical contacts generally are made to each pixel electrode. We assume that no fixed charges (space charge) are present in the detector volume. For simplicity, we also assume that the dimensions of the detector in the x and y directions are large compared with the thickness T, so that edge effects can be neglected in the analysis that follows.

If a common voltage is applied on all the pixel electrodes, then the effects of the small gaps are negligible in determining the actual electric potential and electric field distributions throughout the detector volume. The results of solving the Laplace equation are then equivalent to the case in which a single continuous electrode were to replace the pixelated pattern. The potential changes linearly between the electrode-covered surfaces, and the electric field is essentially uniform throughout the volume. If the pixels are operated at a positive potential relative to the opposite surface, then negative charge carriers (electrons)

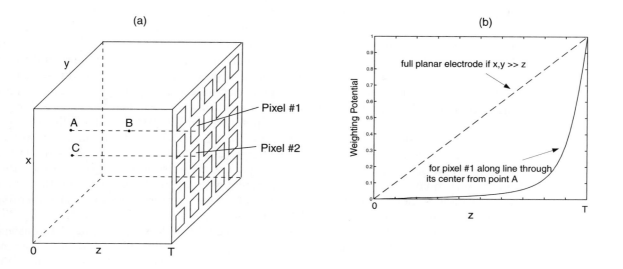

Figure D.1 (*a*) A planar detector configuration with a continuous cathode electrode on the x−y surface at the left and a pixelated anode surface on the right. (*b*) Plot of the weighting potential for a continuous anode (dashed line) and for a single pixel anode (solid line), for the case in which the pixel size is 20% of the detector thickness T.

will be attracted along field lines that are parallel to each other and perpendicular to the detector surfaces. Neglecting diffusion, they will be collected by the pixel electrode that lies directly to the right of the point at which the charges are formed within the volume. The positive charges (holes or positive ions) move in the opposite direction and toward the continuous cathode at the left surface of the detector.

In order to predict the electrical signal expected from a typical single pixel electrode, we must apply the Shockley-Ramo theorem to this case to first find the configuration of the weighting potential. We again solve the same Laplace equation, but now with boundary conditions that set the potential of the pixel of interest to unity, and the potential of all other pixels and the cathode on the opposite surface to zero. The results are plotted in Fig. D.1b, and for different values of detector thickness in Fig. D.2. Notice that the shape of the weighting potential is far from linear, showing a gradient that becomes steeper at distances that are closer to the pixel electrode. Because the induced charge is proportional to the difference in weighting potential between the point of origin and collection, the electron motion will contribute much more than the motion of the positive charge for events over most of the detector volume. Only for charges created very near the pixel surface, a small fraction of the detector volume, will the positive charge motion contribute most of the charge induced on the pixel electrode. For the vast majority of events, the motion of the electron dominates the induced charge. Furthermore, it is the motion close to the pixel surface that contributes most strongly to the induced charge.[†] These consequences are often called the *small pixel effect*[3] and are most pronounced in detectors with pixel dimensions that are small compared with the detector thickness (see Fig. D.2). For detectors in which electrons are much more mobile than positive charges (such as the compound semiconductors described in Chapter 13), deriving the signal from a small pixel rather than from a large-area anode can help improve energy resolution by minimizing sensitivity of the pulse amplitude to the motion of the positive charges that may not be completely collected.

Let us now see how the Shockley-Ramo Theorem can be applied to predict the amplitude of the charge induced on one of the pixel electrodes, for the case in which the detector of Fig. D.1 is a semiconductor (to standardize the terminology). First consider the case of creating n_0 electron–hole pairs from an ionizing event at a point (designated as "A" on Fig. D.1a) that is very near the cathode surface and along an axis that is perpendicular to the center of one of the pixels, call it Pixel #1. The holes will travel only a negligible distance to the cathode and will not make a significant contribution to the signals considered in this example. The induced current observed at the pixel electrode is thus due entirely to the motion of the electrons, and is initially small as the electrons move to the right through a region in which the weighting potential changes slowly (the weighting field is small). As the electrons approach the vicinity of the pixel electrode, the weighting field changes more rapidly, and the induced current rises until the electrons are collected. The electrons travel from a starting point where the weighting potential for the pixel is essentially zero to a point where it is unity (at the pixel surface.) From Eq. (D.5), the total induced charge is given by the product of the total moving charge (equal to $-n_0 e$) multiplied by the difference in the weighting potential between the end and the start of the electron path (unity), and is thus equal to $-n_0 e$. This result is consistent with the prediction based on the simple conservation of energy argument given in Chapter 5 for the pulse from an ion chamber that is also applicable to planar semiconductor detectors.

Next, let us choose a more general case in which the motion of both the electrons and holes contributes to the signal. Assume that the interaction point still lies directly along the axis of Pixel #1, but now at an arbitrary depth (shown as point "B" in Fig. D.1a) where the value of the weighting potential has the value "β" $(0 < β < 1)$. The charge induced on the

---

[†]The effect is somewhat similar to that of the Frisch grid in ion chambers described in Chapter 5, where the output signal is sensitive only to the portion of the electron motion that takes place between the grid and anode, and is free from any contributions from the positive ion motion.

Figure D.2 Plots of the weighting potential for three different pixel electrode sizes. The small pixel effect is enhanced as the pixel size is made small with respect to the detector thickness (top and bottom curves).

pixel electrode will now have two components: one from the motion of electrons, and the other from the motion of holes. Assuming that every charge carrier is collected without loss to recombination or trapping, the total induced charge can again be predicted by applying Eq. (D.5) to both the electrons and holes. The electron motion involves a charge of $-n_0 e$ moving through a difference in weighting potential of $(1 - \beta)$. Thus the electron contribution to the induced charge is equal to $-n_0 e (1 - \beta)$. The hole motion involves an equal but opposite charge of $n_0 e$ moving through a difference in the weighting potential of $-\beta$, so the hole contribution to the induced charge is $-n_0 e \beta$. Combining these two contributions gives the same result as in the first case: the total induced charge is $-n_0 e$. Thus the amplitude of the induced charge is independent of the depth at which the charges are formed, provided all carriers are fully collected.

In detectors with segmented electrodes, such as the pixelated configuration of Fig. D.1, transient signals may be induced on more than one electrode segment even when charges are created at a single point within the detector volume. Now imagine again releasing n_0 electrons very near the cathode, but this time at a position (shown as point "C" on Fig. D.1a) that is along the axis of an adjacent pixel. These electrons follow the actual electric field lines that are horizontal and are collected at the Pixel #2 electrode. The weighting potential for Pixel #1 is essentially zero at the start of this path, and zero at the end, so no net charge is induced when the electrons are collected by the adjacent pixel. But notice from Fig. D.3 that along their path, the electrons pass through a region in which the weighting potential for Pixel #1 rises and passes through a maximum before dropping to zero. Thus a transient signal is induced on the Pixel #1 electrode even though the electrons never reach its surface. The induced current in this case will have a bipolar shape in which the positive and negative lobes balance to produce a net zero integrated charge. If shaped with a long integration time, the resulting pulse amplitude would also be near zero. However, if shorter shaping times are used, a pulse with finite amplitude will result that preferentially responds to the portions of the induced charge with the fastest time characteristics. These

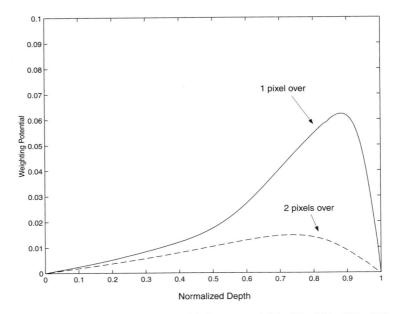

Figure D.3 The solid curve shows the weighting potential for Pixel #1 of Fig. D.1*a* along an axis that is perpendicular to the center of an adjacent pixel. The dashed line shows the same variation along the axis of a pixel that is located two pixels away. Both plots are for the case in which the pixel size is 20% of the detector thickness *T*. Note that the vertical scale is expanded by a factor of 10 compared with the previous two figures.

pulses that are induced on neighboring pixels that do not actually participate in collection of the charge carriers are sometimes given the apt name of "spectator pixel" signals.

In pixel detectors, the size of the induced charge on the primary pixel electrode is the same for all events that occur at positions that are within the projected area of the pixel. Thus simple positioning schemes that are based on sensing the largest pixel signal can locate the event only to within the dimensions of one pixel. The existence of induced signals on neighboring pixels opens the possibility of more precise position-sensing, since these signals will be sensitive to small changes in the exact location of the event. If all the signals in the neighborhood of the primary pixel are read out, an interpolation scheme can be implemented to more accurately position the event to within a fraction of a pixel dimension.

REFERENCES

1. W. Shockley, *Jour. of Appl. Physics* **9**, 635 (1938).

2. S. Ramo, *Proc. of the I. R. E.* **27**, 584 (1939).

3. H. H. Barrett, J. D. Eskin, and H. B. Barber, *Phys. Rev. Lett.* **75**, #1, 156 (1995).

List of Tables

Index